ETHNOMEDICINAL PLANTS
A Biodiversity Treasure

The Editors

Dr. V R Mohan is working as Associate Professor and Head of Botany in V.O. Chidambaram College, Tuticorin. His research areas are Seed Biochemistry and Ethnopharmacology. He did his Ph.D in the field of wild edible legumes. He has nearly 28 years of UG as well as PG teaching experience and 24 years of research experience. During his teaching service he has supervised 38 Ph.Ds on ethnomedicinal plants and 17 M.Phil studies in the same field. He has published more than 400 research articles related to ethnomedicinal plants in various international and national peer reviewed refereed journals. He is a resource person in this field. His other works include documentation of ethnomedicinal plants that are endemic to Southern Tamil Nadu. He had also surveyed and documented several ethnomedicinal plants in the region of Southern Western Ghats, Tamil Nadu. Their pharmacognosy and pharmacological aspects have also been studied. He had completed 3 projects also. He had organized an UGC sponsored seminar and Tamil Nadu State Council for Science and Technology sponsored In Service Training Program. His google scholar citation is 3110 with H-index 26 and I-10 index 97. He also serves as a reviewer in various international journals. Furthermore, he has received "PEARL Foundation Best Senior Scientist Award" conferred by PEARL - A Foundation For Educational Excellence.

Dr. **A Doss** is working as Assistant Professor of Microbiology in Kamaraj College, He has 3 years teaching experience and 10 years research experience in the field of ethnopharmacology. He has received his graduation and post-graduation from St. Joseph's College, Trichy and Doctor of Philosophy from National College affiliated to Bharathidasan University. He has published more than 130 research papers in various international and national peer reviewed refereed journals and 2 books.

Dr. P S Tresina, Assistant Professor of Botany, V.O.Chidambaram College, Tuticorin has achieved her higher degree as Ph.D specializing in seed biochemistry. She holds her UG and PG degree in Botany from St. Mary's College, Tuticorin. She has 3 years teaching and 7 years of research experience in seed biochemistry as well as ethnopharmacology. She has published nearly 77 articles in various international and national peer reviewed refereed journals and 4 book chapters. She have been awarded Tamil Nadu Government stipend for Full- time Scholar for the year 2009-2010 and have worked as UGC Major Research Project Fellow in PG and Research Department of Botany, V. O. Chidambaram College, Tuticorin. Moreover, she have received "Best Young Women Scientist Award in Botany" conferred by PEARL - A Foundation For Educational Excellence.

Dr. **V Sornalakshmi** is working as Assistant Professor of Botany in A.P.C. Mahalaxmi College for Women. She has 10 years teaching experience and 12 years research experience in the field of seaweed chemistry and ethnopharmacology. She has received her graduation, post graduation and master of philosophy from St. Mary's College, Tuticorin and Doctor of Philosophy from V.O.Chidambaram College affiliated to Manonmaniam Sundaranar University. She has published more than 21 research papers in various international and national peer reviewed refereed journals and have completed 5 M.Sc., dissertations. In addition, she have received "Best Young Women Scientist Award in Botany" conferred by PEARL - A Foundation For Educational Excellence.

ETHNOMEDICINAL PLANTS
A Biodiversity Treasure

– Editors –

Dr. V R Mohan

Dr. A Doss

Dr. P S Tresina

Dr. V Sornalakshmi

2018

Daya Publishing House®

A Division of

Astral International Pvt. Ltd.

New Delhi – 110 002

ISBN 9789388173001 (Int. Edition)

Published by : **Daya Publishing House®**
A Division of
Astral International Pvt. Ltd.
– ISO 9001:2015 Certified Company –
4736/23, Ansari Road, Darya Ganj
New Delhi-110 002
Ph. 011-43549197, 23278134
E-mail: info@astralint.com
Website: www.astralint.com

Digitally Printed at : **Replika Press Pvt. Ltd.**

***Prof. Dr. A. Rajendran**, Ph.D.*
Head, Department of Botany,
Dean, Faculty of Sciences,
Bharathiar University,
Coimbatore, Tamil Nadu

Foreword

Medicinal plants products have a long, positive tradition in many countries, especially in India. From a phytochemical stand point, plants were the primary source of medicines for the majority of the world's populations. This is still true today. Plants are not only rich in alkaloids but that some of the alkaloids are unique and could prove useful in the future. The bio-genetic resources present in the medicinal plants are the primarily sources of valuable genes, chemicals, drugs, pharmaceuticals, enzymes or proteins of great health and nutritional importance. The unknown potentials of chemo-diversity found in the plants represent a never ending biological frontier of inestimable value. Comprehensive efforts in research and development, have demonstrated that many plant extract and natural substances are safe and effective drugs. The conventional pharmaceutical companies rely on secondary plant products to provide model chemical structures in the development of new drugs.

In recent years, pharmaceutical companies have become interested in ethnobotanically selected medicinal plants for identification of more effective and marketable drugs. In general, they have a higher hit-rate when compared to plants selected by random screenings. During the present few years, improved analytical methods have enabled studies on the capability and safety of natural products, enhancing the acceptance of those commodities. Many of the plants described in the chapters are easily accessible and they contain chemical compounds that can modify and modulate biological systems, eliciting therapeutic effects. In this book, an assemblage of recent findings on the therapeutic potential of diversified medicinal plants/natural products of the botanists, chemists, microbiologists, ethnobotanists and pharmacologists. The book provides scientific evidence on the use of medicinal plants in the treatment of certain diseases, identified novel plants for treatment of different diseases, explores the mechanisms of actions of the plants and also encouraged the development of plant-based drugs. This informative book will be for students, faculties, scientists, researchers and industry professional in herbal remedies and nutraceuticals. It is hoped that with this book more science student are encouraged to pursue interdisciplinary studies to unlock the secret of plants for the well being of humanity. Definitely it will serve as the engine of more concrete efforts at conserving and propagating our rich plant resources in their natural biodiversity. It is a great pleasure for me to recommend this book without reservation to all researchers in the field of phytomedicine. I wish the book more success and broad distribution.

Prof. Dr. A. Rajendran

Preface

Man's reliance on plant species dates back to the beginning of the human race. In the early days he had limited needs. But with the encroachment of civilization his necessities also grew and the people of each ethnic group developed their own culture, customs, religious rites, legends and myths, taboos, foods and even medicinal practices.

Medicinal plants have played a crucial role in the progression of human culture. Medicinal plants are resources of traditional medicines and many of the modern medicines are produced indirectly from plants. Although medicinal plants have a rich history of utilization in all cultures, no one knows when or where plants first began to be used in the treatment of various ailments. Primitive humans began to distinguish those plants suitable for nutritional purpose from others with an ultimate pharmacological action. This relationship has grown between plants and humans, and many plants have come to be used as drugs.

Plants are sophisticated factories where a variety of chemical compounds are manufactured. These derived products are used as medicines, pesticides, perfumes, fragrances and other utility products. The systematic study of medicinal plants and its claimed use in traditional medicine against a particular ailment not only enables us to discover the active substances responsible for that use but also opens interesting avenues for further research

India has one of the oldest, richest and most assorted cultural traditions associated with the use of medicinal plants. The astonishing fact is that it still remains a living tradition. Over seven thousand and five hundred species of plants are estimated to be used for human and veterinary health care. Nearly all cultures, both ancient and recent, have used plants as source of medicine. We would never be able to say how exactly the ancients discovered the medicinal properties of herbs. Probably, it started with ancient beliefs, myths and lore, got involved with astrology and other occult practices, developed into folk medicine and herbalism, and finally gave rise to traditional systems of herbal medicine. There is at the moment, no exhaustive and reliable inventory available for all the medicinal plants of India.

The utilization of medicinal plants in modern medicine suffers from the fact that although plants are used to cure diseases, scientific evidence in terms of modern medicine is lacking in many cases. Diverse societies use plants according to their own beliefs, knowledge, and previous experiences. Now-a-days, ethnomedicine and pharmacognosy are regarded as a thrust area of research in official agencies like Department of Science and Technology, Department of Environment and Forests, Indian Council of Agricultural Research, Department of Ayurveda, Yoga and Naturopathy, Unani, Siddha and Homoeopathy (AYUSH), Council of Scientific and Industrial Research, *etc.* and have the subject incorporated in syllabi at graduate and post graduate levels or as a special paper in many universities in India.

The principle aim of this book is to provide detailed information about the therapeutic uses of the locally important ethnomedicinal plants. We sincerely hope that this book will serve as a text book and a guide to the readers interested in ethnomedicinal plants in the developing countries and lead to advanced teaching and research work.

Dr. V.R. Mohan

Dr. A. Doss

Dr. P.S. Tresina

Dr. V. Sornalakshmi

Contents

Foreword *v*

Preface *vii*

1. **An Exploration on the Indigenous Ethnomedicinal Plants of Saduragiri Hills, Western Ghats, Tamil Nadu, India** **1**

   P.S. Tresina, K. Paulpriya and V.R. Mohan

2. **Ethnomedicinal Plants of Agasthiarmalai Biosphere Reserve, Tamil Nadu: An Investigation** **67**

   A. Nishanthini and V.R. Mohan

3. **Traditional Knowledge of *Irulas* Tribes in Krishnagiri District, Tamil Nadu, India** **123**

   K. Paulpriya, P.S. Tresina, V. Sornalakshmi and V.R. Mohan

4. **Ethnobotanical Studies on the Medicinal Plants Used by Kani Tribes of Mugilayadi Hills, Kanyakumari Wildlife Sanctuary, Southern Western Ghats, Tamil Nadu, India** **167**

   B. Parthipan, A.V. Babitha, P. Dayana, Y. Jasmine Leena and J.L. Lekshmi

5. **Ethnobotanical Study of Bodha Hills, Southern Eastern Ghats, Namakkal District, Tamil Nadu, India** **189**

   S. Murugesh and P. Deepa

6. **Ethnobotanical Study of Medicinal Plants Used by Kani Tribals in Vamanapuram Block, Nedumangadu Taluk, Kerala, India** **215**

B. Parthipan, C. Biju and M. Johnson

7. **An Ethnobotanical Survey and Investigation of Medicinal Plants from Kiliyur and Pattipadi In Yercaud Hills, Salem** **229**

S. Murugesh and P. Vino

8. **Ethnobotanical Survey of Medicinal Plants Used by Malayali Tribals in Vathalmalai, Dharmapuri District, Eastern Ghats of Tamil Nadu, India** **239**

S. Murugesh, P. Selvi, B. Seelapreethi and R. Rajeswari

9. **Ethnobotanical Survey on Medicinal Plants Used by Malayali Tribals in Palamalai Hills, Salem District, Tamil Nadu, India** **247**

R. Rajeswari, S. Murugesh, P. Selvi and P. Dhandapani

10. **Ethnobotanical Study of Medicinal Plants Used by the *Vetan* Community in Kanyakumari District, Tamil Nadu, India** **263**

J. Celin Pappa Rani, M.S. Kala Swarna and S. Jeeva

11. **Ethnobotanical Studies of Medicinal Plants Commonly Used by the Villagers of Kanyakumari District, Tamil Nadu, India** **293**

Selvaraj Rosemary, Jenisha and S. Jeeva

12. **An Investigation on the Aboriginal Ethnomedicinal Plants of *Kanikkars* of Kalakad-Mundanthurai Tiger Reserve Sanctuary, Tamil Nadu with Special Emphasis on Rheumatism** **307**

E. Daffodil D'Almeida and V.R. Mohan

13. **Therapeutic Uses of Medicinal Plants Used by *Palliyar* Tribe in Sirumalai Hills, Tamil Nadu for the Treatment of Gastro-Intestinal Complaints** **343**

A. Maruthupandian, M. Viji, C. Chithravadivu and V.R. Mohan

14. **Ethnomedicinal Plants used for Skin diseases by *Kanikkars* of Kanyakumari District of Western Ghats, Tamil Nadu, India** **363**

P.S. Tresina, K. Paulpriya, V. Sornalakshmi and V.R. Mohan

15. Ethnomedicinal and Antimicrobial Activities of some Mangrove Species of Godavari Estuary: A Review **381**

G.M. Narasimha Rao

16. *Hybanthus enneaspermus*: A Reliable Herb to Cure different Human Ailments **389**

A. Maruthupandian, M. Viji, B. Perumal, S. Velmani and C. Chithravadivu

17. Medicinal Plants and its Utilization in India **407**

M.S. Rukshana, A. Doss and T.P. Kumari Pushpa Rani

18. An Assessment of Angiosperm Diversity of Manakudy Estuary, Kanyakumari District: A Anthropogenic Polluted Estuariane Ecosystem, Tamil Nadu, India **425**

Parthipan, M. Kalaimathi, M. Mahalekshmi and M. Valarmathi

19. Ethnoveterinary Practices Among the *Malayali* Tribes of Kolli Hills, Eastern Ghats, Tamil Nadu, India **469**

S. Murugesh and P. Deepa

20. Studies on the Indigenous Wild Edible Plants of *Palliyar* Tribes of South India **479**

P.S. Tresina, K. Paulpriya and V.R. Mohan

21. Microscopic Studies of *Cadaba indica* Lam. (Capparaceae) **521**

A. Saravana Ganthi and M. Padma Sorna Subramanian

22. Pharmacognostic and Phytochemical Evaluation of the Whole Plant of *Catharanthus pusillus* (Murr.) G. Don **531**

P. Yokeswari Nithya, S. Mary Jelastin Kala and V.R. Mohan

23. Pharmacognostical, Physico-chemical and Phytochemical Assessment of *Barleria courtallica* Nees (Acanthaceae) **551**

A. Ponmathi Sujatha, R. Michael Evanjaline, S. Muthukumarasamy and V.R. Mohan

24. Pharmacognostic, Physico-chemical Standardization and Phytochemical Analysis of Stem Bark of *Ailanthus excelsa* Roxb. **569**

V. Sornalakshmi, P.S. Tresina, K. Paulpriya and V.R. Mohan

25. Pharmacognostic Study and Establishment of Quality Parameters of Whole Plant of *Beloperone plumbaginifolia* (Jacq.) Nees. **589**

G. Sathiyabalan, K. Paulpriya, P.S. Tresina, S. Muthukumarasamy, V.R. Mohan

26. GC-MS Analysis and Antibacterial Activity of *Myristica fragrans* Seed Extracts against the Lower Respiratory Tract Pathogen *Klebsiella pneumoniae* **607**

T.P. Kumari Pushpa Rani and A. Doss

Index 617

2018, Ethnomedicinal Plants: A Biodiversity Treasure *Pages* **1–65**
Editors: ***V.R. Mohan, A. Doss, P.S. Tresina and V. Sornalakshmi***
Published by: **ASTRAL INTERNATIONAL PVT. LTD., NEW DELHI**

Chapter 1

An Exploration on the Indigenous Ethnomedicinal Plants of Saduragiri Hills, Western Ghats, Tamil Nadu, India

P.S. Tresina, K. Paulpriya and V.R. Mohan

Ethnopharmacology Unit,
PG and Research Department of Botany,
V.O. Chidambaram College, Tuticorin – 628 008, Tamil Nadu
E-mail: vrmohanvoc@gmail.com

ABSTRACT

An ethnobotanical exploration was carried out among the ethnic groups (Palliyars) in Saduragiri hills located in the South Eastern slopes of the Western Ghats, Tamil Nadu. Traditional uses of 165 plant species belonging to 135 genera and 55 families are described under this study. These tribals are using 33 plants to treat rheumatism, 22 plants as to reduce body heat, 20 plants for unknown insect-bites/general poisonous bites, 14 plants for eczema/itches/pimples, 12 plants for treating gastritis and 11 plants for cough and cold. The medicinal plants used by Palliyar tribes are arranged alphabetically, followed by family name, local name, parts used, mode of drug preparation, medicinal uses and dosage etc.

Introduction

The term Ethnobotany was first applied by Harshberger in 1895 to the study of "Plants used by primitive and aboriginal people…." (Anonymous, 1895) and it has been derived from the word, ethnic, which means classification of human beings into social and cultural groups (Singh, 2004). Initially this term referred to the study of plants used by primitive societies; but its scope has widened in recent years. Currently, this term is employed to include the total relationship of plants and people (Jain and Mudgal, 1999) and is often enriched by the rich biodiversity coupled with ethnic diversity. Hough defined it as "the study of plants in their relation to human culture including psychological importance and mythological reference" (Ford, 1978).

Tribals are closely associated with plants and they possess good knowledge about plant resources in their vicinity. With the reach of civilization to the ethnic societies the traditional knowledge on the use of these plants is fast vanishing. There is an urgent need to document this knowledge, as otherwise it will be lost forever. The traditional systems of medicine are being practiced to achieve the elixir of youth and good health along with many indigenous methods. Ethnomedicine or the folk medicine is one of the ways, which is widely practiced among the tribals and aboriginal population of our country for treating ailments. Primitive societies have depended on herbal remedies for the treatment of diseases and disorders since time immemorial. All traditional systems of medicine had their root and origin in folklore medicine and even today large number of rural and tribal populations adopts herbal remedies for primary health care.

According to a survey carried out by the Ministry of Environment and Forests, Government of India, New Delhi for a period of ten years beginning from the late eighties to mid nineties there are over 8000 species of plants used as medicine by around 4600 ethnic groups including tribals, across the various ecosystems in the country.

Millions of the people in the third world still use herbal medicines because they believe in them and regard them as their own system of medicine (Audichya *et al.*, 1983; Austin and Bourne, 1992; Chawdhury, 1992). It has been estimated that 80 per cent of people living in developing countries are almost completely dependent on traditional medical practices, for their primary health care needs. Many higher plants are known to be the main source of drug therapy in traditional medicine (Martini-Bettole, 1980; Farnsworth *et al.*, 1985; Akerele, 1993; Anyinam, 1995; Martinez, 1995).

The various tribal sects of India are repositories of rich knowledge on various uses of plant genetic resources, which have hitherto remained unknown (Khoshoo, 1991). But of late, due to several developmental activities around tribal areas which are after all not related to their welfare, the tribal people are losing their traditional identity resulting in a good deal of loss of such treasure-house of knowledge on plant genetic resources (Shankar, 1995). In view of the harmful developments, the UN declared the year 1993 as the "International year of Indigenous People" based on the recommendations of the Rio de Janeiro Earth Summit. The studies on the relationship between the aboriginal or primitive people and their surroundings including a critical evaluation of some of the important plants used by the tribals have received considerable attention in recent years (Das *et al.*, 1983).

It is a matter of great pride that among the 18 hot spots known for rich flora in the world, two are located in India. They are the Eastern Himalayas and the Western Ghats (Khoshoo, 1996). The hill chain of Western Ghats recognized as a region of high level of biodiversity is under the threat of rapid loss of genetic resources (Gadgil, 1996). A perusal of the available literature reveals that till date there is no comprehensive survey, documentation and enumeration of wild medicinal plants used by the tribe *Palliyars* inhabiting the Saduragiri hills located in the South-Eastern Slope of Western Ghats, Tamil Nadu. And hence, in the present study an attempt is made to survey, document and enumerate the wild medicinal plants present in the study area. The *Palliyars* are the dominant tribal group inhabiting this locality. The present study focuses on the dependence of the *Palliyars* on herbal medicines and attempts at an exhaustive analysis of the therapeutic values of such medicinal plants.

The *Palliyars* have so far asserted their ethnic identity and practiced the traditional ethnomedicine. This is a standing example of the dependence of tribal people on plants for their health care from time immemorial. If it is documented and intellectual rights are guaranteed to them, not only the *Palliyars* but also the whole nation would benefit.

Medicinal Plants and the Palliyars - An Investigation

Palliyars form a tribal group and they one settled in the reserve forest area of the Saduragiri hills located in the South – Eastern slopes of Western Ghats, Tamil Nadu. The present investigation deals with the distribution of the tribe *Palliyars* along the target area, its ecological conditions and the *Palliyars* use of different medicinal plants in therapeutic practices.

The *Palliyar* Tribe

The *Palliyars* belong to the Southern Tribal Zone. The Southern tribals are historically more ancient tribes. There are as many as 36 types of Scheduled tribes in Tamil Nadu. In the serialized list notified by the Government of Tamil Nadu, the *Palliyars* are placed at 32nd position. They live in the low altitude regions of Western Ghats and live in large numbers in the study area (Saduragiri hills, Virudhunagar district). Of the total population of the Scheduled tribes in Tamil Nadu (5, 74, 194), the *Palliyar* tribe accounts for 1890, which ranks 20th among the tribal population (Censes of India, 1991). *Palliyars* can be grouped into three categories based on their life style, namely:

(i) Nomadic *Palliyars*, the hunter-gathers, who live in rock shelters wandering in tracts of forest in search of food and non-timber forest produce. (ii) Semi nomadic *Palliyars*, also the hunter-gatherers, who build huts and do not practice agriculture, go out to collect food and non-timber forest produce and return to their dwellings and (iii) Settled *Palliyars*, who have land holdings, practicing agriculture and living mainly in Kerala (Sankarasivaraman, 2000).

Appearance and Habits of *Palliyar*

The *Palliyars* of the study area at present could be called semi nomadic type. Their ancestors were nomadics. They live as individual families. Like other primitive tribes, the *Palliyars* are short, dark complexioned, curly haired with thick protruding lips and blunt nose with wide nostrils (Plate 1.2). *Palliyars* do not have any established mode of dress. They are scantily dressed but freely wear whatever clothes are available to them. They are non-vegetarians; however they abstain from beef as rigidly as the most orthodox Hindus (Dahmen, 1908). Several species of

Dioscorea provide the basis of *Palliyars* staple food. Besides there are a wide variety of greens, stems, tubers, unripe fruits and ripe fruits, which serve as alternative for food. *Palliyars* also feed on wild animals and birds (like rabbit, rat, deer, hen *etc.*).

The *Palliyars* live in small parties as isolated groups. Generally, a hamlet has about 20 huts. Their small huts are unique with the walls made up of mud or with wiry interwoven stems of *Lantana camara*. Each hut is thatched with the fronds of *Cymbopogon citratus* or *Cymbopogon polyneuros* or with the leaves of *Phoenix pusilla* or *Cocos nucifera*. They sleep on mats woven with the leaves of the above said taxa.

The *Palliyars* as a tribe do not possess much cohesiveness. Each settlement has its Headman whose authority is never challenged and he is solely responsible for settling disputes among the tribals. They are illiterate but in recent years, they have started sending their children to the nearby schools.

The *Palliyars* are adroit in collecting honey. They collelct honey from the branches of towering tall trees and rock caves skillfully using special techniques. *Palliyars* from four to seventy years old are agile on the trees. They also collect Kungillium (resin) from the barks of *Canarium strictum*. They are also good hunters and they trap deer, pig, boar, hare, wild fowl and flying squirrel. The meat obtained is divided among the families within the settlement.

They are good herbalists and often collect medicinal plants from the forests. The knowledge about medicinal plant is rather specialized and is limited to a few members in the community who are recognized as "Vaidyars" or medicine men. They are generally respected the most and considered indispensable members within the tribal society. Each medicine man treats all kinds of illness but some of them are specialized in specific diseases. The remedies of common ailments like cuts, pain, headache, fever, dysentery *etc.* are known to most members of the tribal community. There are several individuals in each locality who though not recognized as medicine men, possess such knowledge and act as reliable informants. They maintain secrecy about the use of certain medicines because they believe that the herbals will lose their healing power if too many people know about them. Some practitioners have inherited the knowledge of certain special remedies.

Rituals and Religious Ceremonies

Palliyars have strong faith in religious customs and practices. They mostly practice Hindu Religion. The deity of their worship is specific to their region. The *Palliyars* in the study area worship Lord Siva and Ayyanar, a village god. They also worship "Forest Goddess" and Goddess "Poomadevi" (Goddess Earth). On special occasions, they stay in the temples in groups for two days and worship offering goats as sacrifice. On these days a large quantity of Mullvalli kizhangu (*Dioscorea pentaphylla* var. *pentaphylla*) are collected and offered to the Goddess Earth, "Poomadevi".

Customs Related to Marriage

Palliyars are monogamous. Elopement is a favorite form of marriage. The elders search for the runaway couples, and then bring them back and get them married.

The marriage of both widowed and divorced persons of either sex is permissible. The dead are buried. On the eighth day after death, they perform the last rites. The *Palliyars* believe in witchcraft. They entertain many curious superstitious beliefs.

The Government of Tamil Nadu has established a medicinal plant conservation area in the forest for preservation and development of herbal wealth. It is maintained and conserved with the assistance of the *Palliyars.*

Materials and Methods

Survey and Collection

Saduragiri hills lies between 9°.42″ and 9°.44″ North latitude and 77°. 37″–77°. 41″ East longitude in Virudhunagar districts of Tamil Nadu in the South - Eastern slopes of Western Ghats of India (Area Map 1.1). The altitude ranges from 200m to 1275m (MSL). The terrain is primarily rocky with sleep slopes, ridges and valley. Many streams cut through Saduragiri hills and act as a perennial source of water. It receives rainfall during the South-West as well as the North–East monsoons. The varied climatic and topographic conditions prevailing in this area present a remarkable diversity of both the flora and fauna. The study area consists of tropical evergreen forests, semi evergreen forests, dry teak forests, southern mixed deciduous forests, tropical thorn forest, tropical moist deciduous forest and dry grass land. (Plate 1.1). In the study area, the *Palliyars* live either in several isolated pockets or in small hamlets.

Survey and collection of medicinal plants used by the *Palliyars* tribe of the Saduragiri Hills, Virudhunagar District, South-Eastern slopes of Western Ghats, Tamil Nadu, were carried out over a period of 24 months (2005-2007). Frequent field trips were undertaken to the areas of study. Information regarding the medicinal plants was gathered by meeting *Palliyars* practicing indigenous medicine, during explorative field trips and by gaining a good rapport and winning over their confidence. Most of the information include in this study was gathered from the elderly and experienced medicine men who have a long acquaintance with the use of medicinal plants. The field notebook delineates all the usage procedures adopted by the tribals. The information thus gathered was cross checked adequately for reliability and accuracy by interacting with different groups of the *Palliyars* from different habitats to confirm the use, mode of administration as well as dosage differences, if any. After eliciting detailed information regarding the wild medicinal plants (Table 1.1) and any plant part(s)/extraction of plant part obtained from local market used in medicinal preparations, as ingredients (Table 1.2), they were carefully brought to the laboratory for identification. Herbaria for all the collected plant specimens except for plants, whose parts or extracts were procured from the local market, were prepared (VOCB No. from 3921 to 4085) and deposited in the Ethnopharmacology Unit, Research Department of Botany, V.O. Chidambaram College, Tuticorin, Tamil Nadu, India.

The collected plants were identified by referring to the following compilations (Fischer and Gamble, 1957; Gamble, 1957a; Gamble, 1957b; Henry *et al.*, 1989; Henry *et al.*, 1987; Nair and Henry, 1983)

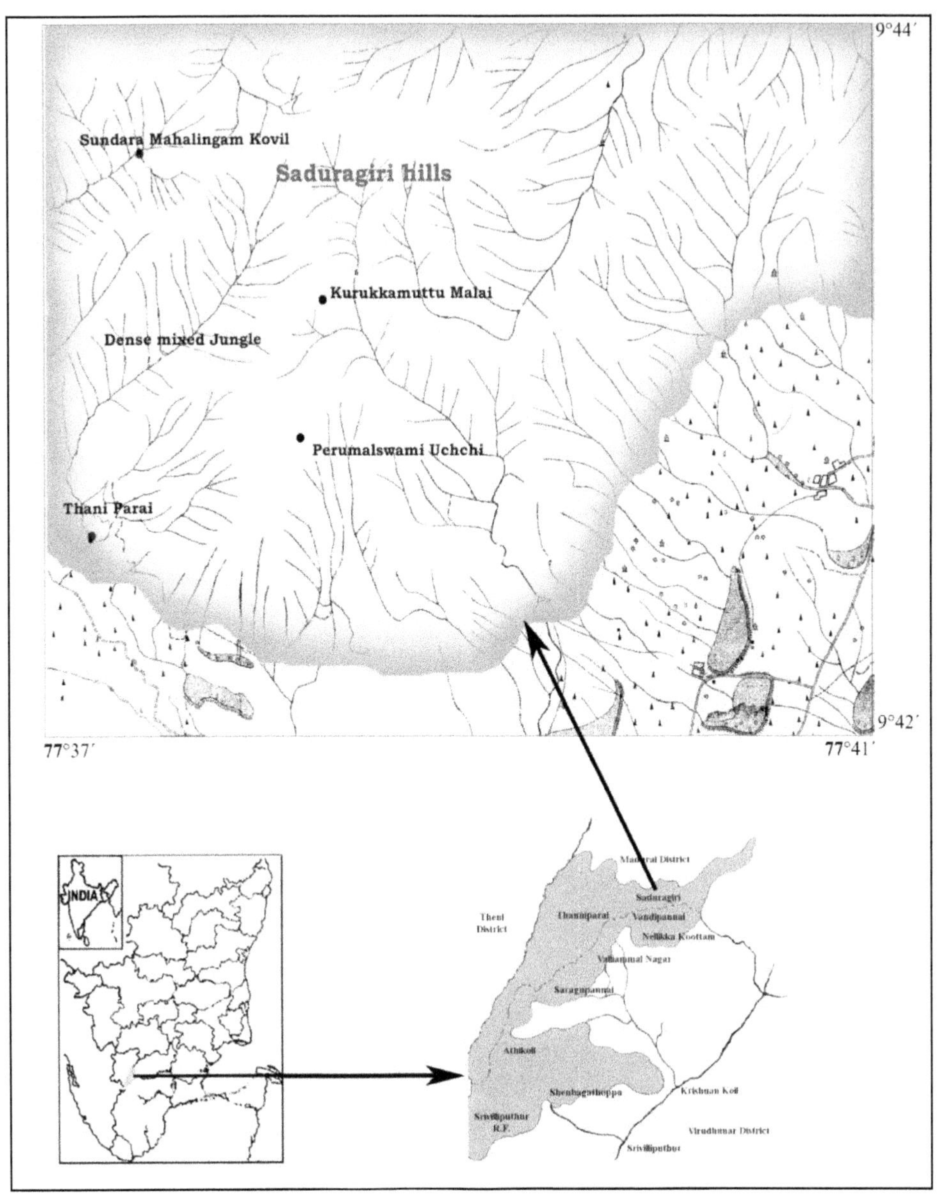

Map 1.1: Area Map.

Plate 1.1: Area Surveyed.

A *Palliyar* huts in Saduragiri hills

A *Palliyar* families

A tribal holding a meidicinal plant

Plate 1.2

Table 1.1: List of Ethnomedicinal Plants Collected and Documented

Sl.No.	Herbarium No.	Botanical Name	Family	Vernacular Name	Habit	Habitat
1	VOCB 3921	*Abrus precatorius* L.	Fabaceae	Kunnimuthu	Climber	Open Wasteland, on fences and bushes
2	VOCB 3922	*Acalypha fruticosa* Forssk	Euphorbiaceae	Sirusinni	Shrub	Dry wasteland, scrub forest
3	VOCB 3923	*Acalypha indica* L.	Euphorbiaceae	Kuppaimaeni	Herb	Wasteland, path sides
4	VOCB 3924	*Achyranthes aspera L.* var.*aspera.*	Amaranthaceae	Nayurivi	Herb	Dry wasteland, path sides
5	VOCB 3925	*Acorus calamus* L.	Araceae	Vasambu	Perennial herb	Semi aquatic places
6	VOCB 3926	*Aegle marmelos* (L) Correa (*Crateva marmelos* L.)	Rutaceae	Vilvam	Thorny tree	Deciduous forest
7	VOCB 3927	*Allium cepa* L.	Alliaceae	Vengayam	Biennial herb	Cultivated in plains
8	VOCB 3928	*Aloe vera* (L.) Burm.f. (*A.perfoliata* L. var. *vera* L.)	Liliaceae	Chotthukathalai	Herb	Dry places
9	VOCB 3929	*Alstonia scholaris* (L.) R. Br. (*Echites scholaris* L.)	Apocynaceae	Ezhilai palai	Tree	Evergreen and deciduous forest
10	VOCB 3930	*Amorphophallus sylvaticus* (Roxb.) Kunth (*Synantherias sylvatica* (Roxb.) Schott)	Araceae	Kattukarunai	Herb	Evergreen forest and deciduous forest
11	VOCB 3931	*Andrographis paniculata* (Burm. f.) Wall.ex Nees (*Justicia paniculata* Burm.f.)	Acanthaceae	Siriyanangai	Herb	Dry forest
12	VOCB 3932	*Anisomeles indica* (L.) Kuntze (*A. ovata* R. Br.)	Lamiaceae	Periya thumbai	Herb	Wasteland and road sides in open forest
13	VOCB 3933	*Anisomeles malabarica* (L.) R. Br.ex Sims (*Nepeta malabarica* L.)	Lamiaceae	Periya thumbai	Herb	Wasteland
14	VOCB 3934	*Argyreia pilosa* Arn.	Convolvulaceae	Thettukkadi	Climber	Deciduous forest
15	VOCB 3935	*Aristolochia indica* L. (*A. lanceolata* Wight)	Aristolochiaceae	Thalaisuruli vaer	Climber	In moist deciduous forest, bushy places
16	VOCB 3936	*Asclepias curassavica* L.	Asclepiadaceae	Kuruthipoo	Herb	Waste land

Contd...

Sl.No.	Herbarium No.	Botanical Name	Family	Vernacular Name	Habit	Habitat
17	VOCB 3937	*Asparagus racemosus* Willd	Asparagaceae	Thanneervittam Kizhangu	Climber	Deciduous forest
18	VOCB 3938	*Azadirachta indica* A. Juss. (*Melia azadirachta* L.)	Meliaceae	Vembu	Tree	Deciduous forest
19	VOCB 3939	*Barleria acuminata* Nees (*Barleria tomentosa* Roth)	Acanthaceae	Vellai Kurinchi	Shrub	Dry forest
20	VOCB 3940	*Bauhinia tomentosa* L.	Caesalpiniaceae	kanchini	Shrub	Dry forest
21	VOCB 3941	*Begonia malabarica* Lam. (*B. fallax* A.DC.)	Begoniaceae	Narayana sanjivi	Herb	Moist deciduous forest, stream sides
22	VOCB 3942	*Bidens pilosa* L. var. *minor* (Blume) Sherff	Asteraceae	Kattu karisalankanni	Herb	Moist shady places
23	VOCB 3943	*Bischofia javanica* Blume (*Microelus roeperianus* Wight and Arn)	Bischofiaceae	Omaviruchu	Tree	Evergreen forest
24	VOCB 3944	*Blepharis maderaspatensis* (L.) Heyne ex Roth (*B. boerhaviaefolia* Pers.)	Acanthaceae	Sadhaiotti	Herb	Dry shady places and path sides
25	VOCB 3945	*Breynia vitis-idaea* (Burm.f.) Fischer (*Breynia rhamnoides* (Retz.) Muell-Arg)	Euphorbiaceae	Kattuniruri	Shrub	Deceduous forest
26	VOCB 3946	*Bupleurum wightii* Mukh.Var.*ramosissimum* (Wight and Arn.) chandrabose comb. nov. (*B. mucronatum* Wight and Arn. Var. *ramosissimum* (Wight and Arn.) Clarke)	Umbelliferae (nom.alter. Apiaceae)	Kattu seeragam	Herb	Moist hill slopes at high altitudes
27	VOCB 3947	*Calotropis gigantea* (L.) R.Br (*Asclepias gigantea* L.)	Asclepiadaceae	Erukku	shrub	Dry land and path sides
28	VOCB 3948	*Canarium strictum* Roxb.	Burseraceae	Kungiliam	Tree	Moist evergreen forest
29	VOCB 3949	*Canthium dicoccum* (Gaertn) Teijsm and Binn	Rubiaceae	Nakkani	Tree	Dry places
30	VOCB 3950	*Capparis sepiaria* L.	Capparaceae	Muruvilikodi	Shrub	Dry forest and Scrub forest.
31	VOCB 3951	*Capsicum annum* L.	Solanaceae	Periya Usimilagai	Herb	Evergreen forest

Sl.No.	Herbarium No.	Botanical Name	Family	Vernacular Name	Habit	Habitat
32	VOCB 3952	*Caralluma adscendens* (Roxb) Haw. var. *attenuata* (Wight) Grav, and Mayuranathan. (*C. attenuata* Wight)	Asclepiadeaceae	sirumankeerai	Fleshy herb	Dry lands and deciduous forest
33	VOCB 3953	*Cardiospermum canescens* Wall	Sapindaceae	Periya Mudakkathan	Climber	Wet place
34	VOCB 3954	*Cardiospermum halicacabum* L.	Sapindaceae	Mudakkathan	Climber	Dry land on fences
35	VOCB 3955	*Carissa carandas* L.	Apocynaceae	Kalakkay	Thorny shrub	Deciduous and Scrub forest.
36	VOCB 3956	*Celtis philippensis* Blanco var. *wightii* (Planch.) Soep. (*Celtis Wightii* Planch)	Ulmaceae	Vakkanai/ Vellaituvarai	Tree	Dry forest
37	VOCB 3957	*Centella asiatica* (L.) Urban (*Hydrocotyle asiatica* L.)	Umbelliferae (nom. alter. Apiaceae)	Vallarai	Prostrate herb	Damb places
38	VOCB 3958	*Chlorophytum heynei* Rottl.ex Baker (*Chlorophytum heyneanum* Wall ex Hook.f.)	Liliaceae	Agathurinji	Herb	Shady areas
39	VOCB 3959	*Cissampelos pareira* L. var. *hirsuta* (Buch. -Ham.ex DC) Forman (*C. pareira* L.)	Menispermaceae	Malaithangi vaer	Climber	Deciduous and scrub forest
40	VOCB 3960	*Cissus quadrangularis* L. (*Vitis quadrangularis* (*L.*) Wall.ex Wight	Vitaceae	Perandai	Climber	Dry scrub forest
41	VOCB 3961	*Cleome viscosa L.*	Cleomaceae	Naaikkadugu	Herb	On path sides
42	VOCB 3962	*Clitoria ternatea L.*	Fabaceae	vellaikakarthan	Climber	On hedges and bushes
43	VOCB 3963	*Cocculus hirsutus* (L.) Diels (*C. villosus* (Lam.) DC.)	Menispermaceae	Vellaikattukodi	Climber	Dry places on scrubs and bushes
44	VOCB 3964	*Cocos nucifera* L.	Arecaceae	Thennai	Tree	Cultivated in plains
45	VOCB 3965	*Commelina benghalensis* L.	Commelinaceae	Amala	Creeping herb	Wet path sides
46	VOCB 3966	*Commelina ensifolia* R. Br. (*C. undulata* var. *setosa* Clarke)	Commelinaceae	Amala	Spreading herb	Wet path sides
47	VOCB 3967	*Commiphora pubescens* (Wight and Arn.) Engler (*Protium pubescens* Wight and Arn.)	Burseraceae	Kodikiluvai	Tree	Dry deciduous forest

Contd...

Sl.No.	Herbarium No.	Botanical Name	Family	Vernacular Name	Habit	Habitat
48	VOCB 3968	*Crotalaria calycina* Schrank	Fabaceae	Nari mirati	Herb	Dry places and bushes.
49	VOCB 3969	*Cryptolepis buchananii* Roem and Schultes	Periplocaceae	Paalkodi/ Karunagatalli	Climber	Deciduous forest
50	VOCB 3970	*Curculigo orchioides* Gaertn. (*C.malabarica* Wight)	Hypoxidaceae	Kuluthupokie/ Nilapanai	Herb	Wet path sides
51	VOCB 3971	*Curcuma longa* L. (*C. domestica* Valeton)	Zingiberaceae	Manjal	Herb	Cultivated in plains
52	VOCB 3972	*Cyanotis tuberosa* (Roxb) Schutles and Schultes (*Tradescantia tuberosa* Roxb.)	Commelinaceae	Palvalikizhangu	Herb	Wetsandy places
53	VOCB 3973	*Cylista scariosa* Roxb.	Fabaceae	Kattumotchi	Climber	Semievergreen forest
54	VOCB 3974	*Cynodon dactylon* (L) Pers (*Panicum dactylon*) L.	Poaceae	Arugampullu	Herb	Wet and dry sandy places
55	VOCB 3975	*Dichrostachys cinerea* (L.) Wight and Arn (*Mimosa cinerea* L.)	Mimosaceae	Vidathalai	Small tree	Dry sandy places
56	VOCB 3976	*Dioscorea bulbifera* L. var. *vera* Prain and Burkill	Dioscoreaceae	Vethalaivalli/ Karuvalli	Climber	At higher elevations in evergreen forest
57	VOCB 3977	*Dioscorea oppositifolia* L. var. *dukhumensis* Prain and Burkill	Dioscoreaceae	Vethalaivalli	Climber	Deciduous forest on bushes
58	VOCB 3978	*Dioscorea pentaphylla* L. var.*pentaphylla* (*D. pentaphylla* L. var. *linnaei* Prain and Burkill)	Dioscoreaceae	Mullvali	Climber	Deciduous forest on bushes
59	VOCB 3979	*Dioscorea tomentosa* Koen. ex Spreng.	Dioscoreaceae	Noolvalli	Climber	Deciduous forest on bushes
60	VOCB 3980	*Dodonaea viscosa* (L) Jacq. (*Ptelea viscosa* L.)	Sapindaceae	Virali	Shrub	shrub forest
61	VOCB 3981	*Drypetes sepiaria* (Wight and Arn.) Pax and Hoffm. (*Hemicyclia sepiaria* Wight and Arn)	Euphorbiaceae	Kalvirai	Tree	Dry evergreen and semi evergreen forests
62	VOCB 3982	*Eclipta prostrata* (L.) L. (*E. alba* (L.) Hassk)	Asteraceae	Karisalankanni	Herb	Wet places and path sides
63	VOCB 3983	*Ehretia ovalifolia* Wight	Boraginaceae	Madavirisu/ kalvari	Tree	Dry evergreen and semi evergreen forests

Sl.No.	Herbarium No.	Botanical Name	Family	Vernacular Name	Habit	Habitat
64	VOCB 3984	*Enicostema axillare* (Lam.) Raynal. (E. littorale auct. non Blume)	Gentianaceae	Vellaragu	Herb	Open waste land
65	VOCB 3985	*Ensete superbum* (Roxb) Cheesman (*Musa superba* Roxb)	Musaceae	Malaivazhai	Tree	Rocky hill sides at high elevations
66	VOCB 3986	*Euphorbia hirta* L. (E. pilulifera sensu Hook. f)	Euphorbiaceae	Amampatcharisi	Herb	Sandy path sides
67	VOCB 3987	*Euphorbia rosea* Retz. (*Chamaesyce rosea* (Retz) Webster)	Euphorbiaceae	Amampatcharisi	Herb	Sandy path sides
68	VOCB 3988	*Evolvulus alsinoides* (L.) L. (*Convolvulus alsinoides* L.)	Convolvulaceae	Vishnukarandi	Herb	Wet and dry path sides and grassy places
69	VOCB 3989	*Ficus benghalensis* L. var. *benghalensis* (*Urostigma benghalense* (L.) Gasp)	Moraceae	Aal	Tree	Deciduous and semi evergreen forests
70	VOCB 3990	*Ficus microcarpa* L. f. (*F. retusa* auct.non L.)	Moraceae	Punniyavirusu	Tree	Deciduous forest and near streams
71	VOCB 3991	*Ficus racemosa* L. (*F. glomerata* Roxb.)	Moraceae	Atthi	Tree	Deciduous forest and near stream
72	VOCB 3992	*Ficus religiosa* L.	Moraceae	Atthi	Tree	Planted in plains
73	VOCB 3993	*Flacourtia indica* (Burm.f.)Merr. (*F. sepiaria* Roxb.)	Flacourtiaceae	Mullumayilai	Shrub	Scrub forest and rocky hill slopes
74	VOCB 3994	*Gardenia resinifera* Roth (*G. lucida* Roxb.)	Rubiaceae	Kattu koiya/ Vetchi	Small tree	Deciduous forest
75	VOCB 3995	*Givotia rottleriformis* Griff.	Euphorbiaceae	Vandalai	Tree	Low altitude grassland and in dry deciduous forest
76	VOCB 3996	*Globba marantina* L. (*G. bulbifera* Roxb.)	Zingiberaceae	Kattumanjal	Herb	Moist grassland and in dry deciduous forest
77	VOCB 3997	*Gloriosa superba* L.	Liliaceae	Karthika kilangu	Herb	Moist grassy hill slopes at high elevations
78	VOCB 3998	*Gymnema sylvestre* (Retz.) R. Br.ex Schultes (*Periploca sylvestris* Retz.)	Asclepiadaceae	Shirukurinjan	Climber	Dry forest on bushes

Contd...

Sl.No.	Herbarium No.	Botanical Name	Family	Vernacular Name	Habit	Habitat
79	VOCB 3999	*Hedyotis puberula* (G.Don)Arn. (*Oldenlandia umbellata* L.)	Rubiaceae	Emburel	Herb	Dry places
80	VOCB 4000	*Hemidesmus indicus* (L.) R. Br. *var.indicus* (*Periploca indica* L.)	Periplocaceae	Nannari	Climber	Open forest on bushes
81	VOCB 4001	*Hibiscus vitifolius* L.	Malvaceae	Periyathutthi	Herb	Path sides
82	VOCB 4002	*Hugonia mystax* L.	Linaceae	Sunguthan Kodi	Shrub	Dry deciduous forest
83	VOCB 4003	*Hybanthus enneaspermus* (L.) F.V.Muell. (*Ionidium suffruticosum* (L.) Ging)	Violaceae	Orithalthamarai	Herb	Path sides and open forest lands
84	VOCB 4004	*Ichnocarpus frutescens* (L.) R.Br. (*Apocynum frutescens* L.)	Apocynaceae	Palvalli	Climber	Lower hills and Hedges and Scrub forest.
85	VOCB 4005	*Indigofera wightii* Graham ex. Wight and Arn.	Fabaceae	Parisa vayittruvalivaer	Pubescent shrub	Grasslands
86	VOCB 4006	*Ipomoea barlerioides* (Choisy) Benth.ex.Clarke (*Aniseia barlerioides* Choisy)	Convolvulaceae	Thoolikkodi	Climber	Grassy places
87	VOCB 4007	*Ipomoea staphylina* Roem and Schultes (*I. racemosa* Roth)	Convolvulaceae	Onaankodi	Climber	On bushes near streams
88	VOCB 4008	*Jasminum roxburghianum* Wall.	Oleaceae	Kattumalli	Climber	Deciduous forest
89	VOCB 4009	*Jatropha glandulifera* Roxb	Euphorbiaceae	Vellaiadalai	Shrub	Open places, waste land
90	VOCB 4010	*Jatropha curcas* L.	Euphorbiaceae	Kattamanaku	Small tree	Open places
91	VOCB 4011	*Jatropha gossypifolia* L.	Euphorbiaceae	Sivappuatalai	Shrub	Dry waste land
92	VOCB 4012	*Justicia adhatoda* L. (*Adhatoda vasica* Nees)	Acanthaceae	Adatoda	Shrub	Open waste land
93	VOCB 4013	*Justicia glauca* Rottl (*Gendarussa tranquebariensis* sensu Wight)	Acanthaceae	Neernotchi/ Kattukurinchi	Herb	Dry forest
94	VOCB 4014	*Kalanchoe pinnata* (Lam.) *Pers.* (*Bryophllum pinnatum* (Lam) Oken)	Crassulaceae	Megasanjeevi	Succulent herb	Open land among bushes
95	VOCB 4015	*Kalanchoe tubiflora* Hamet	Crassulaceae	Sevaka chanjivi	Succulent herb	Dry Waste places.

Sl.No.	Herbarium No.	Botanical Name	Family	Vernacular Name	Habit	Habitat
96	VOCB 4016	*Lantana camara* L.var.*aculeata* (L.) Mold. (*L. aculeata* L.)	Verbenaceae	Unnipoo	Shrub	Scrub forest and path sides
97	VOCB 4017	*Lantana wightiana* Wall. ex Gamble (*L. indica* Roxb var. *albiflora* Wight ex Clarke)	Verbenaceae	Muraikai-chalchedi	Shrub	Scrub forest
98	VOCB 4018	*Lawaonia inermis* L. (L. alba Lam.)	Lythraceae	Marudondi	Shrub	Low plains
99	VOCB 4019	*Leucas biflora* (Vahl) R. Br. var. *biflora* (*Phlomis biflora* Vahl)	Lamiaceae	Siru thumbai	Procumbent herb	Wet path at low elevations
100	VOCB 4020	*Lindernia crustacea* (L.) F.V. Muell.	Scrophulariaceae	Karumsurai-murkontrai/ Sengaththari/ Kodiyanangai	Herb	Wet places
101	VOCB 4021	*Lobelia nicotianifolia* Roth exSchultes var. *nicotianifolia*	Lobeliaceae	Kattupugaiyilai	Herb	Hill slopes at higher elevations
102	VOCB 4022	*Mangifera indica* L.	Anacardiaceae	Maa	Tree	Cultivated in plains
103	VOCB 4023	*Mimosa pudica* L.	Mimosaceae	Thotallsimungi	Herb	Wet and dry waste places, path sides
104	VOCB 4024	*Mimusops elengi* L.	Sapotaceae	Mahilampoo	Tree	Cultivated in plains dry and moist evergreen forest
105	VOCB 4025	*Mitragyna parvifolia* (Roxb.) Korth (*Stephegyne parvifolia* (Roxb.) Korth	Rubiaceae	Neerkadambu	Tree	Dry forest
106	VOCB 4026	*Mollugo cerviana* (L.) Ser. (*Pharnaceum cerviana* L.)	Molluginaceae	Pappadai	Herb	Sandy places of path sides
107	VOCB 4027	*Momordica charantia* L. var. *charantia*	Cucurbitaceae	Kuruvithalai paakakai	Creeping herb	On hedges and fences
108	VOCB 4028	*Morinda pubescens* J.E. Smith var. *pubescens* (*M. tinctoria* Roxb.)	Rubiaceae	Manjanatti	Tree	Dry deciduous forest
109	VOCB 4029	*Moringa concanensis* Nimmo ex. Gibs.	Moringaceae	Kattu moringai	Tree	Deciduous forest on hill slopes

Contd...

Sl.No.	Herbarium No.	Botanical Name	Family	Vernacular Name	Habit	Habitat
110	VOCB 4030	*Mucuna atropurpurea* DC.	Fabaceae	Thellukkai	Woody climber	On bushes and trees near streams
111	VOCB 4031	*Mucuna pruriens* (L.) DC. var. *pruriens* L.	Fabaceae	Punaikali	twinner	Dry deciduous forest
112	VOCB 4032	*Mukia maderaspatana* (L.) M. Roem. (*Melothria maderaspatana* (L.)Cogn.)	Cucurbitaceae	Musumusukkai	Climber	On grasses, hedges and bushes
113	VOCB 4033	*Murraya koenigii* (L.) Spreng.	Rutaceae	Kariveppilai	Small tree	Cultivated in plains
114	VOCB 4034	*Ocimum tenuiflorum* L. (*O. sanctum* L.)	Lamiaceae	Tulsi	Herb	Dry open places
115	VOCB 4035	*Orthosiphon thymiflorus* (Roth.) Sleensen (*O.diffusus* (Benth.) Benth.)	Lamiaceae	Kattutulsi/ Vettukaya patchilai	Herb	Dry open places
116	VOCB 4036	*Pavetta indica* L. var. *indica*	Rubiaceae	Pavattanchedi	Tree	Dry open places
117	VOCB 4037	*Pavonia odorata* Willd.	Malvaceae	Thuthi	Herb	Path sides
118	VOCB 4038	*Pergularia daemia* (Forssk.) Chiov. (*P. extensa* (Jacq.) N.E. Br.)	Asclepiadaceae	Velipparuthi	Climber	On hedges and fences
119	VOCB 4039	*Phyllanthus amarus* Schum and Thonn. (*P. niruri* auct. non L.)	Euphorbiaceae	Kilanalli	Herb	Wet and dry open places
120	VOCB 4040	*Phyllanthus emblica* L. (*Emblica officinalis* Gaertn)	Euphorbiaceae	Nelli	Tree	Dry deciduous forest
121	VOCB 4041	*Phyllanthus maderaspatensis* L.	Euphorbiaceae	Kilanalli	Herb	Dry open places and path sides
122	VOCB 4042	*Phyllanthus reticulatus* Poir (*Kirganelia reticulata* (Poir) Baill)	Euphorbiaceae	Karumpoo-lanchedi	Shrub	Sandy places
123	VOCB 4043	*Phyllanthus urinaria* L. (*Phyllanthus leprocarpus* Wight)	Euphorbiaceae	Kilanalli	Herb	Sandy places
124	VOCB 4044	*Physalis minima* L. *P. minima* L. var *indica* Clarke	Solanaceae	Toppipalam	Herb	Waste place and roadsides
125	VOCB 4045	*Piper argyrophyllum* Miq.	Piperaceae	Kattumilagu	Climber	Evergreen forest

Sl.No.	Herbarium No.	Botanical Name	Family	Vernacular Name	Habit	Habitat
126	VOCB 4046	*Pisonia aculeata* L.	Nyctaginaceae	Murukkalli	Thorny Climbing Shrub	Dry forest
127	VOCB 4047	*Plectranthus amboinicus* (Lour.) Spreng. (*Coleus amboinicus* Lour.)	Lamiaceae	Omavalli	Herb	Moist places among bushes
128	VOCB 4048	*Plectranthus barbatus* Andr. (*Coleus barbatus* (Andr.)Benth.)	Lamiaceae	Omavalli	Herb	Open forest on bushes
129	VOCB 4049	*Polygala javana* DC. var. *angustifolia* Thw. (*P. raoi* R.N. Ban. and L.K. Ban.)	Polygalaceae	Palpiranthai	Herb	Dry open land
130	VOCB 4050	*Pongamia pinnata* (L.) Pierre (*P. glabra* Vent.)	Fabaceae	Pongam	Tree	Near by streams and open land
131	VOCB 4051	*Priva cordifolia* (L.f.) Druce (*P. leptostachya* Juss.)	Verbenaceae	Aadai otti	Sticky herb	On moist rocks and path sides
132	VOCB 4052	*Pterocarpus marsupium* Roxb.	Fabaceae	Vengai	Tree	Deciduous forest
133	VOCB 4053	*Rhinacanthus nasutus* (L.) Kurz var. *nasutus* (*R.communis* Nees)	Acanthaceae	Nagamalli	Shrub	Path sides
134	VOCB 4054	*Sansevieria roxburghiana* Schultes and Schultes (*S.zeylanica* Roxb.)	Agavaceae	Maurl	Fleshy herb	Dry wasteland and among bushes
135	VOCB 4055	*Sarcostemma acidum* (Roxb.) Voigt (*S. brevistigma* Wight and Arn.)	Asclepiadaceae	Kodikkalli	Shrub	Bushes and rocks
136	VOCB 4056	*Scoparia dulcis* L.	Scrophulariaceae	Neer nangai	Herb	Wasteland, path sides at low elevations
137	VOCB 4057	*Senna auriculata* (L.) Roxb. (*Cassia auriculata* L.)	Caesalpiniaceae	Aavarai	Shrub	Wasteland, path sides at low elevations
138	VOCB 4058	*Sesamum indicum* L. (*S. orientale* L.)	Pedaliaceae	Kattu yellu	Herb	Path sides and open forest lands
139	VOCB 4059	*Sesbania sesban* (L.)Merr. (*S. aegyptiaca* (Poir.) Pers	Fabaceae	Chittakatti	Shrub	Waste places

Contd...

Sl.No.	Herbarium No.	Botanical Name	Family	Vernacular Name	Habit	Habitat
140	VOCB 4060	*Sida cordata* (Burm. f.) Borssum (*S. veronicifolia* Lam.)	Malvaceae	Palampasi/ Nilathuti	Herb	Open waste land
141	VOCB 4061	*Sida rhombifolia* L. var.*rhomboidea* (*S. rhomboidea* Roxb. ex Fleming)	Malvaceae	Kurunthatti	Herb	Scrub forest and dry open land
142	VOCB 4062	*Solanum surattense.* Burm. F (*S. xanthocarpurn* Schrader and Wendl.)	Solanaceae	Kandangattiri	Herb	Dry open land, path sides, low hills
143	VOCB 4063	*Solanum trilobatum* L.	Solanaceae	Tuduvalai	Prickly climber	On hedges and bushes.
144	VOCB 4064	*Solanum torvum* Sw	Solanaceae	Sundaikkai	Shrub	Open forest on bushes
145	VOCB 4065	*Sterculia urens* Roxb.	Sterculiaceae	Vennaali/ Senthanakku	Tree	Dry deciduous forest on rocky hills
146	VOCB 4066	*Strychnos nux - vomica* L.	Loganiaceae	Kanjirai	Tree	Deciduous forest
147	VOCB 4067	*Swertia corymbosa* (Griseb.) Wight ex Clarke var. *corymbosa* (*Ophelia corymbosa* Griseb.)	Gentianaceae	Milakainangai/ Malai Siriyanangai	Herb	Grassland at higher elevations
148	VOCB 4068	*Syzygium cumini* (L.) Skeels (*S. jambolanam* (Lam.)DC.)	Myrtaceae	Naval	Tree	Deciduous forest evergreen forest and banks of steams
149	VOCB 4069	*Tabernaemontana divaricata* (L.) R.Br.ex Roem and Schultes	Apocynaceae	Nandiyavattai	Shrub	Open forest on bushes
150	VOCB 4070	*Tamarindus indica* L.	Caesalpiniaceae	Puli	Tree	Open land among bushes
151	VOCB 4071	*Tephrosia purpurea* (L.) Pers. (*T. hamiltonii* Drumm.)	Fabaceae	Kolingi	Herb	Path sides
152	VOCB 4072	*Terminalia bellirica* (*Gaertn.*) Roxb. (*Myrobalanus bellirica* Gaertn.)	Combretaceae	Tani	Tree	Deciduous forest
153	VOCB 4073	*Terminalia chebula* Retz. *Myrobalanus chebula* (Retz.) Gaertn.)	Combretaceae	Kadukai	Tree	Deciduous forest
154	VOCB 4074	*Thespesia populnea* (L.) Soland. ex Correa (*Hibiscus populneus* L.)	Malvaceae	Puvarasu	Tree	Open land among bushes
155	VOCB 4075	*Tinospora cordifolia* (Willd.)Miers ex Hook.f. and Thoms (*Cocculus cordifolius* (Willd)DC.)	Menispermaceae	Chintil/ EnthalKodi	Climber	Dry forest and bushes

Sl.No.	*Herbarium No.*	*Botanical Name*	*Family*	*Vernacular Name*	*Habit*	*Habitat*
156	VOCB 4076	*Trianthema decandra* L.	Aizoaceae	Mookana-charanai	Herb	Fully exposed dry places and hill slopes
157	VOCB 4077	*Tribulus terrestris* L. (*T. lanuginosus* L.)	Zygophyllaceae	Sirunerinji	Herb	Wasteland
158	VOCB 4078	*Trichodesma zeylanicum* (Burm.f.) R.Br. (*Borago zeylanica* Burm.f.)	Boraginaceae	Kavuthumbai	Annual herb	Path sides and open wasteland
159	VOCB 4079	*Tridax procumbens* L.	Asteraceae	Mookuthi elai	Herb	Path sides and Wet place
160	VOCB 4080	*Tylophora indica* (Burm.f.) Merr) (*T. asthmatica* (L.f) Wight and Arn)	Asclepiadaceae	Nanchianuppan palai	Climber	Open forest
161	VOCB 4081	*Vetiveria zizanioides* (L.) Nash	Poaceae	Vetiver	Herb	Along the margins of streams
162	VOCB 4082	*Vigna mungo* (L.) Hepper	Fabaceae	Ulundhu	Herb	Cultivated plains
163	VOCB 4083	*Vitex negundo* L.	Verbenaceae	Vennochi	Shrub	Path sides in plains
164	VOCB 4084	*Wrightia tinctoria* (Roxb.) R.Br. var. *tinctoria* (*Nerium tinctorium* Roxb.)	Apocynaceae	Vetpalai	Tree	Deciduous forest
165	VOCB 4085	*Ziziphus xylopyrus* (Retz.) Willd. (*Rhamnus xylopyrus* Retz.)	Rhamnaceae	Mullukottai	Small tree	Dry deciduous forest

Table 1.2: Botanical Names of Plants whose Parts/Products are Purchased from the Market and Used as Medicine/Ingredients in the Native Medicine Preparations by the *Palliyar* Tribe

Sl.No.	*Botanical Name*	*Family*	*Vernacular Name*	*Plant Part/Extract of Plant Part Procured*
1	*Allium sativum* L.	Allliaceae	Vellai poondu	Bulb
2	*Alpinia calcarata* Roscoe	Zingiberaceae	Chitrattai	Rhizome
3	*Areca catechu* L.	Arecaceae	Pakku	Fruit
4	*Borassus flabellifer* L. (*B.flabelliformis* Murr.)	Recaceae	Panai	Palm candy, Palm jaggery
5	*Cuminum cyminum* L.	Umbelliferae (nom. Alter. Apiaceae)	Shiragam	Fruit
6	*Ferula assafoetida* L.	Umbelliferae (nom.alter.Apiaceae)	Kayam	Root resin
7	*Foeniculum vulgare* Mill.	Umbelliferae (nom.alter.Apiaceae)	Sombu	Fruit
8	*Oryza sativa* L.	Poaceae	Nellu	Grains
9	*Papaver somniferum* L.	Papaveraceae	Kasakasa	Seed
10	*Piper betle* L. (*Chavica betle* (L.) Miq.)	Piperaceae	Vettilai	Leaf
11	*Piper Longum* L. (*Chavica roxburghii* Miq.)	Piperaceae	Tippili	Root
12	*Piper nigrum* L.	Piperaceae	Milagu	Fruit
13	*Ricinus communis* L.	Euphorbiaceae	Amanakku	Castor oil
14	*Trachyspermum ammi* (L.) Sprague (*Carum copticum* Benth.ex Hiern)	Umbelliferae (nom. Alter. Apiaceae)	Omum	Fruit
15	*Zingiber officinale* Roscoe	Zingiberaceae	Sukku	Dried rhizome

Enumeration

Plants used by the *Palliyars* for the treatment of various ailments are enumerated as follows:

Abrus precatorius L.

Two to three grams of fresh leaves or roots of the above plant with white seeds are made into paste and consumed along with cold water or cow's milk two times a day for five to seven days to cure any poisonous bite.

Two to three grams of fresh leaves are chewed carefully without swallowing the saliva for one or two minutes. Then the chewed leaves are spat out and the mouth is rinsed well with water twice or thrice. This procedure is repeated for two times a day for a period of three to four days to get relief from great sensitiveness of teeth (Odonthemedia).

A drop of pure filtered seed extract prepared with clean water is allowed to settle on the eyeballs twice a day for about three days to reduce irritation.

Acalypha fruticosa Forssk. (Plate 1.3, 1)

Four or five fresh leaves are made into paste with a few drops of water and the paste is immediately applied on the spot only once for honeybee-sting.

Two to three grams of fresh leaves are made into juice with hundred ml of water. The filtered juice is taken orally two times a day for one or two days to get relief from headache.

A handful of fresh leaves is made into paste and dissolved in a bucket of lukewarm water. Babies below three years old are bathed with this water to relieve from rickets.

Acalypha indica L

Five grams fresh leaves are made into juice. Small amount of calcium hydroxide is mixed with the juice. This mixture is applied externally on the throat twice a day for five days to get relief from cough.

Achyranthes aspera L. var. *aspera*

A handful of fresh leaves are made into paste with little water. This paste is mixed with a pinch of lime and is applied externally on the spot once in a day for three or four days for treating dog-bite.

Acorus calamus L.

A thread is inserted through a hole made in the dried rhizome and is worn as a necklace to ward off giddiness.

Aegle marmelos (L.) Correa

Five to ten grams of leaves are made into paste with a few drops of water. This paste is applied externally on the affected skin twice a day for a period of two to three days to get relief from itches.

Acalypha fruticosa **Forssk**

Andrographis paniculata **(Burm. f.) Wall.ex Nees**

Aristolochia indica **L.**

Barleria acuminata **Nees**

Breynia vitis-idaea **(Burm.f.) Fischer**

Canthium dicoccum (Gaertn) Teijsm & Binn

Plate 1.3

***Aloe vera* (L.) Burm.f.**

A fresh leaf is taken orally as such after removing the epidermal peel, once in day for a period of three to four days to reduce the body heat.

One fresh leaf is taken per day for about ten days after removing the epidermal peel to cure kidney stones.

***Alstonia scholaris* (L.) R. Br.**

Ten grams of the fresh stem bark is made into paste with a few drops of water. This paste is mixed with hundred ml of cow's milk or rice-fermented water and taken by the feeding mother twice a day for seven to ten days to improve lactation.

One teaspoon of the powder made from the shade dried leaves is orally administered in water or along with a few drops of the stem latex of the same plant, once in a day for a period of one month to get relief from asthma.

***Amorphophallus sylvaticus* (Roxb.) Kunth**

One hundred grams of the tuber is boiled in water with an equal amount of fresh leaves of tamarind (*Tamarindus indica* L.). Then, the peeled tuber is cooked with tamarind. The prepared curry is taken as medicine one time a day for seven to ten days to get relief from bleeding piles.

***Andrographis paniculata* (Burm. f.) Wall.ex Nees (Plate 1.3, 2)**

One teaspoon of fresh juice or one to two grams of the shade dried plant powder is taken twice a day for seven days to treat snake-bite and scorpion-sting.

***Anisomeles indica* (L.) Kuntze**

Twenty to twenty five grams of fresh leaves are boiled in two litres of water in a closed container. The lid of the container is slowly removed and the steam is inhaled (vapour bath) to get relief from headache.

***Anisomeles malabarica* (L.) R. Br. ex Sims**

To get relief from headache twenty to twenty five grams of fresh leaves are boiled along with two liter of water in closed container; the lid of the container is slowly removed and vapour bath is administered.

***Argyreia pilosa* Arn.**

Fifty grams of peeled tubers are roasted or boiled and taken orally two times a day for two days to treat gastric disorders.

***Aristolochia indica* L. (Plate 1.3, 3)**

Five grams of fresh root is made into paste with a few drops of water and the paste is applied externally on the spot or two to three grams of shade dried root powder is taken orally with hundred ml of cold water two times a day for five to seven days to treat snake-bite and scorpion-sting.

Five to ten grams of fresh root is ground into a paste with a little water and the paste is applied on the forehead to get relief from headache.

Juice is prepared from ten to fifteen grams of fresh root with hot water and is taken orally in empty stomach for three days in a single dose to get relief from stomachache.

Asclepias curassavica L.

Ten to twenty grams of shade dried leaf powder mixed with water is taken orally once in a day for four days to treat skin diseases.

Decoctions of root is taken orally once in a day for two days to cure constipation.

Fresh root paste along with the water is taken orally once in a day for four to six days to cure piles.

Asparagus racemosus Willd.

Five to ten grams of the peeled, raw or boiled tuber is given to children in empty stomach in the morning for about one week to stop bed-wetting.

Azadirachta indica A. Juss.

Twenty to thirty grams of fresh leaves are boiled in two litres of water in a closed container. The lid of the container is removed slowly and the steam is inhaled once a day for four days to get relief from cold.

A handful of fresh stem bark is continuously boiled in two hundred ml of water to get fifty ml decoction. This decoction is taken orally once in a day for four days to control fever.

Barleria acuminata Nees (Plate 1.3, 4)

About thirty gram of fresh leaves are made into paste with a few drops of water. This paste is applied externally on the wounds twice a day for a period of five to seven days for healing the wounds.

Bauhinia tomentosa L.

Eight to ten leaves are soaked in two hundred ml of cool water for two to three hours. The leaves are removed and the water is taken orally twice a day for a period of four to five days to reduce the body heat.

Begonia malabarica Lam.

Five or six fresh leaves or five to ten grams of powder prepared from the shade dried aerial part of the plant is taken as an astringent twice a day for one week.

Bidens pilosa L. var. minor (Blume) Sherff

A handful of fresh leaves are ground along with equal quantity of aerial plant part of kilanelli (*Phyllanthus amarus Schum. and Thonn.*) to get a juice. This juice is taken orally as such or along with cow's milk twice a day for a period of seven days to treat jaundice.

Bischofia javanica Blume

Fifty to hundred grams of fresh stem bank is boiled in one litre of coconut oil along with equal amount of fresh leaves of virali. (*Dodonea viscosa* (L.) Jacq) for twenty minutes. The oil is cooled, filtered and the filtrates is used to massage the affected area an hour before taking bath in hot water.This procedure is followed once in a day for one week to get relief from body pain.

Blepharis maderaspatensis (L.) Heyne ex Roth

Forty grams of fresh leaves are made into paste with a few drops of water and it is applied on the wound caused by weapons twice a day for four to five days for healing.

Leaf paste is mixed with powdered black gram (*Vigna mungo* (L.) Hepper), crushed onion (*Allium cepa* L) and white yolk of one egg and the mixture is applied topically over the fractured bones.

Breynia vitis-idaea (Burm.f.) Fischer (Plate 1.3, 5)

Juice is prepared from ten grams of the fresh leaves of the plant is taken orally along with water twice a day for about seven to ten days to treat jaundice.

Bupleurum wightii Mukh. var. _ramosissimun_ (Wight and Arn.) Chandrabose comb. nov.

Five grams of dried fruit is soaked in hundred ml of water for two hours. The water is filtered and the filtrate is taken once in a day for a period of three to four days to reduce the body heat.

Calotropis gigantea (L.) R. Br.

One or two ml of stem latex is mixed with a small quantity of calcium hydroxide. This mixture is applied externally on the throat twice a day for a period of four days to get relief from cold and cough.

Canarium strictum Roxb.

Five grams of powdered stem resin is burnt and the smoke is inhaled (vapour bath) at time of headache. This practice relieves cold when it is administered for two times a day for three to four days.

Canthium dicoccum (Gaertn) Teijsm and Binn. (Plate 1.3, 6)

One teaspoon of the shade dried leaf powder or fruit powder is added to hundred ml of water and taken orally three times a day for a period of two days to get relief from stomach disorders.

Capparis sepiaria L.

Ten grams of the fresh bark of the stem is made into paste with lemon juice and it is applied externally once in a day for a period of one week to treat eczema and dandruff.

The paste prepared from twenty grams of fresh leaves is taken orally along with water two times a day for a period of two days to reduce the body heat.

Capsicum annum L.

Fresh fruits are chewed and the juice is swalled three times a day for two days to get relief from earache.

Caralluma adscendens **(Roxb.) Haw. var.** ***attenuata*** **(Wight) Grav. and Mayuranathan (Plate 1.4, 1)**

Ten grams of clean and fresh rootless plant is taken as refrigerant two times a day for a period of three days to reduce the body heat.

Cardiospermum canescens **Wall (Plate 1.4, 2)**

The paste prepared from ten gram of leaves with water is taken orally along with cow's milk for a period of three days in a single dose a day to reduce bile problems.

Cardiospermum halicacabum **L.**

Two hundred and fifty grams of fresh leaves and five hundred grams of rice are ground well. Required quantity of common salt is added to this paste and a foodstuff called dosai made from this paste is eaten for two to three days to get relief from rheumatic pain.

Five grams of fresh leaves, one gram of jeera (*Cuminum cyminum* L.), five to six leaves of tuduvalai (*Solanum trilobatum* L.), two to three leaves of kandangattiri (*S. surattense* Burm.f.) and hundred mg of asafoetida (*Ferula assafoetida* L.) are continuously boiled in three hundred ml of water to get seventy five ml decoction. This decoction is filtered and taken orally once in a day in the morning for four to five days to relieve cold and cough.

Two or three grams of fresh leaves are continuously boiled along with two to three grams of fresh leaves of tuduvalai (*Solanum trilobatum* L.) and kandangattiri (*S. surattense Burm.* f.), a pinch of asafoetida (*Ferula assafoetida* L.) and one gram of jeera (*Cuminum Cyminum* L.) in three hundred ml of water to get fifty ml decoction. The filtered decoction is taken in empty stomach once in a day for a period of two days to get relief from giddiness.

Carissa carandas **L. (Plate 1.4, 3)**

Fresh fruits are taken, as a coolant twice a day for a period of three days to reduce the body heat.

Celtis philippensis **Blanco var.** ***wightii*** **(Planch) Soep. (Plate 1.4, 4)**

One teaspoon of powder prepared from the shade dried stem is mixed with few drops of water. This paste is applied externally on the wounds twice a day for a period of seven days for healing the wounds.

Centella asiatica **(L.) Urban**

Five to ten grams of fresh leaves or one teaspoon of the shade dried leaf powder is taken along with water or cow's milk two times a day for a week to relieve chest pain, cold and cough.

Caralluma adscendens **(Roxb) Haw. var. attenuata (Wight) Grav, & Mayuranathan.**

Cardiospermum canescens **Wall**

Carissa carandas **L.**

Celtis philippensis **Blanco *var. wightii*** **(Planch.) Soep.**

Chlorophytum heynei **Rottl.ex Baker**

Cissus quadrangularis **L.**

Plate 1.4

Chlorophytum heynei Rottl. ex Baker (Plate 1.4, 5)

Dried root powdered mixed with milk or water is taken orally daily in the morning for a month as a health tonic.

Cissampelos pareira L. var. _hirsuta_ (Buch – Ham. ex DC.) Forman

One teaspoon of shade dried root powder is taken orally in water two times a day for a week for any poisonous bite.

Ten to fifteen grams of fresh root paste is taken along with hundred ml of hot water in empty stomach for three days in a single dose to relieve stomachache.

Cissus quadrangularis L. (Plate 1.4, 6)

One teaspoon of juice prepared from four to five fresh leaves is given to children with a pinch of asafoetida (*Ferula assafoetida* L.) in a single dose at bed time for two days to get relief from tumour like swelling due to wind in the naval region.

Cleome viscosa L.

Five to ten grams of the seeds are soaked in hundred ml of water for ten to twelve hours. The seeds are removed and the water is taken orally or same quantity of seed is made into paste with little water and is taken in empty stomach in the early morning hours for a week as a post natal therapy to arrest over bleeding.

Clitoria ternatea L. (Plate 1.5, 1)

One teaspoon of the powder made from the shade dried root is taken along with hundred ml of cow's milk in the early morning in empty stomach for a period of fifteen to twenty days to remove the kidney stones.

Cocculus hirsutus (L.) Diels (Plate 1.5, 2)

One teaspoon of the shade dried leaf powder is taken along with hundred ml of cow's milk or butter milk as a refrigerant twice a day for a period of one week.

Commelina benghalensis L.

Two to three drops of pure stem sap are applied to the eyes twice a day to remove any foreign particles like dust from the eyes.

Commelina ensifolia R. Br.

To remove any foreign particles like dust from the eyeballs, two to three drops of pure stem sap is applied to the eyes twice a day.

Commiphora pubescens (Wight and Arn.) Engler

The endosperms obtained from four to five fresh or dried seeds are consumed two times a day for a period of two to three days to get relief from stomachache.

Crotalaria calycina Schrank. (Plate 1.5, 3)

Ten grams of dried root made into paste with water is externally applied once a day to treat various skin diseases.

Clitoria ternatea **L.**

Cocculus hirsutus **(L.) Diels**

Crotalaria calycina **Schrank**

Cryptolepis buchananii **Roem & Schultes**

Cylista scariosa **Roxb.**

Dichrostachys cinerea (L.) Wight & Arn

Plate 1.5

Cryptolepis buchananii Roem and Schultes. (Plate 1.5, 4)

One teaspoon of aqueous extract of fresh leaf is taken orally along with water twice a day for four to five days to treat snake-bite.

Curculigo orchioides Gaertn.

One teaspoon of powder obtained from the shade dried root tuber is taken in empty stomach with hundred ml of cow's milk once in a day in the morning for two to three months to increase the sexual vigour in males.

Cyanotis tuberosa (Roxb.) Schultes and Schultes

One or two fresh tubers are warmed by exposing to the flame and the hot tuber is pressed to the tooth for a few minutes to get relief from toothache.

Cylista scariosa Roxb. (Plate 1.5, 5)

Half a cup of decoction prepared by boiling the fresh leaves is taken orally once in a day for seven days to get relief from joint pain.

Cynodon dactylon (L.) Pers.

Prepare a juice containing fifty grams of fresh leaf in hundred ml of water. This juice is taken along with a teaspoon of jeera powder (*Cuminum cyminum* L.) twice a day for three or four days to relieve rheumatic pain.

The leaf juice is taken orally for treating dyspepsia along with one teaspoon of jeera powder (*Cuminum cyminum* L.) in empty stomach in a single dose for a period of three to five days.

Dichrostachys cinerea (L.) Wight and Arn. (Plate 1.5, 6)

The juice prepared from twenty grams of whole plant with 200ml of cow's milk is taken orally for a period of two days in a single dose to obtain relief from lumbago.

The paste prepared from ten grams fresh leaves is taken with two hundred ml of rice fermented water once in a day for a period of two days to arrest dysentery.

Dioscorea bulbifera L. var. _vera_ Prain and Burkill

Fifty grams of fresh or boild tuber is taken orally two times a day for a period of two or three days to arrest dysentery.

Dioscorea oppositifolia L. var. _dukhumensis_ Prain and Burkill

Fifty grams of fresh or boiled tuber is taken orally two times a day for a period of two or three days to arrest dysentery.

Dioscorea pentaphylla L. var. _pentaphyalla_

Fifty to hundred grams of tubers are boiled in water. The boiled tuber is made into curry after removing the skin or roasted and taken two times a day for two weeks to treat piles.

***Dioscorea tomentosa* Koen. ex Spreng.**

Ten grams of the boiled and peeled tuber is given to children once in a day for three days to relieve bowel complaints.

***Dodonaea viscosa* (L.) Jacq.**

Fifty to hundred grams of fresh leaves are boiled in one litre of coconut oil along with an equal quantity of fresh stem bark of omaviruchu (*Bischofia javanica* Blume) for twenty minutes. The oil is cooled and filtered. The filtered oil is used to massage the body one hour before taking hot water bath. This practice is followed once in a day for one week to get relief from the body pain.

***Drypetes sepiaria* (Wight and Arn.) Pax and Hoffm.**

Prepare a paste of ten to twenty grams of fresh stem bark in a few drops of water. The paste is applied on boils externally twice a day for five to seven days.

***Eclipta prostrata* (L.) L.**

Fifteen to twenty ml of aqueous extract of the leaf or whole plant is applied to head an hour before bath. This practice is administered for a few months, once in a week, to produce dark black hair and to arrest the falling of hair. This therapy also reduces body heat.

***Ehretia ovalifolia* Wight (Plate 1.6, 1)**

Five grams of dried leaf powder is taken orally along with cow's milk two times a day for three days to treat ulcer.

***Enicostema axillare* (Lam.) Raynal**

The juice prepared from ten to fifteen grams of whole plant with 200ml of goat's milk or rice fermented water is taken orally three times a day for a period of three days to reduce the body heat, lumbago and leucorrhoea.

***Ensete superbum* (Roxb.) Cheesman**

Hundred ml of filtered juice prepared by grinding the peduncle is taken in empty stomach once in a week for three or four month to remove the kidney stones.

***Euphorbia hirta* L.**

A few drops of stem latex are applied around the breasts ten minutes before feeding the baby to improve lactation.

***Euphorbia rosea* Retz.**

To improve lactation, a few drops of stem latex are applied around the breasts ten minutes before feeding the baby.

***Evolvulus alsinoides* (L.) L.**

Ten grams of fresh leaf is continuously boiled in two hundred and fifty ml of water along with equal quantity of fresh leaves of tulsi (*Ocimum tenuiflorum* L.) and fresh roots of kurunthatti (*Sida rhombifolia* L. var. *rhomboidea*) to get fifty ml decoction. This decoction is taken orally twice a day for two three days to treat/control fever.

Fiscus benghalensis L. var. _benghalensis_

The stem latex is applied on the cracked feet (fissures in foot) twice a day for a week for healing the cracks in the feet.

Ficus microcarpa L.

Applying the stem latex twice a day for seven days is beneficial in the treatment of cracked feet.

Ficus racemosa L.

The stem latex is applied on the cracked feet twice a day for a week for healing the affected feet.

Ficus religiosa L.

Applying the stem latex twice a day for a week can heal the fissures in the foot.

Flacourtia indica (Burm. f.) Merr.

Two or three ripe fruits are taken twice a day for twenty to thirty days to get relief from asthma.One teaspoon of the powder made from the shade dried unripe fruit is taken with hundred ml of cow's milk twice a day for a period of twenty five days as a remedy for asthma.

Foeniculum vulgare Mill

A pinch of powder made from the dried fruit is given to children with water in a single dose for two days to improve digestion.

Gardenia resinifera Roth (Plate 1.6, 2)

Four to five drops of the yellow resin collected by removing the young shoot are taken orally a day for four days to relieve dry cough.

The rind or pericarp of one unripe fruit is taken twice a day for a period of three to four days to get relief from dry cough.

One teaspoon of the powder made from the shade dried immature seeds is added to hundred ml of cow's milk and taken in empty stomach for a period of four days to relieve dry cough.

Givotia rottleriformis Griff.

Ten grams of powdered endosperm is mixed with hundred ml of cow's milk and it is allowed to settle for about half an hour. The filtrate is given orally to children once in a day for three days to improve digestion.

Globba marantina L.

Ten grams of fresh rhizome is made into paste with equal quantity of fresh or shade dried poolankilangu (*Plectranthus barbatus* Andr.) This paste is applied all over the body of the child, as an emollient, a few minutes before bath.

Ehretia ovalifolia **Wight**

Gardenia resinifera **Roth**

Hemidesmus **indicus (L.) R. Br. var.*indicus***

Hugonia mystax **L.**

Kalanchoe pinnata **(Lam.) Pers.**

Kalanchoe tubiflora **Hamet**

Plate 1.6

Gloriosa superba L.

Two to three unripe fruits are cut into small pieces and boiled in two hundred ml of coconut oil for about fifteen to twenty minutes. After cooling, two to three drops of the filtered oil is applied to the eyes twice a day for three to four days to treat any ophthalmic problem.

Gymnema sylvestre (Retz.) R. Br. ex Schultes

One teaspoon of the powder made from shade dried leaves is taken along with hundred ml of water two or three times a day for a period of ten to twelve days to control diabetes.

Hedyotis puberula (G.Don) Arn.

The powder prepared from the shade dried leaves is taken orally along with water for a period of forty eight days in a single dose treatment for diabetes.

Hemidesmus indicus (L.) R. Br. var. _Indicus_ (Plate 1.6, 3)

One teaspoon of shade dried root powder is mixed with two hundred ml of cow's milk or cool water and taken orally twice a day for a period of three to four days to reduce the body heat.

One teaspoon of shade dried root powder is added to two hundred ml of lukewarm water. It is filtered and the filtrate is taken orally two times a day for five to seven days to improve digestion. When the root powder is taken in cool water it can stop thirstiness.

Hibiscus vitifolius L.

Five to ten grams of fresh flowers and two to three Onion bulbs (*Allium cepa* L.) are made into paste. This paste is applied on the boils once in a day for a period of three days for healing.

Hugonia mystax L. (Plate 1.6, 4)

Ten to twenty grams of fresh leaves are made into juice. The juice is taken orally twice a day for a period of five to seven days to get relief from joint pain.

Hybanthus enneaspermus (L.) F.v. Muell.

The shade dried aerial part of the plant is powdered. Men take one teaspoon of this powder with water in empty stomach once in a day for about ten to twenty days to improve the sexual vigour.

Ichnocarpus frutescens (L.) R.Br.

One teaspoon of powder prepared from shade dried root is taken orally along with water three times a day for a period of three days to treat diabetes and to eliminate stones in the gall bladder.

Indigofera wightii Graham ex Wight and Arn.

Half a teaspoon of the shade dried root powder is taken with lukewarm water at the time of stomachache.

Ipomoea barlerioides **(Choisy) Benth. ex Clarke**

About thirty grams of fresh leaves are made into paste with a few drops of water. This paste is applied externally on the wounds twice a day for a period of five to seven days for healing the wounds.

Ipomoea staphylina **Roem. and Schultes**

The stem latex is applied on to the cracked feet (fissures in foot) once in a day at bedtime for a week for healing the cracks.

Jasminum roxburghianum **Wall.**

Five grams of the flower is soaked in hundred ml of gingelly oil for twelve hours. This oil is then boiled for about twenty minutes, cooled and filtered. Two to three drops of the clean filtered oil are applied to the ear twice a day for one week to arrest the oozing pus from the ear.

One gram of fresh root and one gram of fresh turmeric (*Curcuma longa* L.) are made into paste with a little water. This paste is given to children twice a day for two days to get relief from colic pain.

Jatropha glandulifera **Roxb**

The water extract of whole plant is applied externally two times a day for a period of one week to get relief from rheumatic pain.

Jatropha curcas **L.**

The water extract of leaves is applied externally once in a day for a period of one week to obtain relief from rheumatism and pain on knees. The root is introduced into vagina and to be kept for overnight for abortion.

Jatropha gossypifolia **L.**

The water extract of whole plant is applied externally two times a day for a period of one week to get relief from rheumatism.

Justicia adhatoda **L.**

The juice prepared from fresh leaves with 200ml cow's milk is taken orally three times a day for a period of one week to treat cold.

Justicia glauca **Rottl.**

One teaspoon of fresh leaf juice is taken three times a day for five to seven days to treat any poisonous bite.

Kalanchoe pinnata **(Lam.) Pers. (Plate 1.6, 5)**

To reduce the body heat, three to five fresh leaves are taken orally two times a day for three to four days.

Juice from the fresh leaves along with water is taken once in a day for five days to cure sexual disease.

Kalanchoe tubiflora Hamet (Plate 1.6, 6)

The leaf decoction mixed with a teaspoon of *Mucuna pruriens* (L.) DC var. *pruriens* seed powdered is taken orally once in a day for 48 days to cure sexual disorders.

Lantana camara L. var. _aculeata_ (L.) Mold

Three to four grams of fresh flowers, a small onion bulb (*Allium cepa* L.) and ten to twelve fresh leaves of thottalsinungi (*Mimosa pudica* L.) are made into paste with a few drops of water. This paste is applied externally on the spot two times a day for one week to treat any unknown insect-bite.

Lantana wightiana Wall. ex Gamble

Ten to twenty grams of fresh leaves are made into paste with water and the paste is applied on the forehead and the patient is advised to take bath after an hour. This procedure is followed once in a day for two to three days to treat intermittent fever.

Lawsonia inermis L.

Hundred grams of fresh leaves are boiled in one litre of coconut oil for fifteen minutes. The oil is filtered after cooling. This hair oil promotes hair growth and arrests falling of hair, if applied to head regularly.

Leucas biflora (Vahl) R. Br. var. _biflora_

Two or three grams of fresh or air-dried aerial parts of the plant are boiled along with two litres of water in a closed container. The lid is slowly removed and the steam is inhaled once in a day to get relief from headache.

Lindernia crustacea (L) F.v. Muell (Plate 1.7, 1)

Plant paste is applied on the body and taken a bath after a half-hour daily for a week to cure itching, scabies, skin diseases and insect bites.

Lobelia nicotianifolia Roth ex Schultes var. _nicotianifolia_

Four or five fresh leaves are chewed along with betel leaf (*Piper betle* L.) and betel nut (*Areca catechu* L.) to overcome fatigue.

Mangifera indica L.

Twenty to twenty five grams of dried endosperm and dried turmeric (*Curcuma longa* L.) are powdered separately and mixed. One teaspoon of powder is taken orally along with water in empty stomach for four to five days in a single dose to reduce stomachache.

Mimosa pudica L.

Ten to twelve fresh leaves are made into paste with a few drops of water along three to four grams of fresh flowers of unnipoo (*Lantana camara* L. var. *aculeata* (L.) Mold.) and a small onion bulb (*Allium cepa* L.). This paste is applied externally on the spot two times a day for one week to treat any unknown insect-bite.

Lindernia crustacea **(L.) F.V. Muell.**

Mitragyna parvifolia **(Roxb.) Korth**

Mucuna atropurpurea **DC.**

Orthosiphon thymiflorus **(Roth.) Sleensen**

Pavetta indica **L. var.** ***indica***

Phyllanthus reticulatus **Poir**

Plate 1.7

Mimusops elengi L.

Five to ten grams of fresh stem bark is soaked in twenty five ml of water for five to six hours and made into paste or juice. This paste or juice is taken with two hundred ml of rice-fermented water or milk in empty stomach in the early morning once in a day for a period of seven to ten days. The intake of hot drinks or hot food should be avoided for two hours after consumption. This treatment improves the fertility in women.

***Mitragyna parvifolia* (Roxb.) Korth (Plate 1.7, 2)**

Fifty gram of fresh bark of the stem is soaked in 200ml of water for 12hr and the filtrate is applied externally once in a day for one week to get relief from rheumatic pain.

***Mollugo cerviana* (L.) Ser.**

Two or three grams of the entire fresh plant is continuously boiled in one litre of water with equal quantities of the powders made from the shade dried pericarp of nelli (*Phyllanthus emblica* L.), tani (*Terminalia bellirica* (Gaertn.) Roxb.), kadukai (*Terminalia chebula* Retz.), sukku (*Zingiber officinale* Roscoe), chitrattai (*Alpinia calcarata* Roscoe), tippili (*Piper longum* L.) and fresh leaves of tulsi (*Ocimum tenuiflorum* L.), musumusukkai (*Mukia maderaspatana* (L.) M. Roem.), vishnukarandi (*Evolvulus alsinoides* (L.) L.) and hundred grams plam jaggery to get two hundred ml decoction. One teaspoon of this decoction is given to children only once to get relief from cold and cough.

Momordica charantia* L. var. *charantia

As a vermifuge, fifty to one hundred grams of unripe fruit is cooked as curry and taken two times a day for three days.

Morinda pubescens* J.E. Smith var. *pubescens

Ten grams of fresh leaves, two to three small onion bulbs (*Allium cepa* L.) and one teaspoon of turmeric powder (*Curcuma longa* L.) are made into paste. This paste is heated in five to ten ml of coconut oil and the hot paste is applied on the injury caused by weapons once in a day for four to five days to heal the wounds.

***Moringa concanensis* Nimmo ex Gibs.**

Fifteen to twenty grams of fresh leaves are made into paste along with two to three grams of black pepper (*Piper nigrum* L.) and five grams of garlic (*Allium sativum* L.). This paste is taken in the early morning hours in empty stomach once in a day for a period of three to four days to treat jaundice. During this treatment one bucketful of cold water is poured on the head of the patients. One teaspoon of fresh leaf juice is given to children in a single dose to treat bowel disorder.

***Mucuna atropurpurea* DC. (Plate 1.7, 3)**

About five hundred grams of dried seeds are made into a fine powder and boiled in water along with equal quantity of the powder made from the dried seeds of tamarind (*Tamarindus indica* L.) to make a paste. This paste is applied on the fractured area and bandaged tightly using a clean cloth. After one week the bandage is removed and the paste is applied for rebinding. This procedure is repeated for three to four times for proper setting of the fractured bone.

***Mukia maderaspatana* (L.) M. Roem.**

One teaspoon of juice prepared from the fresh leaves is taken orally once in a day for three days to get relief from cold and cough.

Calcium hydroxide is mixed with one to two ml of leaf juice and this mixture is applied externally on the throat thrice a day for a period of three to four days to get relief from cold and cough.

***Murraya koenigii* (L.) Spreng.**

Twenty to twenty five grams of fresh leaves are made into paste along with equal quantity of fresh endosperm of thennai (*Cocos nucifera* L.). The paste is mixed with a little water and applied to head an hour before taking bath. This procedure is practiced once in a week regularly to induce hair growth, to delay the development of silver hairs and to arrest falling of hair.

***Ocimum tenuiflorum* L.**

Ten to twenty grams of fresh leaves are boiled along with two liters of water in a closed container. The lid of the container is removed slowly and the steam is inhaled (vapour bath) only once to get relief from headache.

Twenty to thirty grams of fresh leaves are made into paste. This paste is applied on the wound twice a day for a period of four to five days to heal the wound.

One to two grams of fresh leaf is taken as such twice a day for five to seven days to get relief from cold and cough.

***Orthosiphon thymiflorus* (Roth) Sleensen (Plate 1.7, 4)**

Twenty grams of leaves are made into a paste with a few drops of water. The paste is applied externally on the wounds caused by weapon once in a day for five to six days for healing the wounds.

***Pavetta indica* L. var *indica* (Plate 1.7, 5)**

For a period of three days one teaspoon of powder prepared from the shade dried leaves of the plant is taken orally along with water three times a day to treat poisonous bites.

***Pavonia odorata* Willd.**

Twenty grams of fresh leaves are made into paste with a few drops of water. This paste is applied externally on the wounds twice a day for a period of seven days for healing the wounds.

***Pergularia daemia* (Forssk.) Chiov.**

Ten to fifteen grams of fresh leaves are made into paste along with ten to fifteen grams of garlic (*Allium sativum* L.). This paste is applied externally three times a day, on the stomach and on the dorsal area of the stomach to control gastric trouble.

Phyllanthus amarus Schum. and Thonn.

Prepare a juice using ten grams of the fresh, clean aerial parts of the plant and twenty five ml of water. The filtered juice is taken orally as such or along with hundred ml of cow's milk twice a day for about seven to ten days to treat jaundice.

Phyllanthus emblica L.

Fifty grams of the dried cotyledons are boiled in coconut oil and the oil extract is externally applied thrice a day to treat scabies.

The juice prepared from 10 to 15 grams of fresh fruits and leaves in hot water is taken orally in empty stomach for three days in a single dose to get relief from gastric problems and knee pain.

Phyllanthus maderaspatensis L.

Prepare a juice of ten grams of the fresh and clean aerial parts of the plant using twenty five ml of water. The filtered juice is taken as such or with hundred ml of cow's milk twice a day for seven to ten days to treat jaundice.

Phyllanthus reticulatus Poir. (Plate 1.7, 6)

One fifty ml of juice prepared from the whole plant is taken orally along with water two times a day for a period of three days to reduce the body heat

The juice prepared from twenty grams of leaves is taken orally three times a day for a period of three days to treat skin disease.

Phyllanthus urinaria L. (Plate 1.8, 1)

Prepare a juice using ten grams of the fresh leaves and twenty five ml of water. The filtered juice is taken orally as such twice a day for about seven to ten days to treat jaundice.

Physalis minima L.

Fifty grams of peeled leaves are boiled with salt, asafoetida (*Ferula assafoetida* L), coconut (*Cocos nucifera* L.) and onion bulb (*Allium cepa* L.) are taken orally once in a day for one week to treat ulcer.

Piper argyrophyllum Miq.

Two grams of the shade dried fruit is continuously boiled in two hundred and fifty ml of water along with equal quantities of omum (*Trachspermum ammi* (L.) Sprague), tippili (*Piper longum* L), sikku (*Zingiber officinale* Roscoe) and palm candy to get fifty ml decoction. This decoction is taken once in a day for a period of four to five days to relieve cold and cough.

Pisonia aculeata L. (Plate 1.8, 2)

One hundred gram of dried stem bark powder is boiled in 200ml of coconut oil and the oil extract is externally applied once in a day to treat eczema, psoriasis, rashes, scabies and the ringworm infection.

Phyllanthus urinaria **L.**

Pisonia aculeata **L.**

Plectranthus amboinicus **(Lour.) Spreng.**

Plectranthus amboinicus **(Lour.) Spreng.**

Scoparia dulcis **L.**

Tabernaemontana divaricata **(L.) R.Br.ex Roem & Schultes**

Plate 1.8

Plectranthus amboinicus (Lour.) Spreng

One teaspoon of juice prepared from the fresh leaves is taken twice a day for a period of four to five days to relieve cold and cough.

Plectranthus barbatus Andr. (Plate 1.8, 3)

The paste obtained from ten grams of fresh or shade dried rhizome is applied regularly all over the body of the child, as an emollient, before bath to get glowing skin.

Polygala javana DC. var. _angustifolia_ Thw.

Five to ten grams of fresh leaves are made into paste and applied on the breast twice a day for two to three days to check lactation and to get relief from the pain developed on stopping breast feeding.

Pongamia pinnata (L.) Pierre

A dried seed is tied to a thread by making a hole in the seed and is worn by children, as a necklace, to relieve cold and cough. This is also a part of folk therapy to check abnormal secretion of bile and in the treatment of jaundice.

Priva cordifolia (L. f.) Druce

Twenty five to thirty grams of fresh leaves are made into paste. This paste is applied on the wounds twice a day for five to seven days for healing the wounds.

Pterocarpus marsupium Roxb.

Fifty grams of the stem resin is soaked in hundred ml of water for about twelve hours. It is applied externally on the knee twice a day for five to six days to get relief from the joint pain.

The resin obtained from the stem is soaked in water for a few hours in a small container. A small drop of this resin is placed as a spot (bindhi) on the forehead of the child daily to avoid rickets/rachitis.

Rhinacanthus nasutus (L.) Kurz var. _nasutus_

One teaspoon of aqueous extract of fresh leaf is taken orally or one to two grams of the shade dried leaf powder is taken along with hundred ml of cow's milk twice a day for four or five days as an antidote for snake poison.

Ricinus communis L.

Twenty to thirty ml of castor oil is applied on head before one or two hours of taking bath to reduce the body heat.

Sansevieria roxburghiana Schultes and Schultes

Teeth growth in young children is induced when they bite the tender leaf.

Sarcostemma acidum (Roxb.) Voigt (Plate 1.8, 4)

One teaspoon of powder prepared from the shade dried stem is taken orally along with water three times a day for a period of three days to arrest vomiting.

Four to five drops of stem latex is applied externally on the spot of any insect-bite three times a day for five or six days.

Scoparia dulcis L. (Plate 1.8, 5)

Two to three grams of fresh leaves are taken orally two times a day for five days to treat any poisonous bite. Use of oil and tamarind should be avoided while consuming the leaves.

Senna auriculata (L.) Roxb.

Prepare a paste of five to ten grams of sepals and petals removed flowers by grinding with a small quantity of asafoetida (*Ferula assafoetida* L.) and poppy seeds (*Papaver somniferum* L.) using hot water. The paste is given orally for two days in a single dose to relieve colic pain.

Flowers and young shoots are tied over the stomach at least two times a day for three days continuously to get relief from colic pain even among adults.

Sesamum indicum L.

Twenty five grams of dried seeds are taken in empty stomach only once in a day along with equal quantity of palm jaggery from the thirty-fifth day of menstrual cycle after confirming pregnancy for about seven to ten days to terminate the unwanted pregnancy.

Sesbania sesban (L.) Merr

One hundred grams of fresh leaves is boiled in coconut oil and the oil is applied on hair once in a day to get black, thick and healthy hair.

Sida cordata (Burm.f.) Borssum

Two hundred ml of juice prepared from the whole plant is taken orally along with water two times a day for a period of two days to reduce bile problems.

The paste prepared from the five grams of flower is taken orally along with cow's milk two times a day for three days to treat ulcer.

Sida rhombifolia L. var. _rhomboidea_

Ten grams of fresh root is continuously boiled in two hundred and fifty ml of water along with equal quantities of fresh leaves of vishnukarandi (*Evolvulus alsinoides* (L.)L.) and tulsi (*Ocimum tenuiflorum* L.) to get fifty ml decoction. The filtered decoction is taken orally twice a day for two to three days to control fever.

Solanum surattense Burm.f.

Two to three leaves of the above plant, five to six leaves of tuduvalai (*Solanum trilobatum* L.), one gram of jeera (*Cuminum cyminum* L.) and hundred mg asafoetida (*Ferula assafoetida* L.) are continuously boiled in three hundred ml of water to get seventy five ml decoction. This decoction is taken once in a day in the morning hours for a period of four to five days to get relief from cold and cough.

***Solanum torvum* Sw.**

Cooked unripe fruits used as vermifuge

***Sterculia urens* Roxb.**

Twenty five grams of endosperm obtained from the dried seeds is roasted with ghee and taken twice a day for one week by the feeding mother to improve lactation.

***Strychnos nux-vomica* L.**

A small piece of dried stem is kept inside the house to ward off the evil spirits from the house. The ash obtained by burning the seeds is mixed with a few drops of castor oil and applied on the forehead of the child as a spot (bindhi) daily to avoid rickets/rachitis. A dried seed is worn around the neck of the child as a necklace to avoid rickets/rachitis.

Swertia corymbosa* (Griseb.) Wight ex Clarke var. *corymbosa

One to two grams of the shade dried plant powder is taken orally along with hundred ml of urine of a child of opposite sex twice a day for five to seven days as an antidote for snake-bite.

The shade dried plants are powdered and one teaspoon of powder is taken orally along with water in empty stomach for a period of two or three days to get relief from stomachache.

***Syzygium cumini* (L.) Skeels**

One teaspoon of shade dried seed powder is taken in empty stomach along with hundred ml of cold water daily for a month or two months to control diabetes.

***Tabernaemontana divaricata* (L.) R.Br. ex Roem and Schultes (Plate 1.8, 6)**

Two drops of pure filtered aqueous extract prepared from the flower petal are applied to the eyes for a period of seven days to improve vision.

***Tamarindus indica* L.**

A dried seed is rubbed on a rough floor for a few seconds to remove the outer testa and the rubbed surface of the seed is pressed immediately on the spot of scorpion-sting to relieve pain.

***Tephrosia purpurea* (L.) Pers.**

One teaspoon of juice prepared from the fresh root bark is mixed with a pinch of asafoetida (Ferula assafoetida L.) and given to children in a single dose to get relief from bowel disorders.

***Terminalia bellirica* (Gaertn.) Roxb.**

Two or three grams of the powder made from the shade dried pericarp is continuously boiled in one liter of water along with equal quantities of the powder made form the shade dried pericarp of nelli (*Phyllanthus emblica* L.), kadukai

(*Terminalia chebula* Retz.), sukku (*Zingiber officinale* Roscoe), chitrattai (*Alpinia calcarata* Roscoe), tippili (*Piper longum* L.), fresh leaves of tulsi (*Ocimum tenuiflorum* L.), pappadai (*Mollugo cerviana* (L.) Ser.), musumusukkai (*Mukia maderaspatana* (L.) M. Roem.), vishnukarandi (*Evolvulus alsinoides* (L.) L.) and fifty grams of plam jaggery to get two hundred ml decoction. One teaspoon of filtered decoction is taken orally by the feeding mother twice a day for two days to treat/control her breast-fet baby's fever.

Terminalia chebula **Retz.**

One gram pericarp of the fruit along with equal quantity of turmeric (*Curcuma longa* L.) are made into a paste and taken orally three times a day for a period of two to three days to arrest dysentery.

Thespesia populnea **(L.) Soland.ex Correa**

A few tender shoots are made into paste along with one or two small onion bulbs (*Alliun cepa* L.). This paste is applied immediately on the spot of scorpion-sting to relieve pain.

Tinospora cordifolia **(Willd.) Miers ex Hook.f. and Thoms (Plate 1.9, 1)**

The water extract of whole plant is applied externally two times a day for a period of five days to get relief from rheumatic pain.

Trianthema decandra **L.**

The decoction of whole plant is taken two times a day for a period of three days to get relief from gastric problems.

Tribulus terrestris **L.**

The juice prepared from ten grams of fruits with 200ml of goat's milk is taken orally two times a day for a period of three days to treat leucorrhoea.

Trichodesma zeylanicum **(Burm.f.) R.Br. (Plate 1.9, 2)**

Thirty to forty grams of fresh leaves are roasted along with eight to ten onion bulbs (*Allium cepa* L.) in coconut oil (*Cocos nucifera* L.) and made into curry. This curry is taken one hour before meals for four to five days to treat bleeding piles.

Tridax procumbens **L.**

Fifty grams of fresh leaves are made into juice. This juice is applied externally on shoulder once in a day for a period of three to four days to get relief from shoulder pain.

Ten grams of fresh leaves are made into paste along with a pinch of calcium hydroxide. This paste is externally applied on the eczema affected area or on injury made by weapon once in a day for three or four days for complete cure.

Tylophora indica **(Burm.f.) Merr. (Plate 1.9, 3)**

One to two grams of shade dried leaf powder is taken orally twice a day for five to seven days as for antidote for snake-bite.

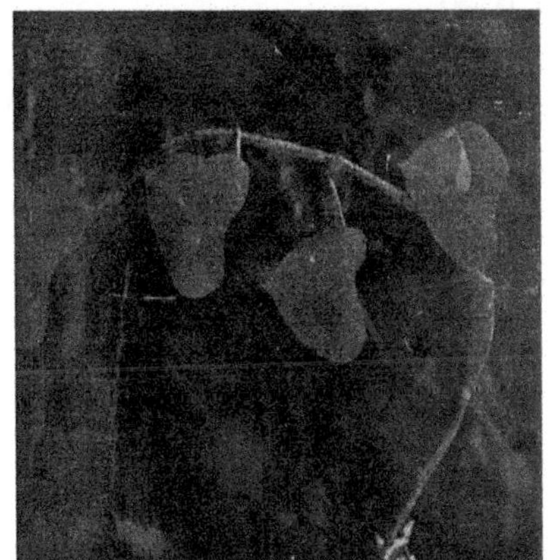

Tinospora cordifolia (Willd.)Miers ex Hook.f.&Thoms

Trichodesma zeylanicum (Burm.f.) R.Br.

Tylophora indica (Burm.f.)Merr)

Vetiveria zizanioides (L.)Nash

Vitex negundo L.

Wrightia tinctoria (Roxb.) R.Br. var. *tinctoria*

Plate 1.9

Vetiveria zizanioides (L.) Nash. (Plate 1.9, 4)

One hundred gram of fresh roots are made into a juice along with *Hemidesmus indicus* (L) R. Br. var. *indicus* root and extire plant of *Hedyotis puberula* (G. Don) Arn. are boiled in coconut oil. The oil extract is externally applied twice a day to treat skin diseases and also is applied to head daily as hair tonic, to induce hair growth and to arrest falling of hair

Vitex negundo L. (Plate 1.9, 5)

Fifty grams of fresh leaves are roasted with equal quantity of sand in a hot pan. The hot mixture is administered in hot fomentation on the forehead for relieving headache.

Wrightia tinctoria (Roxb.) R. Br. var. _tinctoria_ (Plate 1.9, 6)

Ten grams of fresh leaves are cut into small pieces and soaked in five hundred ml of coconut oil for one day.

Then the leaf pieces are removed and the oil is applied on the eczema affected area three or four hour before bath. This practice continued for seven to ten days for complete recovery.

Ziziphus xylopyrus (Retz). Willd

Fifty grams of fresh stem bark is soaked in two hundred ml of water for twelve hours and filtered. This filtrate is taken orally in empty stomach for a period of three days in a single dose to relive stomachache.

Discussion

The tribal's knowledge of indigenous uses of native medicinal plants before their exodus into the urban areas to join the mainstream life needs to be studied and documented. In the present study 180 medicinal plants were collected and documented (Table 1.1). The present study focuses the extensive usage of as many as 180 medicinal plants by the *Pallliyar* tribe inhabitating the study area (Saduragiri hills in the South-Eastern Slopes of the Western Ghats, Tamil Nadu, India).

As an outcome of the present investigation, 180 plants (165 plants collected and 15 plants/extracts procured) belonging to 146 genera and 69 families were recorded (Tables 1.1 and 1.2). Of the recorded plants, a maximum of 16 ethnomedicinal plants belong to Euphorbiaceae, it is followed by Fabaceae (12 plant species). Lamiaceae and Asclepiadaceae (7 plant species each), Rubiaceae, Acanthaceae and Apiaceae (6 plant species each) Apocynaceae, Solanaceae and Malvaceae (5 plant species each).

The documented plants used for the various diseases by the ***Palliyars*** show a great habitual diversity. Among the 180 plants reported as herbal drugs, 77 plants are herbaceous in habit while 26 plants are shrubs, 33 plants are climbers and 42 plants are trees. Among the investigated medicinal plants 63 plants belongs to the polypetalae under 31 families, 60 belong to the gamopetalae under 18 families, 29 belong to the monochlamydeae under 8 families and 28 belong to the monocotyledons under 12 families (Table 1.3).

Table 1.3: Family-wise Distribution of Enumerated Plant Species

POLYPETALAE (DICOTYLEDONS)

Sl.No.	*Family*	*Total No. of Genera*	*Total No. of Species*
1	Aizoaceae	1	1
2	Anacardiaceae	1	1
3	Apiaceae	6	6
4	Begoniaceae	1	1
5	Burseraceae	2	2
6	Caesalpiniaceae	3	3
7	Capparaceae	1	1
8	Cleomaceae	1	1
9	Combretaceae	1	2
10	Crassulaceae	1	2
11	Cucurbitaceae	1	2
12	Fabaceae	10	12
13	Flacourtiaceae	1	1
14	Linaceae	1	1
15	Lytheraceae	1	1
16	Malvaceae	4	5
17	Meliaceae	1	1
18	Menispermaceae	3	3
19	Mimosaceae	2	2
20	Molluginaceae	1	1
21	Moringaceae	1	1
22	Myrtaceae	1	1
23	Papaveraceae	1	1
24	Polygalaceae	1	1
25	Rhamnaceae	1	1
26	Rutaceae	2	2
27	Sapindaceae	2	3
28	Sterculiaceae	1	1
29	Violaceae	1	1
30	Vitaceae	1	1
31	Zygophyllaceae	1	1
	Total	**56**	**63**

GAMOPETALAE (DICOTYLEDONS)

Sl.No.	*Family*	*Total No. of Genera*	*Total No. of Species*
1	Acanthaceae	5	6
2	Apocynaceae	5	5
3	Asclepiadaceae	7	7
4	Asteraceae	2	3
5	Boraginaceae	2	2
6	Convolvulaceae	3	4
7	Gentianceae	2	2
8	Lamiaceae	5	7
9	Lobeliaceae	1	1
10	Loganiaceae	1	1
11	Periplocaceae	2	2
12	Pedaliaceae	1	1
13	Oleaceae	1	1
14	Rubiaceae	6	6
15	Scrophulariaceae	2	2
16	Solanaceae	3	5
17	Sapotaceae	1	1
18	Verbenaceae	3	4
	Total	52	**60**

MONOCHLAMYDEAE (DICOTYLEDONS)

Sl.No.	*Family*	*Total No. of Genera*	*Total No. of Species*
1	Amaranthaceae	1	1
2	Aristolochiaceae	1	1
3	Bischofiaceae	1	1
4	Euphorbiaceae	6	16
5	Moraceae	3	4
6	Nyctaginaceae	1	1
7	Piperaceae	1	4
8	Ulamaceae	1	1
	Total	**15**	**29**

MONOCOTYLEDONS

Sl.No.	*Family*	*Total No. of Genera*	*Total No. of Species*
1	Arecaceae	3	3
2	Araceae	2	2
3	Alliaceae	1	2
4	Asparagaceae	1	1

Sl.No.	Family	Total No. of Genera	Total No. of Species
5	Agavaceae	1	1
6	Commelinaceae	2	3
7	Dioscoreaceae	1	4
8	Liliaceae	3	3
9	Musaceae	1	1
10	Poaceae	3	3
11	Hypoxidaceae	1	1
12	Zingberaceae	3	4
	Total	**22**	**28**
	Grand Total	**146**	**180**

As far as the plant part used is concerned, it is noted that *Palliyars* employed almost all part of plant in ethnomedicine. In terms of percentage of plant part used, the percentages are as follows; Leaf 40 per cent, root, stem/stem bark, root tubers/rhizome/bulb 8 per cent each, fruit/unripe fruit 10 per cent and flower/ inflorescence 3 per cent.

The most prevalent form of administration of medicine is paste (25 per cent). This is followed by juice/extract (23 per cent), plant part as such (20 per cent), powder (19 per cent) decoction (15 per cent), boiled/roasted/burnt plant part (7 per cent) and infusion (7 per cent). Ten plants are used to treat menstrual disorders, to induce lactation in feeding women and as abortifacient. Four plants are used for treating sexual diseases and to improve sexual vigour in men. Six plants are used in child health care.

The enumerated 180 plants used to cure as many as 53 different types 5 human maladies (Table 1.4). A maximum of 22 plants are used to treat cold and cough, followed by 12 plants as refrigerant, 10 plants for treating wounds, 9 plants for treating rheumatism, 9 plants for treating poisonous bites, 7 plants for treating stomachache.

Thirty different maladies of *Palliyars* are grouped into 5 major categories of diseases *viz.*, gastro-intestinal complaints, poisonous bite, body pain, fever, cold and cough and skin ailments. Thirty eight plants belonging to 32 genera and 26 families are used by the *Palliyars* for treating gastro-intestinal complaints like stomachache, dysentery, bowel disorder, bile complaint, jaundice and colic pain *etc.* A few medicinal plants are used as dyspepsia and vermifuge. Fifty plants identified in the present enumeration belonging to 36 genera and 25 families are employed as analgesics to relieve rheumatic pain, headache, general body pain, chest pain and body heat, *etc.* They use twenty seven plants belonging to 24 genera and 17 families are employed to treat fever, cold and cough. The *Palliyars* use 21 plants belonging to 16 genera and 13 families for treating various poisonous bites such as dog bite, snack bite, scorpion sting, honey bee – sting and unknown insect bite. Twenty eight plants belonging to 28 genera and 20 families are used by the *Palliyars* for treating various skin ailments like eczema, boils, wounds, foot crack *etc.*

Table 1.4: Disease Types and Herbal Remedies of *Palliyars*

Sl.No.	*Name of the Ailment*	*Total Number of Herbal Remedies*
1	Abortifacient	3
2	Asthma	3
3	Astringent (Health tonic)	1
4	Bed-wetting	1
5	Boils	2
6	Bone fracture	2
7	Bowel disorder	3
8	Bile complaint	2
9	Chest pain	1
10	Colic pain	2
11	Cold and cough	22
12	Dog-bite	2
13	Dyspepsia (indigestion)	4
14	Dysentery	4
15	Diabetes	4
16	Ear ailments	2
17	Eczema/Itches	5
18	Emollient	2
19	Fever	5
20	Fissures in foot	5
21	General body pain	4
22	Giddiness	2
23	Hair tonic	5
24	Honeybee-sting	1
25	Headache	8
26	Jaundice	6
27	Kidney stone	3
28	Leucorrhoea	2
29	Lumbago	2
30	Menstrual disorder	2
31	Ophthalmic ailments	5
32	Post natal therapy	1
33	Piles	5
34	Refrigerant	12
35	Rheumatism	9
36	Rickets/Rachitis	2
37	Scorpion-sting	4
38	Snake-bite	5
39	Stomachache	8

Sl.No.	Name of the Ailment	Total Number of Herbal Remedies
40	Toothache/tooth ailments	3
41	To induce lactation	4
42	To check lactation	1
43	To increase fertility in women	1
44	To increase sexual vigour in male	2
45	To overcome fatigue	1
46	To stop thirstiness	1
47	Tumour like swelling due to wind in the naval region	1
48	Unknown insect-bite/General poisonous bites	9
49	Vermifuge	2
50	Wound/weapon injury	10
51	Skin diseases/ringworm infection	4
52	Stone in gallbladder	1
53	To arrest vomiting	1
	Total	204

Above 10 per cent of the identified plants possess two fold uses (used to treat two ailments). Eighty five percent of the ethnomedicinal plants are to cure only one particular disease. Some of the plant species recorded in the study exhibit manifold therapeutic uses, *i.e.*, they are used to cure more than two ailments. Five percent of the identified plants possess this type of therapeutic value. For example, the rhizome of *Curcuma longa* is administered in colic pain, stomachache, to heal wounds and dysentery. The seed of *Cuminum cyminum* is used for treating dyspesia, giddiness, rheumatic pain, cold and cough. The leaves of *Cardiospermum halicacabum* is used in curing rheumatic complaints, giddiness, cold and cough. The leaves of *Acalypha fruticosa* is administered in honey bee sting, headache and rickets. The seed of *Abrus precatorius* is used to cure eye irritation; its leaf is used to treat odonthemedia and its leaf and root are used to treat poisonous bites. The root of *Aristolochia tagala* is used for treating snake-bite, scorpion sting, headache and stomachache. The whole plant of *Enicostema axillare* is used to curing body heat, lumbago and leucorrhoea. There above mentioned ethnomedicinal plants are recommended for broad phytopharmacological investigations.

Cross-cultural Ethnomedicobotanical Studies

It is well established that an identical use of same plant by different tribal groups indicates its curative property and therapeutic significance (Jain and Saklani, 1992). The comparison of medicinal claims between different cultural groups in the same region or neighbouring regions has proved very rewarding. The comparative studies on medicinal uses of plants among different cultural groups showed similarities and dissimilarities in uses. In the present investigation, the medicinal claims emanating from the *Palliyars* are compared with claims from three other numerically predominant Indian tribal groups living in the Central and Western zones of the country *viz.*, the *Bhils, Gonds and Santals* and two other predominant tribals of Tamil Nadu namely the *Kanikkars* and *Irulars*. The source of ethnomedicinal information for the above mentioned ethnic groups is selected from the available literature (Jain *et al.*, 1973; Mudgal and Pal, 1980; Ramachandran and Nair 1981a,b; Joshi 1982; 1989, 1992; Tarafder, 1983; 1984; Shah and Gopal, 1985; Saxena, 1986; Sur *et al.*, 1987; 1990; Prasad *et al.*, 1987; Goel and Mudgal, 1988; Pal *et al.*, 1989; Saxena and Tripathi, 1989; Yadav and Bhamare, 1989; Ragupathy and Mahadevan, 1991; Borthakur and Goswami, 1995; Mishra and Naquvi, 1995; Chandra, 1995; Girach and Aminuddin, 1995; Katewa and Arora, 1997; Ignacimuthu *et al.*, 1998; Girach *et al.*, 1999; Viswanathan *et al.*, 2001; Binu *et al.*, 2003; Ayyanar and Ignacimuthu, 2005; Jadhav, 2006a,b).

Comparative Ethnomedicobotany with *Bhils*

Forty one plant species are commonly used by both *Palliyars* and *Bhils*. Both similarities and dissimilarities are discussed.

Sl.No.	*Botanical Name*	*Part Used*	*Palliyars*	*Bhils*
1	*Abrus precatorius*	Root	Poisonous bite	Cough
		Leaf	Poisonous bite, Odonthemedia,	Throat infection
		Seed	Eye irritation	On swelling or inflammation
2	*Aegle marmelos*	Leaf	Skin disease	Skin disease
3	*Asparagus racemosus*	Root tuber	Bed-wetting by children	Piles, constipation, increases lactation in feeding mother, gout, typhoid.
4	*Azadirachta indica*	Stem bark	Fever	Spots on skin caused due to small pox, skin diseases
5	*Calotropis gigantea*	Stem latex	Cold and cough	Cold and cough, asthma, abortifacient, rheumatism
6	*Centella asiatica*	Leaf	Chest pain, cold and cough	As tonic to women after pregnancy, memory tonic
7	*Clitoria ternatea*	Root	Kidney stone	Fever
8	*Cocculus hirsutus*	Leaf	Coolant	Headache,sore, cuts, wounds, toothache, blood purifier
9	*Commelina benghalensis*	Plant sap	To remove dust from eye	Ophthalmic ailments (Swelling in eye)
10	*Curculigo orchioides*	Root tuber	To increase sexual vigour in male	Vermifuge, aphrodisiac
11	*Curcuma longa*	Rhizome	Stomachache	Arthritis
12	*Cissus quadrangularis*	Leaves	Relief from tumour like swelling	Fracture
13	*Eclipta prostrata*	Leaf	Hair tonic, coolant	Hair tonic, on injury as antiseptic, snake-bite
14	*Enicostema axillare*	Whole plant	Body heat, lumbago, leucorrhoea	Gout, headache

Sl.No.	Botanical Name	Part Used	Palliyars	Bhils
15	*Evolvulus alsinoides*	Leaf	Cold and cough, fever	Finger affected with whitlow
16	*Ferula assafoetida*	Root resin	Cold and cough	Toothache
17	*Ficus benghalensis*	Stem latex	Fissures in foot	Rheumatism, muscular pain
18	*Ficus racemosa*	Stem latex	Fissures in foot	Cracked skin
19	*Gymnema sylvestre*	Leaf	Diabetes	Ophthalmic ailments
20	*Hemidesmus indicus*	Root	Coolant, improves digestion, stops thirstiness	Venereal diseases, digestive and reproductive disorders
21	*Jatropha curcas*	Leaves	Rheumatic pain	Toothache
		Root	Abortion	Toothache
22	*Lawsonia inermis*	Leaves	Coolant, hair tonic	To check pregnancy, burning sensation of foot
23	*Lobelia nicotianifolia*	Leaf	To overcome fatigue	Hydrocele
24	*Mimosa pudica*	Leaf	Unknown insect-bite	Skin inflammations
25	*Pergularia daemia*	Leaf	Gastric trouble	Piles
26	*Ricinus communis*	Seed oil	Stomachache, Laxative, coolant	Back-ache, headache, tooth problem
27	*Senna auriculata*	Flower	Colic pain	Jaundice
28	*Sida rhombifolia*	Root	Fever	Cuts, bruises
29	*Sterculia urens*	Stem latex	Increases lactation in feeding mother	Piles, menstrual troubles, back-ache in women, stomach-ulcer
30	*Syzygium cumini*	Seed	Diabetes	Diabetes
36	*Terminalia bellirica*	Fruit	Cold and cough, fever	Cold and cough, jaundice,
37	*Tinospora cordifolia*	Whole plant	Rheumatic pain	Arthritis, leucorrhoea, snake-bite and leprosy.
38	*Tridax procumbens*	Leaf	Shoulder pain, weapon injury, skin disease	Antiseptic and applied an cuts/wound to check bleeding, diabetes
39	*Tribulus terrestris*	Fruit	Leucorrhoea	Leucorrhoea
40	*Vitex negundo*	Leaves	headache	Gout
41	*Zingiber officinale*	Dried rhizome	Cold and cough	After pregnancy for strength and vigour.

Comparative ethnobotanical study between the *Palliyars* and *Bhils* reveals that 11 medicinal plant species (out of 41) exhibit identical uses. They are *Aegle marmelos, Calotropis gigantea, Commelina benghalensis, Eclipta prostrata, Ficus racemosa, Hemidesmus indicus, Syzygium cumini, Terminalia bellirica Tinospora cordifolia, Tribulus terrestris* and *Tridax procumbens.*

Comparative Ethnomedicobotany with the *Gonds*

Twenty nine plant species are commonly used by both the *Palliyars* and *Gonds*. Both similarities and dissimilarities are discussed.

Sl.No.	*Botanical Name*	*Part Used*	*Palliyars*	*Gonds*
1	*Aegle marmelos*	Leaf	Skin disease	Conjunctivitis in eye, febrifuge, jaundice
2	*Alstonia scholaris*	Stem bark	Increases lactation in feeding mother	Gonorrhoea
3	*Andrographis paniculata*	Whole plant	Snake-bite,scorpion-sting	Snake-bite
4	*Asparagus racemosus*	Root tuber	Bed-wetting by children	Piles, constipation, leucorrhoea
5	*Azadirachta indica*	Leaf	Cold	Anthelmintic,otorrhoea
		Stem bark	Fever	Ring worm infections, Anthelmintic
6	*Calotropis gigantea*	Stem Latex	Cold and cough,	Decay of teeth, asthma
7	*Cardiospermum halicacabum*	Leaf	Giddiness,cold and cough, rheumatism	To reduce bulkiness of the body
8	*Cissampelos pareira*	Root	Stomachache, poisonous bite	Cold, cough, fever, stomach pain
9	*Curculigo orchioides*	Root tuber	Increase sexual vigour in male	Blindness and white spot in eye, aphrodisiac
10	*Evolvulus alsinoides*	Leaf	Fever, cold and cough	Finger affected with whitlow
11	*Ficus benghalensis*	Stem latex	Fissures in foot	To treat premature ejaculation,gonorrhoea
12	*Ficus racemosa*	Stem latex	Fissures in foot	Throat pain, stomach pain in children
13	*Gymnema sylvestre*	Leaf	Diabetes	Diabetes, ophthalmic ailments
14	*Hemidesmus indicus*	Root	Improves digestion, coolant and stops thirstiness	Nervous disorders, body pain, cooling agent, alimentary canal and genital disorders
15	*Leucas biflora*	Aerial part	Headache	Skin diseases
16	*Lobelia nicotianifolia*	Leaf	To over come Fatigue	Hydrocele
17	*Mimusops elengi*	Stem bark	Increase fertility in women	Leucorrhoea, bleeding gum, to strengthen teeth
18	*Ocimum tenuiflorum*	Leaf	Cold and cough, headache, wound	Ring worm infection
19	*Phyllanthus emblica*	Fruit	Cold and cough, fever	Diuretic, laxative, refrigerant, epistaxis, gastritis
20	*Senna auriculata*	Flower	Colic pain	Diabetes

Sl.No.	Botanical Name	Part Used	Palliyars	Gonds
21	*Sterculia urens*	Stem Index	Induces lactation in feeding mother	Stomachache, throat infection
22	*Tephrosia purpurea*	Root	Bowel complaint	Diabetes, dyspepsia, Colic pain, Stomachache
23	*Terminalia bellirica*	Fruit	Cold and cough, fever	Bronchitis, cough, diarrhea, purgative, astringent, antipyretics, narcotic
24	*Terminalia chebula*	Fruit	Dysentery, cold and cough, fever	To wash wounds, boils, gastritis, dysentery, conjunctivitis in eye
25	*Tridax procumbens*	Leaf	Skin disease, shoulder pain, weapon injury	Skin disease
26	*Ichnocarpus frute-scens*	Root	Diabetes, elimination of stone in the gall bladder,	Jaundice
27	*Jatropha curcas*	Leaf Root	Rheumatism, abortion	Skin diseases
28	*Mitragyna parvifolia*	Stem bark	Rheumatic pain	Febrifuge and fever
29	*Tinospora cordifolia*	Whole plant	Rheumatic pain	Spermatorrhoea

Comparative ethnobotanical study between the *Palliyars* and *Gonds* reveals that 8 medical plant species (out of 29) exhibit identical uses. They are *Andrographics paniculata, Cissampelos pareira, Gymnema sylvestre, Hemidesmus indicus. Tephrosia purpurea, Terminalia bellirica, Terminalia chebula* and *Tridax procumbens.*

Comparative Ethnomedicobotany with the *Santals*

Thirty plants are commonly used by both the *Palliyars* and *Santals*. Both similarities and dissimilarities are discussed.

Sl.No.	Botanical Name	Part Used	Palliyars	Santals
1	*Abrus precatorius*	Root	Poisonous bite	Abortifacient, snake-bite
		Leaf	Poisonous bite, odonthemedia	Throat pain
		Seed	Eye irritation	Abortifacient
2	*Achyranthes aspera*	Leaf	Dog-bite	Eczema, tetanus, scorpion-sting
3	*Aloe vera*	Leaf	Coolant, to remove kidney stone	Asthma, purgative, vermifuge
4	*Alstonia scholaris*	Stem bark	Induces lactation in feeding mother	Improves secretion of milk in feeding mother
		Stem latex	Asthma	Dysentery
5	*Andrographis paniculata*	Whole plant	Snake-bite, Scorpion-sting	Stomachache, vermifuge, fever

Sl.No.	Botanical Name	Part Used	Palliyars	Santals
6	*Aristolochia indica*	Root	Snake-bite, scorpion-sting, stomachache, headache	Snake-bite, Scorpion-sting, cuts, fever, abortifacient
7	*Asparagus racemosus*	Root tuber	Bed- wetting by children	Abortifacient, fever, syphilis
8	*Calotropis gigantea*	Stem latex	Cold and cough	Abortifacient, rheumatism, eczema, ring worm infection
9	*Centella asiatica*	Leaf	Chest pain, cold and cough	Wound
10	*Cissampelos pareira*	Root	Stomachache, poisonous bite	Tonic in weakness
11	*Curculigo orchioides*	Root tuber	To increase sexual vigour in male	Abortifacient, leucorrhoea, snake-bite, syphilis
12	*Eclipta prostrata*	Leaf	Coolant, hair tonic	General tonic to children, cuts, mycetoma, in growing toe nails
13	*Euphorbia hirta*	Stem latex	Induces lactation in feeding mother	Wounds, headache
14	*Hemidesmus indicus*	Root	Coolant, stops thirstiness, improves digestion	Wounds, malaria
15	*Mimosa pudica*	Leaf	Unknown insect-bite	Ulcer, wounds
16	*Morinda pubescens*	Leaf	Weapon injury	Earache
17	*Ocimum tenuiflorum*	Leaf	Headache, cuts/wounds	Abortifacient
18	*Sesamum indicum*	Seed	Abortifacient	Abortifacient
19	*Sida rhombifolia*	Root	Fever	To treat white discharge in women
20	*Syzygium cumini*	Seeds	Diabetes	Diabetes
21	*Terminalia chebula*	Fruit	Dysentery, cold and cough, fever	Toothache
22	*Tridax procumbens*	Leaf	Shoulder pain, weapon injury, skin diseases	Cuts, wounds, sores, leprosy, scratches
23	*Vitex negundo*	Leaf	Headache	Body ache
24	*Ichnocarpus frutescens*	Root	Diabetes, elimination of stone in the gall bladder	Blood purifier
25	*Jatropha curcas*	Root	Abortion	Abortion
26	*Jatropha gossypifolia*	Whole plant	Rheumatism	Rheumatism
27	*Lantana camera*	Flower	Insect-bite	Cuts
28	*Pavetta indica*	Leaf	Poisonous bite	Boils
29	*Scoparia dulcis*	Leaves	Poisonous bite	Gonorrhoea and headache
30	*Ziziphus xylopyrus*	Stem bark	Stomachache	Fever, cold and cough of infants

Comparative ethnobotanical study between the *Palliyars* and *Santals* reveals that eight medicinal plant species (Out of Thirty) exhibit identical uses. They are *Abrus precatorius, Alstonia scholaris, Aristolochia indica, Jatropha curcas, J. gossypifolia Sesamum indicum, Syzygium cumini and Tridax procumbens.*

Comparative Ethnomedicobotony with the *Kanikkars*

Thirty three plant species are commonly used by both the *Palliyars* and *Kanikkars* both similarities and dissimilarities are discussed.

Sl.No.	Botanical Name	Part Used	Palliyars	Kanikkars
1	*Abrus precatorius*	Root	Poisonous bite	Cough
2	*Acalypha indica*	Leaf	Cough	Scabies, cough and catarrh
3	*Acorus calamus*	Rhizome	Giddiness	Dyspepsia
4	*Allium sativum*	Bulb	Jaundice, gastric problem	Work treatment
5	*Anisomeles malabarica*	Leaf	Headache	Vermicide
6	*Aristolochia indica*	Vegetative part/root	Snake- bite, Scorpion-sting, headache Stomach ache	Snake-bite
7	*Asparagus racemosus*	Root tuber	Bet- wetting by children	A disease in children called "kana"
8.	*Bidens pilosa*	Leaf		Stomachache
9	*Calotropis gigantea*	Stem latex	Cold and cough	Thorn injury
10	*Capsicum annum*	Fruit		Cold and cough
11	*Cardiospermum hali-cacabum*	Leaf	Rheumatism, cold and cough	Health tonic to pregnant women, to reduce labour pain, easy delivery
12	*Centella asiatica*	Leaf	Chest pain, cold and cough	Anti-inflammatory bowel complaints
13	*Clitoria ternatea*	Root	Kidney Stone	Cough
14	*Curculigo orchioides*	Root tuber	To increase sexual vigour in male	To increase sexual vigour in male
15	*Eclipta prostrata*	Leaf	Coolant, hair tonic	Purgative, diuretic, hair tonic, elephantiasis
16	*Hemidesmus indicus*	Root	Coolant, Stops thirstiness, improves digestion,	Antipyretic, sore mouth, to increase eye sight
17	*Hypanthus ennea-spermus*	Shoot	To increase sexual vigour in male	Coolant
18	*Mimosa pudica*	Leaf	Unknown insect-bite	Boils, swellings
19	*Pergularia daemia*	Leaf	Gastric problem	General analgesic
20	*Solanum surattense*	Leaf	Cold and cough	Cough
21	*Solanum trilobatum*	Leaf	Cold and cough	Cough
22	*Terminalia bellirica*	Fruit	Dysentry	Diarrhoea
23	*Tridax procumbens*	Leaf	Shoulder pain, weapon injury	Cuts, wounds, bone fracture
24	*Achyranthes aspera*	Leaves	Dogbite	Eye injury
25	*Allium cepa*	Bulb	Relieve pain	Cold, foot sores
26	*Indigofera wightii*	Root	Stomachache	Ulcer
27	*Mukia madraraspatana*	Leaf	Cold and cough	Giddiness

Sl.No.	Botanical Name	Part Used	Palliyars	Kanikkars
28	*Plectranthus amboinicus*	Leaf	Cold and cough	Giddiness
29	*Evolvulus alsinoides*	Leaf	Cold and cough and fever	Venereal diseases
30	*Alstonia scholaris*	Stem bark	Increase lactation in feeding mother	Swelling
31	*Lantana camera*	Flower	Unknown insects bite	Cold and fever
32	*Gloriosa superba*	Unripe fruit	Ophthalmic problem	Skin diseases
33	*Leucus biflora*	Aerial part	Headache	Skin diseases

Comparative ethnobotanical study between the *Palliyars* and *Kanikkars* reveals that 8 medical plant species (out of Thirty three) exhibit identical uses. They are *Acalypha indica, Aristolochia indica, Curculigo orchioides, Eclipta prostrata, Solanum surattense, Solanum trilobatum, Terminalia bellirica, Tridax procumbens.*

Comparative Ethnomedicobotany with the *Irulars*

Thirty six plants species and commonly used by both the *Palliyars* and *Irulars.* Both similarities and dissimilarities are discussed.

Sl.No.	Botanical Name	Part Used	Palliyars	Irulars
1	*Acalypha furticosa*	Leaf	Honey bee string head ache, rickets in children	Jaundice, stomach pain
2	*Acalypha indica*	Leaf	Cough	Ear pain, snack bite, centipede sting, scabies
3	*Achyranthes aspera*	Leaf	Dog-bite	Fed to new born babies, wheezing, cold and cough, seeds used for snake-bites
4	*Aegle marmelos*	Leaf	Skin disease	Bone fracture
5	*Allium cepa*	Bulb	Scorpion sting unknown insect bite, headache, boils, piles weapon injury	Boils
6	*Andrographis paniculata*	Whole plant	Snake-bite, scorpion-sting	Snake-bite
7	*Anisomeles malabarica*	Leaf	Headache	Paralysis
8	*Aristolochia indica*	Root	Snake bite, scorpion sting, headache, stomachache	Fever, itching on body, snake bite
9	*Asparagus recemosus*	Root tuber	To stop bed -wetting by children	To treat impotency
10	*Azadirachta indica*	Stem bark	Fever	Vermifuge
11	*Bischofia javanica*	Stem bark	Body pain	Bone fracture, dislocation of joints
12	*Calotropis gigantea*	Stem latex	Cold and cough	Abortifacient
13	*Caralluma adscendens*	Stem/ aerial plant part	Coolant	As a tonic for all ailments

Sl.No.	Botanical Name	Part Used	Palliyars	Irulars
14	*Cardiospermum halicacabum*	Leaf	Rheumatism, cold and cough giddiners	Rheumatism, ear pain
15	*Clitoria ternatea*	Root	Kidney stone	Jaundice
16	*Cissus quadrangularis*	Leaves	Relief from tumour like swelling	Dyspepsia
17	*Cocculus hirsutus*	Leaf	Coolant	Body pain
18	*Eclipta prostrata*	Leaf	Coolant, hair tonic	Jaundice, insect and scorpion-bite
19	*Ficus benghalensis*	Stem latex	Fissures in foot	Wound healer
20	*Gymnema sylvestre*	Leaf	Diabetes	Diabetes, Jaundice
21	*Hemidesmus indicus*	Root	Coolant, improve digestion and to over come thirstiness	Snake-bites, to over come thirstiness, coolant, increase blood circulation
22	*Jatropha curcas*	Leaf	Rheumatism	Mouth ulcer
23	*Mimosa pudica*	Leaf	Unknown insect-bite	Reduce body weight
24	*Pergularia daemia*	Leaf	Gastric complaints	Cold and fever in children
25	*Phyllanthus amarus*	Branches/ plant	Jaundice	Jaundice
26	*Phyllanthus emblica*	Fruit	Cold and cough, fever	To prevent acid accumulation in stomach, cold, Diabetes
27	*Pongamia pinnata*	Seed	Cold and cough Jaundice	Joint pain (or) rheumatic pain
28	*Ricinus communis*	Seed oil	Stomachache, laxative, coolant	Redness of eyes
29	*Senna auriculata*	Flower	Colic pain	Small box
30	*Syzygium cumini*	Seed	Diabetes	Diabetes
31	*Tephrosia purpurea*	Root/root bark	Bowel complaints	Headache, stomachache
32	*Terminalia chebula*	Fruit	Dysentery, cold, cough, fever	Indigestion, refrigerant
33	*Tinospora cordifolia*	Whole plant	Rheumatic pain	Fever
34	*Tridax procumbens*	Leaf	Shoulder pain, Weapon injury	Wound injury to stop bleeding antiseptic, wound healer
35	*Vitex negundo*	Leaf	Headache	Body pain, hair lice.
36	*Zingiber officinale*	Dried rhizome	Cold and cough	dyspepsia

Comparative ethnomedicinal study between the *Palliyars* and *Irulars* reveals that 13 medicinal plant species (out of 36) exhibit identical uses. They are *Allium cepa, Andrographis paniculata, Aristolochia indica, Cardiospermum halicacabum, Ficus benghalensis, Gymnema sylvestre, Hemidesmus indicus, Phyllanthus amarus, Phyllanthus emblica, Syzygium cumini, Tephrosia purpurea, Terminalia chebula* and *Tridax procumbens.*

The present study brings to light that all the five tribals in addition to *Palliyars* employ *Tridax procumbens* as an antiseptic to cure wounds/cuts/weapon injury or to check bleeding. *Hemidesmus indicus* with identical medicinal use is being practiced by the *Bhils, Gonds* and *Irulars. Syzygium cumini* is used by the *Bhils, Santals* and *Irulars* to treat diabetes and *Terminalia bellirica* with identitical medicinal use is administered by the *Bhils, Gonds* and *Kanikkars.* Whereas *Andrographis paniculata, Aristolochia indica, Eclipta prostrata, Gymnema sylvestre, Tephrosia purpurea* and *Terminalia chebula* with identical medicinal use are being practiced by at least two ethnic groups under discussion.

Endemic and Endangered Plants of the Study Area

The study area falls in the "hot spot" region in the Western Ghats and the endemic plants are the exclusive biological capital of the nation. Earlier, investigations on endemic and rare plants and their conservation in India are meagre (Jain and Rao, 1983). A large number of endemic plants are at the brink of extinction. And hence the conservation of the threatened endemic species deserves top priority.

Some of the endemic plant species, which were reported to occur in the "hot spot" regions as common or abundant about half-a-century ago, now have become rare or very rate due to over exploitation and are included under the category of endangered species.

The following medicinal plants are identified as the endemic and endangered plant species of the study area (Jain and Rao, 1983; Ahmedullah and Nayar, 1986; Nayar and Sastry, 1987; Nayar, 1996).

Endemic plants of the study area;

1. *Acorus calamus* L.
2. *Argyreia pilosa* Arn.
3. *Ensete superbum* (Roxb.) Cheesman
4. *Hemidesmus indicus* (L.) R.Br. var. *indicus.*
5. *Tabernaemontana divaricata.* (L.) R.Br.ex Roem and Schultes.

Endangered plants of the study area.

1. *Amorphophillus sylvaticus* (Roxb.) Korth.
2. *Gloriosa superba* L.
3. *Swertia corymbosa* (Griseb.) Wight ex Clarke var. *corymbosa*

Conclusion

Since ancient times to the present day as many as 80 per cent of the world's people (WHO) depend on traditional medicine for their primary health care needs in which plants acts as a source of medicines. This great surge of public interest in the use of plants as medicines has been based on the assumption that the plants will be available on a continuing basis. The natural vegetation of the world is disappearing or being altered at an alarming rate; especially, many medicinal plants face extinction or severe genetic loss, but detailed information is lacking. The best means of conservation and ideally all medicinal plant species should be conserved as evolving populations in nature. However, medicinal plant species should be conserved *in situ* by setting aside areas as nature reserves and national parks (collectively termed "Protected Areas") and by ensuring that as many wild species as possible can continue to survive in managed habitats, such as farms and plantation forests and as well as *ex situ* (*i.e.* outside their habitat) which helps to supply plant material for propagation, for re-introduction, for agronomic improvement, for research and for education purposes. Establishment of medicinal gardens and encouraging public; especially, villagers to cultivate medicinal plants can be stimulated by the government at both state and national levels. Despite, putting the information pertaining medicinal plants into computerized databases is also a best approach which has the advantage of enabling retrieval in many different ways yet permitting a constant process of refining and updating.

References

Akerele, O. 1993. Nature's medicinal bounty: don't throw it away. *World Health Forum.* **14**: 390 – 395.

Anonymous. 1895. Some new ideas. *Philadelphia Evening Telegraph* (December 5).

Anyinam, C. 1995. Ecology and Ethnomedicine: Exploring links between current environmental crisis and indigenous medical practices. *Soc.Sci. Med.* **40**: 321 – 329.

Audichya, K.C., Billore, K.V., Joseph, T.G. and Chaturvedi, D.D. 1983. Role of indigenous folk remedies for certain acute illnesses in primary health care. *Nagarjun.* **25**: 199 – 201.

Austin, D.F. and Bourne, G.R. 1992. Notes on Guyana's medical ethnobotany. *Eco. Bot.* **46**: 293 – 298.

Ahmeddullah, M. and Nayar, M.P. 1986. Endemic plants of the Indian Region, Vol. I. Peninsular India, Botanical Survey of India, Culcutta.

Ayyanar, M. and Ignacimuthu, S. 2005. Traditional knowledge of kani tribals in kouthalai of Tirunelveli hills, Tamil Nadu. *J. Ethnopharmacology.* **102**: 246-255.

Binu,S., Shanavaskha, A.K., Santhoskumar, E.S. and Pushpangadan, P. 2003. Plants used as medicine by the Irulars of Palghat district, Kerala, India. *J. Econ. Tax. Bot.* **27**: 808-814.

Borthakur, S.K, and Goswami, N. 1995. Herbal remedies from Dimoria of Kamrup district of Assam in Northeastern India. *Fitoterapia* **LXVI**: 333-340.

Census of India, 1991. Series 1. India, Paper 2 of 1992. Final population totals. Brief analysis of Primary Census abstract.

Chandra, K. 1995. An Ethno-Botanical study on some medicinal plants of district Palamau, (Bihar). *Bull.Med. Eth. Bot. Res.* **XVI**: 11-16.

Chawdhury, R.R. 1992. Why herbal medicines? In: *Herbal Medicine for Human Health.* Regional Publications, SEARO. No.20. WHO: pp 1-3.

Dahmen, F. 1908. The Paliyans, A hill – tribe of the Palni Hills (South India). *Anthropos.* **3**: 19-31.

Das, S.N., Janardhanan, K.P. and Roy, S.C. 1983. Some observations on the ethnobotany of the tribes of Totopara and adjoining areas in Jalpaiguri district, West Bengal. *J. Econ. Tax. Bot.* **4**: 453-474.

Farnsworth, N.R., Akerele, O., Bingal, A.S., Soejarto, D.D. and Guo, Z. 1985. Medicinal plants in therapy. *Bull. WHO.* **63**: 965-981.

Fischer, C.E.C. and Gamble, J.S. 1957. (Rep. Ed.) The Flora of the Presidency of Madras **III**: 1347-2017.

Ford, R.I. (Ed.) 1978. The nature and status of ethnobotany. (Anthropological papers, No. 67. Museum of Anthropology, University of Michigan). The university of Michgan, Ann. Arbor.

Gadgil, M. 1996. Documenting diversity: An experiment. *Curr. Sci.* **70**: 36 – 44.

Gamble, J.S. 1957a. (Rep. Ed.) The Flora of the Presidency of Madras **I**: 1-577.

Gamble, J.S. 1957b. (Rep. Ed.) The Flora of the Presidency of Madras **II**: 578-1346.

Gambel, G.S. 1967. Flora of Presidency of Madras (Rep. Ed.) Botanical Survey of India. Kolkatta.

Girach, R.D. and Aminuddin. 1995. Ethnomedicinal uses of plants among tribals of Singbhum district, Bihar, India. *Ethnobotany.* **7**: 103-107.

Girach, R.D., Singh, S., Brahmam, M. and Misra, M.K. 1999. Traditional treatment of skin diseases in Bhadrak district, Orrissa. *J. Econ. Tax. Bot.* **23**: 499 – 504.

Goel, A.K. and Mudgal, V. 1988. A survey of medicinal plants used by the tribals of Santhal Pargana (Bihar). *J.Econ. Tax. Bot.* **12**: 329-335.

Henry, A.N., Chithra, V. and Balakrishnan, N.P. 1989. Flora of Tamil Nadu, India,

Henry, A.N., Kumari, G.R. and Chithra, V. 1987. Flora of Tamil Nadu, India, Series I: Analysis **II**: 1- 258.

Ignacimuthu, S., Sankarasivaraman, K. and Kesavan, L. 1998. Medico-Ethnobotanical survey among Kanikkar tribals of Mundanthurai Sanctuary, Western Ghats, India. *Fitoterapia.* **LXIX** : 404-414.

Jadhav, D. 2006a. Ethnomedicinal plants used by Bhil of Bibdod, Madhya Pradesh *Ind. J. Trad. Know.* **5**: 263-267.

Jadhav, D. 2006b. Ethnomedicinal survey of Maalgamdi in Ujjain district, Madhya Pradesh, India. *Ethnobotany,* **18**: 157-159.

Jain, S.K. and Mudgal, V. 1999. A hand book of Ethnobotany, Bishen Singh Mahendra Pal Singh Publishers, Dehra Dun, India.

Jain, S.K. and Rao, R.R. 1983. (Eds.) An assessment of threatened plants of India (Proceedings of the Seminar held at Dehra Dun, September 1981), Botanical Survey of India (Department of Environment), Botanic Garden, Howrah.

Jain, S.K. and Rao, R.R. 1983. (Eds.) An assessment of threatened plants of India (Proceedings of the Seminar held at Dehra Dun, September 1981), Botanical Survey of India (Department of Environment), Botanic Garden, Howrah.

Jain, S.K. and Saklani, A. 1992. Cross-cultural Ethnobotanical studies in Northeast India. *Ethnobotany* **4**: 25-38.

Jain, S.K. Banerjee, D.K. and Pal, D.C. 1973. Medicinal plants among certain Adibasis in India. *Bull. Bot. Surv. India.* **15**: 85 – 91.

Joshi, M.C. 1992. Some folk medicines of the tribals of Gujarat. *Bull. Med. Eth. Bot. Res.* **XII**: 115-124.

Joshi, P. 1982. An ethnobotanical study of Bhils – A preliminary survey. *J. Econ. Tax. Bot.* **3**: 257 – 266.

Joshi, P. 1989. Herbal drugs in tribal Rajasthan – from childbirth to childcare. *Ethnobotany* **1**: 77-87.

Katewa, S.S. and Arora, A. 1997. Some plants in folk medicine of Udaipur district (Rajasthan). *Ethnobotany* **9**: 48-51.

Khoshoo, T.N. 1991. Conservation of biodiversity in biosphere. In: *Indian Geosphere Biosphere Programme Some Aspects.* (Eds.) Khoshoo, T.N. and Sharma, M.Nation Academy of Sciences, Allahabad, India. pp. 178-233.

Khoshoo, T.N. 1996. India needs a National biodiversity conservation board. *Curr. Sci.* **71**: 506 – 513.

Martinez, P.H. 1995. Commercialization of wild medicinal plants from Southwest Peubla, Mexico. *Eco. Bot.* **49**: 197-206.

Martini – Bettole, G.B. 1980. Present aspects of the use of plants in traditional medicine. *J. Ethnopharmacology* **2**: 5-7.

Mishra, O.P. and Naquvi, S.M.A. 1995. Ethno-Medico Botany from tribes of Madhya Pradesh. *Bull. Med. Eth. Bot. Res.* **XVI**: 17-26.

Mudgal, V. and Pal, D.C. 1980. Medicinal plants used by tribals of Mayurbhanj (Orissa). *Bull. Bot. Surv. India.* **22**: 59-62.

Nair, N.C. and Henry, A.N. 1983, Flora of Tamil Nadu, India, Series I: Analysis **I**: 1-184.

Nayar, M.P. 1996. "Hot Spots" of endemic plants of India, Nepal and Bhutan, Tropical Botanic Garden and Research Institute, Palode, Thiruvananthapuram, The Ministry of Environment and Forests, New Delhi.

Nayar, M.P. and Sastry, A.R.K. 1987. Red data book of Indian plants, Vol. **I, II**, and **III**., Published by the Director, Botanical Survey of India, **27**: 283-294.

Pal, D.C., Saren, A.M. and Sen, R. 1989. Less known uses of twenty plants from the tribal areas of Bankura district, West Bengal. *J. Econ. Tax. Bot.* **13**: 695-698.

Prasad, P.N., Jabadhas, A.W. and Janaki Ammal, E.K. 1987. Medicinal plants used by the Kanikkars of South India. *J. Econ. Tax. Bot.* **11**: 149-155.

Ragupathy, S. and Mahadevan, A. 1991. Ethnobotany of Kodiakkarai Reserve Forest, Tamil Nadu, South India. *Ethnobotany* **3**: 79-82.

Ramachandran, V.S. and Nair, N.C. 1981a. Ethnobotanical observations on Irulars of Tamil Nadu (India). *J. Econ. Tax. Bot.* **2**: 183-190.

Ramachandran, V.S. and Nair, V.J. 1981b. Ethnobotanical studies in Cannanore district, Kerala State (India). *J. Econ. Tax. Bot.* **2**: 65-72.

Sankarasivaraman, K. 2000. Ethnobotanical wealth of *Paliyar* tribe in Tamil Nadu. Ph.D. Thesis, Manonmaniam Sundaranar University, Tirunelveli, Tamil Nadu.

Saxena, H.O. 1986. Observations on the ethnobotany of Madhya Pradesh. *Bull. Bot. Surv. India* **28**: 149-156.

Saxena, S.K. and Tripathi, J.P. 1989. Ethnobotany of Bundelkhand. Studies on the medicinal uses of wild trees by the tribal inhabitants of Bundlekhand region. *J. Econ. Tax. Bot.* **13**: 381-389.

Shah, G.L. and Gopal, G.V. 1985. Ethnomedicinal notes from the tribal inhabitants of the North Gujarat (India). *J. Econ. Tax. Bot.* **6**: 193 – 201.

Shankar, R. 1995. Tribal community in India and PGR, In: *Farmer's rights and plant genetic resources recognition and reward: A dialogue.* (Ed.) Swaminathan, M.S. Million India Limited, Madras, India. pp.106-111.

Singh, S.K. 2004. Ethnobotany in the present perspective. *Agrobios. News letter.* **2**: p.10.

Sur, P.R., Sen, R., Halder, A.C. and Bandyopadhyay, S. 1987. Observation on the ethnobotany of Malda – West Dinajpur districts, West Bengal – I. *J. Econ. Tax. Bot.* **10**: 395-401.

Sur, P.R., Sen, R., Halder, A.C. and Bandyopadhyay, S. 1990. Observation on the ethnobotany of Malda – West Dinajpur districts, West Bengal – II. *J. Econ. Tax. Bot.* **10**: 453-459.

Tarafder, C.R. 1983. Ethnogynaecology in relation to plants part – II. Plants used for abortion. *J. Econ. Tax. Bot.* **4**: 507-516.

Tarafder, C.R. 1984. Medicinal plants traditionally used by the tribals of Ranchi and Hazaribagh districts, Bihar: Skin diseases and sores. *Bull. Bot. Surv.India* **26**: 149-153.

Viswanathan, M.B., Premkumar, E.H., and Ramesh.N. 2001. Ethnomedicines of Kans in Kalakad – Mundanthurai Tigar Reserve, Tamil Nadu. *Ethnobotany* **13**:60-66.

Yadav, S.S. and Bhamare, P.B. 1989. Ethnomedico-Botanical studies of Dhule forests in Maharashtra State. *J. Econ. Tax. Bot.* **13**:455-460.

2018, Ethnomedicinal Plants: A Biodiversity Treasure *Pages* **67–122**
Editors: ***V.R. Mohan, A. Doss, P.S. Tresina and V. Sornalakshmi***
Published by: **ASTRAL INTERNATIONAL PVT. LTD., NEW DELHI**

Chapter 2

Ethnomedicinal Plants of Agasthiarmalai Biosphere Reserve, Tamil Nadu: An Investigation

A. Nishanthini and V.R. Mohan

Ethnopharmacology Unit,
PG and Research Department of Botany, V.O. Chidambaram College,
Tuticorin – 628 008, Tamil Nadu
E-mail: vrmohanvoc@gmail.com

Introduction

The interest in the therapeutic value of plants is as old as Man's interest in his life. The Roman civilization emphasized the value of Men's sana in corpore sano, which is popularly known as "A sound mind in a sound body". India's contribution to evolving a systematic body of knowledge towards rejuvenation of man precedes most of the established and well-known Western and Euro-centric systems.

In a much influential treatise by Lele (1986) titled, Ayurveda and Modern Medicine, there is a reference to the intimate communion between man and nature from which our knowledge of Herbal therapeutic practices has emerged. In the Atharva Veda Kand 8; Sutra 7.1, the following is recorded: "The boar knows the plant; the Mongoose knows the remedial herb; what ones the serpents, the Gandharvas know; those I call to aid. What herbs of the Angivasas the eagles know, what heavenly ones the Raghatas know, what ones the birds and the swans know, and what all the winged ones, what herbs the wild beasts know, these I call to aid for you. Of how many herbs the inviolable kine partakes, of how many the goats and sheep, let so many herbs, being brought, extend protection to thee."

Popular knowledge of plants which can be used by humans is based on thousands of years' experience. By "trial and error", people learnt how to recognize and use plants including those with a magic religious function (Rodrigues *et al.*, 2003). Traditional medicine or ethnomedicine is a set of empirical practices embedded in the knowledge of a social group often transmitted orally from generation to generation with the intent to solve health problems. It is an alternative to western medicine and is strongly linked to religious beliefs and practices of indigenous cultures. Medicinal plant lore or herbal medicine is a major component of traditional medicine (Bussmann and Sharon, 2006).

Many infectious diseases are known to be treated with herbal remedies throughout the history of mankind. Even today, plant materials continue to play a major role in primary health care as therapeutic remedies in many developing countries (Zakaria, 1991). Plants still continue to be almost the exclusive source of drugs for a majority of the world's population (Hamburger and Hostettman, 1991).

Medicinal plants represent an important health and economic component of biodiversity. It is essential to make the complete inventory of the medicinal component of the flora of any country for conservation and sustainable use. The conservation of the threatened and endangered medicinal species in the wild is indispensable (Rahman *et al.*, 2004).

It is a matter of great pride that among the 16 hot spots known for rich flora in the world, two are located in India. They are the Eastern Himalayas and the Western Ghats (Khoshoo, 1996). The hill chain of Western Ghats recognized as a region of high level of biodiversity is under the threat of rapid loss of genetic resources (Gadgil, 1996). The biodiverse nature of the Eastern Ghats is meagre. A perusal of the available literature reveals that till date there is no comprehensive survey, documentation and enumeration of wild medicinal plants used by the tribe *Kanikkars* inhabiting the Agasthiarmalai Biosphere Reserve, Western Ghats, Tamil Nadu, India. Hence, in the present study an attempt is made for the survey and enumeration of wild medicinal plants in the Agasthiarmalai Biosphere Reserve, Western Ghats, Tamil Nadu. The *Kanikkars* are the dominant tribal group inhabiting this locality. The present study focuses on the dependence of the *Kanikkars* on herbal medicines and attempts at an exhaustive analysis of the therapeutic values of such medicinal plants.

The *Kanikkars* have asserted this ethnic identity and practised the traditional ethnomedicine is a standing example of the symbiotic relationship with man had enjoyed from time immemorial. If it is documented and intellectual rights are guaranteed to them, not only the *Kanikkars* but the whole nation would benefit.

Study Area

The Agasthiarmalai Biosphere Reserve lies between 8° 22′ and 8° 53′ North latitude and between 77° 10′ and 77° 35′ East longitudes in Tirunelveli and Kanyakumari district of Tamil Nadu in the Southern Western Ghats of India (Area Map 2.1). The boundaries are Ambasamudram and Tenkasi taluks of Tirunelveli districts in the north, Ambasamudram and Nanguneri taluks of Tirunelveli districts in the east, Kanyakumari district in the south and Kerala State in the west.

Geographically, it is a part of South Western tip of the Western Ghats, a region that is known for its species richness, diversity and high degree of endemism. The

Area Map
Agasthiarmalai Biosphere Reserve
N
W
E
S
INDIA
08°45'N
08°25'N
Ponmudi Peak
Pothigaimudi Peak
Valayar
Papanasam
Servalar
Ambasamudram
Mundanduri
Kannikatty
Karaiyar
Inchikuzhi
panathirtham
TAMIL NADU
KERALA
Kakachi
Kodayar
MUNDANDURAI SANCTUARY
KALAKAD SANCTUARY
Manimuthar
Muthukuzhivayal
Mahendragiri Peak
Thiruvananthapuram
Tirunelveli
Palayamkottai
Legend
Mountain Peak
Place
Main tourist areas
Reservoirs
Study area
Road
River
Reserve Boundary
77°15'E
77°35'E

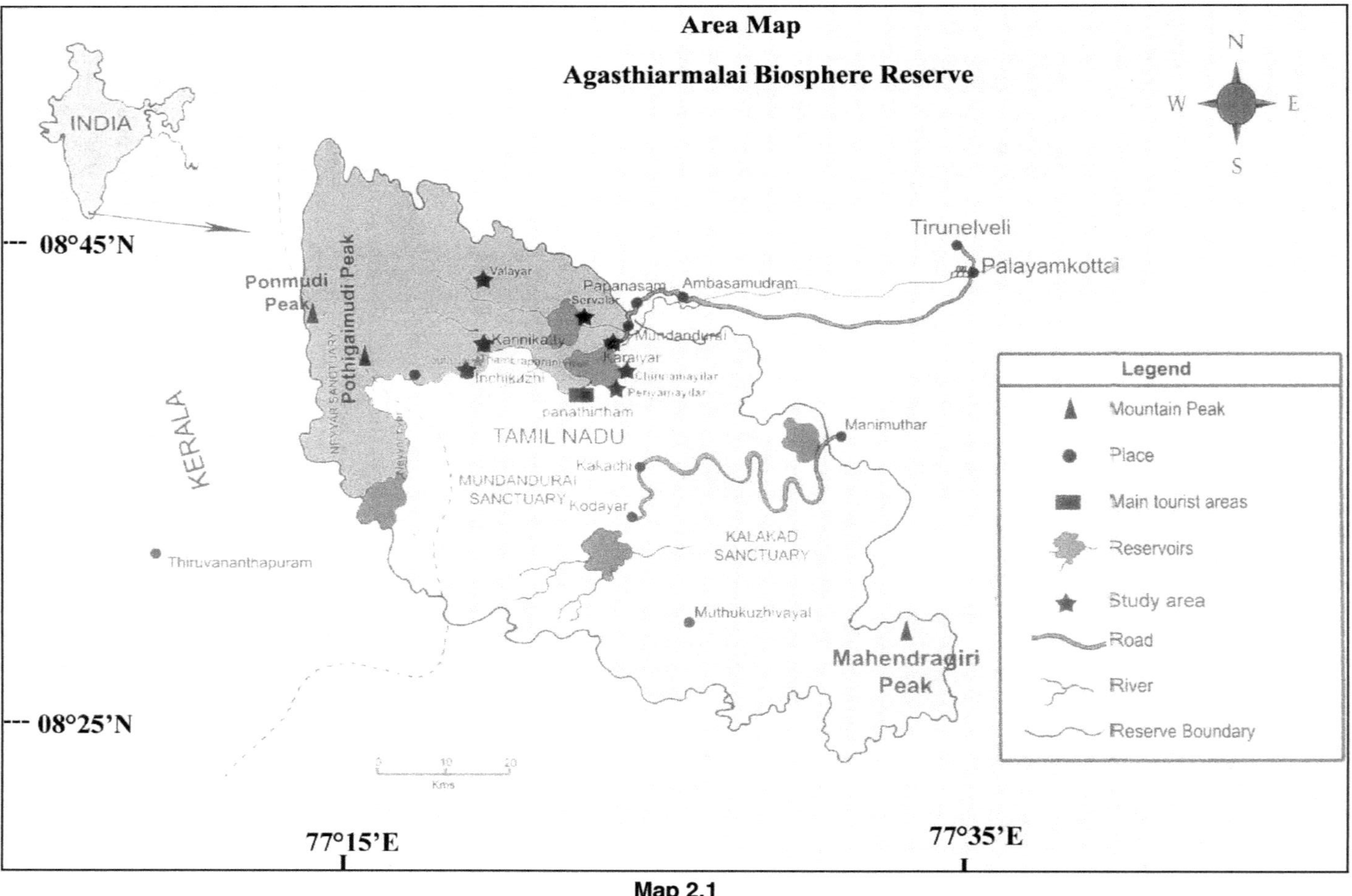

Map 2.1

Agasthiarmalai Biosphere Reserve area has been recognized as one of the 'hot-spots' for Biodiversity conservation by the IUCN (Ayyanar and Ignacimuthu, 2005). The altitude ranges from 100m to 1867m (MSL). It receives rainfall during the South West as well as North East monsoon. The important peaks of the reserve include Agasthiarmalai, Ainthilaipothigai and Nagapothigai. Tamiraparani, the perennial river of Tamil Nadu originates from Agasthiarmalai (Pothigaimalai) and flows through this sanctuary.

The varied climatic and topographic conditions prevailing in the sanctuary present a remarkable diversity of both the flora and fauna. The study area consists of tropical wet evergreen forests, semi evergreen forests, moist deciduous forests, tropical riparian fringe forests, dry teak forests, dry deciduous forests, umbrella thorn forest, wet temperate forests and high and low altitude grass lands.

In the study area (Plate 2.1), the *Kanikkars* live either in isolated pockets or small hamlets. Their habitations are known by the following names.

(i) Ingikuzhi, (ii) Chinnamylar, (iii) Periyamylar, (iv) Agastiyarkanikudiyiruppu, (v) Tharuvattampari Kani Kudiyiruppu (Servalar) (vi) Kouthalai and (vii) Valayar

The Kanikkar Tribe

The Kanikkars belong to the Southern tribal zone. The *Kanikkars* are also known as kanikaran or Kani. The Southern tribals are historically more ancient tribes. There are as many as 36 types of scheduled tribes in Tamil Nadu. In the serialized list notified by the Government of Tamil Nadu, the *Kanikkars* are placed at 7th position. They live in low altitude regions of Western Ghats and live in large numbers. Kanikkars means "hereditary proprietor of land" thus recognizing their ancient rights over the forest lands.

The *Kanikkars* are generally very short in stature and meager in appearance, from their active habits and scanty food. Some have markedly Negroid features. They are simple and straight forward tribals. They are traditionally a nomadic community. They speak in their own dialect, Kanikkar Bhasha or Malampashi, which is close to the Dravidian language Malayalam. (Singh, 1994) They were once lords of the forest and practiced migratory cultivation but *Kanikkars* have now to a large extent abandoned this kind of migratory cultivation because of the forest may not be set fire to or trees felled at the unrestricted pleasure of individuals.

Most of the *Kanikkar* tribals have a general knowledge of medicinal plants that are used for first aid remedies, to treat cough, cold, fever, headache, poisonous bites and some other simple ailments. Kanis still supplement their food by gathering roots and tubers from the near by forest areas. They eat tubers like *Manihot esculenta* and *Dioscorea oppositifolia etc.* They are extremely hard working and can survive without the help of modern facilities. They are socio-economically backward and most of them are very poor. They are also engaged in seasonal collection of honey, bee wax and some minor forest produce. They cultivate edible plants like, tapioca, banana, millets and cash crops such as pepper, areca nut and cashew nut.

Kanikkars in the study area

A view of Western Ghats

A very old kani woman

A view of periyamylar hills

A Kanikkar family of Agastiyarkanikudiyiruppu

A view of servalar

A kanikkar family of servalar

Plate 2.1

Survey and Collection

Survey and collection of medicinal plants used by the *Kanikkars* tribe of the Agasthiarmalai Biosphere Reserve, Western Ghats, Tamil Nadu, were carried out over a period of 24 months. Frequent field trips were undertaken to the areas of study (Vide Area Map). Information regarding the medicinal plants was gathered by meeting *Kanikkars* practicing indigenous medicine, during explorative field trips and by gaining a good rapport and winning over their confidence. Most of the information include in this study was gathered from the elderly and experienced medicine men who have a long acquaintance with the use of medicinal plants. The field notebook delineates all the usage procedures adopted by the tribals. The information thus gathered was cross checked adequately for reliability and accuracy by interacting with different groups of the *Kanikkars* from different habitats to confirm the use, mode of administration as well as dosage differences, if any. After eliciting detailed information regarding the wild medicinal plants (Table 2.1) and any plant part(s)/extraction of plant part obtained from local market used in medicinal preparations, as ingredients (Table 2.2), they were carefully brought to the laboratory for identification. Herbaria for all the collected plant specimens except for plants, whose parts or extracts were procured from the local market, were prepared and deposited in the Ethnopharmacology Unit, Research Department of Botany, V.O. Chidambaram College, Tuticorin, Tamil Nadu, India.

The collected plants were identified by referring to the following compilations (Fischer and Gamble, 1957; Gamble, 1957a; Gamble, 1957b; Hajra *et al.*, 1997; Henry *et al.*, 1989; Henry *et al.*, 1987; Matthew, 1983; Matthew, 1983a; Matthew, 1983b; Matthew, 1999a; Matthew, 1999b; Matthew, 1999c; Nair and Henry, 1983; Sasidharan and Sivarajan, 1996; Sharma *et al.*, 1993; Sharma and Balakrishnan, 1993 and Sharma and Sanjappa, 1993). The medical terms of some diseases were referred from the following volumes (Sambasivam Pillai, 1931; Sambasivam Pillai, 1977; Sambasivam Pillai, 1978.)

After authentication with regional flora, the specimens were matched with the authentic herbarium of Madras Herbarium, Botanical Survey of India (B.S.I), Southern Circle, Coimbatore, Tamil Nadu, India.

Results

As many as 235 (219 plants collected and 16 plant/extracts procured) ethnomedicinal plants were identified as being used by the Kanikkars tribals of Agasthiarmalai Biosphere Reserve, Western Ghats, Tamil Nadu, India. They were tabulated (Tables 2.1 and 2.2) alphabetically with botanical name, family name, vernacular name, plant parts used, mode of administration, dosage *etc.* (Plates 2.2–2.10).

Discussion

The tribal's knowledge of indigenous uses of native medicinal plants before their exodus into the urban areas to join the mainstream life needs to be studied and documented. In the present study 235 medicinal plants were collected and documented (Tables 2.1 and 2.2). The present study focuses the extensive usage of

Table 2.1: List of Ethnomedicinal Plants Collected and Documented

Botanical Name and Family	*Vernacular Name*	*Mode of Administration*
Abrus precatorius L. Fabaceae	Kundumani	Two to three grams of fresh leaves are chewed carefully without swallowing the salva for a minute. Then the chewed leaves are spat out and the mouth is washed well with water. This procedure is repeated for two times a day for a period of three to four day to get relief from sensitiveness of teeth. About fifty grams of warm seed paste is applied over the affected joints twice a day until one experiences relief from stiffness of joints. A drop of pure filtered seed extract prepared with clean water is allowed to settle on the eyeballs twice a day for about three days to reduce irritation. Two to three grams of fresh leaves or roots of the plant with seeds are made into a paste and consumed along with cold water or cow's milk two times a day for five to seven days to cure any poisonous bites.
Abutilon indicum (L.) Sweet Malvaceae	Thuthi	The juice prepared from ten grams of crushed leaves is applied externally two times a day for a period of three days to get relief from earache.
Acacia caesia (L.) Willd (*Mimosa caesia* L.) Mimosaceae	Vellinjal	The dried bark is made into powder and is used like soap to treat scabies, rashes and ringworm infection.
Acalypha racemosa Heyne ex Baill Euphorbiaceae	Sirusinni	The juice prepared from ten to fifteen grams of fresh leaves with fifty ml of water is taken orally two times a day for two days to get relief from indigestion.
Acalypha indica (L.)	Kupaimeni	The paste prepared from 10 gram of fresh leaves with water is given to a children for 3 day in a single dose to arrest cold and cough.
Achyranthes aspera L. Amaranthaceae	Nayurivi	The paste prepared from fifteen grams of fresh leaves with water is applied externally two times a day for a period of ten days to treat dog bites.
Acorus calamus L. Araceae	Vasambu	A thread is inserted through a hole made in the dried rhizome and is worn as a necklace to ward off giddiness.
Actinopteris radiata (Sw) Link Actiniopteridaceae	Mayilkapul	Twenty grams of powder prepared from shade dried leaves is boiled in coconut oil (*Cocos nucifera* L.) and applied externally two times a day for a period of one month to treat fracture.
Aganosma cymosa (Roxb.)G. Don Apocynaceae	Oodhavaly kodie	The juice prepared from twenty grams of stem is boiled with 200ml of water is taken orally two times a day for a period of ten days to treat jaundice.

Contd...

Botanical Name and Family	*Vernacular Name*	*Mode of Administration*
Ageratina adenophora (Spreng.) R. King and H. Robinson	Anandam poochade	The paste prepared from 10 gram of leaves with water is applied externally on the chest to relieve pain when the person suffers a mild heart attack.
Ageratum conzyzoides L.	Narachava Vandana pachilai	The paste prepared from 20 gram of fresh leaves mixed with salt is taken orally two times a day for a period of 2 days to treat indigestion.
Aloe barbadensis Mill. (*A. vera* (L) Burm.F) Liliaceae	Chothukathalai	A fresh leaf is taken orally as such after removing the epidermal peel, once in a day for a period of three to four days to reduce body heat. One fresh leaf is taken per day for about ten days after removing the epidermal peel to cure kidney stone.
Alpinia calcarata Roscoe Zingiberaceae	Sitharathai	The paste prepared from ten grams of the rhizome mixed with water is taken orally twice a day for two days to get relief from stomach disorders.
Alstonia scholaris (L.) R.BR. Apocynaceae	Mukkampalai	Ten grams of the fresh stem bark is made into a paste with a few drops of water. This paste is mixed with one hundred ml of cow's milk or rice fermented water and taken by nursing mothers twice a day for seven to ten days to improve lactation. One teaspoon of the powder made from shade dried leaves is orally administered in water or along with a few drops of the stem latex of the same plant, once in a day for a period of one month to get relief from asthma.
Alstonia venenata R.Br. Apocynaceae	Malaivada modake	The paste prepared from twenty grams of fresh leaves with a few drops of water is applied externally two times a day for a period of one month to get relief from rheumatic pain.
Alysicarpus bupleurifolius (L.) DC.	Tholkanie	The juice prepared from ten to fifteen grams of leaves with two hundred ml of water is taken orally twice a day for a period of three days to reduce heart palpitation.
Amaranthus viridis L. Amaranthaceae	Kuppaikeerai	A handful of the whole plant paste is externally applied in the morning hours to treat scabies.
Amorphophallus paeoniifolius (Dennst.) Nicolson var. *campanulatus* (Blume ex Dence) Sivadasan	Kattukarunai	One hundred grams of the tuber is boiled in water with an equal amount of fresh leaves of tamarind (*Tamarindus indica* L). Then the peeled tuber is cooked in tamarind. The prepared curry is taken as medicine once a day for seven to ten days to get relief from bleecing piles.
Ammannia baccifera L. subsp. *aegyptiaca* (Willd) Koehne Lythraceae	Kallurvi	The paste prepared from ten grams of whole plant with water is mixed with lemon (*Citrus aurantifolia* (Chrishm) Swingle) juice and it is applied externally two times a day for a period of one week to treat eczema.

Botanical Name and Family	*Vernacular Name*	*Mode of Administration*
Anaphyllum beddomei Engler. Araceae	Keerikilangu	About ten grams of the cleaned fresh rhizome made into a paste is externally applied twice a day to treat eczema and scabies.
Anaphyllum wightii Schott.	Kovarakanda mooli	The paste prepared from 10 gram of the bulb is applied externally twice a day for one week to treat snake bite.
Ancistrocladus heyneanus wall.ex Graham Ancistrocladaceae	Malvaduda-madky	The paste prepared from twenty grams of leaves with water is applied externally two times a day for a period of one month to treat rheumatism.
Andrographis paniculata (Burm.f.) Wall.ex Nees Acanthaceae	Neelavembu	The paste prepared from ten grams of leaves along with ten grams of the root of adathoda (*Justicia adathoda* L.) is applied externally two times a day for a period of four days to treat poisonous insect bites.
Anisomeles malabarica (L.) R. Br.ex Sims Lamiaceae	Perunthumbai	Handful of leaves are boiled in two litres of water in a closed container. When the steam emanates the lid of the container is slowly removed and vapour bath is administered to get relief from rheumatic pain
Anisomeles indica (L) Kuntze.	Vaathaneer-patchilai	One hundred gram of leaves are boiled in water and the warm water is used during bath for a period of one week to get relief from rheumatic pain. About one hundred grams of fresh leaves and stem are boiled in one litre of neem oil (*Azadirachta indica* A.Juss) along with ten grams of poppy seeds (*Papaver somniferum* L.) and fifty grams of vellaipoondu (*Allium sativum* L.) in an earthen pot for ten to fifteen minutes. This filtered oil is applied on the joints of legs and hands and gently massaged. This treatment gives ample relief from rheumatic pain.
Argemone mexicana L. Papavaraceae	Kudiyoetti	About ten grams of the leaf paste is externally applied once in a day to treat ringworm infection.
Aristolochia bracteolata Lam. Aristolochiaceae	Aaduthinnapalai	Handful of the whole plant paste is externally applied twice a day to treat eczema, scabies and ringworm infection.
Aristolochia indica L. Aristolochiaceae	Thalaisurulivaer	One teaspoonful of juice prepared from ten grams of fresh root with water is taken orally three times a day for two days to treat snake bite.

Contd...

Botanical Name and Family	*Vernacular Name*	*Mode of Administration*
Aristolochia krysagathra Sivaranjan and Pradeep Aristolochiaceae	Karudakodi	About fifty grams of root and an equal quantity of leaves are boiled in one litre of coconut oil (*Cocos nucifera* L.) for about fifteen to twenty minutes over a low flame. This oil is filtered after cooling and applied on the head once in a day for seven to ten days as the treatment for rheumatism. This therapy is used to reduce excessive body heat.
Aristolochia tagala Cham Aristolochiaceae	Karudakudi	The paste prepared from twenty grams of the root is applied externally twice a day for a period of one week to treat snake bite and insect bite.
Asystasia travancorica Bedd. Acanthaceae	Aathuurinji	About ten grams of a paste made from the leaves and the flowers is mixed with honey and is taken orally, twice a day, for three weeks for the treatment of rheumatism. The use of tamarind, fish and egg is avoided.
Azadirachta indica A Juss Meliaceae	Vembu	The oil extracted from the seeds is massaged over the joints as an embrocation in rheumatism. Neem oil is used in many medicated preparations for the treatment of rheumatism. About twenty to fifty grams of stem bark is boiled in five hundred ml of water for twenty to thirty minutes. This decotion is taken orally for three weeks to treat rheumatic complaints.
Baccaurea courtallensis (Wight). Mullel Arg.	Poovampalam	The juice prepared from 10 gram of fruits with 100 ml of water is taken orally twice a day for one week to reduce the excess of water content in the body.
Bacopa monnieri (L.) Pennell (*Moniera cuneifolia* Michaux) Scrophulariaceae	Nirbrahmi	Fifty to one hundred grams of whole plant is boiled in coconut oil (*Cocos nucifera* L.) and the oil is applied on the knees three times a day for three days to get relief from pain on the knees.
Bauhinia racemosa Lam. Caesalpiniaceae	Aathi	Two to three drops of pure filtered juice prepared from three to five fresh leaves with water is applied on the eyes twice a day to remove particles like dust.
Begonia floccifera Bedd. Begoniaceae	Kalthamare	The juice prepared from ten grams of the fresh leaves with two hundred ml of water is taken orally in empty stomach for a period of two days in a single dose to relieve stomach pain.
Begonia subpeltata Wight Begoniaceae	Oruyelathamare	The paste prepared from twenty grams of leaves with water is taken orally three times a day to get relief from stomachache.
Begonia malabarica Lam	Narayana sanjeevi	The paste prepared from 10 gram of fresh leaves is applied externally once in a day for one week to treat foot sores.

Botanical Name and Family	*Vernacular Name*	*Mode of Administration*
Biophytum sensitivum (L.) DC (*Oxalis sensitiva*) L. Oxalidaceae	Manivatti-patchilai	Twenty grams of leaf is made into a paste with water and is applied externally once in a day to treat skin rashes and eczema.
Bischofia javanica Blume Bischofiaceae	Milachityan	The sap from the bark of the plant is applied externally three times a day for a period of two days to treat inflammation due to an injury.
Blepharis maderaspatensis (L.) Heyne ex Roth Acanthaceae	Sadhaiotti	The paste prepared from fresh leaves with water is applied externally once in a day for a period of one week to heal the wounds.
Boerhavia diffusa L. (*B.* repens L.) Nyctaginaceae	Vethalamai	One hundred grams of leaves is boiled in half a litre of coconut oil (*Cocos nucifera* L.) and the oil extract is applied twice a day to treat scabies and ringworm infection.
Bridelia scandens (Roxb.)Willd Euphorbiaceae	Akakavalie	The paste prepared from twenty five grams of the stem bark with water is taken orally in a single dose for a period of twenty days to treat jaundice.
Bulbophyllum albidum (Wight) Hook. f Orchidaceae	Kalmalpuloravi	The juice prepared from twenty grams of fresh leaves and bulb along with water is taken orally two times a day for a period of one month for strengthening of a weak uterus for conception.
Cadaba trifoliata (Roxb.) Wight and Arn Capparaceae	Mayiladikurutha	The paste prepared from twenty grams of whole plant with a few drops of water is applied externally two times a day for a period of twenty days to treat rheumatism.
Calanthe masuca (D.Don)Lindl Orchidaceae	Kalai kombu	About ten grams of dried tuberous root is powdered and made into a paste with water. This paste is used to treat acnes and sebaceous cysts.
Calophyllum inophyllum L. Clusiaceae	Punnai	Few drops of oil extracted from the seeds are applied externally twice a day to treat ringworm infection.
Calotropis gigantea (L.) R.Br Asclepiadaceae	Erukku	A poultice of roasted leaves is applied on the rheumatic joints for one week or more to soothe pain and to reduce swelling.
Calycopteris floribunda Lam. Combretaceae	Minnarkodi	Ten grams of the leaf paste is applied externally twice a day to treat scabies and skin inflammation.

Contd...

Botanical Name and Family	Vernacular Name	Mode of Administration
Canarium strictum Roxb. Burseraceae	Kungiliam	Fifteen grams of powder prepared from the gum (exudate from the bark) is boiled with coconut oil (*Cocos nucifera* L.) and is applied externally three times a day for a period of two days to get relief from skin allergy and boils.
Canthium parviflorum Lam. Rubiaceae	Kattikarai	The paste prepared from twenty grams of root mixed with coconut oil (*Cocos nucifera* L.) is applied externally three times a day for a period of four days to treat boils.
Capparis sepiaria L. Capparaceae	Muruvilikodi	Ten grams of the fresh bark of the stem is made into paste with lemon juice (*Citrus aurantifolia* (Christhm) Swingle) and is applied externally once in a day for a period of one week to treat eczema and dandruff.
Capsicum annuum L. (*C. frutescens* Sensu Clarke) Solanaceae	Usimilagai	A thick paste made from twenty grams of fruits, along with ten grams of ginger (*Zingiber officinale* Roscoe) and ten grams of vellaipoondu (*A llium sativum* L.) is given orally to women after delivery. This is followed by the administration of fifty milliliters of gingelly oil (*Sesamum indicum* L.) or honey to prevent rheumatic complaints
Carissa carandas L. Apocynaceae	Kalaaka	Fresh fruits are taken as a coolant twice a day for a period of three days to reduce body heat.
Carica papaya L. Caricaceae	Pappali	The juice prepared from twenty five grams of fruit with milk and sugar is taken orally for a period of one week, two times a day to cause abortion.
Carmona retusa (Vahl) Masamune (*Ehretiamicrophylla* Lam) Boraginaceae	Seethavetrilai	The paste prepared from ten grams of fresh leaves and stem along with ten grams of the leaves of keerai (*Phyllanthus lawii* Graham ex Mull-Arg) and applied externally two times a day for a period of six days to get relief from swelling of legs.
Cardiospermum halicacabum L. Sapindaceae	Mudukottan	Five grams of fresh leaves one gram of jeera (*Cuminum cyminum* L.), five to six leaves of thoothuvalai (*Solanum trilobatum* L.), two to three leaves of Kandangattiri (*S. surattense* Burm.f.) and hundred grams of asafoetida (*Ferula assafoetida* L.) are boiled in three hundered ml of water. This decoction is filtered and taken orally once in a day for four to five days to relive cold and cough.
		The paste prepared from ten to fifteen grams of fresh leaves are applied externally once in a day for a period of one month to get relief from rheumatism.

Botanical Name and Family	Vernacular Name	Mode of Administration
		One hundred ml of juice extracted from a sufficient amount of fresh leaves is taken orally with palm sugar (*Borassus flabellifer* L.) for a period of one week to treat rheumatism. The use of tamarind, meat, fish and egg are avoided.
		Two or three grams of fresh leaves are continuously boiled along with two to three grams of fresh leaves of thoothuvalai (*Solanum trilobatum* L.) and kanadangathiri (*S. surattense* Brum.F), a pinch of asafoetida (*Ferula assafoetida* L.) and one gram of jeera (*Cuminum cyminum* L.) in three hundred ml of water to get fifty ml decoction. This filtered decoction is taken in empty stomach once in a day for a period of two days to get relief from giddiness.
Cassia alata L. Caesalpiniaceae	Yanaithavarai	Twenty grams of fresh leaf paste is applied twice a day to treat ring worm infection.
Cassia hirsuta L.	Oolanthavarai	Fifty gram of seeds are powdered and made into a decoction with 200 ml of water and administered orally for 2 days to reduce high blood pressure
Cassia kleinii Wight and Arn Caesalpiniaceae	Mullillathottavadi	Handful of the whole plant is made into a paste and is applied externally twice a day to treat eczema.
Centella asiatica (L.) Urban. Apiaceae	Vallara kerai	The paste prepared from ten grams of fresh leaves with water is taken orally two times a day to increase memory power.
Chlorophytum heynei Rottl.ex Baker Liliaceae	Thengabur	The paste prepared from ten to twenty grams of bulb mixed with coconut milk (*Cocos nucifera* L.) is taken orally for a period of one week in a single dose to treat ulcer.
Chlorophytum laxum R.Br Liliaceae	Pelamanak	The paste prepared from ten grams of fresh leaves is mixed with honey is administered two times a day for period of three days to treat indigestion
Cipadessa baccifera (Roth) Miq Meliaceae	Pullipamcheddi.	The paste prepared from twenty grams of fresh leaves with water is applied externally twice a day for a period of two days to get relief from wasp sting.
Cissus quadrangularis L. Vitaceae	Pirandai	About ten grams of a paste made from the tender shoots is consumed twice a day for three weeks for the treatment of rheumatism
Cissampelos pareira L. Menispermaceae	Malaithankivaer	The paste prepared from leaves with water is taken orally along with cow's milk two times a day for a period of two days to get relief from lumbago

Contd...

Botanical Name and Family	Vernacular Name	Mode of Administration
Clerodendrum inerme (L.) Gaertn (*Valkameria inermis* L.) Verbenaceae	Changukuppi.	The leaf paste is externally applied once in a day for a week to treat psoriasis, scabies and ringworm infection. The leaf paste is also applied on the site of insect bite.
Clerodendrum viscosum Vent. Verbenaceae	Perungilai	The paste prepared from ten to fifteen grams of leaves is applied externally two times a day for a period of three days to prevent excessive bleeding due to injury.
Clitoria ternatea L. Fabaceae	Changu-pushpam	Twenty grams of leaf paste is externally applied twice a day to treat skin inflammation, scabies and ringworm infection
Coccinia grandis (L.) Voig (*C. indica* Wight and Arn) Cucurbitaceae	Kovai	The paste prepared from ten grams of fruit with water is applied externally on the gums two times a day for a period of two days to get relief from toothache.
Commelina benghalensis L Commelinace	Valaipachai	The leaf paste is applied externally twice a day to treat scabies. The leaf paste is also applied once in a day on the wounds for healing and to remove the poisonous spines that had struck accidently on the body parts.
Connarus monocarpus L.	Ponkornadi	Fifty gram of root is boiled in 300 ml of water and is taken orally two times a day to treat general tiredness (as effective as glucose).
Corallocarpus epigaeus (Rottl and willd.) Clarke Cucurbitaceae	Akashagaruden	The paste prepared from twenty grams of bulb with water is mixed with honey and is taken orally two times a day for a period of two day to treat cold and cough in children.
Coscinium fenestratum (Gaertn) Coleb Menispermaceae	Maramanjal	About fifty grams of the dried climbing stem are boiled in one hundred ml of coconut oil (*Cocos nucifera* L.). Few drops of this oil extract are applied on the burns twice a day for healing. The stem paste is used to treat acnes
Costus speciosus (Koen.)J.E. Smith Costaceae	Koshtam	The rhizome and leaf paste are externally applied once in a day to treat ringworm infection.
Crataeva magna (Lour.) DC Capparaceae	Maralingam	The juice prepared from five to ten grams of fresh stem bark with one hundred ml of water is taken orally in empty stomach once in a day for a period of one week during the early stages of pregnancy to cause abortion
Crinum defixum Ker-Gawl Amaryllidaceae	Vishanarayani	About ten bulbs made into a paste with water are applied externally twice a day on the inflamed skin that is formed due to insect bites or allergy.

Botanical Name and Family	Vernacular Name	Mode of Administration
Croton tiglium L. Euphorbiaceae	Nervilankai	The seeds are purified by soaking them in lime water for two days. The oil expelled from the seeds is applied twice a day on the infected areas to treat eczema. The oil is also used to treat alopecia. The leaf paste is externally applied twice a day to treat scabies and ringworm infection.
Curculigo orchioides Gaertn Hypoxidaceae	Nilapenani	The paste prepared from ten to fifteen grams of the root tuber is applied externally three times a day for a period of two days to treat insect bite.
Cyclea peltata (Lam.) Hook.f. and Thomas	Thazhi	One teaspoonful of the juice prepared from 10 grams of fresh leaves with water is given orally twice a day for 2-3 days to reduce excessive body heat, eye burn and stomachache.
Cymbopogon citratus (*DC.*) Stapf (*Andropogon citratus* DC) Poaceae	Chukkuna-aripullu	A sufficient quantity of the whole plant is chopped and boiled in water for thirty minutes and the warm water is used during bath twice a day for twelve days as treatment for rheumatism. Handful of fresh leaves is crushed with water into juice. After filtration, half a cup of this juice is given orally twice or thrice a day for one week as an effective management of rheumatism.
Datura metal L. (*D. fastuosa* L.) Solanaceae	Oomathai	The leaves are soaked in boiling water and they are bandaged over the affected part to get relief from rheumatic pain.The boiled leaves are also used for fomentation on the rheumatic swelling for fifteen to twenty minutes.
Dioscorea bulbifera L. var. *vera* Prain and Burkill Dioscoreaceae	Vethalaivalli	Fifty grams of fresh or boiled tuber is taken orally two times a day for a period of two to three days to arrest Karuvalli dysentery.
Dioscorea pentaphylla L.var *pentaphylla* (*D.pentaphylla* L. var. *linnaei* Prain and Burkill) Dioscoreaceae	Mullvalli	Twenty grams of the tuber is fried in ghee and made into a paste. This paste is taken orally three times a day for a period of four days for general good health.
Dioscorea tomentosa Koen ex Spreng Dioscoreaceae	Noolvalli	Ten grams of boiled and peeled tuber is given to children once in a day for three days to relieve bowel complaints
Diospyros melanoxylon Roxb. Ebenaceae	Vaathabeedi	Fifty grams of fresh leaves is made into a paste. This paste is warmed on low flame and applied over the knee twice a day for five to six days to get relief from joint pain. A hundred to two hundred grams of fresh leaves is boiled in water. A bath with this warm water treats the parts affected by rheumatism.
Diploclisia glaucescens (Blume) Diels (*Cocculus glaucescens* Blume	Erumaithirankodi	Ten grams of dried leaves are soaked in one hundred ml of coconut oil (*Cocos nucifera* L.) for few days. The oil infusion is externally applied thrice a day to treat leprosy and scabies.

Contd...

Botanical Name and Family	Vernacular Name	Mode of Administration
Diplocyclos palmatus (L.) Jeffrey(*Bryonepsis laciniosa* Sensu Naud.) Cucurbitaceae	Malaipusani	Twenty grams of dried tuber made into a paste with water is externally applied twice a day to treat inflammatory sebaceous cysts.
Dipteracanthus patulus (Jacq.) Nees (*Ruellia patula* Jacq.) Acanthaceae	Kanapoodu	The powder prepared from the shade dried whole plant is taken orally along with water three times a day for a period of two days to treat poisonous bites.
Dolichos trilobus L. (*D. falcatus* auct non klein ex willd) Fabaceae	Pirappanelai	The powder prepared from the shade dried leaves is taken orally along with water two times a day for a period of one week to reduce body heat.
Drymoglossum heterophyllum (L.) Trimen Polypodiaceae	Sethavale	The paste prepared from twenty grams of fresh leaves is taken orally two times a day for a period of one month to reduce body heat, reduce excessive white discharge and helps in conceiving.
Drynaria quercifolia (L.) J.Sm Drynariaceae	Aattukkaal malaivahan	The juice prepared from twenty grams of fresh rhizome with water is taken orally in a single dose for a period of one week to treat general weakness of the body.
Eclipta prostrata (L.) L.(*E. alba* (L.) Hassk) Asteraceae	Karisalankanni	Twenty five to fifty grams of the stem and leaves is boiled in one litre of coconut oil (*Cocos nucifera* L). This oil is filtered after cooling. This filtered oil is applied on the head for an hour before bath during the course of treatment. This therapy is used to reduce body heat.
Elephantopus scaber. L. Asteraceae	Aanshovadie	The paste prepared from ten grams of fresh root is applied externally two times a day for a period of three days to get relief from toothache
Eleutherine palmifolia (L.) Merr Iridaceae	Veha Vengayam	The paste prepared from 10 grams of fresh leaves with water is applied externally or 20 gram of leaves boiled with coconut oil and applied externally for 4-5 days to treat burn injury caused by fire.
Emilia sonchifolia (L.) DC Asteraceae	Mouyal Cheviyan	The paste prepared with fifty grams of fresh leaves with water is applied externally two times a day for a period of one week to get relief from sprain.
Ensete superbum (Roxb.) Cheesman Musaceae	Kalvally	The juice prepared from a small soft tissue present in the centre of the fruit with 100 ml of water is taken orally for a period of 10-12 days to dissolve stones in the kidney.

Botanical Name and Family	Vernacular Name	Mode of Administration
Entada pursaetha DC (*E. scandens* auct.non Benth) Mimosaceae	Parandaikodi	A thick paste made from six to ten seeds is applied over the affected and inflamed swellings to reduce pain. To strengthen the joints in infants, this paste is applied over the leg once in a day for ten days. Warm water bath is administered each time.
Erythropalum scandens Bl.Bijdr (*E. populifolium* (Arn.) Mast.) Erythropalaceae	Vaathavallikodi	The chopped tender shoots are boiled with water. When this infusion is comfortably hot bath is taken. This management is continued until there is relief from the rheumatic complaint.A paste made from ten grams of fresh leaves is mixed with one teaspoon of honey is given orally twice a day for seven days as treatment for rheumatism. Consumption of fish, egg, tamarind, liquor, salt and intercourse is prohibited during the course of the treatment.
Erythrina variegata L.(*E.indica* Lam.) Fabaceae	Mullumurungai	Two hundred grams of stem bark is boiled in one litre of neem oil (*Azadirachta indica* A.Juss) together with ten grams of poppy seeds (*Papaver somniferum* L.) and twenty five grams of vellaipoondu (*Allium sativum* L.) for fifteen to twenty minutes. This oil is massaged over the affected joints twice a day until one get relief from rheumatic pain. Fifty to hundred grams of the leaves is boiled in an earthen pot.This preparation is used for fomenting the affected parts after massage.
Eugenia singampattiana Bedd. Myrtaceae	Kaatukorandi	One teaspoon of the powder made from equal quantity of shade dried leaves, flowers and tender fruit is consumed with honey two times a day for twenty one days for the treatment of rheumatism. Consumption of fish, both, fresh and dry, and egg is avoided.
Euphorbia hirta L. Euphorbiaceae	Amampatcharisi	The juice prepared from 10 grams of the whole plant with 100ml of water is taken orally 2-3 times a day to increase lactation
Euphorbia nivulia Buch-Ham Euphorbiaceae	Nagakalli	The paste prepared from ten to fifteen grams of fresh leaves along with ten grams of pineapple leaf (*Ananas comosus* (L.) Merill.) is mixed with a paste of vellaipoondu (*Allium sativum* L.), pepper (*Piper nigrum* L.) and dry ginger (*Zingiber officinale* Roscoe). This paste is taken orally two times a day for a period of two days to get relief from indigestion.
Evolvulus alsinoides (L.) L.	Vishmukarandi	The juice prepared from 10-15 grams of the whole plant is given orally twice a day for a week to reduce body heat.
Fagraea ceilanica Thunb Loganiaceae	Kalalamaram	The paste prepared from twenty grams of fresh leaves with water is applied externally once in a day for a period of one week to reduce swelling in the leg and feet.

Contd...

Botanical Name and Family	Vernacular Name	Mode of Administration
Ficus benghalensis L. var *benghalensis* Moraceae	Aal	The stem latex is applied on the cracked feet (fissures in foot) twice a day for a week for healing cracks in the feet.
Ficus hispida L.f Moraceae	Polavkaye	One teaspoonful of juice prepared from ten grams of fruit with water is taken orally for a period of one week in a single dose to treat prostrate glands.
Ficus microcarpa L.F. Moraceae	Punniyavirusu	Applying the stem latex twice a day for seven days is beneficial for the treatment of cracked feet.
Ficus racemosa L. Moraceae	Atthi	The stem latex is applied on the cracked feet twice a day for a week for healing of the affected part of the feet.
Ficus religiosa L. Moraceae	Arasu	Applying the stem latex twice a day for a week can heal the fissures in the foot.
Gloriosa superba L. Liliaceae	Karthikakilangu	The paste prepared from fifty grams of fresh bulb with water is applied externally to get relief from pain during delivery.
Glycyrrhiza glabra L.	Athimathuram	One hundred grams of shade dried root pieces are powdered with fifty grams of dried ginger (*Zingiber officinale* Roscoe), twenty five grams dried pepper fruit (*Piper nigrum* L.) and twenty five grams of dried long pepper fruit (*Piper longum* L.) About five grams of this powder is mixed with honey and consumed orally twice a day on an empty stomach until one experiences relief from rheumatic pain.
Goniothalamus wightii HK. f. Th. Annonaceae	Kattunarai-patchilai	Five grams of an equal quantity of leaves and tender fruits is made into a paste with a few drops of water. This paste is taken orally twice a day for five to ten days for relief from rheumatic pain.
Gymnema sylvestre (Retz.) R.Br ex Schultes Asclepidaceae	Chakrakoli	The paste prepared from ten to fifteen grams of leaves is mixed with fifteen grams of powder from the seeds of (*Syzygium cumini* (L) Skeels) and boiled with milk is taken orally in a single dose for a period of one month to reduce blood sugar (Diabetes).
Helicteres isora L. Sterculiaceae	Valamburi	The juice prepared from ten grams of fruit with fifty ml of water is taken orally two times a day for a period of three days to treat fever.

Botanical Name and Family	Vernacular Name	Mode of Administration
Hemionitis arifolia (Burm) T. Moore Hemionitidaceae	Nayecheveyan	The juice prepared from fifteen to twenty grams of fresh leaves in one hundred ml of water is taken orally two times a day for a period of ten days to treat dog bite. The leaves are applied as a pack on the wound till it is completely healed.
Hemidesmus indicus (L.) R.Br var. *Indicus* Periplocaceae	Nannari	The paste prepared from 20 gram of root and leaves is mixed with palm jaggary, cooked in coconut milk and taken orally for 3 days to treat ulcer
Hibiscus surattensis L. Malvaceae	Oopanechay	The paste prepared from ten grams of fresh leaves with water is applied externally once in a day for a period of three days to treat boils in children.
Hiptage benghalensis (L.) Kurz Malpighiaceae	Mathavikodi	Twenty grams of leaves and flowers made into a paste with water is externally applied twice a day to treat ringworm infection.
Hugonia mystax L. Linaceae	Motirakanni	One teaspoon of paste made from equal quantity of shade dried leaves and tender fruit is consumed with honey twice a day for twelve days for the treatment of rheumatism. Consumption of fish, both fresh and dry, and egg is avoided.
Indigofera longeracemosa Boiv ex Bail Fabaceae	Poorisa ayittruvali	Handful of leaves is made into a paste along with ten grams of leaves of neelavembu (*Andrographis paniculata* (Burm.f) Wall ex Nees) and ten grams of rhizome of kaccholam (*Kaempferia galanga* L.) are boiled in coconut oil (*Cocos nucifera* L.). This oil extract is externally applied to treat leprosy, leucoderma and scabies. The leaf extract is boiled in coconut oil and the oil infusion is used to treat alopecia.
Indigofera wightii Graham ex Wight and Arn Fabaceae	Neelamari	Half a teaspoonful of shade dried root powder is taken with luke warm water at the time of stomachache.
Isonandra lanceolata Wight Sapotaceae	Milagunari patchilai	A paste is prepared from two grams of fresh leaves and flowers with a few drops of water. This paste is consumed twice a day for ten to twelve days to get relief from rheumatic pain.
Ixora coccinea L. Rubiaceae	Idlipoo	Fifty grams of dried flowers are boiled in coconut oil (*Cocos nucifera* L.) and the oil extract is externally applied twice a day to treat eczema.
Jatropha gossypifolia L. Euphorbiaceae	Aathalai	About fifty grams of the crushed stem bark is boiled in one litre of neem oil (*Azadirachta indica* A.Juss) along with twenty five grams of chopped tuber of nilappanai (*Cucurligo orchioides Gaertn*) for half an hour on a low flame. This oil is massaged on the affected joints helps to relieve rheumatic pain.

Contd...

Botanical Name and Family	*Vernacular Name*	*Mode of Administration*
Justicia adhatoda L.(*Adhatoda vasica* Nees) Acanthaceae	Adathoda	A handful of leaves is cooked and used for fomentation on the affected joint to alleviate rheumatic pain. The juice extracted by crushing fifty grams of fresh leaves together with twenty five grams of leaves of veliparuthi (*Pergularia daemia* (Forssk.) Chiov) is battered with egg white and taken orally for ten to twelve days to treat rheumatic complaint.
Kaempferia galanga L. Zingiberaceae	Kacholam	The paste prepared from ten grams of the fresh rhizome with water is applied externally two times a day for a period of two days to treat insect bite.
Kalanchoe pinnata (Lam) Pers. Crassulaceae	Megasanjeevi	The paste prepared from twenty grams of leaves is applied externally three times a day for a period of four days to treat burnt skin.
Kingiodendron pinnatum (Roxb.ex DC.) Harm. (*Hard wickia pinnata* Roxb.ex DC) Caesalpiniaceae	Kulavu	The resin obtained by piercing the trunk is applied on the affected joints before going to bed along with a soft massaging in circular motion. In the morning lukewarm water is poured over the joints. It is also applied on the fissured foot for five to seven days to get relief.
Knoxia sumatrensis (Retz.) DC. Linearis (Gamble) Bhattacharjee and Deb (Rubiaceae)	Murporade patchilai	Fifty grams of leaves are ground into a paste and applied externally (formentation) on the swelling caused due to injury 2 times a day for one week.
Lantana camara L. var *aculeate*(L.) Mold Verbenaceae		The powder prepared from the shade dried leaves is taken orally along with water two times a day for a period of one week to reduce body heat. A teaspoon of paste made from an equal quantity of leaves and flowers is mixed with honey and taken orally thrice a day for sixteen days as treatment for rheumatism.
Lawsonia inermis L.(*L. alba* Lam) Lythraceae	Unnipoo	The leaves are soaked in coconut oil (*Cocos nucifera* L.) along with the flowers of asokam (*Saraca asoca* (Roxb.) Wilde) for a week and this oil infusion is used as a hair tonic to promote hair growth and to treat dandruff. This oil is also used to treat ringworm infection.
Leea indica (Burm. f.) Merr	Kaatuvalari-patchilai	Twenty grams of tender leaves are boiled with coconut oil and applied externally three times a day for 3 days to treat injury caused by burns
Lepianthes umbellata (L.) Rafin (*Hackeria subpeltata* (Willd.) Kunth Piperaceae	Maruthani	The paste prepared from twenty grams of fresh leaves with water is applied externally twice a day for a period of two days to prevent any swelling caused by wasp bites.
Leucas aspera (Willd) Link Lamiaceae	Tumbai	One teaspoonful of juice prepared from five to ten grams of leaves mixed with honey is taken orally three times a day for a period of two days to get relief from cough in children.

Botanical Name and Family	Vernacular Name	Mode of Administration
Lindernia crustacea (L) F.v. Muell Scrophulariaceae	Manjethumbe	The paste prepared from ten to fifteen grams of fresh leaves with water is applied externally two times a day for a period of two days to get relief from tooth and earache.
Mallotus philippensis (Lam.) Muell-Arg Euphorbiaceae	Kaatuthakadi	One teaspoon of paste is made from an equal quantity of leaves and tender fruit.This is mixed with honey and is taken orally twice a day for thirteen days to get relief from rheumatism. During this period the consumption of salt, spicy food and tamarind is avoided.
Merremia tridentata (L.) Hall. f subsp *hastata* (Desr.) Ooststr Convolvulaceae	Koonthalvalarthi	The entire plant is boiled in coconut oil (*Cocos nucifera* L.) and the extract is used to treat dandruff and to promote hair growth. Ten grams of the leaves made into a paste is applied once in a day to treat various skin infections.
Mimosa pudica (L.)	Thota vadi	The juice prepared from 10 grams of leaves is mixed with 50 ml of cow's milk or coconut milk is taken orally 3 times a day to treat jaundice.
Mirabilis jalapa L. Nyctaginaceae	Anthimantharai.	Twenty grams of dried root tuber made into a paste with water is applied externally twice a day to treat sebaceous cysts
Momordica charantia L. var. *chara ntia* Cucurbitaceae	Paakakai	As a vermifuge fifty to one hundred grams of unripe fruit is cooked as curry and taken two times a day for a period of three days
Mollugo pentaphylla L.	Pulchodie	The paste prepared from 50 grams of whole plant is applied externally six times a day for 12 days to treat blood clot caused by internal injuries.
Morinda pubescens J.E.Smith var *pubescens* Rubiaceae	Manjanatti	Ten grams of fresh leaves, two or three small vengayam (*Allium cepa* L.) and one teaspoon of manjal powder (*Cucurma longa* L.) are made into a paste. This paste is heated in five to ten ml of coconut oil (*Cocos nucifera* L) and the hot paste is applied on the injury caused by weapons once in a day for period of four to five days to heal the wounds.
Moringa concanensis Nimmo ex Gibs Moringaceae	Kattumoringai	Fifteen to twenty grams of fresh leaves are made into a paste along with two to three grams of black pepper (*Piper nigrum* L.) and five grams of vellaipoondu (*Allium sativum* L.). This paste is taken in the early morning hours in empty stomach once in a day for a period of three to four days to treat jaundice. During this treatment one bucketful of cold water is poured on the head of the patients. One teaspoonful of fresh leaf juice is given to children in a single dose to treat bowel disorder.

Contd...

Botanical Name and Family	Vernacular Name	Mode of Administration
Moringa pterygosperma Gaertn (*M. oleifera* acut.non Lam) Moringaceae	Moringai	About two hundred grams of stem bark from a ten year old tree is boiled in neem oil (*Azadirachta indica* A.Juss) along with fifty grams of crushed vellaipoondu (*Allium sativum* L.) and fifty grams of coarsely powdered fenugreek (*Trigonella foenum –graecum* L.) on a moderate flame for about twenty minutes. This oil is applied on the affected part as an embrocation for rheumatic swelling. Hot water bath is administered after few hours.
Mucuna pruriens (L.) DC var *pruriens* Fabaceae	Poonaykali	Powdered hairs on pods are administered with honey for expelling intestinal worms.
Mukia maderaspatana (L.) M.Roem Cucurbitaceae	Musumusukkai	One teaspoon of juice prepared from the fresh leaves is taken orally once in a day for a period of three days to get relief from cold and cough. The juice prepared from ten grams of fresh leaves with two hundred ml of rice fermented water is taken orally two times a day for a period of one week to reduce bile problem.
Murraya koenigii (L.) Spreng Rutaceae	Kariveppilai	Twenty to twenty five grams of fresh leaves are made into a paste along with equal quantity of fresh endosperm of thennai (*Cocos nucifera* L.) The paste is mixed with a little water and applied on the head an hour before taking bath. This procedure is practiced once a week regularly to induce hair growth, to delay the development of grey hair and to arrest falling of hair.
Murraya paniculata (L.) Jack (*M. exotica* L.) Rutaceae	Malaivembu	About fifty grams of fresh leaves is made into a paste with a little water. This paste is warmed for few minutes and bandaged on the affected part in the morning and the patient is advised hot water bath in the evening. This helps to relieve rheumatic pain.
Mussaenda glabrata (Hook.f) Hutchinson ex Gamble Rubiaceae	Vellimadanthai	The leaf paste is applied externally once in a day to treat skin allergy and inflammation. Fifty grams of the fresh leaves are boiled in two hundred ml of coconut oil (*Cocos nucifera* L.) and the extract is applied on the scalp to promote hair growth.
Myxopyrum serratulum A.W. Hill Oleaceae	Saduramullai	Five grams of the leaf paste is applied externally twice a day to treat skin rashes, scabies and ringworm infection.
Naravelia zeylanica (L.) DC	Mookouronjan	Twenty grams of powder prepared from the root is mixed with coconut oil and applied 3 times a day for 2 days to treat treat boils. Fifty grams of powder prepared from the leaves is used as a substitute for snuff to treat allergy cough.
Nilgirianthus heyneanus (Nees) Bremek Acanthaceae	Karunkurinchi	The paste prepared from fifteen grams of leaves with water is applied externally on the chest to get relief from chest pain.

Botanical Name and Family	*Vernacular Name*	*Mode of Administration*
Ocimum gratissimum L. Lamiaceae	Perumtulsi	The paste prepared from one hundred grams of fresh leaves is applied externally three times a day for a period of one week to get relief from muscle pain
Oxalis corniculata L. Oxalidaceae	Ooshnavallie	The paste prepared from fifteen grams of the whole plant is applied externally two times a day for a period of two days to get relief from migraine headache.
Paspalum scrobiculatum L.(*P. commersonii* Lam) Poaceae	Varegu	The paste prepared from equal quantity of leaves and root with water is applied externally to get relief from pain during delivery.
Pavonia odorata Willd Malvaceae	Thuthi	The juice prepared from twenty grams of the whole plant with one hundred ml of water is mixed with five grams of pepper powder (*Piper nigrum* L.) and is taken orally two times a day for a period of five days to treat fever and cough.
Pedilanthus tithymaloides (L.) Poir Euphorbiaceae	Palkalli	The paste prepared from twenty grams of leaves with water is applied externally two times a day for a period of one month to restore loss of hair due to insect bite.
Pergularia daemia (Forssk.) Chiov (*P. extensa* (*Jacq.*) *N.E.Br*) Asclepiadaceae	Veliparuthi	Fifty grams of fresh roots and an equal quantity of fresh leaves is crushed together with twenty five grams of stem bark of mullumurungai (*Erythrina variegata* L.) and ten grams of poppy seeds (*Papaver somniferum* L.). This is boiled for twenty five to thirty minutes on a low flame in an earthen pot or brass vessel. The filtered oil is massaged on the joints and the affected parts two times a day for ten to fifteen days. A hot water bath is administered during the course of the treatment. The use of chicken, pork, fish and having sex is to be avoided.
Phyllanthus amarus Schum and Thonn Euphorbiaceae	Kilanelli	The juice prepared from ten grams of the fresh clean aerial parts of the plant in twenty five ml of water. The filtered juice is taken orally as such or along with hundred ml of cow's milk twice a day for about seven to ten days to treat jaundice.
Phyllanthus gardnerianus (Wight)Baill Euphorbiaceae	Oopanechay	The paste prepared from ten to fifteen grams of leaves and fruits in water are taken orally in a single dose for a period of two weeks to treat jaundice.
Phyllanthus singampattiana (Sebastine and Henry) Kumari and Chandrabose comb. nov Euphorbiaceae	Kattugoiya	One teaspoonful of juice prepared from the fresh leaves is taken twice a day for a period of four to five days to relieve cold and cough.

Contd...

Botanical Name and Family	Vernacular Name	Mode of Administration
Piper nigrum L. Piperaceae	Milagu	One hundred to two hundred grams of leaves with shoot is boiled in water together with tender shoot of bamboo (*Bambusa arundinacea* (Retz.) Roxb) and neem leaves (*Azadirachta indica* A.Juss). Hot water bath is taken twice a day with this water while it is comfortably hot for twelve days or more to get relief from rheumatic pain.
Plectranthus amboinicus (Lour.) Spreng. Labiatae	Nevara Pachala	The juice prepared from ten grams of leaves with one hundred ml of water is taken orally twice a day for a period of three days to arrest cold and cough in children.
Pleiospermium alatum (Wall.ex Wt. and Arn)Swingle (*Limonia alata* wall. ex Wt and Arn) Rutaceae	Malainaarathai	The juice extracted from one hundred grams of fresh leaves and hundred grams of fresh leaves of lemon grass (*Cymbopogon citratus* (DC) Stapf) is boiled in one litre of neem oil (*Azadirachta indica* A.Jess) in a low flame for twenty minutes. This oil is applied on the joints, shoulders and the other affected parts. Hot water is sprinkled to get relief from rheumatic complaints.
Plumbago zeylanica L. Plumbaginaceae	Vellaikkoduvili	The paste prepared from ten to fifteen grams of root with water is applied externally in a single dose for a period of three days to get relief from poisonous insect bites.
Pothos scandens L. Araceae	Paraioutan	Two hundred ml of juice extracted from the leaves and stem of *P.scandens* in hot water is taken orally three times a day for a period of one week to reduce body heat and also helps in conception.
Polygala chinensis L. Polygalaceae	Siriyanangai	The paste prepared from ten grams of the root is applied externally twice a day for a period of three days to treat any poisonous bites.
Polygala javana DC. Polygalaceae	Palpiranthai	Five to ten grams of fresh leaves are made into a paste and applied on the breast twice a day for a period of two to three days to get from pain after breast feeding is stopped.
Pongamia pinnata (L.) Pierre (P.glabra Vent) Fabaceae	Pongan	One hundred grams of dried bark powder is boiled in half a litre of coconut oil (*Cocos nucifera* L.) and the oil extract is applied externally once in a day to treat eczema, rashes and ringworm infection.
Psychotria connata Wall Rubiaceae	Mundantheti	Two teaspoonful of juice prepared from ten grams of root with castor oil(*Ricinus cummunis* L.) is taken orally for three days in a single dose to treat pneumonia in children.
Psychotria nilgiriensis Deb and Gang Rubiaceae	Odaikaapi-patchilai	A ten gram paste made from the tender fruits is consumed along with honey once in a day for twelve days or more for the treatment of rheumatism.

Botanical Name and Family	*Vernacular Name*	*Mode of Administration*
Psychotria nudiflora Wt and Arn Rubiaceae	Kalpoo	A ten gram paste is made from an equal quantity of leaves and flowers which is consumed along with honey once in a day for twelve days for the treatment of rheumatism.
Pterocarpus marsupium Roxb. Fabaceae	Vengai	Fifty grams of the stem resin is soaked in one hundred ml of water for about twelve hours. It is applied externally on the knee twice a day for a period of five to six days to get relief from joint pain.
Rauwolfia densiflora (Wall.) Benth. exHk.f (*Tabernaemontana densiflora* Wall) Apocynaceae	Paarisirunila Patchilai	The paste prepared from ten grams of leaves and flowers is consumed orally two times a day for five days to treat rheumatic complaints.
Rivina humilis L Phytolaccaceae	Molagsar	The paste prepared from ten to fifteen grams of fresh leaves with water is applied externally two times a day for a period of one day to treat insect bites.
Sansevieria roxburghiana Schultes and Schultes (*S. zeylanica* Roxb) Agavaceae	Marul	A hundred grams of chopped leaves, twenty grams of vellaipoondu (*Allium sativum* L.) and twenty grams of vengayam (*Allium cepa* L.), fifty grams of stem bark of moringai(*Moringa pterygosperma* Gaertn.) and ten grams of mustard seeds (*Brassica juncea* (L.) Czerrn) is boiled in one litre of neem oil (*Azadirachta indica* A.Juss) on a low flame. This lukewarm oil is massaged by applying medium pressure on hands and legs for fifteen days or until one gets relief from rheumatic pain.
		The wood paste is applied externally on the skin to treat prickly heat, skin rashes and allergy.
Santalum album L. Santalaceae	Santhanam	One teaspoonful of latex from the plant is applied externally three times a day for two days to treat heat boils in children.
Sarcostemma acidum (Roxb) Voigt Asclepiadaceae	Kodiekalli	One teaspoonful of latex from the plant is applied externally three times a day for two days to treat heat boils in children.
Saraca asoca (Roxb.) Wilde Caesalpiniaceae	Asokam	Fifty grams of the dried flowers along with the leaves of maruthani (*Lawsonia inermis* L.) are boiled in coconut oil (*Cocos nucifera* L.) and the extract is applied externally twice a day to treat eczema and scabies.
Scleria lithosperma (L.) Sw (*Scirpus lithosperma*) Cyperaceae	Kathipul	Twenty grams of dried rhizome made into a paste with water is applied externally twice a day to treat eczema, leucoderma and scabies.

Contd...

Botanical Name and Family	Vernacular Name	Mode of Administration
Scoparia dulcis L. Scrophulariaceae	Chakravembu	One teaspoonful of juice prepared from twenty five grams of the whole plant with water is taken orally two times a day for three days to treat fever.
Senna auriculata (L.) Roxb. (*Cassia auriculata* L.) Caesalpiniaceae	Aavarai	The paste prepared from five to ten grams of sepals and petals from the flowers with a small quantity of asafoetida (*Ferula assafoetida* L.) and poppy seeds (*Papaver somniferum* L.) using hot water. This paste is given orally for two days in single dose to relieve colic pain.
Sida rhombifolia L. var *rhombifolia* Malvaceae	Kurunthatti	The paste prepared from twenty grams of the whole plant with water is applied externally once in a day for a period of one week to reduce body heat.
	Kandargattiri	The decoction prepared from fresh leaves of kandangattiri (*Solanum surattense* Burm.F.) thoothuvalai (*Solanum trilobatum* L.) dried ginger (*Zingiber officinale* Roscoe) coriander seeds.(*Coriandrum sativum* L.) and pepper (*Piper nigrum* L.) is taken orally two times a day for two days to arrest cold, cough and fever.
Solanum torvum Sw Solanaceae	Sundaikai	As vermifuge fifty to one hundred grams of unripe fruits is cooked as curry and taken two times a day for a period of three days.
Solanum trilobatum L. Solanaceae	Thoothuvalai	The decoction prepared from fresh leaves with ginger (*Zingiber officinale* Roscoe) and pepper (*Piper nigrum* L.) is taken oraly two times a day for a period of three days to treat cold and cough.
Sonerila tinnevelliensis Fischer Melastomataceae	Kalpuli	A handful of leaves are boiled with one litre of water that is, further, reduced to a hundred ml and consumed on an empty stomach once in a day for twelve to fifteen days to get relief from rheumatic complaints.
Sphaeranthus indicus L. Asteraceae	Oothuchedi	At the time of giddiness three or four fresh leaves are squeezed using both the palms and then smelled to get relief.
Stephania wightii (Arn.) Dunn Menispermaceae	Koloukone	The paste prepared from ten to fifteen grams of the tuber with water is taken orally to treat cancer.
Syzygium cumini (L.) Skeels Myrtaceae	Naval	One teaspoonful of shade dried seed powder is taken in empty stomach along with hundred ml of cold water daily for a month to control diabetes.
Tabernaemontana heyneana Wall (*Ervatamia heyneana* Wall.) Apocynaceae	Nandhiavattai	Two drops of pure filtered aqueous extract prepared from the flower petal is applied on the eyes for a period of seven days to improve vision.

Botanical Name and Family	*Vernacular Name*	*Mode of Administration*
Tamarindus indica L. Caesalpiniaceae	Puli	A dried seed is rubbed on a rough floor for few seconds to remove the outer testa and the rubbed surface of the seed is pressed immediately on the spot of scorpion sting to relieve pain.
Tephrosia purpurea (L.) Pers Fabaceae	Kattikolingi	One teaspoonful of juice prepared from the fresh root bark is mixed with a pinch of asafoetida (*Ferula assafoetida* L.) and given to children in single dose to get relief from bowel disorder.
Terminalia bellirica (Gaertn.) Roxb.	Tani	The paste prepared from 20 grams of fruit and 10 grams of gooseberry is taken orally 2 times day for 5 days to treat ulcer.
Terminalia chebula Retz. Combretaceae	Kadukai	The juice prepared from ten to fifteen grams of fruit in one hundred ml of water is used as mouth wash and also to get relief from toothache.
Thespesia populnea (L.) Soland ex Correa Malvaceae	Puvarasu	A few tender shoots are made into a paste along with one or two small vengayam (*Allium cepa* L).This paste is applied immediately on the area of scorpion sting to relieve pain
Thottea siliquosa (Lam.) Ding Hou Aristolochiaceae	Kuravan kandi Mooli	Twenty grams of dried roots is powdered with an equal amount of the roots of avalpori (*Rauwolfia serpentina* (L.) Benth ex Kurz) and vaathaneerpatchilai (*Aristolochia indica* (L.) Kuntze) is boiled in coconut oil (*Cocos nucifera* L.) and this extract is applied externally twice a day to treat eczema scabies and ringworm infection.
Tinospora cordifolia (Willd.) Miers ex.Hk.f and Th Menispermaceae	Senthilkodi	Two hundred ml of juice is extracted from a sufficient quantity of leaves without adding water. This juice is boiled in one litre of neem oil (*Azadirachta indica* A.Juss) along with twenty grams of cumin seeds (*Cuminum cyminum* L.), twenty grams of pepper seeds (*Piper nigrum* L.) twenty grams of mustard seeds (*Brassica juncea* (L.)Czerrn) twenty grams of dried ginger (*Zin giber officinale* Roscoe) hundred grams of vellaipoondu (*Allium sativum* L.) and fifty grams of crushed stem bark of murungai (*Moringa pterygosperma* Gaertn) on a low flame for ten minutes. This medicated oil is massaged for a period of ten days which helps to give relief from rheumatic pain.
Toddalia asiatica (L.) Lam.var asiatica Rutaceae	Milagaraniai	The shade dried plants are powdered and one teaspoon of the powder is taken orally along with water in empty stomach for a period of two to three days to get relief from gastric problems. A teaspoon of shade dried stem bark powder is taken after meals with honey three times a day for five to ten days to get relief from rheumatic pain.
Tribulus terrestris L. Zygophyllaceae	Sirunerinji	The juice prepared from ten grams of fruit with two hundred ml of goat's milk is taken orally two times a day for a period of three days to treat leucorrhoea.

Contd...

Botanical Name and Family	*Vernacular Name*	*Mode of Administration*
Trichodesma zeylanicum (*Burm.f*) *R.Br.* (*B ora go zeylanica Burm.f*) Boraginaceae	Kavuthumbai	Thirty to forty grams of fresh leaves are roasted with eight to ten vengayam (*A llium cepa* L.) in coconut oil (*Cocos nucifera* L.) and made into curry. This curry is taken one hour before meals twice a day for a period of four to five days to treat bleeding piles.
Trichopus zeylanicus Gaertn. subsp *travancoricus* (Bedd.) Burkill	Arokya pachilai	Fifty grams of powder prepared from shade dried leaves and a fruit mixed with coconut milk and honey is taken orally to treat frequent abortion and irregular menstrual cycle.
Tridax procumbens L. Asteraceae	Mookuthielai	Fifty grams of fresh leaves are made into juice. The juice is applied externally once in a day for period of three to four days to get relief from shoulder pain.Ten grams of fresh leaves are made into paste along with a pinch of calcium hydroxide. This paste is externally applied on the eczema affected area or an injury made by weapon once in a day for three or four days for complete cure.
Urginea indica (Roxb.) kunth Liliaceae	Kattuvengaum	The paste prepared from ten grams of bulb with water is applied externally once in a day for a period of three days to treat painful corns on the lower surface of the feet.
Ventilago madraspatana Gaertn Rhamnaceae	Vembadan	One hundred grams of the stem bark is coarsely powdered with fifty grams of roots of veliparuthi (*Pergularia daemia* (Forssk.) Chiov) and is mixed with one litre of neem oil (*Azadirachta indica* A.Juss) and heated for twenty minutes on a low flame. This oil is massaged on the joints twice a day.
Vernonia cinerea (L.) Less Asteraceae	Pavankodanthal	The paste prepared from ten grams of leaves with five grams of jeera (*Cuminum cyminum* L.) is mixed with milk and applied externally two times a day for a period of three days to treat eye injury.
Vitex negundo L. Verbenaceae	Notchi	One hundred ml of juice prepared by boiling fresh leaves is taken orally once in a day for a week. Cooked leaves are also used to foment the affected joints to disperse swelling in case of acute rheumatism.
Vetiveria zizanioides (L.) *Nash* Poaceae	Vetiver	One hundred grams of fresh roots are made into a juice along with root of nannari (*Hemidesmus indicus* (L.) B.R. var. indicus) and the entire plant of embural (*Hedyotis puberula* (G. Don) Arn.) are boiled in coconut oil (*Cocos nucifera* L.). The oil extract is applied externally twice a day to treat skin diseases and also is applied on the head daily as hair tonic, to induce hair growth and to arrest falling of hair.
Wedelia chinensis (Osbeck) Merr. Asteraceae	Manjakar-sangani	Fifty grams of fresh leaves is boiled in one hundred ml of coconut oil. The oil extract is applied externally once in a day to arrest hair fall and greying of hair (hair tonic).
Wrightia tinctoria (Roxb.) R. Br. Apocynaceae	Vetpalai	Ten to fifty grams of fresh stem bark is made into paste with water and the paste is taken orally two times a day for period of three days to reduce body heat.

Botanical Name and Family	*Vernacular Name*	*Mode of Administration*
Zanthoxylum rhetsa (Roxb.) DC (*Z. budrunga* (Roxb) DC)	Malvapoou	The paste prepared by rubbing the hard spines on a rock along with water is applied on the breast to give relief from pain and increase lactation in nursing mothers.
Zingiber zerumbet (L.) Rosc.ex J.E. Smith Zingiberaceae	Kolanjee	The paste prepared from ten grams of the rhizome mixed with water is taken orally for a period of one month twice a day for wound healing after child delivery.
Ziziphus xylopyrus (*Retz.*) *willd* Rhamnaceae	Mullukottai	Fifty grams of fresh stem bark is soaked in two hundred ml of water for twelve hours and filtered. This filtrate is taken orally in empty stomach for a period of three days in single dose to relieve stomachache.
Ziziphus rugosa Lam Rhamnaceae	Thodali	The paste prepared from the handful of leaves ground into a paste is applied externally on the skin before the morning bath to treat scabies and ringworm infection.

Table 2.2: Botanical Names of Plants Whose Parts/Products are Purchased from the Market and Used as Medicine/Ingredients in the Preparation by *Kanikkars*

Sl.No.	Botanical Name	Family	Vernacular Name	Plant Part/Extract of Plant Part Procured
1	*Allium cepa* L.	Liliaceae	Vengayam	Bulb
2	*Allium sativum* L	Liliaceae	Vellaipoondu	Bulb
3	*Borassus flabellifer* L.	Arecaceae	Panai	Palm jaggery
4	*Brassica juncea* (L.) Czerrn	Brassicaceae	Kadugu	Seed
5	*Cocos nucifera* L.	Arecaceae	Thennai	Seed oil
6	*Coriandrum sativum.*	Apiaceae	Malli	Seed
7	*Cuminum cyminum* L.	Apiaceae	Shiragam	Fruit
8	*Curcuma longa* L.	Zingiberaceae	Manjal	Dried rhizome
9	*Ferula assafoetida* L.	Apiaceae	Kayam	Root resin
10	*Glycyrrhiza glabra* L	Fabaceae	Athimathuram	Root
11	*Papaver somniferum* L.	Papaveraceae	Kasakasa	Seed
12	*Piper longum* L.	Piperaceae	Tippili	Fruit
13	*Ricinus communis* L	Euphorbiaceae	Amanakku	Castor oil
14	*Sesamum indicum* L.	Pedoliaceae	Yellu	Seed Oil
15	*Trigonella foenum- graecum* L.	Fabaceae	Vendhiyam	Seed
16	*Zingiber officinale* Roscoe	Zingiberaceae	Sukku	Dried Rhizome

Ageratina adenophora (Spreng) R.King & H.Robinson

Ancistrocladus heyneanus Wall ex Graham

Anisomeles malabarica R.Br.

Aristolochia indica L.

Aristolochia tagala L.

Baccaurea courtallensis (Wight) Muell-Arg

Plate 2.2

Begonia floccifera Bedd.

Begonia malabarica Lam.

Begonia subpeltata Wight

Carmona retusa (Vahl) Masamune

Chlorophytum heynei Rottl ex Baker

Chlorophytum laxum R. Br.

Plate 2.3

Connarus monocarpus L.

Corallocarpus epigaeus (Rottle & Willd) Clarke

Cyclea peltata (Lam.) Hook. f. & Thoms

Drymoglossum heterophyllum (L.) Trimen

Drynaria quercifolia (L.)J. Smith

Elephantopus scaber L.

Plate 2.4

Eleutherine palmifolia (L.) Merr.

Fagraea ceilanica Thunb.

Ficus hispida L.F.

Gloriosa superba L.

Hedyotis puberula (G.Don) Arn.

Helicteres isora L.

Plate 2.5

Hemionities arifolia (Burm) T.Moore

Kalanchoe pinnata (Lam.) Pers.

Knoxia sumatrensis (Retz.) DC

Leea indica (Burm. f.) Merr.

Lepianthes umbellata (L.) Rafin.

Lindernia crustacea (L.) F.v. Muell

Plate 2.6

Mollugo pentaphylla L.

Naravelia zeylanica (L.) DC

Oxalis corniculata L.

Pedilanthus tithymaloides (L.)Poir.

Phyllanthus gardnerianus (Wight) Baill

Phyllanthus lawii Graham ex Muell-Arg

Plate 2.7

Piper nigrum L.

Psychotria connata Wall.

Psychotria nilgiriensis Deb & Gang

Rauwolfia densiflora (Wall.) Benth. ex. Hk. f.

Rivina humilis L.

Sarcostemma acidum (Roxb.) Voigt

Plate 2.8

Scoparia dulcis L.

Stephania wightii (Arn) Dunn

Trichopus zeylanicas Gaertn. subsp. *travancoricus* (Bedd) Burkill

Urginea indica (Roxb.) Kunth

Wedelia calendulacea Less

Zingiber zerumbet (L.) J. E. Sm

Plate 2.9

Pothos scandens L.

Bulbophyllum albidum (Wight) Hook

Zanthoxylum rhetsa (Roxb.) DC

***Z. rhetsa* spines**

Plate 2.10

as many as 235 medicinal plants by the *Kanikkars* tribe inhabitating the study area (Agasthiaramalai Biosphere Reserve, Western Ghats, Tamil Nadu, India).

As an outcome of the present investigation, 235 plants (219 plants collected and 16 plants/extracts procured) belonging to 198 genera and 84 families including 4 Pteridiophytes were recorded (Tables 2.1 and 2.2). Of the recorded plants, a maximum of 15 ethnomedicinal plants belong to Euphorbiaceae, it is followed by Fabaceae (13 plant species), Rubiaceae, Asteraceae (9 plants each), Apocynaceae and Caesalpiniaceae, (8 plant species each), Liliaceae (7 species), Rutaceae, Menispermaceae, Acanthaceae (6 plant species each), Malvaceae, Cucurbitaceae, Lamiaceae, Solanaceae, Aristolocaceae, Moraceae, Araceae and Zingiberaceae (5 plant species each).

The documented plants used for the various diseases by the *Kanikkars* show a great habitual diversity. Among the 219 plants reported as herbal drugs, 81 plants are herbaceous in habit while 65 plants are shrubs, 30 plants are climbers, 42 plants are trees and 1 epiphytes. Among the investigated medicinal plants 94 plants belongs to the polypetalae under 37 families, 65 belong to the gamopetalae under 19 families, 35 belong to the monochlamydeae under 9 families and 37 belong to the monocotyledons under 15 families (Table 2.3).

Table 2.3: Family-wise Distribution of Enumerated Plant Species

PTERYDOPHYTE

Sl.No.	Family	Total No. of Genera	Total No. of Species
1	Actniopteridaceae	1	1
2	Polypodiaceae	1	1
3	Drynariaceae	1	1
4	Hemionitidaceae	1	1

POLYPETALAE (DICOTYLEDONS)

Sl.No.	Family	Total No. of Genera	Total No. of Species
1	Ancistrocladaceae	1	1
2	Annonaceae	1	1
3	Apiaceae	4	4
4	Begoniaceae	1	3
5	Brassicaceae	1	1
6	Burseraceae	1	1
7	Caesalpinaceae	6	8
8	Capparaceae	3	3
9	Combretaceae	2	3
10	Connaraceae	1	1
11	Crassulaceae	1	1
12	Cucurbitaceae	5	5
13	Erythropalaceae	1	1
14	Fabaceae	12	13

Sl.No.	Family	Total No. of Genera	Total No. of Species
15	Guttifera (Clusiaceae)	1	1
16	Leeaceae	1	1
17	Linaceae	1	1
18	Lytheraceae	2	2
19	Malpighiaceae	1	1
20	Malvaceae	5	5
21	Melastomataceae	1	1
22	Meliaceae	2	2
23	Menispermaceae	6	6
24	Mimosaceae	3	3
25	Molluginaceae	1	1
26	Moringaceae	1	2
27	Myrtaceae	2	2
28	Oxalidaceae	2	2
29	Papavaraceae	2	2
30	Polygalaceae	1	2
31	Ranunculaceae	1	1
32	Rhamnaceae	2	3
33	Rutaceae	5	6
34	Sapindaceae	1	1
35	Sterculiaceae	1	1
36	Vitaceae	1	1
37	Zygophyllaceae	1	1
	Total	84	**94**

GAMOPETALAE (DICOTYLEDONS)

Sl.No.	Family	Total No. of Genera	Total No. of Species
1	Acanthaceae	6	6
2	Apocynaceae	6	8
3	Asclepiadaceae	4	4
4	Asteraceae	9	9
5	Boraginaceae	2	2
6	Caricaceae	1	1
7	Convolvulaceae	2	2
8	Ebenaceae	1	1
9	Lamiaceae	4	5
10	Loganiaceae	1	1
11	Oleaceae	1	1
12	Pedialaceae	1	1
13	Periplocaceae	1	1

Sl.No.	*Family*	*Total No. of Genera*	*Total No. of Species*
14	Plumbaginaceae	1	1
15	Rubiaceae	7	9
16	Sapotaceae	1	1
17	Scrophulariaceae	3	3
18	Solanaceae	3	5
19	Verbenaceae	3	4
	Total	57	**65**

MONOCHLAMYDEAE (DICOTYLEDONS)

Sl.No.	*Family*	*Total No. of Genera*	*Total No. of Species*
1	Amaranthaceae	2	2
2	Aristolochiaceae	2	5
3	Bischofiaceae	1	1
4	Euphorbiaceae	10	15
5	Moraceae	1	5
6	Nyctaginaceae	2	2
7	Phytolaccaceae	1	1
8	Piperaceae	2	3
9	Santalaceae	1	1
	Total	22	**35**

MONOCOTYLEDONS

Sl.No.	*Family*	*Total No. of Genera*	*Total No. of Species*
1	Amaryllidaceae	1	1
2	Agavaceae	1	2
3	Arecaceae	2	2
4	Araceae	4	5
5	Commelinaceae	1	1
6	Costaceae	1	1
7	Cyperaceae	1	1
8	Dioscoreaceae	2	4
9	Hypoxidaceae	1	1
10	Iridaceae	1	1
11	Liliaceae	5	7
12	Musaceae	1	1
13	Orchidaceae	2	2
14	Poaceae	4	4
15	Zingiberaceae	4	5
	Total	**31**	**37**
	Grand Total	**198**	**235**

As far as the plant part used is concerned, it is noted that *Kanikkars* employed almost all part of plant in ethnomedicine. In terms of percentage of plant part used, the percentages are as follows; Leaf 51 per cent, whole plant 9 per cent, stem 5 per cent, stem bark 6 per cent, root 8 per cent, rootbark 0.4 per cent, tubers/rhizome 8 per cent, seed 6 per cent, fruit 9 per cent and sap/gum/latex 5 per cent. The most prevalent form of administration of medicine is paste (49 per cent). This is followed by juice/extract (18 per cent), powder (16 per cent) decoction (14 per cent) and plant part as such (3 per cent).

The enumerated 235 plants used to cure as many as 52 different types of human maladies (Table 2.4). A maximum of 49 plants are used to treat rheumatism, followed by 22 plants for treating scabies and eczema, 13 plants for treating poisonous bites, 11 plants for treating ringworm infections, 7 plants for treating wounds/injuries, Dyspepsia and to reduce body heat.

Table 2.4: Disease Types and Herbal Remedies of *Kanikkars*

Sl.No.	*Name of the Ailment*	*Total Number of Herbal Remedies*
1	Abortifacient	3
2	Alopecia	1
3	Astringent(Health tonic)	3
4	Boils	4
5	Bone fracture	2
6	Bowel disorder	4
7	Bile complaint	1
8	Cancer	1
9	Chest pain	2
10	Colic pain	1
11	Cold and cough	5
12	Dog-bite	2
13	Dysentery	1
14	Dyspepsia (indigestion)	7
15	Diabetes	2
16	Ear ailments	2
17	Fever	4
18	Fissures in foot	6
19	General body pain	2
20	Giddiness	2
21	Hair tonic	5
22	Headache	2
23	Jaundice	5
24	Kidney stone	2
25	Leucorrhoea	1
26	Lumbago	1

Sl.No.	Name of the Ailment	Total Number of Herbal Remedies
27	Ophthalmic ailments	4
28	Palpitation	1
29	Pneumonia	1
30	Poisonous bites	13
31	Piles	2
32	Rheumatism	49
33	Scabies/eczema	22
34	Scorpion-sting	2
35	Sebaceous cytes	3
36	Skin diseases (Ring worm infection)	11
37	Snake-bite	2
38	Stomachache	5
39	Toothache/tooth ailments	1
40	To induce lactation	1
41	To increase fertility in women	3
42	To increase memory power	1
43	To overcome fatigue	1
44	To reduce body heat	7
45	To reduce high blood pressure	1
46	To treat blood clot	1
47	To treat prostrate gland	1
48	Ulcer	1
49	Urinary infection	1
50	Vermifuge	3
51	Wasp sting	2
52	Wound/weapon/fire injury	7

Fifty two different maladies of *Kanikkars* are grouped into 5 major categories of diseases *viz.*, gastro-intestinal complaints, poisonous bite, body pain, fever, cold and cough and skin ailments. Twenty five plants belonging to 21 genera and 17 families are used by the *Kanikkars* for treating gastro-intestinal complaints like stomachache, bowel disorder, bile complaint, jaundice and colic pain *etc.* A few medicinal plants are used as dyspepsia and vermifuge. Sixty six plants identified in the present enumeration belonging to 64 genera and 44 families are employed as analgesics to relieve rheumatic pain, headache, general body pain, chest pain and body heat, *etc.* They use thirteen plants belonging to 11 genera and 10 families are employed to treat fever, cold and cough. The *Kanikkars* use 21 plants belonging to 20 genera and 19 families for treating various poisonous bites such as dog bite, snack bite, scorpion sting, wasp sting and unknown insect bite. Fifty three plants belonging to 50 genera and 39 families are used by the *Kanikkars* for treating various skin ailments like eczema, boils, scabies, ring worm infections, wounds, acnes and foot crack *etc.*

Above 8 per cent of the identified plants possess two fold uses (used to treat two ailments). Eighty seven percent of the ethnomedicinal plants are to cure only one particular disease. Some of the plant species recorded in the study exhibit manifold therapeutic uses, *i.e.*, they are used to cure more than two ailments. Five percent of the identified plants possess this type of therapeutic value. For example, the rhizome of *Curcuma longa* is administered in colic pain, stomachache and to heal wounds. The root of *Aristolochia indica* is used for treating snake-bite, headache and stomachache. The seed of *Abrus precatorius* is used to cure eye irritation; its leaf is used to treat odonthemedia and its leaf and root are used to treat poisonous bites. The seed of *Cuminum cyminum* is used for treating dyspepsia, eye injury, rheumatic pain, cold and cough. The leaves of *Cardiospermum halicacabum* is used in curing rheumatic complaints, giddiness, cold and cough. The leaves of *Tridax procumbens* is used to treat shoulder pain, skin disease and weapon injury. The above mentioned ethnomedicinal plants are recommended for broad phytopharmacological investigations.

Cross-Cultural Ethnomedicobotanical Studies

It is well established that an identical use of same plant by different tribal groups indicates its curative property and therapeutic significance (Jain and Saklani, 1992). The comparison of medicinal claims between different cultural groups in the same region or neighbouring regions has proved very rewarding. The comparative studies on medicinal uses of plants among different cultural groups showed similarities and dissimilarities in uses. In the present investigation, the medicainal claims emanating from the *Kanikkars* are compared with claims from three other numerically predominant Indian tribal groups living in the Central and Westeren zones of the country *viz.*, the *Bhils, Gonds and Santals* and two other predominant tribals of Tamil Nadu namely the *Palliyars* and *Irulars.* The source of ethnomedicimnal linformation for the above mentioned ethnic groups is selected from the available literature (Jain *et al.*, 1973; Mudgal and Pal, 1980; Ramachandran and Nair 1981; Joshi 1982; 1992; Tarafder, 1983; 1984; Shah and Gopal, 1985; Saxena, 1986; Sur *et al.*, 1987; 1990; Prasad *et al., 1987*; Pal *et al.*, 1989; Saxena and Tripathi, 1989; Yadav and Bhamare, 1989; Ragupathy and Mahadevan, 1991; Borthakur and Goswami, 1995; Mishra and Naquvi, 1995; Chandra, 1995; Girach and Aminuddin, 1995; Katewa and Arora, 1997; Ignacimuthu *et al.*, 1998; Girach *et al.*, 1999; Viswanathan *et al.*, 2001; Binu *et al.*, 2003; Ayyanar and Ignacimuthu, 2005; Jadhav, 2006a,b).

Comparative Ethnomedicobotany with *Bhils*

Twenty two plant species are commonly used by both *Kanikkars* and *Bhils.* Both similarities and dissimilarities are discussed.

Sl.No.	Botanical Name	Part Used	Kannikars	Bhils
1	*Anisomeles indica*	Leaves, Stem	Rheumatism	Rheumatism
2	*Azadirachta indica*	Leaves, Seed	Fever, Rheumatism	Night blindness
3	*Calotropis gigantea*	Latex, Leaves	Rheumatism	Cold and Cough, Asthma

Sl.No.	Botanical Name	Part Used	Kannikars	Bhils
4	*Cissampelos pareira*	Leaves, Root	Lumbago	Sores and Abdominal pain
5	*Clitoria ternatea*	Whole Plant	Skin infection	Fever
6	*Commelina benghalensis*	Leaves	Wounds, Stings and Skin problems	Eye infection
7	*Curculigo orchioides*	Root tuber	Rheumatism, Antidote	Vermifuge Aphrodisiac
8	*Eclipta prostrata*	Stem, Leaves	Reduces body heat, Dandruff, Hair tonic	Hair tonic and Antiseptic
9	*Erythrina variegata*	Stem bark	Rheumatism	Rheumatism
10	*Evolvulus alsinoides*	Leaves	Skin infection, Reduces body heat	Whitlow in fingers
11	*Ficus benghalensis*	Root latex	Fissures in foot	Rheumatism
12	*Helicteres isora*	Fruit, Leaves	Fever	Diarrhoea in children
13	*Hemidesmus indicus*	Roots, Leaves	Ulcer, Stomach pain, Coolant	Venereal disease, Digestive problems
14	*Justicia adhatoda*	Leaves	Joint pain,Bronchitis	Asthma, Pneumonia, Rheumatism
15	*Lawsonia inermis*	Leaves	Hair tonic, Skin infection	Check Pregnancy
16	*Mimosa pudica*	Leaves	Jaundice	Skin disease
17	*Moringa pterygosperma*	Stem bark	Antiinflammatory	Rheumatism
18	*Pergularia daemia*	Roots, Leaves	Joint pain	Piles
19	*Plumbago zeylanica*	Roots	Poisonous insect bites	Rheumatism
20	*Tinospora cordifolia*	Leaves	Rheumatism, Fever	Piles
21	*Tridax procumbens*	Leaf	Shoulder pain weapon injury, skin disease	Antiseptic and applied on cuts/wounds to check bleeding, diabetes.
22	*Wrightia tinctoria*	Leaves, Bark	Skin infections,Reduces body heat	Skin wounds

Comparative ethnobotanical study between the *Kanikkars* and *Bhils* reveals that 7 medicinal plant species (out of 22) exhibit identical uses. They are *Anisomeles indica, Eclipta prostrata, Erythrina variegata, Hemidesmus indicus, Justicia adhatoda, Tridax procumbens and Wrightia tinctoria.*

Comparative Ethnomedicobotany with the *Gonds*

Thiry one plant species are commonly used by both the *Kanikkars* and *Gonds.* Both similarities and dissimilarities are discussed.

Sl.No.	Botanical Name	Part Used	Kannikars	Gonds
1	*Abutilon indicum*	Leaves	Earache	Fever and Bronchitis
2	*Andrographis paniculata*	Leaves	Stings and Antidote	Induce vomitting Stings and Skin infection

Sl.No.	Botanical Name	Part Used	Kannikars	Gonds
3	*Anisomeles indica*	Leaves, Stem	Rheumatism	Nervous disorders, Chronic fever
4	*Aloe barbadensis*	Leaves	Reduce body heat, Cure kidney stones	Paralysis
5	*Argemone mexicana*	Leaves	Skin diseases, Wounds	Jaundice Rheumatism and Skin diseases
6	*Azadirachta indica*	Whole plant parts	Fever,Rheumatism	Skin diseases, Stomaheache, Purgative, Leprosy
7	*Biophytum sensitivum*	Leaves	Skin infection	Headache
8	*Boerhavia diffusa*	Leaves	Skin diseases	Skin diseases and Antidote
9	*Calotropis gigantea*	Leaves, Latex	Rheumatism	Rheumatism, Skin disease, Asthma
10	*Cardiospermum halicacabum*	Leaves	Rheumatism	Reduce body weight
11	*Cissampelos pareira*	Leaves, Root	Lumbago	Fever
12	*Commelina benghalensis*	Leaves	Wounds and Skin diseases, Antidote	Skin diseases
13	*Curculigo orchioides*	Roots	Rheumatism, Stings and Antidote	Skin diseases,Urinary infection and Epilepsy
14	*Datura metal*	Leaves	Joint Pain	Sprain, Lactation
15	*Eclipta prostrata*	Stem, Leaves	Tonic, Reduce body heat	Inflammation Elephantiasis Leucoderma, Spleen disorder
16	*Evolvulus alsinoides*	Whole plant	Reduce body heat, Cure kidney stones	Tonic,Aphrodisiac Asthma and Hair growth
17	*Ficus benghalensis*	Stem and Latex	Fissures in feet	Diarrhoea Abscess Toothache
18	*Helicteres isora*	Leaves and Fruits	Fever, Body pain	Skin diseases Stomach pain
19	*Hemidesmus indicus*	Root and Leaves	Ulcer and Stomach pain	Skin disease, Fever and Rheumatism
20	*Leucas aspera*	Leaves	Fever, Cold and Cough	Snake bites,Stings and Headache
21	*Moringa pteroygosperma*	Stem bark	Rheumatism	Whooping Cough and abortifacient
22	*Murraya paniculata*	Leaves	Hair tonic	Body pain and inflammation
23	*Pergularia daemia*	Roots	Joint Pain	Uterine tonic
24	*Pterocarpus marsupium*	Stem resin	Joint Pain	Urinary disorder, Diabetes
25	*Plumbago zeylanica*	Root	Antidote	Antidote
26	*Scoparia dulcis*	Whole plant	Fever, Rheumatism	Diabetes, Tuberculosis

Sl.No.	Botanical Name	Part Used	Kannikars	Gonds
27	*Tinospora cordifolia*	Leaves	Rheumatism	Diabetes
28	*Tridax procumbens*	Leaf	Skin diseases, Shoulder pain, Weapon injury	Skin diseases
29	*Vernonia cinerea*	Leaves	Eye injury	Skin diseases, fever
30	*Ventilago madraspatana*	Stem bark	Joint Pain	Skin diseases
31	*Vitex negundo*	Leaves	Rheumatism	Mascular pain

Comparative ethnobotanical study between the *Kanikkars* and *Gonds* reveals that 7 medicine plant species (out of 31) exhibit identical uses. They are *Andrographics paniculata, Boerhavia diffusa, Calotropis gigantea, Commelina benghalensis, Argemone mexicana, Tridax procumbens and Plumbago zeylanica.*

Comparative Ethnomedicobotany with the *Santals*

Twenty five plants are commonly used by both the *Kanikkars* and *Santals*. Both similarities and dissimilarities are discussed.

Sl.No.	Botanical Name	Part Used	Kanikkars	Santals
1	Abutilon indicum	Leaves	Earache	Epilepsy
2	Aloe barbadensis	Leaves	Reduce body heat, Cure kidney stones	Abcess, Intestinal worms, Burns,Peptic ulcer
3	Alstonia scholaris	Stem bark, Leaves	Lactation, Asthma	Lactation
4	*Andrographis paniculata*	Leaves	Posionous stings	Stomachache, Fever, Vomitting
5	*Aristolochia indica*	Root	Antidote	Antidote
6	*Azadirachta indica*	Seeds oil, Stem bark	Rheumatism, Joint pain, Fever	Stomachache Skin diseases
7	*Calotropis gigantea*	Leaves, Latex	Rheumatism	Rheumatism, Skin disease
8	*Cardiospermum halicacabum*	Leaves	Rheumatism	Eczema
9	*Cissampelos pareira*	Leaves, Root	Lumbago	Tonic
10	*Clitoria ternatea*	Whole plant	Skin diseases	Skin disease
11	*Curculigo orchioides*	Roots	Rheumatism, Antidote	Abortifacient, Leucorrhoea, Snake bite
12	*Eclipta prostrata*	Stem, Leaves	Reduce body heat, Hair tonic	Tonic,Wounds
13	*Entada pursaetha*	Seed	Antiinflammatory	Antiinflammatory
14	*Helicteres isora*	Leaves, Fruit	Fever, Body pain	Skin diseases, stomach pain
15	*Hemidesmus indicus*	Root, Leaves	Ulcer and stomach pain	Malaria, stomach problems

Sl.No.	Botanical Name	Part Used	Kanikkars	Santals
16	*Jatropha gossypifolia*	Stem bark	Rheumatism	Rheumatism
17	*Justicia adhatoda*	Leaves	Rheumatism	Vermicide
18	*Leucas aspera*	Leaves	Fever, Cold and Cough	Snake bite, Stings and Headache
19	*Mimosa pudica*	Leaves	Jaundice	Skin disease
20	*Moringa pterygosperma*	Stem bark	Antiinflammatory	Epilepsy, Measles, Abortion
21	*Plumbago zeylanica*	Root	Poisonous insect bites	Abortion
22	*Pterocarpus marsupium*	Stem resin	Joint pain	Gonorrhoea, Migraine
23	*Scoparia dulcis*	Whole plant	Fever, Rheumatism	Gonorrhea and headache
24	*Tridax procumbens*	Leaf	Shoulder pain, Weapon injury, Skin diseases	Cuts, Wounds, Sores, Leprosy, Srateches
25	*Vitex negundo*	Leaf	Rheumatism	Tongue sore, Bodyache

Comparative ethnobotonical study between the *Kanikkars* and *Santals* reveals that 9 medicinal plant species (Out of 25) have identical uses. They are *Alstonia scholaris, Aristolochia indica, Calotropis giganitea, Clitoria ternatea, Eclipta prostrata, Entada purseethae, Hemidesmus indicus, Jatropha gossypifolia* and *Tridax procumbens.*

Comparative Ethnomedicobotony with the *Palliyars*

Forty plant species are commonly used by both the *Kanikkars* and *Palliyars* both similarities and dissimilarities are discussed.

Sl.No.	Botanical Name	Part Used	Kannikars	Palliyars
1	*Acalypha indica*	Leaves	Cold, Cough, Fever	Cough
2	*Actinopteris radiata*	Leaves	Fracture, bone setting	Collic pain
3	*Alstonia scholaris*	Stem, bark, Leaves	Lactation, Asthma	Asthma
4	*Andrographis paniculata*	Root	Antidote	Antidote
5	*Anisomeles indica*	Leaves	Rheumatism	Fever, Cold
6	*Aristolochia indica*	Root	Antidote for snake venom	Antidote for Snake and Scorpion venom
7	*Azadirachta indica*	Seeds, Leaves, Stem bark	Rheumatism	Cold, Cough, Fever
8	*Begonia malabarica*	Leaves	Foot Sores, Skin infection	Astringent
9	*Cardiospermum halicacabum*	Leaves	Rheumatism	Cough
10	*Calotropis gigantea*	Leaves	Rheumatism	Cold and Cough
11	*Cissampelos pareira*	Leaves, Roots	Leaves for Lumbago	Roots for Insect bite and stomachache

Sl.No.	Botanical Name	Part Used	Kannikars	Palliyars
12	*Clitoria ternatea*	Leaves, Roots	Leaves for Skin infection	Roots for insect bite and stomachache
13	*Commelina benghalensis*	Leaves	Antidote	Eye infection
14	*Curculigo orchioides*	Root Tuber	Antidote	Increase sexual vigour in male
15	*Diplocyclos palmatus*	Tuber	Skin inflammation Cysts, Boils	For Abortion
16	*Eclipta prostrata*	Stem, Leaves	Reduce body heat, Hair tonic	Hair tonic
17	*Evolvulus alsinoides*	Whole Plant	Reduce body heat	Fever
18	*Ficus benghalensis*	Latex	Foot fissures	Foot fissures
19	*Helicteres isora*	Fruit	Fever	Ear infection
20	*Hemidesmus indicus*	Root	Ulcers	Reduce body heat
21	*Justicia adhatoda*	Leaves	Rheumatism	Asthma, Cold and Cough
22	*Kalanchoe pinnata*	Leaves	Skin burns, Wounds	Reduce body heat
23	*Lawsonia inermis*	Leaves	Hair tonic, Skin infection	Hair tonic
24	*Leucas aspera*	Leaves	Cold and Cough	Headache
25	*Mimosa pudica*	Leaves	Jaundice	Skin problem
26	*Momordica charantia*	Fruit	Vermifuge	Vermifuge
27	*Mukia maderaspatana*	Leaves	Cold and Cough Bile problem	Cold and Cough
28	*Pergularia daemia*	Roots	Joint pain	Gastric problem
29	*Piper nigrum*	Stem, Leaves	Rheumatism, Indigestion	Throat infection
30	*Plumbago zeylanica*	Root	Antidote	Snake bite, Stomachache
31	*Pongamia pinnata*	Bark, Seeds	Bark for Skin infection	Seeds for jaundice
32	*Pterocarpus marsupium*	Resin	Joint pain	Joint pain, Rickets
33	*Sansevieria roxburghiana*	Leaves	Rheumatism	For inducing teeth growth in children
34	*Santalum album*	Wood	Prickles,Rashes, Allergy	Prickles
35	*Scoparis dulis*	Leaves	Fever, Rheumatism	Insect bite
36	*Solanum surattense*	Leaves	Cold and Cough, Fever	Cold and Cough
37	*Thespesia populnea*	Tender shoot	Scorpion Stings	Stings
38	*Toddalia asiatica*	Whole plant	Gastric problems	Stem bark used as tooth powder
39	*Tridax procumbens*	Leaf	Shoulder pain, Weapon injury, skin diseases	Cutes, Wounds, Bone fracture
40	*Wrightia tinctoria*	Stem bark, Leaves	Bark to reduce body heat	Leaves for skin diseases

Comparative ethnobotanical study between the *Kanikkars* and *Palliyars* reveals that 13 medical plant species (out of 40) exhibit identical uses. They are *Acalypha indica, Alstonia scholaris, Andrographis paniculata, Eclipta prostrata, Ficus benghalensis, Lawsonia inermis, Momordica charantia, Mukia maderaspatana, Pterocarpus marsupium, Solanum surattense, Santalum album and Thespesia populnea* and *Tridax procumbens.*

Comparative Ethnomedicobotany with the *Irulars*

Twenty four plants species are commonly used by both the *Kanikkars* and *Irulars.* Both similarities and dissimilarities are discussed.

Sl.No.	*Botanical name*	*Part used*	*Kannikars*	*Irular*
1	*Abutilon indicum*	Leaves	Earache	Bronchitis
2	*Acalypha indica*	Leaves	Cold and Cough	Skin disease, Scabies
3	*Aloe barbadensis*	Leaves	Reduce body heat, Cure kidney stones	Reduce labour pain
4	*Anisomeles malabarica*	Leaves	Rheumatism	Paralysis
5	*Aristolochia indica*	Root	Snake bite, Stings	Fever and skin itching
6	*Azadirachta indica*	Stem bark	Rheumatism	Vermifuge,
7	*Calotropis gigantea*	Leaves	Joint pain	Abortifacient
8	*Cardiospermum halica-cabum*	Seed oil, Leaves	Joint pain, Rheumatism	Contraceptive, Rheu-matism
9	*Clitoria ternata*	Leaves	Skin diseases	Jaundice
10	*Eclipta prostrata*	Stem, Leaves	Reduce body heat,hair tonic	Jaundice
11	*Evolvulus alsinoides*	Whole plant	Reduce body heat	Skin disease
12	*Hemidesmus indicus*	Root, Leaves	Coolant, Ulcers	Snake bite, Coolant, Uurinary disorder
13	*Justicia adhatoda*	Leaves	Rheumatism	Cold and Cough
14	*Lawsonia inermis*	Leaves	Hair tonic, skin infections Dandruff	Dandruff
15	*Leucas aspera*	Leaves	Cold and Cough	Cough and Fever
16	*Mimosa pudica*	Leaves	Jaundice	Reduce body heat
17	*Mukia maderaspatana*	Leaves	Cold and Cough, bile problem	Bile problem
18	*Pergularia daemia*	Leaves	Joint pain	Cold, Fever
19	*Piper nigrum*	Leaves, Stem	Rheumatism, Cold and Cough	Birth control
20	*Plumbago zeylancia*	Root	Antidote	Abortion
21	*Pongamia pinnata*	Bark	Rashes, Scabies, Skin diseases	Rheumatic pain
22	*Scoparia dulcis*	Whole plant	Fever,Rheumatism	Diabetes
23	*Tridax procumbens*	Leaves	Shoulder pain, Weapon injury, Skin diseases	Wound injury to stop bleeding antseptic, wounds healer.
24	*Vitex negundo*	Leaves	Rheumatism	Body pain, Hair lice

Comparative ethnomedicinal study between the *Kanikkars* and *Irulars* reveals that 6 medicinal plant species (out of 24) exhibit identical uses. They are *Cardiospermum halicacabum, Lawsonia inermis, Hemidesmus indicus, Tridax procumbens, Leucas aspera and Mukia maderaspatana.*

The present study brings to light that all the five tribals in addition to *Kanikkars* employ *Tridax procumbens* as an antiseptic to cure wounds/cuts/weapon injury or to check bleeding. *Hemidesmus indicus* with identical medicinal use is being practiced by the *Bhils, Santals* and *Irulars. Eclipta prostrata* is used by the *Bhils, Santals* and *Palliyars* as hair tonic and whereas *Alstonia scholaris, Calotropis gigantea, Lawsonia inermis and Mukia maderaspatana* with identical medicinal use are being practiced by at least two ethnic groups under discussion.

Endemic and Endangered Plants of the Study Area

The study area falls in the "hot spot" region in the Western Ghats and the endemic plants are the exclusive biological capital of the nation. Earlier, investigations on endemic and rare plants and their conservation in India are meagre (Jain and Rao, 1983). A large number of endemic plants are at the brink of extinction. And hence the conservation of the threatened endemic species deserves top priority.

Some of the endemic plant species, which were reported to occur in the "hot spot" regions as common or abundant about half-a-century ago, now have become rare or very rate due to over exploitation and are included under the category of endangered species.

The following medicinal plants are identified as the endemic and endangered plant species of the study area (Jain and Rao, 1983; Ahmedullah and Nayar, 1986; Nayar and Sastry, 1987; Nayar, 1996).

Endemic Plants of the Study Area

1. *Acorus calamus* L.
2. *Asystasia travancorica* Bedd.
3. *Calanthe masuca* (D.Don). Lindl.
4. *Eugenia singampattiana* Bedd.
5. *Goniothalamus wightii* Hk. f. and Th.
6. *Hemidesmus indicus* (L.)R.Br. var. *indicus*
7. *Psychotria nudiflora* Wight and Arn.
8. *Sonerila tinnevelliensis* Fischer

Endangered Plants of the Study Area

1. *Bulbophyllum albidum* (Wight) Hook.F.
2. *Eugenia singampttiana* Bedd.
3. *Psychotria nudiflora* Wight and Arn.
4. *Sonerila tinnevelliensis* Fischer.

Several medicinal plants become threatened because of increasing biotic pressure on natural habitats and unscientific over exploitation. So, their conservation is of paramount importance and that can be made *in situ* and *ex situ*. The establishment of medicinal plant gardens or "germplasm of rare medicinal plants" in suitable agro-climatic regions is the need of the hour at both state and National levels.

References

Ayyanar, M. and Ignacimuthu, S. 2005. Traditional knowledge of kani tribals in kouthalai of Tirunelveli hills, Tamil Nadu. *J. Ethnopharmacology*. **102**: 246-255.

Ayyanar, M. and Ignacimuthu, S. 2005. Traditional knowledge of kani tribals in kouthalai of Tirunelveli hills, Tamil Nadu. *J. Ethnopharmacology*. **102**: 246-255.

Binu,S., Shanavaskha, A.K., Santhoskumar, E.S. and Pushpangadan, P. 2003. Plants used as medicine by the Irulars of Palghat district, Kerala, India. *J. Econ. Tax. Bot.* **27**: 808-814.

Borthakur, S.K, and Goswami, N. 1995. Herbal remedies from Dimoria of Kamrup district of Assam in Northeastern India. *Fitoterapia* **LXVI**: 333-340.

Bussmann, R.W. and Sharon, D. 2006. Traditional medicinal plant use in Loja province, Southern Ecuador. *J. Ethnobiology and Ethnomedicine*. http://www. ethnobiomed. Com/content/2/1/44.

Chandra, K. 1995. An Ethno-Botanical study on some medicinal plants of district Palamau, (Bihar). *Bull.Med. Eth. Bot. Res.* **XVI**: 11-16.

Fischer, C.E.C. and Gamble, J.S. 1957. (Rep. Ed.) The Flora of the Presidency of Madras **III**: 1347-2017.

Gadgil, M. 1996. Documenting diversity: An experiment. *Curr. Sci.* **70**: 36 – 44.

Gamble, J.S. 1957a. (Rep. Ed.) The Flora of the Presidency of Madras **I**: 1-577.

Gamble, J.S. 1957b. (Rep. Ed.) The Flora of the Presidency of Madras **II**: 578-1346.

Girach, R.D. and Aminuddin. 1995. Ethnomedicinal uses of plants among tribals of Singbhum district, Bihar, India. *Ethnobotany*. **7**: 103-107.

Girach, R.D., Singh, S., Brahmam, M. and Misra, M.K. 1999. Traditional treatment of skin diseases in Bhadrak district, Orrissa. *J. Econ. Tax. Bot.* **23**: 499 – 504.

Hajra, P.K., Nair, V.J. and Daniel, P. 1997. Flora of India. **IV**: 1 – 561.

Hamburger, M. and Hostettmann, K. 1991. 7. Bioactivity in Plants: The link between Phytochemistry and Medicine. *Phytochemistry* **30**: 3864-3874.

Henry, A.N., Chithra, V. and Balakrishnan, N.P. 1989. Flora of Tamil Nadu, India, Series I: Analysis **III:** 1- 171.

Henry, A.N., Kumari, G.R. and Chithra, V. 1987. Flora of Tamil Nadu, India, Series I: Analysis **II**: 1- 258.

Ignacimuthu, S., Sankarasivaraman, K. and Kesavan, L. 1998. Medico-Ethnobotanical survey among Kanikkar tribals of Mundanthurai Sanctuary, Western Ghats, India. *Fitoterapia*. **LXIX** : 404-414.

Jadhav, D. 2006a. Ethnomedicinal plants used by Bhil of Bibdod, Madhya Pradesh *Ind. J. Trad. Know.* **5**: 263-267.

Jadhav, D. 2006b. Ethnomedicinal survey of Maalgamdi in Ujjain district, Madhya Pradesh, India. *Ethnobotany,* **18**: 157-159.

Jain, S.K. and Saklani, A. 1992. Cross-cultural Ethnobotanical studies in Northeast India. *Ethnobotany* **4**: 25-38.

Jain, S.K. Banerjee, D.K. and Pal, D.C. 1973. Medicinal plants among certain Adibasis in India. *Bull. Bot. Surv. India.* **15**: 85 – 91.

Joshi, M.C. 1992. Some folk medicines of the tribals of Gujarat. *Bull. Med. Eth. Bot. Res.* **XII**: 115-124.

Joshi, P. 1982. An ethnobotanical study of Bhils – A preliminary survey. *J. Econ. Tax. Bot.* **3**: 257 – 266.

Katewa, S.S. and Arora, A. 1997. Some plants in folk medicine of Udaipur district (Rajasthan). *Ethnobotany* **9**: 48-51.

Khoshoo, T.N. 1996. India needs a National biodiversity conservation board. *Curr. Sci.* **71**: 506 – 513.

Lele, R.D. 1986. Ayurveda and Modern Medicine. Bharatiya Vidhya Bhavan, Mumbai. p. 346.

Matthew, K.M. 1983a. The flora of the Tamil Nadu Carnatic **I**: 1- 688.

Matthew, K.M. 1983b. The flora of the Tamil Nadu Carnatic **II**: 689- 1540.

Matthew, K.M. 1983c. The flora of the Tamil Nadu Carnatic **III**: 1541- 2155.

Matthew, K.M. 1999a. The flora of the Palni Hills **I**: 1- 575.

Matthew, K.M. 1999b. The flora of the Palni Hills **II**: 576- 1196.

Matthew, K.M. 1999c. The flora of the Palni Hills **III**: 1197- 1635.

Mishra, O.P. and Naquvi, S.M.A. 1995. Ethno-Medico Botany from tribes of Madhya Pradesh. *Bull. Med. Eth. Bot. Res.* **XVI**: 17-26.

Mudgal, V. and Pal, D.C. 1980. Medicinal plants used by tribals of Mayurbhanj (Orissa). *Bull. Bot. Surv. India.* **22**: 59-62.

Nair, N.C. and Henry, A.N. 1983, Flora of Tamil Nadu, India, Series I: Analysis **I**: 1-184.

Pal, D.C., Saren, A.M. and Sen, R. 1989. Less known uses of twenty plants from the tribal areas of Bankura district, West Bengal. *J. Econ. Tax. Bot.* **13**: 695-698.

Prasad, P.N., Jabadhas, A.W. and Janaki Ammal, E.K. 1987. Medicinal plants used by the Kanikkars of South India. *J. Econ. Tax. Bot.* **11**: 149-155.

Ragupathy, S. and Mahadevan, A. 1991. Ethnobotany of Kodiakkarai Reserve Forest, Tamil Nadu, South India. *Ethnobotany* **3**: 79-82.

Rahman, M.A., Mossa, S.J. and Al-Said, S.M. 2004. Medicinal plant diversity in the flora of Saudi Arabia 1: a report on seven plant families. *Fitoterapia* **75**: 149-161

Ramachandran, V.S. and Nair, V.J. 1981. Ethnobotanical studies in Cannanore district, Kerala State (India). *J. Econ. Tax. Bot.* **2**: 65-72.

Sambasivam Pillai, T.V. 1931. Tamil-English Dictionary of Medicine, Chemistry, Botany and Allied Sciences. Vols. **I–III**: Published by: Directorate of Indian Medicine and Homeopathy, Madras.

Sambasivam Pillai, T.V. 1977. Tamil-English Dictionary of Medicine, Chemistry, Botany and Allied Sciences. Vol. **IV**: Published by: Directorate of Indian Medicine and Homeopathy, Madras.

Sambasivam Pillai, T.V. 1978. Tamil-English Dictionary of Medicine, Chemistry, Botany and Allied Sciences. Vol. **V**: Published by: Directorate of Indian Medicine and Homeopathy, Madras.

Sasidharan, N. and Sivarajan, V.V. 1996. Flowering Plants of Thrissur Forests: 1-579.

Saxena, S.K. and Tripathi, J.P. 1989. Ethnobotany of Bundelkhand. Studies on the medicinal uses of wild trees by the tribal inhabitants of Bundlekhand region. *J. Econ. Tax. Bot.* **13**: 381-389.

Saxena, S.K. and Tripathi, J.P. 1989. Ethnobotany of Bundelkhand. Studies on the medicinal uses of wild trees by the tribal inhabitants of Bundlekhand region. *J. Econ. Tax. Bot.* **13**: 381-389.

Shah, G.L. and Gopal, G.V. 1985. Ethnomedicinal notes from the tribal inhabitants of the North Gujarat (India). *J. Econ. Tax. Bot.* **6**: 193 – 201.

Sharma, B.D. and Balakrishnan, N.P. 1993. Flora of India **II**: 1-625.

Sharma, B.D. and Sanjappa, M. 1993. Flora of India **III**: 1-639.

Sharma, B.D., Balakrishnan, N.P., Rao, R.R. and Hajra, P.K. 1993. Flora of India **I**: 1- 467.

Singh, K.S. 1994. People of India – The Scheduled Tribes. Anthropological survey of India. Oxford University Press. pp. 460-464.

Sur, P.R., Sen, R., Halder, A.C. and Bandyopadhyay, S. 1987. Observation on the ethnobotany of Malda – West Dinajpur districts, West Bengal – I. *J. Econ. Tax. Bot.* **10**: 395-401.

Sur, P.R., Sen, R., Halder, A.C. and Bandyopadhyay, S. 1990. Observation on the ethnobotany of Malda – West Dinajpur districts, West Bengal – II. *J. Econ. Tax. Bot.* **10**: 453-459.

Tarafder, C.R. 1983. Ethnogynaecology in relation to plants part – II. Plants used for abortion. *J. Econ. Tax. Bot.* **4**: 507-516.

Tarafder, C.R. 1984. Medicinal plants traditionally used by the tribals of Ranchi and Hazaribagh districts, Bihar: Skin diseases and sores. *Bull. Bot. Surv.India* **26**: 149-153.

Viswanathan, M.B., Premkumar, E.H., and Ramesh.N. 2001. Ethnomedicines of Kans in Kalakad – Mundanthurai Tigar Reserve, Tamil Nadu. *Ethnobotany* **13**:60-66.

Yadav, S.S. and Bhamare, P.B. 1989. Ethnomedico-Botanical studies of Dhule forests in Maharahtra State. *J. Econ. Tax. Bot.* **13**:455-460.

Zakaria, M., 1991. Isolation and characterization of active compounds from medicinal plants. *Asia Pacific J. Pharmacology* **6**: 15-20.

2018, Ethnomedicinal Plants: A Biodiversity Treasure Pages 123–166
Editors: V.R. Mohan, A. Doss, P.S. Tresina and V. Sornalakshmi
Published by: ASTRAL INTERNATIONAL PVT. LTD., NEW DELHI

Chapter 3

Traditional Knowledge of *Irulas* Tribes in Krishnagiri District, Tamil Nadu, India

K. Paulpriya[1], P.S. Tresina[1], V. Sornalakshmi[2] and V.R. Mohan[1]

[1]Ethnopharmacology Unit, PG and Research Department of Botany, V.O. Chidambaram College, Tuticorin – 628 008, Tamil Nadu
E-mail: vrmohanvoc@gmail.com
[2]Department of Botany, A.P.C. Mahalaxmi College for Women, Tuticorin – 628 002, Tamil Nadu

Introduction

Plants have been used in traditional medicine for several thousand years (Abu-Rabia, 2005). The knowledge of medicinal plants has been accumulated in the course of many centuries based on different medicinal systems such as Ayurveda, Unani and Siddha. In India, it is reported that traditional healers use 2500 plant species and 100 species of plants serve as regular sources of medicine (Pei, 2001). During the last few decades there has been an increasing interest in the study of medicinal plants and their traditional use in different parts of the world (Lev, 2006; Gazzaneo, 2005; Al-Qura'n, 2005; Hanazaki, 2000; Rossato, 1999). Documenting the indigenous knowledge through ethnobotanical studies is important for the conservation and utilization of biological resources.

Today, according to the World Health Organization (WHO), 80 per cent of the world's people depend on traditional medicine for their primary healthcare needs. There are considerable economic benefits in the development of indigenous medicines and in the use of medicinal plants for the treatment of various diseases (Azaizeh, 2003). Due to less communication, poverty, ignorance and unavailability of modern health facilities, most people especially rural people are still forced to practice traditional medicines for their common day ailments. Most of the tribals have the poorest knowledge in the trade of medicinal plants (Khan, 2002). A vast knowledge of how to use the plants against different illnesses may be expected to have accumulated in areas where the use of plants is still of great importance (Diallo, 1999).

Ethnobotany is not new to India because of its rich ethnic diversity. Jain (1991) pointed out that there are over 400 different tribal and other ethnic groups in India. The tribals constitute about 7.5 percent of India's population. During the last few decades there has been an increasing interest in the study of medicinal plants and their traditional use in different parts of India. There are many reports on the use of plants in traditional healing by either tribal people or indigenous communities of India (Saikia, 2006; Chhetri, 2005; Harsha, 2002; Natarajan, 2000; Maruthi, 2000; Samvatsar, 2000; Kala, 2005; Hebber, 2004). Apart from the tribal groups, many other forest dwellers and rural people also posses unique knowledge about plants (Jain, 1991).

A perusal of available literature reveals that till date there is no comprehensive survey, documentation and enumeration of medicinal plants used by the tribe *Irulas* inhabitating the Krishnagiri district located in the Eastern Ghats, Tamil Nadu. The present study focuses on the dependence of the *Irulas* on herbal medicines and attempts are made to document exhaustive therapeutic values of such medicinal plants.

The *Irulas* belong to the Southern Tribal Zone. The Southern tribals are historically more ancient tribes. There are as many as 36 types of scheduled tribes in Tamil Nadu. In the serialized list notified by the Government of Tamil Nadu, the *Irulas* are placed at 4th position.

Irulas live as individual families. Like other primitive tribes, the *Irulas* are short, dark complexioned, curly haired with thick protruding lips and blunt nose with wide nostrils (Figures 3.1 and 3.2). *Irulas* do not have any established mode of dress. They are scantily dressed but freely wear clothes available to them. They are non-vegetarians; however they abstain from beef as rigidly as the most orthodox Hindus. Besides there are a wide variety of greens, stems, tubers, unripe fruits and

Figure 3.1: A Irulas Family of Sendranakkanoor.

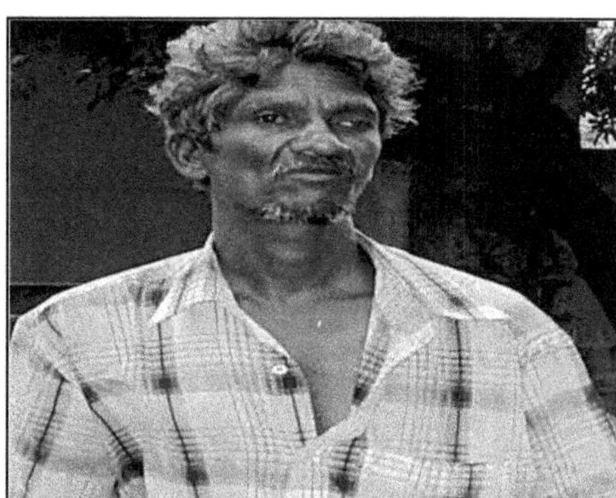

Figure 3.2: A Irulas Man.

ripe fruits, which serve as alternatives for food. *Irulas* also feed on wild animals and birds (like rabbit, rat, deer, hen *etc.*).

The *Irulas* live in small parties as isolated groups. Generally, a hamlet has about 30 huts. Their small huts are unique with the walls made up of mud or with wiry interwoven stems of *Lantana camara*. Each hut is thatched with the fronds of *Cymbopogon citratus* or *Cymbopogon polyneuros* or with the leaves of *Cocos nucifera*. They sleep on mats woven with the leaves of the above said taxa.

The *Irulas* as a tribe do not possess much cohesiveness. Each settlement has its headman whose authority is never challenged and he is solely responsible for settling disputes among the tribals. They are illiterate but in recent years, they have started sending their children to the nearby schools.

The *Irulas* are adroit in collecting honey. They collect honey from the branches of towering tall trees and rock caves skillfully using special techniques. *Irulas* from four to seventy years old are agile on the trees. They are also good hunters and they trap deer, pig, boar, hare, wild fowl and flying squirrel. The meat obtained is divided among the families within the settlement.

They are good herbalists and often collect medicinal plants from the forests. The knowledge about medicinal plant is rather specialized and is limited to a few members in the community who are recognized as "Vaidyars" or medicine men. They are generally respected and considered indispensable members within the tribal society. Each medicine man treats all kinds of illness but some of them are specialized in specific diseases. The remedies of common ailments like cuts, pain, headache, fever, dysentery *etc.* are known to most members of the tribal community. There are several individuals in each locality who though not recognized as medicine men, possess such knowledge and act as reliable informants. They maintain secrecy about the use of certain medicines because they believe that the herbals will lose their healing power if too many people know about them. Some practitioners have inherited the knowledge of certain special remedies.

Materials and Methods

Study Area

The *Irulas* are settled in the forest area of Krishnagiri district located in the Eastern Ghats, Tamil Nadu. Krishnagiri district lies between 11°12´ N and 12° 49´ N latitude and 77° 27´ E and 78° 38´ E longitudes. It occupies an area of 5143 km^2. The district is bounded by Vellore and Thiruvannamalai district of the East, State of Karnataka to the West, State of Andhra Pradesh to the North and Dharmapuri district to the South of Tamil Nadu (Figure 3.5). The Altitude ranges from 300 m to 1400 m (MSL)

The average rainfall is 830 mm per annum is received during the north-east monsoon gaining for 78 per cent and 71 per cent and South-west monsoon for 66 per cent and 56 per cent. The hottest months are April-May (40°C in the open and 37.5°C in the shade).

The vegetation is grouped under two distinct zones: (A) the outer slopes of the hills, (B) the Plateau. The outer slopes consists the following vegetation, scrub forests, dry deciduous forests, dry evergreen forests, and riparian forests. The semi evergreen forests, dry deciduous forests, wet rocky slopes, ponds and stream lets and estates and cultivated fields are surrounding the plateau (Figures 3.3 and 3.4). The cultivated *Annona squamosa* is very famous and popular in Krishnagiri hills.

Figure 3.3: A View of Sendranakkanoor.

Figure 3.4: A View of Ikundam and Pudoor.

Survey and Collection

Survey and collection of medicinal plants used by the *Irulas* tribe of the Krishnagiri district, Eastern Ghats, Tamil Nadu, were carried out over a period of 24 months (2011-2013). Frequent field trips were undertaken to the areas of study (Figure 3.5: Vide Area Map). Information regarding the medicinal plants were gathered by meeting *Irulas* practicing indigenous medicine, during explorative field trips and by gaining a good rapport and winning over their confidence. Most of the information included in this study are gathered from the elderly and experienced medicine men who have a long acquaintance with the use of medicinal plants. The field notebook delineates all the usage procedures adopted by the tribals. The information thus gathered was cross checked adequately for reliability and accuracy by interacting with different groups of the *Irulas* from different habitats to confirm the use, mode of administration as well as dosage differences, if any. After eliciting detailed information regarding the wild medicinal plants (Table 3.1) and any plant part(s)/extraction of plant part obtained from local market used in medicinal preparations, as ingredients (Table 3.2), they were carefully brought to the laboratory for identification. Herbaria for all the collected plant specimens except for plants, whose parts or extracts were procured from the local market, were prepared (VOCB No. from 6101 to 6309) and deposited in the Ethnopharmacology Unit, Research Department of Botany, V.O. Chidambaram College, Tuticorin, Tamil Nadu, India. The collected plants were identified by referring the flora (Fischer, 1957; Gambel, 1957; Matthew, 1983; Nair, 1983)

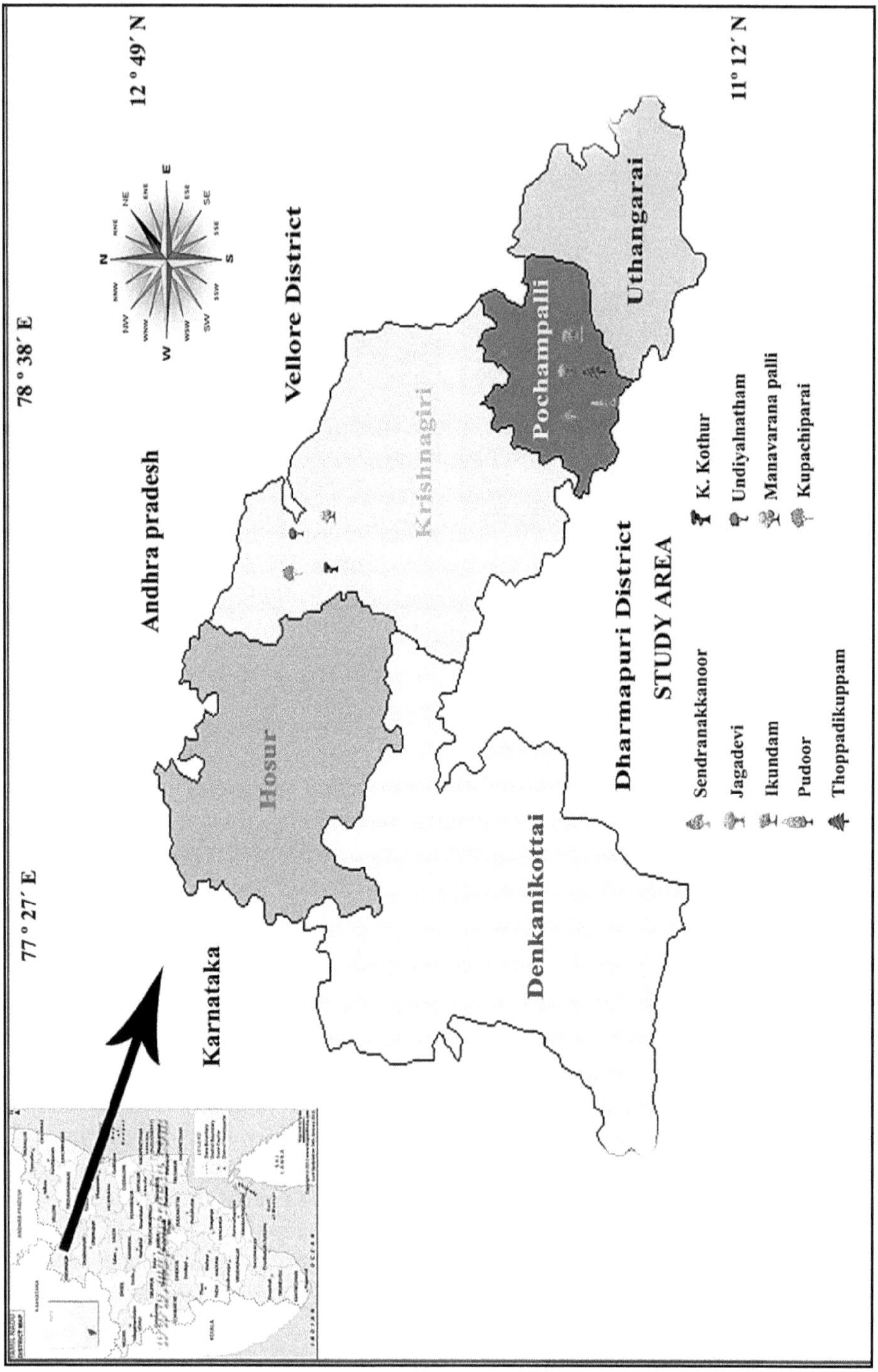

Figure 3.5: Area Map.

Table 3.1: List of Ethnomedicinal Plants Collected and Documented

Sl.No.	Plant Name with Voucher Number	Family with Vernacular Name	Habit	Parts Used	Prepa-rations	Uses/Ailments treated
1	*Abelmoschus esculentus* (L.) Moench,Meth. VOCB 6101	Malvaceae/Vendai	Herb	Unripe fruit	Decoction	Treat gonorrhoea,
2	*Abrus precatorius* L. VOCB 6102	Fabaceae/Gundumani	Climber	Seed, leaves	Paste	Used to cure Stiffness of joint, poisonous bites
3	*Abutilon indicum* (L.) Sweet. VOCB 6103	Malvaceae/Thuthi	Shrub	Leaves, root	Paste	Treat swellings caused piles
4	*Acacia nilotica* (L.) Willd. ex Delile. VOCB 6104	Mimosaceae/Karuvaelam	Tree	Dried resin, leaf, stem bark	Powder	Treat leech bites, control dysentery, treat tooth ache
5	*Acacia pennata* (L.) Willd. VOCB 6105	Mimosaceae/Seengai or indu	Climber	Fresh leaves	Paste	Cures Bleeding gums
6	*Acacia sinuata* (Lour.) Merr. VOCB 6106	Mimosaceae/Mande seengai	Climber	Fresh leaves	Paste	Treat jundice and fever
7	*Acacia torta* Craib. VOCB 6107	Mimosaceae/Seeva keerai	Climber	Leaf	Powder	Expulsion of gas
8	*Acalypha fruticosa* Forssk. VOCB 6108	Euphorbiaceae/Chinni	Shrub	Leaves	Paste	Insect bites
9	*Acalypha indica* L. VOCB 6109	Euphorbiaceae/Kuppameynie	Herb	Leaves	Paste	Psoriasis and allergic dermatitis, treat cold and cough
10	*Achyranthes aspera* L. VOCB 6110	Amaranthaceae/Naai urivi	Herb	Dried leaf	Powder	Tooth powder to whiten the teeth
11	*Acorus calamus* (L). VOCB 6111	Araceae/Vasambu	Shrub	Rhizomes	Paste	Treat skin diseases
12	*Aegle marmelos* (L.) Corr. VOCB 6112	Rutaceae/Vila maram	Tree	Fresh fruit	Extract	treat scabies
13	*Aerva lanata* (L.) Juss VOCB 6113	Amaranthaceae/Siru peelai	Herb	Dried root bark	Powder	Dissolve stones in kidney
14	*Alangium salvifolium* Lamark. VOCB 6114	Alangiaceae/Azhinjal	Shrub	Root bark	Decoction	Treat unknown bites
15	*Albizia amara* (Roxb.) Boivin. VOCB 6115	Mimosaceae/Thurinjil	Tree	Leaf and root bark	Paste	Treat snake bites and scorpion sting

Sl.No.	Plant Name with Voucher Number	Family with Vernacular Name	Habit	Parts Used	Preparations	Uses/Ailments treated
16	*Allium cepa* L. VOCB 6116	Alliaceae/Vengayam	Herb	Bulb	Powder	Treat urinary tract infection
17	*Aloe vera* (L.) Burm.f. VOCB 6117	Liliaceae/Sotrukatrallai	Herb	Fresh leaf	Paste	Treat dysentery
18	*Alternanthera sessilis* (L.) DC. VOCB 6118	Amaranthaceae/Ponaganni	Herb	Fresh leaves, root	Paste	Remedy from night blindness, treat piles
19	*Amaranthus spinosus* L. VOCB 6119	Amaranthaceae/Mullu keerai	Herb	Leaf	Juice	Treat snake bites and kidney stones
20	*Ammannia baccifera* L. VOCB 6120	Lythraceae/Neer mel nerupu	Herb	Whole plant	Paste	Reduces osteoarthritis of the knee joint, treat eczema
21	*Anacardium occidentale* L. VOCB 6121	Anacardiaceae/Munthiri	Tree	Fruit	Juice	Treat wounds
22	*Andrographis paniculata* (Burm.f.) Nees VOCB 6122	Acanthaceae/Nilavembu	Shrub	Whole plant	Decoction	Treat dengu, chikungunya fever and giddiness
23	*Anisomeles malabarica* (L.) R.Br. VOCB 6123	Lamiaceae/Nattaamutti	Shrub	Fresh leaves	Juice extract	Relief from rheumatic pain and stomach pain
24	*Anisomeles indica* (L.) Kuntze VOCB 6124	Lamiaceae/Periya thumbai	Herb	Fresh leaves	Extract	Relief from headache
25	*Annona squamosa* L. VOCB 6125	Annonaceae/seetha	Herbs	Leaf	Paste	Treat Scabies
26	*Areca catechu* L. VOCB 6126	Arecaeae/Kottaipakku	Tree	root	Extract	Treat gum infection
27	*Argemone mexicana* L. VOCB 6127	Papaveraceae/Pirmanthandu	Herb	Fresh leaf, root, whole plant	Paste	Cure skin diseases, cure fungal diseases, cure scabies, ring worm, eczema
28	*Aristolochia bracteolata* Lam. VOCB 6128	Aristolochiaceae/ Aduthinapalai	Climber	Fresh leaves	Paste	Treat wounds and eczema
29	*Artemisia nilagirica* (C.B.Clarke) Pamp. VOCB 6129	Asteraceae/Makkipu	Herb	Leaf	Juice	Treat stomach pain
32	*Azadirachta indica* A. Juss. VOCB 6132	Meliaceae/Veppam maram	Tree	Fresh tender leaves	Paste	Treat psoriasis

Contd...

Sl.No.	*Plant Name with Voucher Number*	*Family with Vernacular Name*	*Habit*	*Parts Used*	*Preparations*	*Uses/Ailments treated*
33	*Bacopa monnieri* (L.) Pennell VOCB 6133	Scophulariaceae/Neerpirami	Herb	Whole plant	Extract	Treat mental disorder and also strengthen bones
34	*Bambusa vulgaris* Schrad. ex J.C.Wendl. VOCB 6134	Poacea/Moongil	Tree like shrub	Leaf, meristem	Juice	Relief from vaginal discharge
35	*Bauhinia purpurea* L. VOCB 6135	Fabacea/Kaatu Mandarai	Tree	Leaf	Paste	Treat wounds caused by injuries, improve digestion
36	*Bauhinia tomentosa* L. VOCB 6136	Fabaceae/Mandhaarai	Shrub	Dried flower, stem bark	Paste	Treat constipation and laxative,induce sweat
37	*Blepharis maderaspatensis* (L.) Heyne ex Roth. VOCB 6137	Acanthaceae/Nethira moolli	Herb	Whole plant	Paste	Treat wounds
38	*Boerhavia diffusa* L. VOCB 6138	Nyctaginaceae/Mookkarattai	Wild herb	Fresh roots	Juice	Treat asthma, jaundice and swellings in legs
39	*Borassus madagascariensis* Bojer and H.Perrier VOCB 6139	Arecaceae/Panai maram	Tree	Endosperm	Paste	Treat pimples
40	*Butea monosperma* (Lam) Taub. VOCB 6140	Fabaceae/Palasu	Tree	Flowers, leaves	Decoction	Treat jaundice, cure eczema
41	*Cajanus cajan* (L.) Millsp. VOCB 6141	Fabaceae/Thuvarai	Herb	Leaf	Decoction	Treat jaundice
42	*Calotropis gigantea* (L.) R.Br. VOCB 6142	Asclepiadaceae/Erukku	Under shrub	Latex, leaves	Ointment	Treat rat and snake bites, relief from rheumatism
43	*Capparis zeylanica* L. VOCB 6143	Capparaceae/Aathondai	Climbing shrub	Leaf	Juice	Treat diarrhoea
44	*Capsicum annum* L. VOCB 6144	Solanaceae/Kaanthari milaghai	Herb	Fruits	Paste	Prevent rheumatic complaints
45	*Carica papaya* L. VOCB 6145	Caricaceae/Papali	Tree	Fruit	Juice	Treat stomach ache, digestive problem
46	*Carissa carandas* L. Mant. VOCB 6146	Apocynaceae/Kalakai or kala	Shrub	Flower	Juice	Treat eye diseases
47	*Cassia alata* L. VOCB 6147	Caesalpiniaceae/Seema agathi	Tree	Leaf	Paste	Treat ringworm infection

Sl.No.	Plant Name with Voucher Number	Family with Vernacular Name	Habit	Parts Used	Preparations	Uses/Ailments treated
48	*Cassia auriculata* L. VOCB 6148	Caesalpiniaceae/Avaram	Shrub	Dried stem bark	Powder	Prevent body odour
49	*Cassia fistula* L. VOCB 6149	Caesalpiniaceae/Sarakonrai	Tree	Flower, stem bark	Paste	Treat eczema and freckles, arrest dysentery, skin diseases
50	*Cassia tora* L. VOCB 6150	Caesalpiniaceae/Oosi thagarai	Small shrub	Leaf	Paste	Treat ringworm
51	*Catharanthus roseus* (L.) G.Don,Gen VOCB 6151	Apocynaceae/Nithiyakalyani	Herb	Leaf	Juice	Treat blood cancer
52	*Celosia argentea* (L). VOCB 6152	Amaranthaceae/Kolikondai poo	Herb	Root	Powder	Treat wounds
53	*Cissampelos pareira* L. VOCB 6153	Menispermaceae/Malai thangi	Shrub vine	Root	Paste	Treat stomach disorders and indigestion
54	*Cissus quadrangularis* L. VOCB 6154	Vitaceae/Perandai kodi	Shrub	Stem, whole plant	Paste	Cure rheumatoid arthritis, bone fracture
55	*Citrullus colocynthis* (L.) Schrader. VOCB 6155	Cucurbitaceae/Peikumatti	Climber	Root	Powder	Treat urinary disorders
56	*Cleome viscosa* L. VOCB 6156	Cleomaceae/Naikkaduku	Herb	Leaf	Juice	Treat wounds and swellings
57	*Clerodendron inerme* (L.) Gaertn. VOCB 6157	Lamiaceae/Sangankuppi	Shrub	Root	Powder	Treat snake bites
58	*Clerodendron phlomoidis* L.f. VOCB 6158	Lamiaceae/Thaluthalai	Shrub	Dried root	Oil	Relief from sprains
59	*Clitoria ternatea* L. VOCB 6159	Fabaceae/Sangupoo	Herb	Leaf	Juice	Cure scabies
60	*Coccinia grandis* (L.) Voigt, Hort. VOCB 6160	Cucurbitaceae/kovakai	Climber	Fruit	Paste	Relief from toothache and mouth ulcer
61	*Cocculus hirsutus* L. Diels VOCB 6161	Menispermaceae/Karum poola	Climber	Leaf	Juice	Cure sexually transmitted diseases like syphilis and gonorrhoea
62	*Cocos nucifera* L. VOCB 6162	Arecaceae/ Thennai maram	Tree	Whole plant	Paste	Treat cuts and wounds, applied on the scalp for thick and long hair
63	*Coldenia pocumbens* L. VOCB 6163	Boraginaceae/Serupad	Herb	Fresh leaves	Paste	Relief from rheumatic pain

Contd...

Sl.No.	Plant Name with Voucher Number	Family with Vernacular Name	Habit	Parts Used	Prepa-rations	Uses/Ailments treated
64	*Colocasia esculenta* (L.) VOCB 6164	Araceae/Karunal kilangu	Herb	Leaf	Juice	Cure skin diseases and rheumatism
65	*Commelina benghalensis* L. VOCB 6165	Commelinaceae/Kanan	Herb	Leaf, stem	Paste	Remove the poisonous spines that stick accidently to body parts, heal wounds
66	*Corallocarpus epigaeus* (Rottl and Willd.) Clarke VOCB 6166	Cucurbitaceae/Aagasa garudan	Climber	Dried root tuber	Paste	Treat goitre and ulcer
67	*Coriandrum sativum* L. VOCB 6167	Apiaceae/Kothamalli	Herb	Leaf	Decoction	Treat indigestion
68	*Crinum asiaticum* L. VOCB 6168	Amaryllidaceae/Nari venkayam	Herb	Leaf	Paste	Treat wounds
69	*Cucumis sativus* L. VOCB 6169	Cucurbitaceae/Vellari	Climber	Seed	Powder	Treat libido
70	*Curculigo orchioides* Gaetrn. VOCB 6170	Hypoxidaceae/Nilappanai	Tree	Root tuber	Powder	Increase sexual libido in males
71	*Curcuma aromatica* Sal. VOCB 6171	Zingiberaceae/Kasthuri manjal	Rhizome	Rhizome	Paste	Cure impetigo
72	*Curcuma longa* L. VOCB 6172	Zingiberaceae/Manjal	Herb	Rhizome	Powder	Treat stomach ache and fever
73	*Cuscuta reflexa* Roxb.Pl.Cort. VOCB 6173	Convolvulaceae/Aagaayakodi	climber	Whole plant	Decoction	Relief from liver disorders and piles
74	*Cymbopogon citratus* (DC.) Stapf. VOCB 6174	Poaceae/Elimichai pul	Grass	Leaf	Juice	Treat Stomach ache
75	*Cynodon dactylon* L. Pers. VOCB 6175	Poaceae/Arugu	Grass	Leaves	Paste	Treat dermatophytosis
76	*Cyperus difformis* L. VOCB 6176	Cyperaceae/Gorai	Grass	Root	Juice	Treat indigestion
77	*Datura discolor* Bernh. VOCB 6177	Solanaceae/Umathai	Herb	Leaf	Juice	Relief from wheezing
78	*Datura innoxia* Mill. Gard.Dict. VOCB 6178	Solanaceae/Vellai umathai	Herb	Leaf	Powder	Treat asthma,wounds

Sl.No.	Plant Name with Voucher Number	Family with Vernacular Name	Habit	Parts Used	Preparations	Uses/Ailments treated
79	*Datura metel* L VOCB 6179.	Solanaceae/Pon umathai	Herb	Fresh leaves	Juice	Treat fever, skin diseases and brain disorders, diabetic mellitus
80	*Delonix regia* (Boj. ex Hook) Rafin. VOCB 6180	Caesalpiniaceae/Mayilkondrai	Tree	Flowers	Paste	Treat ringworm infection
81	*Dichrostachys cinerea* (L.) Wight and Arn. VOCB 6181	*Mimosaceae*/Vadataram	Tree	Seeds, fresh leaves	Juice	Relief from rheumatoid arthritis, dysentry
82	*Diplocyclos palmatus* (L.) C. Jeffrey VOCB 6182	Cucurbitaceae/Malaipusani	Climber	Whole plant	Powder	Treat male infertility
83	*Dodonea viscosa* L. VOCB 6183	Sapindaceae/Virali	Shrub	Leaf, bark	Juice	Treat bronchitis, malaria
84	*Eclipta alba* L.Hassk. VOCB 6184	Asteraceae/Karasala kanni	Herb	Leaf	Juice	Treat snake bites and some poisonous bites
85	*Enicostema axillare* (Lam.) Rayanal VOCB 6185	Gentianaceae/Vellarugu	Herb	Leaf	Paste	To reduce body heat
86	*Erythrina variegata* L. VOCB 6186	Papilionoideae/Mulu murunga	Tree	Fresh Leaves	Juice	Female sterility
87	*Eucalyptus globulus* Labill. VOCB 6187	Myrtaceae/Thaila maram	Tree	Leaf	Powder	Eradicates body pain, fever and cold
88	*Eugenia caryophyllata* Thunb. VOCB 6188	Myrtaceae/Kirambu	Herb	Flower	Powder	Treat toothache
89	*Eupatorium odoratum* DC. VOCB 6189	Asteraceae/Pachai chedi	climbing shrub	Leaf	Paste	Treat wounds
90	*Euphorbia hirta* L. VOCB 6190	Euphorbiaceae/Amman pacharichi	Herb	Leaf	Paste	Treat pimples, asthma
91	*Euphorbia antiquorum* L. VOCB 6191	Euphorbiaceae/Sadurakalli	Herb	Phylloclade	Juice	Treat rheumatism
92	*Euphorbia cyathophora* Murr. VOCB 6192	Euphorbiaceae/Thael thazhai	Herb	Leaf	Juice	Increase lactation in women
93	*Euphorbia tirucalli* L. VOCB 6193	Euphorbiaceae/Thirukalli	Tree	Latex	Paste	Treat arthritis

Contd...

Sl.No.	Plant Name with Voucher Number	Family with Vernacular Name	Habit	Parts Used	Preparations	Uses/Ailments treated
94	*Evolvulus alsinoides* L. VOCB 6194	Convolvulaceae/ Vishnukranthai	Herb	Whole plant	Juice	Reduce body heat
95	*Ficus benghalensis* L. VOCB 6195	Moraceae/Aalam	Tree	Bark	Juice	Treat diabetes and mouth ulcer
96	*Ficus religiosa* L. VOCB 6196	Moraceae/Arasu	Tree	Bark	Powder	Treat poisonous
97	*Ficus racemosa* L. VOCB 6197	Moraceae/Athi	Tree	Bark	Decoction	Treat wounds
98	*Ficus talbotii* King VOCB 6198	Moraceae/Kal ithi	Tree	Bark	Juice	Treat sexually transmitted diseases
99	*Garuga pinnata* Roxb. VOCB 6199	Burseraceae/Karuvemmbu	Tree	Fruits	Juice	Relieve colic pain
100	*Gloriosa superba* L. VOCB 6200	Colchicaceae/Korangu poo	Climber	Root tuber	Powder	Treat snake bites
101	*Gmelina arborea* Roxb. VOCB 6201	Lamiaceae/Perum kumil	Tree	Fresh leaves, Root	Juice	Relief from cough, cardiac diseases
102	*Gmelina asiatica* L. VOCB 6203	Lamiaceae/Mulkumizh	Small tree	Dried root	Juice	Treat rheumatism
103	*Gymnema sylvestre* (Retz.) R.Br. ex chultes. VOCB 6204	Asclepiadaceae/sirukurunjan	Climber	Leaf	Powder	Treat diabetes
104	*Hemidesmus indicus* L. R. Br. VOCB 6205	Apocynaceae/Nannari	Climber	Root	Juice	Act as body coolant
105	*Hibiscus cannabinus* L. VOCB 6206	Malvaceae/Pulicha kerai	Herb	Leaf	Juice	Stop continuous vomiting
106	*Hibiscus rosa-sinensis* L. VOCB 6207	Malvaceae/Semparathai	Shrub	Flower buds	Powder	Treat impotency in males
107	*Hibiscus vitifolius* L. VOCB 6208	Malvaceae/Periathuthi	Shrub	Fresh flower	Paste	Treat healing
108	*Holarrhena antidysenterica* (Roxb. ex fleming) Wall.Cat. VOCB 6209	Apocynaceae/Kasappu vetpallai	Tree	Stem bark	Paste	Treat psoriasis
109	*Hygrophila auriculata* (Schumach.) Heine. VOCB 6210	Acanthaceae/Neermulli	Herb	Dried leaf	Powder	Treat skin diseases
110	*Hyptis suaveolens* (L.) Poit. VOCB 6211	Lamiaceae/Gangathulasi	Herb	Fresh leaves	Juice	Relief from head ache

Sl.No.	Plant Name with Voucher Number	Family with Vernacular Name	Habit	Parts Used	Prepa-rations	Uses/Ailments treated
111	*Indigofera aspalathoides* Vahl.ex DC VOCB 6212	Fabaceae/Civanarvembu	Shrub	Fresh leaf	Paste	Treat skin diseases
112	*Indigofera tinctoria* L. VOCB 6213	Fabaceae/Avuri	Shrub	Whole plant	Juice	Treat leucorrhoea, insect bites
113	*Ixora coccinea* L. VOCB 6214	Rubiaceae/Idly poo	Shrub	Whole flower, flower	Paste	Relief from fever, treat white discharge for women
114	*Jatropha curcas* L. VOCB 6215	Euphorbiaceae/Kattamanaku	Shrub	Leaf, seed	Juice	Treat psoriasis, toothache
115	*Jatropha gossypifolia* L. VOCB 6216	Euphorbiaceae/Seemamanakku	Shrub	Latex	Juice	Treat typhoid
116	*Justicia adhatoda* L. VOCB 6217	Acanthaceae/Aadathoda	Shrub	Fresh leaves	Juice	Cure bronchial asthma
117	*Justicia betonica* L. VOCB 6218	Acanthaceae/ Pachai Kanagamaram	Shrub	Leaf	Extract	Arrest diarrhoea
118	*Justicia glauca* Rottl. VOCB 6219	Acanthaceae/Thavasu murungai	Herb	Leaf	Paste	Relief from knee pain
119	*Lablab purpureus* (L.) Sweet VOCB 6220	Fabaceae/Thattan payaru	Climber	Fresh leaves	Paste	Used as an emollient
120	*Lantana camara* L. VOCB 6221	Verbenaceae/Unimul chedi	Shrub	Leaf	Paste	Treat wounds
121	*Lawsonia inermis* L. VOCB 6222	Lythraceae/Maruthaani	Shrub	Bark, Leaves	Powder	Treat skin diseases, head ache
122	*Leucas aspera* (Willd.) Link. VOCB 6223	Lamiaceae/Thumbai	Herb	Fresh Leaves	Juice	Treat snake bites
123	*Phyla nodiflora* (L.) Greene VOCB 6224	Verbenaceae/Poduthalai	Herb	Leaf	Juice	Cure internal piles
124	*Madhuca longifolia* (Koen.) J.F. Macbr VOCB 6225	Sapotaceae/Illupai	Tree	Stem bark	Decoction	Relief from wounds and skin diseases, fever, cough
125	*Mangifera indica* L. VOCB 6226	Anacardiaceae/Maa	Tree	Seed	Powder	Treat dysentery

Contd...

Sl.No.	Plant Name with Voucher Number	Family with Vernacular Name	Habit	Parts Used	Prepa-rations	Uses/Ailments treated
126	*Manihot esculenta* Crantz. VOCB 6227	Euphorbiaceae/Maravalli kilangu	Shrub	Leaf	Paste	Treat scabies
127	*Manilkara zapota* (L.) P. Royen VOCB 6228	Sapotaceae/Sapota	Tree	Fruits	Paste	Treat gastritis
128	*Marsilea quadrifolia* L. VOCB 6229	Marsileaceae/Aaraikeerai	Herb	Leaf	Powder	Cure wounds and skin diseases
129	*Melia dubia* Cav. Diss. VOCB 6230	Meliaceae/Malai vempu	Tree	Fruits	Paste	Cure from skin diseases
130	*Melochia corchorifolia* L. VOCB 6231	Malvaceae/Pinnakkukkirai	Herb	Fresh leaves	Juice	Relief from urinary disorders
131	*Merremia tridentata* (L.) Hallier f. VOCB 6232	Convolvulaceae/Amaivyar kundal	Herb	Whole plant	Juice	Initiate urination
132	*Michelia champaca* L. VOCB 6233	Magnoliaceae/Sampagam	Tree	Fresh flower buds	Juice	Treat piles, arthritis,fever and leucorrhoea in women
133	*Mimosa pudica* L. VOCB 6234	Mimosaceae/Thottasurungii	Herb	Dried leaves	Powder	Reduce body heat relief from joint pain
134	*Mirabilis jalapa* L. VOCB 6235	Nyctaginaceae/ Anjimalli	Herb	Fresh leaves	Paste	Treat swellings on the skin
135	*Momordica charantia* L. VOCB 6236	Cucurbitaceae/ Pagai sedi	Climber	Leaf, unripe fruit	Paste	Treat external piles, skin diseases
136	*Morinda pubescens* J.E. Smith var. *pubescens* VOCB 6237	Rubiaceae/ Manjanatti maram	Tree	Fresh leaves	Paste	Treat wounds
137	*Morinda tinctoria* Roxb. VOCB 6238	Rubiaceae/Nunaa	Small tree	Leaf	Paste	Applied on the inflammed region of the body
138	*Mukia maderaspatana* (L.) M. Roem VOCB 6239	Cucurbitaceae/Musuku musukai	Climber	Fresh leaves	Juice	Relief from cold cough and piles
139	*Murraya koenigii* (L.) Spreng. VOCB 6240	Rutaceae/Murungai maram	Tree	Fresh leaves	Paste	Increase hair growth
140	*Musa paradisiaca* L. VOCB 6241	Musaceae/Vaalai	Tree	Pseudostem, fruit	Juice	Stop hair loss, increase blood secretion

Sl.No.	Plant Name with Voucher Number	Family with Vernacular Name	Habit	Parts Used	Prepa-rations	Uses/Ailments treated
141	*Myristica fragrans* Houtt. VOCB 6242	Myristicaceae/Jathikai	Tree	Whole plant	Juice	Treat stomach ache
142	*Nervilia aragoana* Gaudich. VOCB 6243	Orchidaceae/Orilaittamarai	Herb	Leaf	Paste	Treat mental disorder and asthma
143	*Nymphaea nouchali* Burm.f. VOCB 6244	Nymphaeaceae/Alli	Herb	Root, fresh flower	Powder	Treat piles, cardiac diseases
144	*Ocimum americanum* L. VOCB 6245	Lamiaceae/Kanjag korai	Herb	Fresh leaves	Paste	Cure from skin diseases
145	*Ocimum basilicum* L. VOCB 6246	Lamiaceae/Karpurathulsi	Herb	Leaf	Juice	Eliminate worms present in the stomach
146	*Ocimum gratissimum* L. VOCB 6247	Lamiaceae/Elumicham Thulasi	Herb	Fresh leaves	Juice	Treat gonorrhea
147	*Ocimum sactum* L. VOCB 6248	Lamiaceae/Thulasi	Herb	Fresh leaves	Leaves	Treat cardiac diseases
148	*Opuntia tehuantepecana* (Bravo) Bravo VOCB 6249	Cactaceae/Sapathikalli	Herb	Stem	Paste	Treat wounds
149	*Pandanus fascicularis* Lam. VOCB 6250	Pandanaceae/Thaalamboo	Shrub	Leaf	Juice	Treat diabetes
150	*Pavetta indica* L. VOCB 6251	Rubiaceae/Kattukarunai	Shrub	Whole plant	Paste	Treat piles
151	*Pavonia zeylanica* (L.) Cav.Diss. VOCB 6252	Malvaceae/Sittamutti	Shrub	Dried roots	Powder	Act as a coolant
152	*Pedalium murex* L. VOCB 6253	Pedaliaceae/Yanai nerunjil	Herb	Leaf	Juice	Treat psoriasis, rheumatism
153	*Pergularia daemia* (Forssk.) Chiov. VOCB 6254	Asclepiadaceae/Veliparuthi	Shrub	Leaf	Decoction	Treat asthma
154	*Phyllanthus acidus* (L.) Skeels VOCB 6255	Euphorbiaceae/Aru nelli	Tree	Leaf	Juice	Treat rheumatism
155	*Phyllanthus amarus* Schum. *and* Thonn. VOCB 6256	Euphorbiaceae/Keezhanelli	Tree	Whole plant	Juice	Treat jaundice and scabies

Contd...

Sl.No.	Plant Name with Voucher Number	Family with Vernacular Name	Habit	Parts Used	Preparations	Uses/Ailments treated
156	*Phyllanthus emblica* L. VOCB 6257	Euphorbiacea/Nelli	Tree	Fresh leaf	Juice	Cure bleeding gums
157	*Phyllanthus reticulatus* Poir VOCB 6258	Euphorbiaceae/Karum poola	Shrub	Leaf	Paste	Treat piles
158	*Piper betle* L. VOCB 6259	Piperaceae/Vetrillai	Climber	Leaf	Powder	Treat stomach disorder
159	*Piper longum* L. VOCB 6260	Piperaceae/Tippili	Shrub	Seed	Powder	Relief from cold and cough
160	*Plectranthus amboinicus* (*Lour.*) Spreng. VOCB 6261	Lamiaceae/Omavalli	Herbs	Leaf	Juice	Dissolve kidney stones
161	*Plumbago zeylanica* L. VOCB 6262	Plumbaginaceae/ Chithiramoolam	Herbs	Root, fruit	Paste	Treat painful joints, inflammation skin
162	*Polyalthia longifolia* (Sonner.) Thw. Enum. VOCB 6263	Annonaceae/ Nettilingam	Tree	Stem bark	Powder	Cure mental problems
163	*Pongamia pinnata* (L.) Pierre. VOCB 6264	Fabaceae/Pungan	Tree	Seed, bark	Powder	Treat skin diseases,ring worm infection, rashes, scabies, eczema and psoriasis
164	*Premna integrifolia* L. VOCB 6265	Verbenaceae/Munnai	Tree	Leaf, root bark	Decoction	Treat lung diseases, arthritis,and nervous disorder
165	*Psidium guajava* L. VOCB 6266	Myrtaceae/Koiya	Shrub	Root	Paste	Treat piles
166	*Punica granatum* L. VOCB 6267	Lythraceae/Maathulai	Shrub	Fruit	Juice	Improve mental relaxation
167	*Randia dumetorum* Lamk. VOCB 6268	Rubiaceae/Kaarai kai	Shrub	Fruit	Paste	Cure skin diseases
168	*Raphanus sativus* L. VOCB 6269	Brassicaceae/Mullangi	Herb	Root tuber	Powder	Treat cancer and to initiate frequent urination
169	*Rauvolfia serpentina* (L.) Benth. Ex Kurz, For. VOCB 6270	Apocynaceae/Pampu kala	Herb	Root	Decoction	Treat blood pressure
170	*Rhinacanthus nasutus* (L.) Kurz VOCB 6271	Acanthaceae/Naagamalli	Shrub	Dried root	Powder	Treat snake bites

Sl.No.	Plant Name with Voucher Number	Family with Vernacular Name	Habit	Parts Used	Prepa-rations	Uses/Ailments treated
171	*Ricinus communis* L. VOCB 6272	Euphorbiaceae/Amanakku	Shrub	Dried leaves	Powder	Treat jaundice
172	*Rosa damascena* Mill. VOCB 6273	Rosaceae/Rose	Shrub	Flower buds	Paste	Induce the growth of hair in black
173	*Ruelia tuberosa* L. VOCB 6274	Acanthaceae/Pattaskai	Herb	root	Juice	Relief from kidney disorders
174	*Sansevieria roxburghiana* Schult VOCB 6275	Liliaceae/Marul	Herb	Leaves	Juice	Treat earache
175	*Santalum album* L. VOCB 6276	Santalaceae/Santhanam	Herb	Wood	Paste	Relief from eczema and acne vulgaris
176	*Schleichera oleosa* (Lour.) VOCB 6278	Sapindaceae/Boovan	Tree	Seed	Paste	Treat rheumatism
177	*Scilla hyacinthina* (Roth) Macbr VOCB 6278	Liliaceae/Kattu vengayam	Tree	Leaf	Juice	Treat cough
178	*Sida acuta* Burm.f. VOCB 6279	Malvaceae/Aruvaamanai poondu	Herb	Leaves	Paste	Treat cuts and wounds
179	*Solanum anguivi* Lam. VOCB 6280	Solanaceae/poondu	Herb	Fresh leaves	Juice	Control vomiting
180	*Solanum nigrum* L. VOCB 6281	Solanaceae/Manathakali	Herb	Fresh leaves	Paste	Treat stomach ulcer,heal burns, wound, and joint pain
181	*Solanum torvum* Sw. Prodr. VOCB 6282	Solanaceae/Sundai	Herb	Fresh root, fruit	Paste	Cure from cracks in heels, heamorroids
182	*Solanum trilobatum* L. VOCB 6283	Solanaceae/Thoodhuvalai	Climber	Fresh leaves	Paste	Strengthen the bones, cure from cough and cold
183	*Solanum virginianum* L. VOCB 6284	Solanaceae/Kandakathiri	Herb	Fresh fruit	As such	Treat throat infection
184	*Solanum xanthocarpum* Schradt. Wendl. VOCB 6285	Solanaceae/Kathirikai	Herb	Flowers	Decoction	Cure bronchial asthma
185	*Stachytrapheta jamaicensis* (L.) Vahl, Enum VOCB 6286	Verbenaceae/Ezhutthaani poondu	Herb	Fresh leaves	As such	Act as coolant
186	*Strebulus asper* Lour. VOCB 6287	Moraceae/Kutti poola	Small tree	Leaf	Latex	Increase hair growth

Contd...

Sl.No.	Plant Name with Voucher Number	Family with Vernacular Name	Habit	Parts Used	Prepa-rations	Uses/Ailments treated
187	*Syzygium cumini* (L.) Skeels. VOCB 6288	Myrtaceae/Naval	Tree	Stem bark	Juice	Treat diarrhoea
188	*Tabernaemontana divaricata* (L.) R. Br. ex Roem and Schultes. VOCB 6289	Apocynaceae/Nanthi vattam	Shrub	Fresh root	As such	Cure from tooth ache
189	*Tamarindus indica* L. VOCB 6290	Caesalpiniaceae/Puli	Tree	Ripened pods	Paste	Treat indigestion
190	*Tephrosia purpurea* (L.) Pers. VOCB 6291	Fabaceae/Kolinji	Herb	Dried root bark	Juice	Treat stroke
191	*Terminalia catappa* L. Syst. Nat. ed. VOCB6292	Combretaceae/Vadhumai	Tree	Leaf	Paste	Applied for skin diseases
192	*Terminalia chebula* Retz. VOCB 6293	Combretaceae/Kadukai	Tree	Fruit	Decoction	Cure eye diseases
193	*Themeda triandra* Forssk., VOCB 6294	Poaceae/Sempulu	Shrub	Whole plant	Powder	Treat wounds
194	*Thespesia populnea* (L.) Sol. ex Correa VOCB 6295	Malvaceae/Poovarasu	Tree	Leaf	Paste	Cure from chronic wounds
195	*Tinospora cordifolia* (Thunb.) Miers VOCB 6296	Menispermaceae/Seendhil	Creeper	Stem	Powder	Treat dysentery
196	*Tragia involucrata* L. VOCB 6297	Euphorbiaceae/Chenthatti	Wild winer	Fresh root	Juice	Arrest dysentery
197	*Trianthema portulacastrum* L. VOCB 6298	Aizoaceae/Baadhar keerai	Creeper	Leaf	Paste	Treat snake bites and scorpion stings
198	*Trichoderma indicum* (L.) R.Br. VOCB 6299	Boraginaceae/Kavil thumbai	Herb	Whole plant	Paste	Treat tumors
199	*Tridax procumbens* L. VOCB 6300	Asteraceae/Thanni poondu	Herb	Whole plant	Paste	Treat cuts and wounds
200	*Tylophora indica* (Burm.f.) Merr. VOCB 6301	Asclepiadaceae/Kakkupaalai	Climbing shrub	Fresh leaves	Juice	Treat bronchial asthma

Sl.No.	Plant Name with Voucher Number	Family with Vernacular Name	Habit	Parts Used	Preparations	Uses/Ailments treated
201	*Vernonia cinerea* (L.) Less. VOCB 6302	Asteraceae/Mukkuthi poondu	Herb	Whole plant	Juice	Treat male infertility
202	*Vigna mungo* (L.) Hepper VOCB 6303	Fabaceae/Ulunthu	Herb	Seed	As such	Treat polio attack
203	*Vitex negundo* L. VOCB 6304	Verbenaceae/Nochi	Tree	Fresh leaves	Vapour	Cure headache, fever, ear infection
204	*Wattakaka volubilis* (L.f) stapf VOCB 6305	Asclepiadaceae/Kodi Palai	Climber	Leaf and root	Decoction	Treat snake bites
205	*Wedelia trilobata* L. VOCB 6306	Asteraceae/Mookuthipoo	Climber	Fresh leaves	Juice	Improve mental abilities
206	*Wrightia laevis* Hook.f. VOCB 6307	Apocynaceae/Vetpalai	Tree	Leaf	Powder	Treat psoriasis
207	*Zingiber officinale* Roscoe VOCB 6308	Zingiberaceae/Sukku	Shrub	Rhizome	Extract	Control nausea
208	*Zizyphus mauritiana* Lam.VOCB 6309	Rhamnaceae/Illanthai	Tree	Root bark	Extract	Improve digestion

Table 3.2: Botanical Names of Plants Whose Parts/Products are Purchased from the Market and Used as Medicine/Ingredients in the Preparation by *Irulas*

Sl.No.	*Botanical Name*	*Family*	*Vernacular Name*	*Plant Part/Extract of Plant Part Procured*
1.	*Allium cepa* L.	Liliaceae	Vengayam	Bulb
2.	*Allium sativum* L	Liliaceae	Vellaipoondu	Bulb
3.	*Arachis hypogaea* L.	Fabaceae	Nillakadalai	Seed
4.	*Cocos nucifera* L.	Arecaceae	Thennai	Seed oil
5.	*Cuminum cyminum* L.	Apiaceae	Shiragam	Fruit
6.	*Curcuma longa* L.	Zingiberaceae	Manjal	Dried rhizome
7.	*Piper nigrum* L.	Piperaceae	Milagu	Seed
8.	*Ricinus communis* L.	Euphorbiaceae	Aamanakku	Seed oil
9.	*Sesamum indicum* L.	Pedaliaceae	Yellu	Seed Oil
10.	*Zingiber officinale* Roscoe	Zingiberaceae	Sukku	Dried Rhizome

After authentication with regional flora, the specimens were matched with the authentic herbarium of Madras Herbarium, Botanical Survey of India (B.S.I), Southern Circle, Coimbatore, Tamil Nadu, India.

Results

The results of the survey are presented in Tables 3.1 and 3.2.

Discussion

The tribals knowledge of indigenous uses of native medicinal plants before their exodus into the urban areas to join the mainstream life needs to be studied and documented. In the present study, 218 medicinal plants were collected and documented (Tables 3.1 and 3.2). The present study focuses the extensive usage of as many as 218 medicinal plants by the tribes inhabiting the area chosen for study (Krishnagiri district, Eastern Ghats, Tamil Nadu, India).

As an outcome of the present investigation, 218 plants (208 plants collected and 10 plants/extracts procured) belonging to 166 genera and 68 families were recorded. Of the recorded plants, a maximum of 16 ethnomedicinal plants belong to Euphorbiaceae followed by Fabaceae (12 plant species each), Lamiaceae (11 plants species), Solanaceae (10 plants species), Malvaceae (9 plant species), Acanthaceae (8 species), Mimosaceae, Apocynaceae and Cucurbitaceae (7 plant species), Caesalpiniaceae, Asteraceae and Moraceae (6 plant species each), Asclepiadaceae, Rubiaceae and Amaranthaceae (5 plant species each).

The documented plants used for various diseases by the Irulas show a great habitual diversity. Among the 208 plants reported as herbal drugs, 36 per cent of

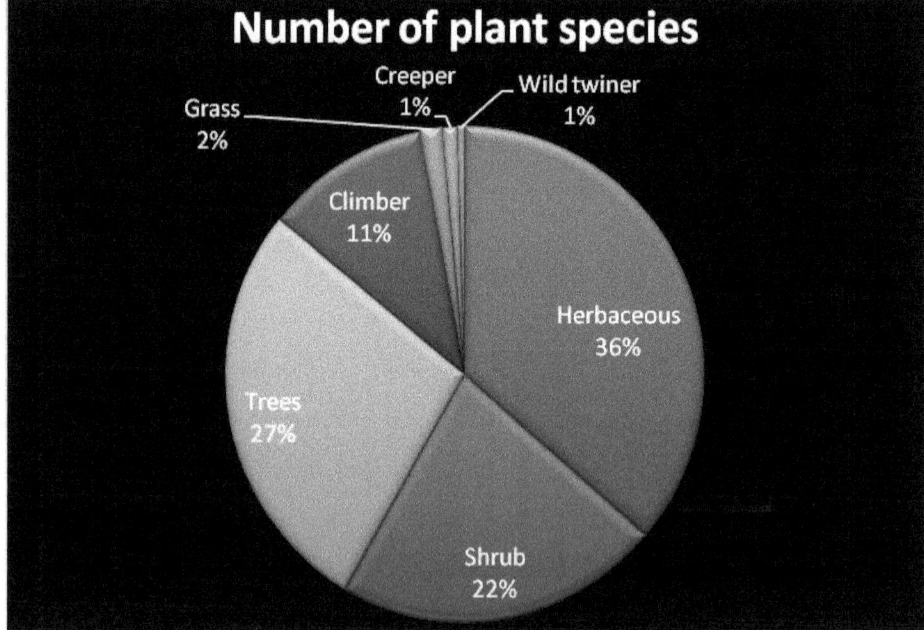

Figure 3.6: Habitual Diversity of the Enumerated Plants.

the plants are herbaceous in habit while 22 per cent plants are shrubs, 27 per cent plants are trees, 11 per cent plants are climbers, 2 per cent plants are grass and 1 per cent plants are creepers and 1 per cent plant is wild twiner (Figure 3.6). Among the investigated medicinal plants 1 plant belongs to pteridophyte under 1 family, 74 plants belong to Polypetalae under 25 families, 71 plants belong to Gamopetalae under 16 families, 35 plants belong to Monochlamydeae under 9 families and 27 plants belong to monocotyledons under 17 families (Figure 3.7).

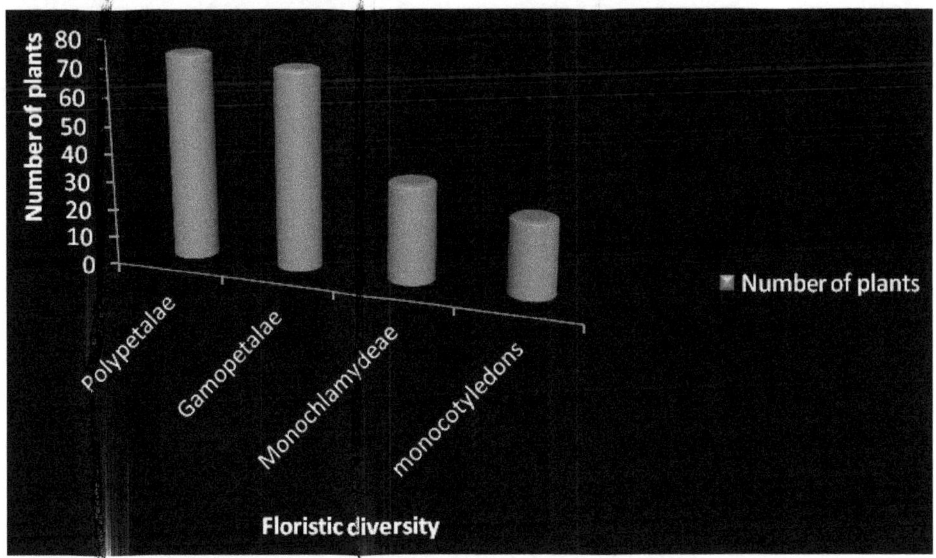

Figure 3.7: Floristic Diversity among the Enumerated Plants.

So far as the plant parts used is concerned, it is noted that almost all part of plant is in employed ethnomedicine. In terms of percentage of plant parts used, the percentages are as follows; Leaf 52 per cent, root bark 14 per cent, fruit and whole plant 9 per cent, stem bark 8 per cent, flower 7 per cent, seed 4 per cent, rhizome and root tuber 2 per cent and latex 1 per cent (Figure 3.8). The most prevalent form of administration of medicine is paste (41 per cent). This is followed by juice/extract (35 per cent), decoction (11 per cent), powder (9 per cent), plant part as such (4 per cent), orally, oil and latex (3 per cent) (Figure 3.9).

The enumerated 218 plants are used to cure as many as 71 different types of human maladies (Table 3.3). A maximum of 25 plants are used to treat cuts and wounds, followed by 18 plants for skin diseases, 17 plants for cold and cough, 12 plants for rheumatism, 11 plants for stomach disorder, 11 plants for piles, 8 plants for asthma and fever, 7 plants for dysentery, urinary disorder and eczema and 6 plants for digestive disorder.

Seventy one different maladives are grouped into 5 major categories of diseases *viz.*, skin elements, gastro – intestinal complaints, body pain, fever, cold and cough and poisonous bites, 64 plants belonging to 57 genera and 40 families are used by the treating various skin ailments like eczema, scabies, ring worm infections, pimples, fungal diseases, leprosy, cuts, wounds, inflammations, acnes, foot crack

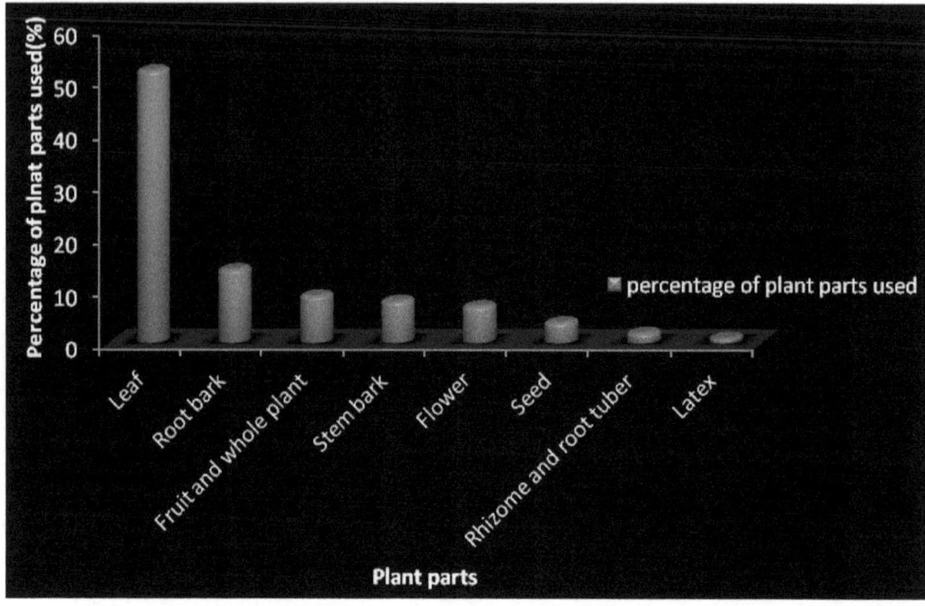

Figure 3.8: Morphology of Plant Parts Used in Phytotherapy.

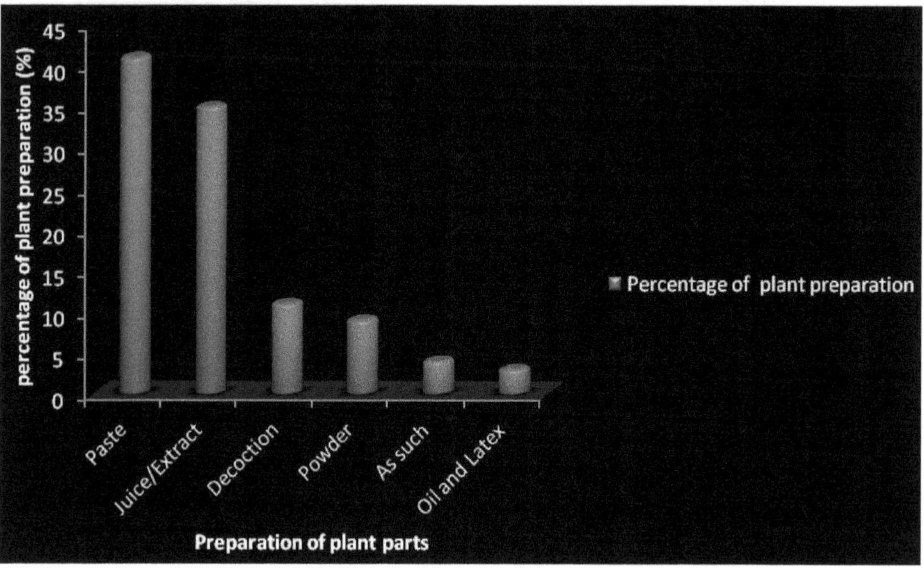

Figure 3.9: Mode of Administration of Herbals by *Irulas.*

and psoriasis. 43 plants belonging to 39 genera and 30 families are used by the *Irulas* for treating gastro-intestinal complaints like piles, jaundice, expulsion of gas, dysentery, stomach ache, indigestion, ulcer, liver disorder, diarrhoea, gastritis, stomach disorder, vomiting sensation and improve digestion *etc.* 33 plants belonging to 33 genera and 28 families are used by the *Irulas* for treating pains like rheumatic pain, arthritis, joint pain, ear ache, tooth ache, head ache, general body pain and body coolant. 22 plants belonging to 19 genera and 15 families are employed to treat fever, cold and cough and related ailments, 15 belonging to 15 genera and 13 families are used by the *Irulas* for treating poisonous bite like snake bites, scorpion stings, rat bites leech bite and insect bites, *etc.*

Table 3.3: Disease Types and Herbal Remedies of *Irulas*

Sl.No.	*Name of the Ailment*	*Total Number of Herbal Remedies*
1	Aphrodisiac	5
2	Arthritis	3
3	Asthma	8
4	Bleeding gums	2
5	Blood cancer	1
6	Blood pressure	1
7	Bone fracture	1
8	Brain disorder	1
9	Bronchial asthma	2
10	Cardiac disorder	3
11	Chikungunya	1
12	Cold and cough	17
13	Colic pain	1
14	Control nausea	1
15	Coolant	6
16	Cracks in heels	1
17	Cuts and wounds	25
18	Dengu	1
19	Diabetes	5
20	Diarrhoea	2
21	Digestive disorder	2
22	Dysentery	7
23	Ear ache	2
24	Eczema	7
25	Eliminate worms	1
26	Expulsion of gas	1
27	Eye disorder	3
28	Fever	8
29	Freckles	1
30	Fungal disease	1
31	Gonorrhea	3
32	Goiter	1
33	Gum infection	1
34	Hair fall	5

Sl.No.	Name of the Ailment	Total Number of Herbal Remedies
35	Headache	4
36	Insect bite	2
37	Indigestion	4
38	Induced sweat	1
39	Jaundice	5
40	Joint pain	3
41	Kidney stones	2
42	Laxative	1
43	Leach bite	2
44	Leucorrhoea	4
45	Mental disorder	3
46	Mouth ulcer	2
47	Night blindness	1
48	Osteoarthritis	2
49	Piles (Haemorrhoids)	11
50	Pimples	2
51	Poisonous bite	1
52	Polio attack	1
53	Psoriasis	8
54	Rheumatism	13
55	Ringworm	4
56	Scabies	6
57	Scorpion sting	3
58	Sexual disorder	5
59	Skin disease	18
60	Sprains	1
61	Stiffness of joint	1
62	Stroke	1
63	Stomach disorder	11
64	Stimulate libido	1
65	Swellings	3
66	Syphilis	1
67	Throat irritation	1
68	Tooth ache	4
69	Tooth paste	1
70	Wheezing	1
71	Urinary disorder	7

Above 28 per cent of the identified plants possess two fold uses (used to treat two ailments). 64 per cent of the ethnomedicinal plants are used to cure only one particular disease. 8 per cent of the plant species recorded in the study exhibit manifold therapeutic uses, *i.e.*, they are used to cure more than two ailments (Figure 3.10). For example, the leaf of *Acalypha indica* is used to cure psoriasis, allergic dermatitis, cold and cough. The leaf of *Argemone mexicana* is used to cure skin diseases; its roots are used to cure fungal diseases; whole plant is used to cure scabies, ring worm and eczema. The flowers of *Butea monosperma* are used to cure leprosy and sexual disorder; its leaf is used to cure eczema. The latex of *Calotropis gigantea* is used to cure rat and snake bite; its leaf is used to treat rheumatism and lesions of eczema. The flower of *Cassia fistula* is used to cure eczema and freckles; its stem bark is used to treat skin diseases. The rhizome of *Curcuma longa* is used to cure stomach ache, fever and wounds. The stem bark of *Madhuca longifolia* is used to cure wounds and skin diseases; its flower buds are used to treat fever and cough. The leaf of *Mukia maderaspatana* is used to cure cold, cough and piles problem. The seed of *Pongamia pinnata* is used to cure skin diseases; its stem bark is used to cure ring worm infection, rashes, scabies, eczema and psoriasis. The leaf of *Solanum nigrum* is used to treat mouth aphthous, stomach ulcer and joint pain. The leaf of *Vitex negundo* is used to treat head ache, fever and ear infections. The above mentioned manifold ethnomedicinal plants are recommended for broad phytopharmacological investigations.

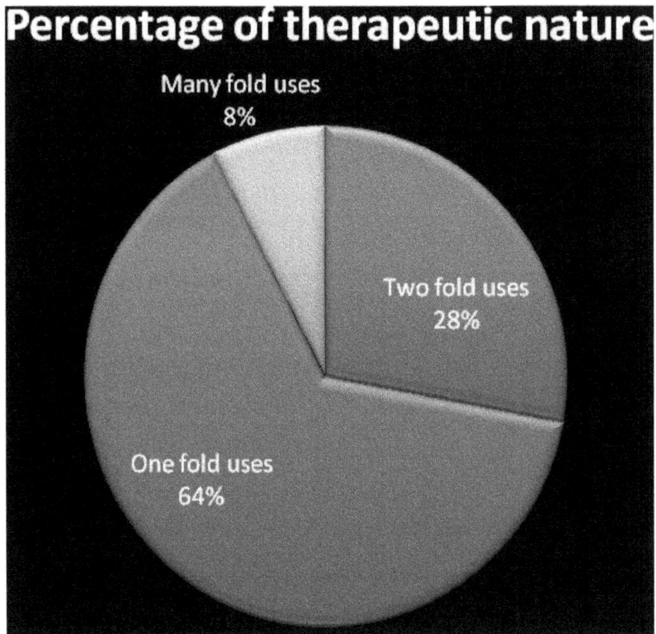

Figure 3.10: Classification of Herbal Therapeutic Nature.

Cross-cultural Ethnomedicobotanical Studies

It is well established that, an identical use of same plant by different tribal groups indicates its curative property and therapeutic significance (Jain and Saklani, 1992). The comparison of medicinal claims between different cultural groups in the same region or neighbouring regions has proved very rewarding. The comparative studies on medicinal uses of plants among different cultural groups showed similarities and dissimilarities in uses. In the present investigation, the medicinal claims emanating from the selected tribal groups are compared with claims from three other numerically predominant Indian tribal groups living in the Central and Western zones of the country *viz.*, the *Bhils, Gonds* and *Santals* and two other predominant tribals of Tamil Nadu namely the *Kanikkars* and *Palliyars*. The source of ethnomedicinal information for the above mentioned ethnic groups is selected from the available literature (Jain *et al.*, 1973; Mudgal and Pal, 1980; Ramachandran and Nair, 1981; Joshi, 1982;1992; Tarafder, 1983; Shah and Gopal, 1985; Sur *et al.*, 1987; 1990; Prasad *et al.*, 1987; Pal *et al.*, 1989; Saxena and Tripathi, 1989; Yadav and Bhamare, 1989; Ragupathy and Mahadevan, 1991; Borthakur and Goswami, 1995; Mishra and Naquvi, 1995; Chandra, 1995; Girach and Aminuddin, 1995; Katewa and Arora, 1997; Ignacimuthu *et al.*, 1998; Girach *et al.*, 1999; Viswanathan *et al.*, 2001; Binu *et al.*, 2003; Das *et al.*, 2006; Ayyanar and Ignacimuthu, 2005; Jadhav, 2006; Senthilkumar *et al.*, 2006; Patil and Patil, 2007; Nazarudeen, 2009; Khanna and Anandkumar, 2009; Sarena *et al.*, 2009; Mitra and Mukherjee, 2009; Lalitharani *et al.*, 2010; Chendurpandy *et al.*, 2010; Jain *et al.*, 2010; Maruthupandian *et al.*, 2011; Murthy, 2012; Nikhil *et al.*, 2013; Divya *et al.*, 2013; Ramkrishna and Saidulu, 2014).

Comparative Ethnomedicobotany with *Bhils*

Thirty one plant species are commonly used by both *Irulas* and *Bhils*. The similarities and dissimilarities are discussed (Table 3.4). Comparative ethnobotanical study between the *Irulas* and *Bhils* reveals that 9 medicinal plant species (out of 30) exhibit identical uses. They are *Achyranthes aspera, Argemone mexicana, Cassia tora, Cissus quadrangularis, Curculigo orchioides, Hibiscus rosa-sinensis, Jatropha* curcas, *Tridax procumbens* and *Vitex negundo*.

Table 3.4: Comparative Ethnomedicobotany with *Bhils*

Sl.No.	Botanical Name	Part Used	Irulas	Bhils
1.	*Abutilon indicum*	Root, Leaf	Piles, Ear ache (Otalgia)	Leucorrhoea
2.	*Acacia nilotica*	Stem bark, Leaf, Dried resin	Leach bites, dysentery	Inflammations and piles
3.	*Achyranthes aspera*	Leaf and Root bark, Latex	Eye disorder, tooth powder to whiten the teeth	Eye disorders, fever, cough, snake bite and Indigestion
4.	*Aloe vera*	Succulent Leaf	Dysentery	Hair fall, Eruptions
5.	*Argemone mexicana*	Leaf, Root, Whole plant	Skin diseases, fungal disease, scabies, ring worm and eczema and wounds	Skin diseases
6.	*Asparagas racemosus*	Root, Leaf	Wounds	Typhoid
7	*Azadirachta indica*	Stem bark Leaf, Seed,	Rheumatic complaints	Ascariasis and skin diseases, Snake bite

Sl.No.	Botanical Name	Part Used	Irulas	Bhils
8.	*Butea monosperma*	Seed, Stem bark, Flowers	Sexual disorder and Eczema	Remove intestinal worms to child and Bodyache
9.	*Calotropis gigantea*	Milky latex, Leaf	Snake bites, Rheumatism, Eczema	Jaundice, To reduce pain and swellings
10.	*Cassia tora*	Seed, Leaf	Ringworm	Asthma, Eczema
11.	*Cissus quadrangularis*	Whole plant, Stem	Rheumatoid arthritis and Bone fractures	Bone fracture, Wound heeling
12.	*Coriandrum sativum*	Seed, Leaf	Digestion	Headache
13.	*Curculigo orchioides*	Root, Tuber	To increase the sexual vigour in males	Vermifuge and Aphrodisiac, Increase potency
14.	*Ficus religiosa*	Leaf, Bark	Psoriasis	Snake bite.
15.	*Hemidesmus indicus*	Root	Body coolant	Dysentery, To increase lactation
16.	*Hibiscus rosa- sinensis*	Flowers, Leaf	To treat impotency in males	Aphrodisiac, Boils and Laxative, To prevent hair fall and to blacken the hairs
17.	*Jatropha curcas*	Twigs, Leaf, Stem, Seed Latex	Psoriasis, Cough, stomach disorder, Toothache and typhoid	Dysentery, Stomachache, toothache, Rheumatism and Poisionous bites
18.	*Lippia nodiflora*	Whole plant, Leaf	Internal piles	To cure kidney stone problem
19.	*Momordica charantia*	Fruit, Leaf	External piles and Skin disease	Diabetes
20.	*Ocimum basilicum*	Leaf	Eliminate worms present in the stomach	To cure ringworm, To keep away the snakes
21.	*Ocimum sactum*	Leaf	Cardiac disease	Cough
22.	*Phyllanthus emblica*	Leaf	To cure bleeding from gums	Dysentery
23.	*Piper longum*	Root, Seed	Cold and Cough	Fever
24.	*Psidium guajava*	Fruits, Root	Piles	Cough and Cold
25.	*Ricinus communis*	Seed, Leaf, Stem	Jaundice	Backache, Headache and Toothache, Hair fall and Purgative.
26.	*Tinospora cordifolia*	Stem	Dysentery	Leucorrhoea, Snake bite and Leprosy
27.	*Tridax procumbens*	Whole plant	To treat cuts and wounds	Antiseptic, Cuts and wounds and Diabetes
28.	*Vitex negundo*	Leaf	To treat fever, Headache and Ear infections.	Hair fall, Headache
29.	*Wrightia tinctoria*	Leaf	Psoriasis	Fish poison
30.	*Zingiber officinale*	Dry rhizome	Nausea	After pregnancy for strength

Comparative Ethnomedicobotany with the *Gonds*

Forty five plant species are commonly used by both *Irulas* and *Gonds.* The similarities and dissimilarities are discussed (Table 3.5). Comparative ethnobotanical study between the *Irulas* and *Gonds* reveals that 14 medicinal plant species (out of 45) exhibit identical uses. They are *Achyranthes aspera, Acacia nilotica, Alangium salvifolium, Azadirachta indica, Cassia fistula, Curculigo orchioides, Datura metel, Dichrostachys cinerea, Jatropha curcas, Mangifera indica, Phyllanthus amarus, Ricinus communis, Syzygium cumini* and *Tridax procumbens.*

Table 3.5: Comparative Ethnomedicobotany with the *Gonds*

Sl.No.	Botanical Name	Part Used	Irulas	Gonds
1.	*Abrus precatorius*	Seed, Leaf	Stiffness of joint and Poisonous bite	Reduce hair fall, Improve hair growth, Natural contraceptive and Fever.
2.	*Achyranthes aspera*	Root, Leaf, Stem	Eye disorder, Tooth powder to whiten the teeth	Lithontriptic, Relieving, Diabetes, Blood dysentery, Leucaemia, Toothbrush and given to pregnant woman for normal delivery
3.	*Acacia nilotica*	Roots, Leaves, Stem, Seeds	Leach bites, Dysentery	Diarrhoea and Dysentery
4.	*Aegle marmelos*	Fruit, Leaf, Stem bark, Fruit	Scabies.	Diarrhoea, Dysentery, Eye infections, Febrifuge and Jaundice
5.	*Alangium salvifolium*	Leaf, Stem, Seed, Rootbark	Bites.	Toothbrush, Malarial fever and Removing poison from the body
6.	*Allium cepa*	Fruit, Bulb	Urinary tract infection	Ear ache, Blood dysentery
7.	*Aloe vera*	Leaf	Dysentery	Skin diseases, Spermatorrhoea
8.	*Andrographis paniculata*	Whole plant	Dengu, Chikungunya Fever and giddiness.	Diarrhoea, Cough, Fever, Malaria, Liver disorder, Typhoid and other Viral fevers
9.	*Aristolochia bracteata*	Whole plant	Wounds and Eczema	Menstrual pains.
10.	*Asparagus racemosus*	Root, Leaf	Wounds	Pregnant lady to control the body heat, Blood dysentery, Aphrodisiac, Dyspepsia, Galactogogue, Nervic tonic, Vigour and Strength

Sl.No.	Botanical Name	Part Used	Irulas	Gonds
11.	*Azadirachta indica*	Stembark Leaf, Seed, Gum, Flower	Rheumatic complaints	Toothache, Swelling, Antidiabetic, Malarial fever, Tooth brush, Rheumatism, Anthelmintic, Diarrhoea in childrens Skin disorder
12.	*Butea monosperma*	Seed, Stem bark, Root, Flowers	Sexual disorder and Eczema	Rheumatism, Diarrhoea Indigestion, Diabetes
13.	*Capparis zeylanica*	Fruits, Leaf Rootbark	Diarrhoea	Immunity, Snake bite
14.	*Carica papaya*	Fruit	Stomach ache, Digestive problems	Stone
15.	*Cassia fistula*	Fruit, Leaf, Bark and Flower	Eczema, Freckles, Dysentery and Skin disease	Tonic, Fever, Dysentery
16.	*Cassia tora*	Seed, Leaf and Root	Ringworm.	Antidiabetic, Asthma
17.	*Catharanthus roseus*	Whole plant	Blood cancer	Blood Pressure
18.	*Cissus quadrangularis*	Whole plant	Rheumatoid arthritis and Bone fractures.	Broken limbs
19.	*Curculigo orchioides*	Root, Tuber	To increase the sexual vigour in males	Aphrodisiac, Blindness, Eye infections
20.	*Curcuma longa*	Flower, Rhizome	Stomach ache (Gastritis), Fever and Wounds.	Jaundice
21.	*Datura metel*	Fruit and Leaf	Fever, Skin diseases, Brain disorders and Diabetes mellitus	Acne and wounds
22.	*Dichrostachys cinerea*	Rootbark, Seed, Leaf	Rheumatoid arthritis and Dysentery.	Joint pains
23.	*Ficus benghalensis*	Aerial root, Bark	Diabetes and Mouth ulcer	Jaundice, Dysentery
24.	*Ficus religiosa*	Bark	Psoriasis	Immunity, Jaundice
25.	*Garuga pinnata*	Roots, Fruits and Stem Bark	Colic pain	Pulmonary infections, Asthma, Cooling, Digestive, Carminative, Blisters Vermifuge and Anthelmintic
26.	*Gmelina arborea*	Bark, Leaf and Root	Cough and Cardiac disease	Fevers
27.	*Hemidesmus indicus*	Root	Body coolant	Tonic, diarrhoea and dysentery.
28.	*Jatropha curcas*	Twigs, Leaf, Stem, Seed, Latex	Psoriasis, Cough, Stomach disorder, Toothache and typhoid	External application, Skin diseases, Dysentery

Sl.No.	Botanical Name	Part Used	Irulas	Gonds
29.	*Mangifera indica*	Seed and Gum	Dysentery	Dysentery, Cough, Eruption of tongue, Dysentery
30.	*Michelia champaca*	Root and Flower	Piles, Arthritis, Fever and Luecorrhoea in women.	Asthma
31.	*Mimosa pudica*	Root, Leaf	Diabetes, Body heat and joint pain (arthralgia)	Liver disorder, Diarrhoea, Aphrodisiac and Strength promoter
32.	*Musa paradisiaca*	Leaf, Stem, Fruit	Hair loss and Increase Blood secretion.	Diarrhoea.
33.	*Ocimum sanctum*	Bark, Leaf	Cardiac disease.	Jaundice
34.	*Phyllanthus amarus*	Entire plant	Jaundice and Scabies	Jaundice
35.	*Ricinus communis*	Root, Leaf, Stem	Jaundice	Jaundice, Strength promoter, Hydrocele
36.	*Sida acuta*	Leaf	Cuts and Wounds	Testicular swellings, Hemorrhoids and impotence
37.	*Solanum nigrum*	Whole plant	Mouth aphthous, Stomach ulcer and joint pain.	Iron Deficiency
38.	*Syzygium cumini*	Seed and Stem bark	Diahhorea	Diabetes, Diarrhoea, and Mouth wash
39.	*Tamarindus indica*	Root, ripened pod	Indigestion	Reducing weight, blood dysentery
40.	*Terminalia chebula*	Fruit, Seed	Initiate urination and Eye diseases.	Nervic tonic, Diarrhoea, Cough.
41.	*Tinospora cordifolia*	Stem, Leaf	Dysentery	Immunity, Swelling and rheumatism.
42.	*Tridax procumbens*	Whole plant	To treat cuts and wounds	Diarrhoea, Cuts and Wounds
43.	*Vernonia cinerea*	Leaf	Male infertility	Regular menstrual cycle in women, Skin diseases and Dysentery
44.	*Wrightia tinctoria*	Stem, bark	Nausea	Typhoid, Bronchitis and Gas trouble
45.	*Zingiber officinale*	Tuber	Nausea	Respiratory problem

Comparative Ethnomedicobotany with the *Santals*

Twenty eight plant species are commonly used by both *Irulas* and *Santals.* The similarities and dissimilarities are discussed (Table 3.6). Comparative ethnobotanical study between the *Irulas* and *sandals* reveals that 6 medicinal plant species (out of 28) exhibit identical uses. They are *Boerhavia diffusa, Cissampelos pareira, Lantana camara, Leucas aspera, Santalum album* and *Tridax procumbens.*

Table 3.6: Comparative Ethnomedicobotany with the *Santals*

Sl.No.	*Botanical Name*	*Part Used*	*Irulas*	*Santals*
1.	*Abrus precatorius*	Leaf and Stem	Stiffness of joint and Poisonous bite	Diabetes, Abortifacient and Throat pain
2.	*Achyranthes aspera*	Leaf and Root bark	Eye disorder, Tooth powder to whiten the teeth	Induce abortion
3.	*Aegle marmelos*	Leaf, Fruit, Stem bark	Scabies	Cold and cough
4.	*Andrographis paniculata*	Whole plant	Dengu, Chikungunya Fever and Giddiness	Malarial fever
5.	*Asparagus racemosus*	Tuber	Wounds	Muscle strain
6.	*Azadirachta indica*	Leaf, Stem, Oil, Gum and Flower	Rheumatic complaints	Stomachache and Skin diseases
7.	*Boerhavia diffusa*	Root and Leaf	Asthma, Jaundice and Swellings in legs	Body swelling
8.	*Butea monosperma*	Stembark, Leaf, Fruit and Seed	Sexual disorder, Eczema	Intestinal worm
9.	*Cajanus cajan*	Leaf	Jaundice	Reddish sores in tongue
10.	*Calotropis gigantea*	Leaf, Root and Flower	Snake bites and Rheumatism	Diarrhoea and Dog bite
11.	*Cassia fistula*	Flower, Leaf, Bark	Eczema, freckles, Dysentery and skin disease.	Ear ache
12.	*Catharanthus roseus*	Flower, Leaf and Root	Blood cancer	Insect sting and Burns
13.	*Cissampelos pareira*	Root and Leaf	Stomach disorders and Indigestion	Stomach pain
14.	*Curculigo orchioides*	Tuber	Increase the sexual vigour in males	White discharge in women
15.	*Ficus racemosa*	Latex, Bark	Wounds	Jaundice
16.	*Hemidesmus indicus*	Root, Leaf	Body coolant	Fever
17.	*Hibiscus rosa -sinensis*	Flowers, Leaf and Root	To treat impotency in males	Excessive bleeding and Whitish discharge in urine of females (leucorrhea).
18.	*Lantana camara*	Leaf	Wounds.	Cuts and Snake bite
19.	*Leucas aspera*	Fruit	Snake bites.	Snake bite and Headache
20.	*Ocimum basilicum*	Leaf	Eliminate worms present in the stomach	Diarrhoea
21.	*Phyllanthus emblica*	Leaf	To cure bleeding from gums	Stomach trouble and Dysentery
22.	*Santalum album*	Wood	Acne	Headache and Skin diseases

Sl.No.	Botanical Name	Part Used	Irulas	Santals
23.	*Solanum tuberosum*	Tuber and Leaf	Strengthen the bones, Cough and Cold	Stomach pain
24.	*Syzygium cumini*	Seed and Stem bark	Diahhorea	Diabetes
25.	*Terminalia chebula*	Fruit	Initiate urination and Eye diseases	Indigestion
26.	*Tridax procumbens*	Leaf	Cuts and Wounds	Cuts and wounds, Sores and Leprosy
27.	*Vernonia cinerea*	Leaf	Male infertility	Oesophagus troubles
28.	*Zizyphus mauritiana*	Leaf, Stem bark and Root bark	Improve digestion	Diarrhoea

Comparative Ethnomedicobotany with the *Kanikkars*

Fifty one plant species are commonly used by both *Irulas* and *Kanikkars*. The similarities and dissimilarities are discussed (Table 3.7). Comparative ethnobotanical study between the *Irulas* and *Kanikkars* reveals that 18 medicinal plant species (out of 51) exhibit identical uses. They are *Abrus precatorius, Aegle marmelos, Albizia amara, Cassia alata, Cassia fistula, Cissus quadrangularis, Cleome viscosa, Clerodendron inerme, Clitoria ternatea, Curcuma longa, Evolvulus alsinoides, Indigofera tinctoria, Jatropha curcas, Momordica charantia, Mukia maderaspatana, Piper longum, Santalum album* and *Vitex negundo.*

Table 3.7: Comparative Ethnomedicobotany with the *Kanikkars*

Sl.No.	Botanical Name	Part Used	Irulas	Kanikkars
1.	*Abrus precatorius*	Leaf, seed, Root and Stem	Stiffness of joint and Poisonous bite	Dandruff, Poisonous bite, Chronic conjunctivitis or inflammation of mucus membrane of eye lids and Skin rashes
2.	*Abutilon indicum*	Root, Leaf	Piles, Ear ache (Otalgia)	Ringworm infection, Cough, Fever, to expel worms and Skin infections.
3.	*Achyranthus aspera*	Stem, Leaf and Root bark	Eye disorder, Tooth powder to whiten the teeth	Antidote and Dog-bite
4.	*Aegle marmelos*	Fruit, Leaf, Stembark, Fruit	Scabies	Rheumatism and Itches.
5.	*Alangium salvifolium*	Leaf, Stem, Seed, Rootbark	Bites	Eye diseases.
6.	*Albizia amara*	Rootbark, Stembark	Snake bites and Scorpion sting	Stomachache, Skin diseases, Snake bite and Scorpion sting

Sl.No.	*Botanical Name*	*Part Used*	*Irulas*	*Kanikkars*
7.	*Aloe vera*	Leaf, Root	Dysentery	Skin diseases, Ulcer, Menstrual problem, Kidney stones and Laxative.
8.	*Andrographis paniculata*	Whole plant	Dengu, Chikungunya Fever and Giddiness	Leprosy, scabies, Ringworm infection, Fever, Body weakness, Skin allergies, Sore throat, Indigestion, Liver problems, Constipation, Snake bite and Scorpion bite.
9.	*Argemone mexicana*	Leaf, Root, Whole plant	Skin diseases, Fungal disease, Scabies, Ring worm and Eczema and Wounds	Wounds and Ringworm
10.	*Asparagus racemosus*	Root, Leaf and Tuber	Wounds	Inflammation, Leucorrhoea, Body heat, Diarrhoea, Tastelessness, Ulcer induced pain, Lumbago and Cracks in the heels.
11.	*Azadirachta indica*	Stembark Leaf, Seed, Gum, Oil, and Flower	Rheumatic complaints	Fever, skin infection, Small pox, chicken pox, Eczema and Skin diseases
12.	*Boerhaavia diffusa*	Leaf and Root	Asthma, Jaundice and Swellings in legs	Scabies and Ringworm infection
13.	*Calotropis gigantea*	Leaf, Root, Latex and Flower	Snake bites and Rheumatism.	Scabies and Ring worm infections
14.	*Cassia alata*	Leaf	Ringworm infection	Ringworm infection
15.	*Cassia fistula*	Flower, Leaf, Bark	Eczema, freckles, Dysentery and skin disease.	Injuries and Eczema
16.	*Catharanthus roseus*	Flower, Leaf and Root	Blood cancer	Tumors, Menorrhagia and Pain due to wasp strings
17.	*Cissampelos pareira*	Root, Leaf	Stomach disorders and Indigestion	Body pain, body heat and Ringworm infection
18.	*Cissus quadrangularis*	Leaf, Stem	Rheumatoid arthritis and Bone fractures	Tumour like swellings caused by wind exposure, Leucorrhoea, Skin diseases and Bone fractures
19.	*Cleome viscosa*	Leaf	Wounds, Swellings and Pain killers in arthritics	Headache, Arthritis, constipation and Worms from the body.

Sl.No.	Botanical Name	Part Used	Irulas	Kanikkars
20.	*Clerodendron inerme*	Root, Leaf	Snake bites and Psoriasis	Insect bite, Psoriasis, Scabies and Ringworm infection
21.	*Clitoria ternatea*	Root, Leaf and Seed	Scabies	Scabies and Skin disease
22.	*Coccos nucifera*	Oil,Tender coconut	Cuts, Wounds and Grow thick and long hair	Intestinal inflammations, Skin rashes, Skin diseases, Itching and Cooking
23.	*Curculigo orchioides*	Tuber, Root	Increase the sexual vigour in males	Energy stimulant
24.	*Curcuma longa*	Flower, Rhizome	Stomach ache (Gastritis), Fever and Wounds	Skin infection, Wounds, Chicken pox, Measles and Scabies
25.	*Evolvulus alsinoides*	Whole plant	Body heat	Leprosy, Scabies, Venereal diseases and Coolant
26.	*Ficus benghalensis*	Bark	Diabetes and mouth ulcer	Dandruff and Cracks on foot
27.	*Gloriosa superba*	Root tuber	Scorpion stings and Snake bites	Skin diseases
28.	*Hemidesmus indicus*	Root	Body coolant	Skin diseases, Semen production and Ulcer
29.	*Hibiscus rosa-sinensis*	Flowers, Leaf	To treat impotency in males	Cough, Cold, Wounds, Hair growth, Herbal shampoo and Breast pain
30.	*Indigofera tinctoria*	Whole plant	Leucorrhoea and Treat insect bites.	Leucorrhoea
31.	*Jatropha curcas*	Leaf and Seed	Psoriasis, Cough, Stomach disorder and Toothache	Skin diseases and Wounds on the lips
32.	*Justicia adhatoda*	Leaf	Bronchial asthma	Body pain
33.	*Lawsonia inermis*	Bark and Leaf	Skin diseases and Headache	Improves hair colour and prevent greying of hair, Mouth ulcers and Small wounds on the mucosa
34.	*Leucas aspera*	Leaf	Snake bites	Ringworm infection and Migraine headache, Cough, Cold and Fever, Antihelmentic, Skin infections
35.	*Mangifera indica*	Root bark, Seed and Gum	Dysentery	Cracks on the foot and Stomachache
36.	*Merremia tridentata*	Whole plant	Initiate urination	Dandruff and Skin diseases

Sl.No.	Botanical Name	Part Used	Irulas	Kanikkars
37.	*Mimosa pudica*	Root, Leaf	Diabetes, Body heat and Joint pain (arthralgia)	Eczema, Cuts and wounds and Jaundice, Body heat or as a laxative for bowel clearance
38.	*Mirabilis jalapa*	Leaf	Swellings in the skin	Sebaceous cyst and Polyps
39.	*Momordica charantia*	Leaf and Fruit	External piles and Skin disease	Scabies and Ring worm infections
40.	*Mukia maderaspatana*	Leaf	Cold, Cough and Piles problem	Cold and Cough.
41.	*Ocimum basilicum*	Leaf	Eliminate worms present in the stomach	Cough, Eczema and Scabies
42.	*Pandanus fasicularis*	Leaf, Root	Diabetes	Antiseptic and Syphilis
43.	*Phyllanthus emblica*	Leaf	To cure bleeding from gums	Scabies
44.	*Piper betle*	Leaf	Stomach disorder	Carminative and stimulant
45.	*Piper longum*	Fruit, Seed	Cold and Cough	Bronchitis, Cough and Cold
46.	*Rauvolfia serpentina*	Root, Oil	Blood pressure (hypertension)	Eczema, Psoriasis and Ringworm infection
47.	*Santalum album*	Wood	Acne	Body heat and Skin diseases
48.	*Solanum nigrum*	Leaf	Mouth aphthous, Stomach ulcer, Heal burns, Wounds and Relieve joint pain	Skin diseases
49.	*Vernonia cinerea*	Whole plant	Male infertility	Leprosy, Scabies, Tumor in breast, Diarrhoea and Stomachache
50.	*Vitex negundo*	Leaf	To treat fever, Headache and Ear infections	Headache
51.	*Wrightia tinctoria*	Stem, bark	Nausea	Skin diseases

Comparative Ethnomedicobotany with the *Palliyars*

Forty four plant species are commonly used by both *Irulas* and *Palliyars*. The similarities and dissimilarities are discussed (Table 3.8). Comparative ethnobotanical study between the *Irulas* and *Palliyars* reveals that 12 medicinal plant species (out of 44) exhibit identical uses. They are *Acalypha fruticosa, Alangium salvifolium, Boerhaavia diffusa, Cajanus* cajan, *Calotropis gigantea, Curculigo orchioides, Ficus benghalensis, Indigofera tinctoria, Rhinocanthus nasutus, Santalum album, Tamarindus indica*, and *Tridax procumbens*.

Table 3.8: Comparative Ethnomedicobotany with the *Palliyars*

Sl.No.	Botanical Name	Part Used	Irulas	Palliyars
1.	*Abrus precatorius*	Leaf, seed and Stem	Stiffness of joint and Poisonous bites	Throat pain, Cough, Head ache and Leucoderma
2.	*Acalypha fruticosa*	Leaf	Insect bites	Poisonous bite and Haemorrhoids
3.	*Achyranthes Aspera*	Stem, Leaf and Root bark	Eye disorder, Tooth powder to whiten the teeth	Scorpion sting, Honeybee sting and Joint pain
4.	*Aegle marmelos*	Fruit, Leaf, Stem bark, Fruit	Scabies	Fever, Headache, Throat pain and Jaundice
5.	*Alangium salvifolium*	Leaf, Stem, Seed, Root bark	Bites	Snake bite and other stings
6.	*Amaranthus Spinosus*	Whole plant	Snake bites and Kidney stones.	Stomach pain and Urinary problems
7.	*Ammannia baccifera*	Leaf and Whole plant	Osteoarthritis of the knee joint	Various stings and Antipyretic
8.	*Anisomeles malabarica*	Leaf	Rheumatic pain and Stomach pain in children	Dysentery for children when a time of teeth growing
9.	*Argemone mexicana*	Leaf, Root, Whole plant	Skin diseases, Fungal disease, scabies, ring worm and eczema and Wounds	Scorpion bite, Anthelmintic and Eye infections
10.	*Asparagus racemosus*	Root, Leaf and Tuber	Wounds	Stomachache, Dysentery and Improve digestion
11.	*Azadirachta indica*	Stem bark Leaf, Seed, Gum, Flower	Rheumatic complaints	Chicken pox, Anthelmintic, Diabetes, Skin diseases and Kidney disorders
12.	*Boerhaavia diffusa*	Leaf and Root	Asthma, Jaundice and Swellings in legs	Asthma, Skin diseases, Fever, Vomiting, Improved digestion, Jaundice and Carminative
13.	*Butea monosperma*	Stem bark, Leaf, Fruit and Seed	Sexual disorder, Eczema	Snake bite and Anthelmintic
14.	*Cajanus cajan*	Leaf	Jaundice	Jaundice
15.	*Calotropis gigantea*	Leaf, Root, Latex and Flower	Snake bites and Rheumatism.	Snake bite, Scorpion sting, Laxative, Anthelmintic, Venereal diseases and Leprosy

Sl.No.	Botanical Name	Part Used	Irulas	Palliyars
16.	*Cassia tora*	Root, Leaf and Seed	Ringworm	Analgesic, Fever and Chronic filiarisis
17.	*Catharanthus roseus*	Flower, Leaf and Root	Blood cancer	Improve digestion, Strengthen to body and Diabetes
18.	*Cissus quadrangularis*	Leaf, Stem	Rheumatoid arthritis. Bone fractures	Colic pain
19.	*Clitoria ternatea*	Root, Leaf and Seed	Scabies	Analgesic, Fever and Chronic filiarisis
20.	*Coccinia grandis*	Fruit	Toothache and mouth ulcer.	Diabetes.
21.	*Curculigo orchioides*	Tuber, Root	Increase the sexual vigour in males	Aphrodisiac
22.	*Curcuma longa*	Dried rhizome	Stomach ache (Gastritis), fever and Wounds.	Bowel complaints
23.	*Datura innoxia*	Leaf	Asthma and Wounds	Anti-rabies medicine
24.	*Dichrostachys cinerea*	Leaf, Seed	Rheumatoid arthritis and Dysentery	Diabetes
25.	*Euphorbia hirta*	Leaf, Flowers and Latex	Pimples and Asthma	Sore in mouth, Stomachache, Urogenital problem, improve lactation and Acne
26.	*Evolvulus alsinoides*	Whole plant	Reduce body heat	Antipyretic
27.	*Ficus benghalensis*	Stem bark and Twig	Diabetes and mouth ulcer	Diabetes and Tooth brush
28.	*Hemidesmus indicus*	Root	Body coolant	Diabetes and Leprosy
29.	*Indigofera tinctoria*	Whole plant	Leucorrhoea and treat insect bites	Diabetes, Leprosy and Antidote for snake bite
30.	*Jatropha curcas*	Leaf and Seed	Psoriasis, cough, Stomach disorder and Toothache	Scabies and Eczema
31.	*Mangifera indica*	Root bark, Seed and Gum	Dysentery	Diarrhoea, Vomiting and Cracks
32.	*Momordica charantia*	Leaf and Fruit	External piles and Skin disease	Anthelmintic, Diabetes, Haemorrhoids and Renal problems

Sl.No.	Botanical Name	Part Used	Irulas	Palliyars
33.	*Myristica fragrans*	Pericarp	Stomach ache	Dysentery,Reduce body heat and Gas trouble
34.	*Ocimum americanum*	Leaf and Pericarp	Skin disease	Dysentery, Cold and Cough, Headache and Piles
35.	*Phyllanthus emblica*	Leaf, Fruit, Stem bark and Seed	Bleeding	Dysentery, Alopecia, jaundice, Throat pain and Aphrodisiac in men
36.	*Pongamia pinnata*	Flower, Stem bark and Seed kernel	Skin diseases, Ringworm infection, Rashes, Scabies, Eczema and Psoriasis	Diabetes, Pneumonia and Lungs disorder
37.	*Ricinus communis*	Leaf, Seed, Root, and Root bark	Jaundice	Jaundice, Wounds, Dysentery, Backache and Laxative
38.	*Rhinacanthus nasutus*	Root, Leaf	Snake bites	Antidote for snake bite
39.	*Santalum album*	Wood	Eczema and Acne vulgaris	Skin diseases, Anti-rabies medicine and Conaria
40.	*Syzygium cumini*	Seed and Stem bark	Diahhorea	Diabetes
41.	*Tamarindus indica*	Stem bark, ripened pod and Leaf	Indigestion	Stomachache, Improve digestion and Urinary troubles
42.	*Tinospora cordifolia*	Leaf, Stem and Flower	Dysentery	Ulcer and Throat pain
43.	*Tridax procumbens*	Whole plant	Cuts and Wounds	Burns, Cuts and wounds
44.	*Vernonia cinerea*	Whole plant	Male infertility	Skin diseases and Dysentery

The present study brings to light that all the four tribals in addition to *Irulas* employ *Tridax procumbens* as an antiseptic to cure wounds/cuts/burns. *Santalum album* with identical medicinal use is being practiced by the *Santals, Palliyars* and *Kanikkars* used as skin diseases. *Curculigo orchioides* is used by the *Bhils, Gonds* and *Palliyars* to treat as sexual disorders. *Achyranthes aspera, Alangium salviforum, Andrographis paniculata* and *Cassia fistula* with identical medicinal use are being practiced by at least two ethnic groups under discussion.

IUCN Red List

The following medicinal plants are identified as the least concern, endangered and vulnerable plants of the study area (Table 3.9). (http://www.iucnredlist.org/).

Several medicinal plants become threatened because of increasing biotic pressure on natural habitats and unscientific over exploitation. So, their conservation is of paramount importance and that can be made *in situ* and *ex situ*. The establishment of medicinal plant gardens or "germplasm of rare medicinal plants" in suitable agro-climatic regions is the need of the hour at both state and National levels.

Table 3.9: The IUCN List of Plant Species

Sl.No.	Botanical Name	Vernacular Name	Status
1.	*Alternanthera sessilis* L. R.Br.	Ponaganni	Least Concern
2.	*Ammannia baccifera* L.	Neer mel nerupu	Least Concern
3.	*Bacopa monnieri* (L.) Pennell	Noorpirami	Least Concern
4.	*Bauhinia purpurea* L.	Kaatu Mandarai	Least Concern
5.	*Borassus madagascariensis* Bojer ex Jum. and H.Perrier	Panai maram	Endangered
6.	*Colocasia esculenta* (L)	Karunal kilangu	Least Concern
7.	*Commelina benghalensis* L.	Kanan	Least Concern
8.	*Cyperus difformis* L.	Gorai	Least Concern
9.	*Dichrostachys cinerea* (L.) Wight and Arn.	Vadataram	Least Concern
10.	*Erythrina variegata* L.	Mulu murunga	Least Concern
11.	*Euphorbia tirucalli* L.	Thirukalli	Least Concern
12.	*Gloriosa superba* L.	Korangu poo	Least Concern
13.	*Hygrophila auriculata* (Schumach.) Heine	Neermulli	Least Concern
14.	*Phyla nodiflora* (L.) Greene	Poduthalai	Least Concern
15.	*Mimosa pudica* L.	Thottasurungii	Least Concern
16.	*Myristica fragrans* Houtt.	Jathikai	Data Deficient
17.	*Nymphaea nouchali* Burm.f.	Alli	Least Concern
18.	*Opuntia tehuantepecana* (Bravo) Bravo	Sapathikalli	Least Concern
19.	*Pongamia pinnata* (L.) Pierre.	Pungan	Least Concern
20.	*Punica granatum* L.	Maathulai	Least Concern
21.	*Santalum album* L.	Santhanam	Vulnerable
22.	*Wrightia laevis* Hook.f.	Vetpalai	Lower Risk/Least concern

Conclusion

In essence, the ethnomedicinal knowledge about the biodiversity reflects many generations of experience and problem solving by the indigenous communities. It represents an immensely valuable database that provides the baseline information for the commercial exploitation of bioresources. Also, the information could be useful for the industry, pharmacologists, physicians, phytochemists, botanists and alike interested in the development of alternative therapies. For lesser known plant species, such a secret treasure trove of information could prove beneficial in phyto-pharmacological research for the discovery of new therapeutic drugs. Also, during the recent past, there has been growing concern among the developing countries about the emerging threats of the biopiracy and the intensity of IPR controversies are increasing day-by-day. Thus, the need of the hour is to speedily document reliable information about the biodiversity and its different uses by the ethnic communities. It is hoped that the information base generated could contribute in filling the knowledge gaps for the compilation of a local biodiversity registers, a key instrument for achieving the regional and global biodiversity conservation and sustainable development goals. The results obtained in the investigation need to be rigorously subjected to pharmaceutical analysis in order to validate their authenticity and future prospects. The paper has only documented the herbal remedies presently in vogue in the region and does not prescribe or recommend for their use till further determination by pharmacologists.

References

Abu-Rabia, A. 2005. Urinary diseases and ethnobotany among pastoral nomads in the MiddleEast. *J.Ethnobiol.Ethnomed.*1:4

Al-Qura'n S. 2005. Ethnobotanical survey of folk toxic plants in southern part of Jordan. *Toxicon.* 46: 119–126.

Ayyanar, M. and Ignacimuthu, S. 2005. Traditional knowledge of Kani tribals in Kouthalai of Tirunelveli hills, Tamil Nadu. *J. Ethnopharmacol.* 102: 246-255.

Azaizeh, H., Fulder, S., Khalil, K. and Said, O. 2003. Ethnomedicinal knowledge of local Arab practitioners in the Middle East Region. *Fitoter.* 74: 98–108.

Binu, S., Shanavaskha, A.K., Santhoskumar, E.S. and Pushpangadan, P. 2003. Plants used as medicine by the *Irulars* of Palghat district, Kerala, India. *J. Econ. Tax. Bot.* 27: 808-814.

Borthakur, S.K, and Goswami, N. 1995. Herbal remedies from Dimoria of Kamrup district of Assam in Northeastern India. *Fitoter.* LXVI: 333-340.

Chandra, K. 1995. An Ethno-Botanical study on some medicinal plants of district Palamau, (Bihar). *Bull. Med. Eth. Bot. Res.* XVI: 11-16.

Chendurpandy, P., Mohan, V.R. and Kalidass, C. 2010. An ethnobotanical survey of medicinal plants used by the *Kanikkars* tribe of Kanyakumari District of Western Ghats, Tamil Nadu for the treatment of skin diseases. *J. Herb. Med. Toxi.* 4: 179-190.

Chhetri DR, Parajuli P, Subba GC. 2005.Antidiabetic plants used by Sikkim and Darjeeling Himalayan tribes, India. *J. Ethnopharmacol.* 99: 199–202.

Das, S., Sheeja, T.E. and Mandal, A.B. 2006. Ethnomedicinal uses of certain plants from Bay Islands. *Ind. J. Trad. Know.* 5: 207-211.

Diallo, D., Hveem, B., Mahmoud, M.A., Berge, G., Paulsen, B.S. and Maiga, A. 1999. An ethnobotanical survey of herbal drugs of Gourma district, Mali. *Pharmaceut. Biol.* 37: 80–91.

Divya, V.V., Karthick, N. and Umamaheswari, S. 2013. Ethnopharmacological studies on the medicinal plants used by Kani tribes of Thachamalai hill, Kanyakumari, Tamil Nadu, India. *Int. J. Adv. Bio. Res.* 3: 384-393.

Fischer, C.E.C. and Gamble, J.S. 1957. (Rep. Ed.) The Flora of the Presidency of

Madras III: 1347-2017.

Gamble, J.S. 1957. (Rep. Ed.) The Flora of the Presidency of Madras I: 1-577.

Gamble, J.S. 1957. (Rep. Ed.) The Flora of the Presidency of Madras II: 578-1346.

Gazzaneo, L.R., Paiva de Lucena, R.F. and Paulino de Albuquerque, U. 2005. Knowledge and use of medicinal plants by local specialists in an region of Atlantic Forest in the state of Pernambuco (Northeastern Brazil). *J. Ethnobiol. Ethnomed.* 1: 9.

Girach, R.D. and Aminuddin, 1995. Ethnomedicinal uses of plants among tribals of Singbhum district, Bihar, India. *Ethnobot.* 7: 103-107.

Girach, R.D., Singh, S., Brahmam, M. and Misra, M.K. 1999. Traditional treatment of skin diseases in Bhadrak district, Orrissa. *J. Econ. Tax. Bot.* 23: 499-504.

Hanazaki, N., Tamashiro, J.Y., Leitao-Filho, H. And Gegossi, A. 2000. Diversity of plant uses in two Caicaras communities from the Atlantic forest coast, Brazil. *Biodiver. Conser.* 9: 597–615.

Harsha, V.H., Hebbar, S.S., Hegde, G.R. and Shripathi, V. 2002. Ethnomedical knowledge of plants used by Kunabi Tribe of Karnataka in India. *Fitoter.* 73: 281–287.

Hebbar, S.S., Harsha, V.H., Shripathi, V. and Hegde, G.R. 2004. Ethnomedicine of Dharwad district in Karnataka, India–plants used in oral health care. *J. Ethnopharmacol.* 94: 261–266.

Ignacimuthu, S., Sankarasivaraman, K. and Kesavan, L. 1998. Medico-Ethnobotanical survey among Kanikkar tribals of Mundanthurai Sanctuary, Western Ghats, India. *Fitoter.* LXIX : 404-414.

Jadhav, D. 2006. Ethnomedicinal plants used by Bhil of Bibdod, Madhya Pradesh. *Ind. J. Trad. Know.* 5: 263-267.

Jain, S.K. 1991. Dictionary of Indian Folk Medicine and Ethnobotany. Deep publications, Paschim Vihar, New Delhi.

Jain, D.L., Baheti, A.M., Jain, S.R. and Khandelwal, K.R. 2010. Use of medicinal plants among tribes in satpuda region of Dhule and Jalgaon Disrict of Maharashtra - An ethnobotanical survey. *Indian J. Trad. Know.* 9: 152-157.

Jain, S.K. Banerjee, D.K. and Pal, D.C. 1973. Medicinal plants among certain Adibasis in India. *Bull. Bot. Surv. India.* 15: 85-91.

Joshi, M.C. 1992. Some folk medicines of the tribals of Gujarat. *Bull. Med. Eth. Bot. Res.* XII: 115-124.

Joshi, P. 1982. An ethnobotanical study of Bhils - A preliminary survey. *J. Econ. Tax. Bot.* 3: 257-266.

Kala, C.P. 2005. Ethnomedicinal botany of the Apatani in the Eastern Himalayan region of India. *J. Ethnobiol. Ethnomed.* 1: 11.

Katewa, S.S. and Arora, A. 1997. Some plants in folk medicine of Udaipur district (Rajasthan). *Ethnobot.* 9: 48-51.

Khan, A.U. 2002. History of decline and present status of natural tropical thorn forest in Punjab. *Pak. Biol. Conserv.* 63: 210–250.

Khanna, K.K. and Kumar, A. 2009. Noteworthy ethnomedicinal uses of plants from the tribals of Betul district, Madhya Pradesh. *J. Econ. Tax. Bot.* 33: 933-939.

Lalitharani, S., Mohan, V.R and Regini, G.S. 2010. Ethnomedicinal plants used by *Kanikkars* of China and Periyamylar regions of Agasthyamalai Biosphere, Tamil Nadu. *J. Non. Tim. For. Prod.* 16: 209-210.

Lev, E. 2006. Ethno-diversity within current ethno-pharmacology as part of Israeli traditional medicine – A review. *J. Ethnobiol. Ethnomed.* 2:4.

Maruthi, K.R., Krishna, V., Manjunatha, B.K. and Nagaraja, V.P. 2000. Traditional medicinal plants of Davanagere district, Karnataka with reference to cure skin diseases. *Environ. Ecol.* 18: 441–446.

Maruthupandian, A., Mohan,V.R. and Kottaimuthu, R. 2011. Ethnomedicinal plants used for the treatment of diabetes and jaundice by *Palliyar* tribals in Sirumalai hills, Western Ghats, Tamil Nadu, India. *Indian J. Nat. Prod. and Resour.*2: 493-497.

Matthew, K.M. 1983. The flora of the Tamil Nadu Carnatic I: 1- 688.

Matthew, K.M. 1983. The flora of the Tamil Nadu Carnatic II: 689- 1540.

Matthew, K.M. 1983. The flora of the Tamil Nadu Carnatic III: 1541- 2155.

Matthew, K.M. 1999. The flora of the Palni Hills I: 1- 575.

Matthew, K.M. 1999. The flora of the Palni Hills II: 576- 1196.

Matthew, K.M. 1999. The flora of the Palni Hills III: 1197- 1635.

Mishra, O.P. and Naquvi, S.M.A. 1995. Ethno-Medico Botany from tribes of Madhya Pradesh. *Bull. Med. Eth. Bot. Res.* XVI: 17-26.

Mitra, S and Mukherjee, S.K. 2009. Some abortifacient plants used by the tribal people of West Bengal. *J. Nat. Prod.* 8: 167-171.

Mudgal, V. and Pal, D.C. 1980. Medicinal plants used by tribals of Mayurbhanj (Orissa). *Bull. Bot. Surv. India.* 22: 59-62.

Murthy, E.N. 2012. Ethnomedicinal plants used by Gonds of Adilabad district, Andhra Pradesh, India. *Int. J Pharm. Life Sci.* 3: 2034-2043.

Nair, N.C. and Henry, A.N. 1983, Flora of Tamil Nadu, India, Series I: Analysis I: 1-184.

Natarajan, B. and Paulsen, B.S. 2000. An ethnopharmacological study from Thane district, Maharashtra, India: Traditional knowledge compared with modern biological science. *Pharmaceut. Biol.* 38: 139–151.

Nazarudeen, A. 2009. Some lesser known wild plants for food and folk use among *Kani* tribes in Southern Kerala, India. *J. Econ. Tax. Bot.* 33:741-745.

Nikhil, B., Rathod, K.S., Rathod and Rafiuddin Naser. 2013. Medicinal plants used by the Andh, Bhil, Pardhi, Thakar communities for treating the hair problems/ disorders in marathwadaregion of Maharashtra. *Golden Res.Thoughts.* 2:1-6.

Pal, D.C., Saren, A.M. and Sen, R. 1989. Less known uses of twenty plants from the tribal areas of Bankura district, West Bengal. *J. Econ. Tax. Bot.* 13: 695-698.

Patil, S.L. and Patil, D.A. 2007. Ethnomedicinal plants of Dhule district, Maharashtra. *Nat. Prod. Radia.*6:148-151.

Pei, S.J. 2001. Ethnobotanical approaches of traditional medicine studies: Some experiences from Asia. *Pharmaceut. Biol.* 39: 74–79.

Prasad, P.N., Jabadhas, A.W. and Janaki Ammal, E.K. 1987. Medicinal plants used by the Kanikkars of South India. *J. Econ. Tax. Bot.* 11: 149-155.

Ragupathy, S. and Mahadevan, A. 1991. Ethnobotany of Kodiakkarai Reserve Forest, Tamil Nadu, South India. *Ethnobot.* 3: 79-82.

Ramachandran, V.S. and Nair, N.C. 1981. Ethnobotanical observations on Irulars of Tamil Nadu (India). *J. Econ. Tax. Bot.* 2: 183-190.

Ramkrishna, N. and Saidulu, C.H., 2014. Medicinal plants used in snakebite and scorpion sting by Gonds and Kolams of Adilabad dist. A. P. *Int. J. Current Pharmaceut. Res.* 2: 39-41.

Rossato, S.C., Leitao-Filho, H. and Gegossi, A. 1999. Ethnobotany of Caicaras of the Atlantic forest coast (Brazil). *Econom. Bot.* 53: 387–395.

Saikia, A.P., Ryakala, V.K., Sharma, P., Goswami, P. and Bora, U. 2006. Ethnobotany of medicinal plants used by Assamese people for various skin ailments and cosmetics. *J. Ethnopharmacol.* 106: 149-157.

Samvatsar, S. and Diwanji, V.B. 2000. Plant sources for the treatment of jaundice in the tribals of Western Madhya Pradesh of India. *J. Ethnopharmacol.* 73: 313–316.

Sarena, A.M., Halder, A.C and Singh, H. 2009. Ethnomedicinal plants of Mahilong forest range in Ranchi district, Jharkhand. *J. Econ. Tax. Bot.* 33: 803-810.

Saxena, S.K. and Tripathi, J.P. 1989. Ethnobotany of Bundelkhand. Studies on the medicinal uses of wild trees by the tribal inhabitants of Bundlekhand region. *J. Econ. Tax. Bot.* 13: 381-389.

Senthilkumar, M., Gurumoorthi, P and Janardhanan, K. 2006. Some medicinal plants used by *Irulars*, the tribal people of Marudhamalai hills, Coimbatore, Tamil Nadu. *Nat. Prod. Radia.* 5:382-388.

Shah, G.L. and Gopal, G.V. 1985. Ethnomedicinal notes from the tribal inhabitants of the North Gujarat (India). *J. Econ. Tax. Bot.* 6: 193-201.

Sur, P.R., Sen, R., Halder, A.C. and Bandyopadhyay, S. 1987. Observation on the ethnobotany of Malda - West Dinajpur districts, West Bengal - I. *J. Econ. Tax. Bot.* 10: 395-401.

Tarafder, C.R. 1983. Ethnogynaecology in relation to plants part – II. Plants used for abortion. *J. Econ. Tax. Bot.* 4: 507-516.

Viswanathan, M.B., Premkumar, E.H., and Ramesh. N. 2001. Ethnomedicines of Kans in Kalakad – Mundanthurai Tiger Reserve, Tamil Nadu. *Ethnobot.* 13:60-66.

Yadav, S.S. and Bhamare, P.B. 1989. Ethnomedico-Botanical studies of Dhule forests in Maharahtra State. *J. Econ. Tax. Bot.* 13:455-460.

2018, Ethnomedicinal Plants: A Biodiversity Treasure Pages 167–187
Editors: V.R. Mohan, A. Doss, P.S. Tresina and V. Sornalakshmi
Published by: ASTRAL INTERNATIONAL PVT. LTD., NEW DELHI

Chapter 4

Ethnobotanical Studies on the Medicinal Plants Used by Kani Tribes of Mugilayadi Hills, Kanyakumari Wild Life Sanctuary, Southern Western Ghats, Tamil Nadu, India

B. Parthipan, A.V. Babitha, P. Dayana, Y. Jasmine Leena and J.L. Lekshmi

P.G. and Research Department of Botany,
South Travancore Hindu College, Nagercoil – 620 002 Tamil Nadu
E-mail: parthipillai64@gmail.com

Introduction

India is recognized as a country that is rich in all aspects of biodiversity like ecosystem, species and genetic diversity mainly due to its tropical location, disparate physical features and climatic types (Jain 1987). The continent has well-documented traditional knowledge, long-standing practice of traditional medicine and the potential for social and economic development of medicinal and aromatic plants in primary health care and industrial scale production. Traditional Medicine (TM) occupies an important place in the health care systems of developing countries. The World Health Organization (WHO) estimates that more than 80 per cent of health care needs in these countries are met through traditional health care practices (Chendurpandy *et al.*, 2010). The people in developing countries depend on TM, because it is cheaper and more accessible than Orthodox Medicine (OM) (Sofowora, 1993; Luoga *et al.*, 2000; World Health Organization, 2002). Traditional medicine is also acceptable than OM because, it blends readily into the people's socio-cultural life (Tabuti *et al.*, 2003).

The various tribal sections of India are repositories of rich knowledge on various uses of plant genetic resources, which have hitherto remained unknown (Khoshoo, 1996). But of later, due to several developmental activities around tribal areas which are after all not related to their welfare, the tribal people are losing their traditional identity resulting in a good deal of loss of such treasures of plant genetic resources (Shankar, 1995). In view of the harmful developments, the UN declared thc year 1993 as the "International Year of Indigenous people" based on the recommendation of the Rio de Janeiro Earth Summit. The studies on the relationship between the aboriginal or primitive people and their surroundings including a critical evaluation of some of the important plants used by the tribes have received considerable attention in recent years (Das *et al.*, 1989).

Many infectious diseases are known to be treated with herbal remedies throughout the history of mankind. Even today, plant materials continue to play a major role in primary health care as therapeutic remedies in many developing countries (Zakaria, 1991). Plants still continue to be almost the exclusive source of drugs for a majority of the world's population (Hamburger and Hostettman, 1991).

It is a matter of great pride that, among the 16 hot spots known for rich flora in the world, two are located in India (Mohan *et al.*, 2010). They are the Eastern Himalayas and the Western Ghats (Khoshoo, 1996). The hill chain of Western Ghats recognized as a region of high level of biodiversity is under the threat of rapid loss of genetic resources (Gadgil, 1996).

It is essential to make the complete inventory of the medicinal component of the flora of any country for conservation and sustainable use. The conservation of the threatened and endangered medicinal species in the wild is indispensable (Rahman *et al.*, 2004).

A review of the literature also reveals that many tribal communities in India are either under explored or unexplored with regard to their floral wealth used in curing disease (Kala, 2006). A few reports on ethnomedicinal uses of plants by kani tribals were available in the Kanyakumari district (Kingston and Raj, 2000; Jeeva *et al.*, 2006; Sukumaran *et al.*, 2008; Chendura Pandy *et al.*, 2010, John De Britto *et al.*, 2010, Kumari subitha *et al.*, 2011, Lalitha Rani *et al.*, 2011 and Divya *et al.*, 2013). However no published research work was available from the present study area. So present study is an attempt to collect information on traditional uses of medicinal plants used in the preparation of herbal drugs by the Kani's settled at Mugilayadi hills under Veerapuli Reserve Forest, Kulasekaram range of Kanyakumari wildlife sanctuary.

The *Kanikkars* belong to the southern tribal zone. They are distributed along the Southeastern slopes of altitude regions of Western Ghats in large numbers. *Kanikkars* means hereditary proprietor of land thus recognizing their ancient rights over the forest lands. The *Kanikkars* are generally very short in stature and meager in appearance. Some have markedly negroid features. They are traditionally a nomadic community. They speak in their own dialect, *Kanikkar Bhasha* or *Malampashi*, which is close to the Dravidian language *Malayalam*. *Kanikkars* once practiced migratory cultivation but have now to a large extent abandoned such cultivation. Most of the *Kanikkar* tribals have a general knowledge of medicinal plants that are used for first aid remedies, to treat cough, cold, fever, headache, poisonous bites and some other simple ailments. *Kanis* still supplement their food by gathering roots and tubers from the nearby forest areas. They eat tubers like *Manihot esculenta* and *Dioscorea oppositifolia, etc.* They are also engaged in seasonal collection of honey, bee wax and some minor forest produce. They cultivate edible plants, like tapioca, banana, millets and cash crops such as pepper, areca nut and cashew nut.

Materials and Methods

About the Study Area

Kanyakumari District, named after the Goddess 'Kanniyakumari', lies at the southernmost tip of peninsular. This is one of the smallest districts in the state of Tamil Nadu. The Kanyakumari district is located in the Southern most part of the Indian peninsula. The study area lies between 77° 05′ to 77° 36′ E longitude and 8° 03′ and 8°35′ N latitude. It occupies an area of 1684 Km2. The district is bounded by Tirunelveli district on the north and east and by the Gulf of Mannar on the south eastern part, the south and southwest are surrounded by the Indian Ocean and the Arabian Sea and the west and northwest by Kerala state (Figure 4.1). The district may be divided into three geographical regions, first the mountainous region with a number of hill tops, next the coastal region and the undulating valleys and finally the plains between the mountainous terrain and the sea coast.

It is predominantly an agricultural region with vast natural resources and variety of geological features. The total area of the forests of Kanyakumari District is 42,762.27 hectare. Kanyakumari forest division was declared as Kanyakumari wildlife sanctuary during 2002 (Divya, *et al.*, 2013). Kanyakumari District forests though small in area have as many as 14 forest types (Champion and Seth, 1968).

Climate

The minimum and maximum temperature is 14.2° C and 32° C during December and May, respectively. Humidity ranges between 65 and 75 per cent. The average monthly rainfall ranges between 20.59 mm and 206.58 mm. The district has a unique advantage of rainfall both during the southwest monsoon between June and September and northeast monsoon between October and middle of December.

Reserve Forests

In Kanyakumari District Forest division there are about 9 Reserve Forests; Therkumalai, Thadakamalai, Poigaimalai, Mahendragiri, Verrapuli, Velimalai, Old Kulasekharam, Kilamalai and Asambu. From the above, Mugilayadi, a part of the Veerapuli Reserved Forest is selected for the present study (Figure 4.1). It is located in the west part of the district between 77°15′-77°20E and 8°30°N latitude in Tamil Nadu. Veerapuli Reserved Forests is a sprawling Reserved Forests spread over Azhiapandipuram and Kulasekaram and ranges with a total area of 24,971 ha (61,650 acres). Wide variation exists in altitude and climate, especially rainfall.

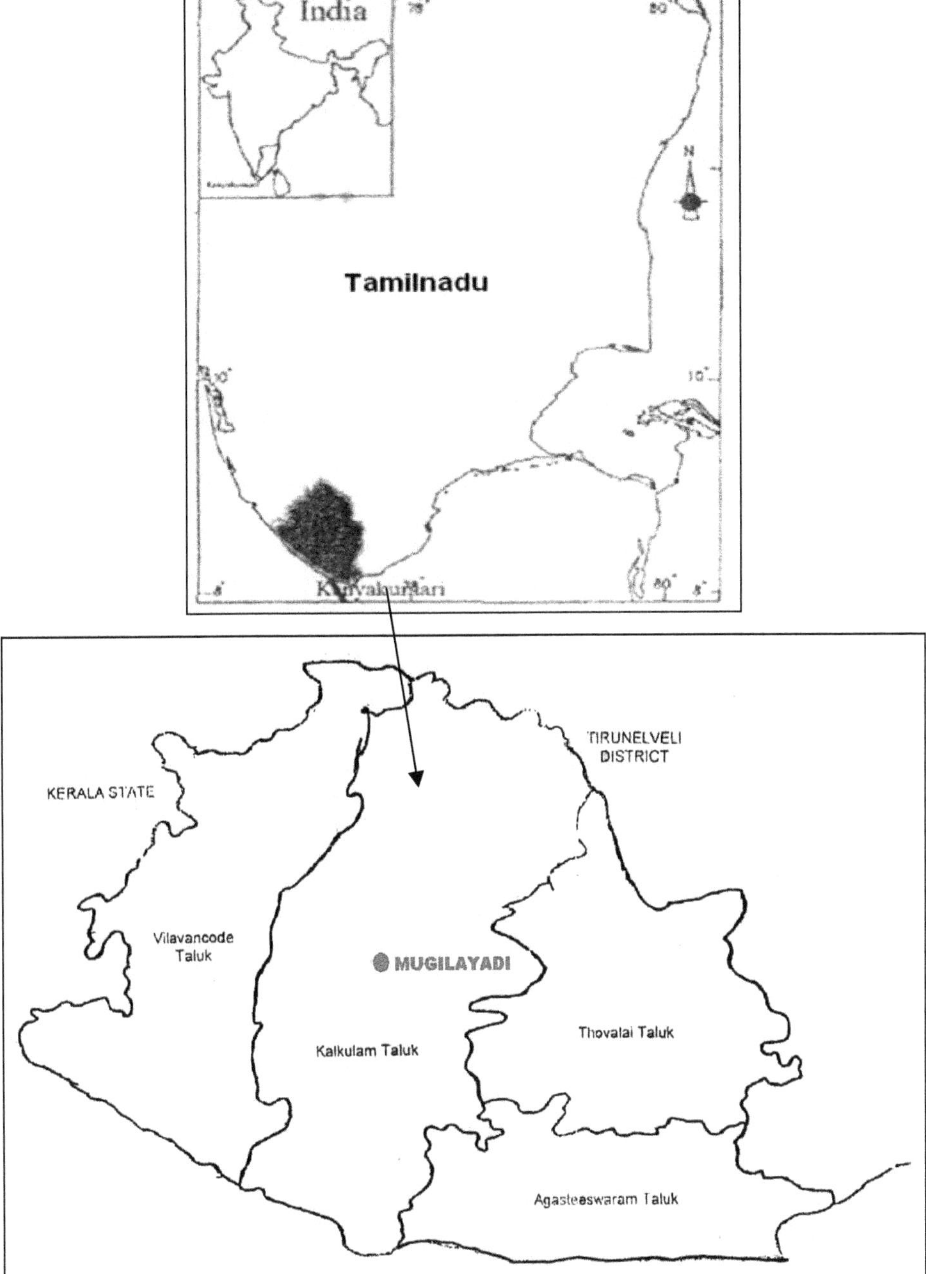

Figure 4.1: Map Showing Study Area.

The important types of forest in this Reserved Forest are Southern hill top tropical evergreen forests, West Coast semi-evergreen forests, slightly moist teak forests, southern moist mixed deciduous forests and Ochlandra reed brakes.

Methodology

The ethnobotanical survey was conducted between January 2013 and October 2013. The success of ethnomedicinal plants study lies in the contact and relationship between the researcher and the tribal people. Hence, more pain is taken to establish such a close relationship with tribes at their settlements and surroundings, which can bring out authentic first hand information about the medicinal uses of plants. Resource persons were identified and selected from all age groups of the tribes and from both sex. In some settlements, elderly women were also contacted.

Plants, which are medicinally used by the *Kanikkars* for treating their various diseases, were collected and their mode of usage was recorded. The majority of illness types are grouped into predefined ethno/economic botany categories (Ragupathy and Newmaster, 2009). Along with that few other illness categories (Table 4.1) that were commonly mentioned during our interviews were also added because they were prevalent in these communities. The use of "general categories" is adopted here as recommended by other ethnobotanical researchers. These 30 illnesses were sorted into usage categories (Table 4.1). All of the illness types were translated as best as possible from the Kani's description of the illness/symptoms to known biomedical terms with few exceptions (*e.g.* spiritualism-keep family wellness (repel evil)). Information on these sources was recorded in video cassettes, which form an authentic and first hand sources of reference. The information thus gathered from one claim was cross checked adequately for reliability and accuracy by interacting with different groups of the *Kanikkars* from different settlements to confirm the usage and dosage differences. Plant specimens were collected for taxonomic identification from different parts of the study area. The photographs of selected plants were also taken during the field trips. The data were recorded in the field note books as mentioned by Jain (1981). Polythene bags were used to keep the collected materials in fresh conditions.

The medicinal plants utilized by these tribal's mostly come from uncultivated area based on the information provided by tribals. Plant specimens were collected; air dried and mounted on herbarium sheets according to the field and herbarium techniques (Mathew, 1981 and Pullaiah, 2007).

All the specimens collected from the fields were identified with the help of regional floras *viz.*, The flora of the presidency of Madras by Gamble (1935) and the Flora of Tamil Nadu Carnatic (Mathew, 1983). Identified plants were finally confirmed by comparing with the authenticated specimens in the Herbarium of Botanical survey of India (Southern Circle), Coimbatore district, Tamil Nadu. Voucher specimens were deposited in the PG and Research Department of Botany, South Travancore Hindu College, Nagercoil, Tamil Nadu.

Table 4.1: Kani's Ailments Grouped by Illness Category

Illness Category	*Biomedical/English Terms*	*Kani's Term*
Dermatological	Skin disease	Thol viyathi
	Pimples	Muhaparu
	Prickly heat	Vaer kuru
	Itching, Psoriasis, Eczema	Arippu, Chori, Chirangu
	Freckle	Thaemel
Diabetes	Diabetes	Neerlivu noie
Fever	Fever	Panie
Gastro intestinal	Digestive power	Cherimana sakthi
	Intestinal worms	Vaitru poochi
	Gastric problems	Vaivu kolaru
	Ulcer	Kudal pun
	Diarrhea	Vaittupoku
General medicine	Hypothermia	Udambu choodu
	Whitlow	Naha chuttu
	Cold	Shzali
General health	Antidotes	Visha kadi
	Snake bite	Paambu kadi
	Anti mosquito (mosquito repellent)	Kosu kadi
	Bone fracture	Yalumbu murivu
Infection	Throat infection	Thondai vali
	Wounds	Vettu kayam
Jaundice	Jaundice	Manja novu
Pain	Head ache	Thala kuthu
	Body ache	Udambu vali
	Ear ache	Chevi kuthu
	Tooth ache	Pal vali
	Rheumatism	Mootu vadham
	Eye pain	Kan vali
Respiratory	Asthma	Izuppu
	Cough	Irummal

Results and Discussion

The present study focuses the extensive usage of as many as 105 ethnotaxa by the Kani tribes inhabiting the Mugilayadi hills, Kanyakumari wildlife sanctuary, southern Western Ghats, Tamil Nadu. As an outcome of the present investigation 105 medicinal plants belonging to 84 genera and 50 families were recorded. Of the recorded plants, a maximum of 7 ethnomedicinal plants belong to Solanaceae and Euphorbiaceae, it is followed by Acanthaceae (6 species), Verbenaceae and Zingiberaceae (5 species each). Among the 105 plants reported as herbal drugs, 51 plants are herbaceous in habit, while 27 plants are shrubs, 10 plants are trees, 15 plants are climbers and 2 plants are twiners (Figure 4.2). This results coincide with the general observations made earlier in relation to ethnomedicinal studies on some of the other tribal communities of India (Sutha *et al.*, 2010., JeyaPrakash *et al.*, 2011., Lalitha Rani *et al.*, 2011., Umapriya *et al.*, 2011., Samydurai *et al.*, 2012., Divya *et al.*, 2013, Pradheeps and Poyyamoli 2013 and Kalaiselvan and Gopalan 2014).

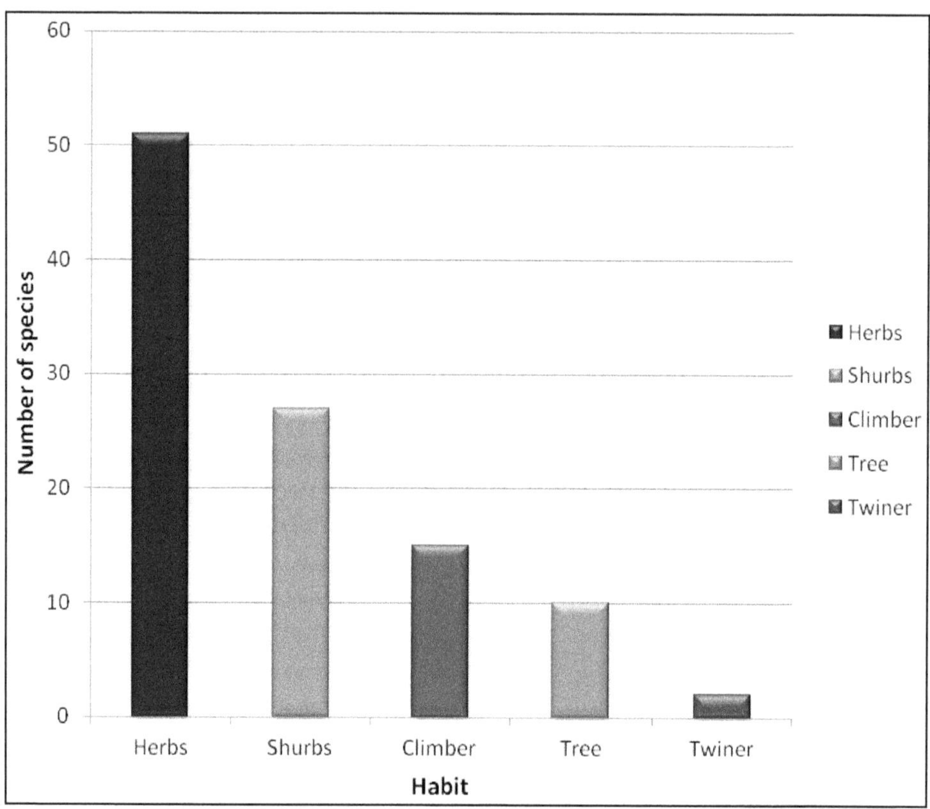

Figure 4.2: Habit-wise Distribution of Medicinal Plants in the Study Area

The percentage of plant parts used and the form of administration is represented in Figure 4.3. Among the different plant parts used in the preparation of medicine (Figure 4.3), leaves (53.3 per cent) were found to be the most frequently used plant parts in the preparation of medicine followed by whole plant, fruits and tuber (8.6 per cent), flower (5.7 per cent), seed (3.85 per cent) root and bark (2.9 per cent) and phylloclade, bulbs(0.95 per cent). Most of the ethnobotanical studies confirmed that leaves are the major portion of the plant used in the treatment of diseases (Muthu *et al.*, 2006., Ragupathy *et al.*, 2008., Ayyanar *et al.*, 2008., Ignacimuthu *et al.*, 2008., Ayyanar 2009., Chenthurpandy 2010 and Fransis Xavier *et al.*, 2011).

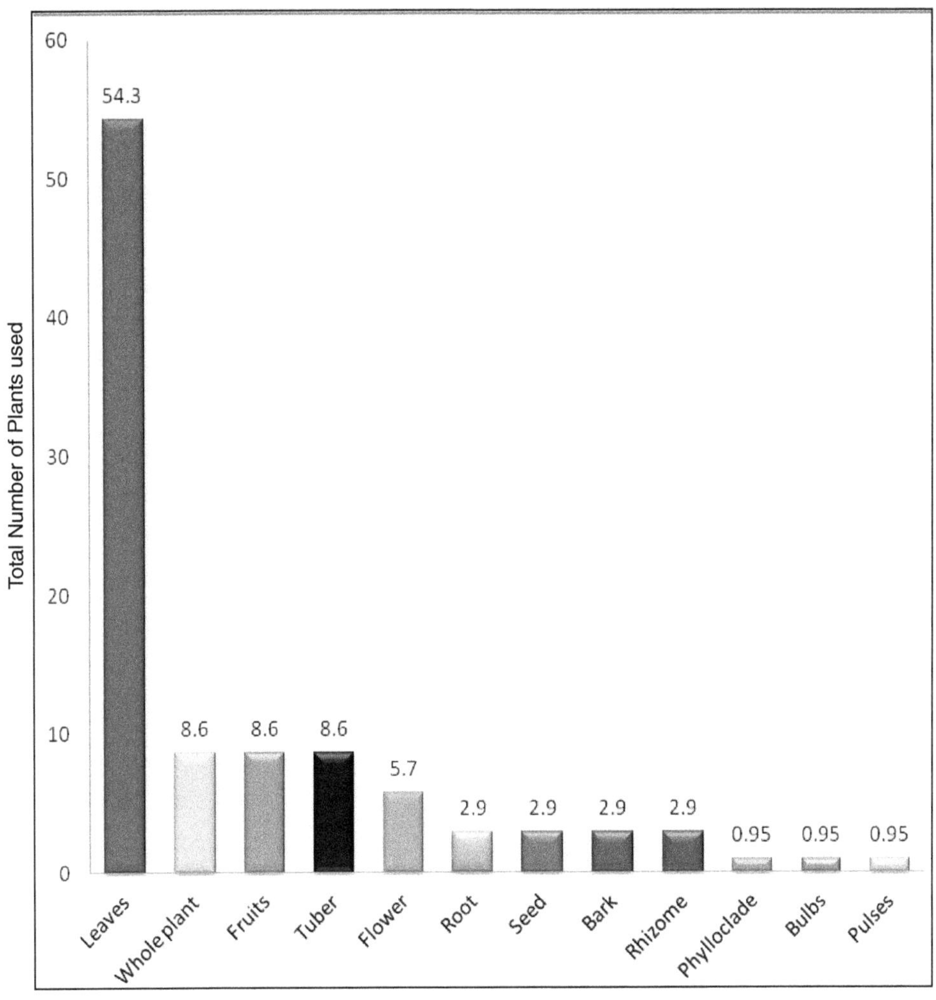

Figure 4.3: Percentage fof Plant Parts Used for the Preparation of Medicine by Kanis in the Present Study Site.

The Kanis revealed ethnotaxa used to treat a variety of illness. Ethnomedicinal plants used to treat 30 illness were sorted into 10 usage categories (Table 4.1) according to Regupathy and Newmaster (2009). Plant remedies were utilized for various illnesses such as skin disease, pimples, prickly heat, itching, psoriasis, eczema, freckle, diabetes, fever, intestinal worms, gastric problems, ulcer, diarrhea, hypothermia, whitlow, cold, antidotes, snake bite, mosquito repellents, bone fracture, throat infection, wounds, jaundice, head ache, body ache, ear ache, tooth ache, rheumatism, eye pain, asthma and cough (Table 4.1) A maximum of 19 plants are used to treat cough, cold and asthma, followed by 17 plants to treat

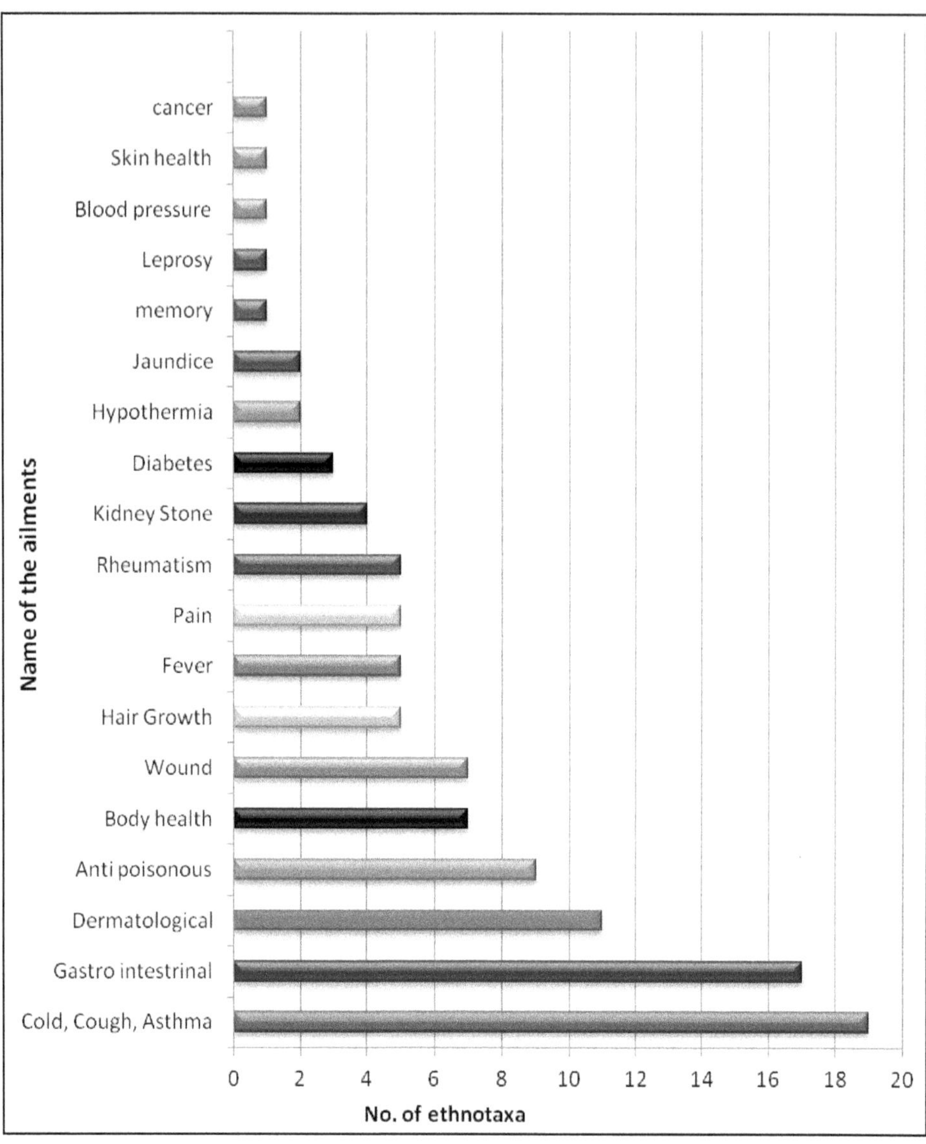

Figure 4.4: Number of Ethnotaxa Used to Treat Various Ailments by Kani Tribes.

gastrointestinal problems, 11 plants for skin diseases, 9 plants for treating poisonous bites and 7 plants to promote body health and wound healing (Figure 4.4).

The Kanis routinely consume plants for their vital well being and good health. Previous research indicates other cultures such as the Malasars and Irulas that put a strong emphasis on daily consumption of plants to promote good health (Regupathy *et al.*, 2008 and Regupathy and Newmaster 2009); which is supported by other studies that have recommended the "general health" category in the standardized illness groupings (Cook, 1995). We found in our survey that 6 plants *viz., Andrographis paniculata, Cardiospermum helicacabum, Cycas circinalis, Dioscorea alata, Trichopus zeylanica* and *Merremia tridendata were* used in the general health category and edible to the Kanis. The great variety of plants that are collected from the wild and distributed among the community or sent off to local markets for trading with Vaidyars or other communities.

The methods of preparation fall into six categories, *viz.*, paste, juice, decoction, powder, infusion and fumes/vapour (Figure 4.5). Plant parts applied as a paste, juice extracted from the fresh parts of the plant, and plant parts used to prepare decoction in combination with water and powder made from fresh or dried material.

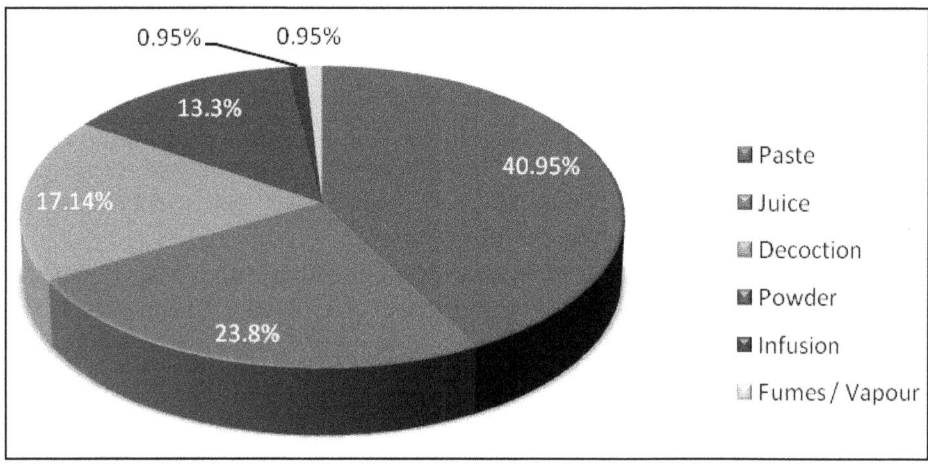

Figure 4.5: Categories of Kanis Mode of Utilization for Various Ailments.

In the present study, 9 ethomedicinal plants are used as antipoisonous by Kanis collected from the selected area of study. *Crinum defixum, Ophiorrhiza mungos, Rauvolfia sarpentina, R. micrantha* are used as antipoisonous plants to treat snake bite poison. *Abrus precatorius, Achyranthus aspera, Acorus calamus* and *Rhinacanthus nastus* are used as antipoisonous to treat poisonous insect bites. The medicinal plants used by the Kanis need to be systematically screened by phytochemists and pharmacologists for the potent active principles. Scientific validation of these remedies may help in discovering new drugs from these medicinal plants.

Enumeration

Plants used by the Kanikkars for the treatment of various ailments are enumerated as follows.

***Abrus precatorius* L** (Fabaceae) (BPN 3425).A juice prepared from ten to fifteen grams of fresh leaves with water is taken orally two times a day for a period of two to three days for curing insect bites.

***Acalypha indica* L.** (Euphorbiaceae) (BPN 3389). A juice is prepared from twenty five grams of leaves with water and salt and is externally applied on the affected part of the skin to treat itching.

Achyranthus aspera **L.** (Amaranthaceae) (BPN 3401). A sufficient quantity of the fresh leaves is made into a paste. This paste is applied over the poisonous insect bite area to reduce the poison.

Acorus calamus **L.** (Araceae)(BPN 3431).A powdered rhizome is used as insect ides in poultry field. The powders rhizome is also having the antipoisonous activity and applied for insect bites.

Aloe vera **(L.) Burm.** (Lilliaceae)(BPN 3414).Handful of leaves is made into paste and applied on the affected face to treat pimples.

Alpinia calcarata **Roscoe.** (Zingiberaceae)(BPN 3388) A decoction prepared from ten to fifteen grams of fresh leaves with fifty ml of water is taken orally once in a day for a period of five days to treat cough.

Alpinia galanga **(L.) SW.**(Zingiberaceae) (BPN 3394) A handful of leaves made into paste is taken orally for two days to cure stomach pain.

Amaranthus viridis **L.**(Amaranthaceae) (BPN 3466).A paste is prepared from fifty grams of fresh leaves and equal quantity of stem is mixed with twenty five grams of fenugreek (*Trigorella fenugreek*). This is taken orally in a single dose for a period of a week to reduce body heat.

Androgarphis paniculata **(Bures.f.)Wall. ex.Ness.**(Acanthaceae).A paste is prepared from equal quantity of leaf, stem and root and is mixed with cow's milk. This is taken orally in a single morning for a period of seven days to improve the health of pregnancy women.

Anisomeles indica **L.** (Lamiaceae),(BPN 3418) A handful of leaves made into paste is taken orally to cure urinary problem.

Annona muricata **L.** (Annonaceae),Fresh fruits are taken daily for a period of two weeks to treat cancer. The paste of the leaves also cure cancer.

Aristolochia indica **L.** (Aristolochiaceae), (BPN 3387).A paste prepared from a sufficient quantity of tuber with water is mixed with milk and is taken orally three times a day for a period of three days to cure fever.

Asparagus racemosus **Willd.**(Lilliaceae),(BPN 3424).A powder prepared from ten grams of stem is consumed orally two times a day for three weeks to treat wound.

Azadirachta indica **Adr. Juss.** (Meliaceae)(BPN 3400).Handful of leaves are boiled with sufficient quantity of water in a closed container. When steam emanates the lid of the container is slowly removed to inhale the vapour to get relief from headache. Handful of the fresh leaf part is mixed with the turmeric powder (*Curcuma longa*, L) and is applied once in a day to treat pimples on the face.

Azima tetracantha **Jum. Lanss.**(Salvadoraceae),(BPN 3383).The leaves are made into paste and is applied externally on the skin to cure the itching.

Bacopa monnieri **(L) Pennell.**(Scorphularaceae), (BPN 3399) About two teaspoon juice extracted from whole plant is applied externally 2 - 3 times a week to get relief from rheumatic pain.

***Bauhinia acuminata* L.**(Fabaceae), (BPN 3434).Take 5 grams of flower bud decoction with water taken orally twice a day for a week to cure stomach pain and kidney problem.

***Biophytum sensitivum* L.DC.**(Oxalidaceae), (BPN 3385).A Paste prepared from leaves is applied externally on the affected part of eye to get relief from eye pain

***Catharanthus pusillus* (Murray) Don.**(Apocyanaceae), (BPN 3433). About two teaspoon of juice is taken orally 2 - 3 times a day to cure ulcer.

***Catharanthus roseus* L. Don.**(Apocynaceae), (BPN 3432). About 10ml of juice from leaves is taken orally twice a day to treat diabetes.

***Capsicum annum* L.**(Solanaceae), (BPN 3482).The fruit is made into a paste and is taken orally for three days to cure blood pressure.

***Caralluma umbellata* Haw.**(Asclepiadaceae), (BPN 3409).One or two drops of juice extracted from phylloclade is applied externally on fore head to treat headache.

***Cardiospermum helicacabum* L.**(Sapindaceae) (BPN 3426).A handful of fresh leaves is boiled in water and the decoction is extracted. The decoction is taken orally morning and evening a day for three months to improve the body health of pregnant women.

***Carica papaya* L.(**Caricaceae), (BPN 3480).One to two teaspoon of juice from the fresh leaves is made is consumed for the treatment of denku fever. Ripened fruit is consumed daily for intestinal worms.

***Cassia alata* L.**(Ceasalpinaceae), (BPN 3460). A paste is made out of leaves is mixed with an equal amount of citrus juice. This is applied to cure skin diseases.

***Cassia klenni* Wight and Arn.**(Ceasalpinaceae), (BPN 3476). Handful leaves made into decoction with water is taken orally for two days to relive cough.

***Cayratia pedata* Var. Glabra.**(Vitaceae), (BPN 3402).A paste prepared from a sufficient quantity of fresh leaves is taken orally two times a day for a period of 3 days to get relief from gas problems.

***Centella asiatica* (L.) Urban.**(Apiaceae), (BPN 3445). Fresh leaves of the plant made into juice mixed with ghee taken orally daily in the morning will increase memory power.

***Chenopodium album* L.** Chenopodiaceae, (BPN 3467). A paste prepared from 25 grams of fresh leaves with water is consumed orally once a day for seven days to reduce body heat.

***Cissampelos pareira* L.** Menispermaceae (BPN 3398).One teaspoon of juice prepared from a sufficient quantity of fresh leaves is taken orally two times a day for a period of 3 days to get relief from stomach problems.

***Cissus quadrangularis* (L.) Wight.**Vitaceae, (BPN 3403).A paste prepared from ten grams of stem is consumed orally two times a day for three weeks to treat rheumatism

***Citrus aurantifolia* (Christas and Pang) Wingle.** Rutaceae (BPN 3434). A juice extracted from the fruit is applied on the face one time a day for a period of three days to treat pimples.

***Clerodendrum inerme* Gaertn.** Fabaceae (BPN 3416).About one teaspoon of juice extracted from fresh leaves is taken orally twice a day for a week to cure cough.

***Clerodendrum phlomidis* L.,** Verbenaceae, (BPN 3413). About 10 grams of leaves made into juice is applied externally for three days on the wounded area to cure wounds.

***Clitoria ternatia* L.** Fabaceae, (BPN 3415).A juice prepared form handful of fresh leaves with water is applied externally 2 times a day for a period of one week to heal the wounds. A juice prepared from handful of fresh leaves and mixed with Ginger extract. This is taken orally 2 times a day for a period of one week to treat fever.

***Cnidoscolus chayamansa* MC Vaugh.** Euphorbiaceae (BPN 3465).A Handful of fresh leaves made into paste is taken orally to cure cough.

***Coccinia indica* Wight and Arn.** Cucurbitaceae (BPN 3485).One teaspoon of juice prepared from the fresh leaves is mixed with half a tea spoon of butter. This is applied externally twice a day to treat skin diseases live skin rashes and scabies.

***Crinum defixum* Ker-Gawl.** Amaryllidaceae (BPN 3381).Sufficient quantity of the fresh bulbs made into a paste in the water is applied externally to treat poisonous snake bites.

***Curcuma angustifolia* Sensu Dalg.**Zingiberaceae(BPN 3463).Fifty grams of the boiled tuber is taken orally three times once in a week to treat diabetes.

***Curcuma longa* L.**Zingiberaceae(BPN 3477).A powder prepared from 10 gms of rhizome is mixed with coconut oil and applied on the affected skin two times a day for a period of one month to treat skin infection.

***Cycas circinalis* L.**Cycadaceae (BPN 3471).Soaked tuber is dried under sunlight for seven days and powdered. The powder mixed with hot water is given to children for starch malnutrition.

***Cymbopogan flexuosus* (Nees ex Steud).** Poaceae (BPN 3404).About 20 gms of leaves are boiled in 200ml water reduced to 50ml. About 25ml of juice is taken orally twice a day for a month to cure mouth infection.

***Datura metel* L.** Solanaceae (BPN 3436).A powder prepared from the shade dried leaves is made into cigrette for smoking two times a day for a period of two to three weeks to get relief from Asthma. A juice prepared from required quantity of fresh unripe fruit is mixed with coconut oil and is applied on the head to control hair fall.

***Dioscorea alata* L.** Dioscoraceae (BPN 3468).Fifty grams of peeled tuber is boiled with water, Boiled tuber is mixed with required amount of garlic pepper and ground it into fine paste. This is consumed to fulfill starch requirement.

***Dioscorea bulbifera* L.** Dioscoreaceae (BPN3478).Tuber paste mixed along with honey will cure dysentry, and also used for conjuctives.

***Eclipta alba* (L.) Hasse.** Asteraceae(BPN 3417).Ten grams of leaves are boiled in one hundred ml of coconut oil and the oil extract is applied on the head twice a day. This helps to improve hair growth.

***Ficus racemosa* L.** Moraceae (BPN 3437).A paste prepared from ten grams of fresh fruit with salt is taken orally once in a day for a period of seven days to treat rheumatism.

***Garcinia gummiguta* L.** Clusiaceae (BPN 3429).Dry fruits are used in traditional recipies. Leaf juice taken in empty stomach can cure ulcer, asthma. Fruits also have the property to reduce obesity and digestion problem.

***Hemidesmus indicus* L.** Asclepiadaceae (BPN 3430).About 10 grams of dried root is powdered and mixed with honey and is taken orally one time a day for a period of one week to get relief from stomach ache.

***Hibiscus rosasinensis* L.** Malvaceae (BPN 3473). A handful of fresh flowers is made into a fine paste with water and mixed with coconut oil. This oil is applied on the head daily to improve hair growth and arrest hair fall.

***Hibiscus surattensis* L.** Malvaceae (BPN 3454).Leaves are cooked as vegetable and the leaf paste is eaten frequently to cure ulcer.

***Indigofera tinctoria* L.**Fabaceae (BPN 3382). A leaf paste mixed with honey is consumed to cure hepatitis

***Indonesiella echinoides* L.** Acanthaceae (BPN 3406). A Handful of leaves made into paste is taken orally for few days to cure leprosy.

***Ixora coccinia* L.** Rubiaceae (BPN 3430).A juice prepared from handful of fresh flowers. This is applied externally twice a day to get relief from body burn.

***Jatropha curcas* L.**Euphorbiaceae (BPN 3456).The oil is made from sufficient quantity of seeds is applied on the head for an hour before bath. This treatment is used to reduce hair fall.

***Justicia adathoda* L.** Acanthaceae(BPN 3481).Mix equal parts of fresh root, bark, leaf and flowers grai it well and extract the juice. This juice is taken orally once in a day for a period of 3 days to get relief from cough and asthma and eye diseases.

***Justicia gendrussa* Burn. F.** Acanthaceae (BPN 3392).Leaves made into a paste is applied on affected part of leg to get relief from rheumatic pain.

***Justicia simplex* L.** Acanthaceae (BPN 3397).About 50 leaves made into paste is applied on the affected area to cure fracture of bone.

***Kaempferia galanga* L.** Zingiberaceae (BPN 3453).Twenty grams of fresh tuber is made in to decoction. This is mixed with cow's milk and taken orally two times a day for a period of three days to control loose motion.

***Lantana camara* L.** Verbenaceae (BPN 3483). About 10 ml of leaf juice taken orally for few days will cure urinary disorders. A juice prepared from the bark is mixed with sugar cake and cow milk. This is taken orally to cure urinary troubles.

***Lawsonia inermis* L.** Lythraceae (BPN 3475).A handful of leaves is made into paste with water. The paste is mixed with coconut oil (*Cocos nucifera,* L.) and boiled. The oil extract is applied on the head before taking bath. This stop hair fall.

***Leucas aspera* (Willd.)** (BPN 3459).One teaspoonful of powder is made from an equal quantity of leaves and flowers. This is mixed with cow's milk is taken orally three times a day for a period of one week to get relief from cough.

***Lippia nodiflora* (L.) A. Riob.** Verbenaceae (BPN 3405).The leaves are soaked in water along with the flowers for a day. The infusion is filtered and half a cup is consumed three times a day for a period of one week to get relief from cough and fever.

***Manihot esculenta* Cranty.** Euphorbiaceae (BPN 3470).The tuber made into paste mixed with pepper is taken orally for two days to cure diarrhea.

***Melia azadirachta* L.** Meliaceae (BPN 3438). Fresh leaves are made into a paste which are applied externally twice a day for 2 - 3 days will cure skin diseases.

***Merremia tridentata* Hallier.** Convolvulaceae (BPN 3411).Decoction taken from the whole plant is taken orally to improve body health.

***Michelia champaca* L.** Magnoliaceae (BPN 3452).Dry flowers of this plant is powdered and made into chutney. This consumed daily at the time of taking food will cure stomach pain and also improves digestion.

***Moringa pterygosperma* Gaertn.** Moringaceae (BPN 3474). Roots and leaves are mixed with some amount of mustard and Garlic and grounded well with water. This is diluted into fluid and drink once in a day for 20 days. This expel helps in kidney stone.

***Mollugo pentaphylla* L.** Molluginaceae (BPN 3393).A juice taken from the leaves taken orally for few days will cure menstruation problems and stomach problems.

***Murraya koengii* (L.) Spreng.** Rutaceae(BPN 3473).The 20grams of leaves made into paste. This is used to cure cough.

***Myristica fragrans* Houtt.** Myristicaceae (BPN 3404).About 10 grams of dried unripened fruit is powdered and made into a paste with honey. This paste is taken orally two times a day for a period of one month will improve digestion.

***Nerium oleander* L.** Apocynaceae (BPN 3439). Flowers are used for ritual purposes.

***Ocimum sanctum* L.** Lamiaceae (BPN 3441). About 5 - 10 fresh leaves are made into a paste. This is consumed daily in the morning and evening to get relief from cough.

***Ophiorrhiza mungos* L.** Rubiaceae (BPN 3446).The tuber is made into a paste. This paste is applied externally to cure snake bites.

***Opuntia dilleni* Ker-Hawl.** Cactaceae (BPN 3447).A decoction is made from the whole plant. This decoction will cure diabetes.

Pavonia zeylanica **(L.) Cav.** Malvaceae (BPN 3422).A decoction prepared using roots will cure nervous disorders. 10 gms of leaves made into paste applied externally will cure swellings.

Pergularia daemia **(Forssen) Chiov.** Asclepiadaceae (BPN 3448). A decoction is prepared from 25 grams of fresh leaves with 100 ml of water. This is taken orally two times a day for a period of one month to treat asthma. A decoction extracted from one hundred grams of fresh leaves is administered with honey and is taken one time for a period of three days to get relief from menstrual problem.

Phyllanthus amarus **L.Schumb.Thonn.** Euphorbiaceae (BPN 3444).A juice prepared from handful of fresh whole plant is mixed with 100 ml of coconut milk. This is taken orally once in a day for a period of three days. This cures jaundice.

Phyllanthus emblica **L.** Euphorbiaceae (BPN 3443).A juice is prepared from twenty grams of fresh leaves. A sufficient quantity of dried ginger, Pepper and clove powder are mixed with 50 ml of leaf juice. This is taken orally along with milk to arrest hair fall.

Phyllanthus rheedi **Wight.** Euphorbiaceae (BPN 3463).A paste made from this whole plant is given in empty stomach for 15 to 20 days. This cures urinary infections.

Physalis angulata **L.** Solanaceae(BPN 3410). A paste prepared from a handful of fresh leaves with water when taken orally in a single dose for a period of two weeks will expel the kidney stones.

Piper betel **L.**Piperaceae (BPN 3461).About six leaves with 5 to 6 pepper boiled with water, one litre of water is further boiled to hundred ml. This reduces the severity in poison due to bites.

Piper longum **L.** Piperaceae (BPN 3390).About half teaspoon of tippli powder and dried ginger powder is mixed with honey and consumed. This helps to get relief from cough.

Piper nigram **L.** Piperaceae (BPN 3457).A sufficient quantity of pepper, seed, dried ginger and palm sugar are mixed well and powdered. Take one teaspoonful of powder boil with water and filter it. After filtration one cup of this decoction is given orally two times a day for period of two to three days to get relief from cough.

Plecranthus ambonicus **(Lour) Spreng.** Lamiaceae (BPN 3443). A handful of fresh leaf juice is boiled with coconut oil (cocos nucifera, L. This oil extract is applied on head for 20 minutes before bath. This treatment helps to get relief from headache and cough.

Pterocarpus santalinus **L.** Fabaceae (BPN 3407).About ten grams of dried bark is powdered and made trade into a paste with rose water (*Rosa indica*, L). This paste is externally applied once in a day to treat facial pimples.

Rauvolfia micrantha **L.** Apocynaceae(BPN 3412). About 10 grams of root is made into paste and applied externally on the snake bitten area to reduce the poisonous effect.

Rauvolfia serpentina **L. Benth ex Kurz.** Apocyanaceae (BPN 3384).A paste prepared using twenty five grams of fresh leaves with water is applied externally twice a day for a period of three days to treat poisonous bites.

Rhinocanthus nasutus **(L.) Karz** Acanthaceae (BPN 3420). About ten grams of fresh leaves are mixed with ten grams of (*Adhathoda vasica*) leaves are made into paste. This paste applied for a period of four days to get relief from poisonous insect bites.

Ruellia tuberosa **L.** Acanthaceae (BPN 3427).A powder is prepared using shade dried tuber and mixed a cup of milk. This is drunk two times a day for a period of three days to treat asthma and cough.

Salacia reticulata **Wight.**Hippocrataceae (BPN 3408).A decoction of bark powder cures early diabetes, gonorrhea, rheumatism.

Sanseveria roxburghiana, **Schultes.** Dracenaceae (BPN 3419).One or two drops of juice is extracted from rhizome and mixed with honey. This is taken orally to cure cough

Solanum nigrum **Sense.Gamble.**Solanaceae (BPN 3469).A decoction prepared from a handful of the leaves with required quantity of water taken once in a day will expel intestinal worms.

Solanum torvum **S.W.** Solanaceae (BPN 3464).About ten grams of unripe fruits is crushed with stone to remove the seeds. The paste prepared from seedless unripe fruits and equal part of coconut is taken orally twice a day for a period of one week will heal the wounds.

Solanum trilobatum **L.** Solanaceae (BPN 3458).A handful of fresh leaves boiled with 100ml of water is allowed to cool. This decoction taken orally twice in a day for a period of 5 days will cure fever and cough.

Solanum virginianum **L.** Solanaceae (BPN 3451).A sufficient quantity of the whole plant powder taken orally along with water twice in a day for a period of one week will cure cough.

Souropsis androgynus **L.** Phyllanthaceae (BPN 3455). The leaves are used as vegetables. The leaf paste is taken orally once in a day for a week to cure Asthma.

Spermacoce hispida **L.** Rubiaceae (BPN 3428).Fresh leaf made into paste with water is mixed with coconut milk and taken orally thrice a day for a few days to cure diarrhoea.

Syzygium aromaticum **L. Merrillberry** Myrtaceae (BPN 3395).The powder prepared from the shade dried flower bud is directly applied on the affected teeth to get relief from tooth ache.

Tinospora cordifolia **(Wild) Hook f. and Thomson.** Menispermaceae(BPN 3421) A handful of fresh leaves, required quantity of small onion (*Allium cepa,* L.) are made into a fine paste with water. This is taken orally once in a day for about seven days to treat jaundice.

***Tribulus terrestris* L.**Zygophyllaceae (BPN 3396).A decoction prepared from 10 to 20 grams of fresh fruits with required amount of water is taken orally once in a day for few days to get relief from kidney problems.

***Trichopus zeylanicus* Gaertn.** Trochopidiaceae (BPN 3450). A sufficient quantity of fresh fruit is consumed daily to improve body vigor.

***Tridax procumbens* L.** Asteraceae (BPN 3484).A juice is prepared using fresh leaves with stem. This is directly applied on the wound once in a day for a period of three days to heal the wounds.

***Vigna radiata* (L.) Wilzek.** Fabaceae (*BPN 3442*).The fruit dried under sunlight is powdered and mixed with manjal (*Curcuma longa* L). This is applied externally on the skin to cure skin diseases.

***Vitex negundo* L.**Verbenaceae (BPN 3480).A handful of leaves boiled with two litres of water in a closed container. Spread the vapours inside the house to repel mosquitoes. Take bath with this water to get relief from body pain.

Conclusion

The tribes have a well- developed system of traditional medicine. They know about number of rare medicinal plants and their applications. But all this knowledge is gradually lost by some superstitious belief of these ethnic groups. They do not reveal the knowledge to others because of the fear that, if they did so the healing power of the plants may be lost. Even though these beliefs have certain advantages, a lot of valuable knowledge has been lost by this way.Another problem with tribal medicine is the absence of recorded data. Numerous ancient knowledge has been lost by the absence of supportive literature. A major reason for this is the illiteracy of the tribes. Further, a large number of medicinal plants are being threatened due to deforestation and urbanization. In these circumstances, ethnobotanical and ethnomedicinal studies have great significance in the collection of traditional knowledge, preparation of recorded data and in the conservation of endangered medicinal plant species. With the help of new technologies the data could be scientifically proved. So that the scientific world will accept the traditional system. Nature is providing what we need and our task is to save nature for posterity.

References

Ayyanar, M and S. Ignacimuthu. 2009. Herbal medicines for wound healing among tribal people in Southern India: Ethnobotanical and Scientific evidences. *International J. Appl.Res. Nat. Products*. 2(3): 29-42.

Ayyanar,M and S. Ignachimuthu. 2005. Traditional knowledge of Kani tribals in Kouthali of Tiruneveli hills,Tamil Nadu, India. *J. Ethnopharm*. 102: 246-255.

Ayyanar,M.,K. Sankarasivaraman and S. Ignacimuth. 2008. Taditional herbal medicines used for the treatment of diabetes among two major tribal groups in south Tamil Nadu, India. *Ethnobotanical Leaflets* 12: 276-280.

Champion, H.C. and S.K. Seth. 1968. A revised survey of forest types of India. Manager of Publication. Govt. India. New Delhi.

Chenthurapandy, P., V.R. Mohan and C. Kalidass. 2010. An ethnoboytanical survey of medicinal plants used by the Kanikkar Tribe of Kanyakumari District of Western Ghats. Tamil Nadu for the treatment of skin diseases. *J. Herbal. Med. Toxicol.* **4**(1): 179-190.

Cook F.E.M. 1995. Economic Botany data collection Standard., Prendergast, Royal Botanic Gardens, Kew.

Das P.C., A.S.Mandal and C.N. Islam. 1989. Antiinflammatory and antimicrobial activity of seed kernels of Mangifera indica. *Fitoterapia* 60: 240-253.

Divya, V.V., N. Karthick and S. Umamaheswari. 2013. Ethnopharmacological studies on medicinal plants used by Kani tribes of Thachamalai hill, Kanyakumari, Tamil Nadu, India. Inter. J. Adv. Biol. Res. **3**(3): 384-393.

Francis Xavier, T., A. Freede Rose and M. Dhivya. 2011. Ethnomedicinal survey of Malayali tribes in Kolli hills of eastern ghats of Tamil Nadu, India. *Indian J. Trad. Knowledge.* **10**: 559-562.

Gadgil,M. 1996. Documenting diversity: An experiment. *Current Science* 70: 1-152.

Gamble J.S., C.E.C. 1915-1936. Flora of the Presidency Madras Vol. I-III. Adlard and Co. London (Reprinted 1956). Botanical Survey of India, Calcutta.

Gamble, J.S., C.E.C. Fisher. 1915-1936.Flora of the Presidency Madras. Vol. I-III. Adlard and Co. London (Reprinted 1956). Botanical Survey of India, Calcutta.

Hamburger, M and K. Hostettman 1991. Bioactivity in plants: The link between phytochemistry and medicine. *Phytochemistry* 30: 3864-3874.

Ignacimuthu,S., M. Ayyanar and K. Sankarasivaraman. 2008. Ethnobotanical study of medicinal plants used by Paliyar tribals in Theni district of Tamil Nadu, India. *Fitoterapia* 79: 562-568.

Jain S.K.1981. Glimpses of Indian Ethno botany. Oxford and IBH Publishing Co., New Dehi, India.

Jeeva S. *et al.*, 2006. Weeds of Kanyakumari District and their value in rural life. *Indian J. Trad. Knowledge* 5(4): 501-509.

Jeyaprakash,K., M. Ayyanar., K.N. Geetha and T. Sekar. 2011. Traditional uses of medicinal plants among the tribal people in Theni District (Western Ghats) Southern India. *Asian Pacific J. Tropical Biomed.* 20-25.

John De Britto, A., R. Mary Sujin, R. Mahesh and K. Dharmar. 2010. Ethnomedicinal wisdom of the Manavalakuruchi people in Kanyakumari District, Tamil Nadu. *Inter. J. Biol. Tech.* **1**(2): 25-30.

Kala, C.P., P.P. Dhyani and B.S. Sajwan. 2006. Developing the Medicinal plants sector in Northern India: Challenges and Opportunities. *J. Ethnobiol. and Ethnomed.* 2: 32-doi: 10.1186/1746-4269-2-32.

Kalaiselvan, M and R. Gopalan. 2014. Ethnobotanical studies on selected wild medicinal plants used by Irula tribes of Bolampatty Valley, Nilgiri Biosphere Reserve (NBR), Southern Western Ghats, India. *Asian J. Pharm.Clinical Res.* 7: 22-26.

Kingston, S. and A.D.S. Raj. 2000. Plants used in ethnomedicine by the habitat people of Kanyakumari District. *Proce. Mano. Ees. Forum*. B20-23.

Koshoo,T.N. 1996. India needs a National biodiversity conservation board. *Current Science* 71: 506-513.

Kumari Subitha., T. Ayyanar, M. Udayakumar and T. Sekar. 2011. Ethnomedicinal plants used by Kani tribals in Pechiparai forests of Southern Western Ghats. Tamil Nadu, India. *Inter. Res. J. Pl. Sci.* **2**(12): 349-354.

Lalitha Rani, S., V. Kanpana Devi, P. Tresina Soris, A. Marughupandian and V.R. Mohan, 2011. Ethnomedicinal plants used by Kanikkars of Agasthiarmalai Biosphere Reserve, Western Ghats. *T. Ecobiotech*. 3(7): 16-25.

Louga E.J.,E.T.F.Witkowski and K.Balkwill. 2000. Differential utilization and ethnobotany of trees in Kitulanghalo forest reserve and surrounding communal lands, Eastern Tanzania. *Economic Botany*. 54:328-343.

Matthew, K.M. 1981. Materials for a flora of the Tamil Nadu Carnatic. The Rapinat Herbarium St. Joseph's College, Tiruchirapalli, India, p. 73-85.

Matthew, K.M. 1983. The flora of the Tamil Nadu Carnatic. Vol. I-III. The Rapinat Herbarium, St. Joseph's College, Tiruchirapalli, India.

Mohan, V.R., C.Kalidass and D. Amisn Abragam. 2010. Ethnomedico botany of the Palliyars of Saduragiri hills., Western Ghats, Tamil Nadu. *J.Economic Taxonomic Botany* 34: 639-658.

Muthu, C., M. Ayyanar., N. Raja and S. Ignacimuthu. 2006. Medicinal plants used by traditional healers in Kancheepuram district of Tamil Nadu. *Indian J. Ethnobiol. Ethnomed*. **2**: 43-52.

Pradeeps, M. and G. Poyyamoli. 2013. Ethnobotany and utilization of plant resources in Irula villages (Sigur plateau, Nilgiri biosphere. Reserve, India. *J. Med. Plants Res*. 7(6): 267-276.

Pullaiah, T. 2007. Field and Herbarium methods. In: Taxonomy of Angiosperms, Regency Publications, New Delhi. pp. 130-145.

Ragupathy, S., S.K. Newmaster., M. Murugesan and V. Balasubramanian. 2008. Consensus of the Malasars traditional aboriginal knowledge of medicinal plants in the Velliangiri holy hills. *Indian J. Ethnobiol. Ethnomed*. 4: 8.

Regupathy, S. and Newmaster, G. 2009. Valorizing the 'Irulas' traditional knowledge of medicinal plants in the Kodiakkarai Reserve Forest. *Indian J. Ethnobiol. Ethnomed*. **5**: 10-22.

Rhaman, M.A., S.J. Mossa and S.M. Al-Said. 2004. Medicinal Plant diversity in the flora of Saudi Arabia-I: areport on seven plant families. *Fitoterapia* 75: 149-161.

Samydurai, P., S. Jagatheshkumar., V. Aravinthan and V. Thangapandian. 2012. Survey of wild aromatic ethnomedicinal plants of Vellingiri hills in the Southern Western Ghats of Tamil Nadu, India. *Int.J. Med. Arom. Plants* 2(2): 229-234.

Shankar, R. 1995. Tribal community in India and PGR, In: Farmer's rights and plant genetic resources recognition and reward: A dialogue.(Ed.) Swaminathan, M.S. Million India Ltd., Madras, India pp. 106-111.

Shanmugam S., K. Rajendran and K. Suresh. 2012. Taaditional uses of medicinal plants among the rural people in Sivagangai district of Tamil Nadu, Southern India. *Asian Pacific J. Tropical Biomed.* 429-434.

Singh K.S. 1994. People of India- the Scheduled Tribes, Anthropological survey of India (Oxford Univ.Press).460-464.

Sofowara, A. 1993. Medicinal plants and traditional medicine in Africa. 2nd ed. Spectrum Books Ltd. Ibadan.

Sudha, S., V.R. Mohan., S. Kumaresan., C. Murugan and T. Athiperumalsami. 2010. Ethnomedicinal plants used by tribals of Kalakkad- Mundanthurai Tiger Reserve (KMTR), Western Ghats Tamil Nadu. *Indian Journal Traditional Knowledge.* **9**(3): 502-909.

Sukumaran, S., S. Jeeva., A.D.S. Raj and D. Kannan. 2008. Floristic diversity. Conservation status and economic value of miniature Sacred Groves in Kanyakumari District, Tamil Nadu, Southern Penisular India. *Turk. J. Bot.* **32**: 185-199.

Tabuti, J.R.S.,S.S.Dhillion and K.A. Lye. 2003. Traditional medicine in Bulamogi country, Uganda: its practitioners, users and viability. *J. Ethnopharmacology* 85: 119-129.

Umapriya T., A. Rajendran., V. Aravindhan,.,B. Thomas and M. Maharajan. 2011. Ethnobotany of Irular tribe in Palamalai hills,Coimbatore, Tamil Nadu. Indian *J. Nat. Products and Resources.* 2(2): 250-255.

World Health Organization (WHO). 2002. WHO traditional medicine strategy 2002. World Health Organization, Geneva. WHO/EDM/TRM/2002.1.

Zakaria, M. 1991. Isolation and characterization of active compounds from medicinal plants. *Asia Pacific. J. Pharmacology* 6: 15-20.

2018, Ethnomedicinal Plants: A Biodiversity Treasure Pages 189–213
Editors: V.R. Mohan, A. Doss, P.S. Tresina and V. Sornalakshmi
Published by: ASTRAL INTERNATIONAL PVT. LTD., NEW DELHI

Chapter 5

Ethnobotanical Study of Bodha Hills, Southern Eastern Ghats, Namakkal District, Tamil Nadu, India

***S. Murugesh*[1] *and P. Deepa*[2]**

[1]*Professor and Head,*
Department of Botany, Periyar University,
Periyar Palkalai Nagar, Salem – 636 011, Tamil Nadu
E-mail: murugeshss@rediffmail.com
[2]*PG and Research Department of Botany,*
Vivekanandha College of Arts and Sciences for Women, Elayampalayam,
Tiruchengode, Namakkal District, Tamil Nadu
E-mail: taanishadeepa@gmail.com

Introduction

Indigenous knowledge is essential. It is oriented and based culturally. It is also integral to the cultural identity of the social group in which it operates and is preserved. Various local communities possess knowledge, innovations and peculiar practices developed from experience gained over centuries and adopted to the local culture and environment. The basic components of any country's knowledge system are its indigenous knowledge. It encompasses the skills, experience and insights of people, applied to maintain or improve their livelihood. World Intellectual Property Office (WIPO) defined that it is tradition-based literacy, artistic, or scientific works; performances; inventions; scientific discoveries; designs; marks; names and symbols, undisclosed information and all other tradition based innovations and creations resulting from intellectual activity in the industrial, scientific, literacy or artistic fields. It can be integrated that traditional knowledge is vast enough to encompass indigenous knowledge related to various categories like agricultural knowledge, medicinal knowledge, biodiversity related knowledge and expressions of folklore in the form of music, dance, song, handicraft, designs, stories and artwork.The term "tradition-based" refers to knowledge systems, creations, innovations and cultural expressions which have generally been transmitted from generation to generations. It is generally regarded as pertaining to a particular people or its territory; and is constantly evolving in response to a changing environment. It tends to be developed in a way that is closely related to the immediate environment in which traditional communities dwell, and to respond to the changing situation of that community.

World Bank (1997) stated that in the emerging global knowledge economy a county's activity to build and mobilize knowledge capital, is equally essential for sustainable developmentas the availability of any physical and financial capital. The indigenous knowledge on utility and utilization aspects of plants has been sensing as suitable tool for botanical and agricultural research owing to its relevance in development and promotional activities of new or less known economic plants (De, 1968; Jain, 1981; Schutes, 1962; Paroda, 2000). An integration of traditional knowledge with modern societies development by value additions is an essential requisite for successful application of indigenous knowledge system for economic welfare of the tribal societies. The indigenous knowledge on botanical data has served as linchpins for theories about the life ways, social and economic history, and health of past peoples (Alcorn, 1995). Modern ethnobotanical data are often used as analogs for interpreting past-human interrelationships (Dimbleby, 1978; Ford, 1979).

Study Area

The area selected for study is Melur of Bodha hills (Latitude: 11.916667; Longitude: 77.55) in Southern Eastern Ghats, Rasipuram taluk, Namakkal District of Tamil Nadu (Figure 5.1). Bodha hills is situated towards the south east of Salem. Rasipuram is the nearest town to Bodha hills. Thengalpalayam, Vadugam and Kullampatti are the villages nearer to the Bodha hills. The village Melur is the highest peak of Bodha Hills.Bodha hills consist of an environment that prevails in a dry deciduous type of forest. The area has a predominant red soil impregnated with organic matter, and granite, bed rock is overlaid with shallow, sandy loam, and glacial soils are moderate to well drain. Temperature begins increasing after March. April is the warmest month with an average temperature of 33.20 °C at noon. January is the coldest with an average temperature of 15.10 °C at night. Temperature drops sharply at night. The temperature is average during the month of February with maximum sunshine. The main source of water for drinking and irrigation comes from wells. It tastes sweet and contains ample amount of minerals. The wells receive water from the seasonal rains.The hill gets rainfall during October and November by the North- East monsoon. This is followed by cyclonic storms in the Bay of Bengal blowing towards the land. There is heavy rainfall due to the south west monsoon also. During summer at times there is shower. This accounts for about 6 per cent of the annual rainfall.

Bodha Hills

Bodhi Malai- Mountain of Enlightenment," a lofty (4,015)' or 1182 meters range (Figure 5.2). It derives its name from the word bodhi (enlightenment, illumination, or awakening) of the religion of Buddhism. The foot print of Budha is further by

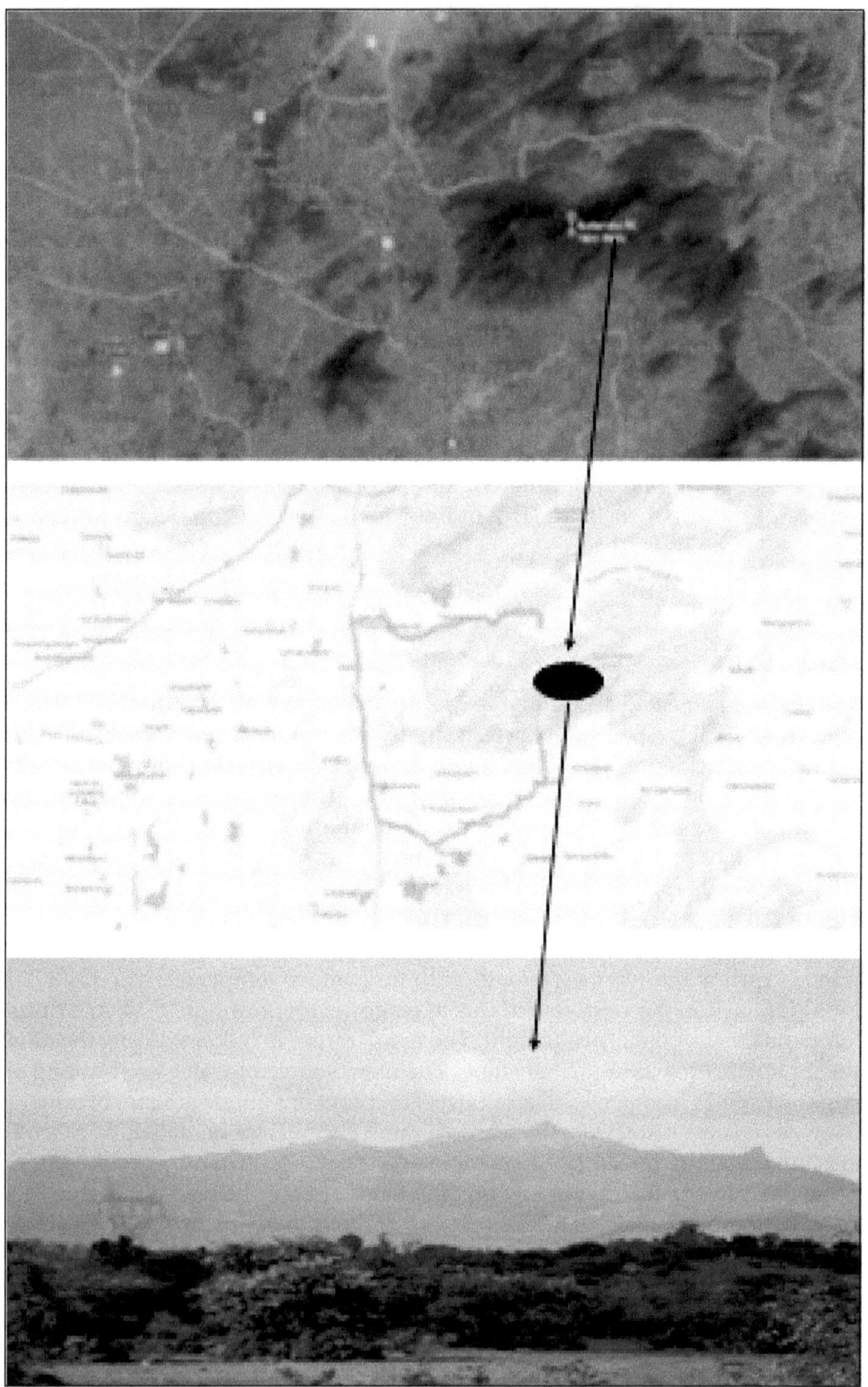

Figure 5.1: Location Map of Bodha Hills in Namakkal District of Tamil Nadu.

Figure 5.2: Panoramic Views of Bodha Hills.

Anandha Malai (Bliss Mountain) the name for one of its peaks. In the thirteen century, it was known as Karungali malai (Red Wood Mountain). Located at the geographical centre of the district, the mountain extends east- west for twelve miles and is separated from the Kollimalai by the Hanuman Ghat. The mountain includes the peaks of Anandha (or Periya) malai (947 meters), Jambuttu malai (3,216′) and Kuttamalai (3,700′). The forests at the eastern end are now abundant with sandal wood trees. The peak near Melur contains a number of Pandukal monuments, on the mountain are the Malayali tribal hamlets of Kizhur, Kutta Malai, Melur (3,861′), and Nadu Valavu. The nearness of the Bodhi Malai Dharmapuri leads one to believe that Salem might have harboured the Buddhist Mahayana Philosopher, Bodhidharma (450-535). According to Chinese tradition, he went to China from South India (Salem) in 480 at the invitation of the Chinese emperor. Buddhists make a claim to this hill calling it Bodhi malai which was later corrupted as Bodha malai (Busnagi Rajannan, 1932).

Melur

Upper Town" is a village (3,861′) on the Bodha malai mountain. It is inhabited by a small group of Malayalis related to the Kolli malayalis. There are more than twenty pandukal tombs (Ancient stone). Pandukals are monuments built of large slabs or blocksbuilt by the earliest inhabitants of South India. They believe in the worship of dead, which was followed by pre-Hindu Brahmin religion that prevailed there. Through the worship or honoring the dead, the generations were bound together in a stabilizing continuity. The living formed a holy union and sequence of blood and flesh stretching far into the past and the future. The ancestral worship held the dead, the living and the unborn together in a sacked unity. They are found on the nearby peak among a grove of sandal wood trees, most of them have been disturbed and the interior dolmens exposed (Figure 5.3). The Malayalis believe that they are remnants of forts and palaces of an extinct race of "Pandiya Raja". They, like the Irish leprechauns, were very small in size, used mushrooms for umbrellas and employed rabbits for ploughing and riding. A heap of Neolithic stone tools, collected from the fields and banks of streams were worshipped by the residents as Pillaiyar (Busnagi Rajannan, 1932).

Malayali Tribal's of Bodha Hills

Malayali (Tamil) "Mountain men" The malayalis of these Bodha malai appear to have migrated during the period between eleventh and thirteenth centuries from the Kolli malai. They cultivate in small plateaus and slopes of the mountain and sell their products by head loads in the weekly markets of the towns in the surrounding plains. Access to and from these hamlets is only by steep and arduous footpaths. They are found in the hamlets of Kutta malai, Kizhur, Kurinjiyur, Melur, Sambutthu, and Naduvalavu. They are grouped into several kulas including Kasakali, Kocha maniyan, Masiyan, and Sirakali. "Sirakali" is the name of a form of mother Kali. It is interesting to notice that a Goddess named Sara Kali (Queen Kali) was worshiped by European Gypsies who were said to have originated from India. They called Sara kali as "their mother, the woman, the sister and the queen. The Bodhi Malayalis worship many minor deities such as Andi Appan, Bodhi Malai

Figure 5.3: ***Megalithic*** **Monuments in Bodha Hills.**

Siddhan, Periya Sadaiyan, Periyasamy and Saelai Katti Karuppu. They also offer prayers at the siddhar kovil and it is now a Vishnu temple, situated at the highest point of the mountain near Melur. The Melur Pillayar temple contains no statue of the elephant-faced God, but a pile of an iconic Neolithic stone tool collected from various places of the mountain (Busnagi Rajannan, 1932).

Materials and Method

The present study is conducted among the Malayali tribal people of Bodha hills, Namakkal District, Tamil Nadu, India. The field trips were arranged for 4 days to a week in the study area once in a month of the year (April 2011- March 2013) to study the Malayali tribal people in Bodha hills. Ethnobotanical data were collected from 90 resource persons (average age of 35-80) of the study area who have much knowledge on plants in semi structured interviews. The interviews were conducted in the local language. *i.e., Tamil.* Ethnobotanical information included the local name of the particular plant, habit and its uses. The collected ethnobotanical information was recorded on field note books and plants were identified using the *Flora of Presidency of Madras* (Gamble. 1935) and *Flora of Tamil Nadu-Carnatic* (Matthew, 1983). The voucher specimens were deposited at Department of Botany, Periyar University, Salem.

Result and Discussion

The present study enumerated 158 plant species under 120 genera belonging to 49 families of ethnobotanical importance based on the observations and interaction with Malayali tribal people of Bodha hills, Southern Eastern Ghats, Namakkal District, Tamil Nadu, India (**Table 5.1**). The Malayali tribal people used the plant species for many different purposes in their day to day life. In the present study 8 plant species are used as agriculture implements formaking cattle yoke, plough-share and handles of axe, hoe, sickle and dagger such as *Anogeissus latifolia, Cassia fistula, Chloroxylon swietenia, Dalbergia latifolia, Dodonaea angustifolia, Ficus benghalensis, Tamarindus indica* and *Tectona grandis*. The similar uses of *Tectona grandis* from Gujarat (Maru and Patel, 2012); *Dalbergia latifolia* from Gujarat (Nirmal Kumar *et al.*, 2007); *Anogeissus latifolia* and *Chloroxylon swietenia* from Orissa (Panda *et al.*, 2005) were already documented.The present study also indicated that the plant species of *Abrus precatorius, Ipomoea staphylina* and *Musa paradisiaca* are used as cordages. The similar use of *Abrus precatorius* and *Musa paradisiaca* were reported by Sahu *et al*,(2013a). Five plant species are also recorded for making basket, net, pillows and cushions. Similar use of *Sansevieria roxburghiana* (Ayyanar and Ignacimuthu, 2009) and *Bambusa arundinacea, Bombax ceiba* (Nirmal Kumar *et al.*, 2007) are also reported.

Tribal peoples are using 19 plant species in crop cultivation for their food and sale. The main crops cultivated by Malayali tribals are; *Oryza sativa, Vigna mungo, Arachis hypogea, Musa paradisiac, Sorghum vulgare, Setaria italica, Pennisetum americanum, Panicum sumatrense, Panicum miliaceum, Capsicum annum, etc.* The species such as *Mangifera indica, Artocarpus heterophyllus* and *Musa paradisiaca* are grown for fruits in the tribal hamlets. Anburajamand and Nandagopalan (2012) also reported that *Eleusine coracana, Oryza sativa, Pennisetum americanum, Setaria*

Table 5.1: Ethnobotanical Study of Bodha Hills

Sl.No.	*Botanical Name*	*Family*	*Part Used*	*Ethnobotanical Uses*
		Agricultural implements		
1	*Anogeissus latifolia* (DC.) Wall. *ex* Guill. and Perr.	Combretaceae	Wood	Wood is used for making handles of harrow, hoe and axe.
2	*Cassia fistula* L.	Caesalpiniaceae	Wood	
3	*Chloroxylon swietenia* DC.	Rutaceae	Wood	Wood is used for making cattle yoke.
4	*Dalbergia latifolia* Roxb.	Fabaceae	Wood	Wood is used for making tools and handles like dagger, hoe and axe.
5	*Dodonaea angustifolia* L.f.	Sapindaceae	Wood	
6	*Ficus benghalensis* L.	Moraceae	Wood	Wood is used for making plough-share.
7	*Tamarindus indica* L.	Caesalpiniaceae	Wood	Wood is used for making handle of sickle.
8	*Tectona grandis* L.f.	Verbenaceae	Wood	Wood is used for making handle of dagger.
		Cordage		
1	*Abrus precatorius* L.	Fabaceae	Stem	
2	*Ipomoea staphylina* Roem. and Schult.	Convolvulaceae	Stem	The wiry stems are used as cordage.
3	*Musa paradisiaca* L.	Musaceae	Leaves	The mid rib of the leaf and leaf sheaths are used as cordage.
		Craft material		
1	*Bambusa arundinacea* (Retz.) Willd.	Poaceae	Culms	The wiry and split stems are used in basket making.
2	*Bombax ceiba* L.	Bombacaceae	Fruit	The fruit floss is used for the stuffing of pillows and cushions etc.
3	*Phyllanthus reticulatus* Poir.	Euphorbiaceae	Stem	The stems are used to make basket.
4	*Sansevieria cylindrica* Boj. *ex* Hook.	Agavaceae	Leaves	
5	*Sansevieria roxburghiana* Schult. and Schult. f.	Agavaceae	Leaves	The leaves are used to make nets.
		Cultivated crops		
1	*Arachis hypogaea* L.	Fabaceae	Seed	The seeds are used as edible.
2	*Capsicum annum* L.	Solanaceae	Shrub	The fruits are edible.

Sl.No.	Botanical Name	Family	Part Used	Ethnobotanical Uses
3	*Citrus medica* L.	Rutaceae	Tree	Fruits are edible.
4	*Coffea arabica* L.	Rubiaceae	Tree	Cultivated for morning drinks.
5	*Coriandrum sativum* L.	Apiaceae	Whole plant	Whole plant is used as greens and seed is edible
6	*Eleusine coracana* Gaertn.	Poaceae	Seed	The seeds are edible and staple food.
7	*Lablab purpureus* (L.) Sweet	Fabaceae	Seed	The seeds are edible.
8	*Luffa cylindrica* (L.) M. Roem.	Cucurbitaceae	Tree	Cultivation for vegetables.
9	*Macrotyloma uniflorum* (Lam.) Verdc.	Fabaceae	Climber	The seeds are edible.
10	*Musa paradisiaca* L.	Musaceae	Herb	Cultivated for vegetables and fruits.
11	*Oryza sativa* L.	Poaceae	Seed	The seeds are used as staple food.
12	*Panicum miliaceum* L.	Poaceae	Herb	
13	*Panicum sumatrense* Roth and Schult.	Poaceae	Herb	
14	*Paspalum scrobiculatum* L.	Poaceae	Herb	
15	*Pennisetum americanum* (L.) Leeke	Poaceae	Herb	
16	*Piper nigrum* L.	Piperaceae	Vine	The seeds are edible.
17	*Setaria italica* (L.) P. Beauv.	Poaceae	Herb	The seeds are used as staple food.
18	*Sorghum vulgare* L.	Poaceae	Herb	
19	*Vigna mungo* (L.) Hepper.	Fabaceae	Seed	
Fiber plants				
1	*Agave americana* L.	Agavaceae	Leaves	These plant parts are used for making fiber.
2	*Corchorus olitorius* L.	Tiliaceae	Stem	
3	*Furcraea foetida* (L.) Haw.	Agavaceae	Leaves	
4	*Hibiscus cannabinus* L.	Malvaceae	Stem	
5	*Sansevieria cylindrica* Bojer	Liliaceae	Leaves	
6	*Sansevieria roxburghiana* Schult. and Schult. f.	Agavaceae	Leaves	

Contd...

Sl.No.	Botanical Name	Family	Part Used	Ethnobotanical Uses
	Fire and Fuel plants			
1	*Acacia nilotica* (L.) Del.	Mimosaceae	Wood	Wood is used as fuel.
2	*Ailanthus excelsa* Roxb.	Simaroubaceae	Wood	
3	*Alangium salviifolium* (L.f.) Wang.	Alangiaceae	Wood	Wood is used as fire wood.
4	*Albizia chinensis* (Osbec.) Merr.	Mimosaceae	Wood	
5	*Bauhinia purpurea* L.	Caesalpinaceae	Wood	Wood is used as fuel.
6	*Bauhinia racemosa* Lam.	Caesalpinaceae	Wood	
7	*Delonix regia* (Hook.) Raf.	Caesalpiniaceae	Wood	Wood is used as fire wood.
8	*Diospyros ebenum* J. Koen. *ex* Retz.	Ebenaceae	Wood	
9	*Diospyros melanoxylon* Roxb.	Ebenaceae	Wood	Wood is used as fuel
10	*Diospyros montana* Roxb.	Ebenaceae	Wood	
11	*Ehretia ovalifolia* Wight	Cordiaceae	Wood	Wood is used as fire wood.
12	*Gyrocarpus americanus* Jacq.	Hernandiaceae	Wood	Wood is used as fuel.
13	*Holoptelea integrifolia* (Roxb.) Plan.	Ulmaceae	Wood	Wood is used as fire wood.
14	*Pithecellobium dulce* (Roxb.) Benth.	Mimosaceae	Wood	
15	*Prosopis juliflora* (Sw.) DC.	Mimosaceae	Wood	
16	*Psidium guajava* L.	Myrtaceae	Wood	Wood is used as fuel.
17	*Syzygium cumini* (L.) Skeels	Myrtaceae	Wood	
18	*Tamarindus indica* L.	Caesalpiniaceae	Wood	Wood is used as fire wood.
19	*Tarenna asiatica* (L.) Kuntz. *ex* Schum.	Rubiaceae	Wood	Wood is used as fuel.
20	*Tectona grandis* L.f.	Verbenaceae	Wood	Wood is used as fire wood.
21	*Terminalia bellirica* (Gaertn.) Roxb.	Combretaceae	Wood	
22	*Terminalia chebula* Retz.	Combretaceae	Wood	

Sl.No.	Botanical Name	Family	Part Used	Ethnobotanical Uses
Fodder plants				
1	*Acacia leucophloea* (Roxb.) Willd.	Mimosaceae	Fruit	Dried fruitsare given to goats as fodder.
2	*Acacia nilotica* (L.) Del.	Mimosaceae	Fruit	The fruit pods are used as fodder for animals
3	*Amaranthus viridis* L.	Amaranthaceae	Whole plant	The whole plant is used as fodder for cattle.
4	*Cassia tora* L.	Caesalpiniaceae	Whole plant	
5	*Cyanotis cristata* (L.) D. Don	Commelinaceae	Whole plant	
6	*Cyperus rotundus* L.	Cyperaceae	Whole plant	
7	*Dactyloctenium aegyptium* (L.) P. Beauv.	Poaceae	Whole plant	
8	*Eleusine coracana* Gaertn.	Poaceae	Whole plant	
9	*Eragrotis tenella* (L.) P. Beauv. ex. Roem and Schult.	Poaceae	Whole plant	
10	*Grewia tiliifolia* Vahl	Tiliaceae	Leaves	The leaves are used as fodder for goat.
11	*Ipomoea hederifolia* L.	Convolvulaceae	Leaves	
12	*Ipomoea marginata* (Desr.) Verdc.	Convolvulaceae	Whole plant	The whole plants are used as fodder for cattle.
13	*Ipomoea staphylina* Roem. and Schult.	Convolvulaceae	Aerial parts	The aerial parts are used as fodder for goat and cow.
14	*Lablab purpureus* (L.) Sweet	Fabaceae	Whole plant	The whole plants and chaffs after threshing are used as fodder.
15	*Pithecellobium dulce* (Roxb.) Benth.	Mimosaceae	Fruit	The pods are given to goats as fodder.
16	*Setaria italica* (L.) P. Beauv.	Poaceae	Whole plant	The whole plants are used as fodder.
17	*Vigna mungo* (L.) Hepper	Fabaceae	Whole plant	
18	*Vigna trilobata* (L.) Verdc.	Fabaceae	Whole plant	
Hairdos for Women				
1	*Barleria buxifolia* L.	Acanthaceae	Flower	Flower is used in doing hairdos for women.
2	*Barleria longiflora* L.f.	Acanthaceae	Flower	
3	*Barleria prionitis* L.	Acanthaceae	Flower	

Contd...

Sl.No.	Botanical Name	Family	Part Used	Ethnobotanical Uses
4	*Celosia cristata* L. Kuntze	Amaranthaceae	Flower	
5	*Jasminum angustifolium* (L.) Willd.	Oleaceae	Flower	
6	*Jasminum cuspidatum* Rottler	Oleaceae	Flower	
Hair Nourisher				
1	*Albizia amara* (Roxb.) Boivin	Mimosaceae	Leaves	The leaves are ground with fruit pods powder of *Acacia concinna* and made a paste. This paste is used as shampoo.
2	*Andrographis echioides* (L.) Nees.	Acanthaceae	Aerial parts	The aerial part is ground to a paste and applied on scalp for hair growth.
3	*Cassia auriculata* L.	Caesalpiniaceae	Leaves	The shade dried leaves are made into powder and used as shampoo.
4	*Cynodon dactylon* (L.) Pers.	Poaceae	Whole plant	The whole plant parts are boiled with castor oil for an hour and applied on scalp to promote the hair growth.
5	*Eclipta prostrata* (L.) L.	Asteraceae	Leaves	The leaves are crushed and boiled with coconut oil and applied in grey hairs.
6	*Indigofera linnaei* Ali	Fabaceae	Leaves	The leaves are ground to a paste and applied on scalp to promote the hair growth.
7	*Lawsonia inermis* L.	Lythraceae	Leaves	The leaves are ground to a paste and applied on hair for grey hair.
8	*Limonia acidissima* L.	Rutaceae	Bark, Leaves	The bark and leaves are dried and powdered and used as shampoo.
9	*Sapindus emarginata* Vahl	Sapindaceae	Seed	The crushed seeds are used as shampoo.
10	*Wedelia chinensis* (Osbec.) Merr.	Asteraceae	Leaves	The fresh leaves are crushed and boiled with castor oil for an hour and applied on scalp to control hair loss and to promote the hair growth.

Sl.No.	Botanical Name	Family	Part Used	Ethnobotanical Uses
Hedge plants				
1	*Acacia caesia* (L.) Willd.	Mimosacea	Whole plant	Whole plant is grown as strong bio fencing around agriculture field and tribal hut.
2	*Acacia pennata* (L.) Willd.	Mimosaceae	Whole plant	
3	*Acacia torta* (Roxb.) Craib	Mimosaceae	Whole plant	
4	*Bambusa arundinacea* (Retz.) Willd.	Poaceae	Whole plant	
5	*Commiphora berryi* (Arn.) Engl.	Burseraceae	Whole plant	
6	*Commiphora caudata* (Wight and Arn.) Engl.	Burseraceae	Whole plant	
7	*Euphorbia antiquorum* L.	Euphorbiaceae	Whole plant	
8	*Lantana camara* L.	Verbenaceae	Whole plant	
9	*Lawsonia inermis* L.	Lythraceae	Whole plant	
10	*Opuntia stricta* Haw.	Cactaceae	Whole plant	
11	*Premna serratifolia* L.	Verbenaceae	Whole plant	
12	*Rhus mysorensis* Don	Anacardiaceae	Whole plant	
Household implement plants				
1	*Acacia sundra* (Roxb.) DC.	Mimosaceae	Wood	Wood is used to make handle of knife
2	*Albizia procera* (Roxb.) Benth.	Mimosaceae	Wood	Wood is used to make household implements.
3	*Artocarpus heterophyllus* Lam.	Moraceae	Wood	Wood is used to make a Churn- staff.
4	*Cassia fistula* L.	Caesalpiniaceae	Wood	Wood is used to make pestle.
5	*Chloroxylon swietenia* DC.	Rutaceae	Wood	Wood is used to make wood grinder.
6	*Dalbergia latifolia* Roxb.	Fabaceae	Wood	Wood is used to make household implements.
7	*Grewia tiliifolia* Vahl	Tiliaceae	Wood	Wood is used to make pounder.
8	*Mangifera indica* L.	Anacardiaceae	Wood	The wood is used to make a wooden grinder.
9	*Terminalia chebula* Retz.	Combretaceae	Wood	Wood is used to make ladle.

Contd...

Sl.No.	Botanical Name	Family	Part Used	Ethnobotanical Uses
House construction and Wooden material				
1	*Adina cordifolia* (Roxb.) Hook. F. ex Brandis	Rubiaceae	Wood	Wood is used to make Hen coops.
2	*Albizia lebbeck* (L.) Benth.	Mimosaceae	Wood	Wood is used in house construction.
3	*Albizia odoratissima* (L.f.) Benth.	Mimosaceae	Wood	
4	*Anogeissus latifolia* (*DC.*) *Wallich ex Guill. and Perr.*	Combretaceae	Wood	
5	*Bambusa arundinacea* (Retz.) Willd.	Poaceae	Stem	The stems are used as cross bars in hut construction. The lengthy slim stems are used in making walls of huts.
6	*Cleistanthus collinus* (Roxb.) Benth. ex Hook. F.	Euphorbiaceae	Wood	The wood is used for making cot, door and window.
7	*Dalbergia lanceolaria* L.f.	Fabaceae	Wood	
8	*Dalbergia latifolia* Roxb.	Fabaceae	Wood	
9	*Dalbergia paniculata* Roxb	Fabaceae	Wood	The wood is used in house construction
10	*Eucalyptus globulus* Labill.	Myrtaceae	Wood	
11	*Gmelina arborea* Roxb.	Verbenaceae	Wood	The timber is used to make furniture.
12	*Gmelina asiatica* L.	Verbenaceae	Wood	
13	*Grewia tiliifolia* Vahl	Tiliaceae	Wood	Wood is used for make dwellings.
14	*Melia azedarach* L.	Meliaceae	Wood	The wood is used for making door, cot and windows.
15	*Shorea roxburghii* Don	Dipterocarpaceae	Wood	
16	*Tectona grandis* L.f.	Verbenaceae	Wood	
17	*Terminalia chebula* Retz.	Combretaceae	Wood	
18	*Ziziphus mauritiana* Lam.	Rhamnaceae	Wood	
Insect/Mosquito repellent				
1	*Aloe vera* (L.) Brum.f.	Liliaceae	Whole plant	Whole plant is used as insect and mosquito repellent.
2	*Leucas aspera* (Willd.) Link	Lamiaceae	Whole plant	
3	*Albizia procera* (Roxb.) Benth.	Liliaceae	Whole plant	

Sl.No.	Botanical Name	Family	Part Used	Ethnobotanical Uses
4	*Azadirachta indica* Adr. Juss.	Meliaceae	Leaves	Leaves are ground with turmeric, water and this mixture along with cow's urine is sprayed inside the hut to keep away insect, mosquito and mites.
5	*Ocimum tenuiflorum* L.	Lamiaceae	Whole plant	The whole plant is used as an insect and mosquito repellent.
		Manure plants		
1	*Crotalaria hebecarpa* (DC.) Rudd.	Fabaceae	Whole plant	The whole plant is used as green manure.
2	*Indigofera suffruticosa* Mill.	Fabaceae	Whole plant	
3	*Melia azedarach* L.	Meliaceae	Leaves	
4	*Tephrosia purpurea* (L.) Pers.	Fabaceae	Whole plant	
5	*Tephrosia strigosa* (Dalz.) Sant. and Mahesw.	Fabaceae	Whole plant	
6	*Tephrosia villosa* (L.) Pers.	Fabaceae	Whole plant	
		Meal plate		
1	*Gyrocarpus americanus* Jacq.	Hernandiaceae	Leaves	Leaves are used as meal plates.
2	*Musa paradisiaca* L.	Musaceae	Leaves	Leaves are used as meal plates.
3	*Ricinus communis* L.	Euphorbiaceae	Leaves	The leaves are used as plates in the ritual for offering.
		Musical Instrument		
1	*Bambusa arundinacea* (Retz.) Willd.	Poaceae	Stem	The stem is used for making musical instruments.
2	*Gmelina arborea* Roxb.	Verbenaceae	Stem	The stem is used for making musical instruments.
		Plants used as Broom		
1	*Cocos nucifera* L.	Arecaceae	Leaves	The leaves are used as broom.
2	*Phoenix loureirii* Kunth	Arecaceae	Leaves	
		Poisonous plant		
1	*Cleistanthus collinus* (Roxb.) Benth. ex Hook. F.	Euphorbiaceae	Leaves and fruit	Poison
2	*Strychnos nux-vomica* L.	Loganiaceae	Fruit	
3	*Thevetia peruviana* (Pers.) Merr.	Apocynaceae	Seed	

Contd...

Sl.No.	Botanical Name	Family	Part Used	Ethnobotanical Uses
Religious plants				
1	*Aegle marmelos* (L.) Corr. Serr.	Rutaceae	Leaves	The leaves are used for offering Poojas in Sivan temples and also this plant is protected and worshiped.
2	*Aerva lanata* (L.) Juss.	Amaranthaceae	Whole plant	At the time of festivals whole herb along with leaves of *Azadirachta indica* and flowering twigs of *Cassia auriculata* are hanged at the entrance of the hamlets.
3	*Asclepias curassavica* L.	Asclepiadaceae	Flowers	The flowers are used in worshiping the God.
4	*Azadirachta indica* Adr. Juss.	Meliaceae	Twigs	Twigs are used in decoration during religious function or festivals and also this whole plant is protected and worshiped.
5	*Bambusa arundinacea* (Retz.) Willd.	Poaceae	Whole plant	The entire plants are considered as sacred.
6	*Barleria prionitis* L.	Acanthaceae	Flowers	The flowers are used in decorative purposes.
7	*Boswellia serrata* Roxb. *ex* Colebr.	Burseraceae	Whole plant	The entire trees are considered as sacred.
8	*Cassia auriculata* L.	Caesalpiniaceae	Flowers	Flowers are used in Pooja.
9	*Cassia fistula* L.	Caesalpiniaceae	Twigs, Flowers	The inflorescence twings are used in decorative purposes.
10	*Cassia montana* Heyne *ex* Roth	Caesalpiniaceae	Flowers	The flowers are used to worship the God.
11	*Catharanthus roseus* (L.) Don	Apocynaceae	Flowers	
12	*Celosia argentea* L.	Amaranthaceae	Flowers	
13	*Celosia cristata* L. Kuntz.	Amaranthaceae	Flower	
14	*Citrus medica* L.	Rutaceae	Fruit	The fruit is used to worship the God.
15	*Crateva religiosa* auct. non Forst.	Capparidaceae	Whole plant	The trees are considered as scared.
16	*Cynodon dactylon* (L.) Pers.	Poaceae	Leaves	The leaves are used to worship the God.
17	*Ficus benghalensis* L.	Moraceae	Whole plant	The plant is protected and worshiped.
18	*Ficus religiosa* L.	Moraceae	Whole plant	
19	*Gomphrena globosa* L.	Amaranthaceae	Flowers	The flowers are used as decoratives and to worship the God.

Sl.No.	*Botanical Name*	*Family*	*Part Used*	*Ethnobotanical Uses*
20	*Holarrhena pubescens* (Buch- Ham.) Wall. *ex* Don	Apocynaceae	Stem	Stem is used in all traditional religious festivals and religious ceremonies.
21	*Mangifera indica* L.	Anacardiaceae	Twigs	A chain of leafy twigs with leaves of *Azadirachta indica* are hanged in front of the huts and entrance of the hamlets during festival time.
22	*Naringi crenulata* (Roxb.) Nicol.	Rutaceae	Whole plant	The whole plant is protected and worshiped.
23	*Nerium oleander* L.	Apocynaceae	Flowers	The flowers are used to worship the God.
24	*Ocimum tenuiflorum* L.	Lamiaceae	Leaves, Flowers	The leaves and flowers are used in Pooja and also this plant is protected and worshiped.
25	*Piper betle* L.	Piperaceae	Leaves	The leaves are used to worship the God.
26	*Plumeria rubra* L.	Apocynaceae	Flowers	The flowers are used to worship the God.
27	*Wrightia tinctoria* (Roxb.) R. Br.	Apocynaceae	Twigs and Stem	Twigs and mature stem are used in all traditional religious festivals.
		Roofing/Thatching plants		
1	*Bambusa arundinacea* (Retz.) Willd.	Poaceae	Leaves	The leaves are used in roof thatching.
2	*Cymbopogon martinii* (Roxb.) Watson	Poaceae	Whole plant	The whole plants are used in roof thatching.
3	*Cyperus rotundus* L.	Cyperaceae	Leaves and branches	Leaves and branches are used in roofing and thatching.
4	*Eleusine coracana* Gaertn.	Poaceae	Whole plant	The plants after threshing are used as roof thatching materials in huts.
5	*Ixora pavetta* Andr.	Rubiaceae	Branches	The branches are used for thatching.
6	*Ophiuros exaltatus* (L.) Kuntz.	Poaceae	Leaves	Leaves are used for roofing and thatching.
7	*Oryza sativa* L.	Poaceae	Leaves and branches	Leaves and branches are used for roofing and thatching.
8	*Sorghum vulgare* L.	Poaceae	Aerial part	Aerial part is used for roofing and thatching.
		Soap/Detergent		
1	*Sapindus emarginata* Vahl	Sapindaceae	Seed	The crushed seeds are used for washing and bathing.

Contd...

Sl.No.	Botanical Name	Family	Part Used	Ethnobotanical Uses
Sorcery-Aphrodisiac				
1	*Evolvulus alsinoides* (L.)L.	Convolvulaceae	Aerial part	Brush up body with aerial part to stimulate sexual desire.
2	*Ocimum canum* Sims	Lamiaceae	Aerial part	
3	*Randia dumetorum* (Retz.) Poiret	Rubiaceae	Twigh	Brush up body with twigs to stimulate sexual desire.
Sorcery- Superstitious belief				
1	*Abrus precatorius* L.	Fabaceae	Root	The root is tied in front of the house to wind off evil spirits.
2	*Aerva lanata* (L.) Juss.	Amaranthaceae	Whole plant	A bunch of herbs are hanged in-front of the huts to ward off evil spirits.
3	*Aloe vera* (L.) Burm.f.	Liliaceae	Whole plant	The entire plant is hanged inside of the huts to wind off evil spirits.
4	*Anisomeles malabarica* (L.) R.Br. ex Sims	Lamiaceae	Whole plant	The leaf twigs are hanged inside of the huts to wind off evil spirits.
5	*Azadirachta indica* Adr. Juss.	Meliaceae	Twig	The twigs are hanged up in-front of their huts to ward off evil spirits.
6	*Calotropis gigantea* (L.) R. Br.	Asclepiadaceae	Root and Twigs	The root is tied around waist of human to wind off evil spirits. The leafy twigs are hanged up in front of the huts to wind off evil spirits. The stem is used to wind off evil spirits from the affected person.
7	*Crotalaria verrucosa* L.	Fabaceae	Root	The root is tied around waist of human to keep off evil spirits.
8	*Pterocarpus marsupium* Roxb.	Fabaceae	Exudates	The stem resin is applied on the forehead of the children to wind off evil spirits.
9	*Sansevieria cylindrica* Bojer ex Hook.	Agavaceae	Leaves	The gently roasted leaf is used by priest to keep wind off evil spirits from affected person.

Sl.No.	*Botanical Name*	*Family*	*Part Used*	*Ethnobotanical Uses*
To away the poisonous insect and snake				
1	*Andrographis paniculata* (Burm.f.) Wallich ex Nees	Acanthaceae	Leaves	The leaves are crushed and applied on body to protect from snake bites.
2	*Moringa oleifera* Lam.	Moringaceae	Stem	The paste of stem bark and *Allium sativum* bulbs is mixed with water and sprayed around the huts to avoid entry of the snakes and poisonous insects.
3	*Plectranthus amboinicus* (Lour.) Spreng.	Lamiaceae	Leaves	The leafy twigs are put inside the huts and used as insect repellent.
4	*Toddalia asiatica* (L.) Lam.	Rutaceae	Leaves	The fresh leaves are crushed and applied on body to protect from snake bites.
Tooth brush/tooth powder				
1	*Acacia nilotica* (L.) Del.	Mimosaceae	Young stem, Bark	The dried bark is powdered and used as tooth powder. Young stem is used as tooth brush.
2	*Achyranthes aspera* L.	Amaranthaceae	Stem, Root	The stem is used as a tooth brush and the root ash is used as tooth powder.
3	*Azadirachta indica* Adr. Juss.	Meliaceae	Young Twigs	Young twig is used as tooth brush.
4	*Ficus benghalensis* L.	Moraceae	Aerial root	Aerial root is used as tooth brush.
5	*Ficus racemosa* L.	Moraceae	Young Twigs	Young twig is used as tooth brush.
6	*Psidium guajava* L.	Myrtaceae	Young Twigs	
7	*Terminalia chebula* Retz.	Combretaceae	Fruit	The powdered dried fruit used as tooth powder.
To rid off lice				
1	*Vitex negundo* L.	Verbenaceae	Leaves	The leaf paste is applied on head upto 30 minutes and taken bath.
Walking sticks				
1	*Bambusa arundinacea* (Retz.) Willd.	Poaceae	Stem	The stem is used as walking sticks.
2	*Cleistanthus collinus* (Roxb.) Benth. ex Hook. F.	Euphorbiaceae	Stem	
3	*Dodonaea angustifolia* L.f.	Sapindaceae	Wood	

*italica*and *Vigna mungo* are cultivated by Malayali tribals in Patchamalai hills. The species such as *Citrus medica, Mangifera indica, Carica papaya, Luffa cylindrical, Artocarpus heterophyllus* and *Musa paradisiaca* were grown for fruits in the *Malayali* tribal hamlets. In the study area, 6 plant species of fiber yielding plants were observed. Their cordage requirements for different purposes, *i.e.,* from the high task of building the houses to the task of just binding the sticks collected for the purpose of fuel were met with the plants locally available. There several species such as *Agave americana, Furcraea foetida, Hibiscus cannabinus, Sansevieria cylindrical and Sansevieria roxburghiana.* The similar use of *Hibiscus cannabinus* (Ayyanar and Ignacimuthu, 2009), *Agave Americana* (Nirmal Kumar *et al.,* 2007) *Corchorus olitorius, Furcraea foetida, Sansevieria roxburghiana* (Sagu *et al.,* 2013) were recorded.Twenty two plant species were used for fire and fuel.Some of the species such as *Acacia nilotica, Albizia chinensis, Alangium salviifolium, Bauhinia purpurea, Diospyros ebenum and Gyrocarpus americanus* were commonly used as fire and fuel wood in the study area. *Acacia nilotica* (Nirmal Kumar *et al.,* 2007; Packiaraj *et al.,* 2014), *Gyrocarpus americanus* (Packiaraj *et al.,* 2014) were also recorded for fire and fuel wood. In the present study, a total of 18 plant species were recored as fodder value plants. Some of the important species are *Acacia leucophloea, Acacia nilotica, Amaranthus viridis, Cyperus rotundus, Ipomoea staphylina, Grewia tiliifolia, Dactyloctenium aegyptium, Lablab purpureus,*etc, were documented in the study area. The similar use of fodder plants such as *Amaranthus viridis* and *Cyperus rotundus* were also reported by Panhwar and Abro (2007). Six plant species were used for hairdos by women in the area selected for study. The species of *Barleria buxifolia, Barleria longiflora, Barleria prionitis, Celosia cristata, Jasminum angustifolium* and *Jasminum cuspidatum* hitherto unknown to hairdos for women were recorded in the study area. Ten plant species are used for hair nourisher. The species such as *Albizia amara, Lawsonia inermis, Eclipta prostrate, Cassia auriculata, Sapindus emarginata, Cynodon dactylon etc.,* were used for hair nourisher in study area. Among them, *Lawsonia inermis* and *Sapindus emarginata* were previously reported as hair nourisher (Punjani and Kumar 2003).

Hedge is the boundary of the open space surrounding the hut or hamlet. In the present study, 12 plant species play their role as hedge for tribal people's huts and agriculture field. Some plant species such as, *Bambusa arundinacea, Premna serratifolia, Lantana camara, Opuntia stricta, Euphorbia antiquorum, Commiphora caudate, Commiphora berry etc.,* were recored in the present study. *Lantana camara* (Ayyanar and Ignacimuthu, 2009) and *Bambusa arundinacea* (Packiaraj *et al.,* 2014) were also recorded for the similar uses. Nine plant species are used for making household implements such as wood grinder, pestle, knife handle, churn- staff, pounder and ladle. Plants such as *Acacia sundra, Albizia procera, Chloroxylon swietenia, Mangifera indica, Dalbergia latifolia, etc.,* were recored for making house hold implements in the present study. Among them, *Albizia procera* (Nirmal Kumar *et al.,* 2007), *Chloroxylon swietenia* (Panda *et al.,* 2005) were also recorded from the previous studies.Eighteen plant species were used for making wooden material and in house construction and 8 plant species were used for roof thatching. The plant species were also used in making other wooden materials like doors, windows and minor furnitures (chairs, stools). These plants were utilized by the tribal people in many ways. Their dwellings, temples, goat-pens and hen-coops were constructed only by using wood materials collected from the under study area. Timber is the most valued forest product, as in most societies, it remains fundamental for construction of temporary shelters of permanent houses.For construction of huts, strong branches of woods are used to erect boles. This is then plastered with red soil to give the appearance of wall. The roof is covered with the aerial parts of *Sorghum vulgare* and poles and rafters from strong plants are preferred for slightly larger constructions. Some of the plant species such as *Bambusa arundinacea, Adina cordifolia, Albizia lebbeck, Dalbergia latifolia, Anogeissus latifolia, etc.*, are used as poles and rafters in the area selected for study. Plants like *Anogeissus latifolia, Bambusa arundinacea, Eucalyptus globulus* and *Terminalia chebula* (Ayyanar and Ignacimuthu, 2009), *Adina cordifolia, Albizia lebbeck* and *Dalbergia latifolia* (Nirmal Kumar *et al.*, 2007; Packiaraj *et al.*, 2014 and Panhwar and Abro, 2007) were also reported.

The following 5 plant species were used as insect/mosquito repellent in the present study: *Aloe vera, Leucas aspera, Albizia procera, Azadirachta indica and Ocimum tenuiflorum.* In the previous studies, *Azadirachta indica* (Pradhan and Rahaman, 2011) and *Leucas aspera* (Pattanayak and Dhal, 2015) were also recorded for the use of mosquito repellents. Six plant species were used as greenmanure. A part of agriculture is done in plantation forms in and around hamlets and also around the surrounding huts of tribal people. Hence, the use of manure in a routine way was not followed in their agricultural practices. The recognition of certain species as a source of manure relates to early agronomic innovations but this practice is extremely rare now and limited to the low elevations and plains. The present study revealed that the species like *Tephrosia strigosa, T. villosa, T. purpurea, Indigofera fruticosa* and *Crotalaria hebecarpa*were used as green manure. In other parts of India, these plants were also recorded for their usage as green manure (Singh, 2013). The leaves of three plant species are used as meal plate.Such plant species are *Gyrocarpus americanus, Musa paradisiaca* and *Ricinuscommunis*. The similar use of *Musa paradisiaca* was also reported by Reddy *et al.* (2008) and Sharma and Pegu (2011).Two plant species were used for making musical instrument. The species of *Bambusa arundinacea* and *Gmelina arborea* are used for making musical instrument. Similarly, *Gmelina arborea* (Ayyanar and Ignacimuthu, 2009; Ajesh and Kumuthakalavalli, 2013) and *Bambusa arundinacea* (Packiaraj *et al.*, 2014) were also reported. In addition to this, 2 plant species were used for making broom. The plant species of *Cocos nucifera* and *Phoenix loureirii*were used for making broom in the present study this observation on was also recorded by Rasigam and Jeeva (2013) and Reddy *et al.* (2008). In the under study, the plant parts of 3 plant species were recorded as poisonous. The present study documented the leaves and fruits of *Cleistanthus collinus*, the fruit of *Strychnos nux-vomica* and the fruit and seed of *Thevetia peruviana* are poisonous in nature. Ranjithkumar *et al.* (2014) reported that the plant species of *Cleistanthus collinus* was highly poisonous.In the present study, 27 plant species were used for religious purposes as decorative, worship, pooja and sacred. The plants flowers and leaves were used during worship for pooja and decoration. The sacred groves were the tracts of virgin forest that were left untouched by the local inhabitants, and were protected by the local people due to their cultural and religious beliefs. Some of the important plant species such as *Ficus benghalensis, F. religiosa, Catharanthus roseus, Azadirachta indica, Mangifera indica, Naringi crenulata, Nerium oleander, Ocimum tenuiflorum, Piper betle, Plumeria rubra, Wrightia tinctoria, etc.*, were recorded in the study area. Plants such as *Mangifera indica, Nerium oleander, Ficus benghalensis, Aerva lanata, Piper betle, Plumeria rubra, Aegle marmelos, Ficus religiosa, Catharanthus roseus, Azadirachta indica, Ocimum tenuiflorum* and *Cynodon dactylon* were also recorded in an ethnobotanical survey amongst the ethnic groups inhabiting in other parts of India (Sarma and Devi, 2015; Sahu *et al.*, 2013b; Sharma *et al.*, 2014; Packiaraj *et al.*, 2014; Panda *et al.*, 2005; Hitesh and Patel, 2013; Panhwar and Abro, 2007 and Kumar *et al.*, 2007). The species *Sapindus emarginata*was used as soap for bathing and washing the cloth in the current study. This was also recorded in Kerala (Singh, 2012). An aphrodisiac is defined as an agent that arouses sexual desire. The species of *Evolvulus alsinoides, Ocimum canum and Randia dumetorum* were recorded for aphrodisiac in the study area. In these, *Evolvulus alsinoides* was also reported in Kerala for similar use (Singh, 2012). Tribes used 9 plant species for ward off evil spirits. Some important species such as *Abrus precatorius, Anisomeles malabarica, Azadirachta indica, Pterocarpus marsupium, Calotropis gigantean, etc.*, were recorded in the study area. Similarly, *Anisomeles malabarica, Calotropis gigantean, Pterocarpus marsupium* were reported by Packiaraj *et al.* (2014) and Ajesh and Kumuthakalavalli (2013) and *Aegle marmelos* was also reported by Kumar *et al.* (2006). The plant species such as *Andrographis paniculata, Moringa oleifera, Plectranthus amboinicus* and *Toddalia asiatica*were used for insect bites and as snake repellent in the study area. The same plant species were also noted for insect and snake repellent in Virudhunagar district (Packiaraj *et al.*, 2014). The plant parts of different species such as *Acacia nilotica, Ficus benghalensis, Azadirachta indica, Psidium guajava, Achyranthes aspera, Psidium guajava,* and *Terminalia chebula* were recorded for their use as tooth brush and tooth powder. These plant species were also recorded for the similar uses by various researchers (Singh *et al.*, 2013; Manikandan, 2013; Pradeep Kumar, 2014; and Deka and Nath, 2014). The leaves of *Vitex negundo*was used to get rid off lice and the stem of *Bambusa arundinacea, Cleistanthus collinus* and *Dodonaea angustifolia* for walking sticks are used by Malayali tribals in the area selected for the study.

Conclusion

Ethnobotanical surveys had developed focus on the relationship between the use of plants and indigenous communities. The ethnobotanical study has also important role in conservation of natural sources, culture and biodiversity of important plant species. Hence, the present study was carried out to document the ethnobotanical knowledge of an unexplored area, Melur of Bodha hills, Namakkal District, Tamil Nadu, India.

Acknowledgement

I thank Mr. S. Kuppusamy, Melur and Kelur panchayat president of Bodha hills for his sincere work and guiding me to visit all tribal hamlets and introducing tribal peoples and traditional healers in Bodha hills, Southern Eastern Ghats, Namakkal District, Tamil Nadu, India.

References

Ajesh, T.P. and Kumuthakalavalli, R. 2012. Ethnic herbal practices for gynaecological disorders from urali tribes of Idukki district of Kerala, India. *International Journal ofPharmacy and Life Sciences* 3(12): 2213–2219.

Alcorn, J.B. 1995. The scope and aims of ethnobotany in a developing world. In: Schultes, R.E. and Reis, S.V. (Eds.) *Ethnobotany – Evolution of Discipline.* Dioscorides Press, Portland, Oregon.

Anburaja, V. and Nandagopalan, V. 2012. Agricultural activities of the Malayali tribal for subsistence and economic needs in the mid elevation forest of Pachamalai hills, Eastern Ghats, Tamil Nadu. *International Journal of Agricultural Research,Innovation and Technology* 2 (1): 32–36.

Ayyanar, M. and Ignacimuthu, S. 2009. Plants used for non-medicinal purposes by the tribal people in Kalakad Mundane Thurai Tiger Reserve, Southern India. *Indian Journal ofTraditional Knowledge* 9(3): 515–518.

Busnagi Rajannan, 1932. Salem cyclopedia (A cultural and historical dictionary of Salem district: Tamil nadu). Institute of Kongu studies, Salem, Tamil Nadu,1-365.

De, J.N. 1968. Ethnobotany a new science in India. *Science Culture.* 34: 326 – 328.

Deka, K. and Nath, N. 2014. Application of local health traditional knowledge in oral health and hygiene among the ethnic tribes of Nalbari and Barpeta districts of western Assam (North East India). *International Journal of Pure and Applied Bioscience* 2 (5): 107–114.

Dimbleby, G.W. 1978. *Plants and archaeology.* Humanities Press, Atlantic Highlands, NJ.

Ford, R.I. 1979. Paleobotany in American archaeology. *Advan. Archaeol. Method Theory* 2: 285 – 336.

Gamble JS. 1915-1935. Flora of the Presidency of Madras, Vols, I-III, Adlard and Co., London.

Hitesh, R.P. and Patel, R.S. 2013. Ethnobotanical plants used by the tribes of R.D.F. Poshina forest range of Sabarkantha District, North Gujarat, India. *International Journal of Scientific and Research* 3(2): 1-8.

Jain, S.K. 1981. *Glimpses of Indian Ethnobotany.* Oxford and IBH Publishing Co., New Delhi.

Kumar, A., Tewari, D.D. and Tewari, J.T. 2006. Ethnomedicinal knowledge among Tharu tribe of Devipatan division. *Indian Journal of Traditional Knowledge*: 310-313.

Kumar, P.P., Ayyanar, M. and Ignacimuthu, S. 2007. Medicinal plants used by *Malasar* tribes of Coimbatore district, Tamil Nadu. *Indian Journal of Traditional Knowledge* 6(4): 579-582.

Manikandan, S. 2013. Ethnomedicinal Flora of Ivanur Panchayat in Cuddalore District, Tamil Nadu, India. *International Journal of Research in Plant Science* 3(2): 39-46.

Maru, R. N. and Patel, R. S. 2012. Certain plants used in house hold instruments and agriculture impliments by the tribals of Jhalod taluka, Dahod district of Gujarat, India. *Life Sciences Leaflets*: 417–426.

Matthew KM. 1981-1984. *The Flora of Tamil Nadu Carnatic*, Vols 1-3, The Rapinant Herbarium, St. Joseph's College, Thiruchirapalli.

Nirmal Kumar, N., Kumar, R.N., Patil, N. and Soni, H. 2007. Studies on plant species used by tribal communities of Saputara and Purna forests, Dangs district, Gujarat. *Indian Journal ofTraditional Knowledge* 6(2): 368-374.

Packiaraj, P., Suresh, K. and Venkadeswaran, P. 2014. Plant species with ethno botanical importance other than medicinal in Paliyars community in Virudhunagar District, Tamil Nadu, India. International Journal on Applied Bioresearch 20: 6–9.

Panda, T., Panigrahi, S.K. and Padhy, R.N. 2005. A sustainable use of phytodiversity by the Kandha tribe of Orissa. *Indian Journal of Traditional Knowledge* 4(2): 173–178.

Panhwar, A.Q. and Abro, H. 2007. Ethnobotanical studies of Mahal Kohistan (Khirthar national park). *Pakistan Journal of Botany* 39(7): 2301–2315.

Paroda, R.S. 2000. Opportunities and challenges for promoting the essential oil beyond; An Indian perspective. *Ind. Perfum.* 44 (3): 93 – 100.

Pattanayak, B. and Dhal, N. K. 2015. Plants having mosquito repellent activity: An ethnobotanical survey. *International Journal of Research and Development inPharmacy and Life Sciences* 4(5): 1760- 1765.

Pradeep Kumar, R. 2014. Ethnomedicinal plants used for oral health care in India. *International Journal of Herbal Medicine* 2 (1): 81–87.

Pradhan, B. and Rahaman, C.H. 2011. Studies on plant wealth associated with folk medicine in Birbhum District, West Bengal, India. *The Socioscan* 3: 17–20.

Punjani, B.L. and Kumar, V. 2003. Plants used in traditional phytotherapy for hair care by tribals in Sabarkantha district, Gujarat, India. *Indian Journal of TraditionalKnowledge* 2(1): 74- 78.

Ranjithkumar, A., Chittibabu, C.V. and Renu, G. 2014. Ethnobotanical investigation on the Malayali tribes in Javadhu hills, Eastern Ghats, South India. *Indian Journal of Medicine and Healthcare* 3(1): 322–332.

Rasigam, L. and Jeeva, S. 2013. Indigenours brooms used by the aboriginal inhabitants of Nilgiri biosphere, Reserve, Western Ghats, India. *Indian Journal of Natural Productsand Resources* 4(3): 312–316.

Reddy, K.N., Chiranjibi Pattanaik, Reddy, C.S., Murthy, E.N. and Raju, V.S. 2008. Plants used in traditional handicrafts in North East Andhra Pradesh. *Indian Journal ofTraditional Knowledge* 7(1): 162- 165.

Sahu, P.K., Kumari, A., Sao, S., Singh, M. and Pandey, P. 2013a. Sacred plants and their Ethno-botanical importance in Central India: A mini review. *International Journal of Pharmacy and Life Sciences* 4(8): 2910–2914.

Sahu, S.C., Pattnaik, S.K., Dass, S. S. and Dhal, N. K. 2013b. Fibre yielding plant resources of Odisha and traditional fibre preparation knowledge- An overview. *Indian JournalNatural Products and Resources* 4(4): 339–347.

Sarma, J. and Devi, A. 2015. Study on traditional worshiping plants in Hindu religion from Nalbari and Sonitpur districts of Assam. *International Journal of Scientific andResearch Publications* 5(5): 1–5.

Schultes, R.E. 1962. The role of ethnobotanists in the search of new medicinal plants. *Lloydia* 25: 257 – 266.

Sharma, M., Sharma, C. L. and Debbarma, J. 2014. Ethnobotnical studies of some plants used by Tripuri tribe of Tripura, NE India with special reference to magico religious beliefs. *International Journal of Plant, Animal and Environmental Sciences* 4(3): 518–528.

Sharma, U.K. and Pegu, S. 2011. Ethnobotany of religious and supernatural beliefs of the Mising tribes of Assam with special reference to the 'Dobur Uie'. *Journal of Ethnobiology and Ethnomedicine* 7: 16.

Singh, A., Satanker, N., Kushwaha, M., Disoria, R. and Gupta, A.K. 2013. Ethnobotany and use of non graminaceous foage species of Chitrakoot region of Madhya Pradesh. *Indian Journal Natural Products and Resources* 4(4): 425–431.

Singh, B., Sinha, B.K., Phukan, S.J., Borthakur, S.K. and Singh, V.H. 2012. Wild edible plants used by Garo tribes of Nokrek Biosphere Reserve in Megalaya, India. *IndianJournal of Traditional knowledge*: 166-171.

Singh, R., Singh, S., Jeyabalan, G. and Ali, A. 2012. An Overview on traditional medicinal plants as Aphrodisiac agent. *Journal of Pharmacognosy and Phytochemistry* 1(4): 43– 56.

2018, Ethnomedicinal Plants: A Biodiversity Treasure Pages 215–227
Editors: V.R. Mohan, A. Doss, P.S. Tresina and V. Sornalakshmi
Published by: ASTRAL INTERNATIONAL PVT. LTD., NEW DELHI

Chapter 6

Ethnobotanical Study of Medicinal Plants Used by Kani Tribals in Vamanapuram Block, Nedumangadu Taluk, Kerala, India

***B. Parthipan*[1], *C. Biju*[1] *and M. Johnson*[2#]**

[1]PG and Research Department of Botany,
South Travancore Hindu College,
Nagercoil – 629 002, Kanyakumari District, Tamil Nadu
[2]Centre for Plant Biotechnology, Department of Botany,
St. Xavier's College (Autonomous), Palayamkottai, Tamil Nadu
[#]Corresponding Author E-mail: ptcjohnson@gmail.com

Introduction

Even from the pre-historic era almost all civilizations have employed plants in the treatment of human sickness (Chah *et al.*, 2006). Over the years, the traditional Indian medicine has become rather well codified in a holistically oriented practice or system of health care as "Ayurvedic medicine". During the last few decades, all over the world they are paying great attention to the studies of a newer branch of science, "Ethnobiology", especially to tribal medicine or Ethnomedicine. In India, the use of different parts of several medicinal plants to cure specific ailments has been in vogue from ancient times. Our knowledge of medicinal plants has mostly been inherited traditionally. Use of plants for curing various ailments are not confined to the doctors only but has been known to several households as well. There are many interesting and sometimes astonishing things to learn from the collectors of medicinal herbs. Spreading and preserving this knowledge on medicinal plants and their uses has become important for human existence.

There is a growing tendency all over the world to shift from synthetic to natural products mainly due to their exponential availability and cheap rate of expenditure. Indian traditional systems of medicines and their practices have been transmitted to the present society because the medicinal aspects of enumerable number of plants lay hidden in the mind of traditional vaidyars. It should be elicited out for the welfare and well-being of modern society. Indigenous knowledge is the collective wisdom of individuals, families, tribes, communities and societies living in the specific geographic locations on a wide spectrum of human activities in relation to their immediate environment. There are altogether 427 tribal communities all over India. According to 1981 Census, the population scheduled tribes population in the country was 53.8 million constituting about 7.5 per cent tribal population (Ragupathy and Mahadevan 1991). "Kanikaran" commonly known as "Kanis" means 'hereditary proprietor of the land'. They are a jungle tribe distributed throughout the mountains of South-Travancore.

During the last few decades there has been an increasing interest in the study of medicinal plants and their traditional use in different parts of India. In the recent years, number of reports on the use of plants in traditional healing by either tribal people or indigenous communities of India is increasing (Savithramma *et al.*, 2007, Pattanaik *et al.*, 2008, Kosalge and Fursule 2009, Namsa *et al.*, 2009, Upadhyay *et al.*, 2010). Several workers reported uses of plants to cure various ailments by rural and tribal people inhabiting various regions of Tamil Nadu (Rajendran *et al.*, 2003, Muthukumarasamy *et al.*, 2003a, Muthukumarasamy *et al.*, 2003b, Ganesan *et al.*, 2005, Ignacimuthu *et al.*, 2006, Kottaimuthu 2008, Shanmugam *et al.*, 2011). Nearly hundred papers have been published and several unpublished reports are also available with ethnomedicinal claims among different tribal communities of Tamil Nadu. A perusal of the literature reveals that a few ethnomedicinal studies among Paliyar tribals have been reported from the various districts of Tamil Nadu (Arinathan *et al.*, 2003, Ayyanar and Ignacimuthu 2005a, Ayyanar and Ignacimuthu 2005b, Ignacimuthu *et al.*, 2008). But there is no report available in Vamanapuram Block of Nedumangadu Taluk, Kerala State. Nature has blessed the Vamanapuram Block of Nedumangadu Taluk, Kerala State with a very rich botanical and ethnomedicinal wealth that has been exploited continuously by the tribals. Kanis are the major inhabitants of this block. As far as our knowledge is concerned, there is no perfect ethnomedicinal survey of Vamanapuram area. Hence an attempt has been made in the present investigation to study the ethno-medico botanical survey of Vamanapuram Block of Nedumangadu Taluk.

Materials and Methods

Geographical Description of the Study Area

The state of Kerala lies along the south-west coast of India, between latitudes 8° 18′ and 12° 48′ N and longitudes 74° 52′ and 77° 22′ E. It is bounded by Karnataka in the North, Tamil Nadu in the East and South and Arabian Sea in the West. Administratively the state consists of 14 districts. The total land area of the state is 38,364 sq.km covering 1.8 per cent of countries geographical area (Ravikumar *et al.*, 2005).

The forest cover of the state is 10336 sq.km which forms 26.6 per cent of the total land area. Topographically, the state can be divided into 3 parallel strips in the north-south direction namely coastal region, mid land region and mountaineous regions. Vamanapuram block of Nedumangadu taluk, Palode and Kulathupuzha range of Thiruvananthapuram forest division, of Western Ghats, Kerala, India, is situated between 8° 49′ north latitude and 77° 3′ and 77° 8′ east longitude (Figures 6.1 and 6.2).

The study area comprises a group of small hills and hillocks. This block is situated at about 35 km from Thiruvananthapuram city and is surrounded by Taluks - Kottarakara on the north-west, Chirayinkeezhu on the west, Thiruvananthapuram on the south-west, Neyyattinkara on the south-east and Pathanapuram on the north-east (Figure 6.2).

Tribal Settlements in Vamanapuram Block

Nedumangadu taluk has three tribal blocks *viz.*, Vamanapuram, Vellanadu and Nedumangadu. Vamanapuram block consists of 9 Panchayaths *viz.*, Peringamala, Nanniyodu, Manickal, Pullampara, Pangodu, Pazhayakunnummel, Polimath, Kazhakkuttom and Sreekaryam. There are 68 settlements reported in this

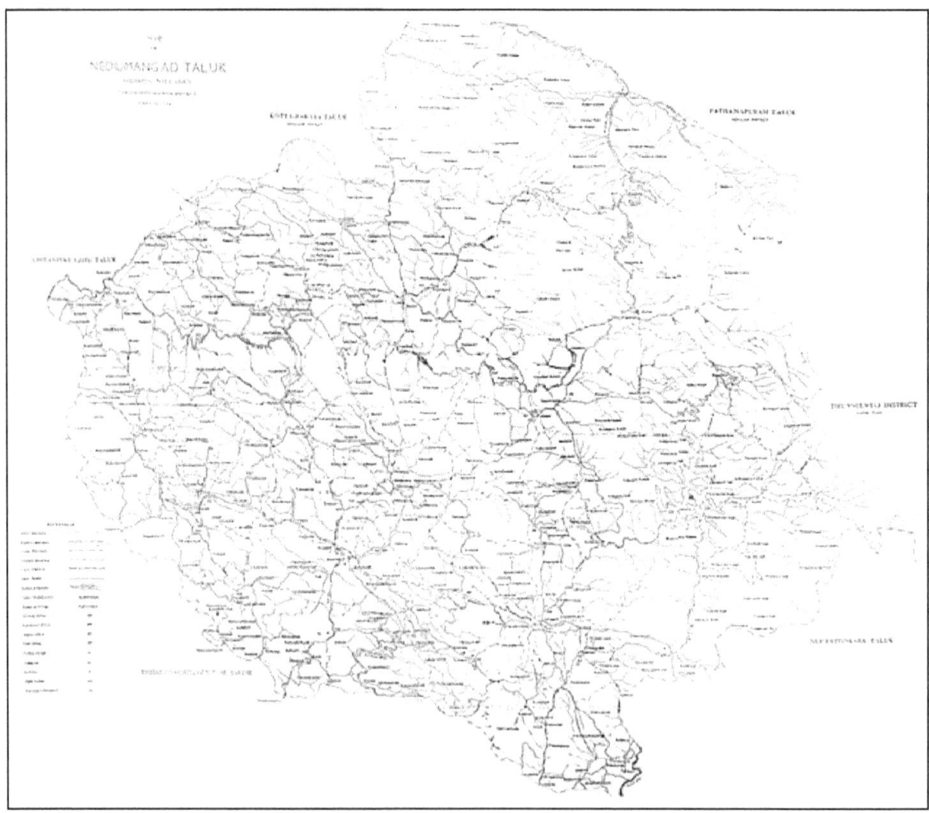

Figure 6.1: Location of the Study Area.

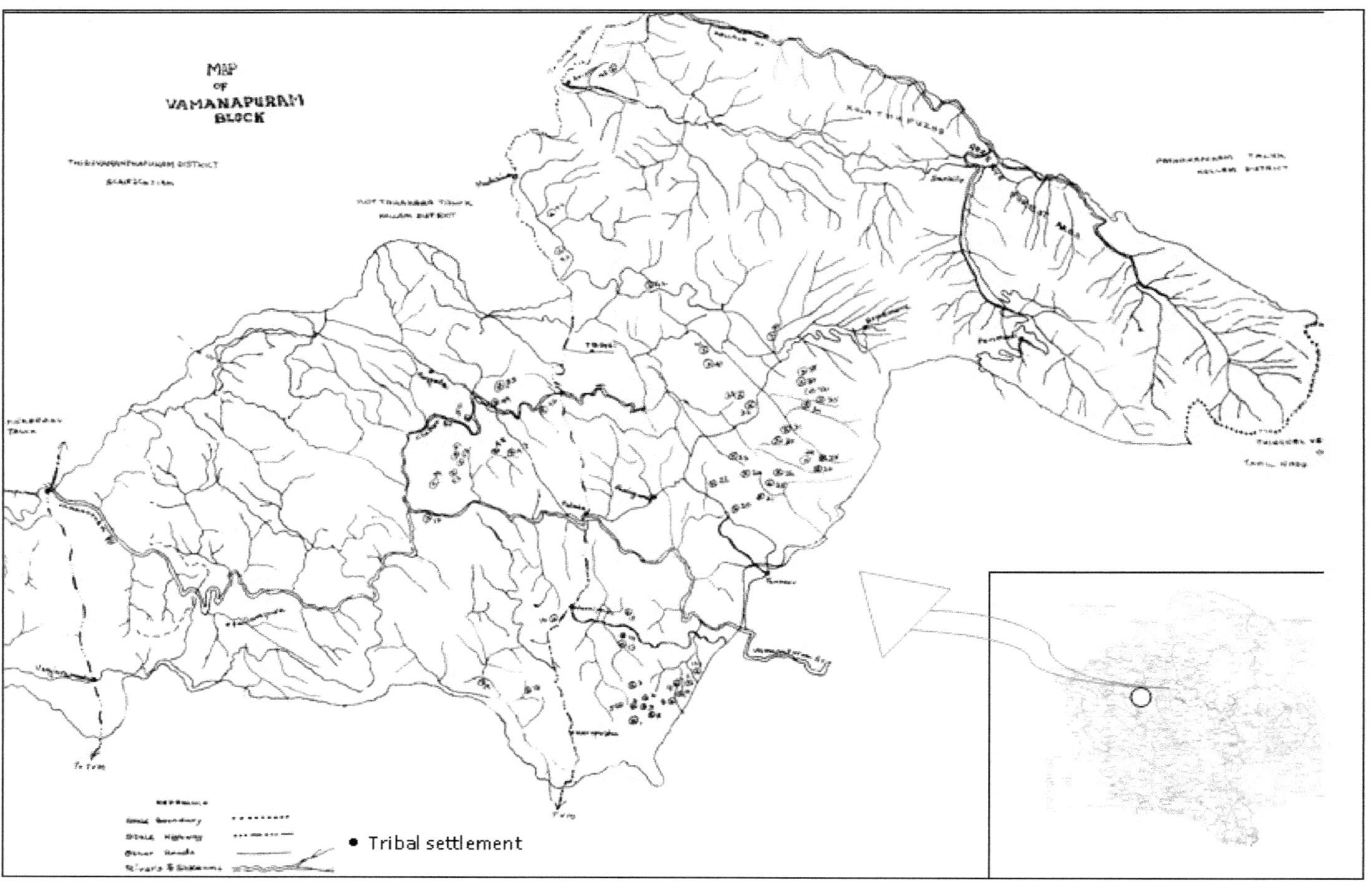

Figure 6.2: Location of the Tribal Settlements in Vamanapuram Block.

Vamanapuram block (Figure 6.2). Of which Peringamala, Nanniyod and Pangodu Panchayaths are the main tribal settlements in Vamanapuram Block.

Plant Collection and identification

The ethno-medico-botanical exploration of Kani tribal pockets of Vamanapuram Block of Nedumangadu Taluk, Kerala was collected and the ethnobotanical survey was carried out among the Kani population residing in this area. Various interviews were carried out with several elder and experienced men and women to get information on medicinal plants used by them. Queries were repeatedly made from the head Ramakrishnan Kani (Muttukani) and different persons (Appukuttan Kani, Kunjukrishnan Kani, Sudarsanan Kani, Pachi, Gopalan Kani, Sundari, Eechan Kani, Mallan Kani and Kuttappan Kani) inorder to verify the accuracy of the information.

The fresh collected plant materials at all times, during the field trips were exhibited in-front of the Kanis to get more information. The photographs of selected plants were also taken during the field trips. The data were recorded in the field note books and audio cassettes using the tape recorder. Polythene bags were used to keep the collected materials in fresh conditions. Hand lens was used for recording the morphological characters in the field note books. The collected plants were brought to the laboratory for identification.

The medicinal plants utilized by these tribals mostly come from uncultivated area based on the information provided by tribals. Plant specimens were collected; air dried and mounted on herbarium sheets according to the field and herbarium techniques. Collected plants were identified correctly and confirmed by referring the various flora *viz.*, The Flora of the Presidency of Madras (Gamble and Fischer 1957) and Excursion Flora of Central Tamil Nadu, India (Matthew 1991).

Identified plants were verified and confirmed by the herbarium of TBGRI, Palode, Thiruvananthapuram. The plant specimens and their medicinal uses, parts used and ailments for which the particular plant is used have been thoroughly verified (Jain 1991). The collected ethnomedicinal plant species from the Vamanapuram block were arranged in alphabetical order. The ethno-medico-botanical data include botanical name of the plant, family, tribal names, medicinal claims/uses and name of the plant collector and field numbers of voucher specimens mentioned.

Mode of Plant Medicine Preparation by *Kanis*

During the survey it was observed that the tribal populations were entirely depending on herbal medicines. The utilization may be in different forms like extract, decoction, paste, infusion, mixture, powder and fumes.

Juice Extract

Cleaned plant parts are ground with water. The extract is filtered and filtrate is used as extract.

Decoction

Decoction is extracted by boiling the plant parts with water at 1:10 ratio and the quantity of the water is reduced to about one fourth of its original volume and the liquid is decanted.

Paste

Both fresh and dried plants and plant parts, in general, are made into paste on a stone and is used both internally and externally.

Infusion

It is partially the soaking of dry medicinal plants or plant parts in fresh water at normal temperature.

Mixture

This is either the combination of the powders of dried materials or the combination of the decoctions and extracts of plant materials at certain ratio.

Powder

The dried materials are cut into pieces and powdered with the help of stones.

Fumes and Vapour

The fumes are obtained after burning the dried material of the specific medicinal plants or plant parts. Vapour can also be obtained by putting fresh medicinal plants in boiling water. In general, it is inhaled as medicine.

Results

The Kani's in Kerala had mainly settled on the Western Ghats region especially Thiruvananthapuram and Kollam Districts. Majority of Kani settlements were located in Thiruvananthapuram District (Figure 6.3). During the present study a total number of 117 ethnomedicinal plants were collected from Kani tribal areas of Vamanapuram Block, Nedumangadu Taluk and Kerala state. Of which herbs contributed 60 species (51.28 per cent), shrubs 24 species (20.51 per cent), trees 25 species (21.36 per cent), climbers 6 species (5.12 per cent) and twiners 2 species with 1.70 per cent (Figure 6.4). Botanical name, medicinally useful parts and the ailments of 117 ethnomedicinally important plants collected from the study area were recorded in Figure 6.4.

All the collected ethnomedicinally important plants in the study area were belongs to fifty five angiospermic families contributed about 104 genera and 117 species. Out of 55 plant families, 33 families such as Alanginaceae, Aristolochiaceae, Anacardiaceae, Begoniaceae, Burseraceae, Bixaceae, Bignoniaceae, Convolvulaceae, Capparidaceae, Clusiaceae, Cyperaceae, Ebenaceae, Goodeniaceae, Hippocrateaceae, Lythraceae, Lecythidaceae, Magnoliaceae, Moringaceae, Moraceae, Myristicaceae, Malvaceae, Menispermaceae, Oleaceae, Orchidaceae, Polygonaceae, Sapotaceae, Sapindaceae, Santalaceae, Scrophulariaceae, Trichopodiaceae, Theaceae, Urticaceae, Vitacae have only one species each.

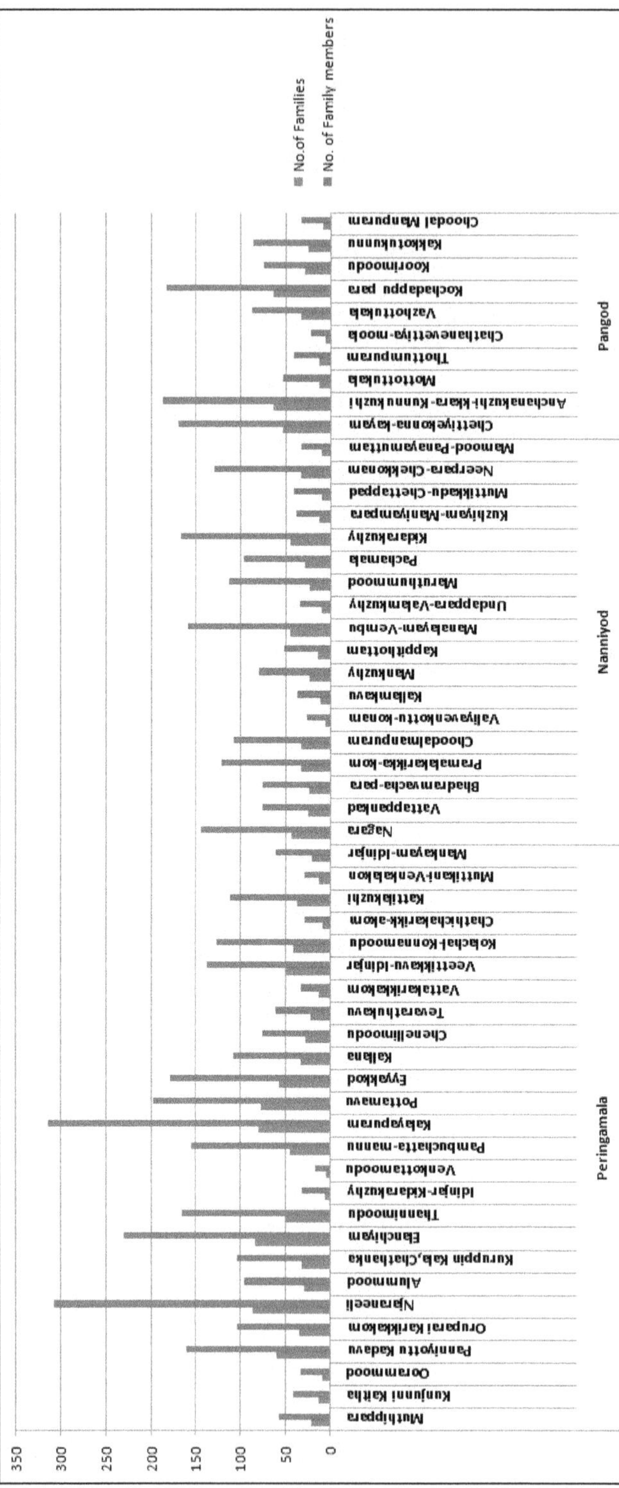

Figure 6.3: Number of Kani Tribal Settlements in Three Panchayaths of Vamanapuram Block.

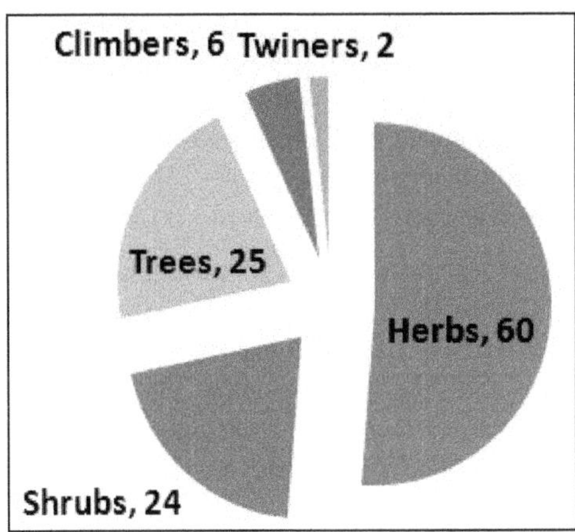

Figure 6.4: Habit-wise Distribution of Ethnomedicinal Plants in the Study Area.

Out of 117 medicinal plants collected from Kani tribal areas of Vamanapuram block, leaves from 46 plants, roots from 16 plants, whole plants, fruits from 16 plants, barks from 11 plants and flowers from 10 plants where used by kani tribes for medicine preparation to cure various ailments (Figure 6.5).

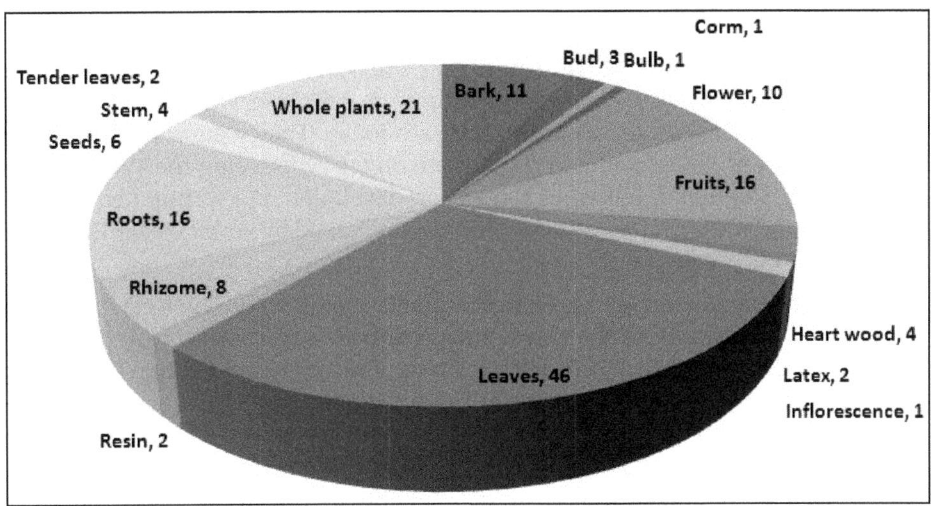

Figure 6.5: Number of Plant Parts Used for the Preparation of Medicine by Kanis.

Antipoisoning Plants Collected from Vamanapuram Block

Twenty plant species were used in the preparation of antipoisoning medicines, which include, *Alangium salvifolium* (Scorpion sting), *Albizzia lebbeck* (Antipoisonous), *Alstonia venenata* (Snake bite), *Aristolochia krisagatra* (Snake bite), *Baliospermum montanum* (Antipoisonous), *Begonia floccifera* (Centepede/Spider poison), *Couroupita guianensis* (Antipoisonous), *Eupatorium triplinerve* (Antipoisonous), *Geodorum densiflorum* (Spider poison), *Jatropha curcas* (Rat poison), *Kaempferia galanga* (Snake bite), *Moringa oleifera* (Antipoisonous), *Pergularia daemia* (Antipoisonous), *Polygonum chinensis* (Antipoisonous), *Pterocarpus santalinus* (Spider poison), *Rauvolfia serpentina* (Snake bite), *Rhinacanthus communis* (Spider poison), *Tamarindus indica* (Antipoisonous), *Tinospora sinensis* (Antipoisonous) and *Clitoria ternatea* (Snake bite).

In the present investigation out of 117 medicinal plants in Kani tribal areas of Vamanapuram block, only 15 species are enlisted in Red Data Book *viz.*, *Acorus calamus, Aegle marmelos, Aristolochia krysagathra, Amorphophallus bulbifera, Baliospermum montanum, Garcinia gummi-gutta, Pterocarpus marsupium, Gymnema sylvestre, Madhuca longifolia, Orthosiphon comosus, Pterocarpus santalinus, Rauvolfia serpentinae, Saraca indica, Trichopus zeylanicus, Michelia champaka* and *Kaempferia galanga.*

Discussion

The Kanis settled in different pockets of Vamanapuram Block, Kerala state, depend on plants growing in their surrounding hereby forests especially for their day to day needs as well as to treat various diseases. In the present study, data about the 117 medicinal plants belonging to 55 angiospermic families and 104 genera used by the Kanis to treat various ailment such as jaundice, bronchitis, leprosy, kidney disoders, poisonous bites, rheumatism, heart disorders, cough, fever, swelling, dental pain, digestive troubles, cancer, throat pain *etc.*

It is evident from the findings that the Kanis have good knowledge about phytomedicines and they are continuing the preservation of vast useful ethnomedicine knowledge on nature care even today. They were well versed with different types of illness and could diagnose only by observing the symptoms of the patients. The herbal remedies have been developed and formulated by their forefathers and these practices are traditionally followed by verbalized means from generation to generation.

The method of using ethnomedicinal plants varied according to the nature of the ailment. In majority of the cases, a decoction of leaves, stem, fruit and root/ tuber is drunk or rubbed on the body to cure ailments. Mostly decoction is extracted by juice crushing the plants in a mortar but sometimes plant parts are boiled with water, and the liquid decanted. Decoction of some plants is applied externally on the wounds or the infected part of the body. In some cases (skin ailments) the patient is given bath of the decoction. Paste of some plant is plastered to set dislocated or fractured bones or for muscular pain. Ailments like body ache, cuts wounds, scabies, boils and skin diseases are treated by external application of the paste. In some cases, combinations of plants are used for best results. The rules of collection time, plant part, storing techniques and method of preparation are largely known by local faith healers. The information collected from this study is in agreement with the previous reports (Pushpangadan and Atal 1984, Anis *et al.*, 2000, Natarajan *et al.*, 2000, Jain 2001, Mahapatra and Panda 2002, Rajan *et al.*, 2002, Ganesan and Suresh 2004, Udayan *et al.*, 2005, Sandhya *et al.*, 2006). Kani tribals in Tirunelveli hills of Tamil Nadu were using 14 plants for the treatment of skin problems (Ayyanar and Ignacimuthu 2005); 52 herbal preparations from 31 plants were used for skin diseases by tribals of Uttar Karnataka district, a nearest state of Tamil Nadu (Harsha 2003) and people of Eastern Cape Province, South Africa used 38 plant species for the treatment of wounds (Grierson and Afolayan 1999).

Majority of the people in the world are relying on the plants rather than commercial products developed from the plant materials (Barret and Kiefer 1996). Most of the ethnobotanical studies confirmed that leaves are the major portion of the plant used for the treatment of diseases. The observations of the present study also coincided with previous observations, out of 117, 46 plants leaves were applied for various ailments (Figure 6.5). The reason why leaves are used mostly is that they are easily accessible and are active in photosynthesis and production of metabolites (Ghorbani 2005). In the present study, the whole plants (21), roots (16), rhizome (8), flower (10), fruits (16) and seeds (6) were found collected in large number of plant species (Figure 6.5). This may lead to the depletion of important ethnomedicinal plants from natural habitats. Due to over exploitation and wholesale collection of rare and endangered plants from wild habitat, their population may decrease rapidly. Kanis exploit the plants broadly for curing 71 ailments covering more or less all major and minor diseases. Analysis of the data indicates that 17 ethnomedicinal plant species are employed for the treatment of stomach diseases. Use of a number of plants for the treatment of stomach diseases indicates that the Kanis face a number of problems associated with sanitary condition, extreme seasonal fluctuation and malnutrition.

The present ethnomedicinal study helps to identify new or less known medicinal herbs of tribal medicine providing clue for new leads, for systematic pharmaco-therapeutic and clinical research, besides listing the local (tribal) names, for the plants. It is not only essential to conserve such a wealth of information found among the tribals but also to enumerate such details and devise a modern biomedical system to meet the ever increasing clinical requirement of mankind. Hence need is often stressed for documenting the knowledge of the tribals, so that the plants can be inspected to intensive clinical pharmacological trials and their efficacy may be evaluated on scientific basis.

Some of the plants used by Kanis in Vamanapuram Block for the preparation of herbal medicine were already investigated for the active principles and for their pharmacological activities and these investigations coincide with the results of the present study. Since the traditional values culture faith and indigenous knowledge related to traditional health care system of Kanis of Vamanapuram Block are facing serious challenges due to acculturation, brought about by migration of the younger generation to cities and these urban migrants show a gap in the cultural beliefs and practices with those of the local inhabitants, the recording of the indigenous knowledge based on traditional health care system becomes increasingly important.

References

Anis, M., M.P. Sharma and M. Iqbal. 2000. Herbal ethnomedicine of the Gwalior forest division in Madhya Pradesh, India. *Pharmaceutical Biology* 38(4): 241-253.

Arinathan, V., V.R. Mohan and A.J. Britto. 2003. Ethnomedicinal survey among

Palliyar tribals of Srivilliputhur Grizzled Giant Squirrel Wild life Sanctuary, Tamil Nadu. *J Econ Tax Bot* 27: 707-710.

Ayyanar, M. and S. Ignacimuthu 2005a. Traditional knowledge of Kani tribals in Kouthalai of Tirunelveli hills, Tamil Nadu, India. *Journal of Ethnopharmacology* 102: 246-255.

Ayyanar, M. and S. Ignacimuthu 2005b. Ethnomedicinal plants used by the tribals of Tirunelveli hills to treat poisonous bites and skin diseases. *Indian J Trad Knowled* 4: 229-236.

Barret, B. and D. Kiefer. 1996. Ethnomedical, biological and clinical support for medicinal plant use on Nicaragua's Atlantic Coast. *Journal of Herbs Spices Medicinal Plants* 4: 77.

Chah, K.F., C.A. Eze, C.E. Emuelosi and C.O. Esimone. 2006. Antibacterial and wound healing properties of methanolic extracts of some Nigerian medicinal plants. *Journal of Ethnopharmacology* 104: 164 -167.

Gamble, J.S. and C.E.C. Fischer. 1957. Flora of the Presidency of Madras, Vol. I-III, Botanical Survey of India, Calcutta.

Ganesan, S., N. Suresh and L. Kesavan 2005. Ethnomedicinal survey of lower Palni Hills of Tamil Nadu. *Indian J Trad Knowled* 3: 299-304.

Ganesan, S., N. Suresh and L. Kesavan. 2004. Ethnomedicinal survey of lower Palni Hills of Tamil Nadu. *Indian Journal of Traditional Knowledge* 3(3): 299-304.

Ghorbani, A. 2005. Studies on pharmaceutical ethnobotany in the region of Turkmen Sahra, North of Iran (Part 1): general results. *Journal of Ethnopharmacology* 102: 58-68.

Grierson, D.S. and A.J. Afolayan. 1999. An ethnobotanical study of plants for the treatment of wounds in the Eastern Cape, South Africa. *Journal of Ethnopharmacology* 67: 327-332.

Harsha, V.H., S.S. Hebber, V. Shripathi and G.R. Hedge. 2003. Ethnomedicobotany of UttatKannada district in Karnataka, India - plants in treatment of skin diseases. *Journal of Ethnopharmacology* 84(1): 37-40.

Ignacimuthu, S., M. Ayyanar and K. Sankarasivaraman. 2006. Ethnobotanical investigations among tribes in Madurai district of Tamil Nadu, India. *Journal of Ethnobiology and Ethnomedicine* 2: 25.

Ignacimuthu, S., M. Ayyanar and K. Sankarasivaraman. 2008. Ethnobotanical study of medicinal plants used by Paliyar tribals in Theni district of Tamil Nadu, India. *Fitoterapia* 79: 562-568.

Jain, S.K. 1991. Dictionary of Indian Folk medicine and Ethno-botany, Deep Publication, New Delhi.

Jain, S.K. 2001. Ethnobotany in modern India. *Phytomorphology Golden Jubilee Issue: Trends in Plant Sciences*. pp. 39-54.

Kosalge, S.B. and R.A. Fursule. 2009. Investigation of ethnomedicinal claims of some plants used by tribals of Satpuda Hills in India. *Journal of Ethnopharmacology* 121: 456–461.

Kottaimuthu, R. 2008. Ethnobotany of the Valaiyans of Karandamalai, Dindigul District, Tamil Nadu, India. *Ethnobotanical Leaflets* 12: 195–203.

Mahapatra, A.K. and P.K. Panda. 2002. Ethno-pharmacological knowledge of Juang and Munda tribes of eastern India. *International Journal of Sustainable Development and World Ecology* 9(2): 151-158.

Matthew, K.M. 1991. An excursion flora of Central Tamil Nadu, India, Oxford and IBH Pub. Company Pvt. Ltd, New Delhi.

Muthukumarasamy, S., V.R. Mohan, S. Kumaresan and V. Chelladurai. 2003a. Herbal medicinal plants used by Paliyars to obtain relief from gastro-intestinal complaints. *Journal of Economic and Taxonomic Botany* 27: 711-714.

Muthukumarasamy, S., V.R. Mohan, S. Kumaresan and V. Chelladurai. 2003b. Herbal remedies of Paliyar tribe of Grizzled Giant Squirrel Wildlife Sanctuary, Western Ghats, Srivilliputhur, Tamil Nadu for poisonous bites *Journal of Economic and Taxonomic Botany* 27: 761-764.

Namsa, N.D., H. Tag, M. Mandal, P. Kalita and A.K. Das. 2009. An ethnobotanical study of traditional anti-inflammatory plants used by the Lohit community of Arunachal Pradesh, India. *Journal of Ethnopharmacology* 125: 234-245.

Natarajan, B., B.S. Paulsen and V. Korneliussen. 2000. An ethnopharmacological study from Kulu district, Himachal Pradesh, India: Traditional knowledge compared with modern biological science. *Pharmaceutical Biology* 38(2): 129-138.

Pattanaik, C., C.S. Reddy and M.S. Murthy. 2008. An ethnobotanical survey of medicinal plants used by the Didayi tribe of Malkangiri district of Orissa, India. *Fitoterapia* 79: 67–71.

Pushpangadan, P. and C.K. Atal. 1984. Ethno-medico-botanical investigations in Kerala I: Some primitive tribal of Western Ghats and their herbal medicine. *Journal of Ethnopharmacology* 11(1): 59-77.

Ragupathy, S. and A. Mahadevan. 1991. Ethnobotany of Kodiakkarai Reserve Forest, Tamil Nadu, South India. *Ethnobotany* 3: 79-82.

Rajan, S., M. Sethuraman and P.K. Mukherjee. 2002. Ethnobiology of the Nilgiri Hills, India. *Phytotherapy – Research* 16(2): 98-116.

Rajendran, S.M., S.C. Agarwal and V. Sundaresan. 2003. Lesser known ethnomedicinal plants of Ayyakarkoil Forest Province of South western Ghats, Tamil Nadu, India-Part I. *J Herbs Spices Med Plants* 10: 103-112.

Ravikumar, K., S. Noorunniza Begum, R. Vijaya Sankar and D.K. Ved. 2005. Profile of Medicinal plants Diversity in Kerala. Pp. 23-28. in *Medicinal Plants of Kerala - Conservation and Beneficiation*. Kerala State Council for Science, Technology and Environment, Trivandrum.

Sandhya, B., S. Thomas, W. Isabel and R. Shenbagarathai. 2006. Ethnomedicinal plants used by the valaiyan community of Piranmalai hills (Reserved forest),

Tamil Nadu, India – A pilot study. *African Journal of Traditional, Complementary and Alternative Medicines* 3(1): 101-114.

Savithramma, N., C. Sulochana and K.N. Rao. 2007. Ethnobotanical survey of plants used to treat asthma in Andhra Pradesh, India. *Journal of Ethnopharmacology* 113: 54-61.

Shanmugam, S., M. Annadurai and K. Rajendran 2011. Ethnomedicinal plants used to cure diarrhoea and dysentery in Pachalur hills of Dindigul district in Tamil Nadu, Southern India. *Journal of Applied and Pharmaceutical Science* 1(8): 94–97.

Udayan, P.S., G. Sateesh, K.V. Thushar and B. Indira. 2005. Ethnomedicine of Chellipale community of Namakkal district, Tamil Nadu. *Indian Journal of Traditional Knowledge* 4(4): 437-442.

Upadhyay, B., Parveen, A.K. Dhaker and A. Kumar. 2010. Ethnomedicinal and ethnopharmaco-statistical studies of Eastern Rajasthan, India. *Journal of Ethnopharmacology* 129: 64–86.

2018, Ethnomedicinal Plants: A Biodiversity Treasure *Pages* **229–238**
Editors: V.R. Mohan, A. Doss, P.S. Tresina and V. Sornalakshmi
Published by: **ASTRAL INTERNATIONAL PVT. LTD., NEW DELHI**

Chapter 7

An Ethnobotanical Survey and Investigation of Medicinal Plants from Kiliyur and Pattipadi In Yercaud Hills, Salem

S. Murugesh and P. Vino

Department of Botany, School of Life Sciences,
Periyar University, Salem, Tamil Nadu

Introduction

India is one of the most important mega biodiversity of hotspots rich in ethnic diversity and traditional knowledge. The tribal people are the ecosystem people who live in harmony with the nature and maintain a close link between man and environment (P Samydurai *et al.*, 2012). The ethnic person residing in different geographical belts of India depends on wild plants to meet their basic requirements and all the ethnic communities have their own pool of secret in ethnomedicinal and ethno pharmacological knowledge about the plants available in their surroundings (M Johnson Gritto *et al.*, 2015).Traditional medicinal practices are an important part of the primary health care system in the developing world (A Cheikhyoussef *et al.*, 2011).India is endowed with a rich wealth of medicinal plants.(Amzad basha kolar, 2013).In India, medicinal plants are widely used by all sections of people either directly as folk remedies in different indigenous systems of medicine or indirectly in the pharmaceutical preparations of modern medicines. Medicinal plants have been used to cure a number of diseases (Algesboopathi, C. 2011). Plants have been used in traditional medicine for several thousand years (Abu Rabia, 2005). The knowledge of medicinal plants has been accumulated in the course of many centuries based on different Indian systems of medicines such as Ayurveda, Unani and Siddha (Pei S.J. 2001).In India it is reported that traditional healers use

2500 plant species and 100 species of plants benefit as usual sources of medicine. (pei sheng ji.2001). In recent years, there has been a tremendous range of interest in the medicinal plants especially in Ayurveda's and other traditional systems of medicines. Drugs obtained from plants are believed to be much safer and exhibit a remarkable efficacy in the treatment of various ailments (Korpenwar, A.N. 2012). Plants have always been the source of medicines and have many uses to mankind (Nadkarni, 2001).Herbal medicines are comparatively safer than synthetic drugs. Plant based information has become a recognized tool to investigate new sources of drugs and neutraceuticals (Algesboopathi,C. 2012). Traditional medicinal scientific investigation is carried specifically when the literature and field work information have been properly evaluated. The documentation of indigenous information on the utility of local plant resources by various ethnic groups or communities is one of the main objectives of ethno botanical research (Haridutta Dandotiya, 2013). Many workers investigated the uses of medicinal plants and analyzed different diseases among rural and tribal people inhabiting different areas of Tamil Nadu. Therefore, the present study aims to identify and document, some of the plant species used for medicinal purposes by the kiliyur and pattipadi local peoples residing at Yercaud hills, Salem district, Tamil Nadu, India.

Materials and Methods

Study Area

The Yercaud Hills range of the Eastern Ghats is situated in Salem district in Tamil Nadu. It is situated at an altitude of 1515 meters (4970 ft) above sea level and the highest point in Yercaud is the Servarayan temple, at 5,236 feet (1.623 m). They are located between 110C 45′ 56″ N latitude and 780C 17′ 55″ E longitude. The temperature ranges from 130 C to 290 C on the peaks and 250 C to 400 C at the foot hills. The average annual rainfall is around 1500 mm – 1750 mm. The soil is deep and non-calcaneal. The types of forest range from ever green to moist deciduous (Alagesboopathi, C. *et al., 1996*).The division of the Yercaud hills by Kiliyur and Pattipadi range of the Eastern Ghats is situated in Salem district in Tamil Nadu. Kiliyur falls is a waterfall in the Servarayan hill range in the Eastern Ghats, Tamil Nadu, India. The waters overflowing the Yercaud lake fall 300 feet (90 m) into the Kiliyur valley. The falls is situated at a distance of 2.5 kilometers (1.6 mi) from the Yercaud lake. The final 500 meters (1,600ft) or so consist only of a rough unpaved steep pathway. But it is a breathtaking sight in the monsoon time when the water levels are the maximum. They are located between 11.795 latitude and 78.2004 longitude. Pattipadi is located between 11.8094 latitude and 78.1877 longitude. It is situated 4km away from sub district head quarter Yercaud and 40km away from district headquarter Salem. The total geographical area of village is 458.23 hectares.

Social Status

The ethnic people of Kiliyur and Pattipadi village from Yercaud hills are farmers and their occupation style is agriculture. Most of them are poor and they are engaged in agriculture as coolies. To supplement their economy rest of the people are doing jobs interrelated to agriculture. They are also involved in hunting, poultry farming,

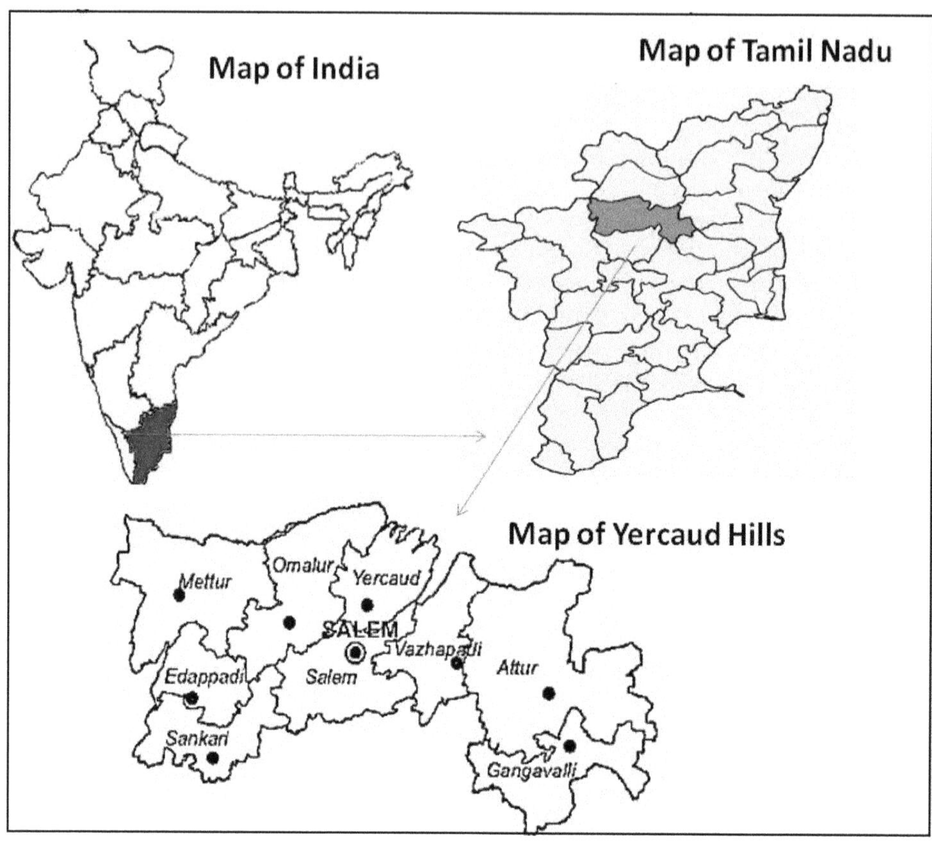

Map 7.1: Showing the Area Selected for Study.

collecting honey and other minor forest products. Most of this area people selected for study are residing in remote and inaccessible forest areas. They donot have any source to get modern medicine for their health care. Hence, they are practicing phytotherapy to treat the common ailments. They also have good knowledge in the herbal treatment for their healthcare (kannan, M. 2016). Enumeration of plants used for several diseases is the part of the ethnobotanical research work carried out by us in the study area.

Data Collection

The present work is an outcome of ethnobotanical studies during December 2015-April 2016). Ethnomedicinal information on the plant species was collected through interviewing local communities. The information was gathered through questionnaires, personal interviews and discussions with them. The interview was conducted with the people who are having the sound knowledge on medicinal plants found in their area and these plants are used by their families and neighbours. The questionnaire contains questions covering the details of the plants, parts used, medicinal uses and mode of preparation of remedies. The ethnobotanical data was collected according to the methodology suggested (Rajendran *et al.*, 2013). In the study, 30 Knowledgeable elders (19 men and 11 women between the ages of 35 to 65) were chosen given by the assistance of local administrators who served as key informants. The collected ethnomedicinal information was recorded on field note books and plants were identified using the Flora of Presidency of Madras (Gamble,1935) and Flora of Tamil Nadu – Carnatic (Matthew,1983). Plant species were identified and kept in our Botany department, Periyar university, Salem.

Results and Discussion

The present study revealed the analysis of plants as herb, shrubs, trees and climbers. The medicinal plants were pre-dominantly found in this hill. In the present study, 57 plant species belonging to 29 families were reported after undertaking the survey (Table 7.1). The 57 plant species were analysed and reported as herbs (26 species), Trees (14 species), Shrubs (11 species), Sub shrubs (3 species), Climbers (2 species) and shrub (1 species); (Figure 7.1). The most dominant plant parts in the study, were leaves (25 species), Whole plant and Barks (7 species), Roots (6 species), Flowers (4 species), Seeds (3 species) other parts with lowest orders are listed: Stems and Roots (2 species); (Figure 7.2). The collected plants have been arranged alphabetically according to their botanical names, followed by their family, local name (Tamil) along with organs used, methods of preparation and ailments treated are provided. Most of the plants used as medicines are either used alone or mixed with other ingredients. (Gilani *et al.*, 2010). The regional people gave some significant information about the species and the suitable period of collection (Alagesaboopathi, 2009). There are considerable economic benefits in the progress of indigenous medications and in the use of medicinal plants for the treatment of several ailments(Francis Xavier *et al.*, 2011). Some important medicinal plants need

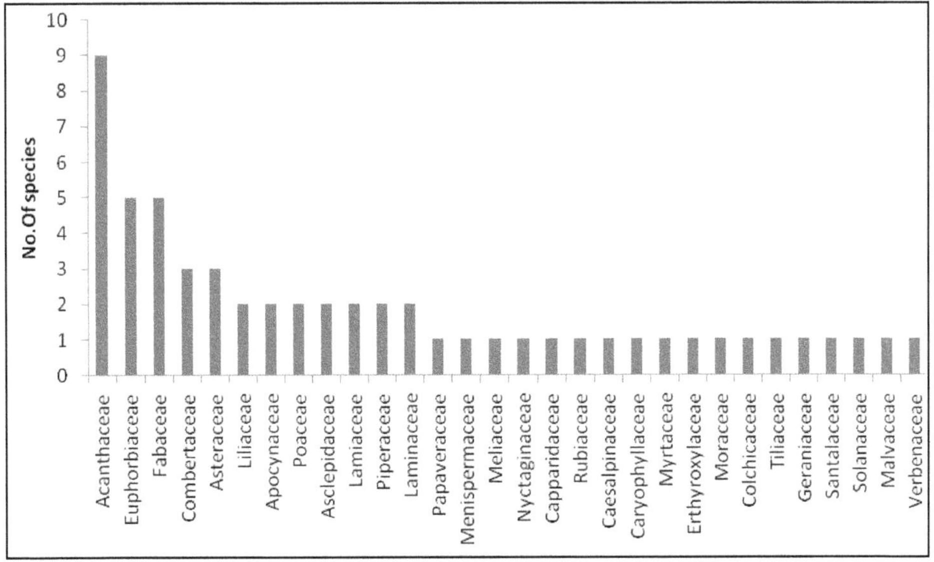

Figure 7.1: Family-wise Distribution of Medicinal Plants in Kiliyur and Pattipadi from Yercaud Hills, Salem District.

Table 7.1: Survey of Medicinal Plants in Kiliyur and Pattipadi from Yercaud Hills

Botanical Name	*Local Name*	*Family*	*Parts of the Plant*	*Medicinal Uses*
Argemone mexicona	Brahmadandu	Papaveraceae	Stem	Antidote poisonous bite
Andrographis affinis	Kodikkurunthu	Acanthaceae	Leaf	Leaf paste mixed with cow's milk used in liver ailments are used to cure
Anamirta cocculus	Kakkaikolli	Menispermaceae	Seed	Seeds epilepsy and as pesticide
Andrographis alata	Nilavembu	Acanthaceae	Leaf	Fever and Diabetes
Andrographis lineata	Periyanangai	Acanthaceae	Leaf	Leaf juice mixed with cow's milk taken for days will cure regularly in liver diseases
Andrographis macrobotrys	Uppali	Acanthaceae	Leaf	Fresh leaf juice is given orally thrice in a day for one week to treat liver disorders
Andrographis paniculata	Nilavembu	Acanthaceae	Whole plant	Jaundice and liver complaints
Baliospermum montanum	Red physic nut	Euphorbiaceae	Root	Root is used to cure abdominal pain,piles.
Aloe barbadensis	Katralai	Liliaceae	Whole plant	Powder of different parts of plant is used as tablet forms
Artemisia nilagirica	Makkippu	Asteraceae	Leaf	The drugs obtained from the leaf,is used to cure malaria
Asparagus racemosus	Shatavari	Liliaceae	Root	The roots are used to cure nervous disorders
Andrographis serpyllifolia	Kaatuppoorankodi	Acanthaceae	Leaf	stomach pain
Azadirachta indica	Vembu	Meliaceae	Bark	Decoction of the bark are taken inside to cure jaundice
Boerhavia diffusa L.	Mukkurattai	Nyctaginaceae	Root	The root powder mixed with cow's milk is used to cure jaundice
Caralluma umbellata	Paraikalli	Asclepidaceae	Leaf	It is used to cure diabetes and fat accumulation.
Ceasalpinia pulcherima	Mayirkonrar	Fabaceae	Seeds	Seeds are used as liver tonic.
Clome viscosa	Naaivalai	Capparidaceae	Leaf	Fresh leaf juice mixed with hot water is used to cure jaundice
Coffea Arabica	Coffee	Rubiaceae	Seed	Seeds are made into powder. This powder is used to make coffee
Cassia fistulata	Sarakondrai	Caesalpinaceae	Flower	Powdered flower is used to cure liver problems.
Cymbopogon citratus	Elumichai	Poaceae	Leaf	Oil is used to treat skin diseases

Contd...

Botanical Name	*Local Name*	*Family*	*Parts of the Plant*	*Medicinal Uses*
Desmodium gangeticum	Oorilai	Fabaceae	Whole plant	Used to overcome headaches and diarrhea
Drymaria cordata	Mudavattukal	Caryophyllaceae	Leaf	Used to cure snake bite
Eclipta alba	Manjalkarisalankanni	Asteraceae	Leaves	Decoction of leaves mixed with hot water is used to treat liver disorders
Eclipta prostrata	Karisalankanni	Asteraceae	Leaf	Used to cure body pain
Emblica officinalis	Nelli	Euphorbiaceae	Fruit	Used to cure bleeding disorders and liver diseases.
Eucalyptus globules	Karpooramaram	Myrtaceae	Stem	Used to cure body aches
Euphorbia hirta	Amman pacharisi	Euphorbiaceae	Leaf	Used to heal wounds
Erythrina indica	Kalyanamurungai	Fabaceae	Leaf	Used to overcome liver troubles,dysentery and joint pain
Erythroxylum monogynum	Sembulichan	Erythroxylaceae	Bark	Bark is made into a chrism and applied on the affected part externally
Ficus glomerata	Attimaram	Moraceae	Whole plant	Powdered and mixed with milk and drank orally
Gloriosa superba	Kanthal	Colchicaceae	Flower	It is used to treat snake bites, open wounds and ulcers
Grewiadi disperma	Uduppai	Tiliaceae	Fruit	Unripe and ripped fruits are eaten
Gymnema sylvestre	Sarakondrai	Asclepiadaceae	Whole plant	Used to treat diabetes, jaundice, cough and asthma.
Hemidesmus indicus	Nannari	Asclepiadaceae	Root	Used as Blood purifier
Indigofera aspalathoides	Sivanar vembu	Fabaceae	Whole plant	Used to treat psoriosis.
Jatropha curcas	Kattamanakku	Euphorbiaceae	Leaf	Leaf and bark juice are mixed with salt. This is applied to cure skin diseases
Justia adhatoda	Adathoda	Acanthaceae	Leaf	Used to cure cough,cold and asthma
Justicia tranquebariensis	Tavashoo Moorunghie	Acanthaceae	Leaf	Used to cure cold and cough. The leaf paste reduces pain in the swellings
Leucas aspera	Thumbai	Lamiaceae	Whole plant	Used to cure Headache and also used as insecticide
Ocimum sanctum	Thulasi	Lamiaceae	Leaves	Used to avoid dry cough
Pelargonium Graveolens	Geranium	Geraniaceae	Leaves	Essential oils are obtained from the leaves
Phyllanthus amarus	Keelanelli	Euphorbiaceae	Root	Used to cure stomach ache and urinary disorder

Botanical Name	Local Name	Family	Parts of the Plant	Medicinal Uses
Piper longum	Tippili	Piperaceae	Fruits	Used to treat asthma and for solving gastric problems
Piper nigrum	Kurumilagu	Piperaceae	Fruit	Used to cure cough
Santalum album	Sandanum	Santalaceae	Bark	Used to treat dipsia and to solve skin diseases
Solanum trilobatum	Thoothuvalai	Solanaceae	Leaf	The leaf is used to treat cough and cold.
Tephrosia purpurea	Kolingi	Fabaceae	Leaves	Used to cure asthma and ulcers.
Terminalia arjuna	Marudam	Combertaceae	Bark	Used as heart remedy
Terminalia bellerica	Thanrikk	Combertaceae	Bark	Beneficial for hair
Terminalia chebula	Kadukai	Combertaceae	Bark	Improves digestion
Thespesia populnea	Poovarasumaram	Malvaceae	Bark	Different parts of plants are taken orally in tablet forms
Vinca rosea	Nithiyakalyani	Apocynaceae	Leaves	Used as anticancer
Vitex agnus	Seemainochi	verbenaceae	Leaves	Used to treat eye diseases and stomachache
Vitex negundo	Nochi	Laminaceae	Leaves	Used to cure malaria
Vitex trifolia	Moovilainochi	Laminaceae	Leaves	Used as pain killer
Vetiveria zizanoides	Vettiver	Poaceae	Root	Used to cure stomach ache and diaphoretic

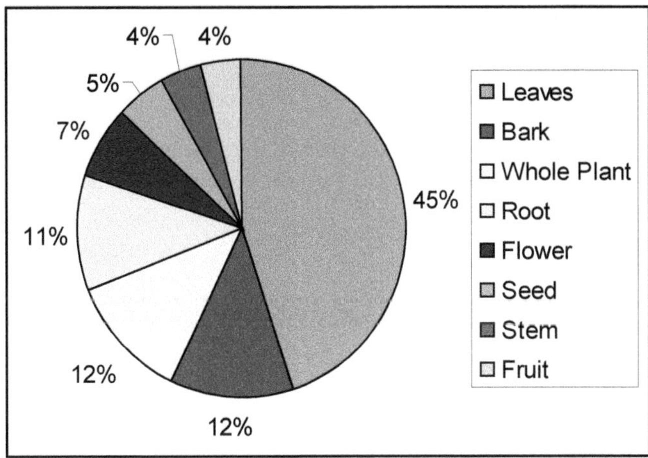

Figure 7.2: Various Parts of Ethnomedicinal Plants Used by Local People of Kiliyur and Pattipadi from Yercaud Hills.

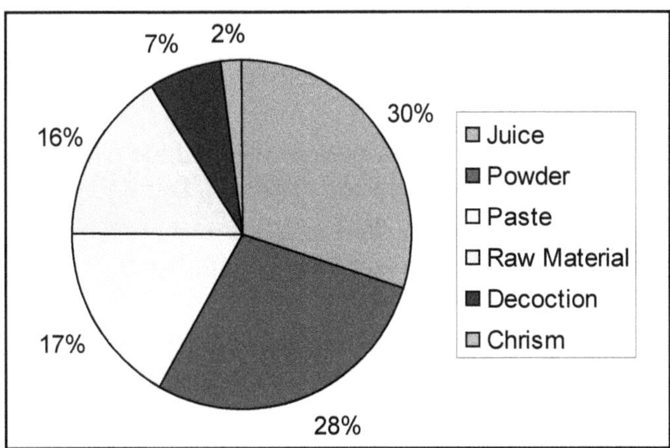

Figure 7.3: Various Modes of Preparations of Medicaments Used to Cure Diseases by Local People of Kiliyur and Pattipadi of Yercaud Hills.

immediate conservation and their cultivation should be encouraged through which their extinction can be prevented and tribal people may also cure their disease at low cost. There is a need to have modern way of identification of plants ensuring correct use of species (Shinwari, 2010).

Conclusion

This study has revealed that medicinal plants play a vital role in the primary healthcare of the people. The information gathered from rural people is useful for further research in the field of ethnobotany and taxonomy. The ethnomedicinal data may provide a base to start the search of phytochemistry, pharmacology and pharmacognosy. Therefore, attention should also be made on proper utilization and conservation of these medicinal plants.

Acknowledgement

Authors are thankful to Periyar university, Salem, Tamil Nadu, India. We specially express our thanks to local people of the study area for sharing their traditional knowledge.

References

Abu-Rabia, A. 2005. Urinary diseases and ethnobotany among pastrol nomands in the middle East. 1:4.

Alagesaboopathi, C. 2012. Ethnobotaical survey of medicinal plants used by Malayali tribals and Rural people in Salem District of Tamil Nadu, India. *Journal of Pharmacy Research*. 5(12).5248-5232.

Alagesboopathi, C. 2009. Ethnobotanical plants and their utilization by villagers in Kumaragiri hills of salem District of Tamil Nadu, India. *Afr. J. Traditional Complementary and Alternative Medicines*, 6: 222-227.

Alagesboopathi, C. 2011. Ethnomedicinal plants used as medicine by the kurumba tribals in pennagaram region, Dharmapuri District of Tamil Nadu, India. *Asian J.Exp.Biol.Sci.* Vol 2(1).

Alagesboopathi, C., Balu, S. and Dwarakan. 1996. Edible fruit, yielding plants of Shevory hills in Tamil Nadu. *Ancient Sciences of Life*. XVI (2):148-151.

Amzad basha Kolar and Ghouse Basha. 2013. Survey of medicinal plants of Pachamalai hills,A part of Eastern Ghats,Tamil Nadu. *International Journal of Current Research*. 5(12).3923-3929.

Cheikyoussef, A., Mapaure, I. and Martin shapi. 2011. The use of some indigenous plants for medicinal and other purposes by local communities in Nambia with emphasis on Oshikoto region: A review. *Research Journal of Medicinal Plants.* 5:406-419.

Francis Xavier, T., Freeda Rosa, A. and Dhivyaa, M.2011. Ethnomedicinal survey of malayali tribes in kolli hills of eastern ghats of Tamil Nadu, India. *Indian.J. Tradit Knowl*, 10:559-562.

Gamble, J.S.1936. Flora of the presidency of madras.Vol I-III. Allard and Co.London. (Reprinted-1956) Botanical Survey of India, Calcutta.

Gilani, S.A., Fuji, Y., Shinwari, Z.k., Adnan, M., Kikuchi, A. and Watanabe, K.N.2010. Phytotoxic studies of medicinal plant species of Pakistan. *Pak. J. Bot.* 42(2).987-996.

Haridutta Dandotiya, Girraj Singh and Kashaw, S.K. 2013. The Galactagogues use by Indian tribal communities to over come poor lactation. *International Journal of Biotechnology and Bioengineering Research.* 3: 243-248.

Johnson Gritto, M., Nanadagopalan, V. and Doss, A. 2015. Ethnobotanical survey of medicinal plants used by Traditional healers in Shobanapuram village of Pachamalai hill, Tamil Nadu. *Advances in Applied Science Aesearch.* 6(3).157-164.

Kannan, M., Senthilkumar, T. and Rao, M.V. 2016. Ethnobotanical plants used for wound healing purposes by Malayali tribes of Kalrayan hills, Salem District, Tamil Nadu, India. *Global J Res. Med. Plants and Indigen. Med.* 5(7).203-216.

Korpenwar, A.N. 2012. Ethnomedicinal plants used to cure skin diseases in Ambabarwa Wild life sanctuary area of Buldhana District (M.S), India. *International Journal of Recent Trends in Science and Technology.* 2(3).36-39.

Mathew, K.M.1981-1983.The flora of the Tamil Nadu Carnatic, Rapinat Herbarium, Tiruchirapalli.

Nadkarni, K.M. 2001. Indian plants and drugs with their medicinal properties and uses. Asiatic Publishing House, New Delhi.

Pei sheng-Ji. 2001. Ethnobotanical approaches of traditional medicine studies: Some experiences from Asia. *Phar. Biol.,* 39:74-79.

Pei, S.J. 2001. Ethnobotanical approaches of traditional medicine studies some experiences from Asia.39:74-79.

Rajendran, A., Aravindhan, V. and Senthilkumar, K. 2013. Ethnobotanical survey of medicinal plants used by Malayali tribes in Yercaud hills of Eastern ghats, India. *Journal of Natural Remedies.* 13(2).

Samydurai, P., Thangapandian, V. and Aravinthan, V. 2012. Wild habits of Kolli hills being staple food of inhabitant tribes of Eastern Ghats, Tamil Nadu, India. *International Journal of Natural Products and Resources,* 3(3): 432-437.

Shinwari, Z.K.2010. Medicinal plants research in pakistan. *Journ. Med. Pl. Res.,* 4(3): 161-176.

2018, Ethnomedicinal Plants: A Biodiversity Treasure Pages 239–246
Editors: ***V.R. Mohan, A. Doss, P.S. Tresina and V. Sornalakshmi***
Published by: **ASTRAL INTERNATIONAL PVT. LTD., NEW DELHI**

Chapter 8

Ethnobotanical Survey of Medicinal Plants Used by Malayali Tribals in Vathalmalai, Dharmapuri District, Eastern Ghats of Tamil Nadu, India

S. Murugesh*, P. Selvi, B. Seelapreethi and R. Rajeswari

Department of Botany, School of Life Sciences,
Periyar University, Periyar Palkalai Nagar, Salem – 636 011 Tamil Nadu
**E-mail: murugeshss@rediffmail.com*

Introduction

Plants are important source of therapeutic drugs and play a significant role in the survival of the tribal and ethnic communities. Ethno botany deals with the relationship between human societies and plants. Medicinal plants are the source of primary health care throughout the world for thousands of years. India has a rich biodiversity of medicinal plants. In the developed countries, 25 per cent of the medicinal drugs are based on plants and their derivatives (Principe *et al.,* 2005). Herbal medicines are also gaining popularity among the western population because they have minor or no side effects if administered properly (Jordan *et al.,* 2010). However, in the middle of 20th century, the use of medicinal plants was reduced to one fourth because researchers encourage the use of synthetic chemicals for curing diseases. But now, the trend is people favour the medicinal plants as they contain natural products which are effective, chemically balanced and have fewer side effects as compared to synthetic chemicals (Sultana *et al.,* 2015). Interest in medicinal plants has been fuelled by the rising cost of prescription drugs in the maintenance of personal health and well-being and the bioprospecting of new plants – derived drugs. In addition, the uses of medicinal plant are widespread, not only because they are easily accessible and affordable, but also due to persistent cultural beliefs and practices, as well as the lack of access to modern health care systems in rural areas (Rokaya *et al.*, 2014). As indigenous cultures are closely maintained by the tribal and other forest dwellers throughout the world, the ethnobotanical investigation is a prerequisite for any developmental planning concerned with the welfare of tribals and their environment. This study aims to document the information of botanical name, family, local name and medicinal uses of the plants.

Materials and Methods

Study Area

The area selected for study is Vathalmalai situated in Dharmapuri district, Tamil Nadu. It is located between latitudes N 11 47′ and 12 33′ and longitudes E 77 02′ and 78 40′. It occupies an area of 4497.77 km² and has a population of 2,856,300 (as of 2001). Temperature in this area ranges from 12°C to 25°C during March-April and average around 12°C during December and 35°C during April-May. Every Malayali village has several hamlets. Hamlets are found in different elevations (1700m).

Data Collection

The field study was carried out from April 2015 to July 2015 in the Vathalmalai in Dharmapuri district, Tamil Nadu, India. The ethnobotanical data was recorded following the standard procedures (Jain, 1981) like participatory research methods such as semi- structure interviews, field inspection, field observation and participation in their social life events. The plant specimens were identified by cross check of literature from 'The flora of presidency of Madras' (Gamble, 1935).

Results and Discussion

The present study revealed that the Malayali tribes of Vathalmalai prefer to use different native plants with medicinal utility. A total of 59 species belonging to 29 families were identified for medicinal purpose during the study. Different parts of plants like leaves, stem, bark, flower, fruits, seeds, roots, rhizome, tuber are used for different ailments. The solanaceae and mimosaceae families contributed the largest number of plant species (each 5), followed by laminaceae (4), acanthaceae, apocynaceae, cesalpineaceae, euphorbiaceae, cucurbitaceae, fabaceae, meliaceae, and verbinaceae families (3 each) and some other huge number of plant species were involved in this present investigation for the treatment of many diseases (Table 8.1; Figures 8.1 and 8.2). Among the different parts used, the leaves were the most frequently used for the preparation of medicine solely or combination with other parts. It was followed by fruit, bark, stem, flower, root, rhizhome, tuber, whole plant, unripe fruit. Many indigenous communities thoughout the world also utilized mostly leaves for the preparation of herbal medicine (Teklehaymanot *et al.*, 2007 and Gonzalez *et al.*, 2010), the reason why leaves were used mostly is that they are collected very easily than underground parts (Giday *et al.*, 2009).

Most of the plants were used to treat common ailments like gastrointestinal disorders, ulcer, debility, pain, skin problem, respiratory problems, bleeding from cuts and woods, urinary tract problem and sexual disorders. In the majority of cases, a single plant part was used for treatment of any given ailment.

Table 8.1: Ethnomedicinal Perspectives of Botanicals Used by Malayali Tribes in Vathalmalai

Botanical Name	*Family*	*Vernacular Name*	*Parts Used*	*Diseases Cured*
Adhatoda vasica Nees.	Acanthaceae	Adhatoda	Leaves	Cold and Cough
Andrographis paniculata Burmf.	Acanthaceae	Nilavembu	Leaves	Skin disorders, Snake bites
Dipterocanthus patula Jacq.	Acanthaceae	Silandi nayagam	Leaves	Cures eye irritation
Achyranthus aspera L.	Amaranthaceae	Nayuruvi	Whole plant	Reduces body weight and skin disorders
Cuminum cyminum, Linn	Apiaceae	Chirakam	Seeds	Anthelmintic, stomach itch, digestion
Centella asiatica (L.) Urban.	Apiaceae	Vallaarai	Leaves	Jaundice
Carissa carandas L.	Apocynaceae	Kilaakkaai	Leaves	Heal wounds
Catharanthus roseus L.	Apocynaceae	Nithya kalyani	Leaves and Root	Cancer disorders
Gymnema sylvestre R.Br.	Apocynaceae	Sirukurinjan	Leaves	Antidiabetes
Aristolochia bractiolata Linn.	Aristolochiaceae	Aduthinna palai	Leaves	Reduces body weight
Tridax procumbens C.	Asteraceae	Vettukkaayathalai	Leaves	Antiulcer
Eclipta prostrata, Linn	Asteraceae	Karisalankannai	Leaves, root	Anti inflammatory
Cassia auriculata Linn.	Caesalpiniaceae	Avarai	Leaves, Flower and Bark	Antidiabetic, Alternative
Cassia occidentalis, Linn	Caesalpiniaceae	Pei-varai	Leaves	Purgative, laxative
Cassia fistula L.	Caesalpiniaceae	Sarakondrai	Flower	Hair growth and colour
Cocinia indica (Linn) voigt.	Cucurbitaceae	Kovi	Leaves, unripe fruit, stem, rhizome	Antidiabetic, diuretic, antispasmodic
Corallocarpus epigaeus HK.f.	Cucurbitaceae	Akashagaruden	Tuber	Jaundice
Momordica charantia, Linn.	Cucurbitaceae	Paagarkaai	Leaves, fruits,seeds	Anthelmintic, diuretic, laxative, tonic
Acalypha indica, Linn.	Euphorbiaceae	Kuppai-meni	Leaves. Seeds, root	Expectorant, Diuretic, anthelmintic
Anisomeles malabarica (L) RBr.Ex.Sims.	Euphorbiaceae	Peyimerutti	Whole Plant	Body Pain
Euphorbia antiquorum L.	Euphorbiaceae	Sadurakalli	Leaves	Lose Motion
Indigofera aspalathoides Vahl.ex.DC	Fabaceae	Sivanar vembu	Whole plant	Skin disorders and Blood purifier

Contd...

Botanical Name	Family	Vernacular Name	Parts Used	Diseases Cured
Macrotyloma uniflorum (Lam) Kerde.	Fabaceae	Kollu	Seeds	Diuretic, tonic
Millettia pinnata L.	Fabaceae	Pungai	Flower	Antidiabetes
Enicostema axillare Lam.	Gentianaceae	Vellarugu	Leaves and flower	Poisonous bite(Beetle larvae)
Coleus aromaticus Benth	Laminaceae	Karppura valli	Leaves	Expectorant, diaphoretic
Leucas aspera Wild.	Laminaceae	Thumbai	Leaves	Ear ache and Cattle disorder
Ocimum basilicum, Linn	Laminaceae	Thiru-neetru Pachai	Whole plant	Expectorant, Dysentery
Ocimum sanctum L.	Laminaceae	Tulsi	Leaves	Cold and Cough
Asparagus racemosus Willd.	Liliaceae	Thanneervittan Kizhangu	Leaves	Body Pain
Cipadessa baccifera Miq.	Meliaceae	Puilipan cheddi	Leaves	Cracks on Foot
Azadirachta indica L.	Meliaceae	Vembu	Whole plant	Antiviral, Jaundice, Anthelmintic
Melia azadirachta L.	Meliaceae	MalaiVembu	Whole plant	Antiviral, Jaundice, Anthelmintic
Prosopis cineraria L.	Mimosaceae	Vannimaram	Bark and Leaves	Snake bites
Acacia catechu (L.f.) Willd.	Mimosaceae	Karungalei	Leaves	Diarrhea
Acacia leucophloea (Roxb.) Willd.	Mimosaceae	Sarai	Bark	Heal Wounds
Plumbago zeylanica L.	Mimosaceae	Chittiramoolam	Bark	Heal Wounds
Mimosa pudica L.	Mimosaceeae	Thottal sinugi	Whole plant	Cattle disorders
Ficus benghalensis, Linn	Moraceae	Alamaram	Leaves, flower, fruit, Bark	Rejuvenation, Laxative
Syzygum cumini (*Linn*) Skeels.	Myrtaceae	Naval	Fruits, bark	Diuretic, Anthelmintic
Sesamum indicum, Linn	Pedaliaceae	Ellu	Leaves, flower, seeds	Demulcent, emollient
Phyllanthus niruri L.	Phyllanthaceae	Kilanelli	Root and leaves	Jaundice
Piper nigram L.	Piperaceae	Milagu	Seeds, stem	Acrid, Antiperiodic, antivatha, antidote
Cynodon dactylon L.	Poaceae	Arugampul	Leaves	Ulcer
Aegle marmelos L.	Rutaceae	Vilvam	Leaves and fruit	Antidiabetic, Digestive, Laxative
Cardiospermuum helicacabum C.	Sapindaceae	Mudakathan	Leaves	Rheumatism

Botanical Name	*Family*	*Vernacular Name*	*Parts Used*	*Diseases Cured*
Dodonaea viscosa L.	Sapindaceae	Virali	Leaves	Blood clot
Solanum anguivi Lam.	Solanaceae	Mullichundai	Fruits	Jaundice, diuretic
Solanum nigram, Linn.	Solanaceae	Milagu thakkali	Fruits, leaves	Ulcer
Solanum trilobatum L.	Solanaceae	*thoodhuvalai*	Leaves	Asthma
Withania somnifera L.	Solanaceae	Asvagantha	Corm	Nerve disorders and Body pain
Solanum xanthocarpum Schrad. and H. Wendl.	Solanaceae	Kandankathari	Whole plant	Nerve disorders and Cough
Gmelina arborea Roxb.	Verbenaceae	Kumalamaram	Bark	Piles
Vitex negundo L.	Verbenaceae	Nochi	Leaves	Cold and Cough
Lantana camara L.	Verbinaceae	Unnimul	Leaves	Heal wounds
Hybanthus enneaspermus Linn.	Violaceae	Oridhazh thamarai	Leaves	Reduces male sterility
*Curcuma longa,*Linn.	Zingiberaceae	Manjal	Rhizome	Antiviral, hepatic tonic
Zingiber officinale, Rose	Zingiberaceae	Inji	Rhizome	Carminative, stomach ache
Tribulus terrestris, Linn.	Zygophyllaceae	Nerunjil	Whole plant	Diuretic, Aphrodisiac

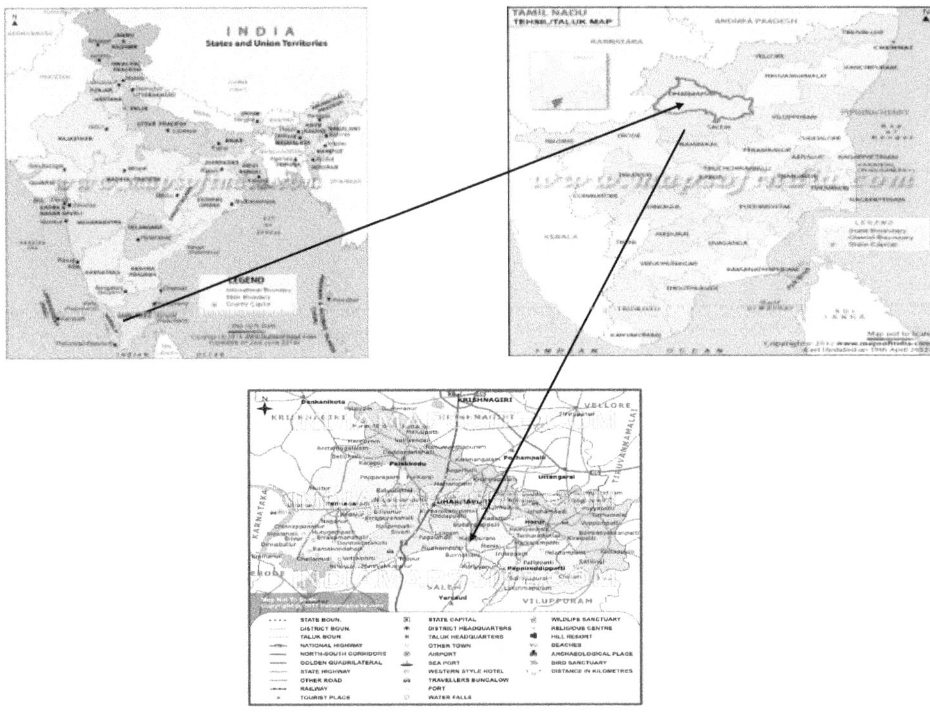

Figure 8.1: The Area Selected for Study with Close View of Dharmapuri District Map.

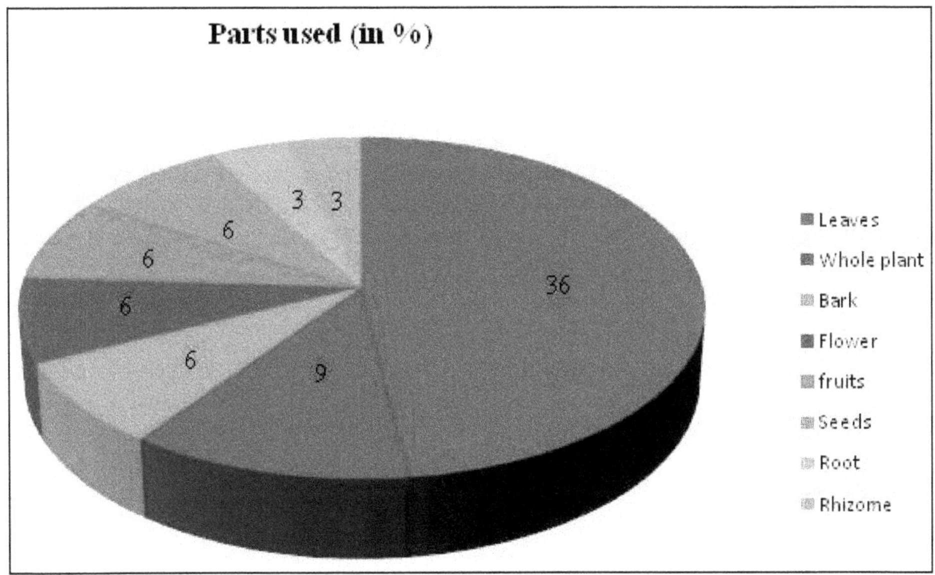

Figure 8.2: Plant Parts Used.

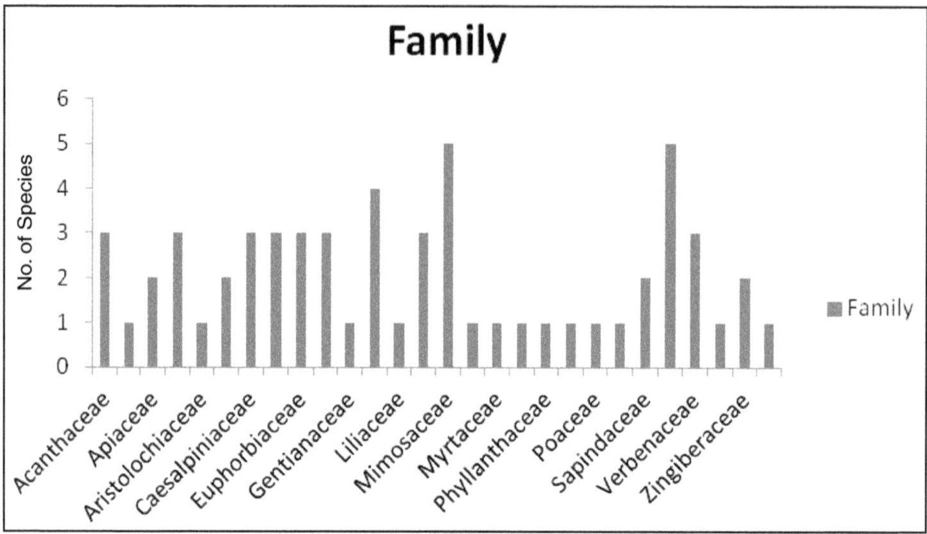

Figure 8.3: Family Used to Cure Various Ailments.

Iyyanar and Ignachimuthu (2005) reported that Andrographis sp is used by tribal people to treat poisonous bites. Similar observation is also seen in the present study. Most of the plants reported in this study gave good evidence of effectiveness and were scientifically validated as significant pharmacological agent. For example, *Gymnema sylvestre* is used in the treatment of diabetics for a long time in Indian traditional medicine and elsewhere in the world and it claimed to has blood glucose lowering activity both *in-vitro* and *in-vivo* by a number of reports (Mukherjee *et al.*, 2006).

The most important aspect of the Malayali tribal medicine is that fresh plant material is used for the preparation of medicine. Alternatively, if the fresh plants parts are not available, dried plant materials are used. For this reason several plants serve as alternative remedy to cure a single disease. From this study it is clear that Malayali tribal possess innate ability to discern the character of plant and exploit the plant resources to meet their health care needs.

Conclusion

The present study reavealed the utilization of 59 species of plants as ethnnomedicine by the Malayali tribes of Vathalmalai, Dharmapuri District of Tamil Nadu. It reveals the knowledge and usage of traditional medicine for the treatment of various ailments among the Malayali tribes. This is still a major part of their life and culture. The present study would be useful in preventing the loss of ethnomedicinal tradition of Malayali tribe community. The new claims which recorded from the study area showed that still much more can be learned from investigating herbals available abundantly in the forest.

Acknowledgment

Authors would like to express thanks to Traditional health practitioners and rural people involved in the interviews for providing information about the medicinal applications of the plants in Vathalmalai, Dharmapuri District of Tamil Nadu.

References

Ayyanar, M., and S. Ignacimuthu. "Traditional knowledge of Kani tribals in Kouthalai of Tirunelveli hills, Tamil Nadu, India." *Journal of Ethnopharmacology* 102.2 (2005): 246-255.

Gamble, J. S., and C. E. C. Fischer. "CEC 1915–1935." *Flora of the Presidency of Madras* 1: 3.

Giday, Mirutse, Zemede Asfaw, and Zerihun Woldu. "Medicinal plants of the Meinit ethnic group of Ethiopia: an ethnobotanical study." *Journal of Ethnopharmacology* 124.3 (2009): 513-521.

González, José A., Mónica García-Barriuso, and Francisco Amich. "Ethnobotanical study of medicinal plants traditionally used in the Arribes del Duero, western Spain." *Journal of ethnopharmacology* 131.2 (2010): 343-355.

Jain, S.K., 1981. Glimpses of Indian Ethnobotany. Oxford and IBH Publishing Co., New Delhi

Jordan, S.A., Cunningham, D.G and marles, R.J. 2010. Assessment of herbal medicinal products: challenges and opportunities to increase the knowledge base for safety assessment. *Toxicology and Applied Pharmacology 243: 198-216*

Mukherjee, Pulok K., *et al.*, "Leads from Indian medicinal plants with hypoglycemic potentials." *Journal of Ethnopharmacology* 106.1 (2006): 1-28.

Principle Monetizing the pharmacological benefit of plants. Washington: US Environmental protection agencies, 2005

Rokaya, M.B., Uprety, Y., Poudel, R.C., Timsina, B., Munzbergova, Z., Asselin, H., Tiwari, A., Shrestha, S.S and Sigdel, S.R. 2014. Traditional uses of medicinal plants in Gastrointestestinal disorders in Nepal. *Journal of Ethnopharmacology.* 158: 221-229.

Sultana, S., Asif, H.M., Akhtar, N. and Ahmad, K. 2015. Medicinal plants with potential antipyretic activity: A review. *Asian Pacific Journal of Tropical Diseases* 5(Suppl 1): S201-S208

Teklehaymanot, Tilahun, and Mirutse Giday. "Ethnobotanical study of medicinal plants used by people in Zegie Peninsula, Northwestern Ethiopia." *Journal of Ethnobiology and Ethnomedicine* 3.1 (2007): 12.

2018, Ethnomedicinal Plants: A Biodiversity Treasure Pages 247–261
Editors: V.R. Mohan, A. Doss, P.S. Tresina and V. Sornalakshmi
Published by: ASTRAL INTERNATIONAL PVT. LTD., NEW DELHI

Chapter 9

Ethnobotanical Survey on Medicinal Plants Used by Malayali Tribals in Palamalai Hills, Salem District, Tamil Nadu, India

R. Rajeswari[1], S. Murugesh[1], P. Selvi[2] and P. Dhandapani[1]

[1]Research Scholar, Professor and Head[2],
Department of Botany, School of Life Sciences,
Periyar University, Periyar Palkalai Nagar, Salem – 636 011, Tamil Nadu
E-mail: rajeswarimsc1988@gmail.com; murugeshss@rediffmail.com

Introduction

Ethnobotany (from ethnology, study of culture, and botany, study of plants) is the scientific study of the relationship that exist between peoples and plants. Ethnobotanists aim to document, describe and explain complex relationships between cultures and (uses of) plants, focusing primarily on how plants are used, managed and perceived across human societies. This includes use for food, clothing, currency, ritual, medicine, dye, construction, cosmetics and a lot more. Richard Evans Schultes is called the "father of ethnobotany". (Schultes 1992).

Malayali Tribes

India is a mega diversity country in the world. A tropical country with rich biological and cultural diversity. There are about 67.37 million tribal people belonging to 573 tribal groups living in different geographic locations with various subsistence patterns (Rekka R, 2014). It is estimated that the predominant tribal areas comprise about 15 per cent of the total geographical area of our country. These tribal groups living in biodiversity rich areas possess a wealth of knowledge on the utilization and conservation of food and medicinal plants.

Tribal Communities

Malayali's were the largest tribal group constituting 47.6 per cent of ST population in the state of Tamil Nadu. It was claimed that they were descents of Vellalas of Kancheepuram and in the following invasion they fled to Shervaroy Hill ranges (Thurston and Rangachari,1909). Malayali's were generally illiterate and speak a local dialect of Tamil, physically they resemble to the Semong of Malaya. It has been well established that tribal communities have survived on their traditional knowledge base. Traditional medicines were the primary healthcare resources for the Malayali tribes to protect and to maintain their health.

According to the World Health Organization (WHO) about 65-80 per cent of the world's population in developing countries depends essentially on plants for their primary healthcare due to poverty and lack of access to modern medicine Awoyemi *et al.* (2012). About 80 per cent of the total population of Ethiopia is depending on traditional medicine to treat different types of human ailments Bekele (2007). The Rig-Veda written during 4500 BC to 1600 BC is believed to be the oldest repository of human knowledge about medicinal usages of plants in Indian subcontinent.

The plants used in ethnomedicine contain a wide range of substances that can be used to teach chronic as well as infectious diseases. Ethnomedical practices are preferred largely because medicinal plants are less expensive, readily available and reliable, and they are considered to have fewer side effects than modern medicines. The objectives of thesis study is to conduct an ethanomedicinal survey of medicinal plants used to cure various ailments in and around Palamalai hill in Salem district, Tamil nadu. The second objective is to intract with local traditional healers and document their knowledge on medicinal plants, their usage and the types of disease treated *etc.* The last objective is to create awareness among tribal people in the conservation of medicinal plants.

Salem is located in the North central part of Tamil Nadu, one of the most botanised areas of southern India. Palamalai hills is one of the famous hills in Salem district. The Palamali is situated in Mettur Tassel and located in Salem district of Tamil Nadu. The Palamalai hills started at the Salem district but ended with Erode district. The Nernjipettai is about 6 km from the heart of Mettur at the foot of Palamalai. Its hills almost reach out of the river with just highway from Mettur to Erode between Kavery river end the Hills. Palamalai is situated at a height of 4920 feet from sea level. Palamalai the shrine of Lord Siddheswara swamy is situated in the North Western foothill, 56 kilometers away from the Periyar university in Salem Nearby Mettur dam 11km from the Palamalai Hills. It is lying in between 11,8385073" North latitude and 77.7508048" East longitude of western Ghats. Palamali village most of the population is from ST (schedule tribes) constitutes 98.65 per cent.

Data Collection

The study was conducted during the period of April 2014 to March 2015. During the survey personal interviews were conducted with tribes of people and other traditional healers. The ethnobotanical medicinal data were collected through discussions among the herbal practioners in and around the area selected for study.

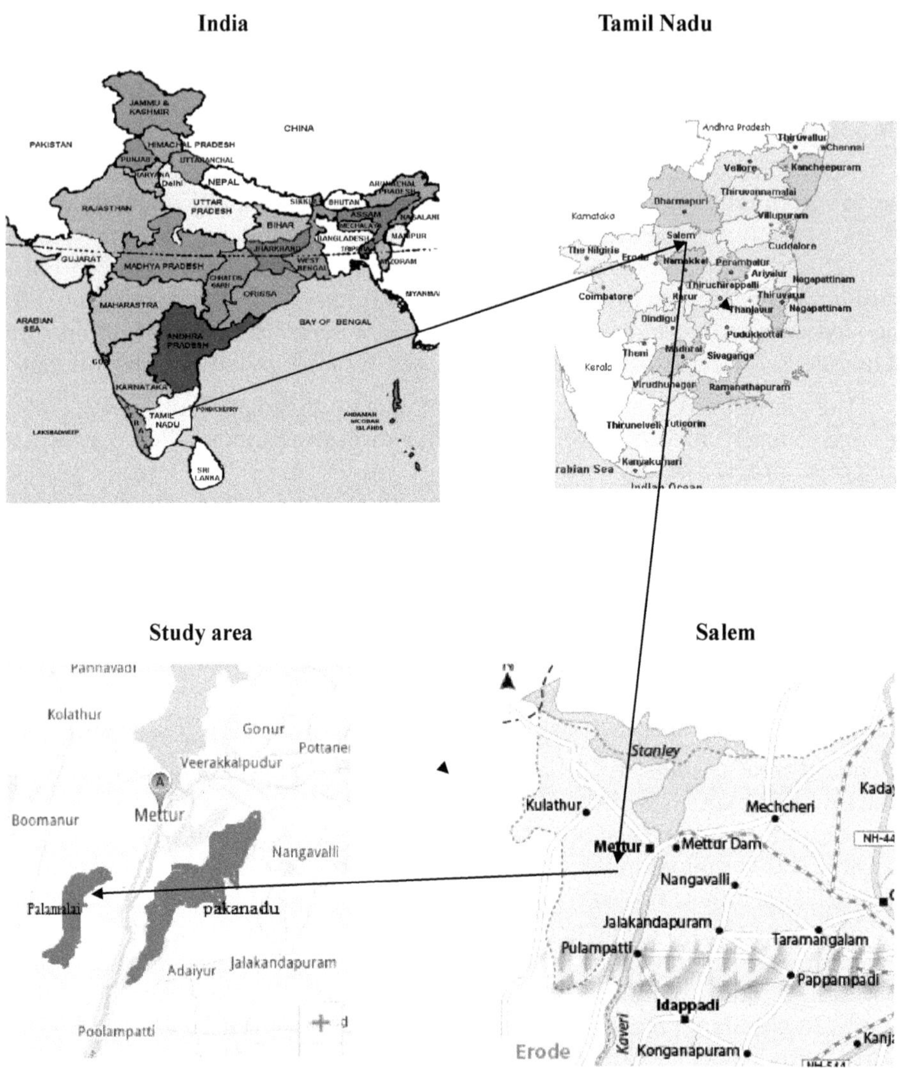

Figure 9.1: Map Showing from Study Area of Palamalai Hills.

Full View of Palamalai.

Temple.

Interaction with Tribals

Tribals involved in work

Figure 9.2: Study Area of Palamalai Hills.

Most of the information was gathered from elderly peoples of malayali tribals, who has very long acquaintance with usage of plants and information gathered from elderly information peoples of malayali tribals. The plant specimens were collected in their flowering and fruiting stage for voucher specimens based on the standard instructions.

The information was gathered through questionnaires, personal interviews and discussions among them. The interview was conducted with the people who are having the sound knowledge on medicinal plants found in their area and used by their families and neighbours in their local language (Tamil). The questionnaire contains the questions covering details of the plants, parts used, medicinal uses and mode of preparation of remedies. Each plant materials were assigned with field book botanical name, local name (Tamil) or vernacular name and medicinal uses. After identification, the families were arranged according to Bentham and Hooker's system of Classification (1883), Taxonomical categories, genera and species within the family are arranged, alphabetically, additional information about habit, life form were collected for each species. Plant species collected were identified with the help of flora books (Gamble, 1963; Hermy *et al.*, 1987; Matthew,1983). Detailed field notes were prepared for each species. Many photographs were taken in different angles with a digital camera. Among many photographs best pictures were chosen, most of the determinations were made using freshly collected plant.

Present investigation provides an ethnobotanical data of the medicinal plants used by the people of Palamalai Hills and nearby villages to cure various ailments. The study comprises survey and documentation of medicinal plants in the area selected for the study. In the present investigation, totally 113 plants belonging to 100 genera and 49 families have been recorded. The documented medicinal plants were tabulated alphabetically with botanical name, family name, common name, and parts used, medicinal uses taxonomical description and their medicinal uses. The plants were tabled with correct botanical name followed by family name (Table 9.1).

Table 9.1: List of Medicinal Plants and their Families

Sl.No.	*Botanical name*	*Family*
1	*Abutilon indicum* (Linn.) Sweet.	Malvaceae
2	*Acalypha indica* L.	Euphorbiaceae
3	*Acacia catechu* (L.f.)willd.	Mimosaceae
4	*Acacia leucophloea* (Roxb.) Willd.	Mimosaceae
5	*Acanthospermum hispidum* DC.	Asteraceae
6	*Acorus calamus* L.	Acoraceae
7	*Achyathus aspera* L.	Amaranthaceae
8	*Aegle marmelos* (L.)Correa.	Rutaceae
9	*Albizia amara* (Roxb.).Boivin.	Fabaceae (Mimosaceae)
10	*Allamanda cathartica* L.	Apocynaceae
11	*Aloe vera* (L.) Burm. f.	Liliaceae
12	*Andrographis serpyllifolia* (Rottler ex Vahl) Wight	Acanthaceae

Sl.No.	Botanical name	Family
13	*Andrographis paniculatata* (Burm.f.) Wall.ex Nees	Acanthaceae
14	*Anisomeles indica*(L.)Kuntze	Lamiaceae
15	*Annona squamosa* L.	Annonaceae
16	*Argemone Mexicana* L.	Papaveraceae
17	*Aristolochia bracteata* Lam.	Aristolochiaceae
18	*Artocarpus heterophllus* Lam.	Moraceae
19	*Azardiracta indica* A.Juss	Meliaceae
20	*Bidens pilosa* L.	Asteraceae
21	*Boerhavia diffusa* L.nom.cons.	Nyctaginaceae
22	*Bambusa arundinacea* (Retz.) Willd.	Graminaceae
23	*Butea monosperma* (Lam.)Taub.	Fabaceae
24	*Borreria verticillata* (L.) G.Mey	Rubiaceae
25	*Blainvillea acmella* (L.) Philipson	Asteraceae
26	*Cassia auriculata* L.	Caesapiniaceae
27	*Cassia fistula* L.	Caesapiniaceae
28	*Cassia occidentails* L.	Fabaceae
29	*Caesalpina sappan* L.	Caesapiniaceae
30	*Canvalia eniformis*(L.)DC.	Fabaceae
31	*Cananga odarata* (Lam.)Hook.f. and Thomson	Annonaceae
32	*Canthium diccoccum*(Gaertn.)Merr	Rubiaceae
33	*Calotropis gigantia* (L.) R. Br. ex Schult.	Asclepidaceae
34	*Centella asiatica* (L.)Urban	Apiaceae
35	*Crotalaria verucosa*	Fabaceae
36	*Ctenolepis garcini* (L.) C.B.Clarke	Cucurbitaceae
37	*Cathranthus roseus* (L.)G.Don.	Apocyanaceae
38	*Commelina clavata* C.B.Clarke	Commelinaceae
39	*Cissus quandragularis* L.	Vitaceae
40	*Cloeme viscose* L.	Capparidaceae
41	*Coleus amboinicus* Lour.	Lamiaceae
42	*Cardiospermum helicacabum* L.	Sapindaceae
43	*Clausena anisata* (Willd.) Hook.f. ex Benth.	Rutaceae
44	*Coccinia indica* (L.) Voigt	Cucurbitaceae
45	*Derris indica* (Lamk) Bennet.	Fabaceae
46	*Dodonea viscose* (L.)Jacq.	Sapindaceae
47	*Enicostemma littorale*	Gentianaceae
48	*Eclipta prostrata*(L.)L.	Asteraceae
49	Evolvulus alsinoides (L.)L.	Convolvulaceae
50	*Euphorbia hirta* L.	Euphorbiaceae
51	*Ficus carica* L.	Moraceae
52	*Gymnema sylvestre* R.Br.	Asclepidaceae

Sl.No.	Botanical name	Family
53	*Gerwia tiliaefolia* Vahl	Tiliaceae
54	*Glycyrrhiza glabra* L.	Fabaceae
55	*Hemidesmus indicus* (L.) R.Br.	Apocynaceae
56	*Heliotropium indicum* L.	Boraginaceae
57	*Justicia adathoda* L.	Acanthaceae
58	*Justicia gendarussa* Burm F	Acanthaceae
59	*Jantropha gossypifolia* L.	Euphorbiaceae
60	*Kyllinga triceps* G. Forst.	Cyperaceae
61	*Leucas aspera* (Wild.)	Lamiaceae
62	*Martyina annua* L.	Pedaliaceae
63	*Melia azedarach* L.	Meliaceae
64	*Mangifera indica* L.	Anacardiaceae
65	*Morinda tinctoria* Roxb.	Rubiaceae
66	*Moringa oleifera* Lam.	Moringaceae
67	*Momordica charantia* L.	Cucurbitaceae
68	*Mimosa pudica* L.	Fabaceae(Mimosoideae)
69	*Murraya paniculata* (L.) Jack.	Rutaceae
70	*Murraya koenigii* (L.)Sprengel	Rutaceae
71	*Myristica fragrans* Houtt.	Myristicaceae
72	*Ocimum tenuiflorum* L.	Lamiaceae
73	*Oldenlandia umbellate* L.	Rubiaceae
74	*Opuntia dillenii* (Ker Gawl.)Haw.	Cactaceae
75	*Parthenium hysterophorus* L.	Asteraceae
76	*Pergularia daemia* (Forssk.) Chiov.	Asclepiadaceae
77	*Plumbago indica* L.	Plumbaginaceae
78	*Plumeria obtuse* L.	Apocynaceae
79	*Pyllanthus niruri* L.	Phyllanthaceae
80	*Phyllanthus emblica* L.	Phyllanthaceae
81	*Physalis minima* L.	Solanaceae
82	*Phyla nodiflora* (L.) Greene.	Verbenaceae
83	*Pterocarpus marsupium* Roxburgh	Fabaceae
84	*Punica granatum* L.	Lythraceae
85	*Portulaca qrandiflora* Hook.	Portulacaceae
86	*Ricinus communis* L.	Euphorbiaceae
87	*Rubia cardifolia* L.	Rubiaceae
88	*Salvadora persica* L.	Salvadoraceae
89	*Sida acuta* Burm.f.	Malvaceae
90	*Sida cordifolia* L.	Malvaceae
91	Solanum atropurpureum L.	Solanaceae

Sl.No.	Botanical name	Family
92	*Solanum nigrum* L.	Solanaceae
93	*Solanum trilobatum* L.	Solanaceae
94	*Solanum torvum* Sw.	Solanaceae
95	*Strobilanthus ciliotus* Wall.	Acanthaceae
96	*Sphaeranthus indicus* L.	Asteraceae
97	*Spilanthes calva* DC.	Asteraceae
98	*Tamarindus indica* L.	Fabaceae
99	*Tribulus terrestris* L.	Zygophyllaceae
100	*Terminalia arjuna* (Roxb.)Wight and Arn	Combretaceae
101	*Terminalia bellirica* (Gaertn.)Roxb.	Combretaceae
102	*Tinospora cordifolia* (Thunb)Miers.	Menispermaceae.
103	*Tephrosia purpurea* (L.) Pers.	Fabaceae
104	*Thevetia peruviana* (L.) Lippold	Apocynaceae
105	*Trichodesma zeylanicum* (Burm.f)R.Br.	Boraginaceae
106	*Urena lobata* L.	Malvaceae
107	*Cyanthillium cinereum* (Carl Linnaeus) H.Rob	Asteraceae
108	*Vitex negundo* L.	Verbenaceae
109	*Vigna mungo* (L.) Hepper	Fabaceae
110	*Waltheria indica* L.	Malvaceae
111	*Wrightia tinctoria* (Roxb.) R.Br.	Apocynaceae
112	*Wrightia arborea*(Dennst.) D.J.Mabberley	Apocynaceae
113	*Zizyphus jujiba* Mill.	Rhamnaceae

Table 9.2: Family-wise Distribution

Sl.No.	Family	No. of Plants	Sl.No.	Family	No of Species
1	Fabaceae	12	13	Cucurbitaceae	3
2	Asteraceae	8	14	Mimosaceae	2
3	Apocyanaceae	7	15	Meliaceae	2
4	Acanthaceae	5	16	Phyallanthaceae	2
5	Rubiaceae	5	17	Annonaceae	2
6	Euphorbiaceae	4	18	Asclepidaceae	3
7	Lamiaceae	4	19	Verbinaceae	2
8	Rutaceae	4	20	Combretaceae	2
9	Solanaceae	5	21	Moraceae	2
10	Malvaceae	5	22	Sapindaceae	2
11	Cesalpinaceae	3	23	Other families	27
12.	Boranginaceae	2			

Table 9.3: Life Forms of Medicinal Plants

Sl.No.	Habit	No. of Plants
1	Herb	**39**
2	Shrub	**24**
3	Tree	**31**
4	Sub shrub	**3**
5	Climber	**16**

Table 9.4: Analysis of Parts Used

Sl.No.	Useful Part	No of Species
1	Leaves	**52**
2	Whole plants	**21**
3	Flower	**5**
4	Fruit	**14**
5	Root	**7**
6	Rhizome	**1**
7	Seeds	**4**
8	Truck wood	**1**
9	Stem	**2**
10	Bark	**6**

The most commonly represented families were Fabaceae-12 species and Asteraceae- 8, Apocyanaceae- 7 species, Euphorbiaceae, Laminaceae, and Rutaceae – 4 species. Solanaceae, malvaceae, Acanthaceae, and Rubiaceae-5 species,Combreataceae, Boranginaceae, Mimosaceae, Meliaceae, phyllanthaceae, Annonaceae, Verbinaceae, Moraceae and Sapindaceae -2species, Cesalpinaceae, Cucubitaceae and Asclepidaceae 3 species,while the remaining 27 families represented by single species (Figure 9.3). Fabaceae was the dominant family with 12 species. The population largely concentrated in five species: *Acalypha indica.*, *Artocarpus heterophyllus* Lam., *Cassia auriculata* L., *Cissus quanddrangularis* L., *lecus aspera* L. Among them most popularly used medicinal plants such as *Andrographis paniculata* Nees, *Solanum trilobatum* L, *Solanum surattense* Burm.f., *Melia azedarach* L., *Phyllanthus amarus* Schult and Thorn, *Andrographis paniculata* Nees, *Cassiafistula* L. *Aristolochia bracteata* L. *Hemidesmus indicus* R.Br. and *Aloe vera* L. plays significant role in the first health care system of tribals and rural people.

The life forms were analysed 39 species were Herbs; Tree species are 31; Shrub species are 24 and 16 species were climbers and Subshrub 3 (Figure 9.4). Analysis of part(s) used in seeds-4, Flowers-5, leaves-52, Bark-6, Fruits-14, Root-7, Rhizome-1,Stem-2,Truckwood-1 (Figure 9.5). Ethnomedicinal uses have been reported and investigation on the medicinal plants among the Palamalai hill. The most prevalent form of administration of medicine was juice/extract. This is followed by paste, decoction, power, plant parts as such and infusion.

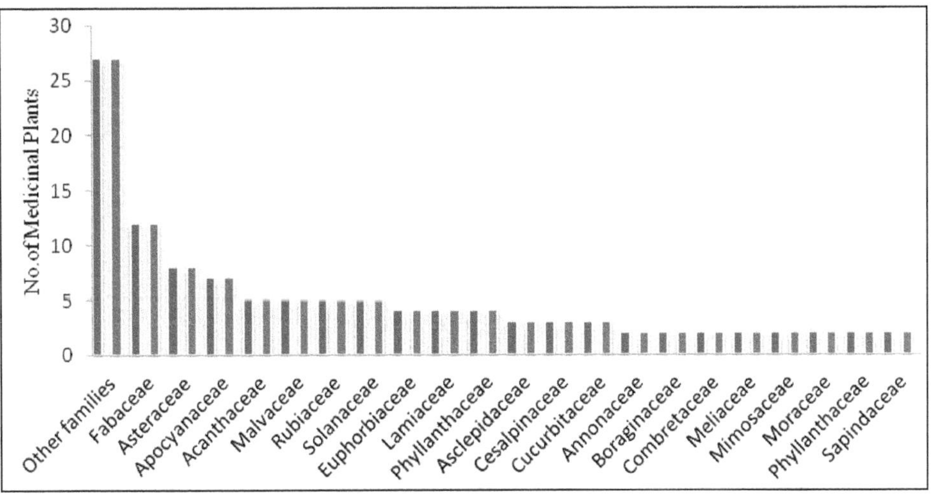

Figure 9.3: Family-wise Distribution of Ethnomedicinal Plants.

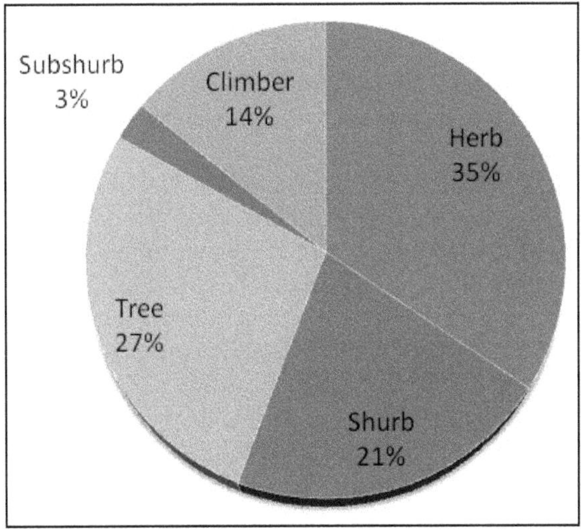

Figure 9.4: Distribution of Medicinal Plants of Various Lifeforms.

In the present study 113 medicinal plants were collected and documented. As an outcome of the present investigation, 113 plants belonging to 100 genera and 49 families were recorded. The documented plants used for the various human diseases by the *Malayali* in Eastern Ghats, Tamil Nadu show a great habitual diversity.

Traditional healers, use their eyes, ear, nose and hands to diagnose the diseases, this way of diagnose is interesting because they live in interior areas. They don't use of modern scientific equipments for treatment, however they treat diseases using medicinal plants (Santhya *et al.*, 2006). The tribal and rural people need plants

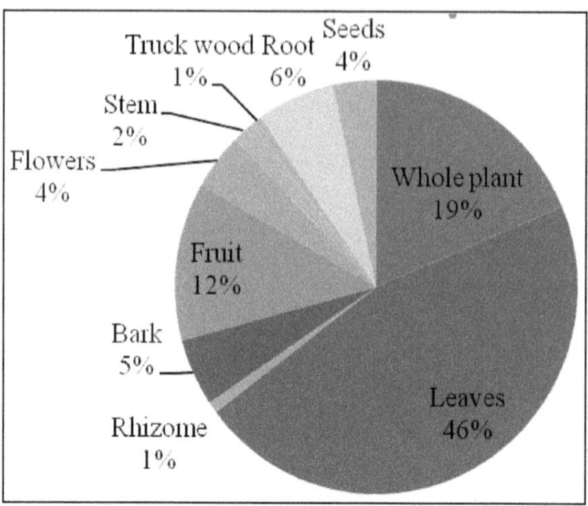

Figure 9.5: Plants Parts Used for of Medicinal Purposes and Percentage of Total Medicinal Species.

for medicine, food, fodder, nuts, craft, agricultural tools, house construction and for other necessities. Plants are utilized to cure snake and scorpion bites, fever, diabetes, jaundice, cough, skin diseases urinary troubles, dysentery, headache and toothache (Rajan *et al.*, 2002; Ayyanar and Ignacimuthu, 2005; Udayan *et al.*, 2006; Venkataswamy *et al.*, 2010; Umapriya *et al.*, 2012; Ramana Naidu *et al.*, 2012). Ethnomedicine is assured and reduced therefore the tribal and rural people of this area still use the traditional ailments. The enumeration has been compared with relevant published literature (Ramana Naidu *et al.*, 2012; Chopra *et al.*, 1956). Comparable studies on medicinal species in similarity is used and conservation has been conducted in different parts of India (Ignacimuthu *et al.*, 2006; Guru Prasad, 2011; Ramana Naidu *et al.*, 2012; Das and Choudhury, 2012). *Andrographis paniculata* plant juice is utilized to cure diabetes, jaundice and dysentery in the study area and the same is used and reported by Jayanthi *et al.* (2012). *Aristolochia bracteolata* plant decoction is given for snake bites and skin diseases.

From our survey of medicinal plants, the results obtained confirm the therapeutic potency of some plants used in traditional medicine. They were using these plants to cure diseases like fever, cough, ophthalmic obligation, intestinal worms, inflammations, indigestion, wounds, rheumatism, ear wound, cold,bone fracture, heal wounds, muscle cramp, anti-inflammatory, diurtic, lithontriptic, tonic, piles, sweet, cooling, stimulant, tonic demulcent, dysentery, scorpion bites, guinea worms, anaemia, fatigue, lassitude, nervous disorders, skin and liver diseases, urethritis and gonorrhea, cholera and haemorrhoids, venereal diseases, heal wounds, foul breath, scurvy, nausea and vomiting, arrest vomiting, diarrhea, ulcer, fits, small pox, rheumatism, skin diseases, strong teeth, *etc.* This was consistent with the general observations made earlier in relation to ethno botanical studies on some of the other tribal communities of Tamil Nadu reported by Karthikeyani (2003)

The leaves of the plant Phyllanthus amarus was combined with white goat milk and consumed in empty stomach in three doses for effective treatment of severe jaundice and liver diseases and it also enhances the appetite (Sankaranarayanan, 2008). *Acalypa indica* for eczema and chest pain, *Achyranthus aspera* extract is used for the treatment of urinary disorder (Jain and Patole 2001). Decoction or juice of the Azadirechta indica is taken for all body problems. It has been claimed to be useful as diuretic, anthelmintic, antidiabetic, expectorant and hepatoprotective in tradition system of medicine. It is also used for antimicrobial and cytotoxicity activity, diuretic, arolithiasis and anti-inflammatory activity (Manokaran *et al.*, 2008).

Certain species of Solanaceae are more important in medicinal field. *Datura metel* leaf paste is mixed with *Curcuma aromatic* rhizomes and applied on the swelling for quick remedy till the swelling reduces. Few drops of leaf juice is poured into ear to treat earache (Jeeva *et al.*, 2007). Leaf paste of *Solanum nigram* is applied externally to treat stomachache. Whole plant parts are taken as food to treat cough (Ramya *et al.*, 2008). *Coleus aromaticus* the vapours from the boiled leaves are inhaled to get relief from cough and cold. In this case the juice is put into the eyes two or three times in a day to relieve burning sensation and redness of eye (Ganesan *et al.*, 2007)) Although considerable research work was being done in India (Alagesaboopathi, 2013; Murthy 2012; Kumar *et al.*, 2010) a lot of important information and indigenous knowledge base are already lost as knowledge hold with older generations could not be transmitted to younger generations and remains unrecorded. Although the literature is replete with general references to ethno-botany for the country as a whole, efforts to document specific details of this knowledge are still limited and several workers have made their efforts on this direction.

The tribal people of Western Madhya Pradesh of India used 13 plants for treatment of Jaundice(Samvaster and Diwanji,2000). In the present study on Phyllanthus amarus and Eclipta prostrate were used for the treatment of Jaundice,It is in agreement with earilar reports in the treatment of oral diseases(tapsoba and Deschampus,2006; Hebbar *et al.*, 2004).Andrographis paniculata,Catheranthus roseus and Gymnema sylvestre were used to treat diabetes by the local traditional healers.The tribal people of Sikkim and Darjeeling Himalayas region in India utilized 37 species of plants belonging to 28 different families as antidiabetic agents.

The forests of Palamalai hills are rich in medicinal plants.The ancestors of tribes had acquired some knowledge about medicinal plants by their experiences. Otherwise technically advanced people need to understand the problems of destruction before conserving the plants. The results of present study reveals that a wide usage of plants by tribes of Palamalai hills. Many medicinal use of plants are endemic to certain tribes. Recently, considerable attention has been given to utilize eco-friendly plant based products for the prevention and cure of different diseases. The tribal knowledge of medicinal use of plants is still an unexploited area. A majority of Non-Timber forest products in Palamalai hills is for medicinal use. At present, tribals depend more on trained doctors from town because it gives immediate relief from pain. Tribals have no ideas about the active principles of the drugs. Palamalai hill is rich in several medicinal plants. Many medicinal plants are no longer used, because people's lack of knowledge about the medicinal values of plants. The ancestors of tribes had acquired some knowledge about plants by their experiences. The knowledge had been passed to the present generation orally. The role of the tribals in the conservation of the plants is a must for preserving plants for the future generation.

Conclusion

The present investigation is mainly focused on the survey and documentation of medicinal plants in Salem district of Palamalai hills Eastern Ghats, Tamil Nadu. 113 medcinal plants were collected and documented. The tribal and rural people of Palamalai hills, Salem district of Tamilandu has been using several plants of therapeutic utilize since time immemorial. Tribals and villagers mainly depend on the plants for all ailments. They are perceptive of the plant remedies for common diseases such as jaundice, diabetes, dysentery asthma, rheumatism, skin diseases, cough, urinary troubles, fever, dyspepsia and constipation. The collected medicinal plants were identified using some standard flora. The ethno botanical information of medicinal plants were gathered and compared with available relevant literature. The number of medicinal plants and their habitual diversity were discussed. The medicinal plants on further analysis, exploration collection of ethnobotanical information, chemical studies and screening gave the following results. It was very clear that the medicinal plants will provide cost effective and reliable source of medicine for the welfare of humanity. This study will promote a practical use of botanicals and must be continued focusing on its scientific validation.

Acknowledgements

The authors are grateful to the tribals from the area selected for study. During the field visit the tribal people were ready to reveal their traditional botanical knowledge. The research scholars are also thankful to the Botanical Survey of India (BSI), Southern Circle, Coimbatore, Tamil Nadu (India) for granting permission.

References

Alagesaboopathi.C (2011).Ethnomedicinal plants used as medicine by the kurumbatribalsin pennagaram, Dharmapuri District of Tamil Nadu.*Asian journal Experimental Biology,* 2 (1).

Anand RM, Nandakumar N, Karunakaran I, Ragunathan M, Murugan V (2006). A survey of medicinal plants in Kollimalai hill tracts, Tamil Nadu.*Nat Prod Rad. 5(2): 139–143.*

Arunachalam.G, Karunanith.M, Subramanian.N, Ravichandran.V and Selvamuthukumar.M(2009). Ethno Medicines of Kolli Hills at Namakkal District in Tamil Nadu and its significance in Indian Systems of Medicine.*Journal. Pharmacological Science and Research. 1(1), 115*

Ayyanar M, Ignacimuthu S.(2011). Ethnobotanical survey of medicinal plants commonly used by Kanitribals in Tirunelveli hills of Western Ghats, India.*J Ethnopharmacol. doi: 10. 1016/j. Jep.*

Ganesan S, Chandhirasekaran M and Selvaraju A (2008). Ethno-veterinary health care practices in Southern districts of Tamil Nadu. *Indian Journal.Traditional. Knowledge.,7*: 347-354.

Harshberger, J.W. (1896). The purpose of ethnobotany. *Bot. Gaz.,* 21:146-148.

Jain, S.K. (2010). Ethno-botany in India: some thoughts on future work. *Ethnobotany,*22: 01 04.

Jeeva s, Kiruba S, Mishra BP, Venugopal n, Das samsukumaran s, Regini GS, Kingston c, Kavitha A, Raj ADS,Laloo RC (2006). Weeds of Kanyakumari district and their value in rural life. *Indian journal of traditional knowledge(4):501-509.*

Kadhirvel (2010).Ethnomedicinal Survey on Plants used by Tribals in Chitteri Hills. *Environ We International Journal of Science. Technology. 5 35-45*

Karthikeyani, T.P. And K.Janardhanan.(2003). Indegenous medicine for snake, scorpion and nsect bites strings in siruvani Hills,Western Ghats, South india, *Asian Jr.of microbial,Biotech. Env,Sc.*Vol.5(4):467-470.

Kumar, V., Sachan, P, Nigam, G, Singh,P.K. (2002). Some ethnomedicinal plant of Chitrakoot district (U.P.).*Biozone International. Journal of Life Science.,*2(1 2): 270283.

Manikandan, AlaguLakshmanan (2014). Ethnobotanical Survey of Medicinal Plants in Kalrayan hills, Eastern Ghats, Tamil Nadu. *International Letters of Natural Sciences* .17 pp111-121

Murthy, E.N.(2012). Ethnomedicinal plants used by Gonds of Adilabad district, Andhra Pradesh, India.*Int. J. Pharm.Life Sci.,* 3(10): 2034-2043.

Parinitha, M., Srinivasa, B.H. and Shivanna, M.B.(2005). Medicinal plant wealth of local communities in ShimogaDistirct of Karnataka, India.*Journal of Ethnopharmacology*.98: 307 – 312.

Rakeshsama, P. N.Shrivastava and manjujain (2015).Ethnobotanical Study of Traditional Medicinal Plants Used By Tribe of Guna District, Madhya Pradesh, India. *Internationalournal Current Microbiology Application Sciences.* 4(7): 466-471.

Ranganathan.R,R.Vijayalakshmi, P.paramaeswari.(2012).Ethnobotanical Survey of JawadhuHills in Tamil Nadu,India.*Asian Journal Pharmacological Clinical Research, 5, Suppl 2,* 45-49.

Raymond Ntume and Godwin Anywar (2015).Ethnopharmacological survey of medicinal plants used in the treatment of snake bites in Central Uganda.*ISSN* 2449-8866.

Rekha, Tamil Selvi S, Bharathidasan, Panneerselvam A, Ilakkiya R and Jayapa. (2013). Study of Medicinal Plants used from Koothanoallur and Marakkadai, Thiruvarur district of Tamil nadu, India. *Hygeia Journal Development Medicinal. 5(1)*: 160-166.

Rekka R., And S.Senthil Kumar. (2014).Indigenous Knowledge On Some Medicinal Plants Among The Malayali Tribals In Yercaud Hills, Eastern Ghats, Salem

District, Tamil Nadu, India. *International Journal Of Pharma And Bio Sciences.* 5(4): 371–376.

SalaiSenthilkumar M. S., D.Vaidyanathan, D. Sivakumar and M. GhouseBasha. (2014). Diversity of ethnomedicinal plants used by Malayali tribals in Yelagiri hills of Eastern ghats, *Asian Journal of Plant Science and Research.* 4(1): 69-80.

Santhya B, Thomas S, Isabel W, Shenbagarathai R,(2006). Ehnomedicinal Plants used by the Valaiyan community of Piranmalai hills (Reserved Forest), Tamil Nadu, India.- A pilot study. *African Journal Traditional CAM,* 3(1): 101-114.

Sasikumar,J.M, karthikeyeni T.P and Janardhan k.(2001).Traditional phytocure of paralysis among Badagas of Nilgiris district. *Advance Lant Science,* 14(1): 517-518.

Schultes, Richard Evans; Robert F. Raffauf (1992). *Vine of the Soul: Medicine Men, Their Plants and Rituals in the Colombian Amazonia.* Oracle, Ariz.: Synergetic Press. *ISBN* 0-907791-24-7.

Vethanarayanan, P., Unnikannan, P.,Baskaran,L.andSundaramoorthy,P (2011). Survey on traditional medicinal plants used by the village peoples of Cuddalore district, Tamil Nadu, India.*Asian J. Bioche.Andpharma. Res.* 3 (1): 351-36.

2018, Ethnomedicinal Plants: A Biodiversity Treasure *Pages* **263–292**
Editors: ***V.R. Mohan, A. Doss, P.S. Tresina and V. Sornalakshmi***
Published by: **ASTRAL INTERNATIONAL PVT. LTD., NEW DELHI**

Chapter 10

Ethnobotanical Study of Medicinal Plants Used by the *Vetan* Community in Kanyakumari District, Tamil Nadu, India

J. Celin Pappa Rani, M.S. Kala Swarna and S. Jeeva*

Department of Botany, Scott Christian College,
Nagercoil – 629 003, Tamil Nadu
**E-mail: solomonjeeva@gmail.com*

Introduction

According to the World Health Organization (WHO) about 65-80 per cent of the world's population in developing countries depend essentially on plants for their primary healthcare due to poverty and lack of access to modern medicine (Calixto 2005). About 85 per cent of the rural population of India depends on varieties of plants for the treatment of various diseases they suffer from (Farnsworth and Jain, 1994). India is rich in cultural and floristic diversity. India is also a storehouse of ethnobotanical knowledge. The indigenous tribal communities possess a broad knowledge about medicinal plants, passed through oral communication from generation to generation (Rajkumar and Shivanna, 2009). The knowledge of the tribes, associated with the traditional healing practices using plants, is now fast disappearing due to modernization, rate of deforestation with the concurrent loss of biodiversity, there is a need for accurate documentation of the tribal knowledge. During the last few decades there has been an increasing interest in the study of medicinal plants and their traditional use in different parts of India. In the recent years number of reports on the use of plants in traditional healing by either tribal people or indigenous communities of India is increasing (Savithramma *et al.*, 2007; Pattanaik, *et al.*, 2008; Kosalge and Fursule, 2009; Namsa *et al.*, 2009; Upadhyay,

2010).There is no report available in the literature about *Vetan* community people. Therefore this study was undertaken to ascertain the detailed information on plants used by *Vetan* community people and their usage based on ethnobotanical knowledge.

Materials and Methods

Study Area and People

Shankaranputhur is a small village situated near Kulasekeranputhur, with 40 acres of land in Kanyakumari district and coversan area of nearly 1684 sq. kms. This village was named after 'Shankaranthambi' who was the divan of Travancore King. In Tamil language the word 'Vettuvan' or 'Vetan' means hunter. Most of the people are farmers. The grains produced were transported to Travancore Samasthanam with the help of bullock carts. Farmers suffered a lot to bring their yield to Samasthanam. The King knew their difficulties and provided land as well a shield (cheppu patayam- shield containing details about *Vetan* community). *Vetan* community people have the knowledge of medicinal plants that are used for first aid remedies and other simple ailments. They still depend on medicinal plants to meet their primary healthcare needs.

Data Collection

An ethnobotanical survey was conducted among the *Vetan* community people who inhabit Shankaranputhur village of Kanyakumari district. Intensive field surveys were conducted between July 2012 and August 2013 according to the methods proposed by Martin (1995), with slight modifications.During the study trips, information were gathered by interviewing the local medicine men, experienced old aged people who have good knowledge about the medicinal uses of plants. Hundred seventy eight informants (98 males and 80 females) between the ages of 35-50 were consulted to gather information in the study area. Ethnobotanical information was collected through questionnaires and discussions among the informants in their local language (Tamil), which include formulations such as mode of preparation (*i.e.*, decoction, paste, powder and juice), diseases it cured and the dosage. Plant specimens were collected, prepared herbariums and identified with the help of local and regional floras (Gamble and Fischer, 1956; Nair and Henry, 1983). Plant names have been checked and updated with the online website (www. the plantlist.org) of the Royal Botanic Gardens, Kew. Herbarium specimens were deposited in the Department of Botany, Scott Christian College, Nagercoil.

Results and Discussion

The present investigation revealed that the *Vetan* community people were using 153 plant species distributed in 134 genera belonging to 60 families for medicinal use. Euphorbiaceae (10 taxa) and Solanaceae (9 taxa) predominates followed by Amranthaceae (8 taxa), Cucurbitacea (8 taxa), Fabaceae (8 taxa), Verbenaceae (7 taxa), Lamiaceae (6 taxa), Asteraceae (5 taxa) Malvaceae (5 taxa), Moraceae (5 taxa), Rutacaeae (5 taxa), Arecaceae (5 taxa), Mimosaceae (4 taxa), Poaceae (4 taxa), Caesalpiniaceae (3 taxa), Asclepiadaceae (3 taxa), Araceae (2 taxa), Apiaceae

(2 taxa), Annonaceae (2 taxa), Convolvulaceae (2 taxa), Nyctaginaceae (2 taxa), Rubiaceae (2 taxa), Sapotaceae (2 taxa), Zingiberacae (2 taxa) remaining families were monospecific. Ahmad *et al.* (2006) and Okgibo *et al.* (2009) reviewed a variety of Euphorbiaceae family contains wide variety of phytochemicals including alkaloids, phenols, flavonoids, saponin, tannins and essential oils and described their origins, characteristics and their therapeutic uses.For each species, botanical names, local names, family, habit, part used, method of preparation, ailments treated and the phytochemical constituents are provided (Table 10.1). *Vetan* community people are using these plants to cure diseases like asthma, blood pressure, chicken-pox, constipation, cough and cold, diabetics, diarrhea, epilepsy, fever, bone fracture, headache, jaundice, malaria, menstural disorder, oral disease, (bleeding gums, mouth blisters, tooth ache), piles, poison bites (dog bite, snake bite), skin diseases (eczema, leprosy, ring worm, burns, cut, wounds), stomach complaints (ulcer). Similar studies were done among *Kani* tribes of Pechiparai forest and documented 58 medicinal plants belonging to 27 families to treat various ailments (Subitha *et al.*, 2011). Herbs (70 species) 44 per cent were found to be the most used plants species followed by shrubs (52 species) 32 per cent, trees (30 species) 19 per cent and climbers (8 species) 5 per cent (Figure 10.1). The frequent use of herbs among the indigenous communities is the result of a wealth of herbaceous plants in their environs (Tabuti *et al.*, 2010; Uniyal *et al.*, 2006). Herbs have been used as medicine since time immemorial in all cultures (Muthu *et al.*, 2006; Barnes *et al.*, 2007; Namukobe *et al.*, 2011; Jeeva and Femila, 2012; Shrestha *et al.*, 2016). Among the different plant leaves (80 species) were most frequently used for the traditional medicine preparation followed by fruits (40 taxa), roots (28 taxa), whole plant(15 taxa), bark, seed and stem (10 taxa each), bulbs (2 taxa), rhizome (5 taxa), tuber (4 taxa).Many indigenous communities elsewhere also utilized mostly leaves for medicine preparation (Mahishi*et al.*, 2005; Srithi *et al.*, 2009). Medicines were prepared in the form of paste, powder, juice, infusion and decoction. Preparing paste for the treatment of ailments was a widespread practice among the tribal communities in India (Raghupathy *et al.*, 2009; Ignacimuthu *et al.*, 2006). In the present study, plants like *Abutilon indicum*, *Phyllanthus amarus* were used for healing wounds, while *Eclipta prostrata*, *Vitex negundo* were used to cure head ache. *Punica granatum* was used to treat abdominal pain. *Terminalia catappa* used to treat asthma, *Cynodon dactylon* used for body ache, *Coccinia grandis* was used as a remedy for cholera, *Acalpha indica* to treat ear ache and constipation, *Ocimum tenuiflorum* leaf paste was used to cure cough, cold and ear ache and *Phyllanthus emblica* fruit was used to heal skin diseases and as a blood purifier. In the present study 4 plant species *Achyranthes aspera*, *Mukia maderaspatana*, *Raphanus sativus* and *Aristolochia bracteolate* were used to treat pneumonia, fever, cuts and wounds. These findings were correlated with the *Kani* tribals in Triunelveli hills (Ayyanar and Ignacimuthu, 2005) and Margala hills national park tribal people (Katewa *et al.*, 2004). In the present study, some of these plants are given for problems such as white discharge (*Tephorsia purpurea*), *Solanum surattense* and *Calotropis gigantea* are applied for tooth infections *Acalpha indica*, *Adhatoda vasica*, *Euphorbia hirta*, *Ocimum tenuiflorum*, *Solanum surattense* and *Leucas aspera* were used for asthma, cold, cough, and fever. The same plants were used by the tribals of Kanjamalai hills of Salem district of Tamil Nadu (Algesaboopathi, 2011). The *Vetan* people used the stem and

Table 10.1: List of Ethnomedicinal Plants used by the *Vetan* Community People

Botanical Name	*Family*	*Local Name*	*Common Name*	*Useful Parts*	*Habit*	*Medicinal Uses/ Mode of Usage*	*Major Phytochemical Constituents**
Abelmoschus esculentus (L.) Moench	Malvaceae	Vendai	Ladies finger	Fruit	Shrub	Daily intake of fruits helps in increasing the memory power and normalizes the blood sugar level.	Catechin, oligomers and hydroxycinnamic
Abrus precatorius L.	Fabaceae	Kunni	Crab's Eye	Leaf	Climber	Leaves with a little sugar chewed to get relieve from throat pain.	Saponins, tannins, triterpene, flavanoids, glycosides
Abutilon indicum (L.) Sweet	Malvaceae	Thutti	Indian mallow	Leaf	Herb	Decoction is taken orally seven days for immediate recovery from bleeding piles	Tannins, alkaloids, asparagines, gallic acid, sesquiterpenes, leucoanthocyanins, flavanoids, triterpenoids, saponins and cardiac glycosides.
Acacia nilotica (L.) Willd	Mimosaceae	Karuvelam	Taruakadam	Bark	Tree	Bark paste applied on infectious parts to cure wounds, ulcer and gum mixture taken orally once a day to cure urinogenital discharges.	Stearic acid, kaempferol-3-glucoside, isoquercetin, leucocyanidin
Acalypha indica L.	Euphorbiaceae	Kupaimeni	Indian copper leaf	Leaf	Herb	Leaf paste applied topically on scabies.	Alkaloids, acalypus and oside and acalyphine, cyanogenetic glucoside and triacetanamine.
Achyranthes aspera L.	Amaranthaceae	Naiyurvi	Prickly chaf flower	Root	Herb	Root paste mixed with honey is given orally twice a day to cure cough.	Achyranthine, betaine, oleanolic acid saponins, tannins and glycosides

Botanical Name	*Family*	*Local Name*	*Common Name*	*Useful Parts*	*Habit*	*Medicinal Uses/ Mode of Usage*	*Major Phytochemical Constituents**
Acmella paniculata (Wall. ex DC.) R.K.Jansen	Asteraceae	Manjal poochedi	Tooth ache plant	Flower	Herb	Flower paste applied externally to treat tooth ache.	Alkaloids, glycosides, flavanoids, tannins, anthraquinones, saponins and cardiac glycosides
Acorus calamus L.	Araceae	Vasambu	Sweet flag	Rhizome	Creeper	Rhizome decoction taken orally once in day to cure dysentery.	Beta-Asarone, isocalamendiol, Monoterpene hydrocarbons and sequestrine ketones
Aegle marmelos (L.) Correa	Rutaceae	Vilvam	Bengal quince	Fruit	Tree	Fruits and leaf decoction are taken for three days used to cure diarrhoea, fever.	Alkaloids, cardiac glycosides
Aerva lanata (L.) A. L.Juss. ex Schultes	Amaranthaceae	Sirukanpilai	Mountain knot grass	Leaf	Herb	Leaf decoction taken orally for 21 days cures kidney stone.	Ervine, aervine, methylaervine,aervoside, ervolanine, aervolanine, kaempferol, quercetin, isorhamnetin, persinol and benzoic acid
Albizia lebbeck(L.) Benth.	Mimosaceae	Vagai	Lebbeck tree, flea tree	Stem	Tree	Stem decoction is used to reduce cholesterol, cures asthma and rheumatism.	Alkaloids, dimethoxy flavone, N- Benzoyl and *L*- phenyl alaninol
Allium cepa L.	Lilliaceae	Vengayam	Onion	Bulb	Herb	Bulb juice and paste used to treat ear ache and head ache.	Acrid volatile oil, albuminoids, soluble carbohydrates and sugar, catechol and protocatechuic acids
Allium sativum L.	Lilliaceae	Vellaipundu	Garlic	Bulb	Herb	Daily intake of bulb cures rheumatism and gas trouble.	Alliin, vitamin C, total phenol and total flavonoid
Aloe vera (L.) Burm. f.	Lilliaceae	Chothukathalai	Aloe	Root and Leaf	Herb	Root used to cure indigestion and leaf gel used to reduce Leucorrhoea.	Aloin, isobarbaloin, emodin, chrysophanic acid and urinic acid

Contd...

Botanical Name	Family	Local Name	Common Name	Useful Parts	Habit	Medicinal Uses/ Mode of Usage	Major Phytochemical Constituents*
Alternanthera sessilis (L.) R. Br. ex DC.	Amaranthaceae	Ponnangani keerai	Ruby alternanthera	Leaf	Herb	Daily intake of cooked leaves taken orally cures night blindness.	α- and β-tocopherols, stigmasterol and β- sitosterol
Amaranthus roxburghianus H.W.Kung	Amaranthaceae	Sirukeerai	Tumble weed	Leaf	Herb	Leaf paste is applied topically on affected places twice a day to heal wounds.	Apigenin, chrysin, isorhamnetin and kaempferol
Amaranthus spinosus L.	Amaranthaceae	Mullukeerai	Prickly amaranth	Leaf	Herb	Cooked leaves taken orally is used to cure anemia and fever.	Flavonoids, spinoside 6, xylofuranosyl uracil, hydroxyl-cinnamates, quercetin, betalains, betaxanthin, betacyanin, amaranthine and isoamaranthine
Amaranthus tricolor L.	Amaranthaceae	Thandankeerai	Chinese spinach	Leaf	Herb	Decoction of leaves drunk to reduce the menstural pain.	Phenols, tannins and flavonoids
Amaranthus viridis L.	Amaranthaceae	Kuppaikeerai	Green amaranth	Leaf	Herb	Root decoction is used twice a day to control excess bleeding during menstruation.	Squalene, trilinolein, polyprenol, phytol and spinasterol
Amorphophallus paeoniifolius (Dennst.) Nicolson	Araceae	Chenaikilangu	Elephant foot yam	Rhizome	Herb	Intake of cooked rhizome cures piles.	Amblyone, Quercetin, steroids
Anisomeles malabarica (L.) R. Br.ex. Sims	Lamiaceae	Perumthumbai	Malabar catmint	Leaf	Shrub	Leaves are boiled with water to cure rheumatic swelling and smoke inhaled to cure cough.	Flavones, echiodinin, echioidin
Annona squamosa L.	Annonaceae	Chethapalam	Custard apple	Leaf and Seed	Tree	Crushed leaf juice taken orally to heal stomach ulcers. Seed paste mixed with oil used to destroy lice from hair.	Kaurenoic acid, taraxerol, β-sitosterol, 16α-hydro-19-al-ent-kauran-17-oic acid, β-hydroxystigmast-4-en-3-one, and 17-acetoxy-16α-ent-kauran-19-oic acid

Botanical Name	*Family*	*Local Name*	*Common Name*	*Useful Parts*	*Habit*	*Medicinal Uses/ Mode of Usage*	*Major Phytochemical Constituents**
Areca catechu L.	Arecaceae	Pakkumaram	Areca nut	Seed	Slender unbran-ched palm	Seed paste is applied on infected parts to cure wounds.	Tannins, flavanoids, lauric acid, palmitic acid, stearic acid, oleic acid, linolinic acid.
Aristolochia bracteolata Lam.	Aristolochiaceae	Karudakodi	Ducthman' s pipe	Leaf	Herb	Leaf paste is applied topically cures wounds and heals skin rashes.	Alkaloid, saponin. Tannins, phenols, sterols
Artocarpus heterophyllus Lam.	Moraceae	Palamaram	Jack fruit tree	Fruit	Tree	Fruits used to cure skin diseases and reduce cough.	Phenolic compounds, flavonoids, stilbenoids, arylbenzofurons, carotenoids, volatile acid sterols and tannins
Artocarpus hirsutus Lam.	Moraceae	Ayenissakai	Anjili	Fruit	Tree	An infusion of the bark is applied to cure small pimples and cracks on the skin.	Alkaloids, flavonoids, saponins, terpenoids
Azadirachta indica A. Juss.	Meliaceae	Veppamaram	Neem	Leaf bark	Tree	Leaf paste used to cure skin diseases and bark powder used to promote tooth strengthening.	Margosine, margosic acid, margosopicrin, glycerides of fatty acids, butyric acid, vatericacid, nimbin, nimbidin, nimbinin, nimbiserin and bakayanin
Azima tetracantha Lam.	Salvodoraceae	Esanku	Needle brush	Leaf	Tree	Leaves are used to cure cough.	Alkaloids, flavanoids, sterols
*Basella alba*L.	Basellaceae	Pasalikeerai	Indian spinach	Leaf	Herb	Cooked leaves are edible and cures fungal infection in stomach.	Alkaloids, Chorogenic acid, Anthocyanin, Steroidal Glycosides, Saponins Glycosides and Coumarin

Contd...

Botanical Name	Family	Local Name	Common Name	Useful Parts	Habit	Medicinal Uses/ Mode of Usage	Major Phytochemical Constituents*
Bauhinia tomentosa L.	Caesalpiniaceae	Eruvachi	Bauhinia	Flower	Tree	Juice of flower is used to cure diarrhoea, dysentery and paste to cure skin diseases.	Beflavone, methylenedioxy flavone, methyldibenzoxepin, flavonolglycoside, triterpene saponin, phenanthraquinone and flavonoids.
Beta vulgaris L.	Chenopodiaceae	Beetroot	Beet root	Root	Herb	Root paste applied on infected area to cure burning sensation.	Glycine betaine, carotenoids, flavonoid anti-oxidants
Boerhavia diffusa L.	Nyctaginaceae	Mukku urataikerai	Pigweed	Whole plant	Herb	Whole plant decoction prepared with jaggery taken once a day used to cure stomach ache and fever.	Boerhaavinone A- F, C-methyl flavone, Kaempferol, Punarnavine-1, Punarnavine-2
Borassus flabellifer L.	Aracaceae	Panaimaram	Palmyra palm	Young fruit	Tall palm	Pulpy kernel water of palmara applied on the eyes to cure eye ache.	Saponins, flavonoids, phenolic compounds and glycosides.
Calotropis gigantea (L.) Dryand.	Asclepiadaceae	Erukku	Crown flower,milk weed	Root and bark	Shrub	Root and bark decoction along with palm sugar taken for 21 days cures elephantiasis.	Alkaloids, Cardiac glycosides, Flavonoids, Saponins, Steroids, Tannins, Terpenoids, Usharin, gigantin, calcium oxalate, alpha and beta-calotropeol
Canna indica L.	Cannaceae	Kalvalai	Indian shot	Rhizome	Shrub	Rhizomes extract acts as diuretic and taken three days to cure fever.	Betulinic acid, Oleonolic acid and Traraxer-14-en-3-one
Capsicum frutescens L.	Solanaceae	Kantharimilagu	Red chilly	Fruit	Shrub	Fruits are edible and reduce blood cholesterol.	Capsaicin, capsaicinoids, oleoresins

Botanical Name	Family	Local Name	Common Name	Useful Parts	Habit	Medicinal Uses/ Mode of Usage	Major Phytochemical Constituents*
Capsicum annum L.	Solanaceae	Milagai	Long chilly	Fruit	Shrub	Daily intake of fruit enhances blood circulation.	Alkaloids, flavanoids, capsaicin, capsaicinoids, oleoresins, dihydrocapsaicin nonivamide. pelargonic acid and vanillylamide
Cardiospermum halicacabum L.	Sapindaceae	Mudakathhan	Balloon vine	Leaf	Climber	Decoction of leaves are used to cure rheumatism.	β-sitosterol, Xalic acid and quebrachitol
Carica papaya L.	Caricaceae	Pappali	Papaya or melon like fruit	Fruit	Tree	Fruits are edible and increases eye vision.	β-sitosterol, Xalic acid and quebrachitol
Cassia auriculata L.	Caesalpinaceae	Avarmpoo	Tannres cassia	Leaf	Shrub	Leaf paste taken orally to cure gas trouble problems.	Alkaloids, glycosides, flavanoids, saponins,tannins, phenols and steroids.
Casuarina litorea (L.) Diss. Forest.	Casurinaceae	Chavukku	Casurina	Bark	Tree	Bark decoction taken orally cures fever.	Not available
Catharanthus roseus (L.) G. Don.	Apocynaceae	Nithayakalyani	Periwinkle	Whole plant	Herb	Whole plant paste taken orally are used to cure leukemia and hypertension.	Alkaloids, Flavonoids, Polyphenols, Saponins, Steroids, Coumarins, Terpenoids, Vit-C Indophenol, Tannins and Cardiac glycosides
Centella asiatica (L.) Urb.	Apiaceae	Vallarai	Water pennywort	Leaf	Herb	Handful of leaves applied on skin to cure skin infection. It increases a memory power.	Vallarine, asiaticoside oxy-asiaticoside, fatty oil, sitosterol, tannin, pectic acid, ascorbic acid and alkaline hydrocotyline

Contd...

Botanical Name	Family	Local Name	Common Name	Useful Parts	Habit	Medicinal Uses/ Mode of Usage	Major Phytochemical Constituents*
Chrysanthemum indicum L.	Asteraceae	Sevanthi	Chrysanthus	Flower	Shrub	Flower extract is drunk to cure fever, head ache and hypertension.	Triterpenoids, Alkaloids, flavonoids
Cissus quadrangularis (L.)	Vitaceae	Pirandai	Adament creeper	Stem	Climber	Cooked stem is taken orally to cure stomach aches.	Ketosteroids, sitosterol, alphaamyrin, α-ampyrone and tetracyclic triterpenoids
Citrullus colocynthis (L.) Schrad.	Cucurbitaceae	Kumattikkai	Bitter apple fruit	Root	Herb	Root paste applied on head before bath used to promote hair growth.	Glycosides, flavonoids, alkaloids, fatty acids essential oils, curcurbitacins A, B, C, D, E, I, J, K, and L and Colocynthosides A and B
Citrus aurantifolia (Christm.) Swingle	Rutaceae	Narthankkai	Bitter orange tree	Leaf, fruit	Shrub	Leaves and fruit extract used to cure vomiting, fever and diabetic problems.	Flavonoids, Glycosides, Anthraquinones and Phlobatannins
Citrus limon (L.) Osbeck	Rutaceae	Elumichai	Lemon	Fruit	Shrub	Fruit juice is taken to cure indigestion and jaundice.	3-Methoxy,3-methyl-1-pentene, Camphor, Limonene, α-Citral, β-Citronellol and α-Pinene
Cleome gynandra L.	Capparidaceae	Thivalai	African spider flower	Leaf	Herb	Juice of leaf used to cure rheumatism, head ache and ear ache.	Phenolics, flavanoids, flavanols, proanthocyanidin
Clerodendrum phlomidis L. F	Verbenaceae	Vathamadaki	Arni	Leaf	Shrub	Leaf paste is applied on the forehead to cure headache and also cures rheumatic pain	D-mannitol, β-D-glucoside of β-sitosterol, β-sitosterol and ceryl alcohol
Clitoria ternatea L.	Fabaceae	Sangupuhpam	Butterfly pea	Leaf	Shrub	Leaf paste applied topically over affected area to cure wounds.	Flavonol glyceride and kaempferol
Coccinia grandis (L.) Voigt.	Cucurbitaceae	Kovaikkai	Ivy-Gourd	Fruit	Climber	Raw fruits taken orally are used to cure stomach ulcer.	Lupeol, Lycopene, Beta-Carotene, Taraxerol Stigmast-7-en-3-one

Botanical Name	Family	Local Name	Common Name	Useful Parts	Habit	Medicinal Uses/ Mode of Usage	Major Phytochemical Constituents*
Cocos nucifera L.	Arecaceae	Thengu	Coconut	Whole plant	Tree	Coconut milk is used to cure administrated for 30 days jaundice, chickenpox.	Phenols, tannins, leucoanthocyanidins, flavonoids, triterpenes, steroids, and alkaloids
Colocasia esculenta (L.) Schott.	Arecaceae	Chem-bukilangu	Taro, Elephant ear	Rhizome	Herb	Daily intake of cooked rhizome cures piles.	Flavones, apigenin
Coriandrum sativum L.	Apiaceae	Kothamalli	Coriander	Leaf	Herb	Leaves are used to cure jaundice and vomiting.	Benzofuran,2,3-dihydro, hexadecanoic acid, methyl ester, benzofuran, 2-methyoxy-4-vinylphenol, dodecanoic acid
Crataeva magna (L.) DC.	Capparidaceae	Mavilingam	Three-lived caper	Root and Bark	Tree	Root and bark are used to treat skin inflammation.	Flavonoids, phenols, saponin, tannin, coumarin and carbohydrates.
Crotalaria retusa L.	Fabaceae	Killukilluppai	Devil bean	Leaf	Shrub	Leaf paste applied topically cures skin diseases.	Pyrrolizidine alkaloid Saponins, Tannins
Cucumis sativus L.	Cucurbitaceae	Vellerikkai	Cucumber	Fruit Seed	Climber	Fruits used as edible and powdered seeds are applied on infected area to cure burning sensation.	Saponins, glycosides, terpenes, phenolics, alkaloids, flavonids
Cucurbita maxima Duchesne.	Cucurbitaceae	Pusanikai	Squash winter	Fruit	Climber	Fruits are edible and acts as coolant.	Lipids, flavonoids, alkaloids and vitamin C
Curcurma longa L.	Zingiberaceae	Manjal	Turmeric	Rhizome	Herb	Rhizome paste applied topically cures skin infection.	Curcumin, desmethoxycurcumin, monodemethoxycurcumin, bisdemethoxycurcumin, dihydrocurcumin and cyclocurcumin

Contd...

Botanical Name	Family	Local Name	Common Name	Useful Parts	Habit	Medicinal Uses/ Mode of Usage	Major Phytochemical Constituents*
Cynbopogon citratus (DC.) Stapf.	Poaceae	Vasanaipullu	Lemon grass	Leaf	Herb	Juice of whole plant is drunk to cure head ache and rheumatism.	Myrcene, Geraniol, Nerol,
Cynodon dactylon (L.) Pers.	Poaceae	Arukampullu	Bermuda grass	Leaf	Herb	Leaf juice taken orally controls blood sugar level.	β-Sitosterol, β-carotene, vitamin C, palmitic acid, triterpenoids, arundoin, friedelin, selenium, alkaloids- ergonovine and ergonovinine, Ferulic, syringic, p- coumaric,
Datura innoxia Mill.	Solanaceae	Umathai	Devil's trumpet	Flower	Herb	Dried flowers are burnt and smoked to cure asthma.	Kaempferol, glucopyran osyl, amnopyranosyl-glucopyranos D-glu-copyranosyl, D-glucopyranosyl, 7-O-a-L-rhamnopyranosyl-and Kaempferol
Dendrocalamus strictus (Roxb.) Cor	Poaceae	Kalmungil	Bamboo	Leaf	Herb	Leaf juice taken for three days cures cough in children.	Not available
Eclipta prostrata (L.) Mant.	Asteraceae	Karisalangani	Trailing eclipta	Root	Herb	Root paste is applied to heal wounds.	Wedelolactone, eclalbasaponin, luteolin and luteolin-7-O-glucoside
Erythrina variegata L.	Fabaceae	Mullu murungaii	Indian coral tree	Leaf	Tree	Leaves are used to cure leprosy.	Alkaloids, Coumarins, Steroids, Flavonoids, Lipids, Triterpenes
Euphorbia heterophylla L.	Euphorbiaceae	Palperuki	Mole plant	Leaf	Herb	Cooked leaves taken orally cures diarrhoea.	Resin, tannins, albuminoids,

Botanical Name	Family	Local Name	Common Name	Useful Parts	Habit	Medicinal Uses/ Mode of Usage	Major Phytochemical Constituents*
Euphorbia hirta L.	Euphorbiaceae	amanpacharsi	Snake weed	Herb	Herb	Herb extract used as a medicine for cold and cough.	Anthocyanins, l- insitol, quercetin, xanthorhamnin, caoutchouc, resin, tannins, albuminoids, gallic acid, shikimic acid and choline
Evolvulus alsinoides L.	Convolvulaceae	Vishnukranthi	Slender dwarf morning glory	Leaf	Herb	Leaf paste applied over head before bathings promote growth of hair.	Phenol, Terpenoids, Chlorogenic acid, Flavonoid
Ficus benghalensis L.	Moraceae	Alamaram	Baniyan tree	Root	Tree	Root boiled with coconut oil applied over head before bath used to cure dandruff and heals cracks on the foot.	Bengalensoide, leucocyanidin, leucopelargonidin and phytosterolin
Ficus racemosa L.	Moraceae	Athimaram	Athi tree	Leaf	Tree	Leaf juice given orally three times per day for three days cures brain fever.	Lupenyl acetate, glutinol, oleanolic acid, pentacyclic triterpenoids, taraxerol and friedelin
Ficus religiosa L.	Moraceae	Arasu	Peepal tree	Bark	Tree	Bark paste applied topically is used to cure rheumatism.	3-0-β-D-Glucopyranoside, leucopelargonidin-3-0-α-*L*-rhamnopyranoside
Gomphrena globosa L.	Amaranthaceae	Vadammali	Globe amaranthus	Root	Herb	Root extract are used to cure cough.	Saponins, alkaloids, reducing sugars and coumarins, stigmasterol, β-sitosterol and isochavicinic acid
Helianthus annuus L.	Asteraceae	suryakanthi	Sun flower	Seeds	Shrub	Seed powdered and taken orally reduce cholesterol and cure diabetics.	Grandifloric acid Paniculoside IV, Helikauranoside
Heliotropium indicum L.	Boraginaceae	Thelkodukai	Indian turn sole	Leaf	Herb	Leaf paste used to cure ring worm infection.	Heliotrine, indicine, retronecine, lasiocarpine

Contd...

Botanical Name	*Family*	*Local Name*	*Common Name*	*Useful Parts*	*Habit*	*Medicinal Uses/ Mode of Usage*	*Major Phytochemical Constituents**
Hibiscus rosa-sinensis L.	Malvaceae	Chembaruthi	Shoe flower	Leaf and flowers	Shrub	Leaf and flower boiled along with the coconut oil applied over head before bath enchances hair growth.	Cyanidin, diglucoside, flavonoids, vitamins, thiamine, riboflavin, niacin, ascorbic acid, Quercetin-3-diglucoside, 3,7-diglucoside, cyanidin-3,5-diglucoside and cyanidin-3-sophoroside-5-glucoside
Ipomoea quamoclit L.	Convulvuaceae	Mayilmanikam	Cypress vine	Leaf	Climber	Leaf paste applied topically cures elephantsis diseases.	Cyanogenetic glycosides, quamoclins I-IV and jalapin
Ixora coccinea L.	Rubiaceae	Thettichedi	Sacred ixora,jungle geranium	Leaf	Shrub	Leaf paste are used to cure skin infection and itching diseases.	Ursolic acid, oleanolicacid, lupeol, Geranyl Acetate, Neryl acetate. α-Terpineol acetate, Borneol acetate, Ethyl cinnamate Cyperene, and α-Copaene
Jasminum angustifolium (L.) Willd.	Oleaceae	kattupichi	Wild jasmine	Leaf	Shrub	Leaf paste applied topically cures skin diseases.	Alkaloids, anthroquinones, coumarins, emodins, fatty acids, flavonoids, glycosides, leucoanthocyanins, lignin's, phenols, saponins, steroids, tannins and triterpenoids.
Jasminum sambac (L.) Aiton.	Oleaceae	Malligai	Jasmine	Flower	Shrub	Flower powder used to cure skin diseases.	Methyl jasmonate, Jasminoside, Jasminol, Jasminolactone, Multiforin, Olueropin, benzyl benzoate, linalool, linalyl acetate, benzyl alcohol, indole, jasmone, methyl anthranilate and verbascoside

Botanical Name	Family	Local Name	Common Name	Useful Parts	Habit	Medicinal Uses/ Mode of Usage	Major Phytochemical Constituents*
Jatropha curcas L.	Euphorbiaceae	Katamanaku	purging nut	Leaf	Shrub	Pasted leaves used to cure skin infection and eczema.	Deoxypreussomerins, palmarumycins, stearic acid, monounsaturated oleic and polyunsaturated inoleic acid
Jatropha glandulifera Roxb.	Euphorbiaceae	Adathalai	Glandular jatropha	Seed and root	Shrub	Seed oil applied topically cures rheumatism and root decoction used to cure abdominal pain.	Alkaloids, flavonoids, cymol, carvacrol, and limonene
Justicia adhatoda L.	Acanthaceae	Adhatoda	Malabar nut	Leaf	Shrub	Leaf juice taken orally cures cough and heals bronchitis problem.	Vasicine, vasicinone, adhatodic acid and 1-pegamine
Lantana camara L.	Verbanaceae	Unnichedi	Lantana weed	Leaf and Root	Shrub	Leaf paste used to cure ulcer and root decoction cures cough and cold.	Triterpene, lantanic acid, lantanine, lantalinilic acid, lantadene A.
Lawsonia inermis L.	Lythraceae	Maruthani	Henna	Leaf	Shrub	Leaf juice is used as cooling agent and it promotes hair growth.	Lawsone, 2-hydroxy-1:4 napthaquinone, gallic acid, glucose, mannitol, fats, resin, hennatannic acid
Leucas aspera (Willd.) Link.	Lamiaceae	Thumbai	Thumbai	Leaf	Herb	Leaf paste applied topically on forehead cures head ache.	Alkaloids, anthroquinones, coumarins, emodins, fatty acids, flavonoids, glycosides, leucoanthocyanins, lignin's, phenols, reducing sugars, saponins, steroids, tannins and triterpenoids.
Limonia acidissima Groff.	Rutaceae	Vilankai	Wood apple	Fruit	Tree	Fruit pulp mixed with honey taken orally cures heart diseases.	Dictamnine, xanthotoxol, scoparone, xanthotoxin, isopimpinellin, isoimperatorin and marmin

Contd...

Botanical Name	*Family*	*Local Name*	*Common Name*	*Useful Parts*	*Habit*	*Medicinal Uses/ Mode of Usage*	*Major Phytochemical Constituents**
Luffa acutangula (L.) Roxb.	Cucurbitaceae	Peerkkangai	Ridged gourd	Fruit	Creeper	Fruits are eaten to cure skin diseases.	β-carotene, flavonoids, luffin, amarin
Lycopersicon esculentum Mill.	Solanaceae	Thakkali	Tomato	Fruit	Herb	Intake of fruits effective in reducing cholesterol levels and lowering blood pressure.	Alkaloids, Flavonoids Proteins, Carbohydrates Tannins, Glycosides Phenols, Saponins Terpenoids
Madhuca longifolia (J.Koenig) J.f.	Sapotaceae	Illupai	South Indian madhca (or) Indian butter tree	Root	Tree	Boil the root along with coconut oil applied topically in legs used to cure rheumatism	Xanthophylls, erthrodiol, palmitic acid, myricetin, 3-O-L-rhamnoside, 3-galactoside
Mangifera indica L.	Anacardiaceae	Mangai	Mango tree	Flower	Tree	Flower decoction taken orally cures chronic dysentery.	Tartaric acid, gallic acid and mallic acids, tannins, β-carotene, kaempferol, myricetin, α- and β-amyrins, gallotanin, glucogallin, indicol, taraxerol, triedelin, lupeol and β-sitosterol
Manihot esculenta Crantz.	Euphorbiaceae	Marchini-kilangu	Tapioca	Tuber	Shrub	Cooked tubers are rich in starch.	Maniesculentins, diterpenoids
Manilkara zapota (L.) P.Royen	Sapotaceae	Sapota	Sapota tree	Fruit	Shrub	Root powder is taken orally for three days to cure cough.	linoleidic acid and linoleic acid
Marsilea quadrifolia L.	Verbenaceae	Aarakeerai	Frog fruit	Leaf	Shrub	Paste of leaf juice applied over head cures dandruff problems.	Steroids, Phenols, Resins, Tannins, Alkaloids, Terpenoids, Cardiac Glycosides, Xantho Protein, Anthocyanidins
Mentha arvensis L.	Lamiaceae	Puthina	Mint	Leaf	Herb	Leaf decoction given orally cures diarrhoea.	Phenols, Steroids, Tannins, Flavonoids

Botanical Name	Family	Local Name	Common Name	Useful Parts	Habit	Medicinal Uses/ Mode of Usage	Major Phytochemical Constituents*
Mimosa pudica L.	Mimosaeceae	Thottal churungi	Touch- me-not	Leaf	Herb	Pasted leaves applied for cuts and wounds.	Mimosine, tyrosine3, 4-dihydroxypiridine, mimosinamine, mimosinic acid
Mirabilis jalapa L.	Nyctaginaceae	Anthimantharai	Four'o clock	Root	Herb	Pasted roots applied topically twice a day to cure sebaceous cysts.	Not available
Momordica charantia L.	Cucurbitaceae	Pavaikkai	Bitter gourd	Fruit	Herb	Fruits taken orally controls blood sugar level.	Momordicolide 10E -3-hydroxyl-dodeca-10-en-9-olide, monordicophenoide A, 4-hydroxyl-benzoic acid, 4-O-beta-D-apiofuranosyl, beta glucopyranoside, dihydrophaseic acid 3-O-beta-D-glucopyranoside, 6,9-dihydroxy megastigman-4, blumenol,
Morinda pubescens Sm.	Rubiaceae	Manjanathi	Morinda tree nuna	Leaf	Tree	Juice of leaves are used to cure diarrhoea, dysentery and fever.	Alkaloid, Glycoside, Steroid, triterpenoid Tannin, Flavonoid, Carbohydrate, and Protein
Moringa oleifera Lam.	Moringaceae	Murungai	Drumstick tree	Leaf	Tree	Leaf extract cures stomach ache and constipation.	Anthraquinone, alkaloids, saponins, steroids, terpenoids, Cardiac glycoside, anthocyanin, tannins and carotenoids

Contd...

Botanical Name	*Family*	*Local Name*	*Common Name*	*Useful Parts*	*Habit*	*Medicinal Uses/ Mode of Usage*	*Major Phytochemical Constituents**
Mukia maderaspatana (L.) M. Roem.	Cucurbitaceae	Musumusukai	Madras pea pumpkin	Leaf	Climber	Roasted leaves cures respiratory problems.	Spinasterol, 23 dihydrospinasterol, 3-O-β-D-glucoside, β-sitosterol, decosaenoic acid and triterpenes
Murraya koenigii (L.) Spreng.	Rutaceae	Kariveppilai	Curry leaf	Leaf	Herb	Leaves boiled with coconut oil applied on the hair to promote hair growth.	Koenigin, β-carotene, coumarin glucoside and scopolin
Musa sapidisiaca L.	Musaceae	Vazhai	Banana	Fruit, flowers and Pseudo-stem	Tall herb	Half cup of Pseudostem juice taken orally for a week used to cure kidney stone. Intake of ripe banana used to reduce constipation. Flowers used to control blood sugar level.	Alkaloids, glycosides, saponins, tannins, flavonoids, polyphenols, and reducing sugars
Nelumbiumnucifera Willd.	Nymphaeaceae	Thammarai	Lotus flower	Flower	Aquatic hydro-phyte	Flower juice used to cure rheumatism and regulate the blood circulation.	Lucenin, flavone-4'-OH, oleic acid trimethylsilyl ester, betulin, à-amyrin-trimethylsilyl ether
Nerium oleander L.	Apocyanaceae	Arali	Indian-oleander	Flower	Shrub	Flower paste applied topically are used to heal foot cracks.	Neriin, α-tocopherol, cardenolides, oleandrin
Nyctanthes arbor-tristis L.	Oleaceae	Pavalamalli	Night jasmine	Leaves	Shrub	Leaf juice are used to cure rheumatism and fever.	β-sitosterole,astragaline, Nicotiflorin, Oleanolic acid, Nyctanthic acid, Tannic acid Methyl salicylate, Friedeline and Lupeol
Ocimum basilicum L.	Lamiaceae	Thiruneetru pachalai	Sweet basil	Leaf	Herb	Powdered leaves taken orally once a day cures giddiness.	Safrole, oimene, cineole, linalool and thymol.
Ocimum tenuiflorum L.	Lamiaceae	Thulasi	Holy basil	Leaves	Herb	Leaf decoction is taken twice a day cures cough and cold.	Eugenol, carvacrol, nerol, eugenol, methylether, apigenin, luteolin, orientin, molludistin and ursolic acid

Botanical Name	Family	Local Name	Common Name	Useful Parts	Habit	Medicinal Uses/ Mode of Usage	Major Phytochemical Constituents*
Opuntia stricta (Haw.) Haw.	Cactaceae	Sapathikalli	Prickly pear	Fruit	Shrub	Fruits are edible. It acts as a coolant, diuretic and laxative.	linoleic, oleic, palmitic and stearic acids
Oryza sativa L.	Poaceae	Nellu	Rice	Grains	Herb	Rice powder is applied to cure inflamed surface.	γ-oryzanol, tocotrienols, tocopherols and tocochromanols
Pandanus fascicularis Lam.	Pandanaceae	Thalampoo	Screw pine	Root	Shrub	Root paste is applied daily on legs to cure rheumatism.	Saponins, tannins, phenolic compounds, alkaloids and flavonoids
Pergularia daemia (Forssk.) Chiov.	Asclepiadaceae	Veliparuthi	Pergularia	Leaf	Herb	Decoction of leaves taken orally once a day cures asthma.	Calactin, calotropin, corotoxigenin and daucosterol
Phoenix sylvestris Roxb.	Arecaceae	Ichanmaram	Wild date palm	Leaf	Palm	Crushed fresh leaves are soaked in water overnight, the water is taken next morning in empty stomach to expel worms, fruits act as nerve tonic	Palmitic acid, β-sitosterol, steroids, alkaloids, terpenoids, and flavonoids
Phyla nodiflora L.	Verbenaceae	Poduthalai	Turkey tangle fog fruit	Leaf	Herb	Paste of leaf is applied on head before bath to cure dandruff problem.	Flavonoids, sterols, and caryophyllene oxide
Phyllanthus acidus (L.) Skeels	Euphorbiaceae	Ari nelli	Carry me seed	Root	Herb	Root extract is drunk once a day in empty stomach to cure jaundice.	Phyllanthusols A and B, aglycon,
Phyllanthus amarus Schumach and Thonn.	Euphorbiaceae	Keelanelli	Keelanelli	Whole plant	Herb	Plant juice given orally once a day for 21 days cures jaundice.	Phyllanthin, hypophyllanthin, niranthin and ursolic acid
Phyllanthus emblica L.	Euphorbiaceae	Nellikkai	Goose berry	Fruits	Tree	Boil the fruits along with coconut oil applied externally cures scabies.	Phyllembin, gallic, ascorbic acid and proanthocyanidin

Contd...

Botanical Name	Family	Local Name	Common Name	Useful Parts	Habit	Medicinal Uses/ Mode of Usage	Major Phytochemical Constituents*
Piper betle L.	Piperaceae	Vetrillai	Betal pepper	Leaf	Creeper	Leaf juice used to cure cough, cold.	α- and β-sitosterol, estragol, caryophyllene, cardinene and allyl catechul
Pithecellobium dulce (Roxb.) Benth	Mimosaceae	Kodukkapuli	Madras thorn	Leaf,	Tree	Leaf extract taken twice a day are used to cure diarrhoea and dysentery.	Tetratriacontanoate,Methyl tricosenoate and tridecatrienoate
Plectranthus amboinicus (Lour.) Spreng	Laminaceae	Karpuravalli	Indian borage	Leaf	Herb	Leaf juice mixed with jiggery taken orally for once in three days used to cure cough.	Thymol, p-cymene, rosmarinic acid
Polyalthia longifolia (Sonn).Thwaites	Annonaceae	Nettilingam	False asoka	Bark	Tree	Bark paste applied on the infected part cures skin diseases.	Polyfothine, noroliveroline, Pendulamine A, Longimide B
Polygala arvensis Willd.	Polygalaceae	Sirianangai	Snake root	Leaf	Herb	Leaves extract is used to cure skin diseases.	Alkaloids, catechism, tannins, saponins, steroids, flavonoids, phenols, silibinin, silidianin
Pongamia pinnata Pierre.	Fabaceae	Punka maram	Indian beech	Bark	Tree	Bark paste is applied externally to heal ring worm infection.	Pongamol, karanjin, glabrin, alkaloid, resin, mucilage, palmitic, stearic, arachidic, lignoceric, linolenic and oleic acid
Psidium guajava L.	Myrtaceae	Koyya	Guava tree	Leaf	Tree	Leaf juice used to cure diarrhoea and diabetics.	β-selinene, β-caryophyllene, caryophyllene, tannins, triterpenes and flavonoid
Punica granatum L.	Punicaceae	Mathulai	Pomegranate	Leaf, flower and fruit	Shrub	Decoction of leaf, flower and fruits drunk orally cures stomach and dysentery.	diglucoside and delfinidin-3-glucoside and tanin

Botanical Name	Family	Local Name	Common Name	Useful Parts	Habit	Medicinal Uses/ Mode of Usage	Major Phytochemical Constituents*
Raphanus sativus L.	Brassicaceae	Mullanki	Radish	Tuber	Herb	Cooked tuber are edible and are used to cure cough and paralysis.	Tannins, saponins
Ricinus communis L.	Euphorbiaceae	Amanaku	Castor seed	Leaf	Tree	Leaf paste applied externally used to cure leg pain, cracks and leaf juice taken once a day cures constipation.	Tannins, alkaloids, cardiac glycosides, terpenoids, flavonoids and steroids
Rosa indica L.	Rosaceae	Rose	Rose	Flower	Shrub	Flowers soaked with honey for 2 weeks. Infusion drunk increases iron content in our body.	Flavonoids, tanins, phenol and steroid
Sesbania grandiflora (L.) Poir.	Fabaceae	Akathikeerai	Sesban	Flowers	Tree	Flowers cooked and taken orally to treat headache.	Arginine, cysteine, histidine, isolcucine, tryptophan, valine, threonine, alanine, aspargine, aspartic acid, oleanolic acid, galactose, rhamnose and glucuronic acid.
Sida rhombifolia L.	Malvaceae	Kurunthottiver	Flannel weed	Root and Leaf	Shrub	Root extract taken orally used to cure gas troubles and leaf paste applied externally to cure headache.	Not available
Solanum melongena L.	Solanaceae	Kaththiri	Brinjal	Fruit	Herb	Fruit extract reduces upper gastrointestinal inflammation such as gastritis, eosinophilic gastroenteritis, peptic ulcer disease,	Pinane

Contd...

Botanical Name	*Family*	*Local Name*	*Common Name*	*Useful Parts*	*Habit*	*Medicinal Uses/ Mode of Usage*	*Major Phytochemical Constituents**
Solanum nigrum L.	Solanaceae	Manathakali	Black	Leaf	Herb	Leaf decoction taken orally for a week to cure ulcer.	Solanosonine, α- and β-solanigrine, α- and β-solamargine, steroidal sapogenins, diosgenin, tigogenin and solasodine
Solanum surattense Burm.f.	Solanaceae	Kandankathri	Yellow-berried-night shade	Roots Fruits	Shrub	Fruits and roots infusion taken orally twice a day cures stomach ache and fever.	Alkaloids, tannins, flavonoids,
Solanum torvum Sw.	Solanaceae	Sundakkai	Thorn apple	Root and Fruits	Shrub	Infusion of root is useful for skin diseases, Fruits sauted with gingelly oil taken orally cures piles	Steroids, steroid saponins, triacontane, Glycoalkaloids
Solanum trilobatum L.	Solanaceae	Thuthuvalai	Purple fruited pea egg plant	Leaf	Shrub	Decoction of leaves taken with jaggery cures cough and cold.	Phenols, flavonoids, alkaloids
Stachytarpheta indica (L.)Vahl.	Verbenaceae	Seemai naiyurvi	Pink rat tail	Root	Shrub	Infusion of root bark is taken twice a day for 3 days to cure diarrhoea and dysentery.	Verbascoside, tannins, flavonoids, saponins, phytosterols
Syzygium cumini (L.) Skeels.	Myrtaceae	Naval	Jambolan, javaplum	Seeds	Tree	Seed powder mixed with milk taken twice a day orally to treat diabetes	Gallic acid, betulinic acid, β-sitosterol, eugenin,
Tabernaemontana divaricata (L.) R. Br. ex Roem. and Schult.	Apocyanaceae	Nanthiavatam	East Indian rose bay wax flower plant	Flower Fruit	Shrub	Flower and fruit extract applied topically cures tooth ache.	Alkaloid– tabernoxidine, coronaridine, vocangine and iboganine
Tamarindus indica L.	Caesalpiniaceae	Puliamaram	Tamarind	Leaf	Tree	Leaf paste used to cure swelling in the body and legs.	Phytic acid

Botanical Name	Family	Local Name	Common Name	Useful Parts	Habit	Medicinal Uses/ Mode of Usage	Major Phytochemical Constituents*
Tectona grandis L.f.	Verbenaceae	Thekku	Teak	Leaf, Bark	Tree	Leaf oil used as a hair tonic and it promote hair growth, bark astringent used in bronchitis.	Alkaloids, anthroquinones, coumarins, emodins, fatty acids, flavonoids, glycosides, leucoanthocyanins, lignin's, phenols, reducing sugars, saponins, steroids, tannins and triterpenoids.
Tephrosia purpurea (L.) Pers.	Fabaceae	Kozhunji	Purple tephorsia	Root	Herb	Decoction of root cures urinary disorders.	Betaphroline
Terminalia catappa L.	Combretaceae	Vallankotai	Indian almond	Seeds	Tree	Powdered seed used to cure cough, ulcer and skin rashes.	β-sitosterol-3-O-β-d-glucoside
Thespesia populnea (L.) Sol. ex Correa	Malvaceae	Poovarasu	Willd indigo	Bark	Tree	Bark paste applied externally which heals leprosy and scabies.	β-Sitosterol, ceryl alcohol, epoxyoleic acid, gossypetin, gossypol and herbacetin
Thevetianeriifolia Juss.ex steud.	Apocyanaceae	Manjalarali	Yellow oleander	Root	Shrub	Root paste is applied topically to cure boils.	Cardenolides- thevetin A and B, anthroquinones, flavonoids, phenolic
Tinospora cordifolia (Willd.) Miers ex Hook. f and Thomson.	Menisperma-ceae	Seenthilkodi	Tinospora, heavenly elixir	Leaf	Climber	Leaf extract is drunk orally to cure jaundice.	Berberine, morphine, psilocin,
Tribulus terrestris L.	Zygophyllaceae	Nerungil	Land-caltrops	Leaf	Herb	Leaf paste is taken orally to cure stomach- ache.	Hecogenin, aurantiamide acetate, xanthosine, fatty acid ester, ferulic acid, vanillin, p-hydroxybenzoic acid and ß-sitosterol,
Trichosanthes cucumerina L.	Cucurbitaceae	Pudalangai	Snake gourd	Fruit	Cimber	Dried fruit powder decoction are given orally with sugar to assist digestion.	cucurbitacin B, cucurbitacin E, isocucurbitacin B, 23,24-dihydroisocucurbitacin B, sterols 2 β-sitosterol and stigmasterol

Contd...

Botanical Name	*Family*	*Local Name*	*Common Name*	*Useful Parts*	*Habit*	*Medicinal Uses/ Mode of Usage*	*Major Phytochemical Constituents**
Tridax procumbens L.	Asteraceae	Muriyan-pachalai	Tridax daisy	Leaf	Herb	Leaf juice is applied over affected place to cure cuts.	Lipids, β- amyrin. Fucosterol, lupeol, sitosterol, luteolin, palmitic and stearic acids
Tylophora zeylanica Decne.	Asclepiadaceae	Palaikeerai	Indian Ipecac	Leaf	Climber	Powdered leaves are used to heal respiratory problem.	Thylophorine.
Vigna mungo (L.) Hepper.	Fabaceae	Ulundhu	Black gram	Seed	Twiner	Seeds are used as a protein food supplement.	Alkaloids, anthroquinones, coumarins, emodins, fatty acids, flavonoids, glycosides, leucoanthocyanins, lignin's, phenols, reducing sugars, saponins, steroids, tannins and triterpenoids.
Vitex negundo L.	Verbenaceae	Nochi	Chaste tree	Leaf	Shrub	Leaf paste applied over affected places cures cuts.	Protocatechuic acid, oleanolic acid, flavanoids, viridiflorol and β-caryophyllene
Zingiber officinale Rosc.	Zingiberaceae	Inchi	Ginger	Rhizome	Herb	Rhizome juice are used to cure indigestion and gas troubles.	Zingiberine, amaldehyde, gingerol, shogaol, and paradol
Ziziphus mauritiana Lam.	Rhamnaceae	Illanthai	Jackal jujube	Fruit Bark	Tree	Fruits are edible and bark decoction taken once in three day is used to cure uterus disorders.	Alkaloids

* The phytochemical constitutes for the plants were extracted from the literatures of 35–37.

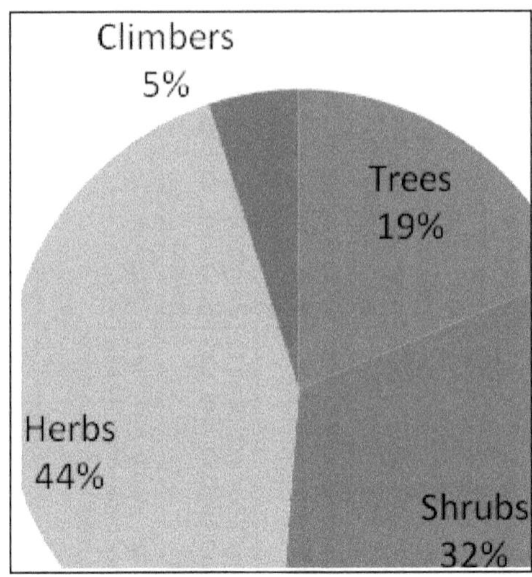

Figure 10.1: Habit-wise Distribution of Ethnomedicinal Plants.

leave for decoction of *Solanum surattense* to cure fever, cough and asthma. These findings were compared with earlier work (Shinwari *et al.*, 2000; Shiva, 2005) Kidney stone and urinary tract disorders are treated with *Beta vulgaris, Aerva lanata, Tridax procumbens, Tribulus terrestris* and *Ricinus communis* (Vijigiri*et al.*, 2013;Jain *et al.*, 2009; Parveen *et al.*, 2007; Katewa and Galav, 2005). In the present study 5 plant species (*Tephrosia purpurea, Acacia nilotica, Aerva lanata, Ipomea quamoclit* and *Tribulus*

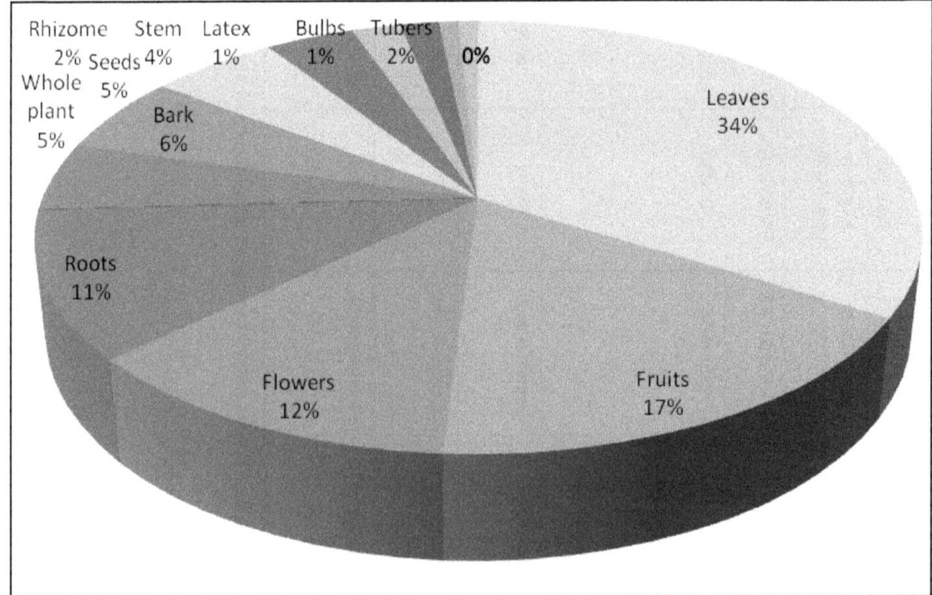

Figure 10.2: The Parts of the Plant Used for the Preparation of Herbal Medicines by Kani Tribal People.

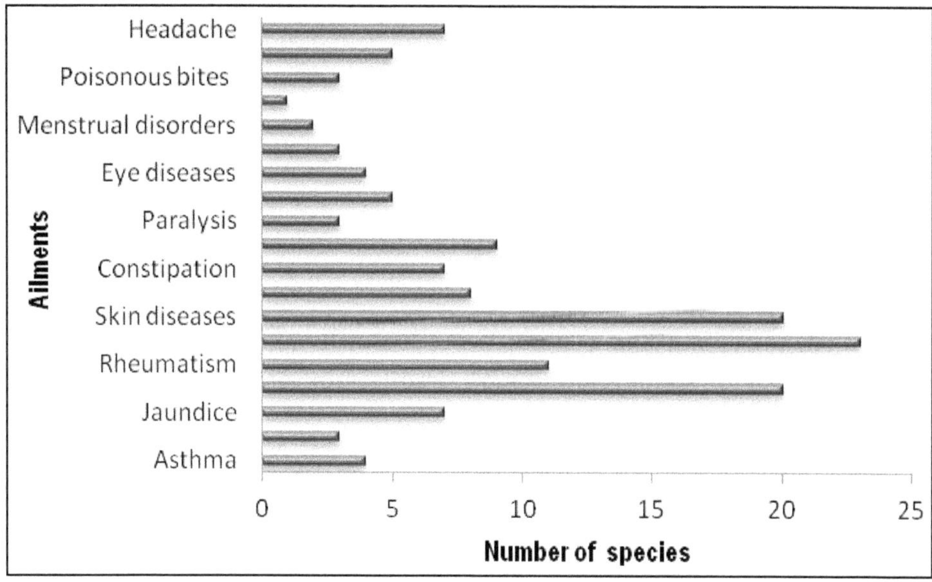

Figure 10.3: Number Medicinal Plant Taxa Used for Various Ailments by the *Vetan community* People.

terrestris) were used to treat urinary tract disorders. The bulb juice of *Allium sativum* and *Allium cepa* plants were used to cure diabetes, hypertension, cooling agent for eyes and also to treat leprosy (Qureshi *et al.*, 2002; Ahamed *et al.*, 2003). Leaf extract of *Phyla nodiflora* was used to reduce dandruff, skin related disease and for joint pains. Similarly, the same plant is used by the rural people of Attock district (Tahrouti *et al.*, 2007) *Cyanodon dactylon* was used to treat wounds and also used for cracks in foot (Shinwari, *et al.*, 2000; Ahmed *et al.*, 2007; Muthu *et al.*, 2006; Natale, 2007). In addition major phytochemical constituents of the ethnomedicinal plants studied in this paper. Of the 153 plants studied, major phytochemical constituents are reported for 150 plants (Table 10.1). This is in concordant with the present study, (Ayyanar and Ignacimuthu, 2005) surveyed the ethnopharmacological plants used by *Kani*tribes of Tirunelveli hills, Tamil Nadu, India.

From this account it was clear that the *Vetan* community people, like other ancient communities have the awareness about the character of various plants and their beneficial properties.

Conclusion

This study shows that usage of plants for treating various ailments is still practiced by *Vetan* community people. Documentation of this knowledge is valuable for the communities and their future generations and for scientific consideration of wider uses of traditional knowledge. The findings of the investigation envisage that the medicinal plants have excellent potentiality to treat various ailments. Their mode of preparation and mode of administration are also easy and suitable and the treatment has no side effects. This study would provide some basic clues of medicinal properties of plants used by native of *Vetan* community of Shankaranputhur village, Kanyakumari district of Tamil Nadu. Ethnomedicinal studies clearly defined that the knowledge of medicinal plants is important not just for *Vetan* themselves but for the entire world.

Acknowledgement

The authors acknowledge the *Vetan* community people for sharing their knowledge about the medicinal plants.

References

Ahmad M, Khan M.A and Qureshi R. A. 2003. Ethnobotanical study of some cultivated plants of Chhuchh Region (District Attock), *Hamdard Medicus*, 46(3) :15-19.

Ahmad, K, Hussain M, Ashraf, M, Luqman, M, Ashraf, M.Y. and Khan, Z.I. 2007. Indigenous vegetation of Soone Valley: At the risk of extinction, *Pakistan Journal of Botany*, 39(3): 679-690.

Ahmad, I., Farrukh, A., Mohammad, O. 2006. Modern Phytomedicine: Turning Medicinal Plants Into Drugs. Wiley-VCH, York, p. 136.

Alagesaboopathi, C., 2011. Ethnobotanical Studies on useful plants of Kanjamalai hills of Salem District of Tamil Nadu, Southern India. *Archives of Applied Science*, 3: 532-539.

Ayyanar M, Ignacimuthu S. 2005. Traditional knowledge of Kani tribals in Kouthalai of Tirunelveli hills, Tamil Nadu India *Journal of Ethnopharmacology*, 102 : 246- 255.

Ayyanar, M, 2013. Traditional herbal medicines for primary healthcare among indigenous people in Tamil Nadu, India.*Journal Homeopathy Ayurvedic Medicine*, 2(5): 1-7.

Ayyanar, M., Ignacimuthu, S., 2005.Ethnomedicinal plants used by the tribals of Tirunelveli hills to treat poisonous bites and skin diseases. *Indian Journal of Traditional Knowledge*, 4(3), 229- 236.

Barnes, J., Anderson, L.A., Phillipson, J.D., 2007. Herbal medicine. 3rd Edition, Pharmaceutical Press, London. pp 1-23.

Calixto, J.B. 2005. Twenty five years of research on medicinal plants in Latin America: a personal review. *Journal of Ethnopharmacology* 100,131-134.

Farnsworth N.R.and Jain S.K. 1994. Ethnopharmacology and drug development,In: *Ethnobotany and search for New Drugs*, edited by DJ Chadwick and March U, (Ciba Foundation Symposium 183, Wiley, Chichester) p. 153.

Gamble J.S. and Fischer CEC. 1956.*Flora of the Presidency of 20.Madras*, Vol. I-III, (Botanical Survey of India, Calcutta).

Ignacimuthu, S., Ayyanar, M., Sankarasivaraman, K., 2006. Ethnobotanical investigations among tribes in Madurai district Tamil Nadu, India, *Journal Ethnobiology and Ethnomedicine*. 11, 1-7.

Jain SC, Jain R.and Singh R, 2009. Ethnobotanical Survey of Sariska and Siliserh regions from Alwar district of Rajasthan, India, *Ethnobotanical Leaflets*, 13:171-188.

Jeeva, S., Femila, V. 2012.Ethnobotanical investigation of Nadars in Atoor village, Kanyakumari District, Tamil Nadu, India. *Asian Pacific Journal of Tropical Biomedicine*. 12, 593-600.

Katewa S. S. and Galav P.K. 2005.Traditional herbal medicines from Shekhawati region of Rajasthan.*Indian Journal of Traditional Knowledge*, 4(3): 237-245.

Katewa, S.S., Chaudhary, B.L., Jain, A., 2004. Folk herbal medicines from tribal areas of Rajasthan, India, *Journal of Ethnopharmacology*, 92(1):41-46.

Kosalge, S.B., Fursule, R.A. 2009. Investigation of ethnomedicinal claims of some plants used by tribals of Satpuda hills in India. *Journal of Ethnopharmacology* 121, 456-461.

Mahishi, P., Srinivasa, B.H., Shivanna, M.B. 2005. Medicinal plant wealth of local communities in some villages in Shimoga district of Karnataka, India, *Journal of Ethnopharmacology*. 98, 307-312.

Martin, G.J. 1995. Ethnobotany: A Methods Manual. London: Chapman and Hall

Muthu C, Ayyanar M, Raja N. and Ignacimuthu S. 2006. Medicinal plants used by traditional healers in Kancheepuram district of Tamil Nadu, India. *Journal of Ethnobiology Ethnomedicine*, 2 (1) : 43- 48.

Nadkarni A.K. 1976. Indian Materia Medica, Popular Prakashan Bombay, India.

Namukobe, J., Kasenene, J.M., Kiremire, B. T., Byamukama, R., Mugisha, M.K.,Krief, S., Dumonte, V., Kadasa, J.D. 2011. Traditional plants used for medicinal purposes by local communities around the Northn sector of Kibale National Park, Uganda.Journal of Ethnopharmacology. 136, 236-245.

Nair N.C., Henry A.N.,1983.Flora of Tamil Nadu, India, Vol. I, Botanical Survey of India, Southern Circle, Coimbatore

Namsa, N.D., Tag, H., Mandal, M., Kalita, P., Das, A.K. 2009. An ethnobotanical study of traditional anti- inflammatory plants usedby the Lohit community of Arunachal Pradesh. India. *Journal of Ethnopharmacology*. 125, 234-245.

Natale, A.D, Pollio, A. 2007. Plant species in the folk medicine of Montecorvino Rovella (inland Compania, Italy), *Journal of Ethnopharmacology*, 109(2): 295-303.

Ogbulie J.N., Ogueke O.O, Okoli I.C., Anyanwu B.N. 2007. Antibacterial activities and toxicological potentials of crude ethanolic extracts of *Euphorbia hirta*.*African Journal of Biotechnology*, 6: 1544-1548.

Parveen B, Roy S, Kumar A. 2007. Traditional uses of medicinal plants among the rural communities of Churu district in the Thar dessert, India. *Journal of Ethnopharmacology*,113(3): 387 - 399.

Pattanaik, C., Reddy, C.S., Murthy, M.S. 2008.An ethnobotanical survey of medicinal plants used by the Diyadi tribe of Malkangiri district of Orissa, India.Fitoterapia. 79, 67-71.

Qureshi, R., Bhatti G.R and Saeed, A. 2002. Obnoxious-Mankind's Need.*Hamdard Medicus*, 45(2): 82-87.

Raghupathy, S., Steven, N.G., Maruthakkutti, M., Velusamy, B., Ul-Huda, M.M. 2009. Consensus of the Malasars traditional aboriginal knowledge of medicinal plants in the Velliangiri holy hills, India, *Journal Ethnobiology and Ethnomedicine* 4, 23- 28.

Rajakumar, N., Shivanna M.B. 2009. Ethnomedicinal application of plants in the eastern region of Shimoga district, Karnataka, India,.*Journal of Ethnopharmacology* 126, 64-73.

Rastogi R.P, Mehrotra B.N, 1990- 1994. Compendium of Indian Medicinal plants Central Drug Research Institute Lucknow and National Institute of Science Communication, New Delhi.

Savithramma, N., Sulochana, C., Rao, K.N.,2007. Ethnobotanical survey of plants used to traet asthma in Andhra Pradesh, India. *Journal of Ethnopharmacology*. 113, 54- 61.

Shinwari Z.K, Gilani S.S, Kohjoma M. and Nakaike, T.2000. Status of medicinal plants in Pakistani Hindu-Kush -Himalayas. In: *Nepal-Japan Joint Symposium on Conservation of Natural Medicinal Resources and their Utilization*, p. 235-242.

Shinwari, M.I., Khan, M.A, 2000. Folk use of medicinal herbs of Margalla Hills National Park, Islamabad. *Journal of Ethnopharmacology*, 69(1): 45-56.

Shrestha,N., Shrestha, S., Koju, L., Shrestha, K.K., Wang, Z. 2016. Medicinal plant diversity and traditional healing practices in eastern Nepal.*Journal of Ethnopharmacology*, 192, 292-301.

Shiva, V. 2005. Traditional knowledge of medicinal plants among rural women of the Garhwal Himalaya, Uttaranchal,*Indian Journal of Traditional Knowledge*, 4(3): 259-266.

Srithi, K., Balslev, H., Wangpakapattanawong, P., Srisanga, P., Trisonthi, C., 2009. Medicinal plant knowledge and its erosion among the Mien (Yao) in northern Thailand. *Journal of Ethnopharmacology*. 123, 335-342.

Subitha, T., Ayyanar, M., Udaya Kumar, M., Sekar, T.2011. Ethnomedicinal plants used by Kani tribals in Pechiparai forests of Southern Western Ghats, Tamil Nadu, India. *Journal of Plant Science* 2, 349-354.

Tabuti, J.R.S., Kukunda, C.B., Waako, P.J. 2010. Medicinal plants used by traditional medicine practitioners in the treatment of Tuberculosis and related ailments in Uganda, *Journal of Ethnopharmacology* 127, 130-136.

Tahraoui A, Hilaly J.E, Israli Z. H, Lyoussi B. 2007. Ethnopharmacological survey of plants used in the traditional treatment of hypertension and diabetes in south–eastern Morocco,*Journal of Ethnopharmacology*, 110(1): 105-117.

Uniyal, S.K., Singh, K.N., Jamwal, P., Lal, B. 2006. Traditional use of medicinal plants among the tribal communities Chhota, Western Himalaya, *Journal of Ethnobiology and Ethnomedicine*, 20, 14-23.

Upadhyay, B., Parveen, Dhaker, A.K., Kumar, A. 2010. Ethnomedicinal and ethnopharmaco- statistical studies of eastern Rajasthan, India. *Journal of Ethnopharmacology*. 129, 64-86.

Vijigiri D, Bembrekar SK and Sharma PP. 2013. Herbal formulations used in treatment of kidney stone by native folklore of Nizamabad District, Andhra Pradesh, India. *Bioscience Discovery*, 4(2): 245-249.

2018, Ethnomedicinal Plants: A Biodiversity Treasure *Pages* **293–306**
Editors: ***V.R. Mohan, A. Doss, P.S. Tresina and V. Sornalakshmi***
Published by: **ASTRAL INTERNATIONAL PVT. LTD., NEW DELHI**

Chapter 11

Ethnobotanical Studies of Medicinal Plants Commonly Used by the Villagers of Kanyakumari District, Tamil Nadu, India

Selvaraj Rosemary, Jenisha and S. Jeeva

Department of Botany,
Scott Christian College (Autonomous),
Nagercoil – 629 003, Tamil Nadu
E-mail: solomonjeeva@gmail.com

Introduction

Plants have been used in the traditional healthcare system from time immemorial particularly among the local and indigenous communities (Laloo, 2006; Balakumar, 2011). The value of medicinal plants to the mankind is very well proven. Many people, especially in the poorer underdeveloped countries rely on wild plants for food, construction materials, fuel wood, medicine and many other purposes. It is estimated that 70 to 80 percent of the people in worldwide rely on herbal medicines (Farnsworth *et al.*, 1985). Out of 18,500 higher plant species recorded in India, about 7,500 species are reported to be medicinal by the rural and tribal communities (Karuppusamy, 2007). In the rural and tribal lives of India, medicinal plant sector has traditionally occupied an important position in the socio-cultural, spiritual and medicinal area (Anon, 2003). The use of plants for medicinal treatment in India dates back to prehistoric time. Today medicinal plants play a great role in human health services worldwide. Many people in the modern world are turning to herbal medicine because they do not have any side effects as like that of modern medicines. Traditional medicines are easy to consume by the rural people and they were obtained from the local plants seen around our homesteads (Jenisha and Jeeva., 2014).

Traditional knowledge on uses of plant parts as medicine used by the tribal and rural people were well documented in many literatures (Subitha, *et al.*, 2011; Jeeva and Femila, 2012; Sukumaran *et al.*, 2014; Sukumaran and Raj, 2010 and Britto *et al.*, 2010), but still most of the areas were unexplored. Thickanamcode village which is rich in plant biodiversity is one among them. Most of the people residing in this area depend on traditional medicine for their primary health care. So, the present investigation was carried out to document the medicinal plants of this area.

Materials and Methods

An ethnobotanical survey was carried out in Thickanamcode village during June 2012 to February 2013. It is situated in Kalkulam taluk of Kanyakumari district, Tamil Nadu, India and is 4.3 km far from the main town Thucklay. It lies between 8.325 North latitude and 77.344 East longitudes. It covers an area of about 130.33 sq. km. Tamil and Malayalam are the main languages spoken by the people of this area. Hindus and Christians form a sizeable percentage of the population. 'Nadar' is the major community seen in this area. Some of the other communities residing in this area are Vellalars, Paravas, Mukthavars and Vilakki. Rice is the staple food of the rich and poor alike in the district. Beverages like tea and traditional coffee (using ginger and palm sugar) are widespread even in the rural areas of the district.

The climate of the district is favourably warm and humid. The forests of this village are coming under traditional agroforesty system with integrated farming practices. Most of the household had lofty trees such as mangoes and tamarind. Ethnomedicinal information on medicinal plants was gathered from knowledgeable experienced people of the study area. The information about plants, their local names, useful parts, mode of administration was documented during the survey. The medicinal uses of species were cross checked through the literature available and were identified with standard floras (Gamble, 1915-1935; Matthew, 1995; Nair and Henry, 1983).

Results and Discussion

A total of 70 plants belonging to 67 genera and 41 families with medicinal value have been reported from the study area (Table 11.1). Family wise distribution of the medicinal plants shows that Euphorbiaceae, Solanaceae, Verbenaceae were the dominant families with 4 species each, the co-dominant position was occupied by Rutaceae, Asteraceae, Malvaceae, Ascelpidaceae, Fabaceae, Amaranthaceae and Moraceae each with 3 species each, Arecaceae, Cucurbitaceae, Lamiaceae, Apocynaceae, Anacardiaceae, and Rubiaceae possess 2 species each, 25 families were monospecific (Figure 11.1).

In the present study plants are used to treat bone fracture, hair fall, diabetes, cold, cough, jaundice, wounds and cuts, poisonous bite, breathing problem, fever, vomiting, body pain, pimple, ulcer, stomach disorders *etc.* Similarly in the present study the various plant parts used as medicines were leaves, whole plant, Roots, flowers, Bark, fruits, nuts, stem, latex and young stem (Figure 11.2). Leaves were

Table 11.1: List of Medicinal Plant Species Recorded from the Study Area

Botanical Name	*Family*	*Local Name*	*Common Name*	*Phenology*	*Useful Parts*	*Mode of Usage*
Abrus precatorius L.	Leguminosae	Kunni	Rosary pea	Throughout the year	Leaves	Leaves juice is used to treat fever and rheumatism.
Abutilon indicum (L.) Sweet	Malvaceae	Thutti	Indian mallow	June - September	Whole plant	Seeds ground into paste and applied on the skin to cure skin diseases. The juice obtained from the whole plant is drunk to cure stomach problems.
Acalypha indica L.	Euphorbiaceae	Kupaimeni	Indian copper leaf	June - November	Leaves	The paste obtained from the leaves is applied on the burns, wounds and it is also applied on forehead to cure headache. The juice extracted from the leaves is mixed with lime and applied on skin to cure skin diseases.
Achyranthes aspera L.	Amaranthaceae	Naiyurvi	Prickly chaff flower	August - February	Whole plant	Leaf paste is applied on the cuts to cure wounds. Juice is drunk by pregnant women for easy delivery. The juice obtained from the root is used to cure snake bites.
Aegle marmelos (L.) Correa	Rutaceae	Vilvam	Bengal quince	Throughout the year	Fruit	Unripe fruit is eaten to treat indigestion, dysentery and kills intestinal parasites.
Aloe vera (L.) Burm. f.	Xanthorrhoeaceae	Chothukathalai	Aloe vera	September - January	Leaves	Leaf paste is used to treat stomachache.
Amaranthus viridis L.	Amaranthaceae	Kuppai keerai	Green Amaranth	August - December	Leaves	The juice obtained from the leaves is used to treat fever.
Anacardium occidentale L.	Anacardiaceae	Mundiri	Cashew nut	December - March	Nuts	The oil taken from the shells of nuts is applied on the cracks.
Ananas comosus (L.) Merr.	Bromeliaceae	Annasi	Pineapple	Throughout the year	Fruit	The fruit is used to promote digestion.
Annona squamosa L.	Annonaceae	Chethapalam	Custard Apple	June - August	Leaves	The juice extracted from the leaves is used to treat cough.

Contd...

Botanical Name	Family	Local Name	Common Name	Phenology	Useful Parts	Mode of Usage
Areca catechu L.	Arecaceae	Pakku	Areca nut	Throughout the year	Leaves	Paste obtained from the leaves is used to cure skin diseases.
Aristolochia indica L.	Aristolochiaceae	Karudakodi	Indian Birth wort	September - January	Leaves	The paste obtained from the leaves is used to cure leg pain.
Artocarpus heterophyllus Lamk.	Moraceae	Pala	Jackfruit	February-May	Fruit	Fruit is used to cure skin diseases and reduces cough.
Azadirachta indica A. Juss.	Meliaceae	Veppamaram	Neem	March - May	Leaves and flower	Paste obtained from the leaf is used to treat skin diseases. Decoction prepared from the leaves is used to treat ulcers. Juice extracted from the leaves and flowers is used to cure stomachache.
Calotropis gigantea (L.) R. Br.	Asclepiadaceae	Erukku	Crown flower	December - March	Leaves and root	The paste obtained from the leaves and root is applied on the knee to cure rheumatism.
Capsium annuum L.	Solanaceae	Kantharimilau	Chilli	Throughout the year	Fruit	The fruit is used to treat blood pressure.
Cardiospermum halicacabum L.	Sapindaceae	Muddakkathan	Wedge leaf rattle pod	December - March	Roots and leaves	Decoction prepared from the roots and leaves is used to treat rheumatism
Carica papaya L.	Caricaceae	Pappali	Melon like fruit	Nearly continuous all year	Fruit and leaves	Juice obtained from the leaves is helpful to cure fever. Fruit is eaten to destroy germs.
Catharanthus roseus (L.)G. Don.	Apocynaceae	Poonaari	Periwinkle	Throughout the Year	Leaves	Decoction prepared from the leaves is used to cure whooping cough.
Centella asiatica (L.) Urb.	Apiaceae	Vallarai	Asiatic pennywort	July - September	Leaves	Fresh leaves are eaten daily to improve memory power.
Chrysanthemum indicum L.	Asteraceae	Sevanthi	Chrysanths	Throughout the Year	flower	Decoction prepared from the flower is used to cure fever and headache.

Botanical Name	Family	Local Name	Common Name	Phenology	Useful Parts	Mode of Usage
Cissus quadrangularis L.	Vitaceae	Pirandai	Hadjora	Throughout the year	Whole plant	The paste obtained from the stem and root of this plant is applied on the bone fractures to cure pain. Juice extracted from the whole plant is helpful to treat asthma, stomach troubles and bleeding of nose.
Citrus limon (L.) Burm. f.	Rutaceae	Elumchai	Lemon	January - March	Leaves and fruit	The decoction of leaves is used to treat fever, headache and cold. Skin of the fruit is used to treat pimples and black dots.
Cleome gynandra L.	Cleomaceae	Thivazai	African spider flower	Throughout the year	Leaves	Juice obtained from the leaves is used to cure cough, headache and rheumatism.
Clerodendrum phlomidis L.	Verbenaceae	Vathamadakki	Ami irun	August - September	Leaves	The paste obtained from the leaves is used to treat rheumatism.
Clitoria ternatea L.	Leguminosae	Sangupuspam	Butterfly pea	July - March	Whole plant	Paste obtained from the whole plant is applied on the cuts to cure wounds.
Coccinia grandis (L.) Voigt	Cucurbitaceae	Kovaikkai	Ivy-Gourd	September - October	Leaves and fruits	Fruit is eaten to cure diabetics. Paste obtained from the leaves is used to cure headache.
Cocos nucifera L.	Arecaceae	Thengu	Coconut tree	Throughout the year	Fruit	Coconut mixed with panner is applied on the pimples and black dots.
Plectranthus amboinicus (Lour.) Spreng.	Lamiaceae	Navarpacchilai	Indian Borage	Throughout the year	Leaves	Used to flavour drinks and also used in herbal medicine.
Crosssandra infundibuliformis (L.) Nees.	Acanthaceae	Kanakambaram	Fire cracker flower	Throughout the Year	root	Juice extracted from the root is used to cure cough.
Crotalaria retusa L.	Leguminosae	Kilukiluppai	Deril bean	November - September	Leaves	The juice extracted from the leaves is used to treat skin diseases.
Cynodon dactylon (L.) Kuntze	Poaceae	Argampul	Bermuda grass	Throughout the year	Leaves	Juice obtained from the leaves is used to cure stomach problems, skin diseases and blood purification.

Contd...

Botanical Name	Family	Local Name	Common Name	Phenology	Useful Parts	Mode of Usage
Datura metel L.	Solanaceae	Umathai	Devil's trumpet	July - December	Leaves	Juice extracted from the leaves is used to cure cough.
Eclipta prostrata (L.) L.	Asteraceae	Karisalangani	False Daisy	Throughout the Year	Roots	The paste obtained from the root is applied on the heel cracks.
Ficus benghalensis L.	Moraceae	Alamaram	Banyan tree	November	Stem latex and yound stem	The stem latex is applied on the heel crackers
Ficus religiosa L.	Moraceae	Arasu	Peepal tree	July-September	Bark	Paste obtained from the bark is used to treat rheumatism.
Hemidesmus indicus (L.) R. Br.	Asclepiadaceae	Nannair	Indian	June - February	Whole plant	The juice obtained from the plant is used to treat fever, rheumatism and urinary disorders.
Hibiscus rosa-sinensis L.	Malvaceae	Sembaruthi	Shoe flower	Throughout the year	Leaves	The paste obtained from the leaves is applied on the head to get rid of dandruff.
Ixora coccinea L.	Rubiaceae	Thettichedi	Jungle geranium	Throughout the year	Leaves	The leaf boiled with coconut oil is used to treat skin infections.
Jasminum angustifolium (L) Willd.	Oleaceae	Kattupichi	Wild Jasmine	November-January	leaves	Paste obtained from the leaves is used to treat skin diseases
Justicia adhatoda L.	Acanthaceae	Adhathodai	Malabar nut	January - June	Leaves	The decoction obtained from the leaves is used to control chest diseases, neurological pain and various skin diseases. Juice extracted is used to treat asthma and cough. The leaves are boiled with coconut oil and applied on the head to cure headache.
Lantana camara L.	Verbenaceae	Unnichedi	Wild sage	September - May	Leaves	The juice extracted from the leaves is used to cure rheumatism.
Lawsonia inermis L.	Lythraceae	Maruthani	Henna	December-June	Leaves	The paste obtained from the leaves is applied on the skin to cure skin diseases.

Botanical Name	*Family*	*Local Name*	*Common Name*	*Phenology*	*Useful Parts*	*Mode of Usage*
Mangifera indica L.	Anacardiaceae	Mau	Mango	November-March	Leaves	Decoction prepared from the leaves is used to treat ulcer.
Mentha arvensis L.	Lamiaceae	Pudina	mint	September – May	Whole plant	Juice extracted from the plant is used to treat diarrhoea.
Mimosa pudica L.	Mimosaceae	Thottal churunki	Touch me not	September - November	Root	Decoction prepared from root is drunk to cure asthma, diarrhea, skin wounds and whooping cough.
Morinda pubescens Sm.	Rubiaceae	Manjanathi	Nuna	Throughout the Year	Leaves	Juice extracted from the leaves is used to treat body pain and dysentery.
Moringa oleifera Lam.	Moringaceae	Murungai	Drumstick tree	March-September	Leaves and Fruit.	Juice obtained from the leaves is used to treat indigestion and eye diseases. Fruit juice is used in hair falling.
Murraya keonigii (L.) Spreng	Rutaceae	Karuveppala	Curry leaf	February - May	Leaves	Juice obtained from the leaves is used in vomiting. Powdered leaves boiled with coconut oil and applied on the hair to promote the hair growth.
Musa paradisiaca L.	Musaceae	Valai	Banana tree	Throughout of the year	Fruit	The fruit is used to cure stomach ache.
Nelumbium speciosum Willd.	Nelumbonaceae	Thammarai	Lotus Flower	August-April	Flower	Juice extracted from the flower is used to cure rheumatism.
Nerium oleander L.	Apocynaceae	Arali	Oleander	April-October	Flower	The paste obtained from the flower is applied on the heel cracks.
Ocimum tenuiflorum L.	Lamiaceae	Thulasi	Sacred basil	November - January	Leaves	The fresh juice extracted from the leaves is drunk to cure cough and fever. Paste obtained from the leaves is used to cure white patches.
Opuntia stricta (Haw.) Haw.var. dillenii (Ker Gawl.) L.D. Benson	Cactaceae	Prickly pear	Sapathikalli	November - April	Whole plant	The juice obtained from the whole plant is used to treat fever, headache and stomachache.

Contd...

Botanical Name	Family	Local Name	Common Name	Phenology	Useful Parts	Mode of Usage
Pergularia daemia (Forsk.) chiov.	Asclepiadaceae	Veliparuthi	Paergularia	November - March	Leaves	The fresh juice extracted from the leaves cures clot blood. The paste obtained from the leaves is used to cure headache.
Phyllanthus amarus Schumach and Thonner	Euphorbiaceae	Keela nelli	Carry me seed	Throughout the year	Leaves	Fresh juice obtained from the leaves is drunk to treat jaundice.
Phyllanthus emblica L.	Euphorbiaceae	Nellikai	Amla	December - January	Fruit	Juice obtained from the fruit is helpful to cure asthma and dysentery. The oil extracted from the fruit is mixed with coconut oil and applied on hair to promote hair growth and prevents hair fall.
Psidium guajava L.	Myrtaceae	Koya	Guava	April - August	Leaves and fruits	The juice extracted from the leaves is used to treat diarrhea. Fruit is used to treat diabetes.
Punica granatum L.	Punicaceae	Mathulai	Pome-granate	March - June	Fruit	The juice extracted from the skin of the fruit is used to treat diarrhea and stomachache. Fruit is used to increase blood level.
Ricinus communis L.	Euphorbiaceae	Amanaku	Caster seed	December - April	Leaves and roots	The fresh juice extracted from the roots is drunk to cure inflammations, fever, cough, skin diseases and rheumatism. Paste obtained from the leaves is used to treat burns.
Santalum album L.	Santalaceae	Santhanam	Sandel wood	August - November	Leaves	Drinking of the juice obtained from the leaves is helpful to receive body heat and blood purification.
Solanum nigrum L.	Solanaceae	Manathakkali	Black	September - February	Whole plant	Juice extracted from the whole plant is drunk in the treatment of liver disorders, fever, dysentery and it promotes urination.

Botanical Name	Family	Local Name	Common Name	Phenology	Useful Parts	Mode of Usage
Solanum trilobatum L.	Solanaceae	Thudavalai	Purple fruited pea egg plant	Throughout the year	Leaves	The fresh juice extracted from the leaves is given to treat cough and itching.
Tamarindus indica L.	Leguminosae	Puliamaram	Tamarind	February - June	Leaves	Paste obtained from the leaves is used to treat swellings.
Tectona grandis L. f.	Verbenaceae	Thekku	Teak	July - December	Leaves	Paste obtained from the leaves is chewed to cure toothache.
Thespesia populnea (L.) Sol. ex. Correa	Malvaceae	Poovarasu	Wild Indigo	Throughout the year	Leaves and flower	Paste obtained from the leaves, flower and the fruit is applied on skin to cure skin disease.
Tribulus terrestris L.	Zygophyllaceae	Nerungil	Puncture vine	Throughout the Year	Leaves	Juice extracted from the leaves is used to treat Stomach ache.
Tridax procumbens L.	Asteraceae	Muriyan pachalai	Tridax daisy	Throughout the year	Leaves	Crushed leaves of this plant are used to stop bleeding from cuts and cure wounds.
Vitex negundo L.	Verbenaceae	Notchi	Chaste tree	Throughout the year	Leaves	Leaf juice is drunk to cure headache, fever, cough and cold. The paste obtained from the leaves is applied on the leg to cure rheumatism.
Ziziphus jujuba Mill.	Rhamnaceae	Illanthai	Jackal Jujube	November-July	Roots	Decoction prepared from the roots is used to cure stomach disorders.

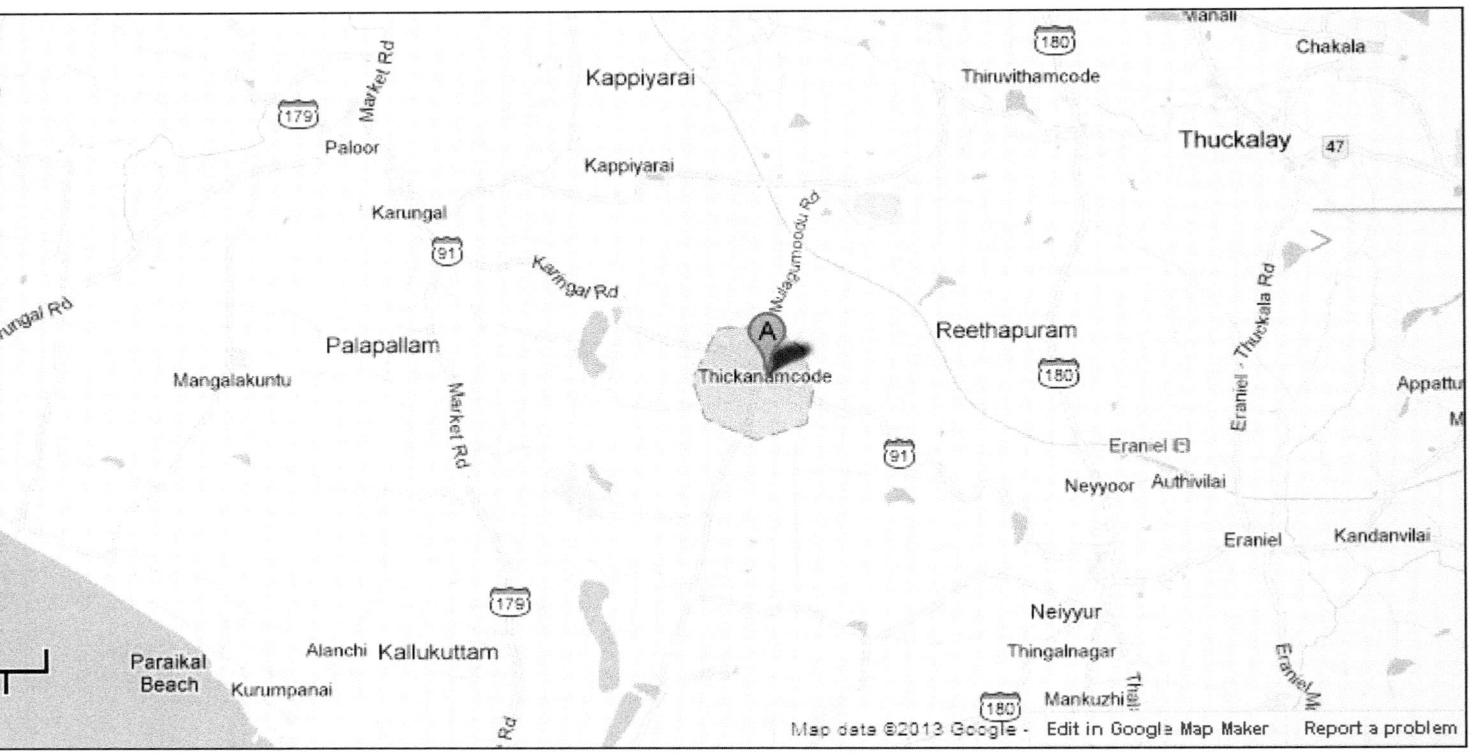

Figure 11.1: Map Showing the Study Area.

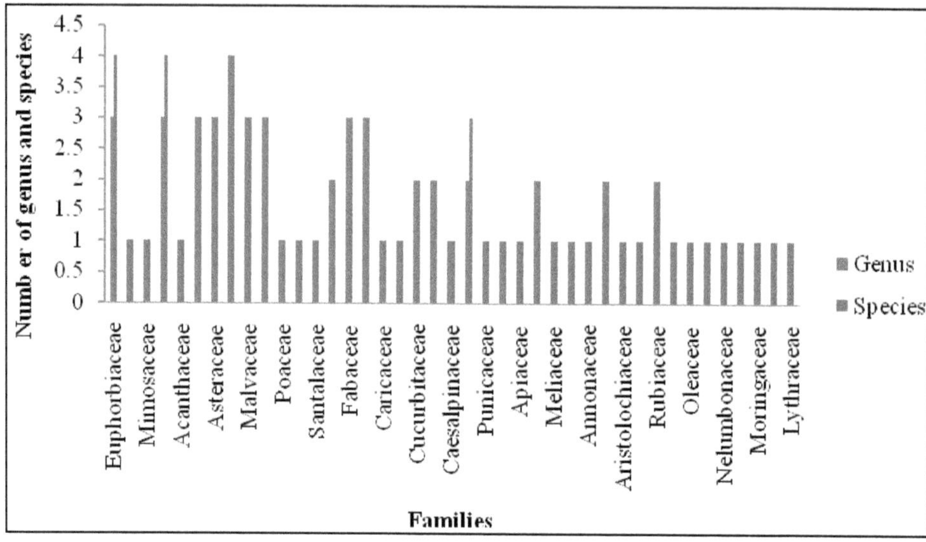

Figure 11.2: Dominant Families Reported from the Study.

predominantly used than other parts of the plants for the medicinal purpose and it was also agreed by other ethnobotanical researches (Bhattacharrya, 2002; Pal, 2003). The plant parts were used as decoction, paste, juice, oil and as raw form. Juice and paste were mostly preferred by people in the study area (Jeeva and Femila, 2012; Sukumaran *et al.*, 2014). Fresh plant parts were commonly used for the medicine preparation. The method of preparation of medicine and use is same or different from place to place.

Hemidesmus indicus the powdered root is given for veneral diseases (Pal, 2003; Singh *et al.*, 2002; Sivaperumal *et al.*, 2010). Decoction of leaves and roots of *Cardiospermum halicacabum* is used to treat rheumatism (Singh *et al.*, 2002). *Centella asiatica* juice is used to cure pyresia, swellings and also used as a vital tonic (Subitha *et al.*, 2011). The fresh juice extracted from the leaves of *Ocimum sanctum* is drunk to cure cough and fever (Ganesan *et al.*, 2008; Karuppusamy, 2007; Sivaperumal *et al.*, 2010). Crushed leaves of *Tridax procumbens* are used to stop bleeding from cuts

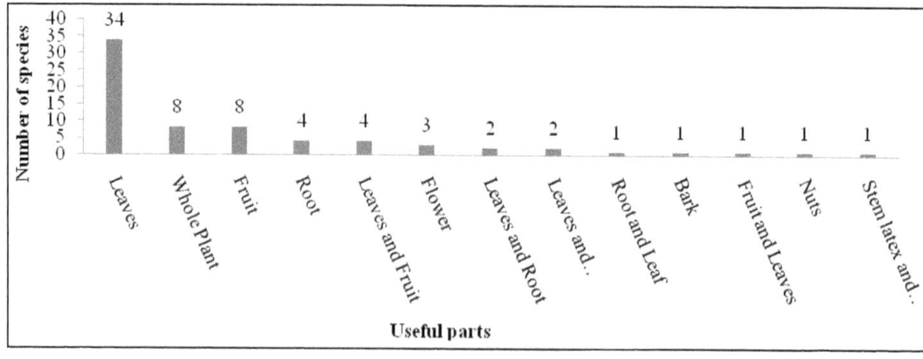

Figure 11.3: Plant Parts Used for Medicinal Purposes.

and cure wounds (Ganesan *et al.*, 2008; Jeeva and Femila, 2012). *Mimosa pudica* leaf juice is used to stop bleeding and dried powder cures diabetes, *Leucas aspera* leaf decoction used as antipyretic (Sukumaran and Raj, 2010) and extracted juice of leaves and young shoots used for gastric disorder, *Aloe vera* fresh leaves are applied on the forehead to cure fever (Sukumaran *et al.*, 2014).

Cissus quadrangularis is ground and the paste to cures fractures and body pain. *Capsicum annum* fruits are given for cold, cough, fever and dyspepsia, cooked plant of *Helitropium indicum* can be used to cure stomachache, *Gynandropis peataphylla* seed oil is used to expel round worms, *Murraya koenigii* leaves are very useful for digestive problems. *Ricinus commuinis* oil is used for rheumatic pain, constipation (Uma and Parthipan *et al.*, 2013). *Abrus precatorius* juice is used in the treatment of rheumatism (Vijayalakshmi *et al.*, 2013).

Conclusion

Findings of the present investigation revealed that, Thickanamcode has a very rich diversity of medicinal plants. The crude drug is obtained from medicinal plants. Due to the influence of modern medicine, the usage of traditional medicine is decreasing day by day. When the people need a small part of the plant they pullout the plant itself. So the wealth of medicinal plants decreases, so we have to conserve the medicinal plants and utilize the crude drugs obtained from medicinal plants. The community should also play a major role to conserve medicinal plants. Today, given the growing demands on the plants and the shrinking of forests, there is no option but to deliberately cultivate medicinal plants and consume of herbal wealth of the study area is the need of their hour indeed.

References

Anon S, 2003. *Science and Technology Policy*, (Ministry of Science and Technology, Govt. of India), pp. 23.

Balakumar S, Rajan S, Thirunalasundari T and Jeeva S, 2011. Antifungal activity of *Aegle marmelos* (L.) Correa (Rutaceae) leaf extract on dermatophytes. *Asian Pacific Journal of Topical Biomedicine*, **1**(4): 309-312.

Bhattacharyya G, 2002. Ethnobotanical studies on some weeds of Gujarat, India. *In*: Recent Progress in Medicinal Plants (Volume 1, *Ethnomedicine and pharmacognosy*), Singh VK, Govil JN and Singh G, (eds.), SCI Tech Publishing LLC, USA, pp. 33-40.

Britto AJD, Sujin MR, Mahesh R and Dharmar K, 2010. Ethnomedicinal wisdom of Manavalakuruchi people in Kanyakumari district, Tamil Nadu. *International Journal of Biological Technology*, **1**(2): 25-30.

Farnsworth NR, Kerele O and Bingei AS, 1985. Medicinal plants in therapy. *Buletin of the World Health Organisation*, **63**: 965-981.

Gamble JS and Fischer CEC, 1915-1935, *Flora of the Presidency of Madras*. Adlord and Sons Limited, London.

Ganesan S, Pandi RN and Banumathy N, 2008. Ethnomedicinal survey of Alagarkoil Hills (Reserved Forest), Tamil Nadu, India. *ejournal of Indian Medicine*, **1**: 1-18.

Jeeva S and Femila V, 2012. Ethnobotanical Investigation of Nadars in Atoor village, Kanyakumari District, Tamil Nadu, India. *Asian Pacific Journal of Tropical Biomedicine*,Pp 593-600.

Jenisha SR and Jeeva S, 2014. Traditional Remedies Used by the Inhabitants of Keezhakrishnanputhoor - A Coastal Village ofKanyakumari District, Tamil Nadu, India. *Medicinal and Aromatic plants*, **3**(4): 175-180.

Karuppusamy S, 2007. Medicinal plants used by Paliyan tribes of Sirumalai hills of Southern India. *Natural product Radiance*, **6**: 436-442.

Laloo RC, Kharlukhi L, Jeeva S and Mishra BP, 2006. Sacred forests of Mehalaya as a treasure house of Medicinal plants: effect of disturbance and population structure of important tree specioes. *Current Science*, **9**(2): 225-232.

Matthew KM, 1995. *An Exclusive Flora of Central Tamil Nadu, India*. The Rapinat Herbarium, Oxford and IBH Publishing Co. Pvt. Ltd, New Delhi.

Nair NC and Henry AN, 1983. *Flora of Tamil Nadu, India*, Botanical Survey of India, Coimbatore.

Pal DC, 2003. Lodha medicine of midnapore district, West Bengal, India. *In* : Recent Progress in Medicinal Plants (Volume 7, Ethnomedicine and Pharmacoghosy II) Singh VK, Govil JN, Hasmi S and Singh G. (eds), Stadium Press, LLC, USA, pp: 443-464.

Singh, Dhan and Pundir YPS, 2002 Wild medicinal plants of Jaun sar-Bawar (Western Himalayas) Uttaranchal, *Indian Forester*, **130**(1): 1529-1271.

Sivaperumal R, Ramya S, Ravi VA, Rajasekaran C and Jayakumararaj R, 2010. Ethnopharmocological studies on the medicinal plants used by tribal inhabitants of Kottur hills, Dharmapuri, Tamil Nadu, India. *Environment and We International Journal of Science and Technology*, **5**: 57-64.

Subitha KT, Ayyanar M, Udayakumar M and Sekar T, 2011. Ethnobotanical plants used by Kani tribals in Pechiparai forest of Southern Western Ghats, Tamil Nadu, India. *International Research Journal of Plant Science*, **2**(12): 349-354.

Sukumaran S and Raj ADS, 2010. Medicinal plants of sacred groove in Kanyakumari district, Southern Western Ghats. *Indian Journal of Traditional Knowledge*, **9**: 294-299.

Sukumaran S, Brintha TSS, Subitha P, Sheeba YA and Jeeva S, 2014. Usage of medicinal plants by two cultural communities of Kanyakumari District, Tamil Nadu, South India. *Journal of Chemical and Pharmaceutical Research*, **6**(8): 67-79.

Uma R and Parthipan B, 2015. Survey of Mredico-botanical climbers in Pazhayaru river bank of Kanyakumari district, Tamil Nadu, *Journal of Medicinal Plant Studies*, **3**(1): 33-36.

Vijayalakshmi N, Anbazhagan M and Arumugam K, 2013. Studies on the Ethnomedicinal plant used by the Irular tribe of Thirumurthi Hill of Western Ghats, Tamil Nadu, India. *International Journal of Research in Plant Science*, **4**(1): 8-12.

2018, Ethnomedicinal Plants: A Biodiversity Treasure Pages 307–341
Editors: *V.R. Mohan, A. Doss, P.S. Tresina and V. Sornalakshmi*
Published by: ASTRAL INTERNATIONAL PVT. LTD., NEW DELHI

Chapter 12

An Investigation on the Aboriginal Ethnomedicinal Plants of *Kanikkars* of Kalakad-Mundanthurai Tiger Reserve Sanctuary, Tamil Nadu with Special Emphasis on Rheumatism

***E. Daffodil D'Almeida*[1] *and V.R. Mohan*[2]**

[1]PG and Research Department of Botany, St. Mary's College (Autonomous), Thoothukudi, Tamil Nadu
[2]Ethnopharmacology Unit, PG and Research Department of Botany, V.O.Chidambaram College, Thoothukudi, Tamil Nadu
Corresponding author's E-mail: vrmohanvoc@gmail.com

Introduction

The tribals are closely associated with plants and possess good knowledge of plant resources in their vicinity. With the reach of civilization to the ethnic societies, the traditional knowledge on the use of these plants is vanishing. There is an urgent need to scientifically authenticate and document this knowledge; otherwise it will be lost forever.

It is a matter of great pride that among the 18 hot spots known for rich flora in the world, two are located in India. They are the Eastern Himalayas and the Western Ghats (Khoshoo, 1996). The hill chain of the Western Ghats recognized as a region of high level of biodiversity is under the threat of rapid loss of genetic resources (Gadgil, 1996). The biodiverse nature of the Eastern Ghats is meager. Keeping these facts in mind, the present study has been carried out in the natural stands of South-Eastern slopes of the Western Ghats, Tamil Nadu, India.

Intensive and extensive survey was made on the literature with particular emphasis on the medicinal plants used for rheumatism. A few available papers give fragmentary information on the medicinal plants used by the *Kanikkars* for treating rheumatism. There seems to be a lacuna in the study of the tribal knowledge for combating this ailment; this fact prompted the present study.

The term "rheumatism" embraces a variety of disorders that have, in common, pain and stiffness referable to the musculoskeletal system. When such symptoms are due to abnormality of the joint itself, the condition can be classified as arthritis. Non articular rheumatism includes those conditions in which the symptoms are produced not by pathologic changes in the joints proper, but in the structures contiguous to, or related to the joints. Although arthritis occurs in a number of different forms, there are essentially two fundamental pathological processes that affect the joints *viz.*, Inflammation, which may be exudative or proliferative or a combination of each and degenerative changes, which are primarily dependent on the limited capacity of articular cartilage to repair itself (Loeb, 1971).

There are many different ways to treat arthritis. Treatments include medications, rest, exercise and surgery to correct damage to the joint. Some medications that offer relief of arthritis symptoms are anti-inflammatory pain killer drugs, such as aspirin, ibuprofen or naproxen, topical pain relievers, corticosteroids and narcotic pain relievers. There are also many strong medications called Disease-Modifying Anti-Rheumatic Drugs (DMARDs) that are used to treat arthritis.

Currently, both steroidal anti-inflammatory drugs and non-steroidal anti-inflammatory drugs (NSAIDs) are used in the relief of inflammation. Steroids have an obvious role in the treatment of inflammatory diseases, but due to their toxicity, can only be used over short periods except in very serious cases where the risks are acceptable. Prolonged use of NSAIDs is also associated with severe side effects, notably gastrointestinal haemorrhage (Robert *et al.*, 1979; Miller, 1983). The newer Cyclooxygenase-2 (Cox-2) selective drugs do not seem to be free of risk (Wallace *et al.*, 1998). Consequently, there is a need to develop new anti-inflammatory agents with minimum side effects (Vane and Botting, 1995; Yesilada *et al.*, 1997).

Inflammatory diseases including different types of rheumatic diseases are very common throughout the world. Although rheumatism is one of the oldest known diseases of mankind and affects a large population of the world, no substantial progress has been made in achieving a permanent cure. The search involving screening and development of drugs for anti-inflammatory activity is an unending problem. There is much hope of finding active anti-rheumatic compounds from indigenous plants, as these are still used in therapeutics despite the progress made in conventional chemistry and pharmacology for producing effective drugs (Handa *et al.*, 1992).

The dried root of *Harpagophytun procumbens* (Devil's claw) (Pedaliaceae), has recently received popular attention for the treatment of painful rheumatic conditions; iridoid glycosides, *e.g.* Harpagoside, are the characteristic constituents. The juice of *Ananas comosus*, the pineapple (Bromeliaceae), contains a mixture of at least five proteolytic enzymes, collectively called bromelin. In western medicine, the enzyme has been introduced for its ability to dissolve fibrin conditions of inflammatory oedema (Evans, 2002).

Modern preparations made from the purified extracts of the resin obtained from the tree *Boswellia serrata* (Burseraceae) are taken internally as well as topically to relieve inflammations associated with rheumatoid arthritis.

Indian systems of medicine have played a highly significant role in combating rheumatism and related ailments. Specifically, the tribal medicines have been enjoying high popularity among the rural folk. Tribal remedies are simple, cost effective and minimal in side effects. In Tamil Nadu, several tribals inhabiting Western Ghats seem to have been endowed with a rich knowledge of herbals, especially ailments like rheumatism. One such tribal community is *Kanikkars* who are popular in healing many ailments including rheumatism. The present study focuses on tapping of the tribal knowledge of anti-rheumatic herbals and to document these data with specific authentication.

The *Kanikkar* Tribe

The *Kanikkars* belong to the Southern tribal zone. The *Kanikkars* are also known as kanikaran or Kani. The Southern tribals are historically more ancient tribes. There are as many as 36 types of scheduled tribes in Tamil Nadu. In the serialized list notified by the Government of Tamil Nadu, the *Kanikkars* are placed at 7th position. They live in low altitude regions of Western Ghats and live in large numbers. *Kanikkars* means "hereditary proprietor of land" thus recognizing their ancient rights over the forest lands.

The *Kanikkars* are generally very short in stature and meager in appearance, from their active habits and scanty food. Some have markedly Negroid features. They are simple and straight forward tribals. They are traditionally a nomadic community. They speak in their own dialect, *Kanikkar* Bhasha or Malampashi, which is close to the Dravidian language Malayalam. (Singh, 1994) They were once lords of the forest and practiced migratory cultivation but *Kanikkars* have now to a large extent abandoned this kind of migratory cultivation because of the forest may not be set fire to or trees felled at the unrestricted pleasure of individuals.

Most of the *Kanikkar* tribals have a general knowledge of medicinal plants that are used for first aid remedies, to treat cough, cold, fever, headache, poisonous bites and some other simple ailments. Kanis still supplement their food by gathering roots and tubers from the nearby forest areas. They eat tubers like *Manihot esculenta* and *Dioscorea oppositifolia etc.* They are extremely hard working and can survive without the help of modern facilities. They are socio-economically backward and most of them are very poor. They are also engaged in seasonal collection of honey, bee wax and some minor forest produce. They cultivate edible plants like, tapioca, banana, millets and cash crops such as pepper, areca nut and cashew nut.

Materials and Methods

Study Area

The Kalakad-Mundanthurai Tiger Reserve (KMTR) lies between 8° 22′ and 8° 53′ North latitude and between 77° 10′ and 77° 35′ East longitudes in Tirunelveli and Kanyakumari district of Tamil Nadu in the Southern Western Ghats of India (Area

Map 12.1). The boundaries are Ambasamudram and Tenkasi taluks of Tirunelveli districts in the north, Ambasamudram and Nanguneri taluks of Tirunelveli districts in the east, Kanyakumari district in the south and Kerala State in the west.

KMTR is India's 17th Tiger Reserve under Project Tiger and the Sanctuary is developed as a National Tiger Reserve from the year 1988 with a total area of 817

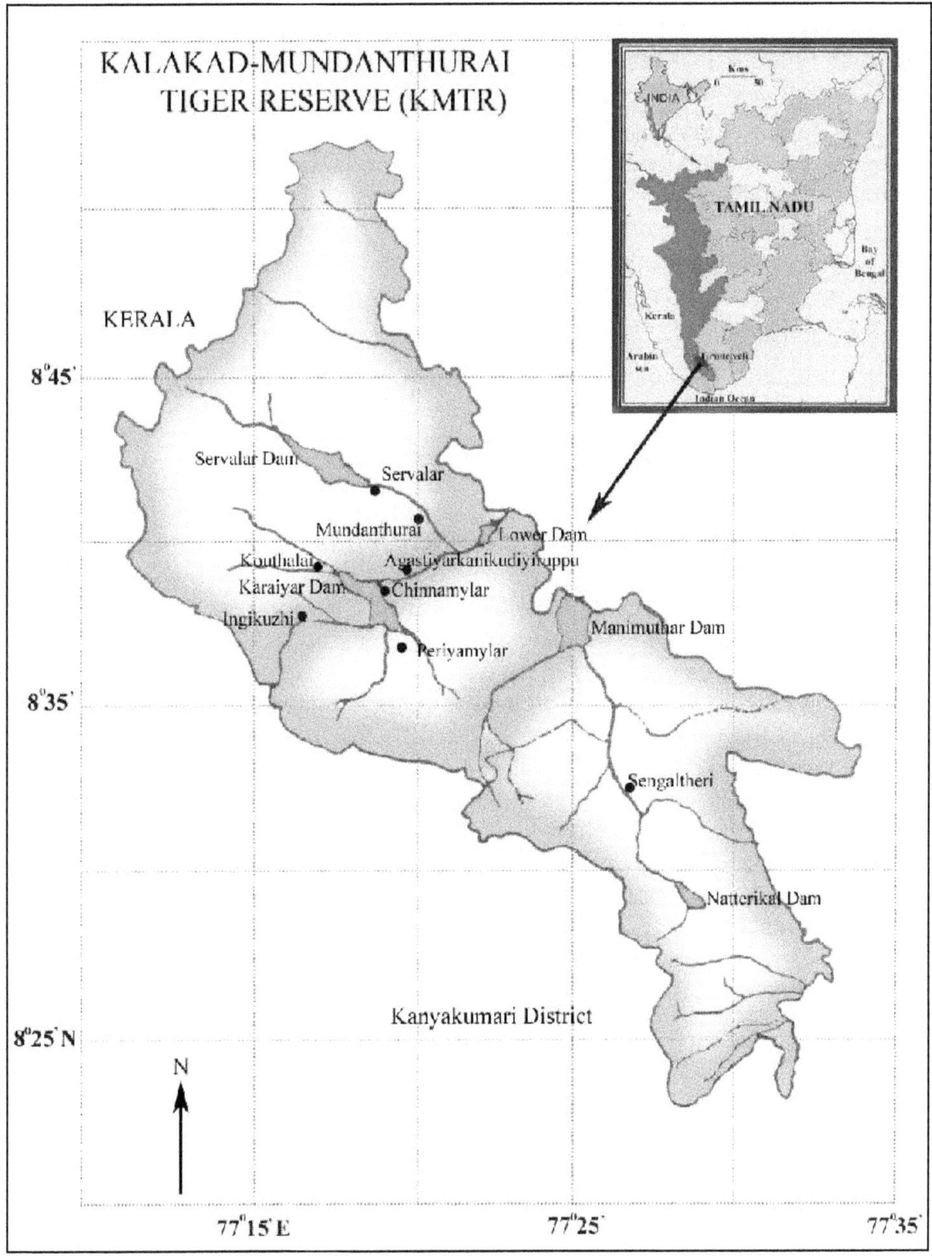

Map 12.1: Area.

Km^2 in the south most Western Ghats ranges. Geographically, it is a part of South Western tip of the Western Ghats, a region that is known for its species richness, diversity and high degree of endemism. The KMTR area has been recognized as one of the 'hot-spots' for Biodiversity conservation by the IUCN (Ayyanar and Ignacimuthu, 2005). The altitude ranges from 100m to 1867m (MSL). It receives rainfall during the South West as well as North East monsoon. The important peaks of the reserve include Agasthiarmalai, Ainthilaipothigai and Nagapothigai. Tamiraparani, the perennial river of Tamil Nadu originates from Agasthiarmalai (Pothigaimalai) and flows through this sanctuary.

The varied climatic and topographic conditions prevailing in the sanctuary present a remarkable diversity of both the flora and fauna. The study area consists of tropical wet evergreen forests, semi evergreen forests, moist deciduous forests, tropical riparian fringe forests, dry teak forests, dry deciduous forests, umbrella thorn forest, wet temperate forests and high and low altitude grass lands.

In the study area (Plate 12.1), the *Kanikkars* live either in isolated pockets or small hamlets. Their habitations are known by the following names.

(i) Ingikuzhi, (ii) Chinnamylar, (iii) Periyamylar, (iv) Agastiyarkanikudiyiruppu, (v) Tharuvattampari Kani Kudiyiruppu (Servalar) (vi) Kouthalai.

Survey and Collection

For survey and collection of medicinal plants used by the *Kanikkar* tribe of the study area (KMTR, Western Ghats, Tamil Nadu), frequent field trips were undertaken. Information regarding the medicinal plants was gathered by meeting *Kanikkars* practicing indigenous medicine, during explorative field trips and by gaining a good rapport and winning over their confidence. Most of the information include in this study was gathered from the elderly and experienced medicine men who have a long acquaintance with the use of medicinal plants. The field notebook delineates all the usage procedures adopted by the tribals. The information thus gathered was cross checked adequately for reliability and accuracy by interacting with different groups of the *Kanikkars* from different habitats to confirm the use, mode of administration as well as dosage differences, if any. After eliciting detailed information regarding the wild medicinal plants (Table 12.1) and any plant part(s)/ extraction of plant part obtained from local market used in medicinal preparations, as ingredients (Table 12.2), they were carefully brought to the laboratory for identification. Herbaria for all the collected plant specimens except for plants, whose parts or extracts were procured from the local market, were prepared (VOCB No. from 3320 to 3371) and deposited in the Research Department of Botany, V.O. Chidambaram College, Tuticorin, Tamil Nadu, India.

The collected plants were identified by referring to the following compilations (Fischer and Gamble, 1957; Gamble, 1957a; Gamble, 1957b; Henry *et al.*, 1989; Henry *et al.*, 1987; Nair and Henry, 1983). After authentication with regional flora, the specimens were matched with the authentic herbarium of Madras Herbarium, Botanical Survey of India (B.S.I), Southern Circle, Coimbatore, Tamil Nadu, India.

A view of Western Ghats

A view of Mylar hills

River Pambanar

A tribal hut in Periyamylar

A Kanikkar family

A Kani tribe holding a mature pod of *Entada pursaetha* DC.

Plate 12.1: Kanikkars in the Study Area.

Table 12.1: List of Ethnomedicinal Plants Collected and Documented for Rheumatism

Herbarium Number	*Botanical Name*	*Family*	*Vernacular Name*	*Part Used*	*Habit*	*Habitat*
VOCB 3320	*Abrus precatorius* L.	Fabaceae	Kundrimani	Seed	Twining shrub	Open waste land on fences and bushes
VOCB 3321	*Actinopteris radiata* (Sw.) Link. *A. dichotama* Kuhn	Actinopteridaceae	Mayilosa	Leaf	Herb	Fully exposed hill slopes
VOCB 3322	*Aloe barbadensis* Mill. *A. vera* (L.) Burm. f.	Liliaceae	Chothukathalai	Leaf	Herb	Hedges in dry zones
VOCB 3323	*Alstonia venenata* R. Br.	Apocynaceae	Malaiarali/ Malaivaathamudakki	Leaf	Shrub	Wet places in ever green forest
VOCB 3324	*Anisomeles indica* (L.) Kuntze *A. ovata* R. Br.	Lamiaceae	Vaathaneerpatchilai	Leaf, stem	Herb	Waste land and road sides in open forest
VOCB 3325	*Anisomeles malabarica* (L.) R. Br. ex Sims *Nepeta malabarica* L	Lamiaceae	Perunthumbai	Leaf	Herb	Waste land
VOCB 3326	*Aristolochia krysagathra* Sivaranjan and Pradeep	Aristolochiaceae	Karudakodi	Root, leaf	Twiner	Wet places in ever green forest
VOCB 3327	*Asystasia travancorica* Bedd.	Acanthaceae	Aathu Urinji	Leaf, flower	Herb	Sandy places of river bank
VOCB 3328	*Azadirachta indica* A. Juss.	Meliaceae	Vembu	Stem bark, leaf, seed	Tree	Cultivated in plains
VOCB 3329	*Begonia malabarica* Lam.	Begoniaceae	Kalsirupuli	Shoot	Herb	Tropical wet evergreen forest
VOCB 3330	*Calotropis gigantea* (L.) R. Br. *Asclepias gigantea* L.	Asclepiadaceae	Erukku	Leaf	Shrub	Dry places and waste land
VOCB 3331	*Capsicum annuum* L. *C. frutescens* Sensu Clarke	Solanaceae	Kaanthaarimilagai	Fruit	Herb	Open land on hill slopes
VOCB 3332	*Cardiospermum halicacabum* L.	Sapindaceae	Mudakkathan	Leaf	Climber	Dry land fences.

Contd...

Herbarium Number	*Botanical Name*	*Family*	*Vernacular Name*	*Part Used*	*Habit*	*Habitat*
VOCB 3333	*Cissus quadrangularis* L.	Vitaceae	Pirandai	Stem	climber	Scrub jungle
VOCB 3334	*Curculigo orchioides* Gaertn. *C. malabarica* Wight.	Hypoxidaceae	Nilappanai	Tuber	Herb	Moist places of evergreen forest
VOCB 3335	*Cymbopogon citratus* (DC.) Stapf. *Andropogon citratus* DC.	Poaceae	Chukkunaaripullu	Whole plant	Herb	Dry areas of open forest
VOCB 3336	*Datura metel* L. *D. fastuosa* L.	Solanaceae	Oomathai	Leaf	Under shrub	Waste land and road sides
VOCB 3337	*Diospyros melanoxylon* Roxb.	Ebenaceae	Vaathabeedi	Leaf	Small tree	Deciduous forest
VOCB 3338	*Drynaria quercifolia*(L.) J. Sm. *Polypodium quericifolium* L.	Drynariaceae	Aattukkaal malaivahan	Rhizome	Herb	Shaded hilly places as epiphyte or lithophyte
VOCB 3339	*Eclipta prostrata* (L.) L. *E.alba* (L.) Hassk.	Asteraceae	Karisalankanni	Shoot	Herb.	Wet sandy places
VOCB 3340	*Elephantopus scaber* L.	Asteraceae	Yanaichuvadi/ Nilanthaangi	Leaf	Subscapige rous herb	Sandy path sides along with grasses.
VOCB 3341	*Entada pursaetha* DC. *E. scandens* auct. non Benth.	Mimosaceae	Parandai kodi	Seeds	Woody climber	Near streams in wet evergreen forest
VOCB 3342	*Erythrina variegata* L. *E. indica* Lam.	Fabaceae	Mullumurungai	Stem bark, leaf	Tree	Deciduous forest
VOCB 3343	*Erythropalum scandens* Bl., Bijdr. *Erythropalum populifolium* (Arn.) Mast.	Erythropalaceae	Vaathavallikodi	Tender shoot	Woody climber	Evergreen forest
VOCB 3344	*Eugenia singampattiana* Bedd.	Myrtaceae	Kaatu korandi	Leaf, flower, tender fruit	Shrub	Semi ever green forest

Herbarium Number	Botanical Name	Family	Vernacular Name	Part Used	Habit	Habitat
VOCB 3345	*Goniothalamuswightii* Hk. f. and Th.	Annonaceae	Kaattunaraipatchilai	Leaf, flower	Tree	Tropical wet evergreen forest
VOCB 3346	*Hemidesmus indicus* (L.) R. Br. var. *indicus* *Periploca indica* L.	Periplocaceae	Nannaari	Root	Climber	Open forest on bushes
VOCB 3347	*Hugonia mystax* L.	Linaceae	Motira kanni	Leaf, tender fruit	Shrub	Tropical dry deciduous forest
VOCB 3348	*Isonandra lanceolata* Wight.	Sapotaceae	Milagunaripatchilai	Leaf, flower	Tree	Tropical dry deciduous forest
VOCB 3349	*Jatropha gossypifolia* L.	Euphorbiaceae	Aathalai	Stem bark.	Shrub	Open forest along road side
VOCB 3350	*Justicia adhatoda* L. *Adhatoda vasica* Nees.	Acanthaceae	Aadathoda	Leaf	Shrub	Open land among bushes
VOCB 3351	*Kingiodendron pinnatum* (Roxb. ex DC.) Harm. *Hardwickia pinnata* Roxb.ex DC.	Caesalpiniaceae	Kulavu	Resin	Tree	Semi evergreen and ever green forests
VOCB 3352	*Leea indica* (Burm. f.) Merr. *Staphylea indica* Burm. f.	Leeaceae	Kaatuvalaripatchilai	Stem, leaf, flower	Shrub	Tropical wet evergreen forest
VOCB 3353	*Mallotus philippensis* (Lam.) Muell. Arg.	Euphorbiaceae	Kaatuthakadi	Leaf, tender fruit	Tree	Tropical dry deciduous forest
VOCB 3354	*Mimosa pudica* L.	Mimosaceae	Thottalvaadi	Leaf	Herb	Path sides in open forest
VOCB 3355	*Moringa pterygosperma* Gaertn. *M. oleifera* acut. non Lam.	Moringaceae	Murungai	Stem bark	Tree	Cultivated in plains
VOCB 3356	*Murraya paniculata* (L.) Jack. *M. exotica* L.	Rutaceae	Malaivembu	Leaf	Medium sized tree	Wet semi ever green forest
VOCB 3357	*Pergularia daemia* (Forssk.) Chiov. *P. extensa* (Jacq.) N.E. Br.	Asclepiadaceae	Veliparuthi	Leaf, root	Climber	On hedges and fence

Contd...

Herbarium Number	Botanical Name	Family	Vernacular Name	Part Used	Habit	Habitat
VOCB 3358	*Piper nigrum* L.	Piperaceae	Milagu	Shoot, fruit	Climber	Cultivated in hills
VOCB 3359	*Pleiospermium alatum* (Wall. ex Wt. and Arn.) Swingle *Limonia alata* Wall. ex Wt. and Arn.	Rutaceae	MalaiNaarathai	Leaf	Small tree	Foot hills of forest
VOCB 3360	*Plumbago zeylanica* L.	Plumbaginaceae	Venkoduveli	Root	Scandent under shrub	Open land among bushes
VOCB 3361	*Psychotria nilgiriensis* Deb and Gang.	Rubiaceae	Odaikaapipatchilai	Tender fruit	Shrub	Tropical wet evergreen forest
VOCB 3362	*Psychotria nudiflora* Wt. and Arn.	Rubiaceae	Kalpoo	Leaf, flower	Shrub	Tropical dry deciduous forest
VOCB 3363	*Pterocarpus marsupium* Roxb.	Fabaceae	Vengai	Resin	Tree	Deciduous forest
VOCB 3364	*Rauwolfia densiflora* (Wall.) Benth. ex Hk. f. *Tabernaemontana densiflora* Wall.	Apocynaceae	Paarisirunila patchilai	Leaf, flower	Shrub	Tropical dry deciduous forest
VOCB 3365	*Sansevieria roxburghiana* Schult and Schult f.	Agavaceae	Marul	Leaf	Herb	Dry waste land among bushes
VOCB 3366	*Scoparia dulcis* L.	Scrophulariaceae	Sakkaraivembu	Leaf	Herb	Waste lands and hill slopes
VOCB 3367	*Sonerila tinnevelliensis* Fischer	Melastomataceae	Kalpuli	Leaf	Shrub	Tropical dry deciduous forest
VOCB 3368	*Tinospora cordifolia* (Willd.) Miers ex Hk. f. and Th. *Menispermum Cordifolium* Willd.	Menispermaceae	Seenthilkodi	Leaf	Climbing shrub	Open forest on bushes
VOCB 3369	*Toddalia asiatica* Lam. var. *asiatica*	Rutaceae	Milagaranai	Stem bark	Prickly scandent shrub	On bushes and hedges
VOCB 3370	*Ventilago madraspatana* Gaertn.	Rhamnaceae	Vembadan	Stem bark	Climber	Deciduous forest
VOCB 3371	*Vitex negundo* L.	Verbenaceae	Notchi	Leaf	Shrub	Path sides in plains

Table 12.2: Botanical Names of Plants whose Parts/Products are Purchased from the Market and Used as Medicine/Ingredients in the Medicinal Preparations by *Kanikkars*

Sl.No.	*Botanical Name*	*Family*	*Vernacular Name*	*Plant Part/Extract of Plant Part Procured*
1	*Allium cepa* L.	Alliaceae	Vengayam	Bulb
2	*Allium sativum* L.	Alliaceae	Vellaipundu	Bulb
3	*Borassus flabellifer* L.	Arecaceae	Panai	Palm candy
	B. flabelliformis Murr.			
4	*Brassica juncea* (L.) Czern.	Cruciferae (nom. alter. Brassicaceae)	Kadugu	Seed
	Sinapis juncea L.			
5	*Cocos nucifera* L.	Arecaceae	Thennai	Seed oil
6	*Cuminum cyminum* L.	Umbelliferae (nom. alter. Apiaceae)	Shiragam	Seed
7	*Curcuma longa* L.	Zingiberaceae	Manjal	Dried rhizome
8	*Glycyrrhiza glabra* L.	Fabaceae	Athimathuram	root
9	*Papaver somniferum* L.	Papaveraceae	Kasakasa	Seed
10	*Piper longum* L.	Piperaceae	Thippili	Fruit
11	*Sesamum indicum* L.	Pedaliaceae	Yellu	Seed oil
	S. orientale L.			
12	*Trigonella foenum-graecum* L.	Fabaceae	Vendhiyam	Seed
13	*Zingiber officinale* Roscoe.	Zingiberaceae	Sukku	Dried rhizome

Enumeration

Plants used by the *Kanikkars* for the treatment of rheumatism are enumerated as follows:

Abrus precatorius L. (Plate 12.2, 1)

About 50 grams of warm seed paste is applied over the affected joints twice a day until one experiences relief from stiffness of joints.

Actinopteris radiata (Sw.) Link. (Plate 12.2, 2)

About hundred grams of fresh leaves are boiled in five hundred millilitres of coconut oil (*Cocos nucifera*) on a low flame for fifteen minutes. This oil is filtered after cooling. The affected joints are massaged with this lukewarm oil. Soft pressure is applied then sprayed with lukewarm water until there is relief from pain.

Aloe barbadensis Mill. (Plate 12.2, 3)

A longitudinal cut is made on a leaf of *Aloe barbadensis* in such a way as to expose the mesophyll. It is heated for a while on a low flame. The mesophyll content is,

Abrus precatorius **L.**

Actinopteris radiata **(Sw.) Link.**

Aloe barbadensis **Mill.**

Anisomeles indica **(L.) Kuntze.**

Aristolochia krysagathra **Sivaranjan & Pradeep**

Asystasia travancorica **Bedd.**

Plate 12.2

then, rubbed on the affected part when it is comfortably hot. This process is repeated for thirteen days to disperse the swelling in the joints.

Alstonia venenata R. Br. (Plate 12.7, 1 and 2)

About hundred to two hundred grams of fresh leaves are boiled in neem oil (*Azadirachta indica*) over a low flame for fifteen to twenty minutes until the oil extracts the complete essence of the drug. The affected part is massaged with this lukewarm oil, specifically, in the direction from the limbs to the upper portion applying soft pressure. Hot water bath is administered after a few hours, until there is relief from the rheumatic complaint.

Anisomeles indica (L.) Kuntze (Plate 12.2, 4)

About hundred grams of fresh leaves and young stem are boiled in one litre of neem oil (*Azadirachta indica*) along with ten grams of poppy seeds (*Papaver somniferum*) and fifty grams of garlic (*Allium sativum*) in an earthen pot for fifteen to twenty minutes on a low flame. The filtered oil is applied on the joints of the hands and legs. A gentle massage is applied for twenty minutes and hot bath is advised after this application. Consumption of fish, egg, meat and sexual intercourse are prohibited for three weeks.

A handful of leaves are boiled in steam and used to foment the affected parts until there is relief from rheumatic complaint.

Anisomeles malabarica (L.) R. Br. ex Sims

Handfuls of leaves are boiled in two litres of water in a closed container. When the steam emanates the lid of the container is slowly removed and vapour bath is administered to get relief from rheumatic pain.

Aristolochia krysagathra Sivaranjan and Pradeep. (Plate 12.2, 5)

About fifty grams of root and an equal quantity of leaves are boiled in one litre of coconut oil (*Cocus nucifera*) for about fifteen to twenty minutes over a low flame. This oil is filtered after cooling and applied on the head once in a day for seven to ten days as the treatment for rheumatism. This therapy is used to reduce excessive heat of the body.

Asystasia travancorica Bedd. (Plate 12.2, 6)

About ten gram of a paste made from the leaves and the flowers of *Asystasia travancorica* is mixed with honey and is taken orally, twice a day, for three weeks for the treatment of rheumatism. The use of tamarind, (*Tamarindus indica*), fish and egg is avoided.

Azadirachta indica A. Juss.

The oil extracted from the seeds is massaged over the joints as an embrocation in rheumatism.Neem oil is used in many medicated preparations for the treatment of rheumatism.

About twenty to fifty grams of stem bark is boiled in five hundred millilitres of water for twenty to thirty minutes. This decoction is taken orally for three weeks to treat rheumatic complaints.

Begonia malabarica **Lam. (Plate 12.3, 1)**

The warm paste of the aerial part is applied externally on the leg once a day for fourteen days to treat rheumatic complaints.

Calotropis gigantea **(L.) R. Br**

A poultice of roasted leaves is applied on the rheumatic joints for one week or more to soothe pain and to reduce swelling.

Capsicum annuum **L. (Plate 12.3, 2)**

A thick paste made from twenty grams of fruits of *Capsicum annuum,* along with ten grams of ginger (*Zingiber officinale*) and ten grams of garlic (*Allium sativum*) is given orally to women after delivery. This is followed by the administration of fifty millilitres of gingelly oil or honey to prevent rheumatic complaints.

Cardiospermum halicacabum **L. (Plate 12.3, 3)**

A hundred milliletres of the juice extracted from a sufficient amount of fresh leaves of *Cardiospermum halicacabum* is taken orally with palm sugar (*Borassus flabellifer*) for seven days for the treatment of rheumatism. The use of tamarind (*Tamarindus indica*), meat, fish and egg is avoided. The use of roasted salt and tamarind is recommended, instead.

A soup is prepared by boiling fifty grams of fresh leaves of *Cardiospermum halicacabum,* half a teaspoon of cumin seeds (*Cuminum cyminum*), four pieces of crushed onion (*Allium cepa*) and a pinch of salt in five hundred millilitres of water. This soup is taken orally twice a day.

About five hundred millilitres of the juice extracted by crushing the leaves of *Cardiospermum halicacabum* is kept in sun light for three days. It is then boiled in one litre of neem oil (*Azadirachta indica*) together with fifty grams of garlic (*Allium sativum*), fifty grams of mustard seeds (*Brassica juncea*), five grams of cumin seeds (*Cuminum cyminum*) and fifty grams of the crushed stem bark of *Moringa pterygosperma* from a ten year old tree, over a low flame for twenty to thirty minutes. It is, then, cooled, filtered and stored. This oil is applied twice a day. A hot water bath is administered mandatoryafter each application. Sexual intercourse is prohibited.

Two hundred and fifty grams of fresh leaves, five hundred grams of rice and a hundred grams of Black gram are ground well with the required quantity of common salt and water to make a paste. A food stuff called 'dosai' (pancake) prepared with this paste is eaten normally for two to three days in a week to prevent the possibility of any rheumatic complaint.

Cissus quadrangularis **L. (Plate 12.3, 4)**

About ten gram of a paste made from the tender shoots of *Cissus quadrangularis* is consumed twice a day for three weeks for the treatment of rheumatism.

1. ***Begonia malabarica* Lam.**
2. ***Capsicum annuum* L.**
3. ***Cardiospermum halicacabum* L.**
4. ***Cissus quadrangularis* L.**
5. ***Curculigo orchioides* Gaertn.**
6. ***Cymbopogon citratus* (DC.) Stapf.**

Plate 12.3

Curculigo orchioides Gaertn. (Plate 12.3, 5)

A hundred grams of chopped roots is boiled in one litre of coconut oil (*Cocos nucifera*) together with fifty grams of cumin seeds (*Cuminum cyminum*) on a moderate flame for fifteen to twenty minutes. Massage with this oil for half an hour is advised. And warm water bath is taken once in a day for fourteen days as treatment for rheumatism.

Cymbopogon citratus (DC.) Stapf. (Plate 12.3, 6)

A sufficient quantity of the whole plant of *Cymbopogon citratus* is chopped and boiled in water for thirty minutes and the warm water is used during bath twice a day for twelve days as treatment for rheumatism.

Handful of fresh leaves is crushed with water in to juice. After filtration, half a cup of this juice is given orally twice or thrice a day for one week as an effective management of rheumatism.

Datura metel L.

The leaves of *Datura metel* are soaked in boiling water and they are bandaged over the affected part to get relief from rheumatic pain.

The boiled leaves are also used for fomentation on the rheumatic swelling for fifteen to twenty minutes.

Diospyros melanoxylon Roxb.

Fifty grams of fresh leaves of *Diospyros melanoxylon* is made into a paste. This paste is warmed on low flame and applied over the knee twice a day for five to six days to get relief from joint pain.

A hundred to two hundred grams of fresh leaves is boiled in water. A bath with this warm water treats the parts affected by rheumatism.

Drynaria quercifolia (L.) J. Sm. (Plate 12.4, 1)

A decoction is prepared by boiling a hundred grams of the chopped rhizome of *Drynaria quercifolia* in five hundred millilitres of water. This decoction is taken along with a teaspoon of cumin (*Cuminumcyminum*) powder twice a day, for seven days to obtain relief from rheumatic complaints.

A paste made from fifty to hundred grams of the rhizome is applied on the part of the body which is affected by rheumatic pain twice a day for a week.

Eclipta prostrata (L.) L.

Twenty five to fifty grams of the stem and leaves of *Eclipta prostrata* is boiled in one litre of coconut oil. This oil is filtered after cooling. The filtered oil is applied to the head for an hour before bath during the course of treatment. This therapy is used to reduce body heat.

Elephantopus scaber L. (Plate 12.4, 2)

One teaspoon of shade dried leaf powder is taken orally with one and a half teaspoon of honey in the morning on an empty stomach, for two weeks to treat

1. ***Drynaria quercifolia*** **(L.) J. Sm.**
2. ***Elephantopus scaber*** **L.**
3. ***Eugenia singampattiana*** **Bedd.**
4. ***Goniothalamus wightii*** **Hk. F. & Th**
5. ***Hemidesmus indicus*** **(L.) R. Br. var.** ***indicus***
6. ***Hugonia mystax*** **L.**

Plate 12.4

pain in the joints.

***Entada pursaetha* DC.**

A thick paste made from six to ten seeds of *Entada pursaetha* is applied over the affected and inflamed swellings to reduce pain.

To strengthen the joints in infants, this paste is applied over the leg once a day for 10 days. Warm water bath is administered each time.

***Erythrina variegata* L.**

Two hundred grams of stem bark of *Erythrina variegata* is boiled in one litre of neem oil (*Azadirachta indica*) together with ten grams of poppy seeds (*Papaver somniferum*) and twenty five grams of garlic (*Allium sativum*) for fifteen to twenty minutes. This oil is massaged over the affected joints twice a day until one get relief from rheumatic pain.

Fifty to hundred grams of the leaves is boiled in an earthen pot. This preparation is used for fomenting the affected parts after massage.

***Erythropalum scandens* Bl., Bijdr. (Plate 12.13, 1 and 2)**

The chopped tender shoots of *Erythropalum scandens* is boiled with water. When this infusion is comfortably hot bath is taken. This management is continued until there is relief from the rheumatic complaint.

A paste made from ten grams of fresh leaves is mixed with one teaspoon of honey is given orally twice a day for seven days as treatment for rheumatism. Consumption of fish, egg, tamarind, liquor, salt and intercourse is prohibited during the course of the treatment.

***Eugenia singampattiana* Bedd. (Plate 12.4, 3)**

One teaspoon of the powder made from equal quantity of shade dried leaves, flowers and tender fruit of *Eugenia singampattiana* is consumed with honey two times a day for twenty one days for the treatment of rheumatism. Consumption of fish, both, fresh and dry, and egg is avoided.

***Glycyrrhiza glabra* L.**

One hundred grams of shade dried root pieces are powdered with fifty grams of dried ginger (*Zingiber officinale*), twenty five grams of dried pepper fruit (*piper nigrum*) and twenty five grams of dried long pepper fruit (*Piperlongum*). About two to three grams of this powder is mixed with honey and consumed orally twice a day on empty stomach until one experiences relief from rheumatic complaint.

***Goniothalamus wightii* Hk. f. and Th. (Plate 12.4, 4)**

Five grams of an equal quantity of leaves and tender fruits of *Goniothalamus wightii* is made into a paste with a few drops of water. This paste is taken orally twice a day for five to ten days for relief from rheumatic pain.

Hemidesmus indicus (L.) R. Br. var. indicus (Plate 12.4, 5)

About fifty grams of fresh roots of *Hemidesmus indicus* are boiled in one litre of coconut oil with ten grams of cumin seeds (*Cuminum cyminum*) on a moderate flame for fifteen to twenty minutes. This oil is filtered after cooling. The filtered oil is applied on the affected part for relief from burning sensation during rheumatism.

One teaspoon of shade dried root powder of *H. indicus* is added to two hundred millilitres of lukewarm water and filtered. This filtrate is taken, orally, twice a day for five to seven days for relief from rheumatic complaint.

***Hugonia mystax* L. (Plate 12.4, 6)**

One teaspoon of paste made from equal quantity of shade dried leaves and tender fruit of *Hugonia mystax* is consumed with honey twice a day for twelve days for the treatment of rheumatism. Consumption of fish both fresh and dry and egg is avoided.

***Isonandra lanceolata* Wight. (Plate 12.5, 1)**

A paste is prepared from two grams of fresh leaves and flowers of *Isonandra lanceolata* with a few drops of water. This paste is consumed twice a day for ten to twelve days to get relief from rheumatic pain.

***Jatropha gossypifolia* L.**

About fifty grams of the crushed stem bark of *Jatropha gossypifolia* is boiled in one litre of neem oil with twenty five grams of chopped tuber of *Curculigo orchioides* for fifteen to twenty minutes over a moderate flame. This oil is massaged on the affected joints as a liniment during rheumatism.

***Justicia adhatoda* L.**

A handful of leaves of *Justicia adhatoda* is cooked and used for a fomentation on the affected joint to alleviate rheumatic pain.

The juice extracted by crushing fifty grams of fresh leaves of *Justicia adhatoda* together with twenty five grams of leaves of *Pergularia daemia* is battered with egg white and taken orally for 10-12 days to treat rheumatic complaint.

***Kingiodendron pinnatum* (Roxb. ex DC.) Harm. (Plate 12.5, 2)**

The resin obtained by piercing the trunk of *Kingiodendron pinnatum* is applied on the affected joints before going to bed along with a soft massaging in circular motion. In the morning lukewarm water is poured over the joints. This oil is also applied on the fissured foot for five to seven days to get relief.

***Leea indica* (Burm. f.) Merr. (Plate 12.5, 3)**

A teaspoon of paste made from an equal quantity of leaves and flowers of *Leea indica* is mixed with honey and taken orally twice a day for sixteen days as treatment for rheumatism.

1. ***Isonandra lanceolata*** **Wight.**
2. ***Resin exudate from the stem of Kingiodendron pinnatum*** **(Roxb. ex DC.) Harm.**
3. ***Leea indica*** **(Burm. f.) Merr.**
4. ***Mallotus philippensis*** **(Lam.) Muell. Arg.**
5. ***Murraya paniculata*** **(L.) Jack.**
6. ***Pergularia daemia*** **(Forssk.)Chiov.**

Plate 12.5

Mallotus philippensis (Lam.) Muell. Arg. (Plate 12.5, 4)

One teaspoon of paste is made from an equal quantity of leaves and tender fruit of *Mallotus philippensis*. This is mixed with honey and is taken orally twice a day for thirteen days to get relief from rheumatism. During this period the consumption of salt, spicy food and tamarind is avoided.

Mimosa pudica L.

About two hundred and fifty grams of fresh leaves of *Mimosa pudica* are crushed and boiled in a mixture of hundred millilitres of gingelly oil (*Sesamum indicum*) and a half litre of neem oil (*Azadirachta indica*) on a low flame for twenty to twenty five minutes. This lukewarm oil is massaged on the affected parts. Hot water bath is administered.

The plant as a whole is made into a paste with a little water. This paste is applied on the inflamed joints once in a day for twelve days. Warm water bath is taken.

Moringa pterygosperma Gaertn.

About two hundred grams of stem bark from a ten year old tree is boiled in neem oil (*Azadirachta indica*) along with fifty grams of crushed garlic (*Allium sativum*) and fifty grams of coarsely powdered fenugreek (*Trigonella-foenum graecum*) on a moderate flame for about twenty minutes. This oil is applied on the affected part as an embrocation for rheumatic swelling. Hot water bath is administered after few hours.

Murraya paniculata (L.) Jack (Plate 12.5, 5)

About fifty grams of fresh leaves of *Murraya paniculata* is made into a paste with a little water. This paste is warmed for few minutes and bandaged on the affected part in the morning and the patient is advised hot water bath in the evening.

Pergularia daemia (Forssk.) Chiov. (Plate 12.5, 6)

Fifty grams of fresh roots and an equal quantity of fresh leaves of *Pergularia daemia* is crushed together with twenty five grams of stem bark of *Erythrina variegata* and ten grams of poppy seeds (*Papaver somniferum*). This is boiled for twenty five to thirty minutes on a low flame in an earthen pot or brass vessel. The filtered oil is massaged on the joints and the affected parts two times a day for ten to fifteen days. A hot water bath is administered during the course of the treatment. The use of chicken, pork, fish and having sex is to be avoided.

Piper nigrum L.

A hundred to two hundred grams of leaves with shoot of *Piper nigrum* is boiled in water together with tender shoots of Bamboo (*Bambusa bamboo*) and neem leaves (*Azadirachta indica*). Hot water bath is taken twice a day with this water while it is comfortably hot for twelve days or more.

Pleiospermium alatum (Wall. ex Wt and Arn.) Swingle.

The juice extracted from a hundred grams of fresh leaves of *Pleiospermium alatum* and a hundred grams of fresh leaves of lemon grass (*Cymbopogon citratus*) is boiled in one litre of neem oil in a low flame for twenty minutes. This oil is applied on the joints, shoulders and the other affected parts. Hot water is sprinkled to get relief from rheumatic complaints.

Plumbago zeylanica L. (Plate 12.6, 1)

Fifty grams of the fresh root of *Plumbago zeylanica* is made into a paste with a few drops of water and boiled in an earthen pot for few minutes. This warm paste is applied on the swollen knee once in a day for 7-12 days for soothing the pain and for reducing swelling.

Psychotria nilgiriensis Deb and Gang. (Plate 12.6, 2)

A ten gram paste is made from the tender fruits of *Psychotria nilgiriensis* is consumed along with honey once a day for twelve days or more for the treatment of rheumatism.

Psychotria nudiflora Wt. and Arn. (Plate 12.6, 3)

A ten gram paste is made from an equal quantity of leaves and flowers of *Psychotria nudiflora* which is consumed along with honey once a day for twelve days or more for the treatment of rheumatism.

Pterocarpus marsupium Roxb. (Plate 12.6, 4)

Fifty grams of the stem resin of *Pterocarpus marsupium* is soaked in hundred ml of water for about twelve hours which is applied externally on the knee twice a day for five to six days to get relief from joint pain.

Rauwolfia densiflora (Wall.) Benth.ex Hk. f. (Plate 12.6, 5)

One gram paste made from an equal quantity of leaves and flowers of *Rauwolfia densiflora* isconsumed two times a day for five days to treat rheumatic complaints.

Sansevieria roxburghiana Schult. and Schult.f.

A hundred grams of chopped leaves, twenty grams of garlic (*Alliumsativum*) and twenty grams of onion (*Allium cepa*), fifty grams of *Moringapterygosperma* bark, and ten grams of mustard seeds (*Brassica juncea*) is boiled in one litre of neem oil on a low flame. This lukewarm oil is massaged by applying medium pressure on hands and legs for fifteen days or until one gets relief from the pain.

Scoparia dulcis L.

The leaves of *Scoparia dulcis* are dried under shade and finely powdered with an equal quantity of shade dried neem leaves (*Azadirachta indica*) and a piece of turmeric (*Curcuma longa*) One teaspoon from this mixture is consumed orally with honey on an empty stomach before breakfast for seventeen days to treat rheumatism. Use of tamarind should be avoided. Cooked rice with little salt and small onion (*Allium cepa*) is recommended as diet.

Sonerila tinnevelliensis Fischer (Plate 12.6, 6)

A handful of leaves of *Sonerila tinnevelliensis*is boiled with one litre of water that is, further, reduced to a hundred millilitres and consumed on an empty stomach once in a day for twelve to fifteen days to get relief from rheumatic complaints.

Plumbago zeylanica L.

Psychotria nilgiriensis Deb & Gang.

Psychotria nudiflora Wt. & Arn.

Pterocarpus marsupium Roxb.

Rauwolfia densiflora (Wall.) Benth. ex. Hk. f.

Sonerila tinnevelliensis Fischer.

Plate 12.6

Tinospora cordifolia (Willd.) Miers ex Hk. f. and Th.

A quarter litre of juice of *Tinospora cordifolia* is extracted from a sufficient quantity of leaves without adding water. This juice is boiled in one litre of neem oil (*Azadirachta indica*) together with twenty five grams of cumin seed (*Cuminum cyminum*) twenty five grams of pepper seeds (*Piper nigrum*), twenty five grams of mustard seeds (*Brassica juncea*), twenty five grams of dried ginger (*Zingiber officinale*), a hundred grams of garlic (*Allium sativum*) and fifty grams of crushed stem bark of *Moringa pterygosperma* on a low flame for fifteen to twenty minutes. The lukewarm medicated oil is massaged on the affected part by applying soft pressure for thirty minutes for seven to twelve days. Hot water bath is administered after the massage.

Toddalia asiatica Lam. var. *asiatica*

A teaspoon of shade dried stem bark powder of *Toddalia asiatica* is taken after meals with honey three times a day for five to ten days to get relief from rheumatic pain.

Ventilago madraspatana Gaertn.

A hundred grams of the stem bark of *Ventilago madraspatana* together with fifty grams of roots of *Pergularia daemia* is coarsely powdered and mixed with one litre of neem oil and heated for fifteen to twenty minutes on a low flame. This oil is massaged on the joints twice a day. This is followed by fomentation with hot boiled leaves of *Justicia adhatoda* to alleviate rheumatic pain.

Vitex negundo L.

Half a cup of decoction prepared by boiling the fresh leaves is taken orally for seven days.

A poultice of cooked leaves is used to foment the affected joints to disperse swelling in the case of acute rheumatism.

Discussion

Survey of Ethnomedicinal Plants for Rheumatism

The tribals' knowledge of indigenous uses of native medicinal plants before their exodus into the urban areas to join the mainstream life needs to be studied and documented. The information collected from field surveys conducted in the tribal pockets of Kalakad-Mundanthurai Tiger Reserve, Western Ghats, Tirunelveli district, Tamil Nadu for plants with medicinal uses to treat rheumatism are summarized in Table 12.1. The listed plants were taxonomically identified with their botanical name, family, local name, habit, habitat and their parts used in treatment.

As an outcome of the present investigation, 65 plants (52 plants collected and 13 plant parts/extracts procured) belonging to 61 genera and 46 families were recorded. Of the recorded plants, 2 plants belonging to Pteridophytes and the remaining 63 plants belonging to 44 families of Angiosperms (Tables 12.1 and 12.2).

Table 12.3: List of Documented Medicinal Plants Used for other than Rheumatism

Botanical Name	*Part used*	*Disease*
Actinopteris radiata (Sw.) Link	Leaf	Bone setting
Aloe barbadensis Mill.	Leaf	Boils, constipation
Anisomeles indica (L.) Kuntze.	Leaf, stem	Astringent
Anisomeles malabarica (L.) R.Br.	Leaf	Fever, indigestion
Azadirachta indica A. Juss.	Stem bark	Anthelmintic, fever
	Leaf	Antiseptic
	Seed	Skin disease
Begonia malabarica Lam.	Leaf	Blisters, boils
Capsicum annuum L.	Fruit	Counter irritant in pain
Cardiospermum halicacabum L.	Leaf	Induce labour pain
Curculigo orchioides Gaertn.	Tuber	Epilepsy, aphrodisiac
Cymbopogon citratus (DC.) Stapf.	Whole plant	Head ache, body pain
Diospyros melanoxylon Roxb.	Leaf	Scabies
Drynaria quercifolia (L.) J. Sm.	Rhizome	Giddiness
Eclipta prostrata (L.) L.	Stem, leaf	Hair tonic
Elephantopus scaber L.	Leaf	Skin disease
Entada pursaetha DC.	Stem bark	Lactation, mumps
Erythrina variegata L.	Leaf	Laxative
Goniothalamus wightii Hk.f. and Th.	Tender fruit	Appetizer
Hemidesmus indicus (L.) R. Br. var. *indicus*	Root	Coolant, mouth sore
Isonandra lanceolata Wight.	Leaf, flower	Piles, stomach problems
Justicia adhatoda L.	Leaf	Bronchitis
Kingiodendron pinnatum (Roxb. ex DC.) Harm.	Resin	Disinfectant
Mimosa pudica L.	Whole plant	Urinary trouble
Moringa pterygosperma Gaertn.	Stem bark	Abortifacient
Pergularia daemia (Forssk.) Chiov.	Leaf	Stomachache
Piper nigrum L.	Fruit	Indigestion, cough and cold
Plumbago zeylanica L.	Root	Piles, antidote
Pterocarpus marsupium Roxb	Resin	Toothache, boils
Sansevieria roxburghiana Schultes and Schultes f.	Leaf	Ear pain, snake bite
Scoparia dulcis L.	Leaf	Diabetes
Tinospora cordifolia (Willd.) Miers ex Hk. f. and Th.	Leaf	Diabetes, fever
	Stem	Aphrodisiac, urinary disorders
Ventilago madraspatana Gaertn.	Stem bark	Skin disease
Vitex negundo L.	Leaf	Headache

The plants, which are routinely used as herbal drugs by the tribal community for the treatment of rheumatism, also have the healing properties for other ailments with different modes of preparation and administration (Table 12.3). They constitute 62 per cent of the recorded plant species, which are used for common diseases like cough, cold, skin disease, diabetes, urinary trouble *etc.* The tribal people for the disease of rheumatism exclusively use only 20 plant species.

The documented plants used for the disease of rheumatism by the *Kanikkars* show a great habitual diversity. Among the plants reported as the herbal drugs, 15 plants are herbaceous in habit, 15 plants are shrubs, 11 plants are climbers and 11 plants are trees. Among the documented medicinal plants, 22 plants belong to Polypetalae under 17 families and 20 plants belong to Gamopetalae under 13 families. 4 belong to Monochlamydeae under 3 families and 4 belong to Monocotyledons under 4 families.

Analysing the plant parts (collected and procured) for treating rheumatism, the leaves are useful in 32 plants. Seeds, flowers and fruits are useful in 7 plants each. Underground modifications and stem barks are useful in 6 plants each. Roots and plant products are useful in 5 plants each. Shoots are used in 3 plants. Stem is useful in 2 plants. An aromatic plant *Cymbopogon citratus* is used as a whole.

The most prevalent form of administration of medicine is oil extracts (33 per cent) and paste (33 per cent). This is followed by fomentation (15 per cent), hot water extracts of plant parts (12 per cent), powder (13 per cent), decoction (8 per cent) and juice (6 per cent)

It is noted that 20 plants have been found to be used by the *Kanikkars* to treat rheumatism (including joint pain and inflammation). The rest of the plants are used to treat rheumatism and also other ailments and 2 plants namely *Azadirachta indica* and *Tinospora cordifolia* have many fold use. *Capsicum annuum* and *Cardiopermum halicacabum* are used as preventive herbal drugs in the case of rheumatism. *Eclipta prostrata* and *Aristolochia karysagathra* are used to reduce excessive body heat during the course of the treatment of rheumatism.

Cross-Cultural Ethnomedicobotanical Studies

It is well established that an identical use of same plant by different tribal groups indicates its curative property and therapeutic significance (Jain and Saklani, 1992). The comparision of medicinal plants between different cultural groups in the same region or neighbouring regions has proved very rewarding. The comparative studies on medicinal uses of plants among different cultural groups showed similarities and dissimilarities in uses. In the present investigations, the medicinal claims emanating from the *Kanikkars* are compared with claims from three other numerically predominant Indian tribal groups living in the Central and Western zones of the country *viz.*, the Bhils, Gonds and Santals and two other predominant tribals of Tamil Nadu namely the Irulars and the Palliyars. The source of ethnomedicinal information for the above mentioned ethnic groups is selected from the available literature (Jain *et al.*, 1973; Saxena and Dutta, 1975; Mudgal and Pal, 1980; Ramachandran and Nair, 1981; Joshi, 1982; Tarafder, 1983, 1984; Shah and Gopal, 1985; Saxena, 1986; Lakshmanan and Narayanan, 1990; Sur *et al.*, 1987, 1990;

Saxena and Tripathi, 1989; Yadav and Bhamare, 1989; Ragupathy and Mahadevan, 1991; Borthakur and Goswami, 1995; Chandra, 1995; Girach and Aminuddin, 1995; Girach *et al.*, 1997; Varma, 1997; Katewa and Arora 1997; Banerjee, 1999; Girach *et al.*, 1999; Sasikumar and Janardhanan, 2002; Singh *at al.*, 2002; Arinathan *et al.*, 2003; Jain *et al.*, 2003; Muthukumarasamy *et al.*, 2003a, b; 2004 a, b; Sahoo and Bahali, 2003; Jadhav, 2006; Pawar and Patil, 2006).

Comparative Ethnomedicobotany with the *Bhils*

The *Kanikkars* and Bhils commonly use 13 plant species. Both similarities and dissimilarities are discussed.

Botanical Name	*Part Used*	*Kanikkars*	*Bhils*
Anisomeles indica	Stem and leaf	Rheumatism, astringent	Rheumatism
Azadirachta indica	Leaf	Rheumatism, antiseptic	Skin disease
Calotropis gigantea	Leaf	Inflammation	Inflammation
Curculigo orchioides	Tuber	Rheumatism, epilepsy, aphrodisiac	Lactation
Eclipta prostrata	Leaf	Rheumatism, hair tonic	Snake bite, antiseptic, hair tonic
Erythrina veriegata	Stem bark	Rheumatism	Rheumatism
Hemidesmus indicus var. *indicus*	Root	Inflammation, mouth sore, coolant	Venereal diseases
Justicia adhatoda	Leaf	Rheumatism, bronchitis	Asthma, pneumonia, rheumatism
Mimosa pudica	Whole plant	Joint pain, urinary trouble	Skin disease
Moringa pterygosperma	Stem bark	Inflammation, abortifacient	Rheumatism
Pergularia daemia	Leaf	Joint pain, stomachache	Piles
Plumbago zeylanica	Root	Rheumatism, piles	Anti rheumatic, headache, cough and cold
Tinospora cordifolia	Stem	Rheumatism, aphrodisiac	Rheumatism
	Leaf	Diabetes, fever, rheumatism	Piles

Comparative ethnobotanical study between the *Kanikkars* and Bhils with special reference to rheumatism reveals that out of 13 plants used by the *Kanikkars* for rheumatism, 7 plants has been found to be used by the Bhils also for the same purpose

Comparative Ethnomedicobotany with the *Gonds*

The *Kanikkars* and Gonds commonly use 17 plant species. Both similarities are dissimilarities are discussed on next page.

Out of 17 plants commonly used by the *Kanikkars* and Gonds, plants such as *Eclipta prostrata, Hemidesmus indicus* var. *indicus* and *Murraya paniculata* have been used for treating rheumatism. Other 14 plants used by *Kanikkars* for treating rheumatism are not used by the Gonds.

Botanical Name	Part Used	Kanikkars	Gonds
Aloe barbadensis	Leaf	Anti-inflammatory, boils, constipation	Paralysis
Anisomeles indica	Stem, leaf	Joint pain, astringent	Nervous disorders, chronic fever
Azadirachta indica	Stem bark	Rheumatic complaint, anthelmintic, fever	Anthelmintic
	Seed oil	Joint pain, skin disease	Leprosy, piles
Cardiospermum hali-cacabum	Leaf	Rheumatism, induce labour pain	Bulkiness of the body
Curculigo orchioides	Tuber	Rheumatism, epilepsy	Leucorrhoea and epilepsy
Datura metel	Leaf	Anti-inflammatory	Sprain, to dry milk in breast
Eclipta prostrate	Leaf	Rheumatism, hair tonic	Inflammation elephantiasis, leucoderma, spleen disorder
Hemidesmus indicus var. indicus	Root	Anti-inflammatory, mouth sore, coolant	Urinary disease, leucoderma, psoriasis, rheumatism, stomach disorders
Moringa pterygosperma	Stem bark	Rheumatism, abortifacient	Whooping cough, abortifacient
Murraya paniculata	Leaf	Anti-inflammatory	Body pain, inflammation
Pergularia daemia	Leaf	Joint pain, stomachache	Uterine tonic
Plumbago zeylanica	Root	Rheumatism, piles	Snake bite
Pterocarpus marsupium	Resin	Joint pain, toothache, boils	Urinary disorders, diabetes
Scoparia dulcis	Leaf	Rheumatism, diabetes	Weakness of semen, tuberculosis
Tinospora cordifolia	Leaf	Rheumatism, diabetes, fever	Diabetes
Ventilago madraspatana	Stem bark	Joint pain, skin disease	Skin disease
Vitex negundo	Leaf	Rheumatism, headache	Muscular pain

Comparative Ethnomedicobotany with the *Santals*

Sixteen plant species are commonly used by both the *Kanikkars* and Santals. Both similarities and dissimilarities are discussed

Botanical Name	Part Used	Kanikkars	Santals
Aloe barbadensis	Leaf	Anti-inflammatory, boils, constipation	Abcess, intestinal worms, burns, peptic ulcer
Azadirachta indica	Stem bark	Rheumatic complaints anthelmintic, fever	Stomachic
	Seed oil	Joint pain, Skin disease	Skin disease
Calotropis gigantean	Leaf	Rheumatism.	Rheumatism, pneumonia, fever, vermicide

Botanical Name	Part Used	Kanikkars	Santals
Cardiospermum halicacabum	Leaf	Rheumatism, to induce labour pain	Eczema
Curculigo orchioides	Tuber	Rheumatism, epilepsy	Boils, ulcers.
Eclipta prostrata	Leaf	Rheumatism, hair tonic	Wounds, dysentry
Entada pursaetha	Seed	Anti-inflammatory	Anti-inflammatory
Hemidesmus indicus var. indicus	Root	Anti-inflammatory, mouth sore, coolant	Stomach trouble, malarial fever
Jatropha gossypifolia	Stem bark	Rheumatism.	Rheumatism
Justicia adhatoda	leaf	Rheumatism, bronchitis	Vermicide
Mimosa pudica	Whole plant	Joint pain, urinary trouble	Cuts, wounds antidote for snake bite.
Moringa pterygosperma	Stem bark	Anti-inflammatory, abortifacient	Epilepsy, measles, abortion
Plumbago zeylanica	Root	Rheumatism, piles	Abortion
Pterocarpus marsupium	Resin	Joint pain, toothache, boils	Diarrhoea
Scoparia dulcis	Leaf	Rheumatism, diabetes	Gonorrhoea, one sided head ache.
Vitex negundo	Leaf	Rheumatism, headache	Tongue sore, bodyache

Out of 16 plants commonly used by *Kanikkars* and Santals, plants such as *Calotropis gigantea, Entada pursaetha* and *Jatropha gossypifolia* have similar use for treating rheumatism. Other 13 plants used by *Kanikkars* for rheumatism are not used by the Santals.

Comparative Ethnomedicobotany with the *Irulars*

The *Kanikkars* and Irulars commonly use 12 plant species. Both similarities and dissimilarities are discussed.

Botanical Name	Part Used	Kanikkars	Irulars
Aloe barbadensis	Leaf	Inflammation	Lessen labour pain.
Anisomeles malabarica	Leaf	Rheumatism, anthelmintic, fever.	Paralysis
Azadirachta indica	Stem bark	Rheumatism, anthelmintic, fever.	Vermifuge
	Seed oil	Joint pain, skin disease	Contraceptive
Cardiospermum halicacabum	Leaf	Rheumatism, to induce labour pain	Rheumatism
Eclipta prostrata	Leaf	Rheumatism, hair tonic	Jaundice.
Hemidesmus indicus var. *indicus*	Root	Inflammation, mouth sore, coolant	Snake bite, coolant, urinary disorder.
Pergularia daemia	Leaf	Joint pain, stomachache	Cold, fever.
Piper nigrum		Rheumatism, indigestion, cough and cold.	Birth control

Botanical Name	Part Used	Kanikkars	Irulars
Plumbago zeylanica	Root	Rheumatism, piles	abortion
Vitex negundo	Leaf	Rheumatism, headache	Body pain, hair lice.

Out of 10 plants commonly used by *Kanikkars* and Irulars, the plant *Cardiospermum halicacabum* has similar use for treating rheumatism. Other 9 plants used by *Kanikkars* for rheumatism are not used by the Irulars.

Comparative Ethnomedicobotany with the *Palliyars*

Sixteen plant species are commonly used by both the *Kanikkars* and Palliyars. Both similarities and dissimilarities are discussed.

Botanical Name	Part Used	Kanikkars	Palliyars
Actinopteris radiata	Leaf	Joint pain, bone setting	Colic pain
Anisomeles indica	Leaf and Stem	Joint pain, astringent	Headache
Anisomeles malabarica	Leaf	Rheumatism, indigestion, fever	Headache
Azadirachta indica	Stem bark	Rheumatism, anthelmintic, fever	Fever, cold
Cardiospermum halicacabum	Leaf	Rheumatism, to induce labour pain	Cough
Curculigo orchioides	Tuber	Rheumatism, epilepsy	Sexual vigour
Hemidesmus indicus var. *indicus*	Root	Rheumatism, coolant, mouth sore	Fever, menorrhagia, stomachache
Justicia adhatoda	Leaf	Rheumatism, bronchitis	Asthma, cold and cough
Mimosa pudica	Leaf	Rheumatism	Insect bite
Pergularia daemia	Leaf	Joint pain, stomachache	Gastric trouble
Piper nigrum	Fruit	Rheumatism, indigestion, cough and cold.	Throat infection
Plumbago zeylanica	Root	Rheumatism, piles, antidote	Stomachache
Pterocarpus marsupium	Resin	Joint pain	Joint pain, rickets
Sansevieria roxburghiana	Leaf	Inflammation, ear pain, snake bite	For inducing teeth growth in children
Scoparia dulcis	Leaf	Rheumatism, diabetes	Poisonous bite
Toddalia asiatica	Stem bark	Rheumatism	Tooth powder

Out of 16 plants commonly used by the *Kanikkars* and Palliyars, the plant *Pterocarpus marsupium* has similar use for treating joint pain (rheumatism). Other 15 plants which are used by the *Kanikkars* for rheumatism are not used by the Palliyars.

This comparative study between the different tribal groups shows some similarities and dissimilarities. The choice of medicinal plants depend on several factors, such as the alternatives available, the reputation of a recipe, past experience in the family, access to the plant and whether it is used singly or in combination. These perceptions vary from community to community and even person to person.

Endemic and Endangered Plants of the Study Area

The study area falls in the "hot spot" region in the Western Ghats and the endemic plants are the exclusive biological capital of the nation. Earlier investigations on endemic and rare plants and their conservation in India are meager (Jain and Rao, 1983). A large number of endemic plants are at the brink of extinction. And hence the conservation of the threatened endemic species deserves top priority.

Some of the endemic plant species, which were reported to occur in the "hot spot" regions as common or abundant about half-a -century ago, now have become rare or very rare due to over exploitation and are included under the category of endangered species.

The following medicinal plants are identified as the endemic and endangered plant species of the study area (Jain and Rao, 1983; Ahmedullah and Nayar, 1986; Nayar and Sastry, 1987; Nayar, 1996).

Endemic Plants of the Study Area

1. *Asystasia travancorica* Bedd.
2. *Eugeniasin gampattiana* Bedd.
3. *Goniothala muswightii* Hk. f. and Th.
4. *Psychotria nudiflora* Wight and Arn.
5. *Sonerilatin nevelliensis* Fischer

Endangered Plants of the Study Area

1. *Eugeniasin gampttiana* Bedd.
2. *Psychotria nudiflora* Wight and Arn.
3. *Sonerilatin nevelliensis* Fischer.

Conclusion

Habitat loss and unchecked commercialization of wild medicinal plants is threatening the future of vital resources, as well as the beauty, diversity, and natural heritage of our planet. As wild lands are destroyed or degraded, we lose unique and precious species, from flowers to frogs to butterflies, and with them potential resources to combat hunger, poverty, natural disasters,and social and economic insecurity. The conservation of our remaining wild species and wild places must be committed so that the loss of these resources is minimized. The best means of conservation is to ensure that the populations of species of plants continue to grow and evolve in the wild - in their natural habitats. For effective recovery of these species *ex-situ* conservation measures should be initiated.

References

1. Ahmeddullah M and Nayar MP (1986). Endemic plants of the Indian Region, Vol. I. Peninsular India, Botanical Survey of India, Kolkatta.

2. Arinathan V, Mohan VR and Britto JD (2003). Ethnomedicinal survey among Palliyar tribals of Srivilliputhur Grizzled Giant Squirrel Wildlife Sanctuary, Tamil Nadu. *Journal of Economic Taxonomic Botany,* 27: 707-710.
3. Ayyanar M and Ignacimuthu S (2005). Traditional knowledge of Kani Tribals in Kouthalai of Tirunelveli hills, Tamil Nadu. *Journal of Ethnopharmacology,* 102: 246-255.
4. Banerjee A (1999). Ethnobotany of some trees in the Santhal villages of Birbhum, West Bengal. *Journal of Economic Taxonomic Botany,* 23: 531-534.
5. Borthakur SK and Goswami N (1995). Herbal remedies from Dimoria of Kamrup district of Assam in Northeastern India. *Fitoterapia,* LXVI: 333-340.
6. Chandra K (1995). An Ethnobotanical study on some medicinal plants of District Palamau (Bihar). *Bulletin of Medico Ethno Botanical Research,* XVI: 11-16.
7. Evans WC (2002). "Trease and Evans' Pharmacognosy", W.B. Saunders Company Limited. London. p. 46.
8. Fischer CEC and Gamble JS (1957). (Rep. Ed.) The Flora of the Presidency of Madras III: 1347-2017.
9. Gadgil M (1996). Documenting diversity: An experiment. *Current Science,* 70: 36-44.
10. Gamble JS (1957a). (Rep. Ed.) The Flora of the Presidency of Madras, I: 1-577.
11. Gamble JS (1957b). (Rep. Ed.) The Flora of the Presidency of Madras, II: 578-1346.
12. Girach RD and Aminuddin (1995). Ethnomedicinal uses of plants among the Tribals of Singhbhum District, Bihar, India. *Ethnobotany,* 7: 103-107.
13. Girach RD, Aminuddin M, Brahmam M and Misra MK (1997). Observations on ethnomedicinal plants of Bhadrak district, Orissa, India. *Ethnobotany,* 9: 44-46.
14. Girach RD, Singh S, Brahmam M and Misra MK (1999). Traditional treatment of skin disease in Bhadrak district, Orissa. *Journal of Economic Taxonomic Botany,* 23: 499-504.
15. Handa SS, Chawla AS and Sharma AK (1992). Plants with anti-inflammatory activity. *Fitoterapia,* LXIII:3-31.
16. Henry AN, Chithra V and Balakrishnan NP (1989). Flora of Tamil Nadu, India, Series I: Analysis, III: 1- 171.
17. Henry AN, Kumari GR and Chithra V (1987). Flora of Tamil Nadu, India, Series I: Analysis II: 1- 258.
18. Jadhav D (2006). Ethnomedicinal plants used by Bhil tribe of Bibdod, Madhya Pradesh, India. *Indian Journal of Traditional Knowledge,* 5: 263-267.
19. Jain SK and Rao RR (1983). (Eds.) An assessment of threatened plants of India. Proceedings of the Seminar held at Dehra Dun, September 1981, Botanical Survey of India (Department of Environment), Botanic Garden, Howrah.

20. Jain SK and Saklani A (1992). Cross-cultural ethnobotanical studies in Northeast India. *Ethnobotany*, 4: 25-38.
21. Jain SK, Banerjee DK and Pal DC (1973). Medicinal plants among certain Adibasis in India. *Bulletin of Botanical Survey of India*, 15: 85-91.
22. Jain SP, Singh SC, Singh J and Kumar S (2003). Ethno-Medico-Botanical survey of Raipur district, Chattisgarh State. *Journal of Economic Taxonomic Botany*, 27: 266-271.
23. Joshi P (1982). An ethnobotanical study of Bhils- A preliminary survey. *Journal of Economic Taxonomic Botany*, 3: 257-266.
24. Katewa SS and Arora A (1997). Some plants in folk medicine of Udaipur District Rajasthan. *Ethnobotany*, 9: 48-51.
25. Khoshoo TN (1996). India needs a national biodiversity conservation board. *Current Science*, 71: 506-513.
26. Lakshmanan KK and Narayanan ASS (1990). Antifertility herbals used by the tribals in Anaikkatty hills, Coimbatore district, Tamil Nadu. *Journalof Economic Taxonomic Botany*, 14: 171-173.
27. Loeb C (1971). Text Book of Medicine. W.B. Saunders Company, Philadelphia – London – Toronto, 1883-1913.
28. Miller TA (1983). Protective effects of prostaglandins against gastric mucosal damage: Current knowledge and proposed mechanisms. *American Journal of physiology*, 245: G601- G623.
29. Mudgal V and Pal DC (1980). Medicinal plants used by tribals of Mayurbhanj. (Orissa). *Bulletin of Botanical Survey of India*, 22: 59-62.
30. Muthukumarasamy S, Mohan VR, and Kumaresan S (2004a). Medico-ethnobotany of Palliyar tribe in Grizzled Giant Squirrel Wildlife Sanctuary, Srivilliputhur, Western Ghats, India. *Journal Medicinal and Aromatic Plant Science*, 26: 507-516.
31. Muthukumarasamy S, Mohan VR, Kumaresan S and Chelladurai V (2003a). Herbal medicinal plants used by Palliyars to obtain relief from gastro-intestinal complaints *Journal of Economic Taxonomic Botany*, 27: 711-714.
32. Muthukumarasamy S, Mohan VR, Kumaresan S and Chelladurai V (2003b). Herbal remedies of Palliyar tribe of Grizzled Giant Squirrel Wildlife Sanctuary, Western Ghats, Srivilliputhur, Tamil Nadu for poisonous bites. *Journal of Economic Taxonomic Botany*, 27: 761-764.
33. Muthukumarasamy S, Mohan VR, Kumaresan S and Chelladurai V (2004b). Traditional medicinal practices of Palliyar tribe of Srivilliputhur in Antenatal and Post-natal care of mother and child. *Natural Product Radiance*, 3: 422-426.
34. Nair NC and Henry AN (1983), Flora of Tamil Nadu, India, Series I: Analysis I: 1-184.

35. Nayar MP (1996). "Hot Spots" of endemic plants of India, Nepal and Bhutan, Tropical Botanical Garden and Research Institute, Palode, Thiruvananthapuram, Ministry of Environment and Forests, New Delhi.

36. Nayar MP and Sastry ARK (1987). Red data book of Indian plants, Vol. I, II and III., Published by the Director, Botanical Survey of India, Kolkatta -700 001.

37. Pawar S and Patil DA (2006). Folk remedies against rheumatic disorders in Jalgaon district, Maharastra. *Indian Journal Traditional Knowledge,* 5: 314-316.

38. Ragupathy S and Mahadevan A (1991). Ethnobotany of Kodiakkarai Reserve Forest, Tamil Nadu, South India. *Ethnobotany,* 3: 79-82.

39. Ramachandran VS and Nair NC (1981). Ethnobotanical observations on Irulars of Tamil Nadu. *Journal of Economic Taxonomic Botany,* 2: 183-190

40. Robert A, Hanchar AJ, Lancaster C and Nezamis JE (1979). Prostacyclin inhibits enteropooling and diarrhea. Prostacyclin. (Eds.), Vane, J.R., and Bergstrom, S. Raven Press, New York, 147-158.

41. Sahoo AK and Bahali DD (2003). Plants used in less known Ethnomedicines and Medico-religious beliefs in Phulbani district of Orissa. *Journal of Economic Taxonomic Botany,* 27: 500-504.

42. Sasikumar JM and Janardhanan K (2002). Ethnomedicinal plants for women folk's health care in Nilgiri Biosphere Reserve, Western Ghats. *Journal of Non-Timber Forest Products,* 9: 138-143.

43. Saxena HO (1986). Observation of the ethnobotany of Madhya Pradesh. *Bulletin of Botanical Survey of India,* 28: 149-156.

44. Saxena HO and Dutta PK (1975). Studies on the Ethnobotany of Orissa. *Bulletin of Botanical Survey of India,***17**: 124-131.

45. Saxena SK and Tripathi JP (1989). Ethnobotany of Bundelkhand. Studies on the medicinal uses of wild trees by the tribal inhabitants of Bundelkhand region. *Journal of Economic Taxonomic Botany,* 13: 381-389.

46. Shah GL and Gopal GV (1985). Ethnomedical notes from the tribal inhabitants of the North Gujarat (India). *Journal of Economic Taxonomic Botany,* 6: 193-201.

47. Singh AK, Raghubanshi AS and Singh JS (2002). Medical ethnobotany of the tribals of Sonaghati of Sonbhadra district, Uttar Pradesh, India. *Journal of Ethnopharmacology,* 81: 31-41.

48. Singh KS (1994). People of India – The Scheduled Tribes. Anthropological survey of India. Oxford University Press.

49. Sur PR, Sen R, Halder AC and Bandyopadhyay S (1987). Observation on the ethnobotany of Malda - West Dinajpur districts West Bengal-I. *Journalof Economic Taxonomic Botany,* 10: 395-401.

50. Tarafder CR (1983). Ethnogynaecology in relation to plants. Part-II. Plants used for abortion. *Journal of Economic Taxonomic Botany,* 4: 507-516.

51. Vane JR and Botting RM (1995). A better understanding of anti-inflammatory drugs based on isoforms of cyclo-oxygenase. *Advances in Prostoglandin, Thromboxane and Leukotriene Research,* 23: 41-48.

52. Varma SK (1997). Comparative studies on folk drugs of tribls of Chotangpur and Santhal Pargana of Bihar, India. *Ethnobotany,* 9: 70-76.

53. Wallace JL, Bak A, cknight W, Asfaha S, Sharkey KA and Mac-Naughton WK (1998). Cyclooxygenase-1 contributes to inflammatory responses in rats and mice: implications for gastrointestinal toxicity. *Gastroenterology,* 115: 101-109.

54. Yadav SS and Bhamare PB (1989). Ethnomedico-Botanical studies of Dhule forests in Maharashtra state. *Journal of Economic Taxonomic Botany,*13: 455-460.

55. Yesilada E, Ustur O, Sezik E, Takaishi Y, Ono Y and Honda G (1997). Inhibitory effects of Turkish folk remedies on inflammatory cytokine: interleukin-1α, interleukin-1β and tumour necrosis factor α. *Journal of Ethnopharmacology,* 58: 59-73.

2018, Ethnomedicinal Plants: A Biodiversity Treasure Pages 343–362
Editors: V.R. Mohan, A. Doss, P.S. Tresina and V. Sornalakshmi
Published by: ASTRAL INTERNATIONAL PVT. LTD., NEW DELHI

Chapter 13

Therapeutic Uses of Medicinal Plants Used by *Palliyar* Tribe in Sirumalai Hills, Tamil Nadu for the Treatment of Gastro-Intestinal Complaints

A. Maruthupandian[1]*, M. Viji[2], C. Chithravadivu[3] and V.R. Mohan[4]

[1]*Department of Botany, School of Life Sciences, Periyar University, Salem – 636 011, Tamil Nadu*
[2]*Department of Botany, Thiagarajar College, Madurai – 625009, Tamil Nadu*
[3]*PG and Research Department of Botany, Vellalar College for Women, Thindal, Erode – 638 012, Tamil Nadu*
[4]*Ethnopharmacology Unit, PG and Research Department of Botany, V.O. Chidambaram College, Tuticorin – 628 008, Tamil Nadu*

Introduction

More than a few millions of years, primary producers of plants are not only used foods by human as well as they are depended on various ailments. In developing countries, 60 per cent -70 per cent of peoples are residing at nearby agricultural lands and forest areas. Regularly those peoples collect various plant parts such as roots, leaves, flowers, fruits and nuts to exercise their daily necessities of food and other needs (Aryal *et al.*, 2009). In this connection, the plants are not only used foods they are also mostly utilized in various human ailments. Such a plant medicine or herbal medicine has fewer side effects when compared to other system of medicines in India. Indigenous Knowledge plays a vital role in plant utilization for various rationales by particular indigenous communities. Indigenous peoples or tribals provide lot of practices to cure variety of diseases based on their traditional knowledge. Recently hundred years of beliefs of tribal's traditional practices are established and extended to modern medicine (Aburjai *et al.*, 2007). Now-a-days 60-70 per cent of peoples turn into herbal medicines to cure their illness. Herbal medicines are fundamental to treat various human ailments including infectious diseases, hypertension etc (Patrick, 2002). Gastro-intestinal complaints is major diseases affected widely in different aged people like stomachache, stomach upset, dysentery, vomiting, constipation *etc.* Rural peoples or tribal peoples are depending on their traditional medicines due to lack of modern medicine facility for this kind of illness.

The survey of literature reveals that not much work has been done on the ethnomedicinal plants with reference to gastrointestinal ailments of *Palliyar* tribals of Sirumalai hills, Dindigul district. Hence, the present work has been undertaken to record the medicinal plants, which are used for the gastrointestinal ailments by the *Palliyar* tribals of this district as a source of medicine.

Materials and Methods

Study Area

The Sirumalai hills (small hills) is located in Dindigul district of Tamil Nadu between 10°07′ -10° 18′ N latitude and 77°55′ -78° 12′ E longitude. They are isolated, compact group of hills 6.5 km south of Dindigul town and 22.5 km north of Madurai city (Area Map 13.1). They are rectangular in outline [19.3 km long (north-south) and 12.8 km broad (east-west)], with an area of 317 km^2. The tract is bounded by the Dindigul-Madurai rail line on the west, the Dindigul – Nattam road on the east and north-east, and the Vadipatti-Nattam road on the south. There is a heavy annual rainfall (132 cm on the plateau, and 120 cm in the plains) during the north-east monsoon gaining for 86 per cent and 76 per cent and South-west monsoon for 77 per cent and 62 per cent. The hottest months are April-May (40°C in the open and 37.5°C in the shade).

Palliyar Tribals

Palliyars are the major ethnic community of the Dindigul district. The *Palliyars* belong to the Southern Tribal Zone. The Southern tribals are historically more ancient tribes. There are as many as 36 types of Scheduled tribes in Tamil Nadu. In the serialized list notified by the Government of Tamil Nadu, the *Palliyars* are placed in 32nd position. They live in the low altitude regions of Western Ghats and live in large numbers in the study area (Sirumalai hills, Dindigul district). Of the total population of the Scheduled tribes in Tamil Nadu (5, 74, 194), the *Palliyar* tribe accounts for 1890, which ranks 20th among the tribal population (Census of India, 1991).

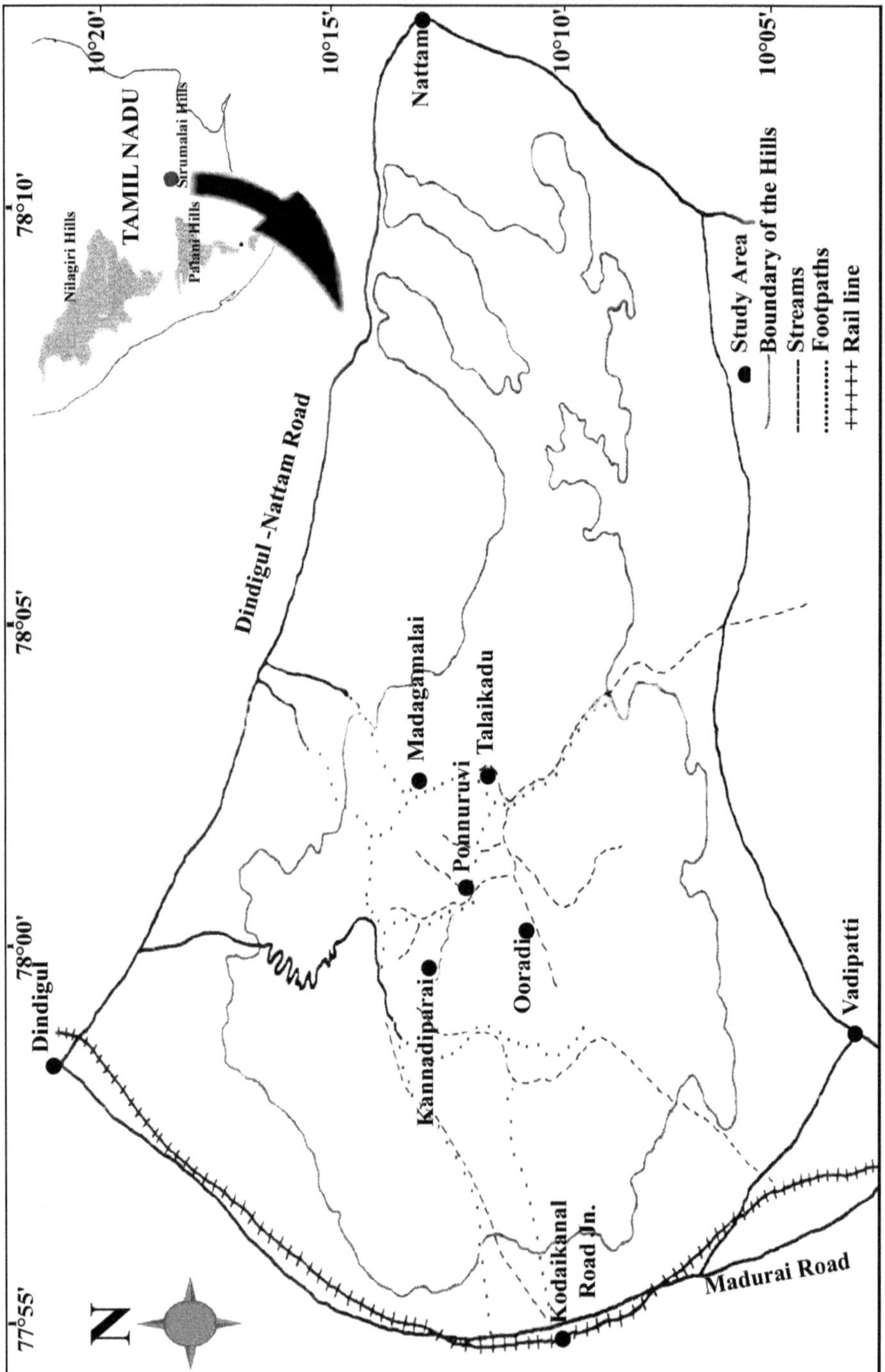

Map 13.1: Study Area.

A old *Palliyar* man

A old *Palliyar* woman

Palliyars Huts

Figure 13.2: *Palliyar* Tribals and their Habitat.

Survey and Enumeration Method

The study focuses on the ethnomedicinal plants used as drugs by *Palliyars* settled in the reserve forest area of Sirumalai hills. Information regarding ethnomedicinal plants was gathered by meeting *Palliyars* during explorative field trips and by gaining a good rapport and winning over their confidence. Most of information was gathered from the elderly people, who have a very long acquaintance with the usage of plants. The information thus gathered was adequately cross checked for the reliability and accuracy by interacting with different groups of *Palliyars* from different habitats to confirm the use, mode of administration as well as dosage differences, if any. Based on the information provided by the *Palliyars*, the plants specimens were collected; air dried, identified and was deposited in the Ethnopharmacology unit and Research Department of Botany, V.O.Chidambaram College, Tuticorin, Tamil Nadu.

Results and Discussion

As many as 114 ethnomedicinal plants were identified from the Sirumalai hills, Dindigul district, Tamil Nadu. They were used by the *Palliyar* tribals for the treatment of gastrointestinal ailments. They were tabulated alphabetically with botanical name, voucher specimen number, vernacular name, family, plant part used and mode of administration (Table 13.1). The investigation reveals that the *Palliyar* tribe of Sirumalai hills, Western Ghats region uses 114 plants species belonging to 100 genera and 51 families for the treatment of gastrointestinal ailments. Among these 114 ethnomedicinal plants, a maximum of 10 plants belong to the

Euphorbiaceae, it is followed by Apocyanaceae, Fabaceae and Rutaceae (6 species each), Mimosaceae and Moraceae (5 species each), Amaranthaceae, Asclepiadaceae and Ceasalpinaceae (4 species each), Cucurbitaceae, Lamiaceae and Malvaceae (3 species each), Acanthaceae, Anacardiaceae, Araceae, Asteraceae, Boraginaceae, Capparaceae, Chenopodiaceae, Combretaceae, Convolvulaceae, Meliaceae, Menispermaceae, Molluginaceae, Nyctaginaceae, Rhamnaceae, Vitaceae and Zingiberaceae (2 species each), Apiaceae, Arecaceae, Asparagaceae, Bombaceae, Brassicaceae, Caryophyllaceae, Commeliniaceae, Cyperaceae, Dioscoriaceae, Flacourtiaceae, Liliaceae, Loganiaceae, Musaceae, Myristicaceae, Myrtaceae, Oxalidaceae, Pappavaraceae, Portulaceae, Punicaceae, Rubiaceae, Sapindaceae, Solanaceae and Zygophyllaceae (each 1 species).

Table 13.1: Ethnomedicinal Plants Used by Palliyar Tribals to Treat Gastrointestinal Complaints

Herbarium No.	*Plant Name*	*Mode of Administration*
VOCB 5402	*Abutilon crispum* (L.) Medicus Siruthutthi Malvaceae	JOne glass of the fresh leaf juice is taken with one teaspoon of ghee twice a day for two to three days to arrest dysentery.
VOCB 5403	*Acacia chundra* (Roxb. ex Pottl) Willd (*Acacia sundra* DC.) Karungali Mimosaceae	One glass of the fresh leaf juice is taken with few drops of honey twice or thrice a day before feeding for a period of one month to treat ulcer complaints.
VOCB 5404	*Acacia ferruginea* DC Parampai Mimosaceae	One teaspoon of the stem bark powder is taken with hot water thrice a day for two days to arrest dysentery.
VOCB 5406	*Acacia nilotica* (L.) Willd. ex Del Karuvelam Mimosaceae	An ounce of the hot water extract of the stem bark is taken twice a day in empty stomach to treat haemorrhoids.
VOCB 5408	*Acalypha fruticosa* Forssk Athathalai Euphorbiaceae	An ounce of the hot water extract of the root is taken twice a day for a period of one month to treat haemorrhoids.
VOCB 5409	*Acalypha indica* L. Kuppaimeni Euphorbiaceae	An ounce of the hot water extract of the root is taken four to five times a day for a period of five days as an anthelmintic (de-wormer).
VOCB 5411	*Acorus calamus* L. Vasambu Araceae	Few shade dried root tubers are powdered, roasted and taken with honey twice a day for four to five days as a carminative
VOCB 5414	*Aerva lanata* (L.) Juss ex Sehultes Sirupulai Amaranthaceae	An ounce of the whole plant extract is taken with honey twice a day for a period of three weeks to dissolve and discharge the kidney stones.

Contd...

Herbarium No.	*Plant Name*	*Mode of Administration*
VOCB 5418	*Albizia lebbeck* (L.) Willd Vagai Mimosaceae	One teaspoon of the stem bark powder is taken thrice a day for a period of one month to treat haemorrhoids. One teaspoon of the powdered gum (bark exudate) is taken with cow's milk to arrest diarrhoea.
VOCB 5420	*Aloe barbadensis* Mill. (*Aloe vera* (L.) Burm) Chotthu kathalai Liliaceae	Thirty grams of fresh leaves are boiled in two hundred ml of water with few bulbs of *Allium cepa* L. (vengayam). The filtrate is taken twice a day to improve digestion and to relieve stomachache.
VOCB 5421	*Alpinia calcarata* Roscoe Sithirathai Zingiberaceae	One teaspoon of the rhizome paste is taken once a day for three days to get relief from colic pain.
VOCB 5423	*Amaranthus spinosus* L. Mullukeerai Amaranthaceae	The paste made from fifty grams of the whole plant with ten grams of the dried fruits of *Carum copticum* Benth and Hook (omam) and few scales of a bulb of *Allium sativum* L. (vellaipoondu) is taken thrice a day to get relief from stomachache and urogenital problems.
VOCB 5425	*Amaranthus viridis* L. Kuppai keerai Amaranthaceae	The leaves are cooked as food and consumed regularly to improve digestion
VOCB 5430	*Anisomeles malabarica* (L.) R.Br ex Sims Aruvasadatchi Lamiaceae	One teaspoon of the paste made from the tender leaves is given to infants with lukewarm water or with coconut oil (oil obtained from the dried copra of *Cocos nucifera* L.) to arrest dysentery at the time of teeth development.
VOCB 5431	*Anogeissus latifolia* (Roxb ex DC) Wall ex Guill and Perr. Vekkali Combretaceae	Two to three ml of the leaf extract is taken with hot water twice a day for a period of one month to treat peptic ulcer.
VOCB 5434	*Argemone mexicana* L. Piramathandu Papavaraceae	An ounce of the leaf extract is taken four to five times for two days as an antidote for scorpion bite. One teaspoon of the root powder is taken with rice fermented water once a day for three days as an anthelmintic (de-wormer).
VOCB 5437	*Artocarpus integrifolia* L.f. Pala Moraceae	An ounce of the juice prepared from ten grams of ripened perianth is taken thrice a day for two days to arrest dysentery.
VOCB 5439	*Asclepias curassavica* L. Kruthipoo Asclepiadaceae	An ounce of the leaf extract is taken with cow's milk twice a day for a period of one month to treat haemorrhoids. The juice prepared from ten grams of the fresh leaves with two grams of the rhizome of *Alpinia galanga* Sw. (perrarati) is taken with cow's milk thrice a day for two days to arrest dysentery.

Herbarium No.	Plant Name	Mode of Administration
VOCB 5440	*Asparagus racemosus* Willd Sathamoolam Asparagaceae	The milk extract of the tubers is taken twice a day to improve digestion, to relieve stomachache and to arrest dysentery.
VOCB 5442	*Azadirachta indica* A.Juss (*Melia azadirachta* L.) Vembu Meliaceae	Twenty grams of the shade dried leaves and the shade dried roots of *Glycyrrhiza glabra* L. (adhimathuram) are jointly powdered. This blend is taken thrice a day for a period of one month to treat chickenpox. An ounce of juice prepared from the tender leaves with an unripe fruit of *Terminalia chebula* Retz. (kadukkai) is taken along with a teaspoon of castor oil (oil obtained from the seeds of *Ricinus communis* L.) twice a day for a period of one week as an anthelmintic (de-wormer). It is also mixed with honey, cow's milk, butter and ghee and taken twice a day for a period of forty five days to treat diabetes. An ounce of the flower infusion is taken once a day in empty stomach for a period of two weeks to improve the performance of kidneys.
VOCB 5446	*Barleria prionitis* L. Kattu Kanagamparam Acanthaceae	An ounce of the fresh leaf juice is taken with water twice a day for two days to get relief from stomachache. It is also taken with honey to treat urogenital problems.
VOCB 5447	*Basella alba* L. (*Basella rubra* L.) Kodi Pasalai Chenopodiaceae	An ounce of the hot water extract of the root is taken twice a day for two to three days to control nausea and to relieve stomachache.
VOCB 5448	*Bauhinia racemosa* Lam. Archi Caesalpiniaceae	An ounce of juice prepared from fifteen grams of the fresh leaves and a few bulbs of *Allium cepa* L. (vengayam) is taken thrice a day for two days to arrest dysentery.
VOCB 5449	*Bauhinia variegata* L. (*Phanera variegata* (L.) Benth.) Segappumantharai Caesalpiniaceae	One teaspoon of the decoction prepared from the flower buds is taken with honey twice a day for a period of one month to treat haemorrhoids.
VOCB 5451	*Boerhavia diffusa* L. (*Boerhavia repens* L.) Mookaratti Nyctaginaceae	Fifteen grams of roots and few black pepper (the dried fruits of *Piper nigrum* L.) (milagu) are gently boiled in one hundred ml of mustard oil (oil obtained from the seeds of *Brassica juncea* (L) Czern and Coss) and filtered. Fifteen ml of the filtrate is given to kids weekly once or twice for a period of six months and twenty ml is taken by adults to treat asthma, dyspepsia, skin diseases, and to control nausea.` One teaspoon of the leaf paste is taken twice a day for a period of one week as a carminative and to treat jaundice.
VOCB 5452	*Bombox ceiba* L. (*Salmalia malabarica* (DC) Schott and Endl.) Mulilavam Bombacaceae	An ounce of the infusion of the stem bark in cow's milk is taken twice a day as a diuretic. An ounce of the milk extract of the fresh leaves is taken twice a day to get relief from stomachache. An ounce of the hot extract of the flowers is taken as a diuretic and laxative.

Contd...

Herbarium No.	Plant Name	Mode of Administration
VOCB 5453	*Brassica juncea* (L.) Czern and Coss Kadugu Brassicaceae	The paste made by grinding fifty grams of the dried seeds with rice fermented water is applied externally on the navel to relieve stomachache.
VOCB 5454	*Breynia retusa* (Dennst.) Alston Peruniruri Euphorbiaceae	An ounce of the fruit juice is taken with a teaspoon of castor oil (oil obtained from the seeds of *Ricinus communis* L.) twice a day for two to three days to arrest dysentery.
VOCB 5455	*Breynia vitis-idea* (Burm.f.) Fischer Kattuniruri Euphorbiaceae	An ounce of the boiled water extract of the fresh leaves is taken with a teaspoon of juice prepared from the fresh rhizome of *Zingiber officinale* Roscoe (inji) once a day for a period of two weeks to improve digestion.
VOCB 5458	*Butea monosperma* (Lam.) Taub (*Butea frondosa* Koeng ex Roxb.) Porasu Fabaceae	One teaspoon of the seed powder is taken with castor oil (oil obtained from the seeds of *Ricinus communis* L.) twice a day for three days as an anthelmintic (de-wormer).
VOCB 5478	*Caesearia tomentosa* Roxb. (*Caesearia elliptica* Willd) Kadakakinchil Flacourtiaceae	An ounce of the boiled water extract of the root bark is taken to treat diabetes and renal problems.
VOCB 5460	*Cajanus cajan* (L.) Millsp (*Cajanus indicus* Spreng) Thuvarai Fabaceae	An ounce of the salted boiled water extract of the fresh leaves is taken in empty stomach for a period of two weeks to treat jaundice.
VOCB 5461	*Calotropis gigantea* (L.) R.Br. Erukku Asclepiadaceae	Five ml of extract prepared from the tender leaves is taken with a teaspoon of castor oil (oil obtained from the seeds of *Ricinus communis* L.) in empty stomach as an anthelmintic (de-wormer) and as a laxative.
VOCB 5463	*Capparis zeylanica* L. (*Capparis horrida* L. f.) Suduthoratti Capparaceae	An ounce of the boiled water extract of the root bark is taken thrice a day for four to five days to control nausea and to improve digestion
VOCB 5465	*Caralluma adscendens* (Roxb.) Haw Kallimulayan Asclepiadaceae	One teaspoon of the paste made from twenty grams of the whole plant, twenty grams of the seeds of *Vigna mungo* (L.) Hepper (black gram), two grams of the fresh rhizome of *Zingiber officinale* Roscoe.(inji), two grams of the fresh leaves of *Coriandrum sativum* L.(malli), few black pepper (the dried fruits of *Piper nigrum* L.(milagu) and few dried fruits of *Cuminum cyminum* L. (shiragam) with ghee is taken to control nausea, to improve digestion and as a coolant.

Herbarium No.	*Plant Name*	*Mode of Administration*
VOCB 5466	*Cardiospermum helicacabum* L. Modakkathan Sapindaceae	One teaspoon of the paste made from the fresh leaves with lemon juice (juice prepared from the fruits of *Citrus aurantifolia* (Christm) Swingle) (elumitchai) is taken with buttermilk thrice a day as a carminative and to arrest dysentery. An ounce of the boiled root infusion is taken twice a day for a period of one month to treat haemorrhoids.
VOCB 5467	*Carissa carandas* L. Kalakkay Apocynaceae	An ounce of the root extract is taken with a glass of cow's milk twice a day for a period of one month to treat peptic ulcer.
VOCB 5468	*Carissa spinarum* L. Kila Apocynaceae	Fruits are consumed regularly to improve digestion.
VOCB 5469	*Cassia fistula* L. (*Cassia rhombifolia* Roxb.) Semaiagathi Caesalpiniaceae	An ounce of the boiled water extract of the flowers is taken twice a day for five days as an anthelmintic (de-wormer) and regularly by diabetic patients to reduce the blood glucose level. One teaspoon of the leaf powder blended with the flower powder is taken with a glass of cow's milk once a day for a period of two weeks to treat jaundice.
VOCB 5472	*Catharanthus roseus* (L.) G.Don (*Vinca rosea* L.) Palaisethai Apocynaceae	An ounce of the boiled water extract of the flowers is taken thrice a day for a period of three weeks to improve digestion and used as an invigorator. An ounce of the boiled water extract is taken thrice a day for a period of one month to treat diabetes.
VOCB 5473	*Cayratia pedata* (Lam.) Juss ex Gagnep Pannikodi Vitaceae	An ounce of the boiled water extract of the leaves is taken with a glass of cow's milk twice a day for a period of one month to treat haemorrhoids. It is also taken with honey twice a day to arrest diarrhoea.
VOCB 5476	*Celosia argentea* L. Pannikeerai Amaranthaceae	One teaspoon of the paste made from the seeds and flowers blended with the seed extract of *Trigonella foenum- graecum* L. (vendhaiyam) is taken with honey twice a day for two days to arrest dysentery.
VOCB 5477	*Centella asiatica* (L.) Urban (*Hydrocotyle asiatica* L.) Vallarai Apiaceae	One teaspoon of the leaf powder is taken with ghee twice a day as a carminative.
VOCB 5479	*Chenopodium ambrosioides* L. Kattasambadam Chenopodiaceae	Twenty five grams of the fresh leaves are made into a paste with coconut milk (the extract of the endosperm of *Cocos nucifera* L.) and taken once for a period of one week to treat ulcer.
VOCB 5481	*Cissampelos pareira* L. var *hirsuta* (Buch-Ham ex DC) Forman Appatta Menispermaceae	One teaspoon of the root powder is taken to relieve stomachache associated with menstruation.

Contd...

Herbarium No.	Plant Name	Mode of Administration
VOCB 5452	*Cissus quadrangularis* L. Pirandai Vitaceae	An ounce of the stem juice is taken once a day for two to three days to improve digestion
VOCB 5483	*Citrus aurantifolia* (Christm) Swingle (*Citrus medica* L.) Elumitchai Rutaceae	In a glass of fruit juice, a handful of dried fruits of *Cuminum cyminum* L.(shiragam) are soaked. Then the shade dried fruits are taken to control nausea and to arrest dysentery.
VOCB 5489	*Colocasia esculenta* (L.) Schott (*Colocasia antiquorum* Schott.) Cheppamkizhangu Araceae	Fresh stem and fresh leaves are cooked with the fruit-pulp of *Tamarindus indica* L. (puli) and taken once a day for a period of one month to treat haemorrhoids.
VOCB 5490	*Commelina benghalensis* L. Kanavazhai Commelinaceae	It is mixed with the leaf extract of *Cynodon dactylon* (L.) Pers (arugampul) and taken twice a day for two days to arrest dysentery and to control nausea.
VOCB 5492	*Cordia obliqua* Willd Naruvili Boraginaceae	One teaspoon of the leaf paste is taken with butter twice a day for a period of one month to treat haemorrhoids. During this period the use of tamarind is avoided in food preparations
VOCB 5493	*Crataeva adansonii* DC Mavilingam Capparaceae	A handful of shade dried leaves are boiled in hundred ml of coconut oil (oil obtained from dried copra of *Cocos nucifera* L.) and the extract is taken with water thrice a day for a period of one month to dissolve and discharge kidney stones. It is also taken to reduce fever and to improve digestion.
VOCB 5495	*Crotalaria verrucosa* L. Salangaisedi Fabaceae	An ounce of the leaf juice is taken with cow's milk twice a day for a period of one week to treat dyspepsia
VOCB 5497	*Cucumis sativus* L. Vellarikai Cucurbitaceae	A glass of sweetened fruit juice is taken regularly as a coolant and to treat urogenital problems.
VOCB 5498	*Cucumis trigonus* Roxb. Kattu thumatti Cucurbitaceae	A glass of sweetened fruit juice is taken twice a day for a period of two weeks to treat bilious.
VOCB 5503	*Cyperus rotundus* L. Kanikorai Cyperaceae	An ounce of the boiled water extract of the tubers is taken with cow's milk twice a day for two to three days to arrest dysentery and to improve digestion. One teaspoon of the powder made from five grams of the epidermal peelings of the tuber and five grams of the dried rhizome of *Zingiber officinale* Roscoe. (sukku) is taken with honey twice a day for five days as an anthelmintic (de-wormer).

Herbarium No.	Plant Name	Mode of Administration
VOCB 5504	*Dalbergia lanceolaria* L.f. (*Dalbergia frondosa* Roxb.) Velangu Fabaceae	An ounce of the boiled water extract of the stem bark is taken twice a day for three days to arrest diarrhoea.
VOCB 5512	*Dioscorea oppositifolia* L. Vettrilaivalli Dioscoriaceae	The fresh or boiled tubers are taken twice a day for two to three days to arrest dysentery.
VOCB 5517	*Eclipta prostrata* (L.) L (*Eclipta alba* (L.) Hassk) Karisalangkanni Asteraceae	Fifty grams of the fresh leaves are continuously boiled in two hundred and fifty ml of water with few fresh leaves of *Leucas aspera* (Willd.) Link (thumbai) and *Phyllanthus amarus* Schumach. and Thonn (keelanelli). The decoction is taken with buttermilk twice a day for a period of one week to treat jaundice.
VOCB 5524	*Euphorbia hirta* L. (*Euphorbia pilulifera* auct. non Linn) Amman patcharisi Euphorbiaceae	An ounce of the boiled water extract of the leaves is taken with cow's milk once a day for five days to get relief from stomachache and urogenital problems.
VOCB 5525	*Euphorbia nivulia* Buch-Ham. Kalli Euphorbiaceae	An ounce of the salted leaf extract is taken for a period of one week to treat jaundice and pneumonia.
VOCB 5528	*Ficus hispida* L.f. Peyathi Moraceae	One teaspoon of the stem bark powder blended with a pinch of the rhizome powder of *Zingiber officinale* Roscoe. (sukku) is taken with hot water twice a day for three days to treat bowel complaints. One teaspoon of the seed powder is taken with honey thrice a day for two days to control nausea.
VOCB 5529	*Ficus microcarpa* L.f. (*Ficus retusa* auct. non Linn.) Kalitchi Moraceae	An ounce of the boiled water extract of the stem bark is taken twice or thrice a day for a period of six weeks to treat liver complaints.
VOCB 5530	*Ficus racemosa* L. (*Ficus glomerata* Roxb.) Vellai athi Moraceae	One teaspoon of the latex is taken with small amount of butter to arrest dysentery
VOCB 5535	*Glinus lotoides* L. (*Mollugo lotoides* (L.) Kuntze. Siruseruppadai Molluginaceae	One teaspoon of the paste made by grinding the whole plant with few fresh leaves of *Acalypha indica* L. (kuppaimeni) and *Eclipta prostrata* (L.) L. (karisalankkanni) is taken with sugar once a day for a period of one month to treat haemorrhoids.

Contd...

Herbarium No.	*Plant Name*	*Mode of Administration*
VOCB 5537	*Glycosmis pentaphylla* (Retz.) DC Kulaparai Rutaceae	One teaspoon of the leaf powder is taken with honey twice a day for a period of two weeks to treat jaundice.
VOCB 5545	*Hibiscus rosa-sinensis* L. Semparithi Malvaceae	An ounce of the boiled extract of the tender leaves is taken with palm jaggery (a sweetener made from the fruit juice of *Borassus flabellifer* L.)once a day for three days as a diuretic. An ounce of the boiled water extract of the flowers is taken with palm jaggery (a sweetener made from the fruit juice of *Borassus flabellifer* L.) once a day for two to three days to relieve chest pain related to gastric complaints.
VOCB 5546	*Holarrhena pubescens* (Buch-Ham) Wall ex G.Don (*Holarrhena antidysenterica* (L) Wall.) Kuthupaalai Apocyanaceae	One teaspoon of the stem bark powder is taken with a half teaspoon of the juice prepared from the fresh rhizome of *Zingiber officinale* Roscoe. (inji) thrice a day to arrest dysentery.
VOCB 5551	*Indigofera tinctoria* L. Kattavuri neeli Fabaceae	A handful of fresh leaves are soaked in two hundred ml of goat's milk and the infusion is taken in the early morning hours for a period of two weeks to treat jaundice.The decoction obtained by boiling a handful of fresh leaves, five grams of the dried seeds of *Cuminum cyminum* L. (shiragam) and few black pepper (dried fruits of *Piper nigrum* L.) (milagu) in two hundred ml of water is taken thrice a day for a period of two weeks to treat jaundice.
VOCB 5553	*Ipomoea obscura* (L.) Ker – Gawl Siruthaali Convolvulaceae	The fresh leaves are cooked with ghee and taken to treat urogenital problems
VOCB 5558	*Jatropha gossypiifolia* L. Adalai Euphorbiaceae	One teaspoon of the root powder is taken with cow's milk to relieve stomachache.
VOCB 5559	*Justicia betonica* L. Pacchai kanakambaram Acanthaceae	An ounce of the boiled water extract of the leaves is taken to arrest diarrhoea.
VOCB 5562	*Lablab purpureus* (L.) Sweet (*Dolichos lablab* L.) Thattam payaru Fabaceae	One teaspoon of the leaf powder is taken with curd twice a day to treat haemorrhoids.

Herbarium No.	*Plant Name*	*Mode of Administration*
VOCB 5563	*Lannea coromandelica* (Houtt.) Merr. Oothian Anacardiaceae	One teaspoon of the powdered resin blended with a pinch of cardamom powder (seed powder of *Elettaria cardamomum* (L.) Maton) (elam) is taken with honey twice a day to treat haemorrhoids.
VOCB 5568	*Mangifera indica* L. Mamaram Anacardiaceae	An ounce of the boiled water extract of the roots is taken thrice a day for two days to arrest diarrhoea and to control nausea.
VOCB 5570	*Melia dubia* Cav. (*Melia composita* Willd.) Malai vempu Meliaceae	An ounce of the boiled water extract of the root bark is taken as an anthelmintic (de-wormer).
VOCB 5571	*Merremia tridentata* (L.) Hallier f. (*Convolvulus tridentatus* L.) Muthiar kunthal Convolvulaceae	The decoction obtained by boiling ten grams of the whole plant, five grams of the stem bark of *Mangifera indica* L. (maa) and two grams of the dried rhizome of *Zingiber officinale* Roscoe. (sukku) are boiled in two hundred ml of water is taken thrice a day to treat intestinal infection leading to diarrhoea.
VOCB 5573	*Mimosa pudica* L. Thottal simungi Mimosaceae	An ounce of the boiled water extract of the leaves is taken thrice a day for a period of thirty days to dissolve and discharge the kidney stones and urogenital problems.
VOCB 5575	*Mirabilis jalaba* L. Anthi mantharai Nyctaginaceae	An ounce of the leaf juice is taken with buttermilk regularly to take care of urogenital problems.
VOCB 5577	*Mollugo cerviana* (L.) Ser Koppada Molluginaceae	The decoction obtained by boiling few whole plants and few fresh leaves of *Adhatoda vasica* Nees. (adatodai) and *Evolvulus alsinoides* L. (vishnu kiranthi) in two hundred ml of water is taken thrice a day to arrest dysentery.
VOCB 5579	*Momordica charantia* L. Pakarkai Cucurbitaceae	The decoction obtained by boiling few whole plants and few fresh leaves of *Adhatoda vasica* Nees. (adatodai) and *Evolvulus alsinoides* L. (vishnu kiranthi) in two hundred ml of water is taken thrice a day to arrest dysentery. An ounce of the leaf juice is taken twice a day for three to four days as an anthelmintic (de-wormer). An ounce of the juice prepared from few fresh fruits and few fresh leaves with pieces of the stem bark of *Syzygium cumini* L. (naval) is taken once in a day to treat diabetes. Dried fruits are roasted and taken with food to treat haemorrhoids and minor renal problems.
VOCB 5582	*Murraya koeinjii* (L) Spreng. Karuvepillai Rutaceae	One teaspoon of the leaf paste blended with small amount of turmeric powder (powder made from the dried rhizome of *Curcuma longa* L.) (manjal) and few dried fruits of *Cuminum cyminum* L. (shiragam) is taken twice a day for two days to arrest dysentery.

Contd...

Herbarium No.	*Plant Name*	*Mode of Administration*
VOCB 5583	*Murraya paniculata* (L.) Jack. Kattu karuvepillai Rutaceae	One teaspoon of the salted paste made from the leaves which are roasted in ghee and few chilies (fruits of *Capsicum frutescens* L.) (milagai) is taken twice a day for two to three days to arrest dysentery and to treat dyspepsia. One teaspoon of the leaf powder is taken with sugar twice a day as a carminative.
VOCB 5584	*Musa paradisiaca* L. Vazhai Musaceae	Flowers cooked with the seeds of *Vigna radiata* (L.) Wilczek (pachai payaru) are regularly taken to treat heamorrhoids.
VOCB 5585	*Myristica fragrans* Houtt. Jathikkai Myristicaceae	One teaspoon of the pericarp powder is taken with cow's milk thrice a day to arrest dysentery.
VOCB 5587	*Ocimum americanum* L. (*Ocimum canum* Sims.) Naithulasi Lamiaceae	An ounce of the leaf juice is taken with cow's milk twice a day to get relief from cold and cough and also to arrest dysentery. The gently roasted pericarp is taken with sugar to relieve headache and to treat haemorrhoids.
VOCB 5589	*Ocimum tenuiflorum* L. (*Ocimum sanctum* L.) Nalla thulasi Lamiaceae	It is taken thrice a day with palm jaggery (a sweetener made from the fruit juice of *Borassus flabellifer* L) to take care of tuberculosis and to improve digestion.
VOCB 5590	*Oxalis corniculata* L. Puliyari Oxalidaceae	A handful of tender leaves are gently boiled in two hundred ml of water with few flowers of *Musa paradisiaca*.L (vazhai). An ounce of the boiled water extract of leaves is given to kids with honey to arrest dysentry.
VOCB 5597	*Pergularia daemia* (Forssk.) Chiov. (*Pergularia extensa* (Jack) N. E. Br.) Veliparuthi Asclepiadaceae	One teaspoon of the root powder is given to kids with cow's milk as an anthelmintic (de-wormer) and as a carminative.
VOCB 5598	*Phoenix loureirii* Kunth. Malai Eecham Arecaceae	An ounce of sap obtained from the peduncle is taken with water twice a day to arrest diarrhoea.
VOCB 5600	*Phyllanthus amarus* Schum and Thonn. Keela nelli Euphorbiaceae	An ounce of the extract obtained by squeezing some tender leaves together with few tender leaves of *Eclipta prostrata* (L.) L. (karisalankanni) and *Leucas aspera* (Willd) Link (thumbai) is taken with cow's milk twice a day for a period of two weeks to treat jaundice.
VOCB 5601	*Phyllanthus emblica* L. Nelli Euphorbiaceae	An ounce of juice prepared from the tender leaves is taken with buttermilk twice a day for two days to arrest dysentery.

Herbarium No.	Plant Name	Mode of Administration
VOCB 5604	*Physalis peruviana* L. Perungkunni Solanaceae	An ounce of the leaf juice blended with a pinch of the gum-resin exuded from incisions in living tap roots of *Ferula assafoetida* L. (perungayam) is given to kids to relieve colic pain.
VOCB 5608	*Polycarpaea corymbosa* (L.) Lam. Palli poondu Caryophyllaceae	One teaspoon of the leaf paste is taken once a day for a period of two weeks to treat jaundice.
VOCB 5611	*Portulaca quadrifida* L. Tharai pasalai Portulacaceae	An ounce of the whole plant juice is taken once a day to treat haemorrhoids.
VOCB 5615	*Psidium guajava* L. Koyya Myrtaceae	The decoction obtained by boiling a handful of leaves with few chillies (fruits of *Capsicum frutescens* L.) (milagai) in one hundred ml of water is taken once a day in empty stomach to control nausea and to treat intestinal infection leading to diarrhoea. The tender leaves are chewed to improve digestion. An ounce of the root infusion is given to kids to arrest dysentery.
VOCB 5517	*Punica granatum* L. Mathulai Punicaceae	One glass of fresh fruit/seed juice is taken to control nausea and to arrest dysentery.
VOCB 5620	*Ricinus communis* L. Amanakku Euphorbiaceae	One teaspoon of the blend made by grinding some shade dried leaves and mixed with dried leaves of *Phyllanthus amarus* Schum and Thonn. (keelanelli) is taken with ghee thrice a day for one week to treat jaundice. One teaspoon of the seed oil is taken with a glass of cow's milk to arrest dysentery. One teaspoon of the root bark powder is taken as a laxative.
VOCB 5622	*Ruta graveolens* L. Pillai valathi Rutaceae	An ounce of the leaf extract is taken twice a day for two to three days to improve digestion and to relieve stomach ache.
VOCB 5627	*Sida cordifolia* L. Nilathuthi Malvaceae	An ounce of the boiled water extract prepared from the shade dried roots is taken thrice a day for a period of one month to treat renal problems. One teaspoon of the seed powder is taken twice a day for a period of one week to treat urogenital problems.
VOCB 5631	*Spermacoce hispida* L. (*Spermacoce articularis* L.f.) Nathaichuri Rubiaceae	A glass of the root juice blended with cow's milk is taken by the feeding mothers to improve lactation. One teaspoon of the powder made from the roasted seeds is taken with cow's milk as a coolant and to dissolve and discharge the kidney stones.
VOCB 5633	*Strebulus asper* Lour. (*Epicarpurus orientalis* Bl.) Pura Moraceae	One teaspoon of the paste made from the tender leaves with ghee and sugar is taken thrice a day to arrest diarrhoea.

Contd...

Herbarium No.	Plant Name	Mode of Administration
VOCB 5635	*Strychnos potatorum* L.f. Thetran kottai Loganiaceae	Few dried fruits are taken to manage nausea and arrest dysentery. One teaspoon of the seed powder is taken with cow's milk to treat urogenital problems. One teaspoon of the seed powder is taken with curd to arrest dysentery.
VOCB 5638	*Tamarindus indica* L. (*Tamarindus officinalis* HK.) Puli Caesalpiniaceae	One teaspoon of the salted stem bark powder roasted and blended with few dried fruits of *Cuminum cyminium* L. (shiragam) is taken twice a day for five days to improve digestion and to relieve stomachache. An ounce of the boiled water extract of the leaves is taken in empty stomach once in a day for a period of two weeks to improve digestion and to treat urogenital problems.
VOCB 5642	*Terminalia chebula* Retz. Kadukkai Combretaceae	An ounce of the decoction obtained by boiling some dried fruits and few cloves (dried unopened floral buds of *Eugenia caryophyllata* Thunb) (kirambu) in one hundred ml of water is taken with ghee in empty stomach to treat haemorrhoids and also as a coolant and laxative.
VOCB 5643	*Thevetia peruviana* (Pers.) Merr. (*Thevetia neriifolia* Juss ex Steud.) Ponnarali Apocynaceae	An ounce of the sweetened leaf juice is taken to control nausea.
VOCB 5644	*Tinospora cordifolia* (Willd.) Miers ex Hook f. and Thorns. Seenthil Menispermaceae	One teaspoon of the leaf powder is taken with water thrice a day for a period of one month to treat peptic ulcer.
VOCB 5645	*Toddalia asiatica* (L.) Lam. (*Toddalia aculeata* Pers.) Milakaranai Rutaceae	An ounce of the hot water extract of the root bark is taken as an antipyretic and to improve digestion.
VOCB 5647	*Tribulus terrestris* L. Nerungi Zygophyllaceae	An ounce of the whole plant extract is taken with buttermilk twice a day for a period of two weeks to treat urogenital problems.
VOCB 5648	*Trichodesma indicum* (L.) R.Br. Kali thumbai Boraginaceae	The paste made by grinding some fresh leaves and little amount of cooked rice (grains of *Oryza sativa* L.) is taken twice a day for three days with palm jaggery (a sweetener made from the fruit juice of *Borassus flabellifer* L.) to arrest diarrhoea.
VOCB 5650	*Vernonia cinerea* (L.) Less. Siru sengalaneer Asteraceae	An ounce of the diluted leaf extract is given to kids to arrest dysentery.

Herbarium No.	*Plant Name*	*Mode of Administration*
VOCB 5654	*Wrightia tinctoria* (Roxb.) R.Br. Veppalai Apocynaceae	An ounce of the stem bark extract is taken in empty stomach for a period of one month to treat hemorrhoids.
VOCB 5655	*Zingiber officinale* Roscoe. Inji Zingiberaceae	An ounce of the extract of the rhizome blended with the extract of few bulbs of *Allium cepa* L. (vengayam) is taken with honey to control nausea.
VOCB 5656	*Zizyphus mauritiana* Lam. (*Zizyphus jujuba* (Lam) Gaertn.) Ilanthai Rhamnaceae	One teaspoon of the paste made from the tender leaves and the stem bark is taken with curd twice a day for three days to arrest diarrhoea. The hot water extract of the root bark is taken in empty stomach thrice in a day for three days to improve digestion.
VOCB 5657	*Zizyphus xylopyrus* (Retz.) Willd. Kottai Ilanthai Rhamnaceae	An ounce of stem bark infusion is taken in empty stomach for a period of one week to improve digestion.

Among the different plant parts used to prepare medicine, the leaves (42 per cent) were used mostly the preparation of medicine followed by root (14 per cent), stem and fruit (9 per cent), whole plant (6 per cent), seed and flower (5 per cent), rhizome and tubers (4 per cent), Latex and resins (1 per cent) (Figure 13.3). Many of the previous studies have been proved that leaves are predominantly used to prepare medicine by different tribal for various human ailments (Rajendran *et al.*, 2002 and 2003; Mahishi *et al.*, 2005; Ganesan *et al.*, 2005; Ayyanar *et al.*, 2005; Ignacimuthu *et al.*, 2006; Jagtap *et al.*, 2006; Ignacimuthu *et al.*, 2008).

The method of preparation and dosage level with period has been provided in mode of administration (Table 13.1). The four methods (Juice/extract, powder, paste and decoction) of preparation have been used as drug to treat following different gastrointestinal complaints such as diabetes, jaundice, dysentery, diarrhea, haemorrhoids, stomachache, nausea, chicken pox, asthma, ulcer, kidney stone, colic pain, vermifuge, dyspepsia, stomachache associated with menstruation and urogenital problems (Figure 13.4).

The ethnomedicinal plants were used frequently to cure diarrhea, dysentery, jaundice, diabetes, stomachache, haemorhoids (piles), ulcer and kidney related diseases. Some of the ethnomedicinal plants were used to treat two or three different types of diseases that plants are categorized. The majority of the ethnomedicinal plants were taken orally in the form of juice/extract, decoction and powder which is used to cure number of stomach disorders. To improve action of herbal medicine some of additives have been used associate with medicinal plants such as bulbs of *Allium cepa*, *Allium sativum*, *fruits of Carum copticum*, dried copra of *Cocos nucifera*, rhizome of *Alpinia galanga*, unripe fruit of *Terminalia chebula*, *etc.*

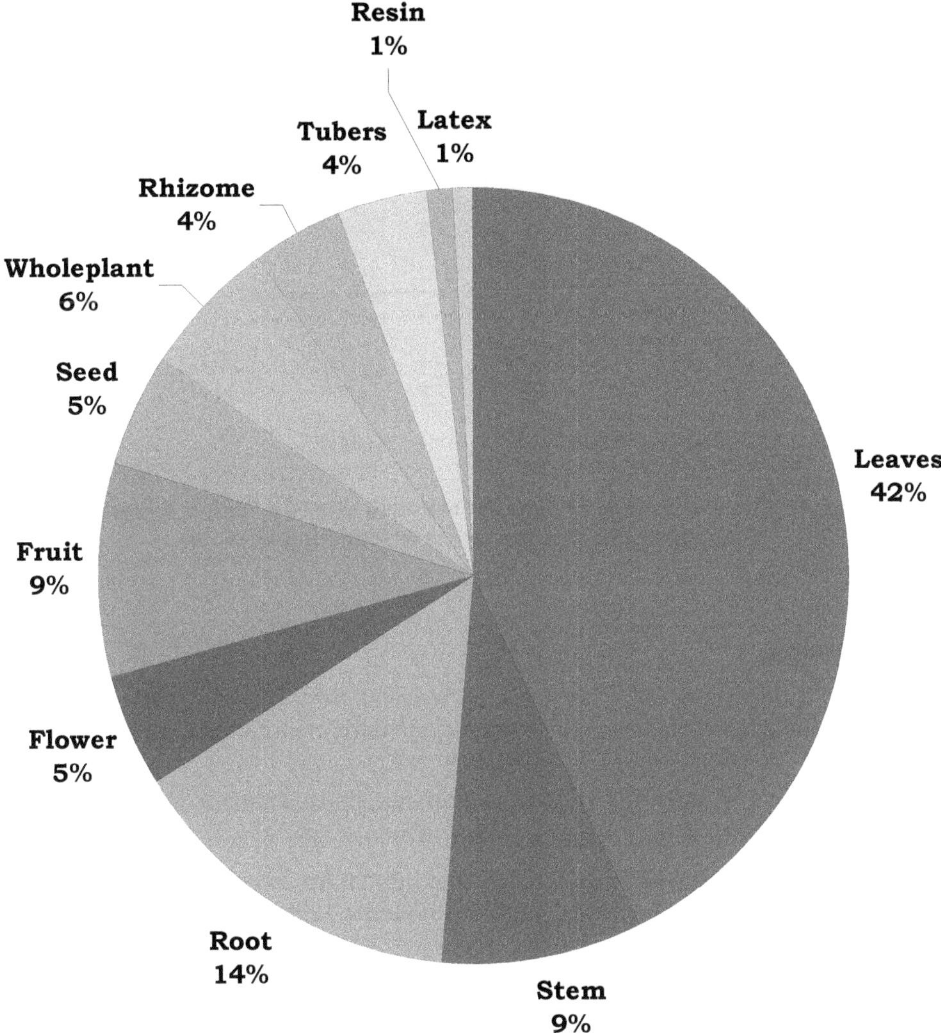

Figure 13.3: Percentage of Plant Parts Used by *Palliyar* Tribals in Sirumalai Hills, Western Ghats in Preparing Medicine.

Traditional Knowledge play a potential role in tribal system of medicine prescribed by traditional healers or vaidhyars establishes in remote areas of *Palliyar* tribals in Sirumalai hills. That indigenous knowledge is very important in the documentation of herbal system of medicine which is directly connected with conservation of biodiversity. In future, this traditional knowledge should validate the medical formulations by using from compound identification to clinical trials.

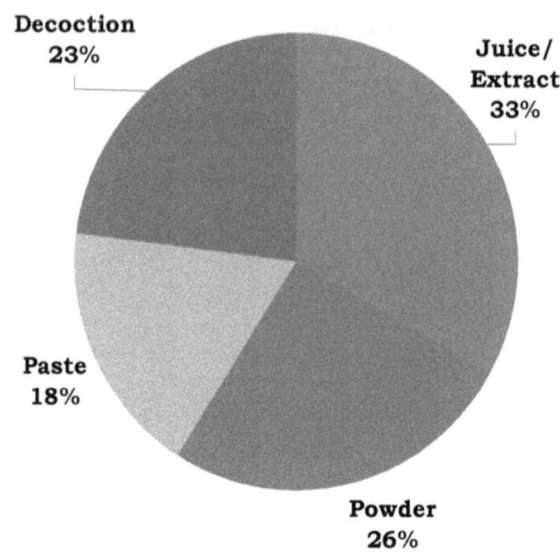

Figure 13.4: Percentage of different Methods Used by *Palliyar* Tribals in Sirumalai Hills, Western Ghats in Preparing Medicine.

References

Aburjai, T, Hudaib, M, Tayyem, R, Yousef and M, Oishawi, M. 2007. Ethnopharmacological survey of medicinal herbs in Jordan, the Ajloum Heights region. *J. Ethnopharmacol.* 110: 294-304.

Aryal, MP, Berg, A, and Ogle, B. 2009. Uncultivated plants and livelihood support – A case study from the Chepang people of Nepal. *Ethnobot. Res. Appl.* 7: 409-422.

Ayyanar, M and Ignacimuthu, S. 2005. Traditional knowledge of kani tribals in Kouthalai of Tirunelveli hills, Tamil Nadu, India. *J Ethnopharmacol.* 102: 246-255.

Ganesan, S, Suresh, N and Kesavan, L. 2005. Ethnomedicinal survey of lower Palni hills of Tamil Nadu, *Indian J Trad Knowl.* 3: 299-304.

Ignacimuthu, S, Ayyanar, M and Sankarasivaraman, K. 2008. Ethnobotanical study of medicinal plants used by Paliyar tribals in Theni district of Tamil Nadu, India. *Fitoterapia.* 79: 562-568.

Ignacimuthu, S, Ayyanar and Sankarasivaraman, K. 2006. Ethnobotanical investigations among tribes in Madurai district of Tamil Nadu, India. *J Ethnobiol Ethnomed.* 2: 25.

Jagtap SD, Deokule, SS and Bhosle, SV. 2006. Some unique ethnomedicinal uses of plants used by the Korku tribe of Amaravati district of Maharashtra, India. *J Ethnopharmacol.*107: 463-469.

Mahishi, P, Srinivasa, BH and Shivanna, MB. 2005. Medicinal plant wealth of local communities in some villages in Shimoga District of Karnataka, India. *J. Ethnopharmacol.* 98: 307-312.

Patrick, OE. 2002. Herbal Medicines: Challenges (Editorial) *Tropical J Pharma Res.* 1: 53-54.

Rajendran, SM, Agarwal SC, and Sundaresan, V. 2003. Lesser known ethnomedicinal plants of Ayyakarkoil Forest Province of Southwestern Ghats, Tamil Nadu, India. Part – I. *J. Herbs Spices Med Plants.*10:103-112.

Rajendran, SM, Chandrasekar, K and Sundaresan, V. 2002. Ethnomedicinal lore of Valaya tribe in Seithur hills of Virudhunagar district, Tamil Nadu, India. *Indian J Trad. Knowl.* 1: 59-71.

2018, Ethnomedicinal Plants: A Biodiversity Treasure Pages 363–379
Editors: V.R. Mohan, A. Doss, P.S. Tresina and V. Sornalakshmi
Published by: ASTRAL INTERNATIONAL PVT. LTD., NEW DELHI

Chapter 14

Ethnomedicinal Plants used for Skin Diseases by *Kanikkars* of Kanyakumari District of Western Ghats, Tamil Nadu, India

P.S. Tresina[1], K. Paulpriya[1], V. Sornalakshmi[2] and V.R. Mohan[1]

[1]Ethnopharmacology Unit, PG and Research Department of Botany, V.O. Chidambaram College, Tuticorin, Tamil Nadu
[2]Department of Botany, A.P.C. Mahalaxmi College for Women, Tuticorin – 628 002, Tamil Nadu

Introduction

Human civilization has evolved as a result of interaction of people with their environment, especially with plants. From the very earliest days of civilization, mankind has turned to plants for healing, a tradition that has survived the arrival of modern medicine and found new strength at the end of 20th century (Sullivan and Shearly, 1997). India is one of the twelve mega-biodiversity countries of the world having rich vegetation with a variety of plants having medicinal values. India accounts for 7-8 per cent of the recorded species of the world. The Botanical Survey of India has recorded over 47,000 species of plants. About 64 per cent of the total global population remains dependent on traditional medicine for their healthcare system (Farnsworth, 1994). Regions with rich biodiversity, with its traditional ethnic people, are the biggest source for the plant resources and its hidden knowledge (Hamilton, 2004). The traditional knowledge of medicinal plants has been recorded in numerous literatures (Chopra *et al.*, 1956; Rajendran and Agarwal, 2007). In Tamil Nadu, a lot of work has been done on the ethnomedicinal plants used for various ailments by different ethnic communities (Ganesan *et al.*, 2004; Maruthupandian and Mohan, 2012).

Tamil Nadu is spotted with tribal pockets, rich in germplasm of medicinal plants. In this regard, the *Kanikkar* tribal pocket constitutes an important source of medicinal plants. The *Kanikkar*- a predominant tribal community of Tamil Nadu is the focus of the present study. The *Kanikkars* are also known as *Kanikkaran* or *Kani. Kanikkars* means "hereditary proprietor of land". Thus, they establish their ancient rights over the forest lands. The *Kanikkars* are generally very short in stature and meager in appearance with Negroid features. They are simple and straight forward tribals. They are traditionally a nomadic community. Total population of the community is 4950 in Kanyakumari district (ASSISI, 2005) and they are only found in adjoining Tirunelveli district. The present study comprises 3184 inhabitants in 48 settlements. Their settlements are composed of low-lying huts, which are built with bamboo and reeds (Thurson, 1998). Most of them are inhabited in hilly areas.

The main aim of the present study is to collect information on traditional uses of medicinal plants used in the preparation of herbal drugs for the treatment of skin diseases by the tribal people living in Kanyakumari district, Tamil Nadu for which no literature is still available. Very few ethnobotanical studies have been carried out about the tribe *Kanikkar* in Tirunelveli district (Ignacimuthu *et al.*, 1998; Viswanathan *et al.*, 2001; 2003; Ayyanar and Ignacimuthu, 2005a,b; Prasad *et al.*, 1987) of Tamil Nadu. The medicinal plants available in Kanyakumari district are not explored well, which deserves a thorough investigation for their pharmacological activities.

Methodology

The *Kanikkars* are distributed along the southeastern slopes of the Western Ghats, Tamil Nadu. The Kanyakumari district is located in the southern most part of the Indian peninsula. The area selectedfor study lies between 77°05′ to 77°36′ E longitude and 8°03′ and 8°35′ N latitude. It occupies an area of 1684 km^2. The district is bounded by Tirunelveli district on the north and east and by the Gulf of Mannar on the south eastern part, the south and southwest are surrounded by the Indian Ocean and the Arabian Sea and the west and northwest by Kerala state (Figure 14.1).

The district may be divided into three geographical regions, first the mountainous region with a number of hill tops, next the coastal region and the undulating valleys and finally the plains between the mountainous terrain and the sea coast. The minimum and maximum temperature is 14°2 C and 32° C during December and May respectively. Humidity ranges between 65 and 75 per cent. The average monthly rainfall ranges between 20.59 mm and 206.58 mm. The district has a unique advantage of rainfall both during the southwest monsoon between June and September and northeast monsoon between October and middle of December.

The ethnobotanical survey was conducted between October 2013 and February 2014. The success of ethnomedicinal plants study lies on the conduct and relationship established between the researcher and the tribal people. Hence, more pain was taken to establish such a close relationship with tribes at their settlements and surrounding, which can bring out authentic first hand information about the medicinal uses of plants. Resource persons were identified and selected from all age

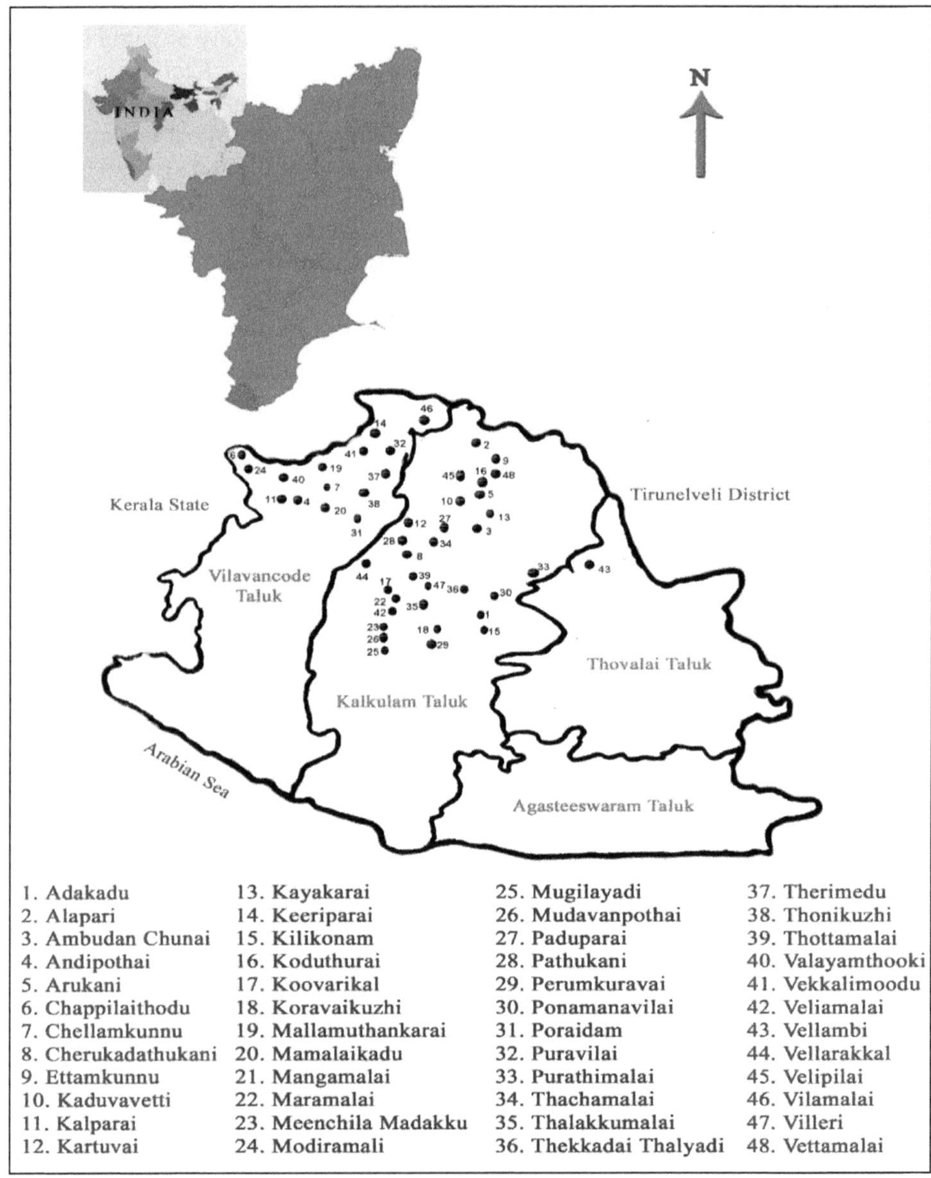

Figure 14.1: Study Area.

groups of the tribes and from both sex. In some settlements, elderly women were also contacted. Plants which are medicinal and used by Kanikkars for treating their skin diseases were collected and their mode of usage was recorded. Information on these sources was recorded in video cassettes, which form an authentic and first hand sources of reference. The information thus gathered from one claim was cross checked adequately for reliability and accuracy by interacting with different groups of the Kanikkars from different settlements to confirm the usage and dosage differences. Plant specimens were collected for taxonomic identification from different parts of the area under study. All the specimens collected from the fields were identified with the help of regional floras and finally confirmed by comparing with the authenticated specimens in the Herbarium of Botanical Survey of India (Southern Circle), Coimbatore, Tamil Nadu. Voucher specimens were deposited in the Ethnopharmacology Unit, Research Department of Botany, V.O.Chidambaram College, Tamil Nadu.

Results

The results of the survey are presented in Table 14.1.

Table 14.1: Medicinal Plants Used by *Kanikkar* Tribe in Kanyakumari District, Tamil Nadu for the Treatment of Skin Diseases

Botanical Name of the Plant with Voucher Number	*Family Name/ Vernacular Name*	*Plant Parts Used/ Preparations/Uses/Ailments Treated*
Abutilon indicum (L.) Sweet VOCB6582	Malvaceae/Thuthi	A handful of fresh leaves is made into a paste and applied externally on skin thrice a day to treat ringworm infection.
Acacia caesia (L.) Willd. VOCB6583	Mimosaceae/ Vellinjal	The dried bark made into powder is used like bathing soap to treat scabies, rashes and ringworm infection.
Acacia pennata (L.) Willd VOCB6584	Mimosaceae/Indu	The stem bark is made into paste and applied topically to get rid of skin rashes and scabies.
Acacia nilotica (L.) Willd VOCB6585	Mimosaceae/ Shikakaai	The paste made from stem bark is externally applied to treat scabies.
Acalypha indica L. VOCB6586	Euphorbiaceae/ Kuppaimeni	A handful of leaf paste is applied externally twice a day to treat eczema, skin rashes, insect bites and ringworm infection.
Ageratina adenomorpha (Spreng) R.King and H.Robinson VOCB6587	Asteraceae/ Aanaivanthan	The leaves are made into a paste and applied externally to treat wounds.
Albizia amara (Roxb.) Boivin VOCB6588	Mimosaceae/Uzil	The paste of leaf is used in the preparation of hair oil to treat dandruff. The stem bark is made into paste and applied externally for healing wounds.
Aloe vera (L.) Burm.f. VOCB6589	Liliaceae/ Chothukathalai	Mucilage taken from the fresh leaf is externally applied thrice a day to treat eczema, leucoderma, scabies and ringworm infection.

Botanical Name of the Plant with Voucher Number	*Family Name/ Vernacular Name*	*Plant Parts Used/ Preparations/Uses/Ailments Treated*
Alstonia scholaris (L.) R. Br. VOCB6590	Apocynaceae/ Mukkampalai	Three drops of milky latex are applied on the infectious wounds twice in a day to clear worms from the wounds.
Amaranthus viridis L. VOCB6591	Amaranthaceae/ Kuppai keerai	A handful of whole plant paste is applied externally in the morning hours to treat scabies.
Anaphyllum beddomei Engler, Pflanzenr VOCB6592	Araceae/Keeri Kilangu	About 10 g of cleaned fresh rhizome made into a paste is applied externally twice in a day to treat eczema and scabies.
Andrographis paniculata (Burm.f.) Wall ex Nees VOCB6593	Acanthaceae/Nila vembu, Kiriyathi	10 g of fresh leaves is made into paste with water and is applied externally twice in a day to treat leprosy, scabies and ringworm infection.
Anisomeles malabarica (L.) R.Br. VOCB6594	Lamiaceae/ Perunthumbai	Handful of leaf is made into paste and applied externally to treat scabies.
Argemone mexicana L. VOCB6595	Papavaraceae/ Ponnumathai	About 10 g of leaf paste is applied externally once in a day to treat ringworm infection. The clean leaf paste mixed with lime water is applied twice a day on the wounds for healing.
Aristolochia bracteata Lam. VOCB6596	Aristolociaceae/ Aaduthinna palai	Handful of whole plant paste is externally applied twice in a day to treat eczema, scabies and ringworm infection.
Aristolochia indica L. VOCB6597	Aristolociaceae/ Karudakodi	10 g of whole plant along with equal amount of rhizomes of *Kaempferiagalangal* and leaves of *Hiptage benghalensis* are boiled in coconut oil and the oil extract is externally applied thrice in a day to treat eczema, scabies, skin rashes and ringworm infection.
Artocarpus hirsutus Lam. VOCB6598	Moraceae/Ayani	The bark infusion is applied externally to treat acnes and cracks on the skin.
Asparagus racemosus Willd. VOCB6599	Liliaceae/Sathavari	20 g of tuberous root paste is applied externally once in a day to treat ringworm infection.
Azadirachta indica A. Juss. VOCB6600	Meliaceae/Veppu	Handful of leaf paste mixed with powdered dried rhizome of *Curcuma longa* is applied externally once in a day to treat all kinds of skin infection, small pox and chicken pox. The oil extracted from the seeds is externally applied twice in a day to treat eczema.
Begonia malabarica Lam. VOCB6601	Begoniaceae/ Narayana Sanjeevi	About 10 g of leaf paste is externally applied twice a day to treat ringworm infection.
Basella alba L. VOCB6602	Basellaceae/ Vasalakkirai	Handful of whole plant is made into paste and applied externally to treat wounds.
Bauhinia purpurea L. VOCB6603	Caesalpiniaceae/ Mantharai	The stem bark is made into paste and externally applied to treat wounds
Biophytum sensitivum (L.) DC. VOCB6604	Oxalidaceae/ Manivatti pachilai	About 10 g of leaf paste is externally applied twice in a day to treat eczema, scabies and skin rashes.

Contd...

Botanical Name of the Plant with Voucher Number	*Family Name/ Vernacular Name*	*Plant Parts Used/ Preparations/Uses/Ailments Treated*
Boerhavia diffusa L. VOCB6605	Nyctaginaceae/ Vethalamai	A handful of leaves are boiled in coconut oil and the oil extract is externally applied twice in a day to treat scabies and ringworm infection.
Calanthe masuca Lind. VOCB6606	Orchidaceae/Kalai Kombu	About 10 g dried tuberous root is powdered and made into a paste with water. The paste is used to treat acnes and inflammatory sebaceous cysts.
Calophyllum inophyllum L. VOCB6607	Clusiaceae/Punnai	Few drops of oil extracted from the seeds are externally applied thricein a day to treat ringworm Infection.
Calotropis gigantea (L.) R.B. VOCB6608	Asclepiadaceae/ Erukku	About 10 ml of latex, rhizomes of *Curcumalonga* and roots of *Aristolochia indica* are boiled in coconut oil and the oil extracts is externally applied thrice in a day to treat eczema.
Calyacopteris floribunda Lam. VOCB6609	Combretaceae/ Pillani	10 g of leaf gall paste is applied externally twice in a day to treat scabies and skin inflammation.
Canarium strictum Roxb. VOCB6610	Burseraceae/ Kungilium	The resin obtained from the plant is made into paste with turmeric and applied externally to treat cracks on heels.
Canthium parvifoloium Lam.	Rubiaceae/Katti karai	Handful of leaf paste is externally applied twice a day to treat scabies and ringworm infection.
Caryota urens L. VOCB6611	Arecaceae/ Koonthappanai	About 5 g of leaf is made into paste and applied externally to treat wounds.
Cassia alata L. VOCB6612	Caesalpiniaceae/ Yanaithavarai	10 g of fresh leaf paste is applied twice in a day to treat ring infection.
Cassia fistula L. VOCB6613	Caesalpiniaceae/ Sarakonrai	The dried bark powder ground into a paste with water is externally applied twice in a day to treat eczema.
Cassia kleinii W. and A. VOCB6614	Caesalpiniaceae/ Mullilla thottavadi	Handful of whole plant made into a paste is externally applied twice in a day for a week to treat eczema.
Cassia occidentalis L. VOCB6615	Caesalpiniaceae/ Peythavarai	The leaves made into paste with water are applied externally once in a day to treat scabies, skin rashes and ringworm infection.
Cissampelos pareira L. VOCB6616	Menispermaceae/ Malaithanki pachilai	Handful of whole plant paste is externally applied on the skin daily in the morning hours for a week to treat ringworm infection.
Cissus quadrangularis L. VOCB6617	Vitaceae/Pirandai	About 10 g of stem is made into paste and applied externally to treat wounds.
Cleome viscosa L. VOCB6618	Capparaceae/ Naikadugau	The leaf paste is externally applied to treat wounds.
Clerodendrum inerme (L.) Gaertn VOCB6619	Verbenaceae/ Changu kuppi	The leaf paste is externally applied once in a day for a week to treat psoriasis, scabies and ringworm infection. The leaf paste is also applied on the site of insect bite.

Botanical Name of the Plant with Voucher Number	Family Name/ Vernacular Name	Plant Parts Used/ Preparations/Uses/Ailments Treated
Clitoria ternatea L. VOCB6620	Fabaceae/Changu pushpam	Handful of leaves made into paste and is applied externally once in a day to treat skin inflammation, scabies and ringworm infection.
Cocos nucifera L. VOCB6621	Arecaceae/Thengu	Oil extracted from the endosperm is used as a base for most of the oil extracts used in the treatment of skin diseases and infections.
Commelina benghalensis L. VOCB6622	Commelinaceae/ Valai pachai	The leaf paste is applied externally twice in a day to treat scabies. The leaf paste is also applied on the wounds once a day for healing and to remove the poisonous spines that had struck accidentally on the body parts.
Coscinium fenestratum (Gaertn.) Coleb VOCB6623	Menispermaceae/ Maramanjal	About 50 g of climbing stem is dried and boiled in 100 ml coconut oil. Few drops of the oil extract are applied on the burns twice in a day for healing wounds. The stem paste is used to treat acnes.
Costus speciosus (Koen.) J.E.Smith VOCB6624	Costaceae/ Koshtam	The rhizome and leaf pastes are externally applied once in a day to treat ringworm infection.
Crinum defixum Ker-Gawl VOCB6625	Amaryllidaceae/ Vishanarayani	About 10 bulbs made into paste with water is externally applied twice in a day on the inflamed skin that is formed due to insect bite or allergy.
Croton tiglium L. VOCB6626	Euphorbiaceae/ Nervilankai	The seeds are ourified by soaking them in lime water for two days. The oil expelled from the seeds is applied twice in a day on the infected areas to treat eczema. The oil is also use to treat alopecia. The leaf paste is externally applied twice in a day to treat scabies and ringworm infection.
Curculigo orchioides Gaertn. VOCB6627	Hypoxidaceae/ Nilapanai	About 10 g of fresh or dried root tuber paste is externally applied twice in a day to treat ringworm infection.
Curcuma aromatic Salisb. VOCB6628	Zingiberaceae/ Kasthuri manjal	About 5 g of powder of dried rhizome made into a paste with water is applied externally on the skin while bathing to treat various skin infections. The powder is also used to treat acnes.
Curcuma longa L. VOCB6629	Zingiberaceae/ Manjal	The rhizome made into a paste with honey is applied externally to treat skin inflammation. The rhizome powder mixed with leaf paste of *Azadirachta indica* is applied topically in the morning hours on the wounds for healing. This paste is also used to treat chicken pox, measles and scabies.
Cyclea peltata (Lam.) Hook.f. and Thomas VOCB6630	Menispermaceae/ Thazhi	The powdered dried leaves is soaked in water and the colloidal paste is applied on the head in the morning and washed after one hour for a week to treat dandruff.

Contd...

Botanical Name of the Plant with Voucher Number	Family Name/ Vernacular Name	Plant Parts Used/ Preparations/Uses/Ailments Treated
Diploclisia glaucescens (Blume) Diels VOCB6631	Menispermaceae/ Erumaithiran Kodi	10 g of dried leaves are soaked in 100 ml coconut oil for few days. The oil infusion is externally applied thrice in a day to treat leprosy and scabies.
Diplocyclos palmatus (L.) Jeffrey VOCB6632	Cucurbitaceae/ Iverelli	5 g of dried tuber made into a paste with water is applied externally twice in a day to treat inflammatory sebaceous cysts.
Eclipta prostrata L. VOCB6633	Asteraceae/ Karisalankanni	Leaves along with the seeds of *Foeniculum vulgare* are boiled in coconut oil and the oil extract is applied on the head daily in the morning hours for a week to treat dandruff. The leaf paste is used to treat ringworm infection.
Elephantopus scaber L. VOCB6634	Asteraceae/Yannai chavuttadi	10 of leaf and rhizome pastes are applied externally twice in a day for a week to treat ringworm infection.
Erythrina variegata L. VOCB6635	Fabaceae/ Mullumurukku	10 g of fresh leaf paste is applied externally to treat leprosy.
Evolvulus alsinoides L. VOCB6636	Convolvulaceae/ Vishnukranthi	50 g of the whole plant extract and 50 g leaves of *Indigofera longeracemosa* are made into paste and boiled in coconut oil. The oil infusion is applied twice in a day for a week to treat leprosy and scabies.
Ficus benghalensis L. VOCB6637	Moraceae/ Alamaram	The dried tender aerial roots are powdered and boiled in coconut oil. The oil extract is used to treat dandruff. Few drops of the latex are externally applied twice in a day for a week to heal cracks on the foot.
Ficus religiosa L. VOCB6638	Moraceae/Arasu	The paste made from fruits is topically applied to treat scabies.
Gloriosa superba L. VOCB6639	Liliaceae/ Senganthal	The rhizome is made into paste and externally applied to treat wounds.
Helicteres isora L. VOCB6640	Sterculiaceae/ Valampuri, Kaivan	10 g of tender leaves made into paste with water is applied externally twice in a day on the wounds for healing.
Heliotropium indicum L. VOCB6641	Boraginaceae/Thel kodukkai	10 g of leaf paste is externally applied twice in a day to treat ringworm infection.
Hemidesmus indicus (L.) R. Br. var. *indicus* VOCB6642	Periplocaceae/ Naruneri	Handful of dried roots are pounded and boiled in 100 ml of coconut oil. Few drops of the oil extract are applied externally twice in a day to treat eczema, scabies and ringworm infection.
Hiptage banghalensis L. Kurz. VOCB6643	Malpighiaceae/ Mathavikodi	20 g of leaves and flowers made into paste with water is externally applied twice in a day to treat ringworm infection.

Botanical Name of the Plant with Voucher Number	Family Name/ Vernacular Name	Plant Parts Used/ Preparations/Uses/Ailments Treated
Indigofera longeracemosa Boivin. VOCB6644	Fabaceae/ Neelamari	Handful of leaves are made into paste along with 10 g leaves of *Andrographis paniculata* and 10 g rhizome of *Kaempferia galangal* and boiled in coconut oil. This oil extract is externally applied to treat leprosy, leucoderma and scabies. The leaf extract is boiled in coconut oil and the oil infusion is used to treat alopecia.
Ixora coccinea L. VOCB6645	Rubiaceae/Idlipoo	50 g of dried flowers are boiled in coconut oil and the oil extract is applied externally twice in a day to treat eczema.
Jatropha curcas L. VOCB6646	Euphorbiaceae/ Kaatamanakku	10 g of leaf paste is applied externally twice in a day to treat eczema, scabies and ringworm infection.
Justicia adhatoda L. VOCB6647	Acanthaceae/ Adhathodai	10 g of leaf paste is applied externally twice in a day to treat scabies and ringworm infection.
Kaempferia galanga L. VOCB6648	Zingiberaceae/ Kaccholam	The extract of the rhizome of *Andrographis paniculata* along with leaves of *Kaempferia galanga* and *Indigofera longeracemosa* in coconut oil is externally applied twice in a day to treat eczema, leprosy, scabies and other skin infections.
Kalanchoe pinnata (Lam.) Pers. VOCB6649	Crassulaceae/ Malaichodakku	10 g of leaf paste is externally applied on cuts and wounds thrice in a day for healing. The leaf paste is also used to treat ringworm infection.
Lawsonia inermis L. VOCB6650	Lythraceae/ Maruthani	The leaves are soaked in coconut oil for a week along with the flowers of *Saraca asoca* and the oil infusion is used as hair tonic to promote hair growth and to treat dandruff. The oil is also used to treat ringworm infection.
Leucas aspera (Wild.) Link VOCB6651	Lamiaceae/ Thumbai	10 g of leaf paste is applied externally twice in a day to treat ringworm infection.
Mangifera indica L. VOCB6652	Anacardiaceae/ Maa	The resin exudates dissolved in water is externally applied twice in a day for healing cracks on the foot.
Merremia tridentata (L.) Hall. F. VOCB6653	Convolvulaceae/ Koonthal valarthi	The entire plant is boiled in coconut oil and the extract is used to treat dandruff and to promote hair growth. 10 g of leaves made into paste is applied once in a day to treat various skin infections.
Mimosa pudica L. VOCB6654	Mimosaceae/ Thotta vadi	10 g of leaf paste is externally applied thrice in a day to treat eczema. A handful of entire plant made into paste and applied on cuts and wounds for healing.
Mirabilis jalapa L. VOCB6655	Nyctaginaceae/ Anthimantharai	10 g of dried root tuber made into a paste with water is externally applied twice in a day to treat sebaceous cysts and polyps.

Contd...

Botanical Name of the Plant with Voucher Number	*Family Name/ Vernacular Name*	*Plant Parts Used/ Preparations/Uses/Ailments Treated*
Momordica charantia L. VOCB6656	Cucurbitaceae/ Pagarkai	Handful of leaves made into paste is externally applied twice in a day to treat scabies and ringworm infection.
Moringa pterygosperma Gaertn. VOCB6657	Moringaceae/ Murungai	About 100 g of bark paste is applied externally once in a day to treat eczema.
Mukia maderaspatana (L.) M. Roem. VOCB6658	Cucurbitaceae/ Kattu vellari	10 g of leaf paste is externally applied twice in a day to treat scabies and ringworm infection.
Mussaenda glabrata (Hook.f.) Hutchinson ex Gamble VOCB6659	Rubiaceae/ Vellimadanthai	The leaf paste is externally applied to treat skin allergy and inflammation. 50 g of fresh leaves are boiled in 200 ml of coconut oil and the extract is applied on the scalp to promote hair growth.
Myxopyrum serratulum A.W.Hill VOCB6660	Oleaceae/ Saduramullai	10 g of leaf paste is externally applied twice in a day to treat skin rashes, scabies and ringworm infection.
Ocimum basilicum L. var. *purpurascens* Benth. VOCB6661	Lamiaceae/ Karunthulasi	10 g of leaves are boiled with equal amount of leaves of *Piper betle* in coconut oil. The oil extract is externally applied twice in a day to treat eczema and scabies.
Phyllanthus emblica L. VOCB6662	Euphorbiaceae/ Nellikai	50 g of dried cotyledons are boiled in coconut oil and the oil extract is externally applied thrice in a day to treat scabies.
Piper nigrum L. VOCB6663	Piperaceae/Nalla milagu	10 g of leaf paste is externally applied twice in a day to treat ringworm infection.
Plumbago zeylanica L. VOCB6664	Plumbaginaceae/ Chithiramoolam	About 10 g of leaves made into paste is externally applied twice a day to treat eczema, scabies and ringworm infection.
Pongamia pinnata (L.) Pierre VOCB6665	Fabaceae/Punga maram	100 g of dried bark powder is boiled in 200 ml of coconut oil and the oil extract is externally applied once a day to treat eczema, psoriasis, rashes, scabies and ringworm infection.
Rauvolfia serpentina (L.) Benth. Ex Kurz VOCB6666	Apocynaceae/ Avalpori	Handful of root along with root of *Thottea siliquosa*, rhizome of *Kaemferia galanga* and fruits of *Helecteris isora* are pounded and boiled in coconut oil. The oil extract is applied externally twice a day to treat leprosy, scabies and ringworm infection.
Rubia cordifolia L. VOCB6667	Rubiaceae/ Manchathi	The root paste and rhizome paste of turmeric is applied topically on the affected portion in the treatment of skin rashes and scabies.
Santalum album L. VOCB6668	Santalaceae/ Santhanam	The wood paste is externally applied on the skin to treat prickly heat, skin rashes and allergy. The essential oil extracted from the wood is used to treat eczema and scabies.

Botanical Name of the Plant with Voucher Number	*Family Name/ Vernacular Name*	*Plant Parts Used/ Preparations/Uses/Ailments Treated*
Saraca asoca (Roxb.) Wilde VOCB6669	Caesalpiniaceae/ Asokam	50 g of dried flowers and leaves of *Lawsonia inermis* are boiled in coconut oil and the extract is applied externally twice a day o treat eczema and scabies.
Scleria lithosperma (L.) Sw VOCB6670	Cyperaceae/ Kathipul	10 g of the dried tuberous rhizome made into paste with water is externally applied twice a day to treat eczema, leucoderma and scabies.
Scoparia dulcis L. VOCB6671	Scrophulariaceae/ Kattu kothamalli	The leaf paste is used to treat ringworm infection.
Smilax zeylanica L. VOCB6672	Smilaceae/ Sathavari	The root tuber paste is used to treat ringworm infection.
Solanum nigrum L. VOCB6673	Solanaceae/ Manathakkali	10 g of leaf paste is externally applied twice a day to treat allergy, rashes and skin inflammation.
Solanum surattense Burn. VOCB6674	Solanaceae/ Kandan Kathiri	The leaf paste is used to treat scabies and ringworm infection.
Sphaeranthus indicus L. VOCB6675	Asteraceae/ Adakkamanian	The leaf extract is mixed with water and used for taking bath to treat scabies.
Spilanthes calva DC VOCB6678	Asteraceae/ Kalamputry	10 g of leaves made into paste is externally applied twice a day to treat scabies and ringworm infection. The leaves are boiled in coconut oil and the oil extract is used to treat dandruff.
Terminalia paniculata Roth VOCB6679	Combretaceae/ Maruthu	Handful of tender leaves is made into paste with lemon juice. The paste kept in a copper vessel for one full day is applied on the scalp in the evening hours to treat alopecia.
Thespesia lampas (Cav.) Dalz. ex Dalz. and Gibs VOCB6680	Malvaceae/Kattu paruthi	The leaf paste is used to treat ringworm infection.
Thespesia populnea (L.) Soland. Ex Correa VOCB6681	Malvaceae/ Poovarasu	100 g of dried bark is boiled in water and the decoction is externally applied thrice a day to treat leprosy and scabies.
Thottea siliquosa (Lam.) Ding Hou VOCB6682	Aristolochiceae/ Kuravan kanda mooli, kuttrilai vayanam	10 g of dried roots pounded with equal amount of roots of *Rauvolfia serpentina* and *Aristolochia indica* is boiled in coconut oil and this extract is externally applied twice a day to treat eczema, scabies and ringworm infection.
Tiliacora acuminata (Lam.) Hook.f. and Thoms VOCB6683	Menispermaceae/ Karpukarasi	The leaf is made into paste and externally applied to treat wounds.
Trichopus zeylanicus Gaertn. subsp. *travancoricus* VOCB6684	Dioscoriaceae/ Arokya pachilai	10 g of fresh leaf paste is externally applied twice a day to treat scabies and ringworm infection.

Contd...

Botanical Name of the Plant with Voucher Number	*Family Name/ Vernacular Name*	*Plant Parts Used/ Preparations/Uses/Ailments Treated*
Tridax procumbens L. VOCB6685	Asteraceae/ Vettukaya poondu	The young leaves is made into paste and applied externally to treat wounds.
Trichodesma indicum (L.) R.Br. VOCB6686	Boraginaceae/ Kavuthumbai	The paste of leaves along with garlic and rhizome of *Acorus calamus* is applied topically to heal wounds.
Vernonia cinerea (L.) Less. VOCB6687	Asteraceae/ Kucharipoo	Handful of leaves are pounded and boiled in coconut oil. The oil extract is externally applied trice a day to treat leprosy and scabies.
Vetiveria zizanioides (L.) Nash VOCB6688	Poaceae/Vetiver	The decoction prepared from root is mixed with water and used for taking bath to treat dandruff.
Wrightia tinctoria (Roxb.) R. Br. VOCB6689	Apocynaceae/ Veppalai	100 g of leaves are pounded and boiled in 200 ml of coconut oil. The red coloured oil extract is applied externally trice a day to treat eczema and scabies. The bark paste is used to treat various skin infections.
Zizyphus rugosa Lam. VOCB6690	Rhamnaceae/ Thodalli	Handful of leaves ground into paste is applied externally on the skin before morning both to treat scabies and ringworm infection.

Discussion

The present study revealed that a total of 109 plants belonging to 59 families have been documented for their therapeutic use against skin diseases. Among them 41 per cent were herbs, 20 per cent were shrubs, 17 per cent were climbers and 22 per cent were trees (Figure 14.2).

Among the documented medicinal plants, 43 per cent belong to Polypetalae, 28 per cent belong to Gamopetalae and 12 per cent belong to Monochlamydeae and 17 per cent belong to Monocotyledons (Figure 14.3). The most commonly represented families are Ceasalpiniaceae (6), Asteraceae (5), Mimosaceae (5), Menispermaceae (5), Euphorbiaceae (4), Fabaceae (4), Apocynaceae (3), Aristolociaceae (3), Cucurbitaceae

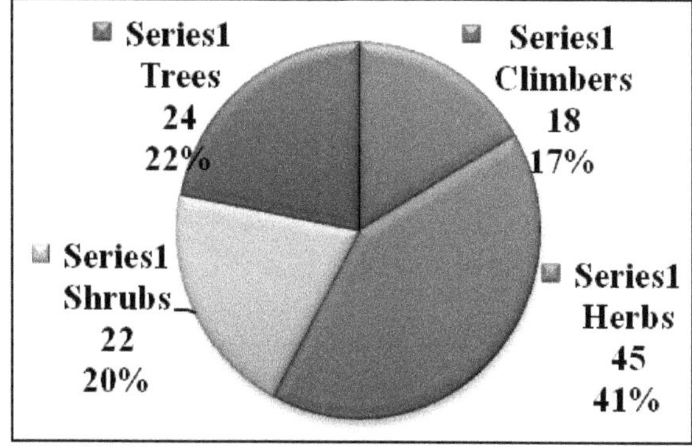

Figure 14.2: Life Forms of Reported Medicinal Plants Used for Various Skin Diseases.

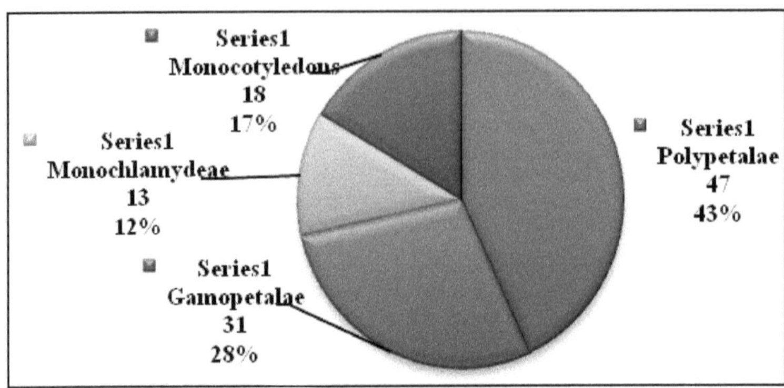

Figure 14.3: Distribution of Reported Medicinal Plants Used for Various Skin Diseases across Sub-class.

(3), Malvaceae (3), Rubiaceae (3), Liliaceae (3), Lamiaceae (3), Araceae (3) and Zingiberaceae (3).

The plant parts used ranged from leaves (52 per cent), root (10 per cent), whole plant (9 per cent), stem bark (9 per cent), rhizome (6 per cent), fruits (3 per cent), flowers (3 per cent), stem (3 per cent), latex (2 per cent), resin (2 per cent) and bulb in one plant (1 per cent) (Figure 14.4). Medicines were prepared in the form of paste, oil extract, powder, decoction and infusion; these plants are used in all forms. The latex is used as such in two plants. External application is involved in the treatment of various types of skin diseases.

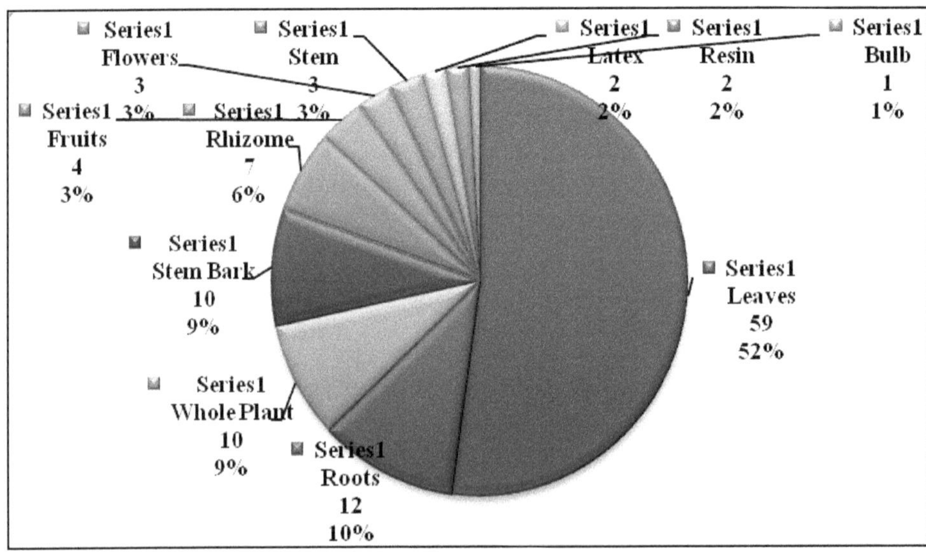

Figure 14.4: Percentage of Plant Parts Used in the Preparation of Medicines.

The investigated 109 plants treat/cure as many as 15 different types of skin diseases. A maximum of 46 plants are used to treat ringworm infection, followed by 48 plants for scabies, 24 plants for eczema, 9 plants for leprosy, 18 plants for cuts and wounds, 7 plants each for dandruff and skin inflammation, 3 plants for alopecia, 4 plants each for healing the cracked foot, sebaceous, cysyts, polyps, acnes and prickle heat. 3 plants are used to treat bites and to promote hair growth. 1 plant is exclusively for treating burns. It is noted that 41 plants are found to be used by the *Kanikkars* to treat only one type of skin disease, 38 plants to treat two skin diseases, 29 plants to treat 3 or 4 skin ailments and 3 plants namely *Azadirachta indica, Curcuma aromatic* and *Curcuma longa* have many fold usage.

The plants which are routinely used as herbal drugs by the tribal community for the treatment of skin diseases also have the healing properties for other ailments with different modes of preparation and administration. The tribal people for skin diseases exclusively use only 35 plant species. In the surveyed plants some plants are used alone and some plants are used in combination. Some of the identified medicinal plants such as *Calanthe masuca, Curculigo orchioides, Hemidesmus indicus* var. *indicus, Kaempferia galangal, Rauvolfia serpentina, Saraca asoca, Trichopus zeylanicus* var. *travancoricus* and *Thottea siliquosa* are endemic and endangered plant species of the study area (Jai and Rao, 1983; Ahmeddullah and Nayar, 1986; Nayar and Sastry, 1987; Pushpagandan and Arogyappacha, 1988; Nayar, 1996). It is well established that an identical use of same plant by different tribal groups indicates its curative property and therapeutic significance (Jain and Saklani, 1992). The comparison of medicinal claims between different cultural groups in the same region or neighbouring regions has proved very rewarding. The comparative studies on medicinal uses of plants among different cultural groups showed similarities and dissimilarities in uses. In the present investigation, the medicinal claims emanating from the *Kanikkars* are compared with claims from three other numerically predominant Indian tribal groups living in the central and western zones of the country *viz.*, the Bhils, Gonds and Santals and two other predominant tribals of Tamil Nadu namely the Palliyar and Irulars. The sources of ethnomedicinal information for the above mentioned ethnic groups are selected from the available literature (Terafter, 1984; Shah and Gopal, 1985; Sur *et al.*, 1987; Sur *et al.*, 1990; Pal *et al.*, 1989; Katewa and Arora, 1997; Girach *et al.*, 1999; Muthukumarasamy, 2003; 2004; Sahoo and Bahali, 2003). This study brings to light that *Wrightia tinctoria* and *Eclipta prostrata* are administrated by the Bhils; *Andrographis paniculata, Anisomeles malabarica, Argemone mexicana, Azadirachta indica, Boerhavia diffusa, Calotropis gigantea, Cassia fistula, Commelina benghalensis, Curculigo orchioides, Heliceres isora, Hemidesmus indicus, Solanum nigrum, Vernonia cinerea* are administered by the Gonds; *Acacia pennata, A. nilotica, Calotropis gigantea, Clitoria ternatea, Eclipta prostrata, Gloriosa superba, Leucas aspera, Mimosa pudica, Santalum album, Sphaeranthus indicus, Thespesia populnea, Wrightia tinctoria* by the *Palliyars* and *Acalypha indica, Aristilochia indica, Azadirachta indica, Cleome viscosa, Lawsonia inermis, Leucas aspera* and *Tridax procumbens* by the Irulars to treat skin diseases. The choice of medicinal plants depend on several factors, such as the alternatives available, the reputation of a recipe, past experience in the family, access to the plant and whether it is used singly or in mixture. These perceptions vary from community to community and even from person to person. Now, for the first time, *Kanikkars* use *Acacia caesia, Calanthe masuca, Calophyllum inophyllum,*

Canthium parvifolium, Cassia klenii, Coscinium fenestratum, Costus speciosus, Cyclea peltata, Diploclisia glaucescens, Ficus benghalensis var. *benghalensis,Heliotropium indicum, Merremia tridentata, Mussaenda glabrata, Saraca asoca* as medicine for the treatment of skin diseases. The medicinal plants used by the *Kanikkars* need to be systematically screened by phytochemists and pharmacologists for the potent active principles. Scientific validation of these remedies may help in discovering new drugs from these medicinal plants.

This study revealed that medicinal plants still play a vital role in the primary healthcare of this tribal community. Traditional medicines also have the potential to form the basis of pharmaceutical drugs for the treatment of a range of diseases. Thus, the loss of these potentially valuable genetic resources ultimately affects the whole society. This could help in creating mass awareness regarding the need for conservation of such plants and also posing ethnic clues for the development of dispensable medicines. There is an urgent need to document the knowledge or otherwise it would be lost forever.

Reference

Ahmeddullah, M. and Nayar, M.P. Endemic plants of the Indian region. Vol.I, Penninsular India, Botanical Survey of India. Calcutta (1986).

ASSISI. ASSISI Forum Survey; Tharankonam, Kanyakumari District, India. (2005)

Ayyanar, M. and Ignacimuthu, S. Medicinal plants used by the tribals of Tirunelveli hills, Tamil Nadu to treat poisonous bites and skin disease. *Ind. J.Trad.Know.* 4:229 (2005).

Ayyanar, M. and Ignacimuthu, S. Traditional knowledge of Kani tribals in Kouthalai of Tirunelveli hills, Tamil Nadu. *J. Ethnopharmacol.* 102:246-255 (2005).

Chopra RN., Nayar,S.L. and Chopra IC. Glossary of medicinal plants of India, Publication and information, Directorate, New Delhi, (1956).

Farnsworth,N.R. Ethnopharmacology and drug development. In: *Ethnobotany and search for new drugs,* (ed.) Chadwick DJ and March V, Ciba Foundation Symposium, 183, Wiley, Chichester, p. 42, (1994).

Ganesan,S., Suresh,N. and Kesavan,L. Ethnomedicinal survey of lower Palani Hills of Tamil Nadu. *Indian J.Traditional Knowledge.* 3:229-304, (2004).

Girach,R.D., Singh, S., Brahman,M. and Misra,M.K. Traditional treatment of skin diseases in Bhadak district, Orissa. *J.Econ.Tax.Bot.* 23:499-504, (1999).

Hamilton, A.C. Medicinal plants, conservation and livelihoods. *Biodiversity Conservation.* 13: 1477-1517, (2004).

Ignacimuthu, S., Sankarasivaraman, K. and Kesavan, L. Medico-ethnobotanical survey among *Kanikkar* tribals of Mundanthurai Sanctuary, Western Ghats, India. Fitoterapia. LXIX: 404-414, (1998).

Jain, S.K. and Saklani, A. Cross-cultural ethnobotanical studies in Northeast India. *Ethnobotany.* 4: 25-38, (1992).

Jain,S.K. and Rao,R.R. An assessment of threatened plants of India, In: Proceedings of the seminar held at Dehradun, September (1981). BSI, Botanical Garden, Howrah (1983).

Katewa, S.S. and Arora, A. Some plants in folk medicine of Udaipur district (Rajasthan). *Ethnobotany*. 9: 48-51, (1997).

Maruthupandian, A. and Mohan, V.R. Ethnomedicinal plants used by Palliyars in Sirumalai Hills, Western Ghats, Tamil Nadu for the treatment of various poisonous bites. *International Journal of Current Trend in Research*. 1: 87-93 (2012).

Muthukumarasamy,S.,Mohan,V.R. and Kumaresan,S. Herbal plants used by Palliyars to obtain relief from gastro-intestinal complaints. *J.Econ.Tax.Bot*. 27: 711, (2003).

Muthukumarasamy,S.,Mohan,V.R. and Kumaresan,S. Herbal remedies of Palliyar tribe of Grizzled Giant Squirrel Wildlife Sanctuary, Western Ghats, Srivilliputhur, Tamil Nadu for poisonous bites. *J.Econ.Tax.Bot*. 27: 761-764, (2003).

Muthukumarasamy,S.,Mohan,V.R. and Kumaresan,S. Medico-ethnobotany of Palliyar tribe in Grizzled Giant Squirrel Wildlife Sanctuary, Western Ghats, Srivilliputhur, Tamil Nadu, India. *J.Medi.Arom.Plant.Sci*. 26: 507-516, (2004).

Nayar, M.P. and Sastry, A.R.K. Red Data Book of Indian Plants. Vol. I, II and III. Published by the Director, Botanical Survey of India, Calcutta, (1987).

Nayar, M.P. Hotspots of endemic plants in India, Nepal and Bhutan. Tropical Botanic Garden and Research Institute, Palode, Thiruvananthapuram. The Ministry of Environment and Forest, New Delhi, (1996).

Pal,D.C., Saren,A.M. and Sen,R. Lesser known uses of twenty plants from the tribal areas of Bankura district, West Bengal. *J.Econ.Tax.Bot*. 13: 695-698.

Prasad, P.N., Jebadhas, A.W. and Janaki Ammal, E.K. Medicinal plants used by the Kannikars of South India. *J.Econ.Tax.Bot*. 11:149-155, (1987).

Pushpagandan, P. and Arogyappacha. The gensing of Kani tribes of Agastyar hills (Kerala) for evergreen health and vitality. *Anc. Sci.Life*. 7:13-16, (1988).

Rajendran,S.M. and Agarwal,S.C. Medicinal plants conservation through sacred forests by ethnic tribals of Virudhunagar district, Tamil Nadu. *Indian Journal of Traditional Knowledge*. 6: 328-333, (2007).

Sahoo, A.K. and Bahali,D.D. Plants used in less known ethnomedicines and medico-religious beliefs in Phulbani district of Orissa. *J.Econ.Tax.Bot*. 27: 500-504, (2003).

Shah,G.L. and Gopal,G.V. Ethnomedicinal notes from the tribal inhabitants of the North Gujarat (India). *J.Econ.Tax.Bot*. 6:193-201, (1985).

Sullivan, K. and Shearly, C.N. *Complete Natural Home Remedies*. Element Books Limited, Shaftesbury.UK. p 3, (1997).

Sur, P.R., Sen, R., Halden, A.C. and Bandyopadhyay, S. Observation on the ethnobotany of Malda –West Dinajpur districts, West Bengal – I. *J.Econ.Tax. Bot*. 10: 395-401, (1987).

Sur, P.R., Sen,R., halden,A.C. and Bandyopadhyay, S. Observation on the ethnobotany of Malda –West Dinajpur districts, West Bengal – II. *J.Econ.Tax. Bot.* 10: 395-401, (1990).

Terafter, C.R. Medicinal plants traditionally used by the tribals of Ranchi and Hazaribagh districts, Bihar. Skin diseases and sores. *Bull. Bot. Surv. India.* 26: 149-153, (1984).

Thurson. E. *Castes and Tribes of Southern India.* Vol. III. Cosmo Publication, New Delhi, India. 25, (1975)

Viswanathan, M.B., Premkumar, E.H. and Ramesh, N. Ethnobotany of the Kanis in the Kalakkad-Mundanthurai Tiger Reserve (KMTR), Tamil Nadu, India-II: Ethnobotany. 13:60-66, (2001).

Viswanathan, M.B., Premkumar, E.H. and Ramesh, N. New ethnomedicine to the literature from the Kanis in (KMTR) of Tamil Nadu in India. *Ind. J.Trop.Med. Pl.* 3:205-212, (2003).

2018, Ethnomedicinal Plants: A Biodiversity Treasure *Pages* **381–387**
Editors: ***V.R. Mohan, A. Doss, P.S. Tresina and V. Sornalakshmi***
Published by: **ASTRAL INTERNATIONAL PVT. LTD., NEW DELHI**

Chapter 15

Ethnomedicinal and Antimicrobial Activities of some Mangrove Species of Godavari Estuary: A Review

G.M. Narasimha Rao

Department of Botany, Andhra University,
Visakhapatnam – 530 003, Andhra Pradesh

Introduction

Mangroves are salt tolerant ecosystems found mainly in the tropical and subtropical intertidal regions of the world. Mangrove ecosystems of the globe are divided into two categories, *viz*., the old world mangroves and the new world mangroves. Chapman (1976) reported the existence of 68 species of mangroves in world wide. Saenger *et al*. (1983), observed 6 distinct mangrove regions in the world mangrove ecosystem. According to FAO (2007), the most extensive mangrove areas are found in Asia, followed by Africa and North and Central America. Duke (1992) has reported 68 true mangrove plant species in the world mangrove species. Mangrove populations of the Godavari estuary have been studied by several investigators (Mathuda,1959; Rao, 1959; Sidhu, 1963, Raju,1968; Blasco,1975; Umamaheswara Rao and Narasimha Rao,1988, Bhaskara Rao *et al*., 1992; Narasimha Rao and Dora, 2009; Narasimha Rao and Subba Rangaiah,2010; Narasimha Rao and Murty,2010 a; 2010b; Narasimha Rao, 2012 and Narasimha Rao et,al, 2012). Narasimha Rao and Murty (2014) presented the ethnomedicinal importance of some important mangrove species present in different estuarine habitats of the Andhra Pradesh. Mangroves and halophytes are of great importance to many people who live along the tropical shorelines. Extracts and chemical compounds from some mangrove species are used mainly in folklore medicine in a wild range of ailments for the mankind.

Mangrove plant extracts have been used for centuries as popular method for treating several health disorders. Plant-derived substances have recently become an area of great interest owing to their versatile applications. Several workers have reported the usefulness of mangrove plants in traditional medicine (Premanathan *et al.*, 1992; Premanathan *et al., 1996*; Kokpal *et al.*, 1990) According to the review of Jun *et al.* (2008) there are about 349 metabolites isolated from mangrove species out of which 200 metabolites are reported exclusively from true mangrove plants. Several mangrove species are used in traditional medicine or have found application as insecticides and pesticides The mangrove plants possess a number of biological activities such as antibacterial, antioxidant, anticancer, cytotoxic, antiproliferative, insecticidal, antimalarial, antifungal, antifeedant, antidiarrheal, central nervous system depressant, antimitotic, antileukemic and anti-plasmodial activities. Information on the biological activities of the mangrove plants have been provided by a number of review and research articles (Bandaranayake, 1998; 2002; Bose and Bose, 2008; Jun *et al.*, 2008; Chandrasekaran *et al.*, 2009; Patra *et al.*, 2009a and 2009b).In the present communication, data is summarized on the ethnomedicinal uses, anti bacterial and anti fungal activities of some important mangrove species.

Ethnomedicinal Importance of Mangroves of Godavari Estuary

Different species of mangrove, associated mangroves and halophytes were utilized by the local inhabitants along the estuarine regions is a regular phenomenon to find out the solutions for their common ailments and disorders of their cattle. Narasimha Rao and Murty (2014) summarized the medicinal applications of some important mangrove species of the Godavari estuary.

Besides, local inhabitants of the Godavari estuary used these mangrove species in various applications such as fodder and in folklore medicine. Extracts from mangroves and associated mangroves have shown the strong remedies against the human, animal, plant pathogens, Skin disorders and leprosy treated with bark of mangrove species such as *Rhizophora mucronata*. Local people used these species in the treatment of asthma, backache, boils, constipation, diarrhea, dysentery, dyspepsia, hemorrhage, inflammation, jaundice, kidney stones, malaria, rheumatism, snakebites and sores.

Antimicrobial Activity of Mangroves Species in Godavari Estuary

Antimicrobial activity of *A. marina* was studied by Bobbarala *et al.* (2009a). In this study antimicrobial activity of aerial parts was evaluated against the pathogens belonging to aquatic, human and plant origin. Among all microorganisms studied *Erwinia caratovara* and *Streptococcus mutans* showed the considerable growth inhibition with chloroform and methanolic extracts.Bobbarala *et al.* (2009b) conducted *invitro* antimicrobial screening of mangrove plant *Avicennia officinalis*. The results showed that *A. officinalis* extracts exhibited antimicrobial activities at a concentration of 20mg/ml. Mature leaf extracts of *A. officinalis* in methanol exhibited promising antimicrobial activity than other solvent extracts. Phytochemical screenings revealed that mature leaf of A. officinalis contain alkaloids, steroids, triterpenoids and flavonoids. Bobbarala *et al.* (2009c) worked on *Sonneratia apetala* antimicrobial activity on selected pathogenic microorganisms. The antimicrobial activity of the mangrove plant extract of *S. apetala* on the various test microorganisms, including clinical multiple antibiotic resistant bacteria and phytopathogens were investigated. *In vitro* screening of *S. apetala* mangrove plant extracts of showed species specific activity in inhibiting the growth of bacteria and fungi. Hexane, chloroform and methanol extracts showed good activity against all the pathogens, where as methanolic extracts were active against most of the pathogens.

Table 15.1: Mangrove Species Used by the Local People for Treatment of various Disorders

Sl.No.	*Name of the species*	*Family*	*Vernacular Name*	*Parts used*	*Benefits*
1	*Acanthus ilicifolius*	Acanthaceae	Aalchi	Leaf fruit pulp	Rheumatism and hair preserver blood purifier, dressing for boils and snake bites
2	*Avicennia officinalis*	Avicenniaceae	Nalla mada	Fruit Leaf	Used to treat boils wounds in cattle
3	*Acicennia marina*	Avicenniaceae	Tella mada	Bark Fruit	Used for curing skin diseases and as fish food
4	*Avicennia alba*	Avicenniaceae	Pasupu mada	Resinous exudates Seed ointment	Used in birth control and to get relief from small pox
5	*Ceriops decandra*	Rhizophoraceae		Bark decoction	Used to arrest hemo-rrhage
6	*Luminitzera racemosa*	Combretaceae	Thanduga	Leaf decoction	To treat piles
7	*Rhizophora mucronata*	Rhizophoraceae	Ponna	Bark	To treat hematoma, diarrhea
8	*Rhizophora apiculata*	Rhizophoraceae		Leaf and Bark	To treat dysentery, leprosy and to cure diabetes.
9	*Sonneratia caseolaris*	*Lythraceae*	Chinna Kalinga	Fruit Leaf Juice	Edible and is used to check hemorrhage.
10	*Sonneratia apetala*	*Lythraceae*	Pedda Kalinga	Fermented juice	To check hemor-rhage
11	*Xylocarpus granatum*	Meliaceae	Kalinga	Bark Seed paste	Used to cure diarrhea and dysentery. Used to treat boils and mumps

Vadlapudi and Naidu, (2009a) worked on bioactivity of mangrove plant *Avicennia alba* on selected plants and oral pathogens. In that study *Avicennia alba* was screened for antimicrobial activity of some clinical and pathogens. The experimental results concluded that plant extracts of *A. alba* have greater potential as antimicrobial compounds against microorganisms and that they can be used in the treatment of infectious diseases caused by resistant pathogenic microorganisms. Vadlapudi and Naidu, (2009b) studied *in vitro* antimicrobial activity *Ceriops decandra* against selected aquatic, human and phytopathogens. Comparative antimicrobial activities studies on hexane, chloroform and methanol extracts of *C. decandra* mangrove medicinal plant was carried out and extracts of *C. decandra* showed prominent antimicrobial activities, while chloroform and hexane extracts show very less or no antimicrobial activity. Vadlapudi and Naidu, (2009c) studied *in vitro* antimicrobial potentiality of mangrove plant *Myriostachya wigtiana* against selected phytopathogens. In their study antimicrobial activity of *M. wigtiana* organic solvent extracts on the various test microorganisms, including bacteria and fungi were investigated. Methanol extracts exhibited promising antimicrobial activity than chloroform and hexane extracts. Among all tested microorganisms highest activity was showed against *B. bicolor*-MTCC 2105 (18mm) with MIC 75mg/ml) followed by *Curvularia lunata* (15mm), *Erwinia carotovora* (15mm) and *Pseudomonas marginales* (16mm), whereas lowest activity was found against *Cladosporum herbarum* and *M. phaseolina* with all extract concentrations (100, 300 and 500mg/ml).

The same researchers vadlapudi and Naidu, (2009d) also evaluated *in vitro* bio efficiency of *Rhizophora conjugata*. In their study antimicrobial activity of *R. conjugata* with organic solvent extracts on the various test microorganisms, including bacteria and fungi were investigated. Methanol extracts exhibited promising antimicrobial activity than chloroform and hexane extracts. Among all tested microorganisms *L. acidophilus* (22mm) showing highest susceptibility followed by *Streptococcus salivarius* (19mm), *Aeromonas hydrophila, Streptococcus mutans* and lowest activity was found with *Cladosporium herbarum, Fusarium oxysporum, Streptococcus anginosus* and *S. mitis* with concentration 100mg/ml. This study, has to some extent, validated the medicinal potential of the mangrove plants. Vadlapudi *et al.* (2009) conducted the comparative screening of selected mangrove and halophytic plant methanolic extracts against clinical and plant pathogens and also the antimicrobial effect of the crude methanol extract of mangrove plants *Tamarix aphylla, Sesuvium portulacastrum* and *Xylocarpus granatum*. Screening revealed that the crude methanol extract of the studied mangrove extracts possess antimicrobial activity against most of the test organisms depending upon the nature of their active ingredients in the extract and capacity of diffusion into the agar medium. Among the test organisms, the extract showed significant antimicrobial activity against *Acremonium strictum, Aspergillus niger, Candida albicans, Ervinia carotovara.* Deepthi Rani, Narasimha Rao, (2013) studied the antimicrobial activity of *Avicennia officinalis* against some clinical and phytopathogens. Plant parts such as leaves and bark of *A. officinalis* were collected from Coringa forest near Kakinada area. The antimicrobial activity of the plant extracts on the various test microorganisms, including multiple antibiotic resistant bacteria, was investigated. The experimental results concluded that the plant extracts of *A.officinalis* have greater potential as antimicrobial compounds against microorganisms and that they can be used in treatment of infectious diseases caused by resistant pathogenic microorganisms. Prasanna Lakshmi and Narasimha Rao (2013) studied the antimicrobial activity of the halophyte *Suaeda monoica* using leaf and shoot extracts on various test microorganisms, including multiple antibiotic resistant bacteria and phytopathogens. The experimental results concluded that the hexane, methanol and water extracts of *S. monoica* leaves have greater potential as antimicrobial compounds against microorganisms and they can be used in the treatment of infectious diseases caused by resistant pathogenic microorganisms.

References

Bandaranayake, W. M. 1998. Traditional and Medicinal Uses of Mangroves. *Mangroves and Salt Marshes,* 2(3): pp. 133-148.

Bandaranayake, W. M. 2002. Bioactivities, Bioactive Compounds and Chemical Constituents of Mangrove Plants. *Wetlands Ecol. and Management,* 10(6): pp. 421-452.

Blasco, F., 1975. *The Mangroves of India,* Pondicherry, French Institute, Fr.Sect.Sci. Tech.14: 1-175.

Bhaskara Rao, V., Narasimha Rao, G.M., Sarma, G.V. S. and Krishna Rao, B.,1992. Mangrove and its sediment characters in Godavari estuary, east coast of India. *Indian J.Mar.Sci.* 21: 64-66.

Bobbarala, V., Vadlapudi, V. and K. Naidu. 2009a. Antimicrobial Potentialities of Mangrove Plant *Avicennia marina. J. Pharm. Res.* (6): pp. 1019-1021.

Bobbarala, V., Vadlapudi, V. and K. Naidu. 2009b. *In vitro* Antimicrobial Screening of Mangrove Plant *Avicennia officinalis. Oriental J. Chem.,* 25 (2): pp. 373-376.

Bobbarala, V., Vadlapudi, V. and K. Naidu. 2009c. Mangrove Plant *Sonneratia apetala* Antimicrobial Activity on Selected Pathogenic Microorganisms. *Oriental J. Chem.* 25 (2): pp. 445-447.

Bose, S. and A. Bose. 2008. *Indian J. Pharma. Sci.,* 70, pp: 821- 823.

Chandrasekaran, M., Kannathasan, K., Venkatesalu, V. and K. Prabhakar. 2009. Antibacterial Activity of Some Salt Marsh Halophytes and Mangrove Plants against Methicillin Resistant *Staphylococcus aureus. World J. Microbiol. Biotechnol.,* 25: pp: 155–160

Chapman, V. J. 1976. Mangrove Vegetation. *J. Cramer, Vaduz,* Liechtenstain. pp. 1-326.

Deepthi Rani S. and G. M. Narasimha Rao. 2013. Antimicrobial Activity of Mangrove Plant *Avicennia officinalis* (Lam. Briqvet) on Selected Pathogens. *Res. J. Pharma. Biol. and Chem. Sci.,* 4(3): pp. 335-341.

Kokpol, U., Chavasiri, W., Chittawong, V. and D. H. Miles. 1990. Taraxeryl Cis Phydroxycinnamate, a novel taraxeryl from *Rhizophora apiculata. J. Natural Products,* 53(4): pp. 953-955.

Jun De Chen, Dan Qin Feng, Zhi Wei Yang, Zhan Chang Wang, Yan Qiu and Yi Ming Lin. 2008. Antifouling Metabolites from the Mangrove Plant *Ceriops tagal. Molecules,* 13: pp. 212-219.

Mathuda, G.S., 1959. Mangrove vegetation of Godavari delta. In: Proceedings of the mangrove symposium, Calcutta, 66.

Narasimha Rao, G. M. 2012. Distribution of Mangroves and Associated Flora of Andhra Pradesh. In: Biodiversity of Aquatic Resources, Daya Publishing House. New Delhi. pp 29-49.

Narasimha Rao, G. M. 2014. Distribution, Density and Conservation of Mangroves in the Godavari Estuary, Andhra Pradesh, India. *J. Biol. Chem. Res.*, 31(2): pp. 614-622

Narasimha Rao, G. M. and G. Subba Rangaiah. 2010. Distribution of Mangroves and Associated Flora of the Pandi Back Waters of Gautami Godavari Estuary. *Anu J. Nat. Sci.*, 2(1): pp. 22-26.

Narasimha Rao, G. M. and P. P. Murthy. 2010a. Mangroves and Associated Flora of Vashista and Vainateyam Estuaries. Andhra Pradesh, India. *J. Not. Sci. Biol.*, 2(4): pp. 40-43.

Narasimha Rao, G. M. and P. P. Murthy. 2010b. Mangrove Populations of Vamsadhara Estuary. *Int. J. Plant Sci.*, 5: pp. 698-699.

Narasimha Rao, G.M. and Pragaya Murty,P. 2014 Survey and documentation of some important medicinal applications of Mangrove plants of Andhra Pradesh, India.The Journal of Ethnobiology and Traditional Medicine. Photon 122. 842-847.

Narasimha Rao, G. M. and S. V. V. S. N. Dora. 2009. An Experimental Approach to Mangroves of Godavari Estuary. *Indian J. Forestry.* 32(3): pp. 263-265. Premanathan, M., K. Chandra, S. K. Bajpai and K. Kathiresan. 1992. A survey of Some Indian Marine Plants for Antiviral Activity. *Bot. Mar.*, 35: pp. 321-324.

Patra, J. K., Mohapatra, D. A., Rath, S. K., Dhal, N. K. and H. N. Thatoi.n2009a. Screening of Antioxidant and Antifilarial Activity of Leaf Extracts of *Excoecaria agallocha* L. *Int. J. Integr. Biol.* 7(1): pp.9–15.

Patra, J. K., Panigrahi,T. K., Rath, S. K., Dhal, N. K. and H. N. Thatoi. 2009b. Phytochemical Screening and Antimicrobial Assessment of Leaf Extracts of *Excoecaria agallocha* L.: a Mangal Species of Bhitarkanika, Orissa, India. *Adv. Nat. Appl. Sci.*, 3(2): pp. 241–246.

Prasanna Lakshmi and G. M. Narasimha Rao. 2013. Antimicrobial Activity of Suaeda monoica (Forssk ex Geml.) against Human and Plant Pathogens. *Res. J. Pharma. Biol. and Chem. Sci.*, Vol. 4(2): pp. 680- 685.

Premanathan, M., H. Nakashima, K. Kathiresan, N. Rajendran and N. Yamamoto. 1996. *In-vitro* Antihuman Immunodeficiency Virus Activity of Mangrove Plants. *Indian J. Medi. Res.*, 103: pp. 278-281.

Raju, D. C. S., 1968. The vegetation on West Godavari. A study of tropical delta. In: *Proceedings of the Symposium on Recent Advances in Tropical Ecology* (Part 1) Banaras Hindu University, Varanasi. 348-358.

Rao, R.S., 1959. Observations on the mangrove vegetation of the Godavari estuary. In: *Proceedings of the mangrove symposium*, Calcutta. 36-44.

Saenger,P., Hegerl. E.J., and Davie, J.D.S.,1983. *Global status of mangrove ecosystems.* Commission on Ecology Papers No.3. Gland, Switzerland, World Conservation Union (IUCN).

Sidhu, S.A.,1963.Studies on the mangroves of India 1. East Godavari region. *Indian Forester,* 86: 337-351.

Vadlapudi, V. and K. Naidu. 2009a. Bioactivity of Marine Mangrove Plant *Avicennia alba* on Selected Plant and Oral Pathogens. *Int. J. Chemtech Res.,* 1(4): pp. 1213-1216.

Vadlapudi, V and K. Naidu. 2009b. *In vitro* Antimicrobial Activity *Ceriops decandra* against Selected Aquatic, Human and Phytopathogens. *Int J. Chem. Tech. Res.,* 1(4): pp. 1236-1238.

Vadlapudi, V and K. Naidu. 2009c. *In vitro* Antimicrobial Potentiality of Mangrove Plant *Myriostachya wigtiana* against Selected Phytopathogens. *Biomedical and Pharmacol. J.,* 2(2): pp. 235- 238.

Vadlapudi, V and K. Naidu. 2009d. *In vitro* Bioefficiency of Marine Mangrove Plant Activity of *Rhizophora conjugata. Int. J. Pharm- Tech. Res.,* 1(4): pp. 1598–1600.

Vadlapudi, V, Bobbarala V. and K. Naidu. 2009. Comparative Screening of Selected Mangrove Plants Methanolic Extracts against Clinical and Plant Pathogens. *J. Pharm. Res.* 2(6): pp. 1062-1064.

Umamaheswara Rao, M, and Narasimha Rao, G.M. 1988.Mangrove populations of the Godavari delta complex. *Indian J. Mar. Sci.,* 326-329.

2018, Ethnomedicinal Plants: A Biodiversity Treasure *Pages* **389–405**
Editors: ***V.R. Mohan, A. Doss, P.S. Tresina and V. Sornalakshmi***
Published by: **ASTRAL INTERNATIONAL PVT. LTD., NEW DELHI**

Chapter 16

Hybanthus enneaspermus: A Reliable Herb to Cure different Human Ailments

A. Maruthupandian[1]*, M. Viji[2], B. Perumal[1], S. Velmani[1] and C. Chithravadivu[3]

[1]*Department of Botany, School of Life Sciences, Periyar University, Periyar Palkalai Nagar, Salem- 636 011, Tamil Nadu*
[2]*Department of Botany, Thiagarajar College, Madurai – 625009, Tamil Nadu*
[3]*PG and Research Department of Botany, Vellalar College for Women, Thindal, Erode – 638 012, Tamil Nadu*

ABSTRACT

Hybanthus enneaspermus Linn. belongs to the family violaceae is a perennial herb or small herb found in the warmer parts, from Uttar Pradesh southward to the Deccan Peninsula in India. The whole plant is used in Ayurveda, Siddha and other traditional systems of medicine in India for curing various human diseases and disorders. The different phytochemicals were separated, identified and characterized by many analytical techniques like TLC, HPLC, HPTLC, GC-MS, LC-MS and FTIR. The pharmacological properties of different parts of Hybanthus enneaspermus have been reported from more than two decades different parts of plants can be acts as herbal drug for antioxidant, antidiabetic, anti-arthritic, cardio and neuroprotective activities etc. In this chapter, therapeutic uses, phytochemical, bioactive potential and pharmacological properties of Hybanthus enneaspermus were compiled and emphasized. This information will be helpful to new drug discovery and formulations relating researches on various health care needs.

Introduction

The therapeutically uses and different pharmacological study of medicinal plants have significantly enhanced during last few decades (Al-qura'n, 2005;

Gazzaneo, Paiva de Lucena and Paulino de Albuquerque, 2005). Uses of herbal medicines in Asia have long history with numerous applications against number of diseases (Duraipandian *et al.*, 2006). Traditional healers have been used more than 2500 plant species as a source of medicine in India (Pei, 2001). Herbal medicines play a key role in different health care system of medicine in India and they are acknowledged by WHO. Diverse traditional system of medicine has been used from ancient days for treating various diseases in developed and developing countries across the world. Medicinal plants are called as chemical factories due to lots of phytoconstituents synthesized from them. From the medicinal plants more than hundreds of bioactive components have been identified and used worldwide (Newman *et al.*, 2007). Plants have been utilized in traditional medicine for thousands of years (Abu-Rabia, 2005). The different folklore medicinal practices are employed besides different herbal traditional medicines such as *Siddha, Ayurveda, Unani, Homeopathy* and *Amchi* (Tibetan Medicine). The great number of ethnomedicinal communication remained endemic to particular areas or people due to need for information. In population, India is the second greatest country in the world. India in various forests for over 550 tribal communities is covered under 227 tribal groups residing about in 5000 villages of vegetation categories through word of mouth and is widely utilized for the treatment of familiar ailments (Mishra *et al.*, 2008). Drugs are supported to be of excellent significance in the initial health care of individuals and communities in several developing countries (Alagesaboopathi, 2014).

Figure 16.1: *Hybanthus enneaspermus.*

H. ennaespermus (L.) is an important medicinal herb used from ancient times. This is a native plant of Himalayan region and warmer parts of India. It is also distributed in Sri Lanka, tropical Asia, Africa and Australia. *Hybanthus* is from the Greek *hypos* (hump-backed) and anthos (flower), referring to the spurred anterior petal; *ennaesperus* is from the Greek *ennea* (nine) and *spermus* (seed), referring to the capsule which contains about nine seeds. *H. ennaespermus* popularly called '**Ratanpurus**' (Sanskrit) and in Tamil it is '**Orithaltamarai**'. The unique ethnomedicinal properties of *Hybanthus enneaspermus* is acquainted by villagers and herbalists (Sudharson *et al.*, 2014).

Table 16.1: Vernacular Names of *Hybanthus enneaspermus*

Common Name	Spade Flower, Pink Ladies slipper
Tamil	Orithalthaamarai
Hindi	Ratanpurush
Malayalam	Orithamatai
Kannada	Purusharathna
Telugu	Ratanpurusha
Sanskrit	Rathnapurusha
Bengali	Nunbora
Marathi	Rathanparas

Therapeutic Uses of *Hybanthus enneaspermus*

The plant is very popular for its various therapeutic uses in diverse herbal based medicinal system. The whole plant is very acrid to relieve strangely painful dysentery, vomiting, burning sensation, wandering of the mind (Tripathy *et al.*, 2009). It is a perennial small herb which is used in Ayurveda, Siddha and other traditional medicines for treating ailments. It is also known as "spade flower" and "pink ladies slipper". In Ayurveda, it is called as "**Sthalakamala**" (Rajsekhar *et al.*, 2016). Recent research work proved that its medicinal potential is used against number of dreadful diseases and disorders. *H. enneaspermus* is the first medicinal plant of violaceae family to elaborate peptide alkaloids (Raveendra Retnam and John De Britto, 2007). It has been used as drug on antimalarial, antirheumatic, emmenagogue, sedative, antispasmodic, aphrodisiac and antiasthmatic (Amuthapriya *et al.*, 2011).

The different traditional therapeutic uses of *Hybanthus enneaspermus* are tabulated from previous literature (Table 16.2). According to the previous literature, the different parts of *Hybanthus enneaspermus* has been widely used in diabetes, asthma, jaundice, poisonous bites, malaria, fever, urinary tract infections and some gastrointestinal complaints. As per the survey of literature, the whole plant is mostly used part of the plant to treat various diseases in the form of extract, decoction and paste (Figure 16.2.). From this information, the fruit is exclusively used for various poisonous bites such as scorpion stings and cobra bites (Raja Reddy *et al.*, 1989; Sudarsanam and Sivaprasand, 1995; Shantha *et al.*, 2001; Dahanukar *et al.*, 2000; Onayade *et al.*, 1990).

A small quantity of the herb it taken with milk, twice a day helps the body gaining strength. If it is taken for 48 days, the internal organ of the body grows healthier and stronger. Scientifically, the organs of reproduction are rejuvenated; get stronger and increase the production of sperm. The peduncle of herb gives a mushy, sticky and smooth, taste like ladies finger when chewed. Daily intake helps in curing ulcers, which occurs due to sexually transmitted diseases and even AIDS. This herb is good for regularizing the monthly periods of women (Sudharson *et al.*,

Table 16.2: Therapeutic Uses of *Hybanthus Enneaspermus*

Sl.No.	*Part Used*	*Medicinal Uses*	*Solvent extract*	*References*
1	Whole Plant	Relieve strangers, painful dysentery, vomiting, burning sensation, wandering of the mint.	Ethanol and Petroleum ether extract	Tripathy *et al.*, 2009.
		Memory power enhancing	Extract	Kanka Rajesham *et al.*, 2013.
		Urinary tract infection, gonorrhea and kidney problems.	Methanol extract	Amutha Priya *et al.*, 2011.
		Aphrodisiac, demulcent, diuretic, urinary infections, diarrhea, leucorrhoea, disarray and sterility.	Extract	Yoganarasimhan, 2000.
		Liver functions	Hydro alcoholic extract	Vetriselvan *et al.*, 2013.
		heart disease, chest pain, increase immune power	Extract	Radjassegarin Arumugam *et al.*, 2012.
		acute and chronic liver injury	Aqueous extract	Subramanian and Pushpangadam, 1999.
		Improve memory, vitality, asthma, fever and leprosy	Hexane, ethyl acetate, ethanol and water extract	Dahanukar *et al.*, 2000; Onayade *et al.*, 1990.
		Arthritis	Extract	Tripathy *et al.*, 2009.
		Aphrodisiac and antipyretic	Extract	Alagesaboopathi, 2014.
		Aphrodisiac activity in sexually inactive male rats	Aqueous extract	Narayanswamy *et al.*, 2007.
		Malaria, diabetes, male sterility, gonorrhea, urinary treat infection, jaundice and cholera	Aqueous extract	Sarita *et al.*, 2004; Patel *et al.*, 2011; Kheraro and Bouquet, 1950; Pushpangadam and Atal, 1984; Gopal and Shah, 1985.
		Improve memory and to treat asthma, tuberculosis, leprosy and eye diseases	Aqueous extracts	Udayan and Indira, 2009.

Sl.No.	Part Used	Medicinal Uses	Solvent extract	References
2	Root	Relieve sugar complaints.	Extract	Aswini Kumar Dixit and Sudurshan, 2011.
		Epilepsy and hysteria	Ethanol and Petroleum ether extract	Khandelwal, 2005.
		Diuretic and administered in infections.	Methanol extract	Amutha *Priya et al.*, 2011.
		Bowel complaints of children	Hexane, ethyl acetate, ethanol and water extract	Dahanukar *et al.*, 2000; Onayade *et al.*, 1990.
		Diuretic and gonorrhea	Ethanol extract	Prakash *et al.*, 1999.
		Diuretic, gonorrhea and urinary infections	Ethanol extract	The Wealth of India, 1959; Nagaraju and Rao, 1996.
3	Leaf	Decoction or electuary: mixed with oil.	Methanol extract	Amutha Priya *et al.*, 2011.
		Diabetes as well as development of macrosomnia in fetus	Methanol extract	Awobajo and olatunji-Bello, 2010.
		Asthma	Extract	Shantha *et al.*, 2001.
		Demulcent and decoction or electuary	Hexane, ethyl acetate, ethanol and water extract	Dahanukar *et al.*, 2000; Onayade *et al.*, 1990.
		Antidotes for scorpion stings and cobra bites	Extract	Raja Reddy *et al.*, 1989; Sudarsanam and Sivaprasand, 1995.
4	Fruit	Antidotes for scorpion stings and cobra bites	Extract	Raja Reddy *et al.*, 1989; Sudarsanam and Sivaprasand, 1995.
		Snake and Scorpion sting	Extract	Shantha *et al.*, 2001.
		Scorpion sting	Hexane, ethyl acetate, ethanol and water extract	Dahanukar *et al.*, 2000; Onayade *et al.*, 1990.
5	Flower	Urinary tract infection and water retention	Methyl Chloride extract	Pushpangadan and Atal, 1984.
		jaundice	Aqueous extract	Gopal and Shah, 1985.
		Venomous snake bites	Methanol extract	Majumder *et al.*, 1979.

Contd...

2014). It is used for the treatment of various disorders including urinary infections, diarrhea, leucorrhoea, dysuria, inflammation and male sterility (Dinesh *et al.*, 2013). It has been reported to have anti-inflammatory (Boominathan *et al.*, 2004) antitussive (Boominathan *et al.*, 2003) antiplasmodial (Weniger *et al.*, 2004) anticonvulsant and free radical scavenging activity (Hemalatha *et al.*, 2003).

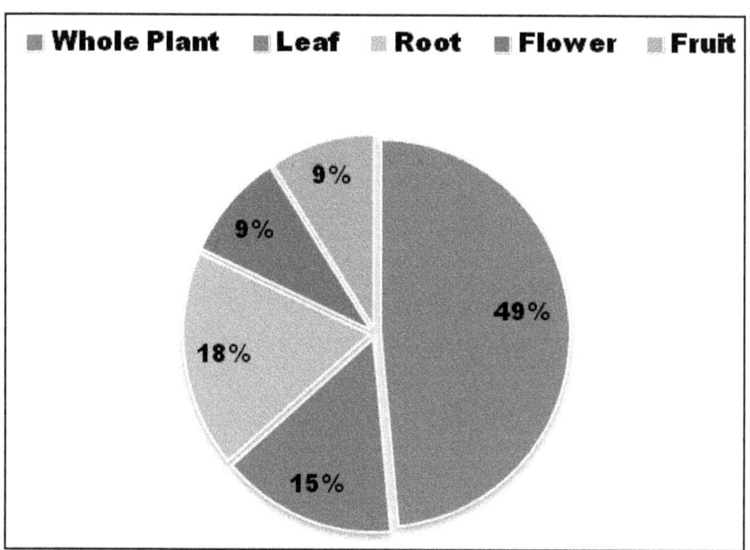

Figure 16.2: Diagrammatic Representation of Therapeutic Uses of *Hybanthus enneaspermus*.

The root sandals are employed for the bowel complaints of children. The leaves and tender stalks are demulcent and used as a decoction or electuary. They are employed in preparing a cooling liniment for head ache. An infusion of the plant is given in the case of cholera and decoction or powder of the whole plant is taken to improve memory, vitality and as a remedy for tuberculosis, asthma, fever and leprosy. Its infusion is good for all diseases of eyes. Fruit is used to treat scorpion sting (Thyagaraju *et al.*, 2015). *H. ennaespermus* is of great importance in traditional system of medicine. The plant is bitter and acrid; easily digested, removes kappa and pita, urinary calculi, stangury, pain, dysentery, vomiting, burning, wandering of the mind, urethral discharge, blood troubles, asthma, epileptic fits, cure cough, give stone to the breast, alexetric (Ayurveda) (Baviya Sampath *et al.*, 2016). The roots of this plant have been indigenously used in epilepsy and hysteria (Khandelwal, 2005). In some part of India, the plant is used to treat diabetes and is also having anti-oxidant activity (Das *et al.*, 2004).

Phytochemicals and Pharmacological Studies on *Hybanthus enneaspermus*

The results of various phytochemical studies of *Hybanthus enneaspermus* has been presented in Table 16.3. The table comprises types of phytochemicals have been reported in different solvent extracts of *Hybanthus enneaspermus*.

Table 16.3: Phytochemicals in *Hybanthus enneaspermus*

Sl.No.	*Extracts*	*Phytochemical compounds*	*References*
1.	Ethanol	Alkaloids, Flavonoids, Tannins, Saponins and Reducing sugar	Dinesh Kumar Patel *et al.,* 2011.
		Alkaloids, Terpenoids, Flavonoids, Tannins, Steroids, Phenols and Reducing sugar	Krishnamoorthy *et al.*, 2014.
		Phytosterols, Saponins, Terpenoids, Alkaloids, Carbohydrates, Flavonoids, Polyphenols and Tannins	Thyaga Raju *et al.,* 2014.
		Alkaloids, Flavonoids, Tannins, Glycosides, Steroids, Phenols, Terpenoids, Carbohydrate, Phlobatannins, Protein and Amino acid	Vennila and Pavithra, 2015.
		Steroids, Flavonoids, Alkaloids, Saponins, Carbohydrates and Terpenoids	Baviya Sampath *et al.*, 2016.
2.	Acetone	Alkaloids, Terpenoids, Tannins and Glycosides	Krishnamoorthy *et al.*, 2014.
		Tannins	Remya Krishnan *et al.*, 2012.
		Alkaloids, Tannins, Glycosides, Steroids, Phenols, Terpenoids, Carbohydrate, Phlobatannins, Resins, Protein and Amino acid	Vennila and Pavithra, 2015.
3.	Aqueous	Terpenoids, Flavonoids, Tannins and Phenols	Krishnamoorthy *et al.*, 2014.
		Phytosterols, Terpenoids, Alkaloids, Carbohydrates, Flavonoids and Polyphenols	Thyaga Raju *et al.,* 2014.
		Alkaloids, Flavonoids, Tannins, Glycosides, Steroids, Phenols, Terpenoids, Saponins, Resins, Carbohydrate, Anthroquinones, Protein and Amino acid	Vennila and Pavithra, 2015.
		Alkaloids, Flavonoids, Steroids, Phenols, Terpenoids, Anthroquinones and Sugar	Remya Krishnan *et al.*, 2012.
		Steroids, Flavonoids, Alkaloids, Saponins, Tannins, Carbohydrates and Terpenoids	Baviya Sampath *et al.*, 2016.
4.	Ethyl acetate	Tannins	Remya Krishnan *et al.*, 2012.
		Phytosterols, Terpenoids, Alkaloids, Carbohydrates and Polyphenols	Thyaga Raju *et al.,* 2014.
		Steroids, Flavonoids, Alkaloids, Carbohydrates and Terpenoids	Baviya Sampath *et al.*, 2016.
5.	Hexane	Phytosterols, Terpenoids, Alkaloids, Carbohydrates and Tannins	Thyaga Raju *et al.,* 2014.
		Steroids, Flavonoids, Alkaloids, Terpenoids and Carbohydrates	Baviya Sampath *et al.*, 2016.
6.	Petroleum ether	Tannins	Remya Krishnan *et al.*, 2012.
		Alkaloids, Flavonoids, Tannins, Glycosides, Steroids, Phenols, Terpenoids, Carbohydrate, Anthroquinones, Protein and Amino acid	Vennila and Pavithra, 2015.

Sl.No.	Extracts	Phytochemical compounds	References
7.	Methanol	Alkaloids, Flavonoids, Tannins, Steroids, Phenols, Terpenoids, Saponins, Anthroquinones and Sugar	Remya Krishnan *et al.*, 2012.
		Alkaloids, Flavonoids, Tannins, Steroids, Terpenoids	Snehal Singh *et al.*, 2015.
8.	Chloroform	Steroids, Flavonoids, Alkaloids, Carbohydrates and Terpenoids	Baviya Sampath *et al.*, 2016.
9.	Hydro-alcoholic	Alkaloids Phytosterols, Flavonoids, Glycosides, Carbohydrates, Saponins, Protein and Amino acid	Vetriselvan *et al.*, 2013

The collected data of phytochemical analysis of *Hybanthus enneaspermus* exhibits alkaloids, flavonoids, terpenoids, tannins, saponins, phenols, steroids and carbohydrates are predominantly presented in different extracts of *Hybanthus enneaspermus*. Generally, medicinal plants are used to cure various human ailments due to the presence of secondary metabolites (Nostro *et al.*, 2000). All the primary and secondary metabolites or phytochemicals are exhibits unique activities. From the previous literature survey, eight phytochemicals were primarily present in different parts of *Hybanthus enneaspermus* such as alkaloids, flavonoids, terpenoids, tannins, saponins, phenols, steroids and carbohydrates. Different biological activity of these eight phytochemicals was compiled from previous reports and showed in Table 16.4.

Table 16.4: Phytochemical's Activity of *Hybanthus enneaspermus*

Sl.No.	Name of the Phytochemical Compund	Activity	References
1	Alkaloids	Anaesthetic agent	Herouart *et al.*, 1988
		Antimicrobial, antiparasitic and anti HIV	Bouayad *et al.*, 2011
		Antiarrhythmic, anticholinergic stimulant, adenosine receptor antagonist cough medicine, analgesic, remedy for gout, antiprotozoal agent, sympathomimetic, vasodilator, antihypertensive, analgesic stimulant, antiarrhythmic, antipyretics, antimalarial antihypertensive muscle relaxant, antitumor vasodilating, antihypertensive stimulant, aphrodisiac	Neha Babbar, 2015
2	Flavonoids	Antimicrobial, Antiinflammatory and Antioxidant	Anand and Gokulakrishnan, 2012
3	Terpenoids	Antiviral, antibacterial, anti-inflammatory, anticancer, antimalarial, inhibition of cholesterol synthesis	Mahato and Sen, 1997
4	Tannins	Anti-inflammatory, gastrointestinal complaints	Cheng *et al.*, 2002
		Antiviral	Lin *et al.*, 2004
		Antibacterial	Akiyama *et al.*, 2001
		Antiparasitic	Funatogawa *et al.*, 2004

Sl.No.	*Name of the Phytochemical Compund*	*Activity*	*References*
5	Saponins	Antifungal, antioxidant, antimicrobial, antiprotozoal, antiviral, absorption of minerals and vitamins, immunostimulatory effects, increase permeability of intestinal mucosa cells, reduction in fat absorption, ruminal ammonia concentration and still birth's in swine	Guclu Ustundag and Mazza, 2007
		Antiinflammatory	Balandrin, 1996
		Hypocholesterolomic	Oakenfull, 1996
		Immune stimulating	Klausner, 1988
		Control blood cholesterol levels, bone health, cancer, and building up the immune system.	Jan Chaboya – Hembree, HTTPS://www.medicalmarijuana.com/author/askjan)
		Antibacterial, antileishmanial, antifungal, antimalarial, antiplasmodial, antiviral and antitumoral	Dinda *et al.*, 2010
		hemolytic, expectorative, antiinflammatory and immunestimulating activity	Ray Sahelian, 2016
6	Phenols	Antimicrobial and antioxidant activity	Prashith *et al.*, 2010; Sahu and Mahato 1994
7	Steroids	Arthritis, asthma, autoimmune diseases, skin diseases and some kind of cancer	Josette Govington, 2015
8	Carbohydrates	Anticancer, cardiovascular diseases, anti-inflammatory, diabetes, heart disease, stroke, kidney damage, retinopathy and diabetic foot	David Platt, 2013

The early reports established on eight phytochemicals encompass different biological activities. This bioactive potential of *Hybanthus enneaspermus* could make the useful herbal drug for treating various human ailments. Medicinal plants are rich source of bioactive compounds which are playing a vital role in the human health maintenance. There are so many types of methods are available for identification and characterization of bioactive compounds studies. However, from the previous literature medicinal plant *Hybanthus enneaspermus* were subjected in to the following few kinds of studies only such as TLC, HPLC, HPTLC, GC-MS and FTIR. Gas Chromatography Mass Spectroscopy is a very simple and widely used technique to identify and quantify the unknown organic compounds from a complex mixture with interpretation and also by matching the spectra with reference spectra (Rohald Hites, 1997).

The ethanolic extract *H. ennaespermus* was subjected in to GC-MS analysis. Interpretation on mass spectrum GC-MS was conducted by using the database of National Institute Standard and Technology (NIST) which is having more than 62,000 patterns. The name, molecular weight and structure of the components of the test materials were ascertained. GC-MS results shown 13 compounds were present in ethanolic extraction of *H. ennaespermus*. They were reported that the plant extract contain propane,1,1,3-triethoxy-phenol,4,6-di(1,1-dimethyl)-2-methyl, 1,14-tetradecanediol, phytol, 2-pepridione, N-4[4-bromo-nbutyl]-, cedarn-diol, 8S, 14-and 2H-pyran, 2-(7-heptadecynylox)tetrahydro possess anticancer, hepatoprotective, anti-inflammatory, antimicrobial and inhibition of parasitic growth (Anand and Gokulakrishnan, 2012).

The GC-MS analysis showed that 39 compounds were present in the methanol extract of *H. ennaespermus*. These compounds were identified through mass spectrometry attached with GC. Among these 39 compounds only three compounds have bioactivity such as n-Hexadecanoic acid (Hyperlipidemic, Nematicide and Antioxidant), Dodecanoic acid, 1,2,3 propa (Antibacterial, Antioxidant and Antiviral) and Octadecanoic acid (Hypocholesterolemic) (Suman *et al.*, 2016). The qualitative and quantitative phytochemical studies as well as characterization of bioactive compound studies are very important to the herbal drug standardization. The presence of phytocompounds is responsible for the particular bioactivity which may be helpful to the pharmacological studies on medicinal plants for new drug development. Several pharmacological studies have shown different parts and extracts of *Hybanthus enneaspermus* have therapeutic significance.

The antibacterial and antifungal activity of *Hybanthus enneaspermus* showed maximum inhibitory effect on both bacterial and fungal strains (Sahoo *et al.*, 2006; Raveendra Retnam and John De Britto, 2007; Arumugam *et al.*, 2012; Nepolean *et al.*, 2011 and 2012; Anand and Gokulakrishnan, 2012). Recently anti-HIV (antiviral) activity has been reported from the leaf extract of *H. enneaspermus* by reverse transcriptase inhibition assay method. The methanol and hexane extracts showed positive result and ethanol, chloroform and petroleum ether extracts showed no inhibition. The hydroethanolic extract of *H. enneaspermus* along with paracetamol exhibited significant antioxidant activity and decreased lipid peroxidation activity in rats (Deepika Thenmozhi and Premalashmi, 2011). The alcoholic extract of H. *enneaspermus* showed significant antioxidant activity with well reducing power against reactive oxygen species (DPPH radical, nitric oxide, hydrogen peroxide and deoxyribose) Patel *et al.*, 2011. The ethyl acetate extract of *H. enneaspermus* showed strong antioxidant activity due to the presence of phenolic compounds (Anish and Rajesh, 2014). The ethanol extract of *H. enneaspermus* act as a natural antioxidant agents instead of synthetic toxic antioxidants (Mohanapriya *et al.*, 2016; Rex Dab, and Ragavan, 2014).

The ethanol extract of *H. enneaspermus* showed maximum inhibitory activity of alpha-amylase and alpha-glucosidase enzymes (Vennila and Pavithr, 2015). The antidiabetic potential of alcoholic extract of *H. enneaspermus* in rat hemi-diaphragm was analysed. Blood glucose level was reduced in streptozotocin-induced diabetic rat model. Moreover, improvement in body weight was observed which indicated its ability to reduce hyperglycemia (Patel *et al.*, 2011). The alcoholic extract of *H. enneaspermus* is exhibited very good antiarthritic potential against Freund's adjuvant-induced arthritis in albino rats (Tripathy *et al.*, 2009). Antinociceptive activity of *H. enneaspermus* has been proved against albino rats (Afolabi *et al.*, 2014). The *invitro* study of ethanol and petroleum ether extracts of aerial parts of

H. enneaspermus showed high level inhibition against albino mice by milk induced leucocytosis and eosinophilia methods (Thamizhmozhi *et al.*, 2013). In addition, Anticonvulsant activity (Hemalatha *et al.*, 2003), Antihyperlipidemic activity (Vetriselvan *et al.*, 2013), Nephroprotective activity (Manjunath Setty *et al.*, 2007), Cardioprotective activity (Radhika *et al.*, 2011), Neuroprotective activity (Kar *et al.*, 2010) and antioxidant activity (Dinesh Kumar Patel *et al.*, 2011) of *H. enneaspermus* were reported previously.

Conclusion

Medicinal plants are used as herbal medicine or phytomedicine to cure various human health improvements based on their therapeutic values. Due to the lesser side effects of phytomedicine, the medicinal plants are widely used as a medicine to cure various ailments. According to survey of literature, the medicinal plant *H. enneaspermus* is extensively used as a folk medicine to treat kind of diseases. *H. enneaspermus* have produced number of active constituents which are responsible for many biological activities. The different extracts of *H. enneaspermus* were used as a potential herbal drug for the treatment of several diseases like diabetes, arthritis, analgesic, allergic, epilepsy, kidney disorder and cardiovasular. Finally, it is concluded that the medicinal plant *H. enneaspermus* have high medicinal properties due to their nature of therapeutic uses, phytochemical composition and their bioactivity with proved different pharmacological properties. Through this review it is suggested that the medicinal plant *H. enneaspermus* is a reliable drug for treating human ailments. In the conservation point of view, the regeneration potential of *H. enneaspermus* is very poor due to low seed viability (Praksah *et al.*, 1999), over harvesting, sporadic distribution and germination are major threats to this plant (Arunkumar *et al.*, 2009). Several plant tissue culture methods may be helpful in conservation of this medicinal plant.

References

Abu-Rabia, (2005). Urinary diseases and ethanobotany among pastoral nomads in the Middle East. *J. Ethnobiol. Ethnomed.* 1, 4.

Akiyama H, Kazuyasu F, Yamaski O, Oono T and Iwatsuki K, (2001). Antibacterial action of several tannins against *Staphylococcus aureus. J. Antimicrobial Chemotherapy*, 48 (18): 487-491.

Alagesaboopathi C, (2014). Medicinal Plants used by Tribal and Non-Tribal People of Dharmapuri District, Tamil Nadu, India. *International Journal of Current Research in Bioscience and Plant Biology*, Vol 1(2): 64-73.

Alagesaboopathi C, (2014). Medicinal plants used by Tribal and Non-Tribal People of Dharmapuri District, Tamil Nadu, India, *International Journal of Current Research in BioScience and Plant Biology*, Vol 1(2): 64-73.

Amutha priya D, S Ranganayaki and P Suganya Devi (2011). Phytochemical Screening and antioxidant potential of *Hybanthus ennneaspermus*: Arare Ethano botanical herb. *Journal of Pharmacy research*, 4(5): 1497-1502.

Anand T and K Gokulakrishnan, (2012). GC-MS Analysis and antimicrobial Activity of Bioactive components of *Hybanthus enneaspermus. International journal of Pharmacy and Pharmaceutical Sciancia,* (2014) vol, 4 (3): 646-650.

Anish NP, MG Rajesh, (2014). *International Pharmacutica sciencia,* 4(1): PP 14-20.

Arunkumar BS and M Jayaraj, (2011). Rapid *In vitro* callogenesis and phytochemical screening of leaf callus of *Ioonidium suffruticosum,* Ging-a seasonal multipotent medicinal herb. *World j Agri Sci.* vol. 7: . 55-61.

Aswini Kumar Dixit and Sudurshan. M, (2011). Review of flora of anti-diabetic plants of Puducherry ut. *International Journal of Applied Biology and Pharmaceutical Technology.* Vol.2(4): 455-462.

Awobajo. FO and I I Olatunji-Bello, (2010). Hypoglycemic activities of aqueous and methanol leaf extract of *Hybanhus enneaspermus* on normal and allaioxan induced diabetic female Sprague dawley rats. *Journal of Phytology,* vol, 2(2): 1-9.

Balandrin, MF, (1996). Commercial Utilization of Plant-derived Saponins: An Overview of medicinal, Pharmaceutical and Industrial Applications, In: Saponins Used in Food and Agriculture, G. R. Waller and K. Yamasaki (Eds). Plenum Press. pp. 1-14.

Baviya Sampath R, Radha and R Vadivu, (2016). Phytochemical studies of *Hybanthus enneaspermus* Linn, F Mull., *World Journal of Pharmaceutical Sciences,* 4(3): 440-447.

Boominathan RB, Parimaladevi, SC, Mandal and SK Ghoshal, (2004). Anti-inflammatory evalution of Ionidium Suffruticosam Ging. In rets, *Journal of Ethnopharmacol,* 91 (2-3): 364-70.

Boominathan RB, Parimaladevi, SC Mandal, (2003). Evaluation of antitussive potential of Ionidium suffruticosam Ging, (Violaceae) exteact in albino mice, Phytotherapy Res., 17: 838-9.

Bouayad N, Rharrabe K, Lamhamdi M, Nourouti NG, Sayah F, (2012). Dietary effects of harmine, a b-carboline alkaloid, on development, energy reserves and a amylase activity of Plodia interpunctella Hubner [Lepidoptera: Pyralidae]. Saudi *J Biol Sci,* 19(1): 73-80.

Cheng Y, VM Canuto and AM Howard, (2002). An improved model for the turbulent PBL. *J. Atoms. Sci,* 59: 1550-1565.

Dahanukar SA, (2000). The Beneficial values of plants. *Indian Journal of Pharmacological.* 32: 81-118.

Das S, Dash SK, Padhy SN, (2004). Ethno botanical information from Orissa state, India: a review, *Journal of Human Ecology,* 14: 165-227.

Deepika Thenmozhi C and V Premalashmi, (2011). *Int. J. Res. Pharm.Biomed. Sci.,* 2(3): 1285-1287.

Dinda B, Debnath S, Mohanta BC, Harigaya Y, (2010). Naturally occurring triterpenoid saponins. *Chem Biodivers* 7: 2327–2580.

Dinesh kumar Patel, Rajesh Kumar, Damiki Laloo and Sive Hemalatha, (2011). Evalution of Phytochemical and antioxidant activities of the different fractions of *Hybanthus ennneaspermus* (Linn.). *Asian Pacific journal of Tropical Medicine,* pp. 391-396.

Dinesh K, Patel, Sairam Krishnamurthy, S Hemalatha, (2013) Evalution of Glucose utilization Capacity of Bioactivity guided Fractions of *Hybanthus ennneaspermus* and Pedalium murexin isolated rat hemidiaphragm, *Journal of Acute Disease,* pp. 33-36.

DJ Newman, GM Cragg, J. Nat. Prod., 2007, 70: 461-477.

Duraipandiyan V, Ayyanay M and Lgnacimuthu S, (2006). Antimicrobial activity of some ethnomedicinal plants used by Paliyar tribe from Tamil Nadu, India. BMC Complimentary and Alternative Medicine, 6: 35-41.

Funatogawa K, Hayashi S, Shimomura H, Yoshida T, Hatano T, Ito H and Irla Y, (2004). Antibacterial activity of hydrolysable tannins derived from medicinal plants against Helicobacter pylori. Microbiol, Immunol, 48 (4): 251-261.

Gopal GV and GL Shah, 1985. Some folk medicinal plants used for jaundice in Gujarat, India. *Journal of Research and Education in Indian in Indian Medicine* 4: 45.

Guclu-Ustundag O and Mazza G (2007). Saponins: properties, applications and processing, Crit Rev Food Sci Nutr, 47: 231-538.

Hemalatha S, Wahi AK, Singh PN, Chansouria, (2003). Anticonvulsant and free radical scavenging activity of *Hybonthus enneaspermus:* A preliminary screening, *Indian Journal Traditional Knowledge,* 2(4): 389.

Herourt D, Sangwan RS, Fliniaux MA and Sangwan-Norreel BS, (1988). Variations in the leaf alkaloids content of androgenic diploid plants of Datura innoxia. *Planta Medica,* 54: 14-17.

Kanaka Rajesham Ch, N Narasinga Rao, M Venkateshwarlu, D Sammaiah, U Anitha and Ugandhar, (2013). Studies on the Medicinal plant Biodiversity in Forest Ecosystem of Mahadevpur Forest of Karimnagar (AP) India. Biosience Discovery, 4(1): 82-88.

Khandelwal KR, (2005). Practical Pharmacognosy Techniques and Experiment. 13th ed. Pune: Nirali Prakashan: pp. 149-153.

Kheraro J and A Bouquet, (1950). In-Plants medicinal et toxiques de la cote d' lvoire-Haute-Volta. Vigot Freres, Paris, pp. 170.

Kirshnamoorthy BS, N Nattuthurai, R Logeshwari, H Dhaslim Nasreen and I Syedali Fathima. (2014). Phytochemical study of *Hybanthus enneaspermus* (L.) F. Muell, Journal of Pharmacognosy and Phytochemistry, 3(1): 6-7.

Klausner RD, (1998). From Receptors to Gene –Insights from molecular Iron metabolism. *Clin. Res.* 36: 494-500.

Lin LU, Shu-Wen L, Shi-bo J and Shu-guang W, (2004). Tannins inhibits HIV-1 entry by targeting gp 41. Acta Pharmacol Sin, 25 (2): 213-218.

Luiz Rodrigo Saldanha Gazzaneo, Reinaldo Farias Paiya de Lucena and Ulysses Paulino de Albuquerque, (2005). *Journal of Ethnobiogy and Ethnomedicine,* 1:9, DoI: 10.1186/1746-4269-1-9.

Mahato SB and Sen S, (1997). Advances in triterpenoids research-1990-1994. *Phytochemistry,* 44: 115-123.

Majumder PL, A Basu and D Mal, (1979). Chemical constituents of *Hybanthus enneaspermus* F. Muel. (Violaceae). *Indian Journal of Chemistry Section* B 17: 297-298.

Mohanapriya M, J Jayanthi, D Devakumar and MG Ragunathan, (2016). *In Vitro* Evaluation of Free Radical Scavening Activity of the Flower *Hybanthus enneaspermus* (Linn.) F. Muell. Extract, International Journal of Phytopharmacology. 7(4): 159-163.

Nagaraju N and KN Rao, (1996). A survey of plant crude drugs of Rayalseema, Andhara Pradesh, India. *Journal of Ehno Pharmacol* 29: 137-158.

Napoleon Arumugam, SasiKumar Kandasamy, M. Sekar (2012). *In- Vitro* Antifungal Activity of *Hybanthus enneaspermus* F Muell, *International journal of Pharmacy and Pharmaceutical Sciences,* Vol 4, Issue 2: 594-596.

Napoleon A, K Sasikumar, M Himaja, M SeKar, (2011). *Int. J. Res.* Ayurveda Pharm. 2(4): 1184-1185.

Narayanswamy VB, M Manjunatha Setty, S Malini, Annie Shirwaikar, (2007). Preliminary Aphrodisiac Activity of *Hybanthus Enneaspermus* in Rats. *Pharmacologyonline,* 1: 152-161.

Neha Babbar, Sandra Van Roy, Marc wijnants, Winnie Dejonghe, Augusta Caligiani, Stefano Sforza and Kathy Elst, (2015). Effect of extraction condition on the saccharide (Neutral and Acidic) Composition of the crude pectic extract from various Agro-industrial Residues, *Journal of Agricultural* and *Chemistry,* 64: 268-276.

Neha Babbar, (2015). an introduction to alkaloids and their applications in pharmaceutical chemistry, *The Pharma Innovation Journal,* 4(10): 74-75.

Nostro A, Germanò MP, D'angelo V, Marino A and Cannatelli MA (2000). Extraction methods and bioautography for evaluation of medicinal plant antimicrobial activity. *Lett. Appl. Microbiol.* 30: 379-384.

Oakenfull, G and Glib B, (1996), Research-Based Advertising to Preserve Brand Equity but Avoid Genericide, *Journal of Advertising Research,* 36, (September/ October), pp. 65-72.

Onayade OA, JJ Scheffer and AB Svensen, (1990). The importance of phytotherapy and screening of plants used medicinally in Africa women studies. *Planta Medica,* 56: 503-504.

Patel DK, Kumar R, Prasad SK, Sairam K, Hemalatha S, (2011). Antidiabetic and *In vitro* antioxidant potential of *Hybanthus enneaspermus* (Linn) F. Mull in streptozotocin-induced diabetic rats, *Asin Pacific Journal of Tropical Biomedicine,* vol. 1: 316-322.

Pei SJ, (2001). Ethnobotanical approaches of traditional medicine studies: Some experiences from Asia. Pharmaceutical Biology 39: 74-79.

Prakash E, PS Sha Valli khan P Sairam Reddy, (1999). Regeneration of plants from seed-derived callus of *Hybanthus enneaspermus* L. Muell., a rare ethnobotanical herb. *Plant Cell Reports,* 18: 873-878.

Pushpangadam P and CK Atal, (1984). Ethno–Medico - Botanical Investigations in Kerala. I. Some primitive tribal's of Western Ghats and their herbal medicine. *Journal of Ethnopharmacology* 11: 811-816.

Radhika S, KH Smila, R Muthezhlian, (2011). *Int. J. Fund.Appl. Life Sci.* 1(3), PP 90-97.

Raghav Kumar Mishra, Shio Kumar Singh, (2008). Safety assessment of Syzygium aromaticum flower bud (clove) extract with respect to testicular function in mice. *Food Chem Toxicol,* 46: 3333-3338.

Rajesekhar PS, RS. Arvind Bharani, K Jini Angel, Maya Ramachandran and Sharadha Priya Vardhini Rajesekhar, (2016). *Hybanthus enneaspermus* (L) F. Muell: A Phytopharmacological review on herbal medicine, vol 8(1): 351-355.

Ranandeep Singh, Ashraf Ali, Gaurav Gupta, Alok Semwel, Jeyabalan G, (2013). Some medicinal plants with aphrodisiac potential: A current status. *Journal of Acute Dieses,* pp. 179-188.

Raveenadra Retnam and John De Britto, (2007) Antibacterial activity of medicinal plants *Hybanthus enneaspermus* (L.) F Mull. *Journal of Pharmacy and Pharmaceutical Sciences,* 6(6): 567-572.

Reddy K, G Sudarsanam and PG Rao, (1989). Plant drugs of Chittoor District, Andhara Pradesh, India," *Indian Journal of Crude Drug Research,* 27(1): 41-54.

Remya Krishnan V, Shirmila Jose G and Radhamany PM, (2012). Pharmacognostic and Phytochemical Investigation of *Hybanthus enneaspermus* Linn. *International Journal of Pharmacy and Pharmaceutical Sciences,* 4(4): 77-80.

Rex Dab and B Ragavan, (2014) Studies on Phytochemicals, Antioxidant and Cytotoxicity Effect of *Hybanthus enneaspermus. International Journal of Pharmacy* and *Pharmaceutical Science,* 6(6): 567-572.

Rey Sahelian, MD, (2016). Natural Health News letter www.rays sahelian.com/ natural health news letter.html.

Ronald Hites A, (1997). Gas Chromatography Mass Spectroscopy: Handbook of Instrumental Techniques for Analytical Chemistry. pp. 609–611.

Sahoo S, DM Kar, S Mohapatra, SP Rout and SK Dash, (2006). antibacterial activity of *Hybanthus enneaspermus* against Selected Urinary Tract Pathogen. *Indian Journal of Pharmaceutical Science,* 68(5): 653-655

Sahu NP and Mahato SB, (1994). Phytochemistry, 37: 1425.

Sarita D, SK Dash and SN Padhy, (2004). Ethno botanical information from Orissa State, India. A Review. *Journal Human Ecology,* 14(3): 227.

Shantha TR, Saraswathi Pashupathy, JK Shetty, B Vijayalakshmi, P Kandavel and T BIkshapathy. (2001). *Ancient Sci.Life,* XXI (1): 1-12.

Subramoniam A and P Pushpangadam, (1999). Development of phytomedicines for liver disease. *Indian Journal of Pharmacological,* 31: 66-75.

Sudarsanam G and G Sivaprasand, (1995). Medical Ethnobotany of Platsused as Antidotes by Yanadi Tribes in South India. *Journal of Herbs, Spices and Medicinal Plants.* 3(1): 57-65.

Sudharson. S, M Anbazhgan, B Balachandran and K Arumugam, (2014). Effect of Bap on in vitro propagation of *Hybanthus enneaspermus* (L.) F Muell, an Important Medicinal Plant. *International Journal of Current Microbiology and Applied Science,* 8: 397-402.

Suman, TY, SR Radhika Rajasree, C Jayaseelan, R Regina Mary, S Gayathri, L Aranganathan, R and R Remya, (2016). GC-MS analysis of bioactive compounds and biosynthesis of silver nanoparticles using *Hybanthus enneaspermus* at room temperature evaluation of their stability and its larvicidal activity. *Enuiron Sci Pollut Res,* 23: 2705-2714.

TR Prashith Kekuda, N Mallikarjun, D Swathi, KV Nayana, Meera B Aiyar and TR. Rohini, (2010). Antibacterial and Antifungal efficacy of steam distillate of Moringa oleifera Lam. *J. Pharm. Sci. and Res.* 2(1): 34-37.

Tamizh mozhi M, Swarnalatha S, Sakthivel P, Manigandan LS, Jayabharath A and Suresh Kumar P. (2013). Anti-Allergic and Analgesic activity of Aerial parts of *Hybanthus enneaspermus, International Research Journal of Pharmacy,* 4(6): 243-248.

The Wealth of India, 1959. A dictionary of Indian raw materials and industrial products. Vpl 5. Publication and Information Directorate, Council of Scientific and Industrial Research, New Delhi India. pp. 139.

Thyaga Raju K, B Namratha Rani, K Kamala and G Sujatha, (2014). Studies on the Phytochemical, Antioxidant, Proximate and Elemental Analysis of *Hybanthus enneaspermus* (L.) F.Muell Leaf, *World Journal of Pharmaceutical Research,* 4(1): 1244-1253.

Thyaga Raju K, B Namratha Rani, K Kamala and G Sujatha, (2015). Studies on the Phytochemical, Antioxidant, Proximate and Elemental Analysis of *Hybanthus enneaspermus* (L.) F.Muell Leaf, *World Journal of Pharmaceutical Research,* 4(1): 1244-1253.

Tripathy S, SP Sahoo, D Pradhan, S Sahoo and DK Satapathy (2009). Evalution of anti arthritic potential of *Hybanthus enneaspermus* (L.) *African Journal of Pharmacy and Pharmacology,* 3(12): 611-614.

Udayan PS and B Indira, (2009). Medicinal plants of Aryavaidyasala herbs garden. Aryavaidya sala, kpttakkal pub, pp. 199.

Vennila V and Pavithra V, (2005). *In-vitro* alpha amylase and alpha glucosidase inhibitory activity of various solvent extract of *Hybanthus enneaspermus* linn. *World Journal of Pharmacy* and *Pharmaceutical Science,* 4(4): 1425-1437.

Vetriselvan S, V Suganya and T Muthuramu, (2013). Anti hyperlipidemic activity of hydrochloric extract of *Hybanthus enneaspermus. Asian Journal of Phytomedicine* and *Clinical Research.* 1(1): 27-33.

Weniger B, L Lagnika, C Vonthron-Senecheau, D Adjabimey, J Gbenou, M Moudachirou, R Brun, R Anton and A Sanni, (2004). Evalution of ethnobotanically Selected Benin Medicinal Plants for their *in vitro* antiplasmodial activity, *J. Ethnopharmacol.* 90(2-3): 279.

Yoganarasimhan S. N, (2000). *Medicinal Plants of India – Tamil Nadu,* Vol. 2. (Bangalore: Cyber Media), pp. 276.

2018, Ethnomedicinal Plants: A Biodiversity Treasure Pages 407–423
Editors: V.R. Mohan, A. Doss, P.S. Tresina and V. Sornalakshmi
Published by: ASTRAL INTERNATIONAL PVT. LTD., NEW DELHI

Chapter 17

Medicinal Plants and its Utilization in India

M.S. Rukshana, A. Doss and T.P. Kumari Pushpa Rani

[1]PG and Research Department of Microbiology,
Kamaraj College, Tuticorin, Tamil Nadu
E-mail: rukshanasana94@gmail.com

Introduction

Plants which have one or more of its parts having substances that can be used for treatment of diseases, are called "**MEDICINAL PLANTS**". Medicinal plants are the "backbone" of traditional medicine. It has been identified and used throughout human history. Plants make many chemical compounds that are for used thier biological functions. Before the introduction of chemical medicines, man relied on the healing properties of medicinal plants. Some people value these plants due to the ancient belief which says plants are created to supply man with food, medical treatment, and other effects. It is thought that about 80 per cent of the 5.2 billion people of the world live in the less developed countries and the World Health Organization estimates that about 80 per cent of these people rely almost exclusively on traditional medicine for their primary healthcare needs. There are nearly 2000 ethnic groups in the world, and almost every group has its own traditional medical knowledge and experiences. About 250,000 higher plant species on Earth, more than 80,000 species are reported to have at least some medicinal value and around 5000 species have specific therapeutic value. The use of herbs to treat disease is widespread in non-industrialized societies. Because medicines derived from plants are widely famous due to their safety, easy availability and low cost. Many of the herbs and spices used by humans to season food also yield useful medicinal compounds.

Herbs are staging a comeback and herbal 'renaissance' is happening all over the globe. The herbal products today symbolize safety when compared to the synthetics that are considered as unsafe to human and environment. Even though herbs had been priced for their medicinal, flavoring and aromatic qualities for centuries, the synthetic products of the modern age surpassed their importance, for a while. But nowadays people are returning to the herbals with hope of safety and security. Medicinal herbs are more significant to the health of individual and community. Over three-quarters of the world population relies mainly on plants and plant extracts for health care. More than 30 per cent of the entire plant species were used for medicinal purposes. The medicinal value of these plants lie in bioactive phytochemical constituents that produce definite physiological action of human body. Some of the most important bioactive phytochemical constituents are alkaloids, essential oils, flavonoids, tannins, terpenoids, saponins, phenolic compounds, and many more. These natural compounds formed the foundations of modern prescription drugs as we are using today. Use of plants based drugs for curing various ailments is as old as human civilization using in all cultures throughout history. The primitive man started to distinguish between useful and harmful plants by trial and errors. A well defined herbal pharmacopoeia was developed by tribal people, which was based on the information collected from local flora, religion and culture. The knowledge of medicinal plants was gradually developed and passed on from one individual to other, which laid foundation for traditional medicine throughout the world.

Medicinal Plants in India

India is a country known for its ancient natural medicine products from plants. Use of plants as a source of medicine has been an early practice and is an important component of the health care system in India. It is the largest producer of medicinal herbs and is called the botanical garden of the world. It is one of the world's 12 biodiversity centres with the presence of over 45000 different plant species. Among these, about 15000-20000 plants have good medicinal value. However, only 7000-7500 species are used for their medicinal values by traditional communities. Medicines in India are used by about 60 percent of the world's population. These are not only used for primary health care not just in rural areas in developing countries, but also in developed countries as well where modern medicines are predominantly used. While the traditional medicines are derived from medicinal plants, minerals, and organic matter, the herbal drugs are prepared from medicinal plants only.

India has a rich culture of medicinal herbs and spices, which includes about more than 2000 species and has a vast geographical area with high potential abilities for Ayurvedic, Unani, Siddha traditional medicines but only very few have been studied chemically and pharmacologically for their potential medicinal value. Among these systems, Ayurveda is most developed and widely practiced in India. The term comes from the Sanskrit root Au (life) and Veda (knowledge). As the name implies it is not only the science of treatment of the ill but covers the whole gamut of happy human life involving the physical, metaphysical and the spiritual aspects. There are currently about 250 000 registered medical practitioners of the Ayurvedic system, as compared to about 700,000 of the modern medicine system.

In rural India, 70 per cent of the population depends on the traditional type of medicine, the Ayurveda. Ayurvedic form of medicine is believed to be existent in

India for thousands of years. It employs various techniques and things to provide healing or relief to the ailing patients. One of the thing that ayurveda using is medications of plant origin. It is estimated that about 80,000 species of plants are utilized in some form or other by the different systems of Indian medicine. Many herbs and spices are used in Indian cooking, such as onion, garlic, ginger, turmeric, clove, cardamom, cinnamon, cumin, coriander, fenugreek, fennel, ajowan (ajwain), anise, amchur, bay leaf, hing (asafoetida) *etc.* Ayurvedic medicine uses all of these either in diet or as medicine. Some of these medicinal plants have been featured on Indian postage stamps. The first set of stamps showing medicinal plants came out in 1997. The set had four stamps showing four different medicinal plants - Tulsi (*Ocimum sanctum*), Haridra (*Curcuma longa*), Sarpagandha (*Rauvolfia serpentina*), and Ghritkumari (Aloe barbadensis). From this it shows the great importance of medicinal plants and its value in India.). The codified traditions have about 25,000 plant drug formulations that have emerged from such studies. In addition to this, over 50,000 formulations are believed to be existing in the folk and tribal traditions. All these point to the deep passion for exhaustive knowledge about medicinal plants that have existed in this land from time immemorial. Uncurable disease can also get cured by treating with medicinal plants and the major type of medicinal plan was in India. All medicinal plants have its own medicinal value. Each and every part of the plant was used as a medicine. It can be used as anti-diabetic, anti-hepatotoxicity, anti-cancer, anti-bacterial, anti-viral, anti-inflammatory, antioxidant and so on.

Indian Medicinal Plants in Treatment of Severe Diseases

Anti-diabetic Plants

Among the challenging health conditions that had been increasingly affecting human beings is diabetes. This is a disease condition caused by a metabolic disorder of the body systems as a result of chronic high blood sugar (hyperglycemia). This sort of health challenge is often associated with disturbances in the metabolism of fat, carbohydrate and protein thereby causing defects in insulin activities and secretion due to the inability of pancreas to produce enough insulin. This can also be as a result of failure of the body cells to respond to already-produced insulin. There are several means of managing and treating diabetes however, many authors reveal that natural remedies are more viable unlike the synthetic drugs and oral medications that may pose undesirable side effects to the body. Thank goodness for nature because several medicinal plants have been identified and proven very effective for treating diabetes mellitus. Some key active components present in certain medicinal plants play significant roles in treating diseases and such powerful components continue to show promising effects in treating diabetes mellitus and any complications associated with it. Discussed below are some powerful medicinal plants for treating diabetes.

Gymnema sylvestre

Gymnema is a woody climbing shrub native to India and Africa. It belongs to the family Apocynaceae and genus *Gymnema*. The leaves are used to make medicine. *Gymnema* has a long history of use in India's Ayurvedic medicine. The Hindi name,

Plate 17.1: *Gymnema sylvestre.*

gurmar, means "destroyer of sugar." Today, *Gymnema* is used for diabetes, metabolic syndrome, weight loss, and cough. It is also used for malaria and as a snake bite antidote, digestive stimulant, laxative, appetite suppressant, and diuretic. *Gymnema* contains substances such as dihydroxy gymnemic, triacetate, gymnemic acids I-VII, conduritol A and triterpenoid saponins (gymnemosides A-F and gymnemoside W1-2) that decrease the absorption of sugar from the intestine. Gymnema may also increase the amount of insulin in the body and increase the growth of cells in the pancreas, which is the place in the body where insulin is made. Leaves contain glycoside designated as Gymnemic acid (I) a bitter principle; tartaric acid and calcium oxalate and hentriacontane ($C_{31}H_{64}$) have been isolated from leaves and demonstrated antidiabetic activity. The leaves of *Gymnema sylvestre* are reported to possess many medicinal properties in the ayurvedic literature (Chopra *et al.*, 1928). The hypoglycemic action of *Gymnema* leaves was first documented in the late 1920s (Mhasker and Caius, 1930). The alcoholic extract of leaves yields water soluble acidic fraction, which show good hypoglycemic activity (Plate 17.1).

Momordica charantia

Momordica charantia, known as bitter melon, bitter gourd, bitter squash, or balsam-pear in English, has many other local names. Goya from Okinawan and kerela from Sanskrit are also used by English-language speakers and haagalakaayi in Kannada and pavakai in Tamil. Bitter melon is a tendril-bearing vine used mainly for treating diabetes especially in India. It belongs to the family of Cucurbitaceae. The plant contains several biologically active compounds, chiefly momordicin I and II, and cucurbitacin B. The plant also contains several bioactive glycosides (including momordin, charantin, charantosides, goyaglycosides, momordicosides) and other terpenoid compounds (including momordicin-28, momordicinin, momordicilin, momordenol and momordol) (Kimura *et al.*, 2005, Chang *et al.*, 2008, Akihisa *et al.*, 2007). It also contains cytotoxic (ribosomeinactivating) proteins such as momorcharin

Plate 17.2: *Momordica charantia.*

and momordin (Ortigao and Better, 1992). Both the leaf extracts, fruits and seeds of bitter melon have hypoglycemic effects on diabetic patients. However, as many studies report, there has been substantial emphasis on the anti-diabetic compounds and their hypoglycemic properties (Islam *et al.*, 2011, Hazarika *et al.*, 2012). The components of bitter melon which are responsible for hypoglycemic effects are momordicin, stearic acid, charantin, eleostearic acid, insulin-like peptide, oleanolic acids and cucurbutanoids. They are in use since a very long time in Ayurveda medicine. Bitter melon has been used in various Asian and African herbal medicine systems for a long time. In traditional medicine of India different parts of the plant are used to relieve diabetes, as a stomachic, laxative, antibilious, emetic, anthelmintic agent, for the treatment of cough, respiratory diseases, skin diseases, wounds, ulcer, gout, and rheumatism. *M. charantia* has a number of purported uses including cancer prevention, treatment of diabetes, fever, HIV and AIDS, and infections (Plate 17.2).

Anti-hepatotoxicity Plants

Liver is the very important part of our body responsible for the maximum metabolic and secretory activities and therefore appears to be a sensitive target site for substances modulating biotransformation. Liver injury or liver disfunction is a major health problem that challenges not only medical professionals but also the pharmaceutical companies and drug regulatory authorities. Liver cell injury is caused by various toxic chemicals like certain antibiotics, chemotherapeutic agents, carbon tetrachloride, thioacetamide, excessive alcohol consumption and microbes. Herbal medicines have been applied for the treatment of liver disorder for a lengthy period in India. Many herbal preparations are available to cure liver disorder. Phytochemicals from medicinal plants have been used to treat liver disorder from ancient Indian medical history.

Zingiber officinale

Ginger (*Zingiber officinale*) is a flowering plant whose rhizome, ginger root or simply ginger, is widely used as a spice or a folk medicine. It comes under Zingiberaceae family and genus *Zingiber*. Oral or topical uses of ginger to treat various disorders, such as nausea or arthritis pain, hepatoprotective, seasickness, chemotherapy, dyspepsia *etc.* Presence of alkaloids and/or nitrogenous bases, carbohydrates and/or glycosides, tannins, flavonoides, polyphenols, saponins and unsaturated sterols and/or triterpenes in both ginger. Ginger act as a hepatoprotective by decreasing the liver enzymes and bilirubin in plasma concentration. This is the traditional medicine which was used in treating liver disorders and it would not cause any side effects as synthetic medicines. It not only used for hepatoprotective treatment, many diseases can also be get cured by this rhizome (Plate 17.3).

Plate 17.3: *Zingiber officinale*.

Andrographis paniculata

Andrographis paniculata is an annual herbaceous plant in belonging to the family *Acanthaceae*, native to India and Sri Lanka. It is widely cultivated in Southern and

Plate 17.4: *Andrographis paniculata.*

Southeastern Asia, where it has been traditionally used to treat infections and some diseases. Mostly, the leaves and roots were used for medicinal purposes. Since ancient times, *A. paniculata* has been used in traditional Siddha and Ayurvedic systems of medicine as well as in tribal medicine in India.The whole plant is also used in some cases. As an Ayurveda herb, it is known as *Kalmegh* or *Kalamegha*, meaning "dark cloud". It is also known as *Nila-Vembu*, meaning "neem of the ground", since the plant, though being a small annual herb, has a similar strong bitter taste as that of the large Neem tree (*Azadirachta indica*). It is used to treat various diseases such as upper respiratory tract infection and so on. But it is mainly used to treat liver damage. It also works same as ginger by reducing the liver enzymes to prevent the liver damage (Plate 17.4).

Anti-cancer Plants

The traditional Indian medicine – Ayurveda, describes various herbs with anticancer properties. Various plant products have been used in treatment of cancer over the years. India is a peninsula of herbal hub, in which Ayurvedic system of medicine has flourished as an enlightment in the field of Medicine. Currently medicinal plants have become the paramount source of drug discovery in research for treating diverse form of disease including cancer. The synthetic anticancer remedies are beyond the reach of common man because of cost factor. Herbal medicines have a vital role in the prevention and treatment of cancer and medicinal herbs are commonly available and comparatively economical. Herbal medicines will boost immune cells of the body against cancer.

Piper longum

Long pepper (*Piper longum*), sometimes called Indian long pepper, is a flowering vine in the family Piperaceae which was used to in traditional medicine as well as in the Ayurvedic system against various disorders and infection. Long pepper has a taste similar to, but hotter than, that of its close relative Piper nigrum - from

Plate 17.5: *Piper longum.*

which black, green and white peppers are obtained. The first reference to long pepper comes from ancient Indian textbooks of Ayurveda, where its medicinal and dietary uses are described in detail. Long pepper is known to contain the chemical compound piper longumine. Unripe, dried fruits are used as an alternative to tonic. Various plant preparations like decoction of young fruits and roots are used for treating chronic bronchitis, cough and cold. It is also used as antidote in snake bites and scorpion stings. Mainly it involves in curing cancer by stopping the cancerous cell multiplication. It has been in use of traditional Indian medicine which was not known to the world for many years. It is very effective when compared to synthetic medicines. And the synthetic medicines will cause many side effects and will be very painful to the patients. But herbal medicines will not cause any of these things. It will act mainly in the control of multiplication of cancerous cells (Plate 17.5).

Vitex negundo

Vitex negundo, commonly known as five-leaved chaste tree, or horseshoe vitex, is a large aromatic shrub with quadrangular, densely whitish, tomentose branchlets. It is widely used in folk medicine, particularly in South and Southeast Asia. Vitex negundo is native to tropical Eastern and Southern Africa and Asia. It is widely cultivated and naturalized elsewhere. The principal constituents of the leaf juice are casticin, isoorientin, chrysophenol D, luteolin, p–hydroxybenzoic acid and D-fructose. The main constituents of the oil are sabinene, linalool, terpinen-4-ol, β-caryophyllene, α-guaiene and globulol constituting 61.8 per cent of the oil. The chemicals isolated from the plant have potential anti-inflammatory, anti-cancer, antibacterial, antifungal and analgesic activities. Roots and leaves are used in eczema, ringworm and other skin diseases, liver disorders, spleen enlargement, rheumatic pain, gout, abscess, backache; seeds are used as vermicide. Many experimental evidences are there for the plant activity against cancer cells. It was also one of the traditional herbs used to treat various diseases and later it was found that it can also treat cancer (Plate 17.6).

Plate 17.6: ***Vitex negundo.***

Anti-bacterial Plants

A total of 82 Indian medicinal plants traditionally used in medicines were subjected to preliminary antibacterial screening against several pathogenic and opportunistic microorganisms. Since time immemorial, the traditional medicinal practices have been known for the treatment of various ailments in India. The medicinal plants were selected in this study based on their common uses in Indian traditional systems of medicine. Plants are prospective source of antimicrobial agents in different countries. About 60 to 90 per cent of populations in the developing countries use plant-derived medicine. Traditionally, crude plant extracts are used as herbal medicine for the treatment of human infectious diseases. Plants are rich in a variety of phytochemicals including tannins, terpenoids, alkaloids, and flavonoids which have been found *in vitro* to have antimicrobial properties. Global prevalence of infectious diseases caused by bacteria is a major public health problem. The bacterial agents including *Staphylococcus aureus, Escherichia coli, Pseudomonas aeruginosa, Bacillus subtilis,* and *Proteus vulgaris* cause several human infections. Recent emergence of antibiotic resistance and related toxicity issues, limit the use of antimicrobial agents and is prompting a revival in research of the antimicrobial role of plants against resistant strains due to comparable safety and efficacy.

Vachellia nilotica

Vachellia nilotica is widely known by the taxonomic synonym *Acacia nilotica*. The common names gum arabic tree, Babul/Kikar, Egyptian thorn, Sant tree, Al-sant or prickly acacia called thorn mimosa or prickly acacia in Australia, lekkerruikpeul or scented thorn in South Africa, karuvela maram in South India is a species of *Vachellia* native to Africa, the Middle East and the Indian subcontinent. It comes under the family fabaceae. The whole plant is used as a medicine. Its bark is used to treat

Plate 17.7: *Vachellia nilotica.*

cough, acute gonorrhoea dysentery, diarrhoea, cancers, syphilitic affections and genitourinary affections which are the bacterial diseases. It has many antimicrobial activity which cures many diseases (Plate 17.7).

Vachellia leucophloea

Vachellia leucophloea is a moderate sized tree found in southern India. It comes under the family Fabaceae. The bark extracts of *V. leucophloea* are used as traditional medicine in India as an astringent, a bitter, a thermogenic, a styptic, a preventive of infections, an anthelmintic, a vulnery, a demulcent, an expectorant, an antipyretic, an antidote for snake bites and in the treatment of bronchitis, cough, vomiting, wounds, ulcers, diarrhea, dysentery, internal and external hemorrhages, dental caries, stomatitis, and intermittent fevers and skin diseases. An ethanolic extract ointment has shown marked wound healing activity in trials. Whole plant has anti-microbial activity. The bark of plant is used as antimicrobial, anti-helmentic, expectorant and blood purifier. It is also used to treat skin diseases (leprosy), ulcer, gum bleeding, mouth ulcer, dry cough, dysentery, diabetes and fever. It is also a traditional Indian medicine used to treat many diseases (Plate 17.8).

Plate 17.8: *Vachellia leucophloea.*

Anti-viral Plants

Medicinal plants have been traditionally used for different kinds of ailments including infectious diseases. There is an increasing need for substances with antiviral activity since the treatment of viral infections with the available antiviral drugs often leads to the problem of viral resistance. There is an increasing need for search of new compounds with antiviral activity as the treatment of viral infections with the available antiviral drugs is often unsatisfactory due to the problem of viral resistance coupled with the problem of viral latency and conflicting efficacy in recurrent infection in immune compromised patients. Ethno-pharmacology provides an alternative approach for the discovery of antiviral agents, namely the study of medicinal plants with a history of traditional use as a potential source of substances with significant pharmacological and biological activities. The Indian subcontinent is endowed with rich and diverse local health tradition, which is equally matched with rich and diverse plant genetic source. A detailed investigation and documentation of plants used in local health traditions and ethno pharmacological evaluation to verify their efficacy and safety can lead to the development of invaluable herbal drugs or isolation of compounds of therapeutic value.

Swertia chirata

Swertia chirata is known as Chirayata in India. This annual herb is found all across India and grows up to the height of 1.5 meters. It is known to have leaves in opposite pairs which are about 10 cms in length with no stalks. The tonic made from this plant is used to relieve general weakness and also during convalescence. The infusion prepared from the *S. chirata* used for relieving fever. The plant contains glycoside chiratin which yields on hydrolysis. The ophidic acid in the plant is brown in colour and is identified as a hydroscopic substance. It contains tannin, resin and ash. The chemical compounds present in the herb are amarogentin, chiratin, chiratogenin, enicoflavine, gentianine, swertianin and swerchirin. These act as an anti-viral plant. It corrects the nutrition disorders in the body and helps in bringing normality into the system. The herb is used widely to stimulate the appetite of people

Plate 17.9: *Swertia chirata.*

suffering from anorexia and other such problems. It helps in relieving acidity, nausea and biliousness. It is used as a laxative, vermifuge, sedative and alterative. It has the properties to relieve cough, bronchial infections, malaria and asthma (Plate 17.9).

Leucas lavandulifolia

Leucas lavandulifolia is a genus of plants in the Labiatae family. It is widespread over much of Africa, and southern and eastern Asia, Iran, India, China, Japan, Indonesia with a few species in Queensland and on various islands in the Indian Ocean. It is commonly known as Halkusha, is a well-known plant in Indian traditional medicine. On the basis of its traditional use and literature references, this plant was selected for evaluation of its wound healing potential. It also treats cough, cold, fever, loss of appetite, skin diseases, headache, snake bites and scorpion stings. A decoction of leaves is used as a sedative in nervous disorders, as expectorant, carminative, vermifuge and stomachic. Flowers are stimulant, expectorant and diaphoretic. Juice of flower with honey and a few grains of borax mixed together are very much useful for nasal and laryngeal coughs and colds. The aerial parts have a strong characteristic odour and are used as sedative, laxative, anthelmintic, inflammation, jaundice, dyspepsia, vermifuge, stomachic, scabies, psoriasis, dermatosis, migraine, glaucoma, asthma, anthelmintic, urinary discharge, fever and paralysis. It mainly acts as hepatoprotective, hypoglycemic, antipyretic, anti-diarrahoel, anti-viral, **wound healing** and psychopharmacological, antimicrobial properties (Plate 17.10).

Plate 17.10: *Leucas lavandulifolia.*

Anti-Inflammatory Plants

Most of the medicinal plant parts are used as raw drugs and they possess varied medicinal properties Plants have a great potential for producing new drugs and used in traditional medicine to treat chronic and even infectious diseases. The phytomedicine are more import in the treatment of inflammation. In recent years, there is an increasing awareness about the importance of medicinal plants. Many medicinal plants have shown to exhibit potent anti-inflammatory effect in the treatment of inflammation by using various models. Large group of medicinal plants are used as traditional medicine, which have potential to cure various ailments. These medicinal plants which containing potential sources of phytocompounds which are used for many therapeutics. Almost all parts of the medicinal plant are used as medicines. Commonly available many medicinal plants are used in Indian traditional medicine.

Lawsonia inermis

Lawsonia inermis is commonly known in India as Henna comes under the family Lythraceae. It is indigenous to northern Africa, western and southern Asia and northern Australasia. It grows well in temperatures ranging from 35 to 45°C. Leaves, seeds and bark of the plant are used for medicinal purposes. *L. inermis* (Lythraceae) is a perineal plant commonly called as Henna, having different vernacular names in India *viz.,* Mehndi in Hindi, Muruthani in Tamil, Mailanchi in Malayalam and Mehedi in Bengali Henna leaves are very popular natural dye used to colour hand, finger, nails and hair. The dye molecule, lawsone is the chief constituent of plants, its highest concentration is detected in the petioles (0.5-1-5 per cent). In Folk medicine, henna is used as astringent, anti-haemorrhagic, anti-neoplastic, cardioinhibitory, sedative and also as therapeutic against amoebiosis, headache, jaundice and leprosy. They contain arachidic, behenic, oleic, stearic and linoleic acids. Jaundice and enlargement of the liver can be effectively relieved with the bark of the plant. The seeds of the plant are useful in relieving dysentery. These also cause allergic reactions and chronic inflammatory reaction (Plate 17.11).

Plate 17.11: *Lawsonia inermis.*

Centella asiatica

Centella asiatica, commonly known centella and gotu kola as is a small, herbaceous, frost-tender perennial plant of family Apiaceae and is native is Asia. It is used as a medicinal herb in Ayurvedic medicine. It is otherwise called as sarswathi aku in Telugu, kudangal in Malayalam, vallaarai in Tamil, brahmi booti in Hindi, brahmi in Marathi and so on. In traditional herbal medicine, centella has been used for varicose veins, anti-inflammatory, chronic venous insufficiency, psoriasis, and minor wounds. It act as anti-inflammatory by the depression of inflammation.

Plate 17.12: *Centella asiatica.*

Biochemical compounds such as alkaloids, flavonoids, glycosides, triterpenoids, saponins, amino acids, inorganic acids, vitamins, sterols and lipid compounds are present in centella. Due to the presence of active principles such as flavonoids and tritrepenoids asiaticoside, madecassoside and related polyphenols may responsible for this activity. Centella can be used as a potent anti-inflammatory agent (Plate 17.12).

Antioxidant Plants

Since the beginning of human civilization, a number of Indian medicinal plants have been used in the traditional system of medicine (Ayurveda), aiming to maintain health and to cure diseases. As plants produce a lot of antioxidants they can represent a source of novel compounds with promising antioxidant activity. Nowadays there has been an increasing interest worldwide to identify antioxidant compounds that are pharmacologically active with less or no side effects. Many indigenous herbal plants of regional interest have been used popularly as folk medicines in India. Among them many plants are used as antioxidant. Medicinal plants for their potential antioxidant activities. Each and every part can be act with medicinal properties. Natural antioxidants represent a potentially side effect-free alternative to synthetic antioxidants in the food processing industry and for use in preventive medicine.

Lantana camara

Lantana is native to India and grows in tropical and subtropical regions, wild and cultivated in India like Madhya Pradesh, Gujarat, Maharashtra, Andhra Pradesh and Tamil Nadu *etc.* It also grows also in Mexico, Central America, Bahamas, Colombia and Venezuela. *L. camara* is a low erect or vigorous shrub which can grow to 2-4 meters in height. It comes under the family Verbenaceae. They contain chemical constituents such as theveside, theviridoside, oleanolic acid,

Plate 17.13: *Lantana camara*.

lantaiursolic acid, naphthoquinones: diodantunezone, triterpenoids: lantic acid, camaric acid, camarinic acid, camaryolic acid, lantanolic acid, lantabetulic acid, oleanolic acid, betulonic acid, betulic acid, lantalonic acid, lantoic acid, lantanone, flavonoids, lantanoside, linaroside, phenylethanoid glycosides, acteoside. All these compounds have many activities against diseases. Extract of the plant exhibited potent antioxidant activity and has decreased the extent of lipid peroxidation in the kidneys (Plate 17.13).

Vaccinium oxycoccos

V. oxycoccos a very hardy and drought tolerant plant, comes under the family Ericaceae, it thrives well throughout the tropical and subtropical climates, wild and cultivated in India (like Madhya Pradesh, Gujarat, Maharashtra, Andhra Pradesh, Tamil Nadu *etc.*). It grows also in Japan, Central America, Srilanka, Myanmar, and Malacca. Heavy rainfall and waterlogged conditions are not desirable. It can be grown on a wide range of soils including saline and sodic soils. It is also called as Karmard in Sanskrit, Karaunada in Hindi, Cranberry in English, Kalakai in Tamil, Karaka in Malayalam, Vakkay in Telugu *etc.* The *Vaccinium* fruit is a rich source of iron and contains a fair amount of vitamin C. It is a source of polyphenol

Plate 17.14: *Vaccinium oxycoccos*.

antioxidants, phytochemicals under active research for possible benefits to the cardiovascular system and immune system, and as anti-cancer agents. Cranberry contains tannin and flavonoids like Proanthocyanidins, flavonols and quercetin. *Vaccinium* ranks high among fruit in both antioxidant quality and quantity because of its substantial flavonoid content and a wealth of phenolic acids. The crude extracts inhibit oxidative processes including oxidation of low-density lipoproteins and oxidative inflammatory damage to the vascular endothelium. Dried cranberry juice showed significant decrease in serum level of advanced oxidation protein products determine the antioxidant activity of cranberry fruits and reported decrease risk of heart disease and cancer in healthy human (Plate 17.14).

Conclusion

Above mentioned are some of the diseases and some of the medicinal plants that are widely used in India. But there are many medicinal plants and their uses have not been discovered. Uncurable diseases itself can be cured by using herbal medicines. And the main aim of using herbal medicines will not cause any side effects as the synthetic medication. The phytochemical-analysis of these medicinal plants have revealed a large number of compounds including flavonoids and phenolic which have shown potent antioxidant activity. These compounds have been used as anti-oxidant, ant-carcinogenic, antifungal, antibacterial, anti-spasmodic, anti-inflammatory and anti-diabetic. Plants have been used for medicinal purposes long before prehistoric period. Traditional systems of medicine continue to be widely practiced on many accounts. Population rise, inadequate supply of drugs, prohibitive cost of treatments, side effects of several synthetic drugs and development of resistance to currently used drugs for infectious diseases have led to increased emphasis on the use of plant materials as a source of medicines for a wide variety of human ailments. Among ancient civilizations, India has been known for its rich repository of medicinal plants. The forest in India is the principal repository of large number of medicinal and aromatic plants, which are largely collected as raw materials for manufacture of drugs and perfumery products. About 8,000 herbal remedies have been codified in AYUSH systems in INDIA. Ayurveda, Unani, Siddha and Folk (tribal) medicines are the major systems of indigenous medicines. Among these systems, Ayurveda and Unani Medicine are most developed and widely practised in India. Recently, WHO (World Health Organization) estimated that 80 percent of people worldwide rely on herbal medicines for some aspect of their primary health care needs. According to WHO, around 21,000 plant species have the potential for being used as medicinal plants.

As per data available over three-quarters of the world population relies mainly on plants and plant extracts for their health care needs. More than 30 per cent of the entire plant species, at one time or other was used for medicinal purposes. It has been estimated, that in developed countries such as United States, plant drugs constitute as much as 25 per cent of the total drugs, while in fast developing countries such as India and China, the contribution is as much as 80 per cent. Thus, the economic importance of medicinal plants is much more to countries such as India than to rest of the world. These countries provide two third of the plants used in modern system of medicine and the health care system of rural population depend on indigenous systems of medicine. Treatment with medicinal plants is considered very safe as there is no or minimal side effects. These remedies are in sync with nature, which is the biggest advantage. The golden fact is that, use of herbal treatments is independent of any age groups and the sexes. The ancient scholars believed that herbs are the only solutions to cure a number of health related problems and diseases. They conducted thorough study about the same, experimented to arrive at accurate conclusions about the efficacy of different herbs that have medicinal value. Most of the drugs, thus formulated, are free of side effects or reactions. This is the reason why herbal treatment is growing in popularity across the globe. These herbs that have medicinal quality provide rational means for the treatment of many internal diseases, which are otherwise considered difficult to cure.

References

1. Myung, H.K., Min, S.L., Mi-Kyeong Choi, Kwan, S.M. and Takayuki, S., 2012. Hypoglycemic activity of *Gymnema sylvestre* extracts on oxidative stress and antioxidants status in diabetic rates. *J. Agri. Food Chem.,* 60(10): 2517-2524.

2. Chippada, S.C., Volluri, S.S., Bammid, S.R. and Meena Vangalapati, 2011. *In-vitro* anti-inflammatory activity of methanolic extract of *Centella asiatica* by HRBC membrane stabilization. *Rasayan J. Chem.,* 4(2): 457-460.

3. Sunila, E.S. and Kuttan, G., 2004. Immunomodulatory and antitumour activity of *Piper longum* Linn. and piperine. *J. Ethnopharmacol.,* 90(2-3): 339-46.

4. Mahdi-Pour, B., Jothy, S.L., Latha, Ly., Yeng Chem and Sasidharan, S., 2012. Antioxidant activity of methanol extracts of different parts of *Lantana camara. Asian Pac. J. Trop. Biomed.,* 2(12): 960-965.

2018, Ethnomedicinal Plants: A Biodiversity Treasure Pages 425–468
Editors: V.R. Mohan, A. Doss, P.S. Tresina and V. Sornalakshmi
Published by: ASTRAL INTERNATIONAL PVT. LTD., NEW DELHI

Chapter 18

An Assessment of Angiosperm Diversity of Manakudy Estuary, Kanyakumari District: A Anthropogenic Polluted Estuariane Ecosystem, Tamil Nadu, India

Parthipan, M. Kalaimathi, M. Mahalekshmi and M. Valarmathi

P.G. and Research Department of Botany,
S.T. Hindu College, Nagercoil-2, Kanyakumari District, Tamil Nadu
E-mail: parthipillai64@gmail.com

Introduction

An estuary may be defined as "a semi-enclosed coastal body of water which has a free connection with the open sea and within which sea water is measurably diluted by fresh water from land drainage" Pritchard (1967). It is an abode for unique flora and fauna and considered to be an usually productive ecosystem. According to Khear (1998) estuaries have a unique combination of physical features, associated with their shape, catchment area, connection to the sea and tidal regime Odum (1988) reported that water salinity is considered as the dominant factor determining the distribution of plants in estuarine marshes and marine marshes (Partridge and Wilson, 1989). Due to urbanization, estuarine wetlands are increasingly being replaced by residential/industrial areas throughout the world (Rosa *et al.*, 2003). Kennish (2001) reported that pollution due to industrial and residential developments, recreation and other activities in both the estuary and its catchment area greatly affect these sensitive habitats, as well as living resources of the estuary.

The Manakudy estuary and its surrounding terrestrial environment is marked with various human activities such as salt industry, coir industry, lime shell, sand mining and waste disposal on the banks of the estuary. The Manakudy estuary acts as a sink for sewage discharge in Manakudy and near by area, which adversely affect its self - purification capacity. There are two ways to reverse the quality and loss of any habitat: i) Conservation of currently viable habitat and ii) restoration of degraded habitats. Conservation properties and restoration measures must be decided based upon the inventorisation of biological diversity (Kunte *et al.*, 1999). World Conservation Monitoring Centre (WCME, 1992) reported that botanical assessments, such as floristic composition and structure are essential to understand the extract of phyto-diversity of any ecosystem. Kanyakumari District is a favourite spot for taxonomists for over centuries as it includes a part of diversity rich southern Western Ghats. A number of people have explored the flora of Kanyakumari District from 18th centuary onwords. (Lawson, 1894., Rama Rao, 1914., Jacob, 1942., Lawrence, 1959 and 1960., Nayar, 1959., Govindarajalu, 1967; 1970, 1972 and 1973., Vasudevan, 1967., Henry and Subramanyam, 1972 and Sharma *et al.*, 1973) Most of the studies have focused on the hill ranges of Kanyakumari District (Rao, 1975 and Henry and Swaminathan, 1979, 1980 and 1981).

Kanyakumari District which is occupied by aquatic and shore vegetation establishes strong association between aquatic and terrestrial ecosystems. This plays a significant role in the primary production, nutrient cycling, and serves as bio indicators for eutropication process (Reginold, 2010., Rejini Balasigh, 2010., Sathya and Sangeetha 2010., Thangam *et al.*, 2010 and Vasantha, 2010). Only a few studies have documented plant diversity of costal regions of Kanyakumari District (Rao and Sastry, 1972., Rao *et al.*, 1974., Rao *et al.*, 1975 and Ravikumar, 2010).

However, little attention has been paid to the systematic study on aquatic and wetland plants of Kanyakumari District (Sukumaran and Raj, 2009., Sukumaran *et al.*, 2010 and Sukumaran and Jeeva, 2010). Consequently botanical explorations of wetlant plants are necessary to gain more knowledge on species richness as well as their geographical disturbance. In view of this fact the present research work is carried out to document the angiosperm diversity of the Manakudy estuary, Kanyakumari District, Tamil Nadu, India as well as to identify the threats that previal in the estuarine region and to suggest necessary conservation strategies.

Review of Literature

Indian sub continent has a coastline of about 7516 Km long with 2.1 million Km^2 exclusive economic zone and 0.13 million Km^2 continental shelf (Khoshoo, 1996) and spread over 9 states and the coastline of Tamil Nadu has a length of about 1076 Kms, constitutes about a 15 per cent of the total coastal length of India. The coastal zone is one among the 10 geographically important habitats of the Indian sub continent (Rodgers and Panwarm, 1988).

According to Champion and Seth (1968) and Rao and Sastry (1972 and 1974) the Indian coastal vegetation type can be categorized into the following two subtypes: Strand vegetation and Estuarine borderland vegetation. The estuarine border land vegetation is characterised by dense and gregarious growth of woody plants, shrubs and succulent herbs in varying proportions dispersed on a relief lying under constant influence of tidal and fresh water resources. In India this type of vegetation is predominant in the deltatic regions and reverine mouths along the east coast whereas it is confined to seainlets, small river mouths, lagoons, bays and back water systems along the west coast.

The nearly 7516 Km long coast line in India is subjected to the wave actions of the Bay of Bengal, the Arabian Sea and the Indian Ocean on the east and the west coast and at the southern most end respectively. The west coast line is long and more or less straight starting from.Cape Comerin in the south to the 20° parallel N and includes two peninsulas, Saurashtra and Kutch and is spread into the states of Gujarat, Maharashtra, Karnataka (Mysore) and Kerala. The east coast runs in wide curves changing directions form north to north east from 16° parallel N and is covered in the state Tamil Nadu, Andra Pradesh, Orissa, and West Bengal. In general the seaboards are wider along the east coast rather than in the west coast. Between the coastline and the adjacent plains lie the following two major inshore ecosystem; strand and Eastuary. Strand vegetation of the Tamil Nadu coastal belt which extends south of Pulicat lake up to cape comarin exhibits similarities of vegetation to the Andra Pradesh Coast was reported by Sastry and Rao (1973) and Rao (1974).

Anarda Rao and Sastry (1974) studies of the coastal vegetation of India is described in detail with regard to its sub - types, the taxa components in each and also its phytogeographical affinities. Temperate coastal dunes as compared to studies on tropical coastal dunes were well studied and documented (Koske and Gemma, 1997), (Kulkarni *et al.,* 1997). Coastal vegetation contains many species of specific flora and thus it is an ecological storehouse rich in Biodiversity and also has high ecological values (Untawale, 1994 and Banerjee, 1944). Coastal sand dune comprise a variety of flora and fauna, which play a vital role in provisioning ecological and economical services to the coastal communities (Martinez *et al.,* 1997).

Rao, (2000) made an extensive studies on coastal and marine plant diversity of Andaman and Nicobar Islands. He reported that 3552 plant species forming 76.5 per cent of estimated species. Angiospermic flora constitutes about 2000 species of which 80 per cent are indigenous. The endemism in Angiosperms is about 14 per cent. The flora in general has affinities to the Indo - Malayan region.

Temperate coastal dunes as compared to studies on tropical coastal dunes were well studied and documented (Sridhar and Bhagya 2007). Survey of common folk medicinal plants as AVM Canal used by fisherman community in Kollencode Revenue village segment of Kanyakumari district was made by Sundar *et al.* (2009). About 85 plant species belonging to 71 genera from 34 families on AVM canal were used as medicines by fishermen community to treat various ailments.

Padamavathy *et al.* (2010) made a detailed vegetation survey of Naliavadu village, Puducherry, India. They reported that a total of 41 species belonging to 35 genera and 20 families were identified at different distances from the shoreline towards inland where various edaphic factors decline facilitating more floral colonization.

Traditional herbal medicines of the coastal diversity in Tuticorin District, Tamil Nadu India were reported by Muthukumar and Selvin Samuvel (2010). They reported that the coastal plant species of Tuticorin district bears high medicinal and ecological values. A total of 41 medicinal plants have been collected and their popular uses were recorded.

Antimicrobial activity of medicinal plants along Kanyakumari coast, Tamil Nadu, India was reported by *Ravikumar et al.* (2010). In vitro antibacterial and antifungal activity of the *chloroform* extracts of the seventeen different coastal medicinal plants were screened against different gram possitive and gram negative and fungal ornamental fish pathogens of the selected plant. *Datura innoxia* showed wide range of antimicrobial activity, against many of the fish pathogens.

Skumanan and Jeeva (2011) made qualitative floristic surveys in the wetland ecosystem of Kanyakumari district, Tamil Nadu, India. They reported 124 species of Angiosperms belonging to 31 families and 81 genera from the study area. Dominant families in the wetlands of Kanyakumari district were Poaceae in 39 species followed by Cyperaceae (24), Scrophelariaceae (9) Commelinaceae (5) Acanthaceae and Convolvulaceae (4 species each) and Hydrocaritaceae and Verbenaceae (3 species each). Out of 31 families 10 families were represented by 2 species each, where as 13 species were monospecific. They also reported that of the 124 species there are 21 dominant pantropical species 15 subdominant Asiatic species and 11 Codominat Indian species 5 species endemic to Southern Western Ghats.

Bala Krishnan *et al.* (2012) studied the biodiversity of plant resources during the end of summer season in Rameshwanm Coastal Island. Nearly 5 angiosperm plant species were recorded in the Rameswaram Island. In Dhanuskodi region they observed the xerophytic plant species. *Spinix sporosis* belonging to the family Poaceae.

Selvakumari and Raja Kumar (2010) made the floristic and phytogeographical analysis of Kudiraimozhi Theri in Tuticorin District, Southern India. Plant diversity on coastal sand dune flora, Tirunelveli Dist. Tamil Nadu was investigated by Ramarajan and Murugesan (2014). They investigated current vegetative status of sand dunes at Tirunelveli District and the communities living close to the coastal sand dunes. They reported that 55 sp belonging to 46 genera and 26 families were identified at different distances from the share line. Poaceae was the most common dominant family followed by Cyperaceae, Fabaceae, Molluginaceae, Euphorbiaceae, Amaranthaceae, Asclepiadaceae, Astraceae and Malvaceae.

Enilda Rexy *et al.* (2014) studied the plant diversity and their uses of selected coastal villages of Tirunelveli district Tamil Nadu, India. A total of 61 plant species belonging to 54 genera representing 31 families were collected for the coastal villages *viz.*, Chettikulam, Ethankadu, Kootapuli and Perumanal of Tirunelveli District. Among these four sites Chettikulam had 51 species, Ethankadu had 19 species, Kootapuli had 11 species. About 51 medicinal plants were used by fishermen community, from the study area.

Materials and Methods

Description of Study Area

The present study is carried out in the Manakudy estuary situated between 8° 05' E latitude; 77° 32' N longitude located in Kanyakumari district of Tamil Nadu State and lying along the South West Coast of India (Figure 18.1). Kanyakumari district (longitude 77° 15' - 77° 36' E, latitude 8° 03'- 8° 35' N) is located in the Southernmost tip of the peninsular India. Regarding the boundry, on north – east is bounded by Tirunelveli; north – west is Thiruvananthapuram district; Southern-Eastern is Gulf of Mannar (Bay of Bengal); south and west is Indian ocean and Arabian sea respectively. It covers an area of about 1,675 sq. Km and it is a thickly populated district of South Tamil Nadu. It is administratively divided into 4 Taluks namely Agastheeswaram, Thovalai, Kalkulam and Vilavancode, 9 blocks and 88 villages. Normally, annual rainfall over this district varies from about 103 to 310 cm under the influence of both south – west and north – east monsoons and elevation is 1829 Msl (Raj, 2002 and Dharmaraja *et al.*, 2012). The rivers Pazhayar, Ponnivaikal, Valliar, Pompoori vaikal and Thamiraparani are enriching this district and form smaller and larger estuaries along the south – west coast, Thengapattnam and Manakudy are the two major estuaries. Rajakamangalam, Kadiyapattnam and Colachel are the prominent minor estuaries of this district (Figure 18.1).

According to the census record the populations of Kanyakumari district is 11, 37, 181 (Sukumaran *et al.*, 2008). Tamil and Malayalam are the main languages used by the population. Hindus and Christians form a sizable percentage of the population of the district and there are a number of Muslim dominated belts. The cast system in the society has weakened to a great extent especially after independence because of the growth of education and improvements in transports and communication. 'Nadar' is the major community of this district. Some of the other communities in the district are Nangil Nadu Vellars, Parvavas, Mukthavas, Vilakki Thalanayar, Asari, Chackarevars, Kerala Mudalis *etc.* Rice (*Oryza sativa*) is the stable food of the rich and poor alike in the district. Some poor people use Tapioca (*Manihot esculenta*) as food. Beverages like tea and traditional coffee (using dried ginger and palm sugar = sukku coffee) are widespread even in the rural areas of the district (Raj, 2003).

Topographically this district may be broadly classified as mountainous region, middle region and coastal region.

The Mountainous Region

The mountainous region of the Western Ghats provides a continuous wall along the northern side of the district. Fourteen types of forests from luxurient tropical wet evergreen to tropical thorn forests occur in this district because of the diverse nature of the landscape. The total forest cover is estimated to be 30.2 per cent of the total geographical area of the district and the forests of the district are verdant and virgin forests and are said to be 75 million years old (Champian and Seths, 1968).

In this district Mahendragiri is the highest peak reaching an altitude of about 1800 m. Muthukuzhivayal (C.1500 m) is floristically the richest region of the district (Henry and Swaminathan, 1981). The southern end of the Western Ghats is seen near Aralvaimozhi pass. Southward, the pass rises into a perfectly detached Kathadimalai, a rocky spur extending on its south with two breaks for a distance of about 6 Km and terminating its Maruthuvalmalai, about 5 Km north west of Cape Comorin. In many places hill ranges radiate into the Mirdland Region east to west, enclosing fertile valleys.

The Middle Region

This is full of hills and dales that stretch slantingly from east to west, approximately 60 meters above sea – level. The soil of this area is graveal mixed with pebbles. The Thiruvananthapurum – cape road runs along this strip touching Kuzhithurai, Thuckalay and Nagercoil. Presence of a large number of perennial and ephemeral ponds and irrigation canals promote rich aquatic vegetation. Coconut, paddy, banana, mango and jack are some of the commonly cultivated plants in this region.

The Coastal Region

Kanyakumari district has a long seashore of about 68 Km extending from Rasthakaadu in the south – east to Neerodicolony in the west (Figure 18.1). Most of the beaches are erosional in nature and are enriched with workable deposits of placer minerals (Angusamy and Rajamanickam, 2000). The Sandy belt is however broken at Thengapattnam by the Thambaraparani delta. The soil is alluvial and very fertile. Many of the important townships and the Colachal port are located on this strip. There is no bay or gulf on the western shore. The coast is rocky at cape comerin and Muttom. A chain of rocks locally known as Crocodile Rock, projects a few hundred meters into the sea at Muttom point. A light house has been constructed here to give warning signals to passing ships. There is a sea cliff at Colachel. According to a local Tamil legend, there was a land beyond Cape Comorin with a hill named Kumaricode and a river known as Paruliyar. It is believed to have been immersed under the sea by a deluge thousands of years ago.

Manakudy Estuary

Manakudy estuary is unique and specific as it is the first estuary located on the south west coast of Peninsular India (8° 05' E latitude; 77° 32' N longitude). It is also the second largest estuary in Kanyakumari district (Figure 18.1).

This estuary has area of about 150 hectare is situated about 8 Km northernwest of Cape Comorin in Kanyakumari District (Figure 18.1). It is the confluence of river pazhayar which has its origin from the Western Ghats. Pazhayar, one of the main source of irrigation for the plains of Kanyakumari district originated from Mahendragiri while flowing through the Nanjil Nadu plain the water is diverted for irrigation all along its stretch of 37 km (Figure 18.1).

The Manakudy estuary is bound with varied habitats that include shallow open water, sandy beaches, muddy flats, mangrove forests, river belt and sea grass. This estuary is a typical bar built estuary, which remains land locked during non-monsoon season (January, February, March, April, May and September). It opens during southeast monsoon (June to August) and northwest monsoon seasons

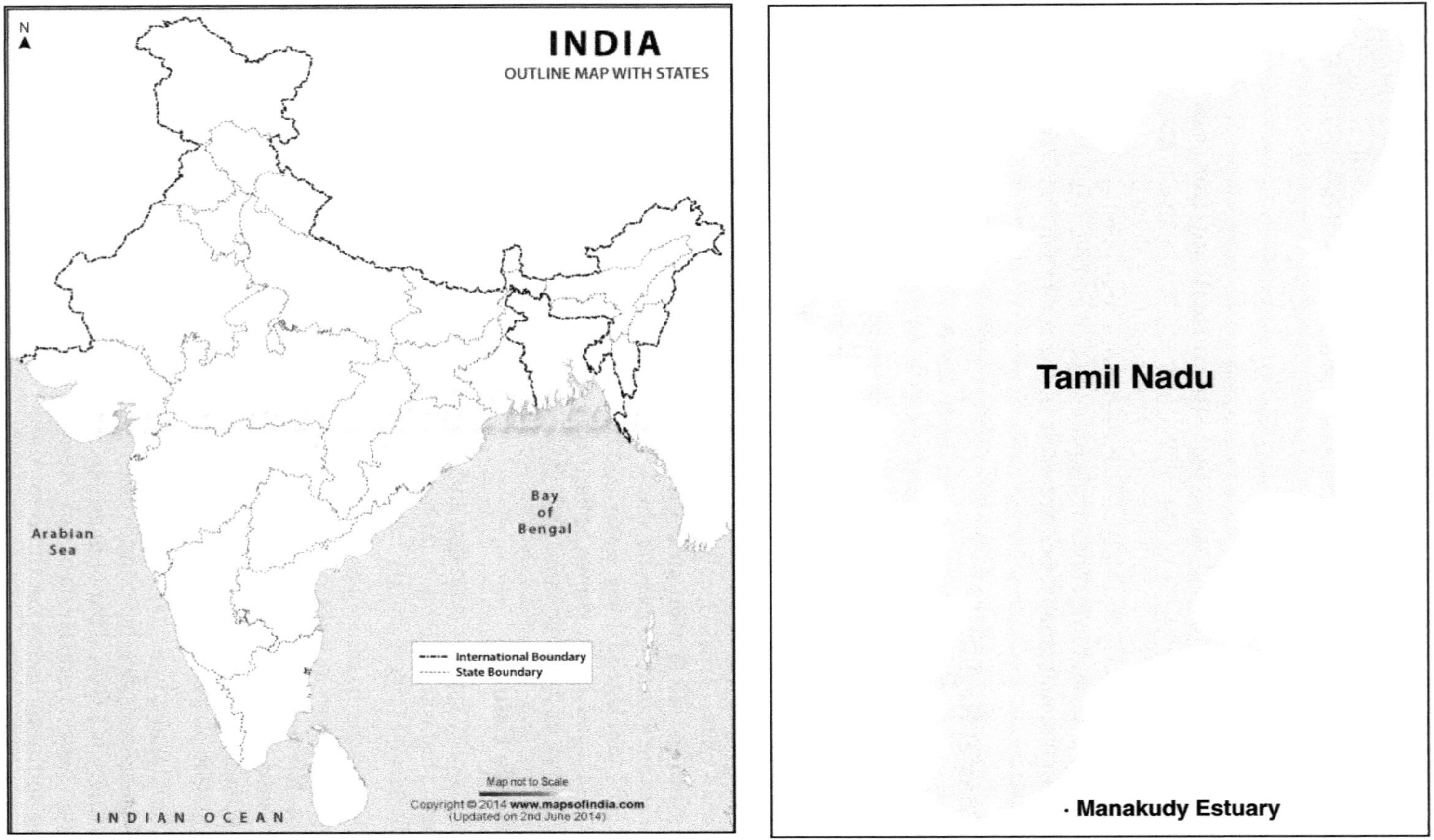

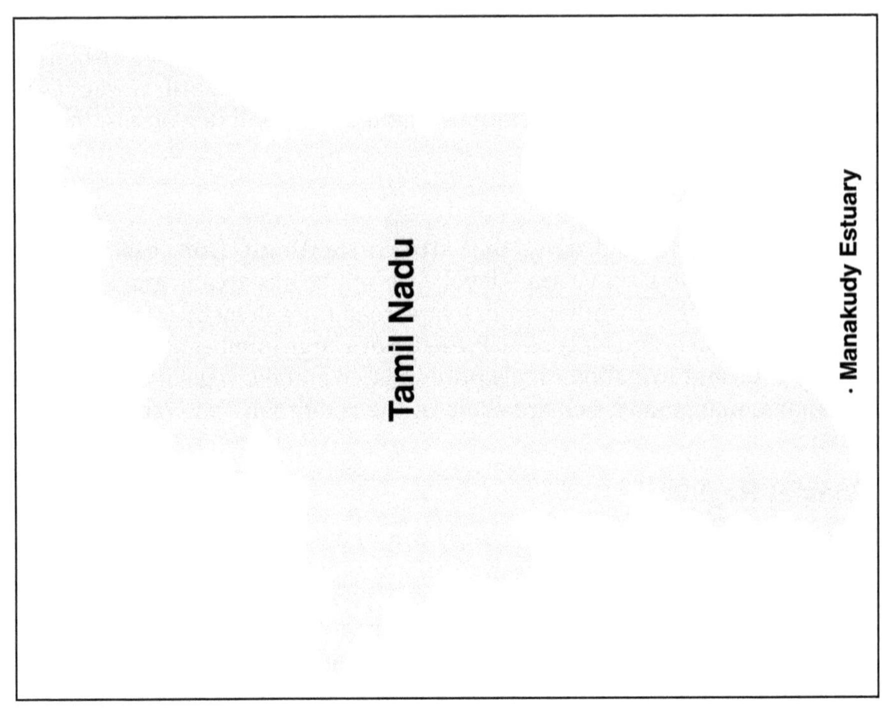

Figure 18.1a: Map Showing the Manakudy Estuary in Kanyakumari District, Tamil Nadu, India.

Figure 18.1b

(October to December) (Figure 18.2A). Mangrooves are a significant ecosystem in the estuary with a luxuriant growth on the mud flats (Figure 18.2B). Mangroove plant species of *Rhizophora mucoronata* and *Avicinnia officinalis* (Figure 18.2C) are planted in the area, which has enhanced prawn breeding. The surrounding terrestrial environment is marked with various human activities such as salt industry (Figure 18.2D), coir industry (Figure 18.2E), lime shell (Figure 18.2F), sand mining (Figure 18.2G), waste disposal and frequent defecation on the banks of the estuary. Some other interesting observations made during the present study is provided in (Figures 18.2H-L).

Data Collection

Regular field surveys were under taken in and around the Manakudy estuary between June 2014 to October 2014. Plants either with flowers or fruits were collected and photographed. The collected plants were tagged and brought to the laboratory for examination. Plants were collected according to the field and herbarium techniques (Pullaiah, 2007). A field note book was maintained during the course of investigation for writing the observable characters in the field. Polythene bags were used to keep the collected materials in fresh condition for short periods. For spot identification and to record their morphological characters, hand lenses were used. Many of the plants were photographed. The collected plants were treated with standard solution of 0.3 per cent mercuric chloride dissolved in 95 per cent alcohol. The poisoned specimens were pressed and after drying, they were pasted on standard sized herbarium sheets using fevicol as adhesive (Fosberg and Sachet, 1965). The identified plants were confirmed with available regional floras (Gamble. 1915 – 1936; Matthew,1982; 1983; 1988., Nair *et al.*, 1983 and Nayer *et al.*, 2006).

Plant families are arranged according to Angiosperm Phylogeny Group III classification (APG III, 2009). Abbrevations of author's names of plant names strictly

Figure 18.2A-L: Interesting Observations made in the Manakudy Estuary.

A) Manakudy Estuary

B) Open bar built Manakudy Estuary

C) Mangrove forest

D) Rhizophora and Avicinia found in mangrove forest

E) Salt pan

F) Coir Industry

G) Lime shell collection

H) Sand collection

I) Waste disposal

J) Sewage discharge

K) Grazing of cattle

L) Migratory birds in the Estuary

followed according to Burmmitt and Powel (1992). Voucher specimens have been deposited at STHC herbarium, South Trivancore Hindu College, Nagercoil for reference. The current nomenclature of all taxa was further determined by reffering to authentic databases, such as International Plant Names Index (IPNI), The plant List and Tropicos.

Results and Discussion

A total of 143 species belonging to 109 genera distributed in 40 families from 20 orders and 7 clades/groups according to Angiosperm Phylogeny Group III Classification (2009) were recorded during the present investigation from Manakudy estuary. These taxa are represented in Table 18.1. Lamids (45 spp.), Fabids (40 spp.), Malvids (34 spp.), Commelinids (17 spp.) and Campanulids (5 spp.) are the major clades/groups representing a total of 136 taxa that constitute 95 per cent of the flora (Figure 18.3).

Table 18.1: Enumeration of Angiosperms of Manakudy Estuary According to APG III (2009)

Sl.No.	*Name of the Clades/ Order/Families*	*Name of Species/Varieties*	*Life Form*	*Worldwide Distribution@*	*V.N*
	COMMELINIDS				
	Arecales Bromhead				
1	Arecaceae Bercht. and J.Presl,nom.cons.	*Borassus flabellifer* L.	T	HT,ST,WT	3635
2		*Cocos nucifera* L.	T		3641
3		*Phoenix sylvestris* (L.) Roxb.	T		3654
	Commelinales Mirb. ex. Bercht., and J. Presl				
4	Commelinaceae Mirb, nom,cons.	*Commelina benghalensis* L.	H	TR,TEM	3628
	Poales Small				
5	Cyperaceae Juss., nom. cons.	*Cyperus rotundus* L.	H	WW	3666
6		*Cyperus tenuiculmis* Boeckeler.	H		3637
7	Poaceae Barnhart, nom. cons.	*Aristida setacea* Retz.	H	WW	3658
8		*Chloris barbata* SW.	H		3566
9		*Cynodon dactylon* (L.) Pers.	H		3681
10		*Dactyloctenium aegyptium* (L.) Willd.	H		3638
11		*Digitaria bicornis* (Lam) Romer and Schultes	H		3595
12		*Eragrostis tenella* (L.) Roem. and Schult.	H		3642
13		*Eragrostis viscosa* (Retz.) Trin.	H		3655
14		*Echinochloa colona* (L.) Link.	H		3611

Contd...

Sl.No.	Name of the Clades/ Order/Families	Name of Species/Varieties	Life Form	Worldwide Distribution@	V.N
15		*Saccharum spontaneum* L.	H		3598
16		*Sporobolus tenuissimus* (Schrank) Kuntye	H		3579
17		*Sporoblous virginicus* (L.) Kunth	H		3639
	EUDICOTS				
	Ranunculales Juss. ex. Bercht. and J.Presl.				
18	Papaveraceae Juss., nom. cons.	*Argemone mexicana* L.	H	WW	3575
	ROSIDS				
	Vitales Juss.ex Bercht. and J.Presl				
19	Vitaceae Juss.,nom.cons.	*Cissus quadrangularis* L.	Cl	PAN	3640
	FABIDS				
	Zygophyllales Link.				
20	Zygophyllaceae R.Br., nom. cons.	*Tribulus terrestris* L.	H	TEM,TR	3577
	Malpighiales Juss.ex. Bercht. and J. Presl.				
21	Calophyllaceae J.Agardh.	*Calophyllum inophyllum* L.	T	THO,TR	3682
22	Euphorbiaceae Juss., nom.cons	*Acalypha indica* L.	H	PAN,WT	3684
23		*Corton sparsiflorus* Morong.	H		3545
24		*Euphorbia antiquorum* L.	S		3686
25		*Euphorbia cyathophora* Murray.	H		3543
26		*Euphorbia hirta* L.	H		3620
27		*Euphorbia tirucalli* L.	S		3552
28		*Jatropha glandulifera* Roxb.	S		3600
29		*Pedilanthus tithymaloides (L.)* Poit.	H		3685
30	Passifloraceae Juss.ex. Roussel.,nom.cons.	*Passiflora foetida* L.	Cl	TRO, esp AM, AF,WT	3547
31	Phyllanthaceae Martinov., nom. cons.	*Phyllanthus amarus* Schumach and Thonn	H	Pan, some TEM	3619
32		*Phyllanthus debilis* Klein ex Willd.	H		3562
33	Rhizophoraceae Pers., nom.cons.	*Rhizophora mucronata* Lam	T	PAN	3679
34	Violaceae Batsch., nom. cons.	*Hybanthes enneaspermus* (L.) F. Muell	H	WW	3604
	Cucurbitales Juss.ex Bercht. and J. Presl				

Sl.No.	Name of the Clades/ Order/Families	Name of Species/Varieties	Life Form	Worldwide Distribution@	V.N
35	Cucurbitaceae Juss., nom. cons.	*Citrullus coloncynthis* (L.) Schrad	Cl	TRO, ST	3571
36		*Coccinia grandis* (L.) Voigt	Cl		3550
37		*Ctenolepis garcini* (L.) C.B	Cl		3676
38		*Mukia maderaspatana* (L.) M.Roem	Cl		3565
39		*Trichosanthes cuspidata* Lam.	Cl		3683
	Fabales Bromhead				
40	Fabaceae Lindl.,nom.cons.	*Acacia planiforns* Wright.	T		3572
41		*Acacia nilotica* (L.) Delile.	T		3673
42		*Caesalpinia bonduc* (L.) Roxb.	S	WW	3687
		Canavalia cathartica Thouars	H		3546
43		*Cassia auriculata* L.	S		3675
44		*Cassia occidentalis* L.	S		3558
45		*Cassia surattensis* Burm.f.	T		3674
46		*Clitoria ternatea* L.	Cl		3602
47		*Crotolaria globosa* Wight and Arm	H		3574
48		*Crotolaria Juncea* L.	S		3588
49		*Crotolaria pallida* Aiton	H		3677
50		*Crotolaria verrucosa* L.	H		3680
51		*Indigofera aspalathoides* Dc	S		3607
52		*Mimosa pudica* L.	H		3560
53		*Prosopis juliflora* (S.W) Dc.	T		3670
54		*Rhynchosia minima* (L.) Dc.	CL		3583
55		*Tamarindus indica* L.	T		3624
56		*Tephrosia purpurea* (L.) Pers	S		3573
	Fagales Engl.				
57	Casuarinaceae R. Br., Nom. cons.	*Casuarina equisetifolia* L.	T	SEA, M, SWP, AUS	3586
	Rosales Bercht. and J.Presl				
58	Rhamnaceae Juss.nom. cons.	*Ziziphus oenoplia* (L.) Mill.	S	WW, esp, TRO, WT	3672
	MALVIDS				
	Myrtales Juss. ex Bercht. and J.Presl				
59	Lythraceae J.St.-Hil.,nom. cons.	*Lawsonia inermis* L.	S	TRO but some TEM	3678
	Brassicales Bromhead				
60	Cleomaceae Bercht. and J. Presl	*Cleome gynandra* L.	H	TRO WT, AM	3569

Contd...

Sl.No.	*Name of the Clades/ Order/Families*	*Name of Species/Varieties*	*Life Form*	*Worldwide Distribution@*	*V.N*
61		*Cleome viscosa* L.	H		3559
	Malvales Juss.ex Bercht. and J.Presl				
62	Malvaceae Juss., nom. cons.	*Abutilon hirsutum* (Vell.) K. Schum.	S	Largly TRO also TEM	3668
63		*Abutilon pannosum* (G.Forst)	S		3564
64		*Corchorus aestuans* L.	S		3585
65		*Hibiscus virtifolius* L.	S		3605
66		*Melochia conchorifolia* L.	H		3601
67		*Pavonia odorata* Willd.	S		3667
68		*Sida acuta* Burm f.	H		3671
69		*Sida cordifolia* Roxb.	H		3688
70		*Thespesia populnea* (L.) Sol.ex.	T		3596
71		*Waltheria americana* L.	H		3690
72	Muntingiaceae C.Bayer, M.W.Chase and M.F.Fay	*Muntingia calabura* L.	T	TRO, AM, IN	3669
	Sapindales, Juss. ex Bercht and J. Presl.				
73	Meliaceae Juss., nom. cons.	*Azadirachta indica* A. Juss	T	PAN	3597
	Caryophyllales Juss. ex Bercht. and J. Presl				
74	Amaranthaceae Juss., nom.cons.,	*Amaranthus hybridus* L.	H		3576
75		*Achyranthes aspera* L.	H	WW, TEM, ST, esp, SA	3689
76		*Aerva lanata* (L.) Juss. ex Shult	H		3581
77		*Aerva persica* (Burm) f. Merr.	H		3645
78		*Alternanthra pungens* Kunth	H		3647
79		*Alternanthra sessilis* (L) RBr. Ex. DC.	H		3662
80		*Alternanthra tenella* Colla	H		3665
81		*Alternathera paronychzioides* St. Amara	H		3663
82		*Amaranthus viridis* L.	H		3548
83		*Gompherena globosa* L.	H		3594
84		*Gompherena serrata* L.	H		3599
85	Cactaceae Juss., nom. cons.	*Opuntia stricta* (Haw.)	S	All NW	3646
86	Gisekiaceae Nakai	*Gisekia pharnaceoides* L.	H	AF, AS	3664
87	Molluginaceae Bartl. nom. cons.	*Mollugo oppositifolia* L.	H	TRO, TEM, largely SA	3615

Sl.No.	Name of the Clades/ Order/Families	Name of Species/Varieties	Life Form	Worldwide Distribution@	V.N
88	Nyctaginaceae Juss., nom. cons.	*Boerhavia diffusa* L.	H	TRO to WT	3606
89		*Boerhavia erecta* L.	H		3554
90	Phytolaccaceae R.Br. nom. cons.	*Rivinia humilis* L.	H	TRO, WT, esp. AM	3613
91	Plumbaginaceae Juss. nom.cons.	*Plumbago zeylanica* L.	H	MED, CA	3609
92	Polygonaceae Juss., nom cons.	*Polygonum barbatum* L.	H	WW	3610
	LAMIDS				
93	Borangiaceae Juss.,nom. cons.	*Heliotropium curassavicum* L.	H	TEM, TRO	3660
	Gentianales Juss. ex Bercht. and J. Presl.				
94	Apocynaceae Juss., nom. Cons.	*Calotropis gigantea* (L.) W. T. Aiton	S	Largely TROP, TEM	3644
95		*Catharanthus roseus* (L.) G.Don	H		3555
96		*Cryptostegia grandiflora* Roxb. ex. R. Br	Cl		3652
97		*Oxystelma esculentum* **(L. f) R. Br. ex. Schultes**	H		3616
98		*Pergularia daemia* (Forssk) Chiov.	Cl		3603
99		*Sarcostemma intermedium* Decne.	S		3656
100		*Thevetia peruviana* (Pers.) K. Schum	S		3618
101		*Tylophora indica* (Burm. F.) Mera	Cl		3648
102		*Wattakaka volubilis* (L.F) Stapf.	S		3551
103	Rubiaceae Juss., nom. cons.	*Canthium coromandelicum* (Burm. f.) Alston	S	WW, TRO	3659
104		*Oldenlandia umbellate* L.	H		3580
105		*Spermacoce hispida L*	H		3561
	Lamiales Bromhead				
106	Acanthaceae Juss., nom. cons.	*Asystesia gangetica* (L.) T. Anderson.	H	Mostly TRO	3627
107		*Avicennia marina* (Forssk) Vierh	T		3657
108		*Avicennia officinalis* L.	T		3557
109		*Barleria buxifolia* L.	S		3612
110		*Barleria cristata* L.	S		3661
111		*Ecobolium ligustrinum* (Vahl) Vollesen	H		3608
112		*Hygrophila auriculata* (Schumach) Heine	S		3590

Contd...

Sl.No.	*Name of the Clades/ Order/Families*	*Name of Species/Varieties*	*Life Form*	*Worldwide Distribution@*	*V.N*
113		*Justicia glauca* Rottler	H		3643
114		*Justicia procumbens* L.	H		3617
115		*Justicia simplex* D. Don	H		3593
116		*Justicia prostrata* (Nees) T. Anderson.	H		3591
117		*Ruellia prostrate* Poir	H		3582
118		*Ruellia tuberosa* L.	H		3626
119	Lamiaceae Martinov, nom. cons.	*Anisomeles indica* (L.) Kuntze	H	WW	3567
120		*Anisomeles malabarica* (L.) R Br. Ex Sims	H		3556
121		*Hyptis suaveolens* (L.) Poit	H		3651
122		*Leucas aspera* (Willd) Link	H		3549
123		*Ocimum americanum* L.	H		3563
124		*Ocimum filamentosum* Forssk	H		3649
125		*Ocimum tenuiflorum* L.	H		3592
126	Pedaliaceae R. Br. nom. cons.	*Pedalium murex* L.	H	Mostly TROP, CO or EH	3629
127	Verbenaceae J. St - Hil. nom. cons.	*Clerodendrum inerme* (L.) Gaerth.	S	PAN, WT, NW	3542
128		*Lantana camara* L.	S		3625
129		*Stachytarpheta jamaicensis* (L.) vahl	H		3587
	Solanales Juss. ex. Bercht. and J. Presl				
130	Convolvulaceae Juss., nom. cons.	*Evolvulus alsinoides* L.	H	WW	3633
132		*Ipomoea pes-caprae* (L.) R.Br.	H		3650
133		*Ipomoea indica* (Burm.) Merr.	H		3623
134	Solanaceae Juss., nom. Cons.	*Datura metel* L.	S	WW	3622
135		*Physalis minima* L.	H		3631
136		*Solanum nigrum* L.	H		3630
137		*Solanum surattense* Burm F.	H		3621
138		*Solanum trilobatum* L.	S		3634
	CAMPANULIDS				
	Asterales Link				
139	Asteraceae Bercht. and J. Presl., nom. cons,	*Parthenium hyterophorus* L.	H		3570
140		*Syndrella nodiflora* (L.) Gaertner	H	WW	3584
141		*Tridax procumbens* L.	H		3544

Sl.No.	*Name of the Clades/ Order/Families*	*Name of Species/Varieties*	*Life Form*	*Worldwide Distribution@*	*V.N*
142		*Vernonia cinerea* (L.) Less	H		3553
143		*Xanthimun indicum* J. Koing ex Roxb	S		3632

H: Herb; S: Shrub; T: Tree; Cl: Climber; HT: Humid Tropics; ST: Sub Tropical; WT: Warm temperate; TR: Tropical; TEM: Temperate; WW: World Wide; PAN: Pantropical; Tho: Throughout; Esp: Especially; AM: America; AF: Africa; SEA: South East Asia; MA: Malaysia; SWP: South West Pacific; AU: Australia; I: India; SA: South Africa; AS: Asia; MED: Mediterranean; SH: Saline habitats; NW: New World; CA: Central Asia; CO: Coastal; EH: Erid Habitats.

@ According to Raviprasad Rao and Prasanna (2010).

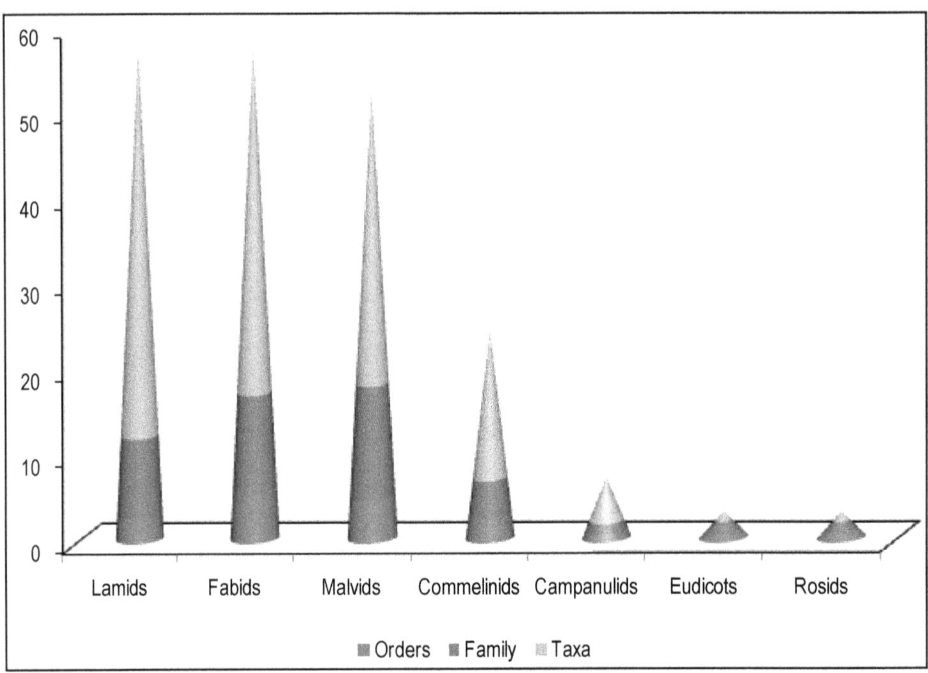

Figure 18.3: Diversity of Orders, Families and Taxa in each Clade.

The most diverse families in the Manakudy estuary includes Fabaceae with 18 species and 12 genera, Acanthaceae with 13 species and 7 genera, Poaceae and Amaranthaceae (11 species each with 9 species and 6 genera respectively), Malvaceae (10 species and 8 genera), Apocynaceae (9 species and 9 genera), Euphorbiaceae (8 species and 5 genera), Lamiaceae (7 species and 4 genera), Asteraceae and Cucurbitaceae (5 species with 5 genera each), Solanaceae (5 species with 3 genera), Convolvulaceae (3 species with 2 genera), Arecaceae, Rubiaceae, Verbenaceae (3 species with 3 genera), Cyperaceae, Phyllanthaceae, Cleomaceae and Nyctaginaceae (2 species with 1 genera), Commelinaceae, Papavaraceae, Vitaceae, Zygophyllaceae, Calophyllaceae, Passifloraceae, Rhizophoraceae, Violaceae, Casuarinaceae, Rhamnaceae, Lythraceae, Muntingiaceae, Meliaceae, Cataceae,

Gisekiaceae, Molluginaceae, Phytolocaceae, Plumbaginaceae, Polygonaceae, Boranginaceae and Pedaliaceae were monospecific (Table 18.2).

Table 18.2: Family-wise Distribution of Plant Species in the Manakudy Estuary

Family	*Genus*	*Species*
Acanthaceae	7	13
Amaranthaceae	6	11
Apocyanaceae	9	9
Arecaceae	3	3
Astraceae	5	5
Boranginaceae	1	1
Cactaceae	1	1
Calophyllaceae	1	1
Casuarinaceae	1	1
Cleomaceae	1	2
Commelinaceae	1	1
Convolvulaceae	2	3
Cucurbitaceae	5	5
Cyberaceae	1	2
Euphorbiaceae	5	8
Fabaceae	12	18
Gisekiaceae	1	1
Lamiaceae	4	7
Lythraceae	1	1
Malvaceae	8	10
Meliaceae	1	1
Molluginaceae	1	1
Muntingiaceae	1	1
Nyctaginaceae	1	2
Papaveraceae	1	1
Pasifloraceae	1	1
Pedaliaceae	1	1
Phyllanthaceae	1	2
Phytolocaceae	1	1
Plumbuginaceae	1	1
Poaceae	9	11
Polygonaceae	1	1
Rhamnaceae	1	1
Rhizophoraceae	1	1
Rubiaceae	3	3
Solanaceae	3	5

Family	*Genus*	*Species*
Verbenaceae	3	3
Violaceae	1	1
Vitaceae	1	1
Zygophyllaceae	1	1
Total	109	**143**

The dominant genera of the flora are *Alternanthera, Crotalaria, Euphorbia* and *Justicia* (4 spp. each), *Cassia, Ipomea, Ocimum* and *Solanum* (3 Spp. each) and Abutilon, Acacia, Aerva, Anisomelas, Boerhavia, Cleome, Cyperus, Ecbolium, Eragrosis, Phyllanthus, Ruellia, Sida and Sporobolus (2 spp. each) (Table 18.2).

An analysis of the floristic diversity denotes that the family Fabaceae dominates the flora with 18 species, followed by Acanthaceae with 13 species, Amaranthaceae and Poaceae with 11 species each, Apocynaceae 9 species, Euphorbiaceae 8 species and Lamiaceae 7 species (Figure 18.4). Fabaceae dominates the flora with 37 species in floristic diversity assessments of Adayar Estuary, Chennai was already reported by Karthigeyan *et al.* (2013).

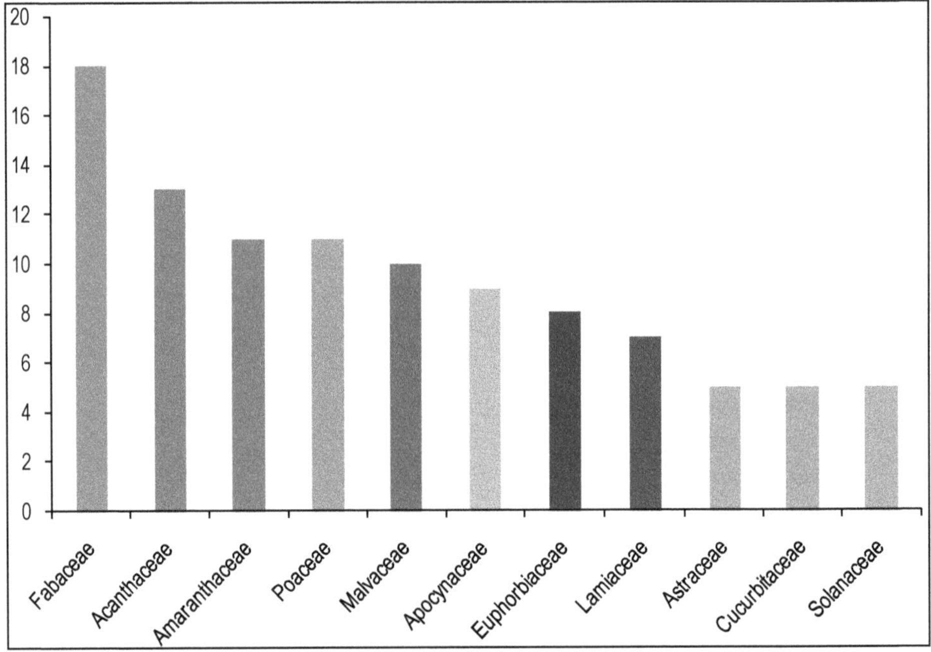

Figure 18.4: Diversity of Dominant Families of Plants Species from the Area under Study

Habit wise distribution analysis shows that herbs dominant the flora of Manakudy estuary with a total of 85 species (59.44 per cent), followed by trees with 30 species (20.97 per cent), trees with 16 species (11.18 per cent) and Climbers with 12 species representing 8.39 per cent of the flora (Figure 18.5).

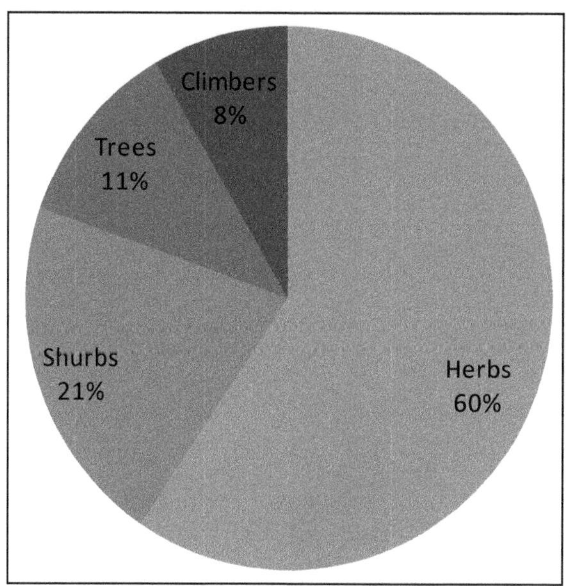

Figure 18.5: Habit-wise Distribution of Plant Species in the Manakudy Estuary.

Life form classification of plants species shows that contribution of herbaceous plants to total diversity to the maximum i.e 85 species belonging to 60 genera and 28 families, followed by shrub by plants 30 species belonging to 27 genera and 13 families, trees with 16 species belonging to 14 genera and 8 families and the remaining species was climbers 12 species belonging to 12 genera and 5 families. (Figure 18.6).

Three species of mangroove and halophytes were reported, which include *Avicennia officinalis, Avicennia marina* and *Rhizophora micronata*. These plants are used as medicine by local inhabitats of Manakudy village, Kanyakumari District. Similar findings were reported by Ravindran *et al.* (2005).

One interesting observation is made in the present study area. That is a threatened medicinal plant species *Calophyllum inophylum* L. belonging to Calophyllaceae was collected from the region of fresh water from the Manakudy estuary. This threatened species (Rangit Daniels *et al.*, 1995) possess potential threat due to decline in the population because of various biotic (Wain House *et al.*, 1998) and abiotic factors being a littoral species (Thengare *et al.*, 2006) the seeds are taken away in the tidal water thereby limiting the propagation rate.

A large number of exotic flora (25 species) were reported from the Manakudy estuary (Table 18.3). Which include *Ruellia tuberosa, Alternanthera paronichyoides, A. pungens, Gomphrena serrata, Catharanthus roseus, Thevitia peruviana, Cryptostegia grandiflora, Synedrella nodiflora, Tridax procumbens, Opuntia stricta, Cassia Occidentalis, Casuarina equisetifolia, Ipomea indica, Croton sparsiflorus, Euphorbia cyathophora, Euphorbia tirucalli, Pedilanthus tithymaloides, Phyllanthus amarus, Hyptis suaveolens, Mimosa pudica, Prosopsis juliflora, Argemone*

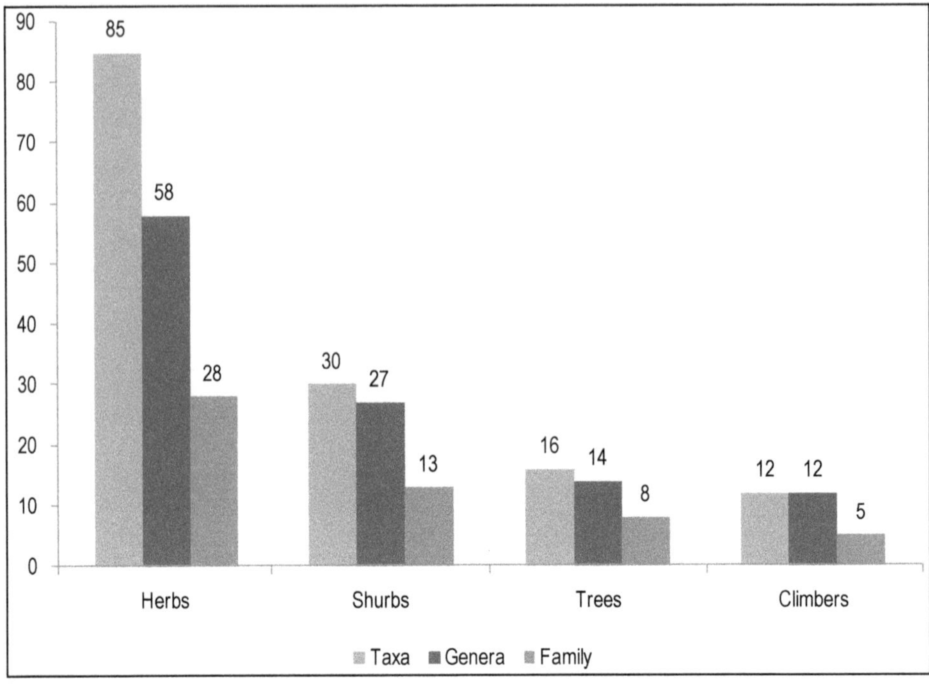

Figure 18.6: Life-form Classification of Plant Species Collected from the Manakudy Estuary.

amaricana, *Corchorus aestuans*, *Muntingia calabura* and *Stachytarpheta jamaicensis* (Nair *et al.*, 2006).

Table 18.3: List of Plant Species Recorded from Manakudy Estuary and their Economic Uses

Sl. No.	*Botanical Name*	*Family Name*	*Habit*	*Uses*	*Flowering and Yielding Fruits*
1	*Abutilon hirsutum* (Vell.) K.Schum.	Malvaceae	H	M, F, Fi, O	Throughout the year
2	*Abutilon pannosum* (G.Forst.) Schltdl.	Malvaceae	US	Fi	October-March
3	*Acacia planifrons*	Fabaceae	T	Fo, Ta	February-May
4	*Acacia nilotica* (L.) Delile	Mimosaceae	T	M, F, Ti	May-October
5	*Acalypha indica*	Euphorbiaceae	H	M	Throughout the year
6	*Achyranthes aspera* L.	Amaranthaceae	H	M, F, D	Throughout the year
7	*Aerva lanata* (L.) Juss.	Amaranthaceae	H	M, F	Throughout the year
8	*Aerva persica* (Burm.f.) Merr.	Amaranthaceae	US	M	Throughout the year
9	*Alternanthera paronichyoides* A. St.-Hil #	Amaranthaceae	Prostrate H	-	December-August.
10	*Alternanthera pungens* Kunth #	Amaranthaceae	Prostrate H	M	Throughout the year

Contd...

Sl. No.	Botanical Name	Family Name	Habit	Uses	Flowering and Yielding Fruits
11	*Alternanthera sessilis* (L.) R.Br. ex DC.	Amaranthaceae	Prostrate H	M, F, Fo	Throughout the year
12	*Alternanthera tenella* Colla #	Amaranthaceae	ST	-	Throughout the year
13	*Amaranthus hybridus* L.	Amaranthaceae	H		Novemeber-December
14	*Amaranthus viridis* L.	Amaranthaceae	H	M, F, Fo	Throughout the year
15	*Anisomeles indica* (L.) Kuntze	Lamiaceae	USH	M,O	Throughout the year
16	*Anisomeles malabarica* (L.) R.Br. ex Sims	Lamiaceae	US	M	Throughout the year
17	*Argemone mexicana* L. #	Papaveraceae	US	M	Nov ember– may
18	*Aristida setacea* Retz.	Poaceae	PH	Fo	Throughout the year
19	*Asystasia gangetica* (L.) T.Anderson	Acanthaceae	H	M,Fo	Throughout the year
20	*Avicennia marina* (Forssk.) Vierh.	Acanthaceae	T	M, Fo	March-July
21	*Avicennia officinalis* L.	Aviciniaceae	T	M, F, Fo, Ti	April-November
22	*Azadirachta indica* A.Juss.	Meliaceae	T	M,F	Flower: March-May, Fruit:April-July
23	*Barleria buxifolia* L.*	Acanthaceae	AU	M	Throughout the year
24	*Barleria cristata* L.	Acanthaceae	U	M	December- March
25	*Boerhavia diffusa* L.	Nyctaginaceae	H	M,F,Fo	Through out the year
26	*Boerhavia erecta* L. #	Nyctaginaceae	H		Throughout the year
27	*Borassus flabellifer* L.	Arecaceae	T	M, F, Fi, Ti	December-January
28	*Caesalpinia bonduc*	Fabaceae	S	M, O	August-February
29	*Calophyllum inophyllum* L.	Calophyllaceae	T	M,F, Ti, O	March-October
30	*Calotropis gigantea* (L.) Dryand.	Apocynaceae	S	M, Fi, Gm	Throughout the year
31	*Canavalia cathartica* Thouars	Fabaceae	CH	M,F	September-December
32	*Canthium coromandelicum* (Burm.f.) Alston	Rubiaceae	S	M	Flower: June-Aug, Fruit: Throughout the year
33	*Cassia surattensis*	Caesalpiniaceae	T	M	Throughout the year
34	*Cassia auriculata* L.	Caesalpiniaceae	S	M, F, Fi	Flower.Jan-May Fruit.February-July
35	*Cassia occidentalis* L. #	Caesalpiniaceae	S	M, F, O	Throughout the year
36	*Casuarina equisetifolia* L. #	Casuarinaceae	T	M,D, Ti, Or	December-April
37	*Catharanthus roseus* (L.) G.Don #	Apocynaceae	H	M	Throughout the year
38	*Chloris barbata* Sw.	Poaceae	PH	Fo	Throughout the year

Sl. No.	Botanical Name	Family Name	Habit	Uses	Flowering and Yielding Fruits
39	*Cissus quadrangularis*	Vitaceae	Cl	M, F, Fi	August-December
40	*Citrullus colocynthis* (L.) Schrad.	Cucurbitaceae	TH	M,F,O	August-March
41	*Cleome gynandra* L.	Cleomaceae	H	M,F,O	Throughout the year
42	*Cleome viscosa* L.	Cleomaceae	H	M,F,O	Throughout the year
43	*Clerodendrum inerme* (L.) Gaertn.	Verbenaceae	S	M	Throughout the year
44	*Clitoria ternatea* L. #	Fabaceae	CH	M, F, Fo, D	Through out the year
45	*Coccinia grandis (L.)* Voigt	Cucurbitaceae	CH	M,F	Apr-Aug
46	*Cocos nucifera* L.	Arecaceae	T	M, F, Ti, D, O	Throughout the year
47	*Commelina benghalensis* L.	Commelinaceae	H	M,F	Throughout the year
48	*Corchorus aestuans* L. #	Malvaceae	H	M, Fi	September-January
49	*Crotalaria globosa* Wight and Arn. *	Fabaceae	H	-	January
50	*Crotalaria juncea* L.	Fabaceae	US	M,Fi	June-December
51	*Crotalaria pallida* Aiton	Fabaceae	S	M, Fo, Fi, O	Throughout the year
52	*Crotalaria verrucosa* L.	Fabaceae	H	M	Oct-April
53	*Croton sparsiflorus* Morong #	Euphorbiaceae	US	Gm,O	Throughout the year
54	*Cryptostegia grandiflora* Roxb. ex R.Br. #	Apocynaceae	Cl	L, Fi	Throughout the year
55	*Ctenolepis garcini* (L.) C.B.Clarke	Cucurbitaceae	Cl	M	December-January
56	*Cynodon dactylon* (L.) Pers.	Poaceae	PH	M, F, Fo, Or	Throughout the year
57	*Cyperus rotundus* L.	Cyperaceae	PH	M,F,O	Throughout the year
58	*Cyperus tenuiculmis* Boeckeler	Cyperaceae	PH	-	October-January
59	*Dactyloctenium aegyptium (L.)* Willd.	Poaceae	AH	M,F, Fo	Throughout the year
60	*Datura metel* L.	Solanaceae	US	M	Throughout the year
61	*Digitaria bicornis* (Lam.) Roem. and Schult.	Poaceae	H	Fo	July-December
62	*Ecbolium ligustrinum* (Vahl) Vollesen	Acanthaceae	U	M	Throughout the year
63	*Echinochloa colona* (L.) Link	Poaceae	H	-	July-February
64	*Eragrostis tenella* (L.) Roem. and Schult.	Poaceae	AH	-	March-Dec
65	*Eragrostis viscosa* (Retz.) Trin.	Poaceae	AH		December-March
66	*Euphorbia antiquorum* L.	Euphorbiaceae	S	M	December-March

Contd...

Sl. No.	Botanical Name	Family Name	Habit	Uses	Flowering and Yielding Fruits
67	*Euphorbia cyathophora* Murray #	Euphorbiaceae	H	-	Throughout the year
68	*Euphorbia hirta* L.	Euphorbiaceae	H	M,F	Throughout the year
69	*Euphorbia tirucalli* L.	Euphorbiaceae	S	M,Fo,L	April-June
70	*Evolvulus alsinoides* (L.) L.	Convolvulaceae	H	M	Throughout the year
71	*Gisekia pharnaceoides* L.	Gisekiaceae	H	M,F, Fo	August-February
72	*Gomphrena globosa* L.	Amaranthaceae	H	M, F, Or	Throughout the year
73	*Gomphrena serrata* L. #	Amaranthaceae	Dece-mber H	-	Throughout the year
74	*Heliotropium curassavicum* L.	Boraginaceae	H	-	November-March
75	*Hibiscus vitifolius* L.	Malvaceae	US	-	January-September
76	*Hybanthes enneaspermus* (L.) F.	Violaceae	H	M	Throughout the year
77	*Hygrophila auriculata* (Schumach.) Heine	Acanthaceae	H	M	August-May
78	*Hyptis suaveolens* (L.) Poit. #	Lamiaceae	S	M, F, Fo, Gm, O	Throughout the year
79	*Indigofera aspalathoides* DC.	Fabaceae	H	M	July-December
80	*Ipomoea pescarpae* (L.) R. Br. Muell.	Convolvulaceae	H	M, F, D, Fo, O	Throughout the year
81	*Ipomoea indica* (Burm.) Merr. #	Convolvulaceae	TH	-	July-January
82	*Jatropha glandulifera* Roxb.	Euphorbiaceae	S	M, Fi, D, O	Throughout the year
83	*Justicia glauca* Rottler*	Acanthaceae	US	-	October-February
84	*Justicia procumbens* L.	Acanthaceae	H	M,F	June-December
85	*Justicia prostrata* (Nees) T. Anderson	Acanthaceae	Pros-trate H	-	Throughout the year
86	*Justicia simplex* D. Don	Acanthaceae	H	-	August-May
87	*Lantana camera* L.	Vernemaceae	S	M, F, O, or	Throughout the year
88	*Lawsonia inermis* L.	Lythraceae	S	M, D, O	February-August
89	*Leucas aspera* (Willd.) Link	Lamiaceae	H	M,F	Throughout the year
90	*Melochia corchorifolia* L.	Malvaceae	H	M, F, Fi	July-February
91	*Mimosa pudica* L. #	Mimosaceae	H	M, Fo, O	Throughout the year
92	*Mollugo oppositifolia* L.	Molluginaceae	H	M,F	Throughout the year
93	*Mukia maderaspatana (L.)* M.Roem.	Cucurbitaceae	CH	M	Throughout the year
94	*Muntingia calabura* L.	Muntingiaceae	T	M, F, Fi, Or	Throughout the year

Sl. No.	Botanical Name	Family Name	Habit	Uses	Flowering and Yielding Fruits
95	*Ocimum americanum* L.	Lamiaceae	Aromatic H	M, F, O	October-February
96	*Ocimum filamentosum* Forssk.	Lamiaceae	H	-	May-June
97	*Ocimum tenuiflorum* L.	Lamiaceae	H	M,F,O	Throughout the year
98	*Oldenlandia umbellata* L.	Rubiaceae	H	M	September-March
99	*Opuntia stricta* (Haw.) Haw. #	Cactaceae	S	M, F, Fi, Fo	Throughout the year
100	*Oxystelma esculentum* (L. f.) Sm	Apocynaceae	TH	M,F,L	November-March
101	*Parthenium hysterophorus #*	Asteraceae	H	M	Throughout the year
102	*Passiflora foetida* L.	Passifloraceae	CH	M, F, Or	Throughout the year
103	*Pavonia odorata* Willd.	Malvaceae	US	M, F, Fi, or, O	Throughout the year
104	*Pedalium murex* L.	Pedaliaceae	H	M,F	July- December
105	*Pedilanthus tithymaloides* (L.) Poit. #	Euphorbiaceae	H	M,L, Or	February-June
106	*Pergularia daemia* (Forssk.) Chiov.	Apocynaceae	Cl	M,F,Fi	October-April
107	*Phoenix sylvestris* (L.) Roxb.	Arecaceae	T	M, Fi, F, Fo	October-April
108	*Phyllanthus amarus* Schumach. and Thonn. #	Phyllanthaceae	H	M	Throughout the year
109	*Phyllanthus debilis* Klein ex Willd.	Phyllanthaceae	H	-	January-October
110	*Physalis minima* L.	Solanaceae	H	M, F	Throughout the year
111	*Plumbago zeylanica* L.	Plumbaginaceae	US	M	November-March
112	*Polygonum barbatum* L.	Polygonaceae	H	-	September-February
113	*Prosopis juliflora* (Sw.) DC. #	Mimosaceae	T	-	Flower:January-April,Fruit:February-July
114	*Rhizophora mucronata* Lam.	Rhizophoraceae	T	M, F	Throughout the year
115	*Rhynchosia minima* (L.) DC.	Fabaceae	CH	M,Fo	Throughout the year
116	*Rivina humilis* L.	Phytolaccaceae	US	-	Throughout the Year
117	*Ruellia prostrata* Poir.	Acanthaceae	H	M	Throughout the year
118	*Ruellia tuberosa* L. #	Acanthaceae	H	M, Or	March-November
119	*Saccharum spontaneum* L.	Poaceae	PH	M, Fo, Fi	November-January
120	*Sarcostemma intermedium* Decne.*	Apocynaceae	Cl	M	August-December
121	*Sida acuta Burm.f.*	Malvaceae	H	M, Fi	Throughout the year
122	*Sida cordifolia* L.	Malvaceae	US	M, Fi	Throughout the year
123	*Solanum nigrum* L.	Solanaceae	H	M, F	Throughout the year

Contd...

Sl. No.	Botanical Name	Family Name	Habit	Uses	Flowering and Yielding Fruits
124	*Solanum surattense* Burm. f.	Solanaceae	H	M,F, Fo	September-April
125	*Solanum trilobatum* L.	Solanaceae	CLS	M,F	August-April
126	*Spermacoce hispida* L.	Rubiaceae	H	M, F, Fo	August-February
127	*Sporobolus tenuissimus* (Schrank) Kuntye	Poaceae	H	-	September-December
128	*Sporobolus virginicus (*L.*)* Kunth	Poaceae	PH	Sand binder	July-September
129	*Stachytarpheta jamaicensis* (L.) Vahl #	Verbenaceae	W	M, F, Fo	August – May
130	*Synedrella nodiflora (*L.*)* Gaertner #	Asteraceae	H	M,F	Throughout the year
131	*Tamirindus indica #*	Fabaceae	T	M, F	March-May
132	*Tephrosia purpurea* (L.) Pers.	Fabaceae	US	M, Fi, Gm, D,O	Throughout the year
133	*Thespesia populnea* (L.) Sol. ex Corrêa	Malvaceae	US	M, F, Fi, Ti, Gm	Throughout the year
134	*Thevetia peruviana* (Pers.) K.Schum.#	Apocynaceae	S	M, Or, L	Throughout the year
135	*Tribulus terrestris* L.	Zygophyllaceae	H	M, F, Fo	Throughout the year
136	*Trichosanthes cuspidata* Lam.	Cucurbitaceae	CS	M	July-March
137	*Tridax procumbens* (L.) L. #	Asteraceae	H	M	Throughout the year
138	*Tylophora indica* (Burm. f.) Merr.	Apocynaceae	TS	M,Fi	August-December
139	*Vernonia cinerea* (L.) Less.	Asteraceae	H	M,F	Throughout the year
140	*Waltheria americana* L.	Malvaceae	H	-	Throughout the year
141	*Wattakaka volubilis* (L.f.) Stapf	Apocynaceae	L	M,Fi,F	Throughout the year
142	*Xanthium indicum* Roxb.	Asteraceae	W	M,F	Throughout the year
143	*Ziziphus oenopolia* (L.) Mill.	Rhamnaceae	S	M, F	Flower:October-November, Fruit:November-December

M: Medicine; F: Food; Or: Ornamental; Fo: Fodder; Fi: Fiber; L: Latex; Ti: Timber; O: Oil; Gm: Gum; R: Resin; H: Herb; CS: Climbered; TS: Trees; W: Wine; S: Shrub

*: Endemic to India; #: Exotic.

During the present study, typical coastal sand dune plants like *Borassus flabellifer*, *Casuarina equisetifolia*, *Gisekia pharanoides* and *Thespecia populnea* were collected from Manakudy estuary. The primary stabilizing plants *viz.*, *Gisekia pharanoides*, *Thephrosia purpurea*, coastal tree species like *Borassus flabellifer* and the introduced *Casuarina equisetifolia* were collected from the study area.

Economically Important Plants

Most of the plants recorded from Manakudy estuary of Kanyakumari district are economically important. The medicinal plant ranked first with 108 species,

60 species having food value, 28 species having fibre, 21 species having fodder, 22 species having oil, 8 species having ornamental and 7 species having timber value and 5 species were gum yielding plants (Figure 18.7). Many multipurpose species were also reported from the present study area.

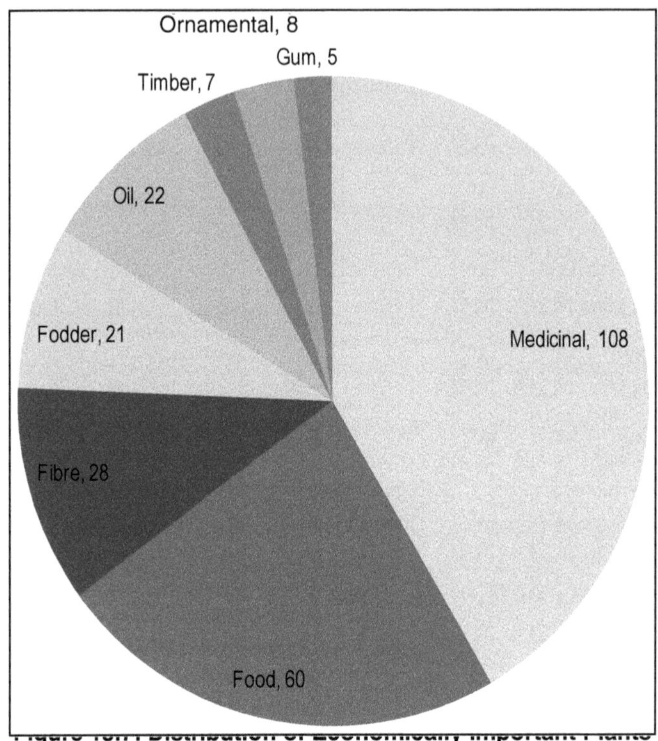

in the Area Selected for Study.

Plants enumerated from the Manakudy - estuary and their economic uses are presented in Table 18.3. Many species collected from the present study area are medicinally valuable, which include *Aerva lanata, Achyranthes aspera, Coccinia grandis, Datura metel, Calophyllum inophyllum, Evolvulus alsinoides,* Mukia maderaspatana, *Indigofera aspalatho ides, Mollugo oppositifolia, Pedalium - murex, Plumbego zeylanica, Physalis minima, Solanum trilobatum, Clerodendrum inerme, Tribulus - terrestris etc.* Plants used by the local people for medicine preparation to treat various diseases are enumerated and provided (Table 18.4).

About 108 medicinal plants are used by local community to cure various diseases. Out of 108 medicinal plants whole plant parts of 47 (43.51 per cent) plant species, roots and leaves of 41 (37.96 per cent) plant species each, fruits from 17 species (15.74 per cent), bark from 12 species (11.1 per cent), oil and stem from 3 species each, resins and tender shoots from 2 species each and gum, tuber, root bark, and inflorescence from 1 plant species each were used for the medicine preparation to treat various ailments by local people from the present study area (Figure 18.8).

Table 18.4: Name of the Medicinal Plants Used to Cure different Ailments by the Local People of the Manakudy Estuary

Sl.No.	*Name of the Species*	*Name of the Plant Parts Used as Medicine*	*Medicinal Values*
1	*Abutilon hirtum* (Vell.) K. Schum.	Root	Cough, dysuria, diabetes fever, thirst, dysmenorrheal, piles, diarrhoea, worm troubles.
2	*Acacia nilotica* (L.) Detile	Bark, Le, Se, GM	Urino - genital disease, wounds, haemorrhage, diarrhea, dysentery, piles, worm infestiation, skin disease, ulcers, cough, dysentery, polyuria
3	*Acalypha indica*	R, L, Wp	Worm infestation, Burns, Piles, Cough, Constipation, Ear diseases, Urinary diseases.
4	*Achyranthes aspera* Linn.	L, R, Se	Insect bites, Scorpion stings
5	*Aerva lanata* (L). Juss. ex Schult	WP	Diuretic, bladder stones, haemetemesis, diabetes, Lithiasis.
6	*Aeva persica*	Wp, R	Dioretic, diarrhoea
7	*Alternanthera pungens* Kunth	WP	Diuretic, gonorrhoea.
8	*Alternanthera sessilis* (L.) R. Br. ex. Dc	WP	Night blindness, burning sensation, diarrhoea, leprosy, skin diseases, fever and snake bites.
9	*Amanranthus viridis*	Wp	Body heat
10	*Anisomeles indica* (L.) Kuntze	WP	Gastric catarrh, itches, intermittent fevers, convulsions, uterine affections.
11	*Anisomeles malabarica* (L) R.Br. ex simes	Le	Dyspepsia, fever rheumatism. fever, haematological disorders, infertile, catarrh, cough, coilic
12	*Argemone mexicana* L.	R, Se, La	Opthalmia, blisters, ulcers, leprosy, cough, catarrhal affections asthma, ulcers
13	*Asystasia gangetica* (L.) *T. Anderson*	WP	Swellings, worms infestation rheumatism
14	*Avicinia officinalis*	R, Fr, Se, Re	Tumours, boils, abscesses, small pox sores, contraceptive, ulcers, snake bites poison
15	*Azadirachta indica* A. Juss.	B, Le, Fl, Fr, Se	Skin diseases, antiseptic boils ulcer, eczema, stomachic, tonic purgative, emollient, skin troubles, rheumatism, intestinal worms, impurity of blood, eye diseases, diabetes, fever, dysponea, cough, liver disorder.
16	*Barleria buxifolia* L.	L, R	Cough and inflammations
17	*Barleria cristata* L.	L, R	Cough, snake bites, inflammations
18	*Boerhavia diffusa* L.	WP, R	Jaundice, anaemia, cardiac diseases, hydrocele, stomach disorder, pain, piles, urinary calculi, diabetes, diuretic
19	*Borassus flabellifer* L.	R, Le, In, Ba, Fr, Se	Bleeding, thirst, oedema, fainting, burning sensation, anorexia, cardiac disease, amenorrhoea, dysuria, skin diseases.
20	*Caesalpinia bonduc*	R, L, Fr	Hydrocele, urinary calculi, abdominal disorders

Sl.No.	Name of the Species	Name of the Plant Parts Used as Medicine	Medicinal Values
21	*Calophyllum inophyllum* L.	Ba, Fl, Re, Se, O	Ophthalmitis, internal haemorrage, rheumatism, skin affections
22	*Calotropis gigantea* (L.) R. Br	R, Ba, Le, Fl and La	Skin diseasse, promotes digestive power, leprosy, pruritus, ulcers, liver diseases, abdominal disorders, piles, cough, asthma catarrh, worms
23	*Cassia auriculata* Linn	Ba, R, Le, Fr, Se	Urinary discharge, Tumors, skin diseases, asthma, chronic fever, tooth ache, diabetes, spermatorrhoea.
24	*Cassia occidentalis* (L) Link	R, Le, Se	Skin troubles, eczema, filariasis cough, dyspnoea, polyuria, ringworm, purifies blood, improves digestion
25	*Cassia surrattensis* (Burm f) Irwn and Barnehy	Se	Skin disorders.
26	*Casuarina equisitifolia*	Ba, Le	Diarrhoea, dysentery
27	*Catharanthus roseus* (L) G. Don	WP	Insommia, blood cancer, diabetes, dysentery, hypertension, anxiety, neurosis and cardiac tonic
28	*Cissus quadrangularis*	WP, R. St.	Piles, abdominal disorders, diarrhoea, dysentery, venereal diseases, worm infestation.
29	*Citrullus colocynthis* (L) Schrad	Fr, R	Laxative, diuretic, antiperiodic, haemorrhages, alcohol poisoning
30	*Cleome gyhandra* (Linn.)	Se, R, Le	Snake bites, anticancer, fever, skin disease, rheumatism muscular pain, head ache
31	*Cleome viscosa* L	WP	Wounds, ulcers, ear ache, antiseptic, painful joints
32	*Clerodendrum inerme* (L.) Gaertn.	R, Le	Fever, rheumatism
33	*Clitoria ternate* Linn.	R, B, Se, Le	Purgative, diuretic, absortion, diuretic, laxative, chronic bronchitis.
34	*Coccinia grandis* (L). voigt	Le, R, Fr	Ear pain, antidiabetic, antipyretic, head ache, skin eruptions, bloody dysentery.
35	*Cocos nucifera* Linn.	Fl, R, Fr, O	Diuretic, astringent, fever, urinary disorders.
36	*Commelina benghalensis* L.	WP	Haemorrhage, leprosy, rheumatism.
37	*Corchorus aestuans* L.	Le, Se	Pneumonia, wounds, stomach disorders, dysentery, haemorrhage, leprosy, itching, rat poisioning.
38	*Critalaria Pallida* Ait	R, Le	Diarrhea, dysentery, bleeding disorders, swelling leprosy, other skin diseases, ulcers.
39	*Critalaria, verrucosa* L.	WP	Vomiting, diarrhoea, dysentery, bleeding, swellings, leprosy, skin diseases
40	*Crotalaria Juncea* L.	Le, Se	Anorexia, blood disorders, amenoorhoea, skin diseases, obsesity, fever, poisioning, dysentery
41	*Ctenolepis garcini*	Wp	Ulcers

Contd...

Sl.No.	*Name of the Species*	*Name of the Plant Parts Used as Medicine*	*Medicinal Values*
42	*Cynodon dactylon* (L.) Pers	WP	Fever, burning sensation, chronic diarrhea, dysentery, dropsy, wounds, catarrhal ophthalmia, haemorrhage skin diseases, vomiting piles, menorrhagia
43	*Cyperus rotundus* Linn	Tu	Indigestion, diarrhea, demulcent, diuretic, vermifuge, diaphoretic
44	Dactyloctenium aegyp-ticum	Wp	Worm infestation, wounds
45	*Datura metel* L.	WP	Scorpion sting poisoning, psychosis, epilepsy, fever, delirium, burning sensation, dysponoea, leprosy, dysuria, skin diseases, asthma, cough, piles, ear ache, eye diseases
46	*Ecbolium ligustrinum* (Vahl) *vollesen*	WP	Dysuria, jaundice, menorrhagia, rheumatism
47	*Euphorbia antiquorum* L.	R, S, Le, La	Nervine troubles, dropsy, kill maggots
48	*Euphorbia hirta* Linn	WP	Antiasthmatic, worms, cough, breast pain, cardiovascular pain
49	*Euphorbia tirucalli* Linn.	TB, Le, S, R, Ba	Leprosy, leucorrhoea, Whooping cough, asthma dropsy, leprosy, jaundice tumours, scorption stings
50	*Evolvulvus alsinoides* Linn.	WP, Le, O	Fever, anthelmintic, haemorrhage
51	*Gisekia pharnaceoides*	Le	Chest disorders, worm infestation mental disorders
52	*Gomphrena globose*	R	Cough
53	*Hybanthus enneas permus*	Wp	Diarrhoea, Polyuria, Leucorrhoea
54	*Hygrophila auriculata (schumach.) Heine*	Se, R, WP	Dysuria, urinary calculic arthritis, amavata, eye disease, bladder stones, arrest abortion
55	*Hyptis Suaveolens* (L.) Poit.	WP	Worm, infestation
56	*Indigofera aspalathoides*	Wp	Leprosy, cancer, oedema, abscess, Skin diseases
57	*Ipomoea pes-caprae*	L, Se,	Rheumatism, Stomach ache
58	*Jatropha glandulifera* Roxb	Se	Skin disease, snake bite poisoning, leprosy, oedema.
59	*Justicia procumbens* L.	WP	Laxative, diaphoretic, diuretic, expectorant, asthma, cough, rheumatism.
60	*Lantana camere*	L.	Haemorrhage, rheumatism, malaria.
61	*Lawsonia inermis*	L, Fl, Se, Ba	Rheumatoid arthritis, headache, ulcers, leprosy, fever, diarrhoea.
62	*Leucas aspera* (Willd.) Link	WP	Worm problems, fever, intestinal disorders, catarrh in children, cough, dyspepsia, jaundice psoriasis, respiratory disease, skin diseases
63	*Melochia conchorifolia* L.	Le	Dysentery, bites of water snakes.

Sl.No.	Name of the Species	Name of the Plant Parts Used as Medicine	Medicinal Values
64	*Mimosa pudica* L.	WP, R	Urinary complaints, sinus, sores, piles, diarrhea, leprosy, uterine disorders, haemorrhage, wounds, oedema, skin diseases, burning sensation
65	*Mollugo oppositifalius* L.	WP	Stomachic, antiseptic, skin diseases
66	*Mukia maderaspatana* (L) Roem	WP	Burning sensation, dyspsia, flatulence, constipation, ulcers, cough, neuralgia, odontalgia, vertigo polyuria, tuberculosis
67	*Ocimum americanum* L.	WP	Helminthiasis, anorexia, dyspepsia, flatulence, dysentery, poisonous affections, haemolysis, migraine
68	*Ocimum tenuilylorum* L.	WP	infantile cough, cold, bronchitis, catarrh, dysentery, improve appetite, skin disease, itches, ringworm, leprosy, intestinal worms, ulcers, poisonous affection fevers
69	*Oldenlantia umbellata*	Le, R, B	Asthuma, bronchitis, tuberculosis, haemoptysis
70	*Opuntia striata* (Haw.) Haw.	Ba and Fr	Whooping cough, spasmodic, cough gonorrhoea and opthalmia.
71	*Oxystelma esculentum*	Wp, R	Antiseptic, jaundice
72	*Parthenium hysterophorus*	Wp	Febrifuge, emmenagogue analgestic, dysentery.
73	*Passiflora foetida* L.	Fr	Asthma, biliousness, hysteria, itches.
74	*Pavonia odorata* Willd.	R	Rheumatism, fever
75	*Pedalium murex*	Wp	urinary disorders, leucoderma, Urinary calculi
76	*Pedilanthus thymaloides*	Le	Venereal diseases warts, leucoderma patches
77	*Pergularia daemia* (Forssk) Chiov.	WP	Diarrhoea, asthma, rheumatism, amenorrhoea, dismenorrhoea, snake bites, anaemia leprosy, piles
78	*Phoenix sylvestris*	Fr, Gu, Se, Ts	Dystentery, ulcer, wheezing, digestive disorders, urinary disorders, eye diseases.
79	*Phyllanthus amarus* Linn.	WP	Diuretic troubles in genito urianary tract, febrifuge, stomach troubles, fever, dysentery.
80	*Physalis minima* L.	WP, R	Diabetes, heat sores of mouth, burning sensation, ulcers, cough, bronchitis and erysipelas
81	*Plumbago zeylanica* L.	R	Leprosy, oedema, piles, worm infestation, anaemia, anorexia, fever, bronchial asthma, leucoderma, skin diseases, dyspepsia, anaemia, diabetes, inflammations, bronchitis, helminthiasis elephantiasis
82	*Rhizosphora mucronata*	B	Haemorrhage, diabetes
83	*Rhyncosia minima*	Le	Abortifacient
84	*Rivina humilis* Linn.	Le, F, Fr	Febrifuge
85	*Ruellia prostrata* poir	L	Gonorrhoea, ear disease.
86	*Ruellia tuberosa* L.	WP	Diarrhoea

Contd...

Sl.No.	*Name of the Species*	*Name of the Plant Parts Used as Medicine*	*Medicinal Values*
87	*Saccharum spontaneum*	R, Wp	Vomiting, mental disorders, abdominal disorders, dyspnoea, anaemia, obesity, urinary disorders, diarrhoea, haemorrhage, urinary calculli, eye diseases, burning sensation
88	*Sarcostemma inter-medium*	R, S, La	Rabies, digestive disorders, eczema, scabies, poisoning, skin diseases, abdominal disorders
89	*Sida acuta* Burm. F.	R	Arthritis, leucorrhoea, gonorrhea, diarrhea promote strength.
90	*Sida cordifolia* L.	R	Astringent, diuretic, tonic, urinary troubles, cystitis, stranguary, haematuria in hemiplegia, sciatica, facial paralysis, dysentery, rheumatism, neurological disorders, headache, tuberculosis ophthalmia.
91	*Solanum nigrum* L.	WP, Fr	Leprosy, piles, fever urinary disease, eye diseases, vomitting, heart aliments inflammatory swelling, enlargement of liver and spleen, increases semen.
92	*Solanum surattense* Burm. F.	WP, R	Dyspnoea, anorexia, cough, worm infestation, blood disorders, oedema, influenza, amenorrhoea, coryza, dysuria, bladder stones, skin diseases, toothache.
93	*Solanum trilobatum* L.	R, Fr	Cough and bronchial disorders
94	*Spermacoce hispida* L.	WP	Haemorrhoids, dyspepsia, colic, flatulence, general desability, gallstones, diarrhea, dysentery
95	*Stachytarpheta jamicensis* (L.) Vahl.	WP	Worms, venereal disease, ulcers, erysipelas, dropsy, stomach ailments, vomiting, fever rheumatic inflammations.
96	*Synedrella nodiflora*	Le	Laxative, rheumatism
97	*Tamirindus indica*	Ba, L, Fl, Fr, Se	Ulcers, anaemia, oedema, leucorrhoea, pain.
98	*Tephrosia purpurea* (L.) Pers.	WP	Inflammations, skin diseases, elephantiasis, dyspepsia, stomach ache, flatulence, asthma, bronchitis, anaemia, verminosis, fever, boils, pimples, syphilis, gonorrhea, rat poisoning
99	*Thespesia populnea* (L.) Sal. ex correa	WP	Psoriasis, ring worm, guinea worm, leprosy, dysentery, cholera, diabetes
100	*Thevetia perceviana* (Pers.) K. Schum	Rb, Le, Se, La	Diuretic, cardio tonic, cures oedema, tumours, fever, body cooling.
101	*Tribulus terrestris* L.	R, F	Dysuria, oedema, bronchial asthma, cardiac disease, nervous disorders, urinary disorder, cough, diabetes, piles, dropsy, rheumatism, burning sensation.
102	*Trichosanthus cusbidata*	Wp	Antimicrobial
103	*Tridax procumbens* L.	Le	Bronchial catarrh, dysentery, diarrhoea, haemorrhages, bruises and wounds
104	*Tylophora indica*	Wp	Snake bites, food poisoing, syspnoea, asthma, bronchial, asthuma, spider poisoning

Sl.No.	Name of the Species	Name of the Plant Parts Used as Medicine	Medicinal Values
105	*Vernonia cinerea* Less	R, Le	Purifies blood, bile and semen, excessive bleeding, chronic skin disease, bladder stones, piles, worms.
106	*Wattakaka volubilis*	Le	Boils, abscesses
107	*Xanthium indicum*	R, Le, Fr, Se	Cancer, ulcers, boils, herpes, inflammation, urinogenetal disorders
108	*Ziziphus oenoplia* (L.) Mill.	R, Fr	Anthelmintic, digestive, anti septic, hyperacidity, ascaris infection stomachalgia, healing of wounds.

WP: Whole plant; R: Root; Fr: Fruit; Se: Seed; Le: Leaves; LA: Latex; F: Flower; B-Bark, Gm: Gum; Re: Resinous exuadtes; St: Stem; Fl: Flower; Tu: Tuber; L: Leaves; Rb: Root bark.

The plant parts used for the preparation of medicine, whole plants were found to be the most frequently used morphological parts of the medicinal plants collected from the study area (Figure 18.8). The name of illness, number of medicinal plants used to treat respective diseases are given (Table 18.5).

Plant species such as *Alternanthera sesilis, A. tenella, Amaranthus viridis* and *Cleome gynandra* leaves are used as green vegetables by the poor people selected from the study area. Unripe fruits of *Coccinia grandis* are used as vegetables by the local people from the study area.

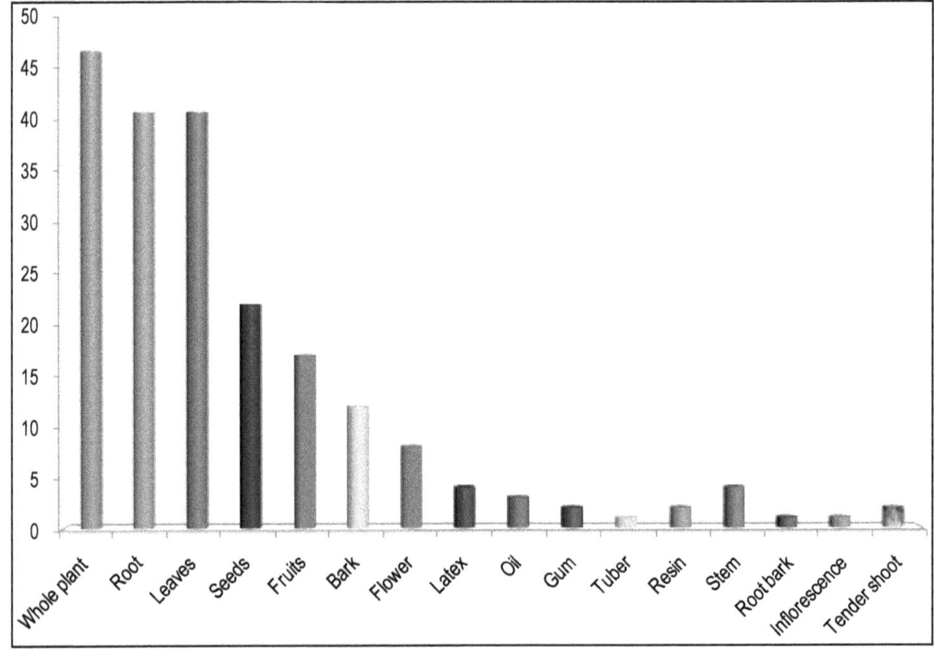

Figure 18.8: Number of Medicinal Plants Parts Used by Local Peoples of the Present Study Area.

Table 18.5: Name of the Disease, Total Number of Plant Used and Name of the Medicinal Plants Used by the Local People from the Area taken under Study

Sl.No.	Name of the Disease	Total Number of Plants Ysed	Name of the Species
1	Asthma	10	*Calatropis gigantea, Justicia procumbens, Pergularia daemia, Cassia auriculata, Euphorbia tirucalli, Ephrosia purpurea, Argemone mexicana, Passiflora foetida, Plumbago zeylanica, Datura metal, Tribulus terresteris.*
2	Anxiety	1	*Catharanthus roseus*
3	Swelling	5	*Asystasia gangetica, Cocos nucifera, Crotataria pallida, Crotalaria. Verrucosa, solanum nigrum*
5	Infestation worms	10	*Asystasia gangetica, Calotropis gigntea, Vernonia cinerea, Euphorbia hirta, Leucas aspera, Stachytarpheta jamaicensis, Hyptis suaveolens, Acacia hilotica, plumbago zeylanica, Solanum surattense*
6	Intestinal worms	2	*Azadirachta indica, Ocimum tenuiflorum*
7	Cough	20	*Cassia occidentalis, Barleria buxifolia, Barleria cristaeta, Justicia procumbens, Calocropis gigantea, Opuntia striata, Mukia madraspatana, Euphorbia hirta, Euphorbia trucalli, Anisomeles malabarica, Leucas aspera, Ocimum tenuiflorum, Abutilon hirstum, Azadirachta indica, Argemone Mexicana, Datura metal, Physalis minima,. Solanum nigrum, Solanum surattense, Solanum crilobatum, Tribulus terrestris*
8	Inflammation	7	*Stachytarpheta jamaicensis, Barleria buxifalia, Barleria cristata, Tephorsia purpurea, Hyptis suavelolens, Plumbago zeylanica, Tribulus terrestris*
9	Dysuria	6	*Ecbolium ligustrinum, Abutilon hirtum, Datura metal, Tribulus terrestris, Hygrophila auriculata, Heine Solanum surattense*
10	Jaundice	5	*Ecbolium ligustrinum, Vollesen Euphorbia tirucalli, Leucas aspera, Boerhavia diffusa*
11	Menorrhagia	3	*Echolium Ligustrinum, Vollesen Cynoden dactylone*
12	Rheumatism	11	*Ecbolium ligustrinum, Vollesen Justicia procumbens, Pergularia daemia, Cleome gynandra, Calophyllum inophyllum, Commelina benghalensi, Pavonia odorata, Sida cordifolia, Azadirachta indica, Stachytarpheta jamaicensis, Vahl. Tribulus terrestris*
13	Diaphoretic	2	*Justicea procumbens, Cyperus rotundus*
14	Diuretic	14	*Justicia procumbens, Aerva lanata, Alternanthera pungens, Achyranthes aspera, Thevetia peruviana, Colos nucifera, Citrullus colocynthis, Schrad cyperus rotundus,. Phyllanthus amarus, Clitoria ternate, Sida cardifolia, Boerhavia diffusa, Cocos nucifera, Solanum nigrum*
15	Urinary Calculic	6	*Hydrophila auriculata, Cassia auriculata, Phyllanthus amarus, Mosa pudica,. Boerhavia diffusa, Tribulus terestris*
16	Arthritis	1	*Hydrophila auriculata*
17	Amavata	1	*Hydrophila auriculata*

Sl.No.	Name of the Disease	Total Number of Plants Ysed	Name of the Species
18	Eye disease	5	*Alternanthera sessils* *Hygrophila auriculata, Heine Azadirachta indica, Juss Datura metel, Solanum nigrum*
19	Bladder stones	2	*Aerva lanata, Hydrophila auriculata*
20	Laxative	2	*Citrullus colocynthis, Justicia procumbens*
21	Diarrhoea	11	*Ruellia tuberose, Alternanthera sessils, pergularia daemia, Crotalaria Pallid, Crotalaria verrucosa, Abutilon hirtum, Sida acuta, Acacia nilotica, Mimosa pudica, Cynoden dactylon, Spermacoce hispida*
22	Gonorrhoea	3	*Ruellia prostrate, Alternanthera pungens, Sida acuta*
23	Ear disease	5	*Ruellia prostrate, Cleome viscose, Coccinia grandis, Ocimum tienui florum, Datara metal*
24	Haemetemesis	1	*Aerva lanata*
25	Diabetes	9	*Aerva lanata,Catharanthus roseus, Cassia auriculata, Abutilon hirstum, Azadirachta indica, Boerhavia diffusa, Plumbago zeylanica, Physalis minima, Tribulus terrestris*
26	Lithiasis	1	*Aerva lanata*
27	Leprosy	12	*Alternanthera sessils, Calotropis gigntea, Pergularia daemia, Commelina benghalensis, Euphorbia tirucalli, Jatropha glandulifera, Crotalaria pallid, Crotalaria verrucosa, Ocimum tenuiflorum, Mimosa pudica, Datura metal, Corchorus aestuans*
28	Skin disease	22	*Alternanthera sessils, Borassus flabellifer, Calotropis gigntea, Vernonia cinerea, Cassia auriculata, Cassia occidentalis, Cassia surrattensis (Burm f), Coccinia grandis, Jatropha glandulifera, Crotalaria Juncea, Crotalaria pallid, Crotalaria verrucosa, Tephrosia purpurea, Leucas aspera, Ocimum tenuiflorum, Azadirachta indica, Acacia nilotica, Delile Mimosa pudica, Mollugo oppositifolices, Plumbago zeylanic, Cynodon dactylon, Solanum surattense*
29	Fever	18	*Alternanthera sessils, Evolvulvus alsinoides, Cocos nucifera, Oleome gyandra, Phyllanthus amarus, Tephrosia purpurea, Anisomeles malabarica, Leucas aspera, Ocimum tenuiflorum, Abutilon hirtum, Plumbago zeylanica, Cynodon dactylon, Datura metel, Solanam nigrum, Azadirachta indica, Clerodendrum inerme, Stachytarpheta Jamaicensis*
30	Snakebite	4	*Alternanthera sessils, Pergularia daemia, Cleome gynandra, Jatropha glandurlifera*
31	Insommia	1	*Catharanthus roseus*
32	Blood cancer	1	*Catharanthus roseus*
33	Dysentery	12	*Catharanthus roseus, Tridax procumbens, Crotalaria Juncea, Crotalaria pallid, Crotalaria verrucosa, Ocimum americanum, Ocimum tenuiflorum, Thespesia populnea, Correa Melochia corchorifolia, Corchorces aestuans, Cynodon dactylon, Acacia nilotica*

Contd...

Sl.No.	Name of the Disease	Total Number of Plants Ysed	Name of the Species
34	Hypertension	1	*Catharanthus roseus*
35	Vomiting	4	*Crotalaria verrucosa, Cynodon dactylon, Solanum nigrum, Stachytarpheta Jamaicensis*
36	Heurosis	1	*Catharanthus roseus*
37	Cardio tonic	3	*Catharanthus rosesus, Thevetia peruviana, Boerhavia diffusa*
38	Oedema	5	*Thevetia peruviana, Jatropha glandulifera, Borassus Flabellifer, Mimosa pudica, Solanum surattense*
39	Tumours	1	*Thevetia peruviana*
40	Body coaling	2	*Thevetia peruviana, Citrullus colocynthis*
41	Bleeding	3	*Borrassus Flabellifer, Crotalaria pallid, Crotalaria verrucosa*
42	Ulcers	9	*Calotropis gigntea, Cleome viscose, Mukia madera spatana, Crotalaria pallid, Ocimum tenuiflorum, Azadirachta indica, Physalis minima, Stachytarpheta Jamaicensis*
43	Fainting	1	*Borassus flabellifer*
44	Burning	1	*Borassus flabellifer*
45	Anorexia	1	*Borassus flabellifer*
46	Amenorrhoea	3	*Borassus flabellifer, Pergularia daemia, Solanum surattense*
47	Liver disease	2	*Calocropis gigntea, Solanum nigrum*
48	Piles	11	*Calotropis gigntea, Pergularia daemia, Vernonia cinera, Abutilon hirtum, Acacia nilitica, Mimosa pudica, Boerhavia diffusa, Plumbago zeylanica, Cynodon dactylon, Datura metal, Solanum nigrum*
49	Rheumatism	10	*Pergularia daemia, Cleome gynandra, Calophyllum inophyllum, Commelina benghalensis, Anisomeles malaharica, Sida cordifolia, Azadirachta indica, Clerodendrum inerme, Stachytarpheta Jamaicensis, Tribulus terrestris*

Leaflets obtained from *Phoenix sylvestris* are split and woven into mats, brooms, baskets. Toddy is obtained from the inflorescence. Inflorescence of *Aristida setacea* is used for making brooms by the local people and the brooms are commercially available in the local markets or street vendors. *Thevetia peruviana* kernel taken from their poisonous seeds, are used by the local people for sucide purpose.

The people of the area, selected for study use 10 medicinal plants to treat poisonous bites (Table 18.6). Out of 9 plants, 5 plant species are used to treat snake bites, 2 species are used to treat rat bites and 1 species namely *Achyranthus aspera* was used to treat insect bites and scorpion stings. One interesting observation made by the researchers of the present study is the antipoisonous plant *Melochia corchorifolia* leaf is used by people to treat water snake bite poison.

Table 18.6: List of Antipoisonous Plants Collected from the Study Area

Sl.No.	Name of the Plants	Useful parts	Poison
1	*Achyranthes aspera*	Leaves, Root, Stem	Insect bite, Scorpion stings
2	*Alternanthera sessilis*	Whole plant	Snake bite
3	*Avicinia officinalis*	Root, Fruit, Stem, Resin	Snake bite
4	*Cleome gynandra*	Stem, Root, Leaves	Snake bite
5	*Corchorus aestuans*	Leaves, Stem	Rat poisoning
6	*Jatropha glandulifera*	Stem	Snake bite
7	*Melochia corchorifolia*	Leaves	Water snake bite
8	*Perguleria daemia*	Whole plant	Snake bite
9	*Tephrosia purpurea*	Whole plant	Rat poisoning
10	*Tylophora indica*	Whole plant	Spider poisoning

One of the exotic plant species reported from the Mankudy estuary, *Parthenium hysterophorus* is found to be harmful to native flora. This American flora (Sukumaran and Parthipan, 2014) has spread very fast in the past two decades in the area selected for study and posing a threat to as biological pollutants due to their destructive effects on natural and man - managed ecosystem (Westbrooks, 1991). Serious ecological effects of the fast - spreading introduced flora have been already reported by Di Castri *et al.* (1990), Punalekar *et al.* (2010). Non - indigenous plant species are considered a major threat to bio - diversity was reported by various researchers (Mooney, 1988; Lodge, 1993; Huston, 1994 and Mc Geoch *et al.*, 2006). Plants either with flowers or fruits were collected and photographed (Figures 18.9A–R).

During the present investigation, contamination of estuary water by plastics, thermocol, aluminium foil and other waste materials are observed to cause adverse effects on the mangrove habitats (Figures 18.2A-L). The surrounding terresterial environment is marked with various human activities such as newly constructed bridge, Coir industry, lime shell and sand mining, waste disposal and cattle grazing. These activites are identified as threats to the estuarine vegetation of Manakudy.

Environmental laws should be strictly implemented to prevent or to reduce the sewage discharge both from industries and residents. Similarly, dumping of non - biodegradable materials into the estuary should be prohibited. It is suggested to increase the vegatation cover by planting fast growing native littoral tree species along the estuary. In addition to this, environmental awareness should be created among the people residing near by the village *viz.*, Keezha Manakudy, Mela Mankudy and people residing on the banks of the Pazhayar river regarding the importance of ecosystems, biodiversity and conservation of nature and their values.

Conclusion

The present study reveals that a number of valuable plant species are found in the Manakudy estuary of Kanyakumari district. These plant species taken from the area of present study are extremely important, because they play a vital role in the medicinal and social life of people. Conservation and Judicious utilization of

Figure 18.9: List of Medicinal Plants Collected from the Manakudy Estuary.

A) *Borassus flabellifer* L.

B) *Phoenix sylvestris* (L.) Roxb.

C) *Cynodon dactylon* (L.) Pers.

D) *Rhizophora mucronata* Lam

E) *Citrullus coloncynthis* (L.) Schrad

F) *Caesalpinia bonduc* (L.) Roxb.

G) *Crotolaria globosa* Wight and Arm

H) *Cleome viscosa* L.

I) *Mollugo oppositifolia* L.

J) *Heliotropium curassavicum* L.

K) *Oxystelma esculentum* (L. f) R. Br. ex. Schultes

L) *Thevetia peruviana* (Pers.) K. Schum

M) ***Wattakaka volubilis*** **(L.F) Stapf.**

N) ***Ocimum tenuiflorum*** **L.**

O) ***Pedalium murex*** **L.**

P) ***Clerodendrum inerme*** **(L.) Gaerth.**

Q) ***Ipomoea pes-caprae*** **(L.) R.Br.**

R) ***Solanum trilobatum*** **L.**

this plant wealth is important because they have become threatened due to over exploitation, clearing of mangrove forest for fire wood, industrialization, rapid urbanization human settlements, *etc.* If conservation measures are not initiated in the near future there may be a great loss of plant diversity of the Manakudy estuary. Thus, there is ample scope for further research on plant diversity, community attributes and natural regeneration. Elaborate bio diversity explorations from the area slected from the present study could be helpful in developing suitable measures for conserving precious plant wealth. The findings from the above mentioned studies will help in using appropriate plant species for plantation. The reorientation of local indigenous knowledge through environmental awareness programmes by the Government of India or Tamil Nadu will be an effective strategy for conservation of plant resources in the Manakudy estuary, Kanyakumari district. These conservation efforts can help to improve the life of the local people and also to maintain ecological balance of the Mankudy eastuary, Kanyakumari district.

References

Ananda Rao, T. and Sastry, A.R.K., 1974. *Bull. Bot. Sur. India,* 16(1 and 2): 101-115.

Angusamy, N and G.V. Rajamanikam.2000. *J.Geol. Soc. India.* 56(8): 199-211.

APG III.2009. An update of Angiosperm Phylogeny Group Classification for the orders and Families of Flowering Plants. *Bot. J. Linn. Soc.* 16:105-121.

Balakrishnan, V., Prabhu, V. and Sundari, T., 2012. *Asian J. Plant Science* and *Res.,* 2(4): 539-545.

Banerjee, L.K., 1994. *Bull. Bot. Surv. India.,* 36(1-4): 160-165.

Burmmitt,R.K and C.E.Powell.1992. Authors of Plant Names. Kew. Royal Bot. Gardens. p. 732.

Champion H.G. and S.K. Seth.1968. A Revised Survey of Forest types of India. Government of India. Press, Delhi, India.

D' Antonio C.M. and P. Vitousek 1992. *Ann. Rev. Ecol. System* 23: 63-87.

Dhavmaraja,J.,S. Shobana.,T.C. Pillai and J. Balamurugan. 2012. *J. Sci.*1(2): 133-137.

Dicastri, F. *et al.,* 1990. Biological invasion in Europe and Mediterranean basin. Kulwer Academic Publisher, Fordretht.

Gamble, J.S.1915-1936. Flora of the presidency of Madras.11 Parts (Parts 1-7,by Gamble and 8-11 by C.E.C. Fisher). London. Adlard and Son Co.Repr. 1957. Calcutta. Botanical Survey of India. 2017 p.

Henry, A.N and M.S.Swaminathan. 1981. *Bull. Bot. Soc. Indian* 23:135-139.

Henry, A.N and M.S.Swaminathan. 1989. *J. Bombay Nat. His. Soc.* 75 (2):462-464.

Henry, A.N. and M.S.Swaminathan.1980. *J. Bombay Nat. His. Soc.* 76(2):373-376.

Hobbs, R.J. and L.F. Huenneke 1992. Conser. Biol. 6: 324-337.

Kennish, M.J. 2001. *J. Coastal Research.* Special issue No.32:243-273.

Karthigeyan, K. *et al.,* 2013. Check List, 9: 920-940.

Khedr, B.A. 1998 Vegetation zonation and management in the Damiette estuary of the River Nile. *J. Coastal. Conser.* 4(1): 79 – 86.

Khoshoo, T. N., 1996. Curr. Sci., 71: 506-513.

Koske, R.E. and Gemma, J.N., 1997. *J. Bot.* 84: 118-130.

Kulkarni, S.S., Raviraja, N.S. and Sridhar, K.R., 1997. *J. Coastal Res.,* 13: 931-936.

Kunte, K.A.Joqlekar.,G.Utkarsh and P.Padmarabhan. 1999. *Cur. Sci.* 77(4):577-586.

Lawrence, C.A 1959. *J. Bombay. Nat. His. Soc.* 57:95-100.

Lawrence, C.A 1960. *J. Bombay Nat. His. Soc.* 57: 184-195.

Martinez, M.L., Moreno-Casasola, P and Vazquez, G, 1997, *Can, J. Bot.* 75: 2005-2004.

Mathew, K.M. 1983. The flora of the Tamil Nadu, Carnatic. Vol.3 (Parts 1 and 3), Madras: The Diocesan Press, pp. 2154.

Mathew, V.M,1988. Further Illustrations of the Flora of the Tamil Nadu Carnatic. Vol.4.Madras : The Diocesan Press, pp.995.

Mathew,K.M.1982. Illustrations on the Flora of the Tamil Nadu.Cartatic.Vol.2 Madras. The Diocesan Press. 1027 p.

Mc Geoch, M.A. *et al.,* 2006. *Conser. Biol.* 20(6): 635-664.

Muthukumar, K. and Selvin Samuel, A., 2011. *Journal of Threatened Taxa,* 3(11): 2211-2216.

Nair, N.C. and A.N. Henry.1983. Flora of Tamil Nadu India (3 Vols.) BSI, Southern Circle, Coimbature.

Nayar, M.P.1959. *Bull. Bot. Surv.* 1: 122-126.

Nayar. J.S., A.R.Beegam., N.Mohanan and G.Rajkumar.2006.

Nayar. T.S.,A.R.Beegam.,N.Mohan and G. Rajkumar.2006. Flowering Plants of Kerala. A Hand Book. TBGRI, Palode, Tvm. Kerala, India P.1069.

Odum, W.E. 1988. *Ann. Rev. Ecol. Systematics.* 19: 147 – 176.

Padmavathy, K., Poyamoli, K. and Balachandran, N., 2010, Checklist.

Partridge, T.R and J.B. Wilson. 1989. *Newzeyland J. Bot.* 27(1): 35-47.

Pritchard, D.W. 1967. What is an estuary; Physical view point In Estuaries (ed.) G.H. Lauff. Publ. no. 83, Washington. DC.: American association for the Advancement of Science.

Pullaiah. T. 2007. Field and Herbarium Methods.In Taxonomy of angiosperms, Regency Publication, New Delhi, p. 130-145.

Punalekar, S. *et al.,* 2010. *Indian J. For.* 33(4): 549-554.

Raj A.D.S. 2002. Profile of Kanyakumari Dist. http//www.Kanyakumari. tn.nicing

Raj, A.D.S-2003. Profile of Kanyakumari district. Directorate of Information Nagercoil, T.N. India.

Ramarajan, S. and Murugesan, A.G., 2014. *Indian J. Plant Sci.* **3**(2): 42-48.

Ranjit Daniels, D.J. and V. Patil 1995. Curr. Sci. 68: 243-244.

Rao, J.A., A.R.K. Sastry, P. Base and N.R. Mandal. 1975. *Indian Forester* 101: 460-475.

Rao, J.A. and Sastry, R.K., 1972. *An Indian Forest*. 98: 594-607.

Rao, P.S.N., 2000. *Bull. Bot. Surv.* India 43: 1-40.

Rao, T.A. and Sastry, R.K., 1974. *Ibid* 100: 438-452.

Rao,T.A.,L.K. Banergee and A.K. Mukhergee.1974. *J. Bombay Nat. His. Soci.* 71: 346-349.

Rao,T.B.and A.R.K. Sastry. *Indian Forester* 98: 594-607.

Ravikumar, S. 2010. *Africa J.Basic and Appl.Sci*. **2:** 153-157.

Ravikumar, S. Palani Selvan, G. and Anitha, N., Anandha Gracelin, 2010. *Afric. J. Basis and Applied Sciences*, 2(5-6): 153-157.

Rejini, Balasingh, G.S. 2010. *J. Basic and Applied Biol.* 4(3): 188-193.

Rejinold, M. 2010. *J. Basic and Appl. Biol.* 4(3): 204-206.

Revendran, K.C. *et al.*, 2005. *Indian J. Trad. Know.* 4: 409-411.

Rodgers, W.A. and Panwar, H.S., 1998. http://www.irrd./org19/16/srid 19084html.

Rosa, S., J.M, Palmeirim and F. Moreira 2003. *The Inter.J. Waterbird Biology* 26(2): 226-232.

Sastry, A.R.K. and Ananda Rao, T., 1973. *Bull. Bot. Surv. India.*, 15(1 and 2): 92-102.

Sathia Geetha V, M. Rejinoad Appavoo and S. Jeeva. 2010. J. Basic Appl. Biol. 4(3):69-85.

Seklvakumari, R. and Rajakumar, J.J.S., 2010. *Indian J. Forestry.*, 33: 253-256.

Sridhar, K.R. and Bhagya, R. 2007. http://www.irrd. org/1rrd 19/61 Srid, 19084. html.

Sukumaran, S and A.D.S Raj. 2009. *J. Economic and Taxonomic Bot.* 33(1): 26-31.

Sukumaran, S. and B. Parthipan 2014. *Bio. Sci. Discovr.* 5(2): 204-217.

Sukumaran, S. and Jeeva, S., 2011. *Checklist*, 7: 486-495.

Sukumaran, S.,S.Jeeva., A.D.J. Raj and D. Kannan. 2008. *Turk. J. Bot.* 32: 185-199.

Sukumaran, S.Uma Devi and C.Kingston.2010. Westland Medicinal Plants of Vilavancode Talluk, Kanyakumari,T.N,India; P.23. In K.Paul Raj, P.D. Samuvel and S. Jeeva(ed.)National Seminar On Conservation and Management of Vetlands in an area of climate change, city:Publisher.

Sunder, S.K., Bency and Parthipan, B., 2009. Survey of Common folk medicinal plants on AVM canal used by fisherman community in kollencode. Revenue village segment of Kanyakumari District, Tamil Nadu, paper presented in National Seminar on Marine Resources, Sustainable Utilization and Conservation, 11 and 12th Dec., pp. 71-74.

Thangam, R.T.R. Meena and H. Prabhavathy. 2010. *J. Basic Appl. Biol.* 4(3): 194-198.

Thengane, S.R. *et al.*, 2006. *Curr. Sci.* 90: 1393-1397.

Untawale, A.G., 1994. (WWF India, IIRR SIDA), pp.3-10.

Vasantha, R. 2010. *J. Basic and Appl. Biol.* 4(3): 123 – 128.

WCMC (World Conservation Monitoring Centre) 1992. Global Biodiversity : Status of the Earths Living Resources. London : Chapman and Hall. 585 p.

Westbrooks, R. 1991. *Weed. Tech.* 5: 232-237.

2018, Ethnomedicinal Plants: A Biodiversity Treasure *Pages* **469–478**
Editors: ***V.R. Mohan, A. Doss, P.S. Tresina and V. Sornalakshmi***
Published by: **ASTRAL INTERNATIONAL PVT. LTD., NEW DELHI**

Chapter 19

Ethnoveterinary Practices Among the *Malayali* Tribes of Kolli Hills, Eastern Ghats, Tamil Nadu, India

S. Murugesh[1] and P. Deepa[2]

[1]Professor and Head, Department of Botany,
Periyar University, Periyar Palkalai Nagar, Salem – 636 011, Tamil Nadu
E-mail: murugeshss@rediffmail.com
[2]PG and Research Department of Botany,
Vivekanandha College of Arts and Sciences for Women,
Elayampalayam, Tiruchengode, Namakkal District, Tamil Nadu
E-mail: taanishadeepa@gmail.com

Introduction

India is one of the twelve mega-biodiversity countries of the world having rich vegetation with wide variety of plants with medicinal values. Current level of assessment puts our country's total count of flora at more than 40,000 species and that of fauna at more than 8,900 species. The current course of flowering plant species in India estimated at 17,500 represent approximately 7 per cent of the global count. It is also one among the 25 hot spots of the richest and highly endangered eco-regions of the world. It contains over 5 per cent of the world's diversity though it covers only 2 per cent of the Earth surface. About 20,000 plant species are known from this region.

The areas rich in biodiversity are also an abode of diverse ethnic groups possessing a valuable reservoir of indigenous knowledge system acquired and developed during a long period of time. It is also a band of physical, cultural, social and linguistic diversity which is endowed with an ecosystem and tremendous biodiversity, genetic as well as of species. In India, there are over 400 tribal and other ethnic groups. The tribal constitute about 7.5 per cent of Indian population.

Apart from the tribal groups, many other forest dwellers and rural populations also possess unique knowledge about plants. Some of such folklores and traditions had also survived among urban societies.

The communities over the years through trial and error have established the indigenous knowledge on the application of different sporadic, their specific part, their specific combinations and a specific way of formulations and applications. The traditional knowledge on applications of plants for different medicinal uses that has evolved and is currently maintained are largely determined by the locally available biodiversity lost in the past and the present. The indigenous knowledge on utility and utilization aspects of plants has been sensing as suitable tool for botanical and agriculture research owing to its relevance in development and promotional activities of new or less known economic plants (Jain and De 1964 ;1966).

Plants are important source of therapeutic drugs and play a significant role in the survival of the tribal and ethnic communities. The tribal people are the ecosystem people who live in harmony with the nature and maintain a close link between man and environment. The observations concerning the human use of beliefs and other interactions with people are the foundation of ethnoveterinary which is profound understanding and appreciation of the richness and intimacy of relationships between humans and nature.

Study Area

Kolli hills of Eastern Ghats of Kolli hills is situated in NamakkalDistrict of Tamil Nadu. Its Longitude: 78° 17′05″E to 78° 27′45″E and Latitude: 11° 55′05″N to 11° 21′10″N.The total block area is 441.41 sq. kms. It stretches 29 km from north to south and 19 km from east to west.Physiographically it is a hilly region with altitude ranging from 180 m at the foot hill to 1415 m at the plateau. Geologically the study area occupied by the hill is highly undulating, cut by a network of streams and most of them are semi-perinealand seasonal flowing in all the directions but mostly in the eastern and southern directions. Annual rainfall is 1324 mm which is received largely between May to December. Annual mean temperatures of maximum and minimum are 35° C and 18° C respectively. The type of soil is red loamy and black soil. As per the 2011 census, the total population of Kollihills is about 42,200 in 14 village panchayats and 275 hamlets. Four veterinary dispensaries and one mobile veterinary service unit are available for the veterinaryservice to the tribal people of Kolli hills. Kolli hill is ethnobotanically rich in a wide variety of medicinal plants.

Materials and Method

The present study was conducted among the Malayali tribal people of Kolli hills,Namakkal District, Tamil Nadu, India. The people of the study area were basically agriculturists and most of them were having domestic animals such as cow, goat, sheep, buffalo and pigs. The ethnoveterinary healers of the study area offered some necessary indigenous treatments with medicinal plants to treat their animals. Field trips ranging from 4 days to a week were made once in a month in the study area in every year of the study (April 2015- March 2016) among the Malayali tribal people in Kollihills. The major livelihood of these Malayali tribes were cattle farming, agriculture, collection of fuel-wood and counting forest resources such as herbal medicines, honey and some edible fruits and tubers in Kolli hills. Malayali tribes are one of the 36 scheduled tribal communities of Tamil Nadu with rich population when compared to the other major tribal peoples of Tamil Nadu. Ethnoveterinary data were collected from 40 resource persons (average age of 40-65) from the area under study who have much knowledge on medicinal plants with semi structured interviews. The interviews were conducted in the local language. *i.e.*, Tamil. Ethnoveterinary information included the local name of the particular plant, habit, parts used and medicinal uses. The collected ethnoveterinary information was recorded on field note books and plants were identified using the Flora of Presidency of Madras (Gamble. 1935) and Flora of Tamil Nadu- Carnatic (Matthew, 1983). The vouchers specimens were deposited at Department of Botany, Periyar University, Salem.

Results and Discussion

The present study enumerated 29 plant species belonging to 24families of ethnoveterinary importance based on the observations and interaction with local people, the herbal medicine practitioners and traditional healers of Kolli hills, Eastern Ghats, Tamil Nadu, India (Table 19.1 and Figure 19.1). For each reported species, the botanical name of the plant, followed by family, parts used, method of preparation, application and mode of administration were provided. Herbs are the primary source of medicine (44 per cent) followed by trees (28 per cent), shrubs (11 per cent), stragglers (8 per cent), twiners, straggling shrub and climbers (3 per cent) (Figure 19.2). The data proved that herbs are used as major sources of medicine among the tribal people in the study area. The different plant parts are used to cure various ailments in the present study. The leaves (42 per cent) are most frequently used followed by seeds (22 per cent), whole plant, fruit and aerial parts (8 per cent), rhizome (6 per cent), root and latex (3 per cent) (Figure 19.3). The preparation and utilization of plant parts were grouped into fivecategories *viz.*, paste, raw parts, powder, decoction, oil and latex. Of these, most commonly used method of preparation is paste (38 per cent) followed by decoction (19 per cent),

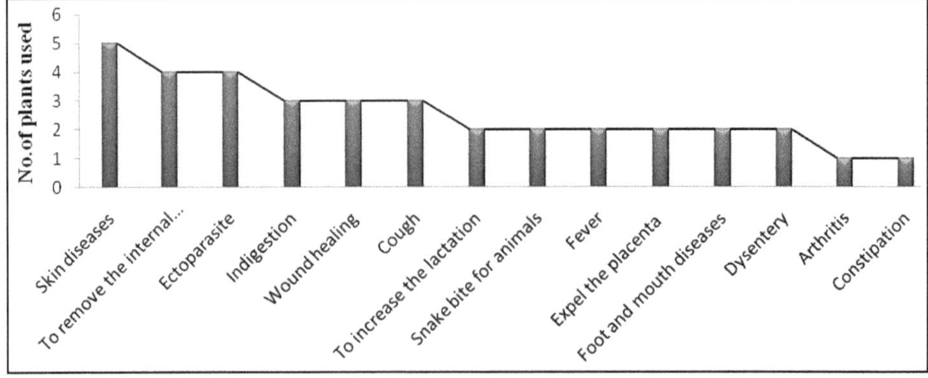

Figure 19.1: Ethnoveterinary Ailments in the Study Area.

Table 19.1: Ethnoveterinary Practices among the *Malayali* Tribes in Kolli Hills

Sl.No.	*Botanical Name/ Family*	*Vernacular Name*	*Habit*	*Part used/ Preparation*	*Application*	*Mode of administration*
			To increase lactation			
1	*Acacia nilotica* (L.) Del./ Mimosaceae	Karuvelamaram	Tree	Fruit pods/Raw part	Oral ingestion	The tender fruit pods are given to animals to increase lactation.
2	*Tribulus terrestris* L./ Zygophyllaceae	Nerungi	Herb	Whole plant/Raw part	Oral ingestion	The whole plant is given to cattle to increase lactation.
			Snake bites			
1	*Accacia caesia* Willd./ Mimosaceae	Karuindu	Straggler	Leaves/Paste	Oral ingestion	The paste is given to animals for snakebite
2	*Plumbago zeylonica* L./ Plumbaginaceae	Chitramoolam	Herb	Root/Paste	Oral ingestion	The root paste is given to animals for snake bites.
			To remove Internal worm			
1	*Acorus calamus* L./Acoraceae	Vasambu	Herb	Rhizome/Paste	Oral ingestion	The rhizome paste is given to animals to remove the internal worms.
2.	*Azadirachta indica* Adr. Juss./ Meliaceae	Vembu	Tree	Seed/Oil	Oral ingestion	The oil is given to animals to remove the internal worms.
3	*Pergulari adaemia* (Forsskal) Chiov./Asclepiadaceae	Veliparuthi	Straggler	Leaves/Paste	Oral ingestion	The leaf paste is given to animals removing ring worms.
4	*Vitex negundo* L./Verbenaceae	Notchi	Tree	Leaaves/Decoction	Oral ingestion	The leaf decoction is given to animals for removing internal worms.
			Ecto parasite			
1	*Acalypha indica* L./Euphorbiaceae	Kuppaimeni	Herb	Leaves/Paste	Topical	The leaf paste is applied externally for cattle removing the external parasite.
2	*Curcuma domestica* Valeton/Zingiberaceae	Manjal.	Herb	Rhizome/Paste	Topical	The rhizome paste is applied on externally to animals.

Sl.No.	*Botanical Name/ Family*	*Vernacular Name*	*Habit*	*Part used/ Preparation*	*Application*	*Mode of administration*
3	*Leucas aspera* (Willd.) Link/Lamiaceae	Thumbai	Herb	Aerial part/Paste	Topical	The paste is applied on externally to the cattle.
4	*Ocimum sanctum* L./Lamiaceae	Thulasi	Herb	Leaves/Paste	Topical	The leaf paste is appliedcattle.
				Indigestion		
1	*Piper nigrum* L./Piperaceae	Milagu	Herb	Seed/Paste	Oral ingestion	The seed paste is given to animals for indigestion.
2	*Tinospora cordifolia* (Willd.) Hook.f. and Thoms./Menispermaceae	Seenthilkode	Climber	Leaves/Decoction	Oral ingestion	The leaf decoction is given to animals for indigestion.
3	*Tylophora indica* (Burm.f.) Merr./Asclepiadaceae	Nappalai	Twiner	Leaves/Decoction	Oral ingestion	The leaf decoction is given to animals to treat indigestion
				Fever		
1	*Adhatoda zeylanica* Medikus/Acanthaceae	Adathodai	Shrub	Leaves/Decoction	Oral ingestion	The leaf decoction is given to animals to treat fever.
2	*Andrographis paniculata* (Burm.f.) Wall. *Ex* Nees/Acanthaceae	Siriyanangai	Herb	Leaves/Decoction	Oral ingestion	The leaf decoction is given to animals to treat fever.
				Wound healing		
1	*Achyranthes aspera* L./ Amaranthaceae	Nayuruvi	Herb	Whole plant/Paste	Topical	Paste is applied on the affected parts
2	*Calotropis gigantean* (L.) R. Br./Asclepiadaceae	Erukku	Shrub	Latex/Latex	Topical	The stem latex is applied on affected parts of the animals.
3	*Tamarindus indica* L./ Caesalpiniaceae	Puliyamaram	Tree	Fruit/Paste	Topical	The fruit paste is applied on the affected parts.

Contd...

Sl.No.	*Botanical Name/ Family*	*Vernacular Name*	*Habit*	*Part used/ Preparation*	*Application*	*Mode of administration*
			Cough			
1	*Adhatoda zeylanica* Medikus/Acanthaceae	Adathodai	Shrub	Leaves/Decoction	Oral ingestion	The leaf decoction is given to animals for treating the cough.
2	*Andrographis paniculata* (Burm.f.) Wall. *Ex* Nees/ Acanthaceae	Siriyanangai	Herb	Leaves/Decoction	Oral ingestion	The leaf decoction is given to animals to get relief from cough and cold.
3	*Ocimum sanctum* L./Lamiaceae	Thulasi	Herb	Aerial part/Raw part	Oral ingestion	The aerial part is given to animals to get relief from cold.
			Arthritis			
1	*Dodonaea viscosa* L./ Sapindaceae	Virali	Shrub	Leaves/Paste	Topical	Paste is applied on fractured bones and tied tightly.
			Expel the placenta			
1	*Cissus quadrangularis* L./ Vitaceae	Pirandai	Straggling shrub	Aerial part/Raw part	Oral ingestion	The aerial part is given to animals to expel the placenta after delivery.
2	*Oryza sativa* L./Poaceae	Nellu	Herb	Whole plant/Raw part	Oral ingestion	The whole plant is given to animals to expel placenta.
			Foot and mouth disease			
1	*Azadirachta indica* Adr. Juss./ Meliaceae	Vembu	Tree	Seed/Oil	Topical	The seed oil is applied on the affected parts.
2	*Cocos nucifera* L./Arecaceae	Thennai	Tree	Seed/Oil	Topical	The seed oil is applied on affected parts.
			Constipation			
1	*Coriandrum sativum*L./ Apiaceae	Kothamalli	Herb	Seed/Decoction	Oral ingestion	The seed decoction is given to animals for constipation.
			Dysentery			
1	*Mangifera indica* L./ Anacardiaceae	Mamaram	Tree	Seed/Paste	Oral ingestion	The seed paste is given to animals to treat dysentery.

Sl.No.	*Botanical Name/ Family*	*Vernacular Name*	*Habit*	*Part used/ Preparation*	*Application*	*Mode of administration*
2	*Terminalia chebula* Retz./ Combretaceae	Kadukai	Tree	Fruit/Powder	Oral ingestion	The fruit powder is given to animals for dysentery.
				Skin diseases		
1	*Acalypha indica* L./Euphorbiaceae	Kuppaimeni	Herb	Leaves/Paste	Topical	The leaf paste is applied on the affected parts.
2	*Aloe vera* (L.)Burm.f./Liliaceae	Sothukathalai	Herb	Leaves/Paste	Topical	The leaf paste is applied on the affected parts of animals.
3	*Cocos nucifera* L./Arecaceae	Thennai	Tree	Seed/Oil	Topical	The seed oil is applied on affected parts.
4	*Pergularia daemia* (Forsskal) Chiov./Asclepiadaceae	Veliparuthi	Straggler	Leaves/Paste	Topical	The leaf paste is applied on the affected parts of animals.
5	*Pongamia pinnata* (L.) Pierre/Fabaceae	Pungamaram	Tree	Seed/Oil	Topical	The seed oil is applied on affected parts.

powder (17 per cent), oil and raw part (12 per cent) and latex (2 per cent) (Figure 19.4). The oral applications (58 per cent) of remedies are predominating over topical applications (42 per cent) (Figure 19.5).

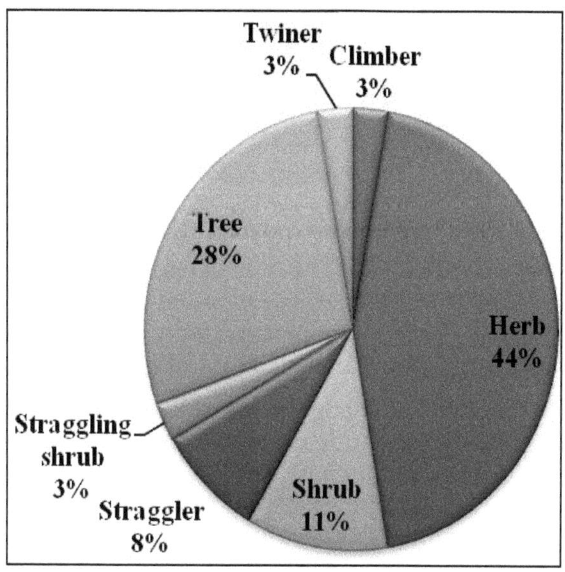

Figure 19.2: Habit-wise Analysis of Ethnoveterinary Plants.

Most of the reported plants in the present study are also used by the different types of tribal people in India for the treatment of various diseases in livestock (Girach *et al.*, 1998; Reddy *et al.*, 2006; Mini and Sivadasan, 2007; Harsha *et al.*, 2008; Satya and Solanki, 2009; Yadhav, 2009; Rahman *et al.*, 2009). Geetha *et al.* (2006) reported that, the plants such as *Aloe vera*, *Azadiracta indica* and *Cissus quadrangularis* were used by the Malaiyali tribals in Kolli hills of Namakkal district.

Traditional veterinary practices reports from Dindigul district (Rajan and Sethuraman, 1997) and some southern districts of Tamil Nadu (Ganesan *et al.*, 2008) showed some resemblance with the present study but most of the uses

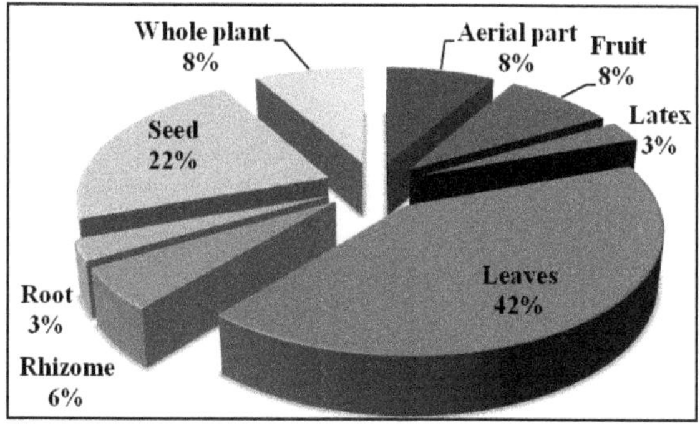

Figure 19.3: Plant Parts Used for Curing the Ailments.

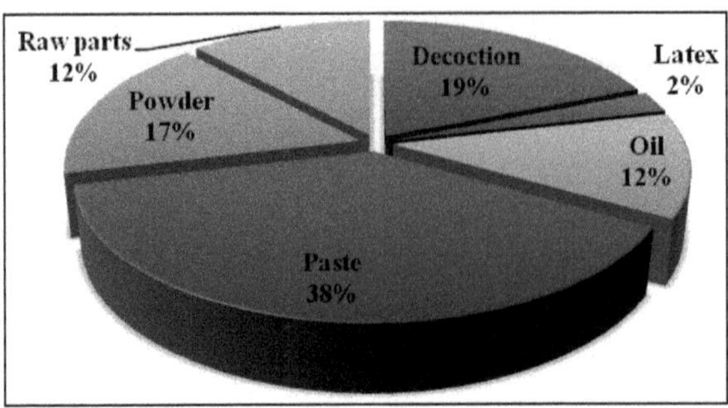

Figure 19.4: Preparations Used for Curing the Ailments.

found to be different. Similarly, Kiruba *et al.* (2006) reported that *Acalypha indica, Aloe vera, Andrographis paniculata, Azadirachta indica, Calotropis gigantea, Cissus quadrangularis, Pergularia deamia* and *Vitex negundo* are used by the indigenous people of Kanyakumari district for the treatment of different types of diseases in livestock.

Documentation of this knowledge is valuable for the communities and their future generations and for scientific consideration of wider uses of traditional knowledge in treating domestic animals. The low cost and almost no side effects of these traditional preparations with medicinal plants make them adaptable by the local community. The wealth of this tribalknowledge of medicinal plants points to a great potential for research and the discovery of new drugs to cure the diseases of animals.

Conclusion

The present investigation espied that the countries indigenous knowledge and wisdom on ethnoveterinary medicine cumulated in the minds of people rooted on

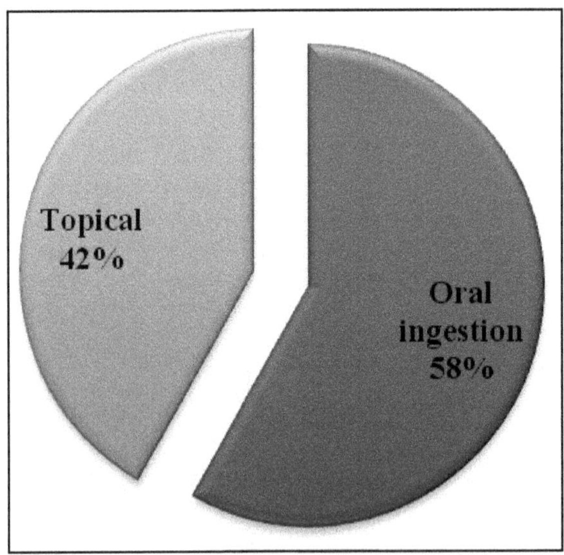

Figure 19.5: Application of Curing the Ailments.

the trials and errors of generations were gone by threatened, slowly but surely. With this information, scientific experiments can be performed which may provide potent and proliferate from herbs against diverse diseases.

References

Gamble JS. Flora of the Presidency of Madras, Vols, I-III, Adlard and Co., London, 1915-1935.

Ganesan, S., Chandhirasekaran, M. and Selvaraju, A. 2008. Ethno-veterinary health care practices in Southern districts of Tamil Nadu. *Ind. J. Trad.Knowl.*7: 347-354.

Geetha, S., Lakshmi, G. and Ranjithakani, P. 2006. Ethnoveterinary medicinal plants of Kollihills, Tamil Nadu.*J. Econ. Taxon. Bot.* 12: 284-291.

Girach, R.D., Brahman, M. and Misra, M.K. 1998.Folk veterinary herbal medicine of Bhadrak district, Orissa, India.*Ethnobot.*10: 85-88.

Harsha, V.H., Shripathi, V. and Hegde, G.R. 2008.Ethnoveterinary practices in Uttara Kannada districts of Karnataka. *Ind. J. Trad.Knowl.*4: 253-258.

Jain, S.K. and De, J.N. 1964. Some less known plant foods among the tribals of Purulia District, West Bengal.*Sci. and Cult.*30: 285-286.

Jain, S.K. and De, J.N. 1966. Observations on ethnobotany of Purulia District, West Bengal.*Bull. Bot. Surv. India* 8: 237-251.

Kiruba, S., Jeeva, S. and Dhas, S.S.M. 2006. Enumeration of ethnoveterinary plants of Cope Comorin, Tamil Nadu. *Ind. J. Trad.Knowl.*7: 576-578.

Matthew KM. The Flora of Tamil Nadu Carnatic, Vols 1-3, The Rapinant Herbarium, St. Joseph's College, Thiruchirapalli, 1981-1984.

Mini, V. and Sivadasan, M. 2007. Plants used in Ethno-veterinary medicine by Kurichya tribes of Wayanad district in Kerala India. *Ethnobot.*19: 94-99.

Rahman, C.H., Ghosh, A. and Mandal, S. 2009.Studies on the Ethnoveterinary medicinal plants used by the tribes of Birbhum district, West Bengal.*Ind. J. Trad.Knowl.*33: 333-338.

Rajan, S. and Sethuraman, M. 1997.Traditional veterinary practices in rural areas of Dindigul, Tamil Nadu, India.*Indigen.Knowl. Dev. Mon.* 5: 709.

Reddy, K.N., Subbaraju, G.V., Reddy, C.S. and Raju, V.S. 2006.Ethnoveterinary medicine for treating live stock in Eastern Ghats of Andhra Pradesh, India. *Ind. J. Trad.Knowled.*,5: 368-372.

Satya, V. and Solanki, C.M. 2009.Indigenous knowledge of veterinary medicines among tribes of West Nimar, Madhya Pradesh.*Ind. J. Trad.Knowl.*33: 896-902.

Yadav, D. 2009. Ethno veterinary plants from tribes in habited localities of Ratlam district Madhya Pradesh, India. *Ind. J. Trad.Knowl.*33: 64-67.

2018, Ethnomedicinal Plants: A Biodiversity Treasure Pages 479–520
Editors: ***V.R. Mohan, A. Doss, P.S. Tresina and V. Sornalakshmi***
Published by: **ASTRAL INTERNATIONAL PVT. LTD., NEW DELHI**

Chapter 20

Studies on the Indigenous Wild Edible Plants of *Palliyar* Tribes of South India

P.S. Tresina, K. Paulpriya and V.R. Mohan

Ethnopharmacology Unit, PG and Research Department of Botany, V.O.Chidambaram College, Tuticorin – 628 008, Tamil Nadu
E-mail: vrmohanvoc@gmail.com

Introduction

Plant Genetic Resources for Food and Agriculture (PGRFA) are the biological basis of world food security and directly or indirectly support the livelihood of every person on Earth (FAO, 1996). In general, plants provide 65 per cent of the global requirement of edible protein (Young and Pellet, 1994) and in particular about 80 per cent of the protein consumed by the humanity in developing countries accrues from plants (Singh and Singh, 1992).

The study on the diversity and distribution of plant wealth vis-à-vis its utilization in this context is of great concern to mankind. Among the various kinds of plants, food plants received the earliest attention of mankind (Burkill, 1952) and reflect man's search for knowing more and more about the nutrient qualities of food plants. Against this back-drop, studies on the diversity and distribution of such edible plants, both of the wild and the cultivated types are relevant to codify exploitation of bioresources.

Such plants sustain them and remain their stable diet as plants are the predominant harvesters of solar energy and they constitute primary sources of carbohydrates, vitamins, proteins, essential fatty acids and provide enough calories for consumers.

The wild relatives of cultivated crops are actively considered for genetic improvement because they have beneficial genes for tolerance to environmental stress and resistance against pests and insects and higher levels of nutrients; while cultivated species often have a very narrow genetic base and lack genes for resistance to certain diseases. Wild species of several cultivated plants act as reservoirs for crop improvement including developing resistance to pests and pathogens (Ignacimuthu and Babu, 1987; Babu *et al.*, 1988; Burdon and Jarosz, 1989; Doney and Whitney, 1990; Lenne and Wood, 1991; Rajaram and Janardhanan, 1991a; Hodgkin and Debuock, 1992; Mary Josephine and Janardhanan, 1992; Siddhuraju *et al.*, 1992; 1994; Vijayakumari *et al.*, 1993; Mohan and Janardhanan, 1993; 1994a; 1994b; Frankel *et al.*, 1995; Vadivel and Janardhanan, 2000a). Hence, the identification and introduction of such wild crops, little known grain legumes, including the tribal pulses throughout the tropical regions as well as their genetic improvement of the quality and quantity would, indeed, be a great contribution (Maikhuri *et al.*, 1991; Rajaram and Janardhanan, 1993).

In India, certain wild tubers, root-types, green leaves, flowers, unripe as well as ripe fruits and grain legumes including tribal pulses are known to be consumed by different tribal sects (Jain, 1981; 1991; Gunjatkar and Vartak, 1982; Janardhanan, 1990; Maikhuri *et al.*, 1991; 2000; Rajaram and Jarnardhanan, 1993; Siddhuraju *et al.*, 1993; Mohan and Janardhanan, 1994a; Borathakur, 1996; Radhakrishna *et al.*, 1996; Sahu, 1996; Singh, 1996; Sinha, 1996; Vadivel and Janardhanan 2000b; 2001). Tribals in India are, in fact, the indigenous and autochthonous people of the land. They had settled in different parts of India, in Kabul and Indus Valley long before the Aryans penetrated into India to settle down (Roy Burmann, 1986). There are more than 550 tribal communities under 227 ethnic groups spread throughout India (Anonymous, 1994). Tribal population of India is 67.76 million, constituting about 8.08 per cent of total population (Census of India, 1991). The tribal sects live in the relatively less disturbed forest areas.

The various tribal sects of India are repositories of rich knowledge on various uses of plant genetic resources, which have hitherto remained unknown (Khoshoo, 1991). But of late, due to several developmental activities around tribal areas which are, after all, not related to their welfare, the tribal people are losing their traditional identity resulting in a good deal of loss of such treasure house of knowledge on plant genetic resources (Shankar, 1995). In view of these harmful developments, the UN declared the year 1993 as the "International Year of Indigenous People" based on the recommendations of the Rio de Janerio Earth Summit.

The studies on the relationship between the aboriginal or primitive people and their surroundings including a critical evaluation of some of the important plants used by the tribals have received considerable attention in recent years (Das *et al.*, 1983).

It is a matter of great pride that among the 18 hot spots known for rich flora in the world over, two are located in India. They are the Eastern Himalayas and the Western Ghats (Khoshoo, 1996). The hill chain of Western Ghats recognized as a region of high level of biodiversity is under the threat of rapid loss of genetic resources (Gadgil, 1996). The biodiverse nature of the Eastern Ghats is meager.

Keeping these facts intact in mind, in the present study, the wild edible food plants have been surveyed and documented from the natural stands of the South-Eastern slopes of Western Ghats, Tamil Nadu, India. The Palliyars, the dominant tribal group, inhabit the locality of the study area. The present study focuses on the dependence of the Palliyars on edible plants and attempts at an exhaustive analysis of the nutritional qualities of such edible plants.

Wild Edible Plants and the *Palliyars* – An Investigation

The present investigation studies the distribution of the tribe Palliyars, along the target area, and the ecological conditions, consumption pattern of different food plant and their nutritional potential.

The Palliyars are distributed along the South-Eastern slopes of the Western Ghats, Tamil Nadu (adjoining Virudhunagar and parts of Madurai and Tirunelveli districts – Area Map 20.1).

The present study focuses on the wild edible food plants consumed by the Palliyars settled in the reserve forest area of the Grizzled Giant Squirrel Wildlife Sanctuary. This area lies mostly in Virudhunagar and parts of Madurai district. This wildlife sanctuary is established in the year 1989. It encompasses an area of 480 sq.km. The study also covers tribal hamlets around Puliangudi, Tirunelveli district.

The area of investigation lies between 70°.3′ E and 77 °.9′ E longitude and 9 °.1′N and 9 °.8′N latitude. This altitude varies from 100 m to 2210 m (MSL). It receives rainfall from both South-West and the North-East monsoons. The varied climatic and topographic conditions prevailing in the Sanctuary present remarkable diversity of both the flora and fauna.

Variations in the altitude and rainfall always have a bearing on the vegetation in general. The study area consists of tropical evergreen forests, semi evergreen forests, dry teak forests, southern mixed deciduous forests and dry grassland (Plate 20.1).

In the study area the Palliyars, live in several isolated pockets or in small hamlets. Their habitations are known by the following names.

1. Saduragiri
2. Thanniparai
3. Vandipannai
4. Nellikka koottam
5. Valliammal Nagar
6. Saraupannai
7. Petchikeni Koottam
8. Athikoil
9. Shenbagathoppu
10. Ayyanarkoil
11. Thallaianai
12. Shelimbuthoppu
13. Kulirattu
14. Manapadai
15. Rakkamal Koil Parai
16. Kottamalai

Life Style of the *Palliyar* Tribals

Tribals are a distinct ethnic group who are usually confined to definite geographical areas. They speak a common dialect, and are culturally homogenous.

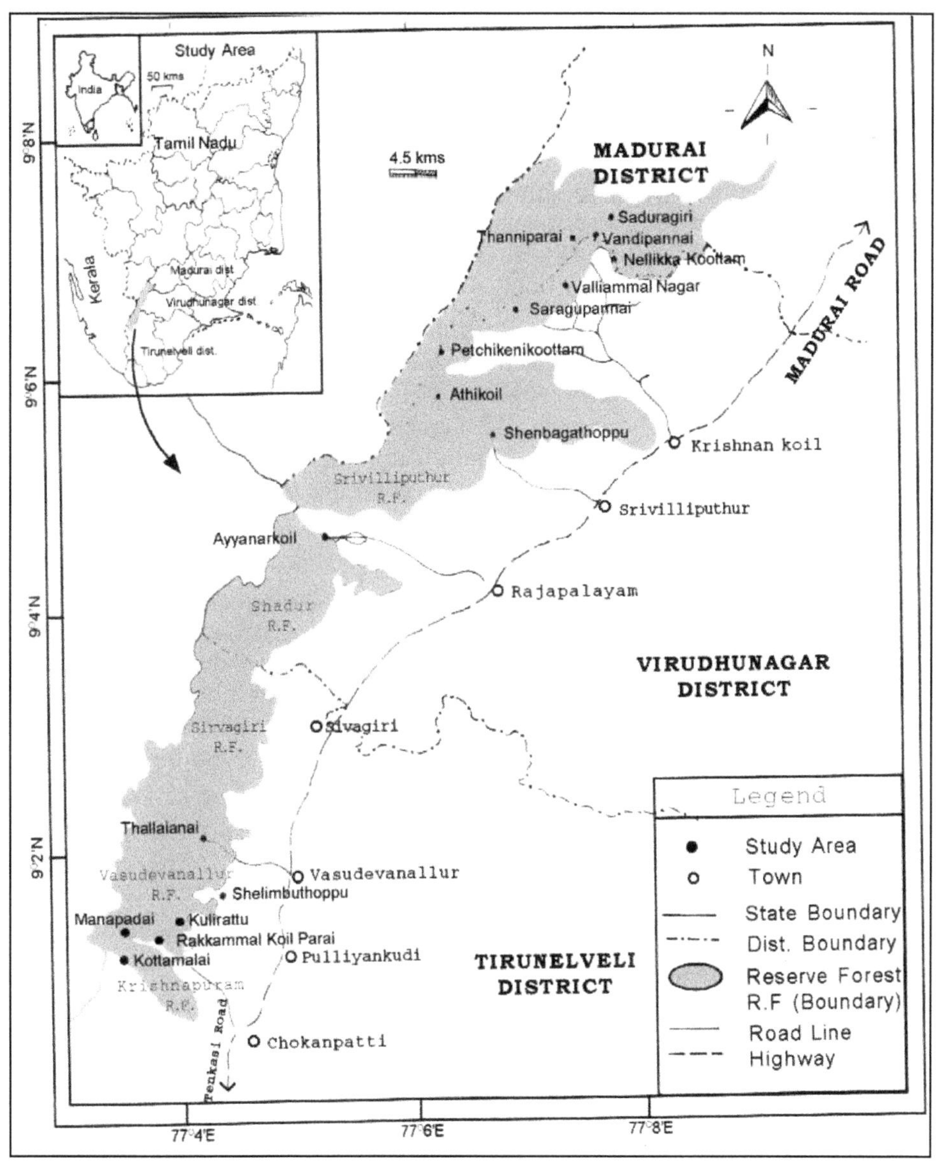

Map 20.1: Study Area.

The tribal people of India live mostly in the forests, hills, plateaus and regions of natural isolation. They are differently named as Adivasi (original settlers), Adim Niwasi (oldest ethno-logical sector of the populations), Aboriginal (indigenous), Vanavasi (forest inhabitants) and several such other names symbolizing either their ecological, economic or historical or cultural characteristics (Jain, 1987).

The Indian sub-continent is divided into three major tribal zones on the basis of the tribal demography and geography of India: (i) the North and North-Eastern Zone (ii) The central or middle zone and (iii) The Sothern zone (Ajit Raizada, 1984). The southern tribals are historically the most ancient tribes. There are as many as 36 types of scheduled tribes in Tamil Nadu. In the serialized list notified by the Government of Tamil Nadu number 32 denotes the Palliyars. The Palliyars live in the low altitude of Western Ghats. They live in large number in Virudhunagar district, parts of Madurai and Tirunelveli districts.

Of the total population of the scheduled tribes in Tamil Nadu, 5,74,194 the Palliyar tribe accounts for 1,890 which ranks 20th among the total tribal population (Census of India, 1991). At present the population of Palliyars in the study area is 1,070. Of this, 373 are men, 342 women and 355 being children (Authors investigation). The Palliyars live as individual families. They are short, dark complexioned, curly haired with thick protruding lips and blunt nose with wide nostrils (Plate 20.2).

Generally, a hamlet has about 20 huts. Each hut is thatched with the fronds of Tharagu pull (*Cymbopogon polyneurus*) or lemon grass (*Cymbopogon citratus*) or leaves of Iechamaram (*Phoenix pusilla*) or coconut leaves (*Cocos nucifera*) (Plate 20.2). They use sand and stone for wall construction. They decorate their houses with the same leaves. They sleep on mats woven with these leaves. They eat tubers, greens, unripe fruits, ripe fruits *etc.* they also feed on wild animals and birds like rabbits, rats, deer, hens *etc.* They roast besides tubers, the flesh of these animals and birds and eat.

The Palliyars have strong faith in religious customs and practices. They worship "Forest Goddess" and "Goddess Poomadevi" (Goddess Earth). On special days the stay in the 'temples' in groups for two days and worship by offering goats as sacrifice. On those days a large quantity of Mullvallikizhlangu (*Dioscorea pentaphylla* var. *pentaphylla*) are collected and offered to the Goddess Earth, 'Poomadevi'.

The Palliyars are monogamous. Elopement is the favorite form of marriage. The elders search for the eloped couples; them bring them back and get them married. Widow remarriage is common. The dead are buried. On the eighth day after death, they perform the last rites. The Palliyars believe in witchcraft. They entertain many curious superstitious beliefs.

The Palliyars are adroit in collecting wild honey. They collect it from the branches of towering tall trees and rock caves skillfully using special techniques. They also collect Kungillium (resin) from the barks of *Canarium strictum.*

They use many herbs to treat diseases. Vaithiyar is a tribal person who is an expert in administering herbal medicines. He collects medicinal plants from various remote parts of the hills. People respect the vaithiyar and hold in good esteem. The government of Tamil Nadu has established a medicinal plant conservation area in the forest for preservation and development of herbal wealth. It is maintained and conserved with the assistance of the Palliyars.

At present, a free residential school is run by the Government of Tamil Nadu at Maharasapuram for the exclusive educational development of the tribal children.

Materials and Methods

Survey and Collection

Survey and collection of wild edible plants of the Palliyars of the South-Eastern slopes of the Western Ghats, Tamil Nadu, India (adjoining Virudhunagar, parts of Madurai and Tirunelveli districts) were carried out over a period of 24 months. Frequent field trips were undertaken to the areas of study. Information regarding the wild edible food plants was gathered from the elderly people who have a very long acquaintance with the usage of plants. The field note book delineates the usage procedures adopted by the tribals. The information thus gathered was adequately cross checked for reliability and accuracy by interacting with different groups of the Palliyars from different habitats to confirm the usage pattern as well as differences, if any, in the mode of consumption.

After eliciting detailed information regarding the wild edible food plants and their useful parts, they were carefully brought to the laboratory for identification and chemical analysis. Herbaria for all the plant specimens were prepared and deposited in the PG and Research department of Botany, V. O.Chidambaram College, Tuticorin, Tamil Nadu, India.

The collected plants were identified by referring the following compilations (Fischer, 1980; Gamble, 1957a, 1957b; Henry *et al.*, 1989, 1987; Mathew, 1983a,b,c; Mathew, 1999a,b,c; Nair and Henry, 1983; Sasidharan and Sivarajan, 1996; Sharma *et al.*, 1993a, b; Sharma and Sanjappa, 1993; Singh, 2001).

After authentication with the regional flora, the specimens were matched with the authentic herbarium of Madras Herbarium, Botanical Survey of India (BSI), Coimbatore, Tamil Nadu, India.

Results

Survey of Wild Edible Plants

As many as 171 wild edible plants were identified as being eaten by the Palliyar tribals of the South-Eastern slopes of the Western Ghats, Tamil Nadu, India. They were tabulated alphabetically with Botanical name, Family name, Vernacular name (used by the Palliyars) plant part/s used and based on the nature of plant part/s consumed by the Palliyar tribals, the plants were grouped into different categories (Tables 20.1–20.7 and Plates 20.3–20.13). 171 wild edible plants belong to 67 families. Of which one belongs to Gymnosperms and the rest belong to Angiosperms.

Table 20.1: Edible Tubers, Rhizomes, Corms and Root Types Consumed by the Palliyar Tribals

Sl.No.	*Botanical Name*	*Family*	*Vernacular Name*	*Plant Parts Used*	*Mode of Use**
1	*Abelmoschus moschatus* Medicus	Malvaceae	Kattuvendai	Fleshy root	Raw/Boiled/Roasted
2	*Amorphophallus sylvaticus* (Roxb.) Kunth	Araceae	Kattukarunai Keerai	Corm	Boiled in Tamarindus water and consumed next day
3	*Argyreia pilosa* Arn.	Convolvulaceae	Thettukkadi	Root	Boiled/Roasted
4	*Asparagus racemosus* Wild.	Liliaceae	Neervalli	Tubers	Raw/Boiled/Roasted
5	*Borassus flabellifer* L.	Arecaceae	Panai	Root like fleshy cotyledons	Boiled/Roasted
6	*Canna indica* L.	Cannaceae	Kalvazhai	Rhizome	Boiled/Roasted
7	*Colocasia esculenta* (L.) Schott.	Araceae	Kattuchembu	Corm	Boiled/Roasted/Cooked with chilli and salt as curry
8	*Curculigo orchioides* Gaertn.	Hypoxidaceae	Kuluthupokie	Tubers	Raw/Boiled/Roasted
9	*Curcuma Neilgherrensis* Wt.	Zingiberaceae	Kattukalvazhai	Rhizome	Raw/Boiled/Roasted
10	*Dioscorea bulbifera* L. var. *vera* Prain and Burkil	Dioscoreaceae	Karuvalli	Tubers	Boiled/Roasted
11	*Dioscorea oppositifolia* L. var. *dukhumensis* Prain and Burkill	Dioscoreaceae	Vethalaivalli	Tubers	Boiled/Roasted
12	*Dioscorea oppositifolia* L. var. *oppositifolia*	Dioscoreaceae	Thallavalli/ Kavalavalli	Tubers	Boiled/Roasted
13	*Dioscorea pentaphylla* var. *pentaphylla*	Dioscoreaceae	Mullvalli	Tubers	Boiled/Roasted
				Bulbils	Boiled/Roasted
14	*Dioscorea tomentosa* Koen. Ex Spreng.	Dioscoreaceae	Noolvalli	Tubers	Tender part of the tubers are boiled/roasted and eaten by Palliyar tribals in Ayyanarkoil
					Entire tubers boiled/roasted and eaten by the Palliyar tribals in Shenbagathoppu
15	*Dolichos trilobus* L.	Fabaceae	Minna	Tubers	Boiled/Roasted

Contd...

Sl.No.	Botanical Name	Family	Vernacular Name	Plant Parts Used	Mode of Use*
16	*Hemidesmus indicus* (L.) R. Br. var. *indicus*	Periplocaceae	Nannari	Root	Juice extracted, mixed with water and sugar
17	*Hemidesmus pubescens* (L.) R. Br. *pubescens* (Wight and Arn) Hook.f.	Periplocaceae	Nannari	Root	Juice extracted, mixed with water and sugar
18	*Ipomea staphylina* Roem. and Sch.	Convolvulaceae	Oanakodi	Root	Raw/Boiled
19	*Sterculia urens* Roxb.	Sterculiaceae	Vennaali	Root	Raw/Boiled

* All items are skinned off before use.

Table 20.2: Edible Stems, Piths and Apical Meristems Consumed by the Palliyar Tribals

Sl.No.	*Botanical Name*	*Family*	*Vernacular Name*	*Plant Parts Used*	*Mode of Use*
1	*Amaranthus tricolor* L.	Amaranthaceae	Thandan keerai	Stem	Boiled with chilli and salt
2	*Arenga wightii* Griff.	Arecaceae	Ala panai	Pith	Powdered pith used in the preparation of dosai/bread. Powdered pith mixed with sugar is used in the preparation of sweets.
				Apical meristem	Raw/Cut into pieces and boiled
3	*Basella alba* L. var. *alba*	Chenopodiaceae	Kattupasali	Stem	Boiled along with salt and chilli
4	*Borassus flabellifer* L.	Arecaceae	Panai	Pith	Raw
				Apical meristem	
5	*Caralluma adscendens* (Roxb.)Haw.	Asclepiadaceae	Periyasiruman keerai	Stem	Raw
6	*Caralluma lasiantha* (Wight)N.E.Br.	Asclepiadaceae	Siruman keerai	Stem	Raw
7	*Caryota urens* L.	Arecaceae	Koonthal panai	Pith	Raw
8	*Cissus quadrangularis* L.	Vitaceae	Pirandai	Stem	Peeled stem mashed and used in preparation of chutney and side dishes
9	*Phoenix pusilla* Gaertn.	Arecaceae	Iechamaram	Apical meristem	Raw
10	*Sarcostemma acidum* (Roxb.) Voigt	Asclepiadaceae	Thannikkodi	Stem	Raw
11	*Sterculia urens* Roxb.	Sterculiaceae	Vennaali	Exudates from the stem	Mixed with water and salt – kept overnight and used
12	*Streblus asper* Lour.	Moraceae	Pira	Latex from the stem	Mixed with milk and sugar and used

Table 20.3: Greens Consumed by the Palliyar Tribals

Sl.No.	Botanical Name	Family	Vernacular Name	Plant Parts Used	Mode of Use
1	*Acacia grahamii* Vajravelu	Mimosaceae	Indu	Tender leaves	Boiled and mashed with salt and chilli
2	*Allamania nodiflora* (L.)R.Br. ex.Wight var. *aungustifolia* Hook f.	Amaranthaceae	Chengkumaati keerai	Leaves	Boiled and mashed with salt and chilli
3	*Allamania nodiflora* (L.)R.Br. ex.Wight var. *procumbens* Hook f.	Amaranthaceae	Kumattikerai	Leaves	Boiled and mashed with salt and chilli
4	*Aloe vera* (L.) Burm.f.	Liliaceae	Chothukathalai	Succulent leaves	Peeled and cooked with dhal, tomato and salt as curry
5	*Alternanthera bettzickiana* (Regel) Nicholson	Amaranthaceae	Ponnankanni keerai	Leaves	Fried along with chilli, coconut and salt
6	*Amaranthus roxburghianus* Nevski	Amaranthaceae	Araikeerai	Leaves	Boiled and mashed with salt and chilli
7	*Amaranthus spinosus* L.	Amaranthaceae	Mullukkeerai	Tender leaves	Boiled and mashed with salt and chilli
8	*Amaranthus tricolor* L.	Amaranthaceae	Thandankeerai	Tender leaves	Boiled and mashed with salt and chilli
9	*Amaranthus viridis* L.	Amaranthaceae	Kuppaikeerai	Tender leaves	Boiled and mashed with salt and chilli
10	*Amorphophallus sylvaticus* (Roxb.) Kunth	Araceae	Kattukkarunai keerai	Leaves	Boiled nad mashed along with more tamarindus water
11	*Basella alba* L. var. *alba*	Chenopodiaceae	Kaatupasali	Leaves	Boiled and mashed with salt and chilli
12	*Begonia malabarica* Lam.	Begoniaceae	Naarayana sanjeevi	Leaves	Raw
13	*Boerhavia diffusa* L.	Nyctaginaceae	Mookkana charana	Leaves	Boiled and mashed with salt and chilli
14	*Boerhavia erecta* L.	Nyctaginaceae	Kuthucharana	Leaves	Boiled and mashed with salt and chilli
15	*Borassus flabellifer* L.	Arecaceae	Panai	Young tender leaves	Raw
16	*Brassica juncea* (L.) Czern. and Coss.	Cruciferae	Kattukadugu	Leaves	Boiled and mashed with salt and chilli
17	*Canthium parviflorum* Lam.	Rubiaceae	Periyakaarai/ Malukkaarai	Tender leaves	Raw/Boiled and mashed with salt and chilli
18	*Capsicum frutescens* L.	Solanaceae	Kanamillakukeerai	Leaves	Boiled/Fried with salt and chilli

Sl.No.	Botanical Name	Family	Vernacular Name	Plant Parts Used	Mode of Use
19	*Cardiospermum halicacabum* L.	Sapindaceae	Mudakkathan	Leaves	Boiled along with salt and chilli
20	*Celosia argentea* L.	Amaranthaceae	Mavulikkeerai	Leaves	Boiled and mashed with salt and chilli
21	*Cissus quadrangularis* L.	Vitaceae	Perandai	Leaves	Boiled and mashed with salt and chilli
22	*Cleome gynandra* L.	Cleomaceae	Thaivalaikeerai	Leaves	Boiled and mashed with salt and chilli
23	*Cleome viscosa* L.	Cleomaceae	Naaikkadugu	Leaves	Boiled and mashed with salt and chilli
24	*Coccinia grandis* (L.) Voigt.	Cucurbitaceae	Kovai	Leaves	Boiled and mashed with salt and chilli
25	*Cocculus hirsutus* (L.) Diels.	Menispermaceae	Vellakattukkodi	Leaves	Extract poured into sugar solution and the semisolid form consumed
26	*Colocasia esculenta* (L.) Schott	Araceae	Kattuchembu	Leaves and petiole	Boiled and mashed with salt, chilli and coconut
27	*Commelina benghalensis* L.	Commelinaceae	Amala	Leaves	Boiled and mashed with salt and chilli
28	*Commelina ensifolia* R.Br.	Commelinaceae	Amala	Leaves	Boiled and mashed with salt and chilli
29	*Cycas circinalis* L.	Cycadaceae	Paereenji	Tender leaves	Decanted leaves mashed with chilli and salt
30	*Digera muricata* (L.) Mart.	Amarathaceae	Kattukkeerai	Leaves	Boiled and mashed with salt and chilli
31	*Diplocyclos palmatus* (L.) Jeffrey	Cucurbitaceae	Aattupudal/ Malaipusanni	Tender leaves	Boiled and mashed with salt and chilli
32	*Gisekia pharnaceoides* L.	Molluginaceae	Manalkeerai	Leaves	Boiled and mashed with salt and chilli
33	*Heracleum rigrens* Wall.ex.DC.var. *rigens*	Umbelliferae	Kattu kothamalli	Leaves	Fried and mashed along with chilli, tamarindus and salt
34	*Hybanthus enneaspermus* (L.) F.v.Muell.	Violaceae	Orithalthamarai	Leaves	Raw
35	*Jasminum auriculatum* Vahl.	Oleaceae	Kattumullai	Leaves	Boiled and mashed with salt and chilli
36	*Jasminum calophyllum* Wall. ex.A.DC.	Oleaceae	Kattumullai	Leaves	Boiled and mashed with salt and chilli
37	*Kalanchoe pinnata* (Lam.) Oken.	Crassulaceae	Rannakalli	Leaves	Raw/Boiled/Fried along with salt, chilli and coconut

Contd...

Sl.No.	Botanical Name	Family	Vernacular Name	Plant Parts Used	Mode of Use
38	*Moringa concanensis* Nimmo ex Gibs.	Moringaceae	Kattu moringai	Leaves	Boiled/Fried along with salt, chilli and coconut
39	*Mukia maderaspatana* (L.) M.	Cucurbitaceae	Musumusukkai	Tender leaves	Raw
40	*Murraya koenigii* (L.) Spreng.	Rutaceae	Kariveppilai	Leaves	Mashed with chilli, salt and a piece of coconut (chutney)
41	*Murraya paniculata* (L.) Jack	Rutaceae	Kattu kariveppilai	Leaves	Mashed with chilli, salt and a piece of coconut (chutney)
42	*Oxalis corniculata* L.	Oxalidaceae	Puliyotharakeerai	Leaves	Boiled and mashed with salt and chilli
43	*Physalis minima* L. var. *indica* Clarke	Solanaceae	Kuttythakkali	Leaves	Boiled and mashed with salt and chilli
44	*Portulaca oleracea* L. var. *oleracea*	Portulacaceae	Udambukolup-pukeerai	Leaves	Boiled and mashed with salt and chilli
45	*Portulaca quadrifida* L.	Portulacaceae	Paruppukeerai	Leaves	Boiled and mashed with salt and chilli (taken only by ladies)
46	*Premna corymbosa* (Burm.f.) Rottl. and Willd.	Verbenaceae	Minnakkeerai	Leaves	Boiled and mashed with salt and chilli
47	*Psilanthus wightianus* (Wight and Arn)J.	Rubiaceae	Kattumalli	Leaves	Fried along with chilli and salt
48	*Sarcostemma acidum* (Roxb.) Voight	Asclepiadaceae	Thannikkodi	Leaves	Raw
49	*Sesbania grandiflora* (L.) Poir.	Fabaceae	Agathi keerai	Leaves	Fried along with chilli, coconut and salt
50	*Solanum anguivi* Lam. var. *multiflora*	Solanaceae	Kattuthudhuvalai	Leaves	Boiled and mashed with salt and chilli
51	*Solanum nigrum* L.	Solanaceae	Milaguthakkali	Leaves	Boiled and mashed with salt and chilli
52	*Solanum trilobatum* L.	Solanaceae	Thudhuvalai	Leaves	Boiled and mashed with salt and chilli
53	*Tamarindus indica* L.	Caesalpiniaceae	Puli	Tender leaves	Raw
54	*Trianthema portulacastrum* L.	Aizoaceae	Vattachanathikeerai	Leaves	Fried along with chilli, coconut and salt

Table 20.4: Edible Flowers Consumed by the Palliyar Tribals

Sl.No.	*Botanical Name*	*Family*	*Vernacular Name*	*Plant Parts Used*	*Mode of Use*
1	*Arenga wightii* Griff.	Arecaceae	Ala panai	Peduncle	Toddy taped from the peduncle
2	*Borassus flabellifer* L.	Arecaceae	Panai	Peduncle	Toddy taped from the peduncle
3	*Caryota urens* L.	Arecaceae	Koonthalpanai	Peduncle	Toddy taped from the peduncle
4	*Enseta superbumm* (Roxb.) Cheesman	Musaceae	Malai vazhai	Entire flower	Cut into pieces and boiled with chilli and coconut
5	*Ipomea alba* L.	Convolvulaceae	Mukkuthikkay	Swollen pedicle	Fried
6	*Moringa concanensis* Nimmo ex Glbs.	Moringaceae	Kattu moringai	Entire flower	Fried in oil along with coconut
7	*Murraya koenigii* (L.) Spreng.	Rutaceae	Kariveppilai	Entire flower	Fried along with salt, chilli and coconut oil
8	*Murraya paniculata* (L.) Jack	Rutaceae	Kattu kariveppilai	Entire flower	Fried along with salt, chilli and coconut oil
9	*Phoenix pusilla* Gaertn.	Arecaceae	Lechamaram	Tender infloresence	Fried along with salt, chilli and coconut oil
10	*Tamarindus indica* L.	Caesalpiniaceae	Puli	Entire flower	Raw/Mashed along with chilli and salt

Table 20.5: Edible Unripe Fruits Consumed by the Palliyar Tribals

Sl.No.	Botanical Name	Family	Vernacular Name	Plant Parts Used	Mode of Use
1	*Abelmoschus moschatus* Medicus	Malvaceae	Kaatuvendai	Entire unripe fruit	Raw/Fried in oil along with chilli and salt
2	*Atrocarpus heterophyllus* Lam.	Moraceae	Pala	Unripe perianth	Cut into pieces and boiled with chilli and coconut
3	*Atalantia racemosa* Wight and Arn.	Rutaceae	Kattuelumichai	Entire unripe fruit	Pickled in salt and chilli powder/Juice extracted, mixed with water and sugar
4	*Canavalia gladiata* (Jacq.)DC.	Fabaceae	Thampattai	Tender fruit	Boiled with chilli and salt
5	*Capparis zeylanica* L.	Capparaceae	Kaathatikaai	Entire unripe fruit	Raw/Boiled/Fried
6	*Capsicum frutescens* L.	Solanaceae	Kanamillaku keerai	Entire unripe fruit	Raw
7	*Carissa carandas* L.	Apocynaceae	Kilakkay	Fleshy part of the unripe fruit	Pickled in salt and chilli powder
8	*Coccinia grandis* (L.) Voigt	Cucurbitaceae	Kovai	Entire unripe fruit	Unripe fruits sun dried and fried
9	*Commiphora caudata* (Wight and Arn) Engler.	Burseraceae	Mangkiluvai	Fleshy part of the unripe fruit	Raw
10	*Commiphora pubescens* (Wight and Arn) Engler.	Burseraceae	Kadikiluvai	Fleshy part of the unripe fruit	Raw
11	*Dolichos trilobus* L.	Fabaceae	Minna	Tender pod	Raw/Boiled
12	*Enseta superbum* (Roxb.) Cheesman	Musaceae	Malai vazhai	Skinned off unripe fruit	Boiled unripe fruit fried along with coconut, salt and chilli
13	*Heracleum rigens* Wall.ex.DC.var. *rigens*	Umbelliferae	Kattu kothamalli	Entire unripe fruit	Boiled and mashed along with chilli and salt
14	*Hibiscus lobatus* (Murr.) Kuntze	Malvaceae	Kaatuvendai	Entire unripe fruit	Raw
15	*Hibiscus lunarifolius* Willd.	Malvaceae	Vendai	Entire unripe fruit	Raw
16	*Hibiscus ovalifolius* (Forsk.) Vahl	Malvaceae	Theingaapottu	Entire unripe fruit	Raw
17	*Lablab purpureus* (L.) Sweet var. *lignosus* (Prain) Kumari comb.	Fabaceae	Kattumochai	Entire tender pod	Boiled
18	*Luffa actungula* (L.) Roxb. var. *amara* Clarke	Cucurbitaceae	Kattu pirkku	Skinned off unripe fruit	Boiled along with chilli and salt

Sl.No.	Botanical Name	Family	Vernacular Name	Plant Parts Used	Mode of Use
19	*Magifera indica* L.	Anarcardiaceae	Maa	Fleshy part of the unripe fruit	Raw
20	*Momordica charantia* L. var. *charantia*	Cucurbitaceae	Kuruvithalai paakakai	Entire unripe fruit	Cut and oil fried with chilli, onion and salt
21	*Momordica dioica* Roxb.ex.Willd.	Cucurbitaceae	Palupaakakai	Entire unripe fruit	Cut and oil fried with chilli, onion and salt
22	*Moringa concanensis* Nimmo ex Gibs.	Moringaceae	Kattu moringai	Entire unripe fruit	Fried with salt, chilli and coconut
23	*Pavetta indica* L. var. indica	Rubiaceae	Pavattai	Tender unripe fruit	Fried in oil with salt and chilli
24	*Phoenix pusilla* Gaertn.	Arecaceae	Iechamaram	Fleshy part of the unripe fruit (red in colour)	Raw
25	*Phyllanthus emblica* L.	Euphorbiaceae	Kattunelli	Fleshy part of the unripe fruit	Raw
26	*Psidium guajava* L.	Myrtaceae	Kattu koyya	Entire unripe fruit	Raw
27	*Secamone emetica* (Retz.) R. Br.ex Schultes	Asclepiadaceae	Karuppattikodi	Entire unripe fruit	Raw
28	*Senna occidentalis* (L.) Link.	Caesalpiniaceae	Ponnavarai/Tagarai	Tender pod	Fried along with salt, coconut and chilli
29	*Solanum anguivi* Lam.var.*multiflora* (Roth ex Roem. and Sch.) Chithra comb.nov.	Solanaceae	Kattu thuduvalai	Entire unripe fruit	Fried
30	*Solanum erianthum* D. Don	Solanaceae	Kattuchundai	Entire unripe fruit	Fried in oil
31	*Solanum melongena* L. var. *insanum* (L.) Prain	Solanaceae	Mullukathari	Spine removed unripe fruit	Fried along with dry fish, salt and chilli
32	*Solanum pubescens* Willd.	Solanaceae	Kattusundai	Entire unripe fruit	Sun dried/Roasted used in curry
33	*Solanum torvum* Sw.	Solanaceae	Kattusundai	Entire unripe fruit	Sun dried/Roasted used in curry
34	*Solanum trilobatum* L.	Solanaceae	Thudhuvalai	Entire unripe pod	Fried in oil
35	*Tamarindus indica* L.	Caesalpiniaceae	Puli	Fleshy part of the unripe fruit	Raw

Contd...

Sl.No.	Botanical Name	Family	Vernacular Name	Plant Parts Used	Mode of Use
36	*Vigna bourneae* Gamble	Fabaceae	Kattu payaru	Entire tender pod/ Unripe seed	Raw/Boiled
37	*Vigna radiata* (L.) Wilczek var. *sublobata* (Roxb.) Verdc.	Fabaceae	Kattu pasipayaru	Entire tender pod	Raw/Boiled
38	*Vigna trilobata* (L.) Verdc.	Fabaceae	Kalpayaru	Entire tender pod	Raw/Boiled
39	*Vigna unguiculata* (L.) Walp. subsp. *cylindrica* (L.) Eselt.	Fabaceae	Panni minnappayaru	Unripe pod	Raw/Boiled
40	*Zizyphus mauritiana* Lam. var. *fruticosa* (Haines) Seb. and Balakr.	Rhamnaceae	Periya ilanthai	Fleshy part of the unripe fruit	Raw
41	*Zizyphus xylopyrus* (Retz.) Willd.	Rhamnaceae	Mullukottai	Fleshy part of the unripe fruit	Raw

Table 20.6: Edible Fruits Consumed by the Palliyar Tribals

Sl.No.	Botanical Name	Family	Vernacular Name	Plant Parts Used	Mode of Use
1	*Aegle marmelos* (L.) Correa.	Rutaceae	Vilvam	Fleshy part of the fruit	Raw
2	*Atrocarpus heterophyllus* Lam.	Moraceae	Pala	Fleshy perianth	Raw
3	*Atalantia racemosa* Wight and Arn.	Rutaceae	Kattu elumichai	Fleshy part of the fruit	Raw
4	*Azadirachta indica* A. Juss.	Meliaceae	Vembu	Fleshy part of the berry	Raw
5	*Borassus flabellifer* L.	Arecaceae	Panai	Pulp of the tender fruit	Raw
6	*Bridelia retusa* (L.) Spreng.	Euphorbiaceae	Kadukaipalam	Fruit	Raw
7	*Canthium parviflorum* Lam.	Rubiaceae	Periyakaarai/ Malukkaarai	Entire fruit	Raw
8	*Capsicum frutescens* L.	Solanaceae	Kanamillaku keerai	Dry fruit	Powdered and used in curry preparation
9	*Carissa carandas* L.	Apocynaceae	Kilallay	Fleshy part of the berry	Raw
10	*Chomelia asiatica* O.Kze var. *rigida* Gamble	Rubiaceae	Therani	Fleshy part of the fruit	Raw
11	*Coccinia grandis* (L.) Voigt	Cucurbitaceae	Kovai	Entire fruit	Raw
12	*Commiphora caudata* (Wight and Arn) Engler.	Burseraceae	Mangkilluvai	Fleshy part of the fruit	Raw
13	*Commiphora pubescens* (Wight and Arn) Engler.	Burseraceae	Kodikilluvai	Fleshy part of the fruit	Raw
14	*Cordia obliqua* Willd. var. *obliqua*	Boraginaceae	Virusu	Fleshy part of the fruit	Raw
15	*Cordia obliqua* Willd. var. *tomentosa* (Wall.) Kazmi	Boraginaceae	Kal virusu	Fleshy part of the fruit	Raw
16	*Diospyrus ferrea* (Willd.) Bakh.var. *neilgherrensis* (Wight) Bakh.	Ebenaceae	Karunthuvarai	Entire fruit	Raw
17	*Diospyrus foliolosa* Wall.ex.A.DC.	Ebenaceae	Thumla	Entire fruit	Raw
18	*Diplocyclos palmatus* (L.) Jeffrey	Cucurbitaceae	Attupudal/ Malaipusanni	Entire fruit	Raw
19	*Elaeocarpus tectorius* (Lour.) Poir.	Elaeocarpaceae	Kotla	Fleshy part of the fruit	Raw
20	*Enseta superbum* (Roxb.) Cheesman.	Musaceae	Malai vazhai	Fleshy part of the fruit	Raw

Contd...

Sl.No.	Botanical Name	Family	Vernacular Name	Plant Parts Used	Mode of Use
21	*Erythtoxylon monogynum* Roxb.	Erythroxylaceae	Chemmana	Fleshy part of the fruit	Raw
22	*Ficus benghalensis* L.var.*benghalensis*	Moraceae	Aal	Entire fruit	Raw
23	*Ficus racemosa* L.	Moraceae	Atthi	Entire fruit	Raw/Dipped in honey and dried
24	*Ficus religiosa* L.	Moraceae	Arasu	Entire fruit	Raw
25	*Flacourtia indica* (Burm.f.) Merr.	Flacourtiaceae	Mullumayilai	Fleshy part of the fruit	Raw
26	*Gardenia gummifera* L.f.	Rubiaceae	Karadivetchi	Fruit	Raw
27	*Gardenia resinifera* Roth	Rubiaceae	Vetchi	Fleshy part of the fruit	Raw
28	*Glycosmis pentaphylla* (Retz.) DC.	Rubiaceae	Panam palam/ Pannichedi	Fleshy part of the fruit	Raw
29	*Grewia flavescens* Juss.	Tiliaceae	Odaachu	Fleshy part of the fruit	Raw
30	*Grewia heterotricha* Mast.	Tiliaceae	Periyaachhu	Fleshy part of the fruit	Raw
31	*Grewia hirsuta* Vahl	Tiliaceae	Chinnaachu	Fleshy part of the fruit	Raw
32	*Grewia laevigata* Vahl	Tiliaceae	Karuachu	Fleshy part of the fruit	Raw
33	*Grewia tiliifolia* Vahl	Tiliaceae	Valukkaimaram	Fleshy part of the fruit	Raw
34	*Grewia villosa* Willd.	Tiliaceae	Vattachi	Fleshy part of the fruit	Raw
35	*Heracleum rigens* Wall.ex.DC.var.*rigens*	Umbelliferae	Kattu kothamalli	Entire dry fruit	Roasted fruit fried mashed along with coconut, salt and chilli
36	*Lantana camara* L. var. *aculeata* (L.) Mold.	Verbenaceae	Uni	Entire fruit	Raw
37	*Mangifera indica* L.	Anarcardiaceae	Maa	Fleshy part of the fruit	Raw
38	*Miliusa eriocarpa* Dunn.	Annonaceae	Nedunaarai	Fleshy part of the fruit	Raw
39	*Mimusops elengi* L.	Sapotaceae	Karuvithalai paakakai	Fleshy part of the fruit	Raw
40	*Momordica charantia* L. var. *charantia*	Cucurbitaceae	Mahilam	Fleshy part of the fruit	Raw
41	*Momordica dioica* Roxb.ex.Willd.	Cucurbitaceae	Palupaakakai	Fleshy part of the fruit	Raw
42	*Mukia maderaspatana* (L.) M.	Cucurbitaceae	Musumusukkai	Entire fruit	Raw

Sl.No.	Botanical Name	Family	Vernacular Name	Plant Parts Used	Mode of Use
43	*Murraya koenigii* (L.) Spreng.	Rutaceae	Kariveppilai	Fleshy part of the fruit	Raw
44	*Murraya paniculata* (L.) Jack	Rutaceae	Kattukariveppilai	Fleshy part of the fruit	Raw
45	*Opuntia dillenii* (Ker-Gawl.) Haw.	Cactaceae	Sappathikalli	Fleshy part of the fruit	Raw
46	*Osyris quadripartite* Salzm.ex.Decne. var.*puberula* (Hook.f.)Kumari.	Santalaceae	Sundaravalli	Entire fruit	Raw
47	*Passiflora foetida* L.	Passifloraceae	Poonakkaali	Entire fruit	Raw
48	*Phoenix pusilla* Gaertn.	Arecaceae	Iechamaram	Fleshy part of the fruit	Raw
49	*Phyllanthus emblica* L.	Euphorbiaceae	Kattunelli	Fleshy part of the fruit	Raw
50	*Phyllanthus reticulata* POir.	Euphorbiaceae	Poola	Entire fruit	Raw
51	*Physalis minima* L. var. *indica* Clarke	Solanaceae	Kutty thakkali	Fleshy part of the fruit	Raw
52	*Polyalthia cerasoides* (Roxb.) Bedd.	Annonaceae	Nedunarai	Fleshy part of the fruit	Raw
53	*Polyalthia suberosa* (Roxb.) Thw.	Annonaceae	Kodinaaval	Fleshy part of the fruit	Raw
54	*Psidium guajava* L.	Myrtaceae	Kattu koyya	Fleshy part of the fruit	Raw
55	*Rubus niveus* Thunb.var.*niveus* Gamble	Rosaceae	Maekattu illanthai	Entire fruit	Raw
56	*Secamone emetica* (Retz.) R. Br.ex Schultes	Asclepiadaceae	Karuppattikodi	Fleshy part of the fruit	Raw
57	*Solanum nigrum* L.	Solanaceae	Milaguthakali	Entire fruit	Raw
58	*Solanum trilobatum* L.	Solanaceae	Thudhuvalai	Entire fruit	Raw
59	*Syzygium cumini* (L.) Skeels.	Myrtaceae	Naval	Fleshy part of the fruit	Raw
60	*Tamarindus indica* L.	Caesalpiniaceae	Puli	Fleshy part of the fruit	Raw
61	*Uvaria rufa* Blume.	Annonaceae	Thevakodi	Fleshy part of the fruit	Raw
62	*Zizyphus mauritiana* Lam. var. *fruticosa* (Haines) Seb. and Balakr.	Rhamnaceae	Periya ilanthai	Fleshy part of the fruit	Raw
63	*Zizyphus oenoplia* (l.) Mill.	Rhamnaceae	Pulichi	Fleshy part of the fruit	Raw
64	*Zizyphus xylopyrus* (retz.) Willd.	Rhamnaceae	Mullukottai	Fleshy part of the fruit	Raw

Table 20.7: Edible Seeds and Seed Components by the Palliyar Tribals

Sl.No.	*Botanical Name*	*Family*	*Vernacular Name*	*Plant Parts Used*	*Mode of Use*
1	*Atrocarpus heterophyllus* Lam.	Moraceae	Pala	Kernel	Boiled/Roasted
2	*Atylosia scarabaeoides* (L.) Benth.	Fabaceae	Kattuthuvarai	Seed	Raw/Boiled/Roasted
3	*Borassus flabellifer* L.	Arecaceae	Panai	Endosperm	Raw
4	*Bupleurum wightii* Mukh.var. *ramosissium* (Wight and Arn.) Chandrabose comb. nov.	Umbelliferae	Kattuseeragam	Seed	Side dish is prepared along with salt, chilli and coconut
5	*Canarium strictum* Roxb.	Burseraceae	Kungilium	Kernel	Raw/Roasted
6	*Canavalia gladiata* (Jacq.)DC.	Fabaceae	Thampattai	Seed	Seeds consumed after repeated boiling
7	*Capparis zeylanica* L.	Capparaceae	Kaathatikaai	Kernel	Raw/Roasted
8	*Celtis philippensis* Blanco var. *wightii* (Planch.) Soep.	Ulmaceae	Vellaithuvarai	Seed	Raw/Roasted
9	*Chamaecrista absus* (L.) Irwin and Barneby.	Caesalpiniaceae	Kattukanam	Seed	Boiled/Roasted
10	*Cycas circinalis* L.	Cycadaceae	Paereenji	Kernel	Powdered decoated dry seeds used in the preparation of various items like 'dosai'/'puttu'
11	*Dolichos trilobus* L.	Fabaceae	Minna	Seed	Raw/Boiled/Roasted
12	*Drypetes sepiaria* (Wight and Arn.) Pax and Hoffm.	Euphorbiaceae	Kalvirai	Seed	Raw
13	*Elaeocarpus tectorius* (Lour.) Poir.	Elaeocarpaceae	Kotla	Kernel	Raw/Boiled
14	*Eleusine coracana* (L.) Gaertn.	Poaceae	Kattu kepai	Seed	Raw/Boiled
15	*Ensete superbum* (Roxb.) Cheesman.	Musaceae	Malai vazhai	Seed	Roasted
16	*Entada rheedi* Spreng.	Mimosaceae	Malam thelluka	Kernel	Kernels roasted and soaked for 3 days
17	*Givotia rottleriformis* Griff.	Euphorbiaceae	Vandalai	Kernel	Milky juice extracted from the powder of raw kernels

Sl.No.	Botanical Name	Family	Vernacular Name	Plant Parts Used	Mode of Use
18	*Heracleum rigrens* Wall.ex.DC. var. *rigens*	Umbelliferae	Kattu kothamalli	Seed	Fried seeds are ground along with coconut, chilli and kari leaf
19	*Lablab purpureus* (L.) Sweet var. *lignosus* (Prain) Kumari comb.	Fabaceae	Kattumochai	Seed	Raw/Boiled
20	*Moringa concanensis* Nimmo ex Gibs.	Moringaceae	Kattu moringai	Kernel	Raw kernels cooked with chilli and salt as curry
21	*Mucuna atropurpurea* DC.	Fabaceae	Thelluka	Kernel	Roasted
22	*Neonotonia wightii* (Wight and Arn.) Lackey.var.*coimbatorensis* (Ajita Sen) Karthik.	Fabaceae	Kattumochai	Seed	Raw/Boiled/Roasted
23	*Ocimum gratissimum* L.	Labiatae	Kattuthulasi	Seed	Boiled
24	*Oryza meyeriana* (Zoll. and Mor.) Baill var. *granulata* (Nees and Arn. ex.Watt) Duist.	Poaceae	Nell	Grain (Rice)	Raw/Boiled with salt
25	*Pithecellobium dulce* (Roxb.) Benth.	Mimosaceae	Kodukkapuli	Aril	Raw
26	*Rhychosia cana* DC.	Fabaceae	Kattuthuvarai	Seed	Raw/Boiled/Roasted
27	*Rhychosia filipes* Benth.	Fabaceae	Kattuthuvarai	Seed	Raw/Boiled/Roasted
28	*Rhychosia rufescens* (Willd.) DC.	Fabaceae	Kattuthuvarai	Seed	Raw/Boiled/Roasted
29	*Rhychosia suaveolens* (L.f.) DC.	Fabaceae	Kattuthuvarai	Seed	Raw/Boiled/Roasted
30	*Sapindus emarginatus* Vahl	Sapindaceae	Pullichi	Seed	Raw mostly children prefer
31	*Sesamum indicum* L.	Pedaliaceae	Kaatu yellu	Seed	Raw/Roasted ground along with salt and red chilli/with jaggery
32	*Sterculia foetida* L.	Sterculiaceae	Kongatti	Kernel	Raw/Boiled/Roasted
33	*Sterculia guttata* Roxb.ex.DC.	Sterculiaceae	Kattu iluppai	Kernel	Raw/Boiled/Roasted
34	*Sterculia urens* Roxb.	Sterculiaceae	Vennaali	Kernel	Raw/Boiled
35	*Strychnos nux-vomica* L.	Loganiaceae	Thaethankottai	Kernel	Raw/Roasted
36	*Tamarindus indica* L.	Caesalpiniaceae	Puli	Kernel	Raw/Boiled/Roasted

Contd...

Sl.No.	Botanical Name	Family	Vernacular Name	Plant Parts Used	Mode of Use
37	*Teramnus labialis* (L.F.) Spreng.	Fabaceae	Kattukanam	Seed	Raw/Boiled/Roasted
38	*Terminalia bellirica* (Gaertn.) Roxb.	Combretaceae	Tani	Kernel	Raw/Roasted
39	*Terminalia chebula* Retz.	Combretaceae	Kadukkai	Kernel	Soaked, dried and powdered kernels used in the preparation of 'dosai' and 'kali'
40	*Vigna bourneae* Gamble	Fabaceae	Kattu payaru	Seed	Raw/Boiled/Roasted
41	*Vigna radiata* (L.) Wilczek var. *sublobata* (Roxb.) Verdc.	Fabaceae	Kattu pasipayaru	Seed	Raw/Boiled/Roasted
42	*Vigna trilobata* (L.) Verdc.	Fabaceae	Kalpayaru	Seed	Raw/Boiled/Roasted
43	*Vigna unguiculata* (L.) Walp. subsp. *cylindrica* (L.) Eselt.	Fabaceae	Panni minnppayaru	Seed	Raw/Boiled/Roasted
44	*Vigna unguiculata* (L.) Walp. *unguiculata*	Fabaceae	Kattukanam	Seed	Raw/Boiled/Roasted
45	*Zizyphus xylopyrus* (Retz.) Willd.	Rhamnaceae	Mullukottai	Kernel	Raw/Roasted

The South Eastern slopes of Western Ghats,Ayyanarkoil.

Shenbagathoppu Hills.

Thanniparai Hills.

Saduragiri Hills,Saptoor.

The upper Hills of Shenbagathoppu.

A view of Shenbagathoppu Falls.

Plate 20.1: Area Surveyed.

Tribal men of Valliammmal Nagar.

A young tribal woman of Valliammal Nagar with her children

A tribal man of Thaniparai holding *Abelmoschus moschatus* Medicus. with the edible root.

An old Palliyar tribal man of Athikoil sector of Grizzled Giant Squirrel Wildlife Sanctuary.

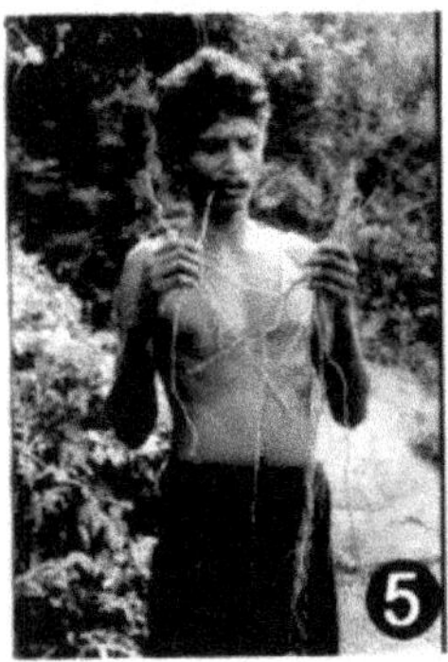

Mayandi,a tribal man of Thanniparai eating *Sarcostemma acidum*(Roxb.) Voigt.

A Tribal family of Rakkammalkoil Parai.

Plate 20.2: The Tribal People of the Area Surveyed.

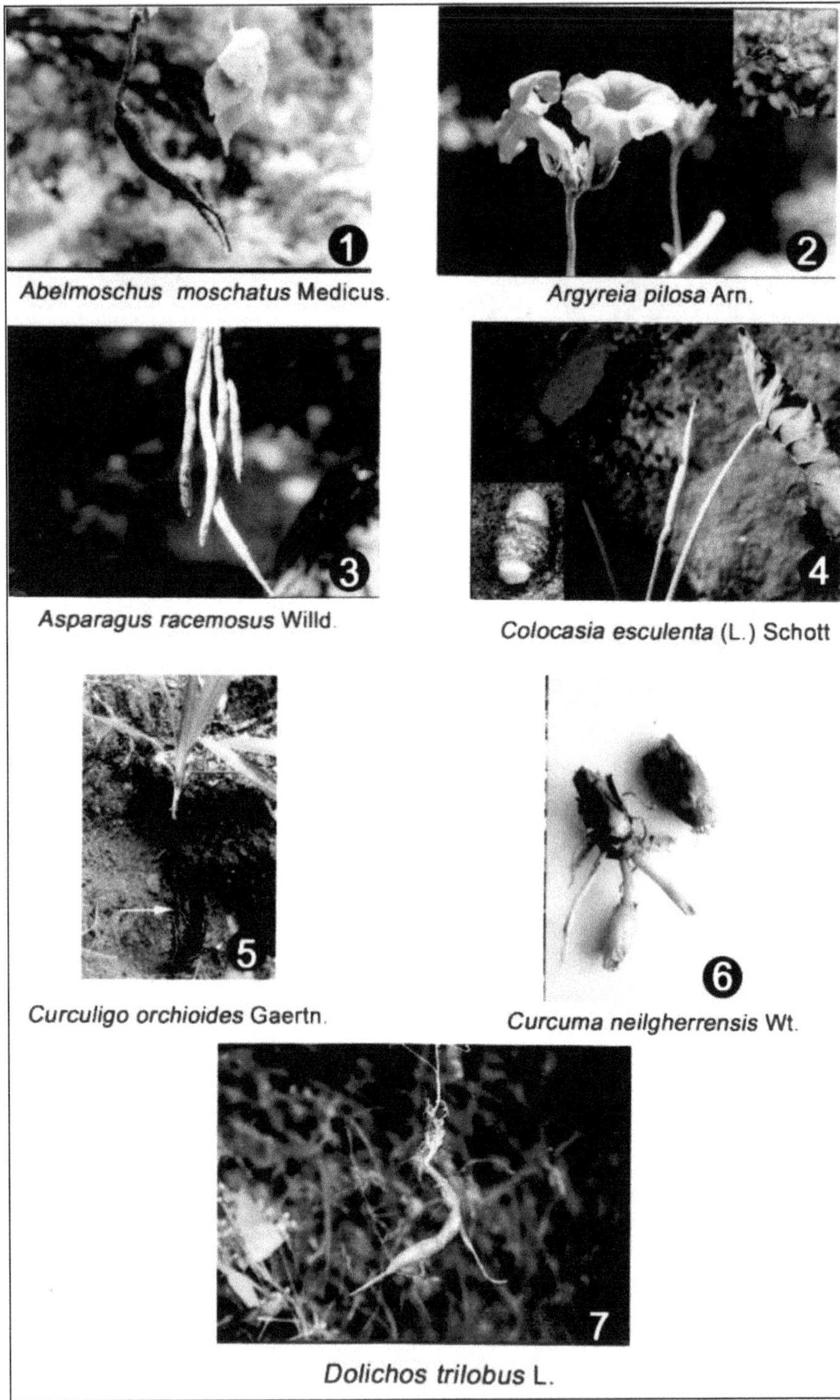

Plate 20.3: Edible Tubers, Rhizome, Corm and Root-Types.

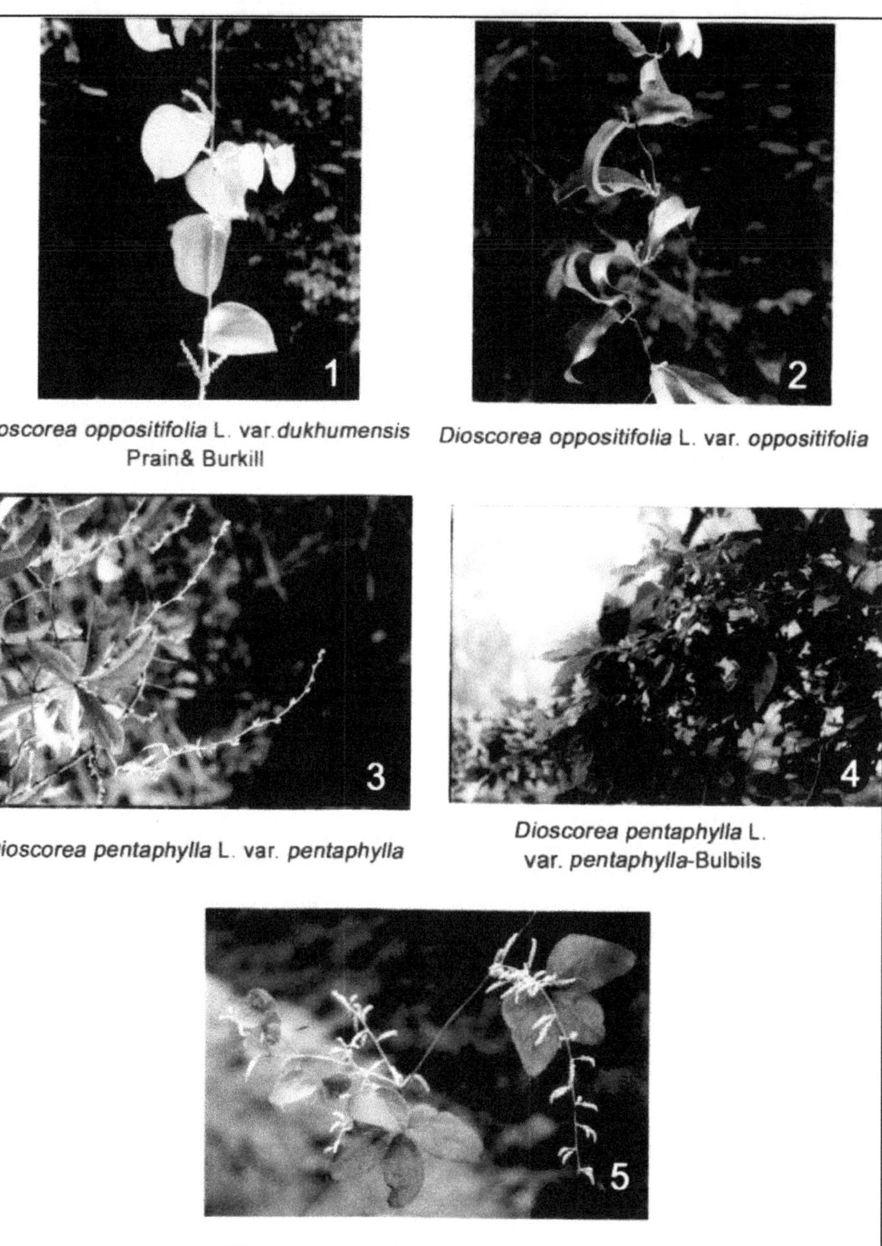

Dioscorea oppositifolia L. var.*dukhumensis* Prain& Burkill

Dioscorea oppositifolia L. var. *oppositifolia*

Dioscorea pentaphylla L. var. *pentaphylla*

Dioscorea pentaphylla L. var. *pentaphylla*-Bulbils

Dioscorea tomentosa Koen.ex Spreng.

Plate 20.4: Edible Tubers and Bulbils in their Natural Habitat.

Dioscorea oppositifolia L. var. *dukhumensis* Prain& Burkill

Dioscorea oppositifolia L. var. *oppositifolia*

Dioscorea pentaphylla L. var. *pentaphylla*

A tribal man holding a tuber of *Dioscorea tomentosa* Koen.ex Spreng.

Plate 20.5: Edible Tubers of *Dioscorea* Species.

1. *Basella alba* L. var. *alba*
2. *Cardiospermum halicacabum* L.
3. *Celosia argentea* L.
4. *Commelina ensifolia* R.Br.
5. *Heracleum rigens* Wall.ex DC. var. *rigens* Clarke
6. *Jasminum calophyllum* Wall. ex A. DC.

Plate 20.6: Greens in their Natural Habitat.

Ipomoea alba L. - flower

Ipomoea alba L. with swollen pedicel

Moringa concanensis Nimmo ex Gibs.

Phoenix pusilla Gaertn.

EDIBLE STEM AND APICAL MERISTEM

Caralluma lasiantha (Wight) N.E. Br.

Phoenix pusilla Gaertn.

Plate 20.7: Edible Fruits in their Natuaral Habitat.

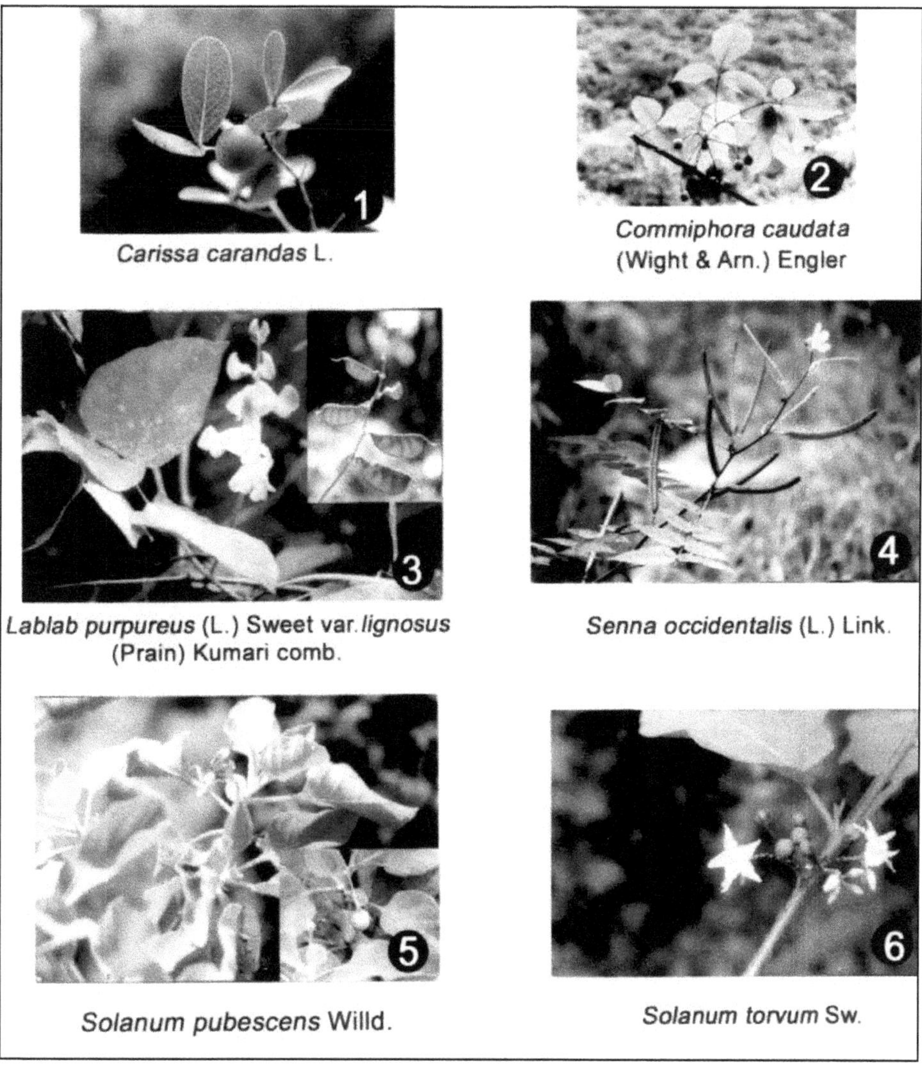

1. *Carissa carandas* L.
2. *Commiphora caudata* (Wight & Arn.) Engler
3. *Lablab purpureus* (L.) Sweet var. *lignosus* (Prain) Kumari comb.
4. *Senna occidentalis* (L.) Link.
5. *Solanum pubescens* Willd.
6. *Solanum torvum* Sw.

Plate 20.8: Edible Unripe Fruits and Pods in their Natural Habitat.

Carissa carandas L.

Ficus racemosa L.

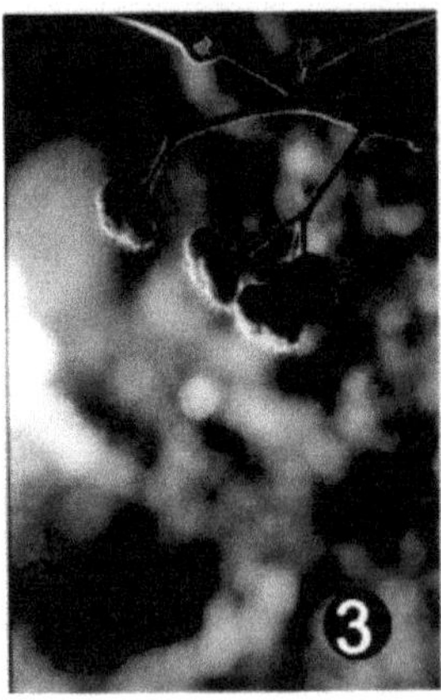

Grewia heterotricha Mast.

Grewia hirsuta Vahl.

Lantana camara L. var. *aculeata* (L.) Mold.

Opuntia dillenii (Ker-Gawl.) Haw.

Plate 20.9: Edible Ripe Fruits.

Passiflora foetida L.

Phyllanthus emblica L.

Polyalthia cerasoides (Roxb.) Bedd.

Syzygium cumini (L.) Skeels

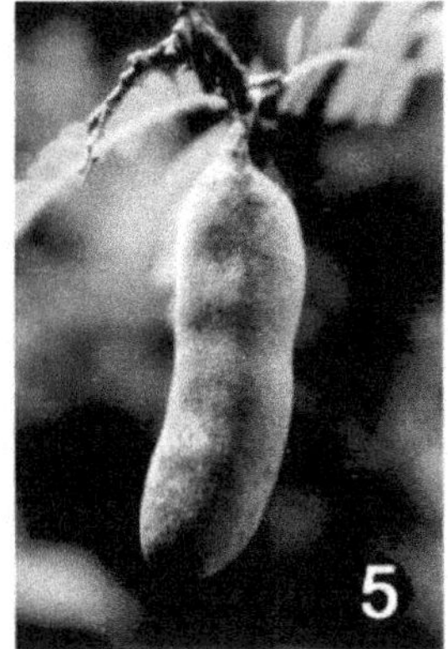

Tamarindus indica L.

Uvaria rufa Blume.

Plate 20.10: Edible Ripe Fruits.

Atylosia scarabaeoides (L.) Benth.

Canavalia gladiata (Jacq.) DC.

Givotia rottleriformis Griff.

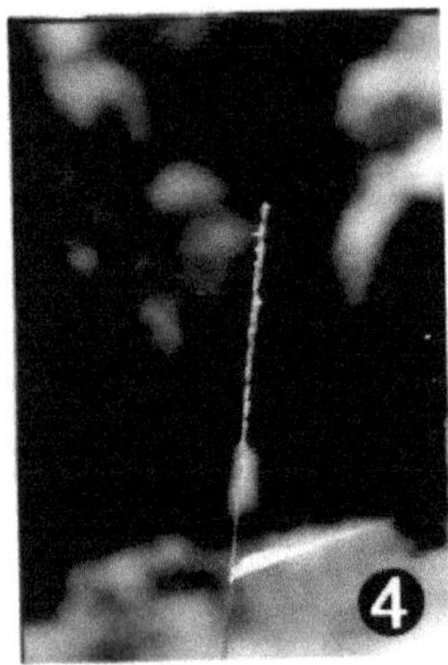

Oryza meyeriana (Zoll.& Mor.) Baill. var.*granulata* (Nees & Arn.ex Watt) Duist.

Rhynchosia filipes Benth.

Sterculia guttata Roxb. ex DC.

Plate 20.11: Edible Seeds in their Natural Habitat.

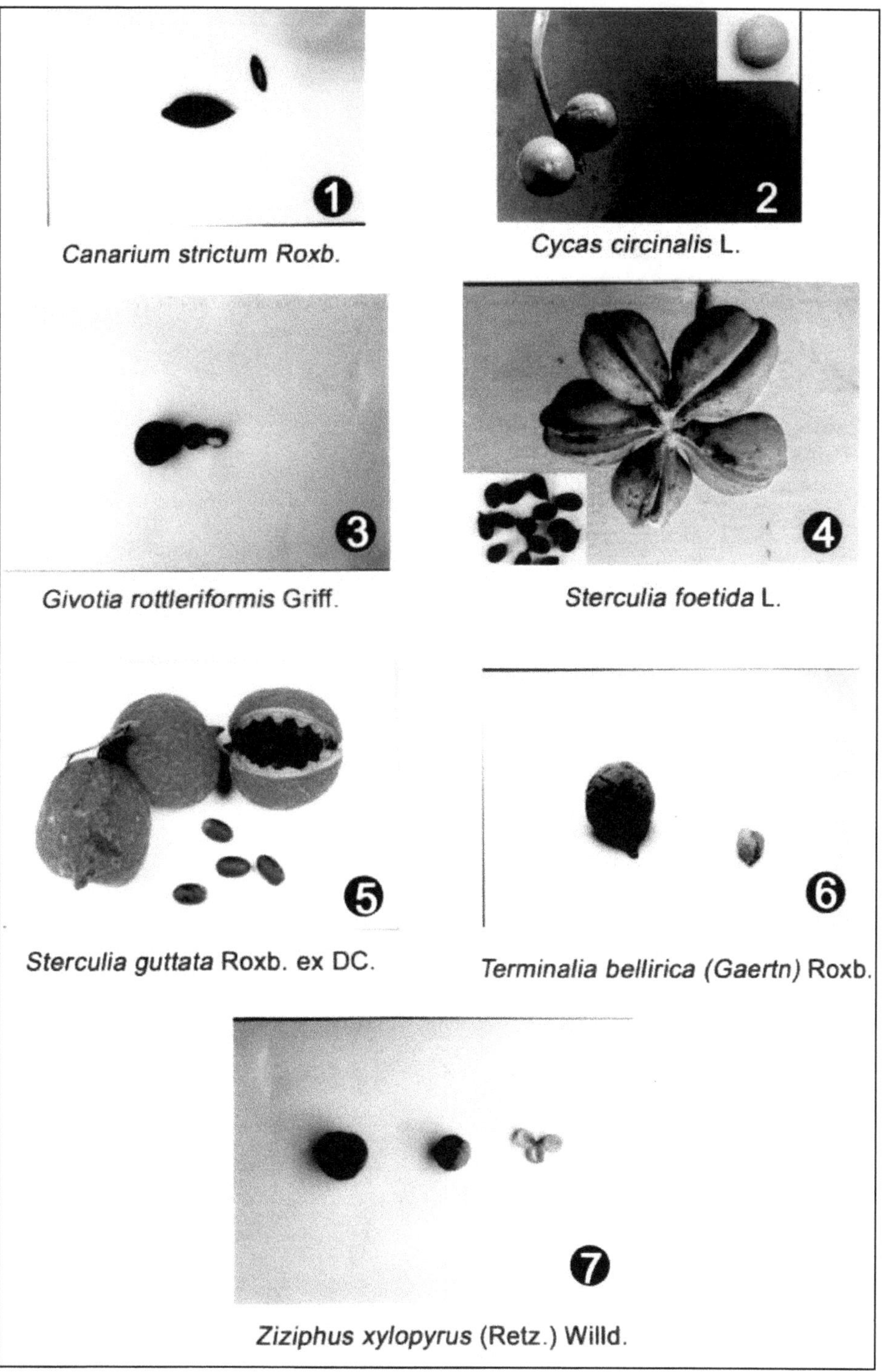

Canarium strictum Roxb.

Cycas circinalis L.

Givotia rottleriformis Griff.

Sterculia foetida L.

Sterculia guttata Roxb. ex DC.

Terminalia bellirica (Gaertn) Roxb.

Ziziphus xylopyrus (Retz.) Willd.

Plate 20.12: Edible Seeds.

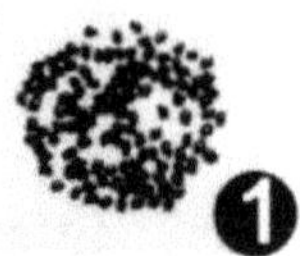

Chamaecrista absus (L.) Irwin & Barneby

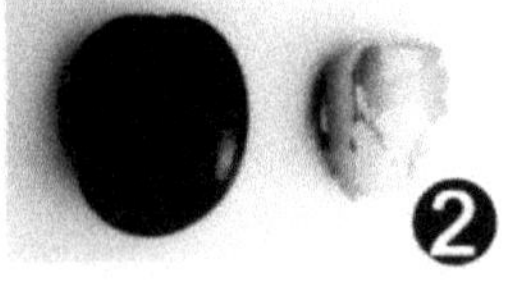

Entada rheedi Spreng.

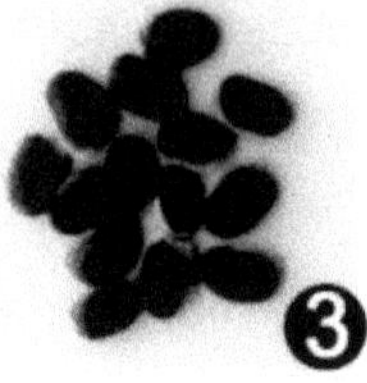

Lablab purpureus (L.) Sweet var.*lignosus* (Prain) Kumari comb.

Mucuna atropurpurea DC.

Rhynchosia cana DC.

Rhynchosia filipes Benth.

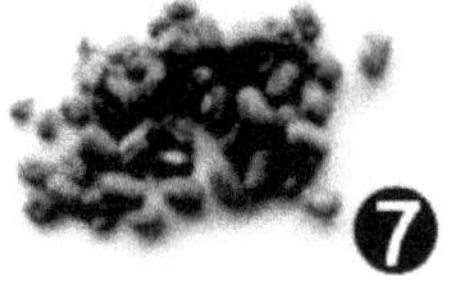

Teramnus labialis (L.f.) Spr.

Vigna trilobata (L.) Verdc.

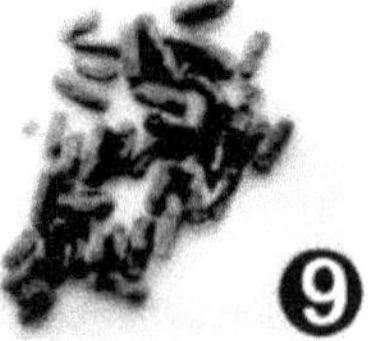

Vigna unguiculata (L.) Walp.subsp.*cylindrica* (L.) Eselt.

Vigna unguiculata (L.) Walp.subsp. *unguiculata*

Plate 20.13: Edible Seeds.

Discussion

Of the various parts of the edible plants consumed by the Palliyar tribals, rhizomes, corms, tubers, bulbils and root-types of 19 plant species, stem, pith and apical meristems of 12 plant species, leaves of 54 plant species, flowers of 10 plant species, unripe fruits/pods of 41 plant species, ripe fruits of 64 plant species and seeds, endosperm and kernels of 45 plant species are consumed. The plant parts are either eaten raw or cooked and eaten as shown in Tables 20.1 to 20.7. Tubers of *Dioscorea pentaphylla* var. *pentaphylla* are over-exploited by Palliyars, living in Shenbagathoppu. Similarly tribes over-exploit both the varieties of *Dioscorea oppositifolia*. Tubers of *Dioscorea bulbifera* var. *vera* are used only by the Palliyar tribe living in Athikoil and Shenbagathoppu. The plant parts are either eaten raw or cooked and eaten. Palliyars rarely use the roots of *Argyreia pilosa* and tubers of *Curculigo orchioides* as food. Tubers of *Dioscorea tomentosa* are used as food by Palliyars settled in Shenbagathoppu.

Among the 54 plant species as greens, the leaves of different species of the genus *Amaranthus, viz., A. roxburghianus, A. spinosus, A. tricolor* and *A. viridis* are used in large quantities. These greens are known to be consumed not only by the Palliyars but also by several other India tribals. The leaves of *Amaranthus viridis* and *A. spinosus* are used as greens by the tribal people of Rajasthan (Singh and Singh, 1981); the Bhoxa tribe of Bijnor and Pauri Garhwal districts, Uttar Pradesh (Maheswari and Singh, 1984); the tribals of Pune and neighbourinng districts, Maharashtra (Vartak and Kulkarni, 1987; Vartak and Suryanarayana, 1995). The Totos, the Mech, the Modesla and the Napalese of Jalpaiguri district, West Bengal uses the leaves of *Amaranthus spinosus* as greens (Das *et al.*, 1983). The Kondareddies, the Valmiki, the Koyas and the Kongakammars of East-Godavari district, Andhra Pradesh uses the leaves of *A. spinosus* as greens (Prasad *et al.*, 1999). So the documentation and conservation of these plants is significant in the current era of documenting nutritional content of the plants to provide an update on indigenous know-how.

The ripe fruits of various species of *Grewia, viz., G. flavescens, G. heterotricha, G. hirsuta, G. laevigata, G. tiliifolia* and *G. villosa*, which are sweet in taste, are eaten raw by the Palliyars like several other tribes (the ripe fruits of *G. tiliifolia* are eaten raw by the Gonds of Madhya Pradesh – Jain, 1963; the tribal people of Rajasthan – Singh and Singh, 1981; the tribals of Susale Island, Pune district, Maharashtra – Vartak and Suryanarayana, 1995; the ripe fruits of *G. villosa* are eaten by the Irulars of Coimbatore district, Tamil Nadu – Ramachandran and Nair, 1981; the ripe fruits of *G. hirsute* are eaten by the tribal people of Orissa – Aminuddin and Girach, 1991).

The ripe fruits of *Flacourtia indica* and *Syzygium cumini* which are sweet in nature, are eaten raw by the Palliyars as in the case of several other tribals in India (the ripe fruits of *Flacourtia indica* and *Syzygium cumini* are eaten raw by the Maldharis of Gujarat – Shah *et al.*, 1981; the Kurumbars, the Kurichyas and the Paniyans of Cannanore district, Kerala – Ramachandran and Nair, 1981; and the tribes like the Minas and the Bamanwas of the Todabhim and Nadoti tehsils of Karauli and Sawai Madhopur districts, Rajasthan – Das, 1997).

The seeds of *Canavalia gladiata, Vigna radiata* and *V. trilobata* consumed by the Palliyars are also consumed by other tribals in different parts of India. The seeds of *Canavalia gladiata* were originally consumed by the tribals of India (Rajaram and Janardhanan, 1992). The seeds of *Vigna radiata* are consumed by the Kathodi, the Varli, the Thakur and the Mahadevs kolis of Pune district, Maharashtra (Gunjatkar and Vartak, 1984). The seeds of *V. trilobata* are consumed by the Indian tribal sect Kurumba (Siddhuraju *et al.*, 1992a) and other poor sections people (Mittre, 1981).

The seed kernels of *Cycas circinalis* consumed by the Palliyars are also consumed by other tribals like the Nayadis of North Kerala (Prasad and Abraham, 1984) and the Gonds of Orissa (Girach and Aminuddin, 1992). The seed kernels of *Entada rheedi* are consumed by the Palliyars as in the case of their counter parts in other parts of India. The soaked seed kernels of *Entada phaseoloides* are roasted and eaten by the Great Andamanes and the Onges of Andaman and Nicobar Islands (Parkinson, 1972; Thathathri, 1980); the Kanikkars of Tirunelveli district, Tamil Nadu (Mohan and Janardhanan, 1993).

The seed kernels of *Mucuna atropurpurea* are consumed by the Palliyars, while the seed kernels of other species of *Mucuna* are consumed by various other Indian tribals (Jain, 1981; Mohan and Janardhanan, 1995). The seed kernels of *sterculia* species namely *S. guttata* and *S. urens* which are consumed by the Palliyars are also taken by other tribals (Ramachandran and Nair, 1981; Singh and Singh, 1981; Vartak and Suryanarayana, 1995; Prasad *et al.*, 1999).

Among the documented wild edible seed plants of the Southeastern slopes of Western Ghats; 7 species (*Atylosia scarabaeoides, Canavalia gladiata, Entada rheedi, Mucuna atropurpurea, Vigna bourneae, Vigna radiata* var. *sublobata* and *V. trilobata*) are endemic. Of 7 endemic plant species, *Atylosia scarabaeoides, Canavalia gladiata, Entada rheedi, Mucuna atropurpurea* and *Vigna bourneae* are listed in threatened category, so suitable conservation method may be carried out to preserve them. Because of urbanization, the traditional knowledge on the use of plants by tribals is fast vanishing. There is an urgent need to document the knowledge or otherwise it will be lost forever.

References

Ajit Raizada. 1984. Historical background. In: The tribal development in Madhya Pradesh. A planning perspective. Inter-India publications, New Delhi, pp. 13-20.

Aminuddin and Girach, R.D. 1991. Ethobotanical studies on Bondo tribe of district Koraput (Orissa), India. *Ethnobotany*. 3: 15-19.

Anonymous, 1994. *Ethnobiology in India, a status report*. (All India coordinated research project in Ethnobiology), Ministry of Envirionment and Forest, Government of India, New Delhi, India.

Arora, R.K. 1985. *Genetic resources of less-known cultivated food plants*. NBPGR Sci. Monogr. No.9, ICAR, New Delhi, India.

Babu, C.R., Sharma, S.K., Chatterjee, S.R. and Abrol, Y.P. 1988. Seed protein and amino acid composition of wild *Vigna radiata* var. *sublobata* (Fabaceae) and two cultigens. *V. mungo* and *V. radiata*. *Econ. Bot.* 42: 54-61.

Borthakur, S.K. 1996. Wild edible plants in markets of Assam, India – An Ethnobotanical investigation. In: *Ethnobiology in human welfare.* (Eds.) Jain, S.K. Deep Publications, New Delhi, India. pp. 31 – 34.

Burdon, J. J. and Jarosz, A.M. 1989. Wild relatives as sources of disease resistance. In: *The use of plant genetic resources.* (Eds.) Brown, A.H.D., Frankel, O.H., Marshall, D.R. and Williams, J.T. Cambridge Univ. Press, Cambridge, UK. pp. 280 – 296.

Burkill, I.H. 1952. Habits of man and the origins of the cultivated plants of the old world. *Proc. Limm. Soc.London.* 164: 12-42.

Census of India, 1991. Series1, India, Paper 2 of 1992. Final population totals. Brief analysis of primary census abstract.

Das, S.N. 1997. A study on the Ethnobotany of Karauli and Sawai Madhopur Districts, Rajasthan. *J. Eco. Taxon. Bot.* 21: 587- 605.

Das, S.N., Janardhanan, K.P. and Roy, S.C. 1983. Some observations on the Ethnobotany of the tribes of Totapara and adjoining areas in Jalpaiguri district, West Bengal. *J. Econ. Taxon. Bot.* 4: 453-474.

Doney, D.L. and Whitney, E.D. 1990. Genetic enhancement of Beta for disease resistance using wild relative. A strong case for the value of genetic conservation. *Econ. Bot.* 44: 445 – 451.

FAO, 1996. *The state of the world's plant genetic resources for food and agriculture.* FAO, Rome, Italy.

Fischer, C.E.C. 1980. The Flora of the Presidency of Madras J.S. Gamble III: 1347-2017.

Frankel, O.H., Brown, A.H.D. and Burdon, J.J. 1995. *The conservation of plant biodiversity.* Cambridge Univ. Press, Cambridge, UK. pp. 29 – 33.

Gadgil, M. 1996. Documenting diversity: An experiment. *Curr. Sci.* 70: 36-44.

Gamble, J.S. 1957a. The Flora of the Presidency of Madras I: 1-577.

Gamble, J.S. 1957b. The Flora of the Presidency of Madras II: 578-1346.

Girach, R.D. and Aminuddin. 1992. Some little known edible plants from Orissa. *J. Econ. Taxon. Bot.* 16: 61-68.

Gunjatkar, N. and Vartak, V.D. 1982. Enumeration of wild edible legumes from Pune district, Maharastra State. *J. Econ. Taxon. Bot.* 3: 1-9.

Henry, A.N., Chithra, V. and Balakrishnan, N.P. 1989. Flora of Tamil Nadu, India, Series I: Analysis III: 1-171.

Henry, A.N. Kumari, C.R. and Chitra, V. 1987. Flora of Tamil Nadu, India, Series I: Analysis II: 1-258.

Hodgkin, T. and Debuock, D.G. 1992. Some possible applications of molecular genetics in the conservation of wild species for crop improvement. In: *Conservation of plant genes, DNA banking and in vitro biotechnology.* (Eds.) Adams, R.P. and Adams, J.B. Academic Press, California, USA. pp. 153 – 182.

Ignacimuthu, S. and Babu, C.R. 1987. *Vigna radiata* var. *sublobata* (Fabaceae): Economically useful wild relative of urd and mung bean. *Econ. Bot.* 41: 418-422.

Jain, S.K. 1963. Studies in Indian Ethnobotany.Less known uses of fifty common plants from the tribal areas of Madhya Pradesh. *J. Bull. Bot. Surv. India.* 5: 223-226.

Jain, S.K. (Ed.) 1991. *Contributions to Ethnobotany of India.* Scientific Publishers, Jodhpur, India.

Jain, S.K. (Ed.) 1981. *Glimpses of Indian Ethnobotany.* Oxford and IBH Publishing Co., New Delhi, India.

Jain, S.K. 1987. Ethnobotany – its concepts and relevance. Tenth Botanical Conference, Indian Bot. Soc., Department of Botany, Univ. of Patna, India. pp. 3-12.

Janardhanan, K. 1990. Germplasm resources of pulses of tribal utility in India. In: *Proceedings of the national seminar on advances in seed science and technology.* (Eds.) Shetty H.S. and Prakash, H.S., Univ. of Mysore, Mysore, India. pp. 407-409.

Khoshoo, T.N. 1991. Conservation of biodiversity in biosphere, In: *Indian geosphere biosphere programme* – some aspects. (Eds.) Khoshoo, T.N. and Sharma, M. National Academy of Sciences, Allahabad, India. pp: 178 – 233.

Khoshoo, T.N. 1996. India needs a national biodiversity conservation board. *Curr. Sci.* 71: 506-513.

Lenne, J.M. and Wood, 1991. Plant diseases and the use of wild germplasm. *Ann. Rev.Phytopathol.* 29: 35-63.

Maheswari, J.K. and Singh, J.P. 1984. Contribution to the Ethnobotany of Bhoxa tribe of Bijnor and Pauri Gharwal districts, U.P. *J. Econ. Taxon. Bot.* 5: 253-259.

Maikhuri, R.K., Nautiyal, M.C. and Khali, M.P. 1991. Lesser – known crops of food value in Garhwal Himalaya and a Strategy to conserve them. *FAO/IBPGR Plant Genet. Resour. Newslett.* 86: 33 – 36.

Maikhuri, R.K., Nautiyal, S., Rao, K.S. and Semwal, R.L. 2000. Indigenous knowledge of medicinal plants and wild edibles among three tribal subcommunities of the Central Himalayas, India. *J. Indigenous Know. Devt. Mon.* 8: 7-13.

Mary Josephine, R. and Janardhanan, K. 1992. Studies on chemical composition and antinutritional factors in three germplasm seed materials of the tribal pulse, *Mucuna pruriens* (L.) DC. *Food Chem.* 43: 13 – 18.

Mathew, K.M. 1983a. The Flora of Tamil Nadu Carnatic I: 1-688.

Mathew, K.M. 1983b. The Flora of Tamil Nadu Carnatic II: 689-1540.

Mathew, K.M. 1983c. The Flora of Tamil Nadu Carnatic III: 1541-2155.

Mathew, K.M. 1999a. The Flora of the Palani Hills. I: 1-575.

Mathew, K.M. 1999b. The Flora of the Palani Hills. II: 576-1196.

Mathew, K.M. 1999c. The Flora of the Palani Hills. III: 1197-1635.

Mohan, V.R. and Janardhanan, K. 1993. Chemical and nutritional evaluation of raw seeds of the tribal pulses *Parkia roxburghii* G. Don. and *Entada phaseoloides* (L.) Merr. *Inter. J. Food Sci. Nutr.* 44: 47-53.

Mohan, V.R. and Janardhanan, K. 1994a. Chemical and nutritional evaluation of two germplasms of the tribal pulse, *Bauhinia racemosa* Lamk. *Plant Food Hum. Nutr.* 46: 367-374.

Mohan, V.R. and Janardhanan, K. 1994b. The biochemical composition and nutrient assessment of less known pulses of the genus *Canavalia. Inter. J. Food Sci. Nutr.* 45: 255-262.

Mohan, V.R. and Janardhanan, K. 1995. Chemical analysis and nutritional assessment of lesser known pulses of the genus *Mucuna. Food Chem.* 52: 275-280.

Mittre, V. 1981. Wild plants in Indian folk life- a historical perspective. In: *Glimpses of Indian Ethnobotany.* (edited by S.K.Jain) Oxford and IBH Publishing Co., New Delhi. pp. 37-58.

Nair, N.C. and Henry, A.N. 1983. Flora of Tamil Nadu, India Series I AnalysisI: 1-184.

Prasad, P.N. and Abraham, Z. 1984. Ethnobotany of the Nayadis of North Kerala. *J. Econ. Taxon. Bot.* 5: 41-48.

Prasad, V.K., Rajagopal, T., Kant, Y. and Badarinath, K.V.S. 1999. Food plants of Konda Reddies of Rampa Agency, East Godavari district, Andhra Pradesh – a Case Syudy. *Ethnobotany.* 11: 92-96.

Radhakrishnan, K., Pandurangan, A.G. and Pushpangadan, P. 1996. Ethnobotany of wild edible plants of Kerala, India. In: *Ethnobiology in human welfare.* (Eds.) Jain, S.K. Deep Publications, New Delhi, India. pp. 48 – 51.

Rajaram, N. and Janardhanan, K. 1991a. Studies on the underexploited tree pulses, *Acacia catchu* Willd, *Parkinsonia aculata* L. and *Prosopis chilensis* (Molina) Stinz. Chemical composition and antinutritional factors. *Food Chem.* 42: 265 - 273.

Rajaram, N. and Janardhanan, K. 1992. Nutritional and chemical evaluation of raw seeds of *Canavalia gladiata* (Jacq.) DC and *C.ensiformis* DC: The underutilized food and fodder crops in India. *Plant Foods. Hum. Nutr.* 42: 329-336.

Rajaram, N. and Janardhanan, K. 1993. *Ex situ* conservation of genetic resources of tribal pulses and their wild related species. *FAO/IBPGR Plant Genet. Resour. Newslett.* 91/92: 29-32.

Ramachandran, V.S. and Nair, V.J. 1981. Ethnobotanical observations on Irulars of Tamil Nadu (India). *J. Econ. Taxon. Bot.* 2:183-190.

Roy Burmann, B.K. 1986. Tribal demography: A preliminary appraisal. In: *The tribal situation in India.* (Eds.) Suresh Singh, K. Motital Banarsidass, New Delhi, India.

Sahu, T.R. 1996. Life support promising food plants among aboriginals of Bastar (M.P.), India. In: Etnobiology in Human Welfare. (Ed.) Jain, S.K. Deep Publications, New Delhi, India. pp. 26-30.

Sasidharan, N. and Sivarajan, V.V. 1996. Flowering plants of Thrissur forests. 1- 579.

Shah, G.L., Menon, A.R. and Gopal, G.V. 1981. An account of the ethnobotany of Gujarat State (India). *J. Econ. Taxon. Bot.* 2:173-181.

Shankar, R. 1995. Tribal community in India and PGR. In: Farmer's rights and plant genetic resources recognition and reward: A dialogue. (Ed.) Swaminathan, M.S. MacMillion India Limited, Madras, India. pp. 106 – 111.

Sharma, B.D., Balakrishnan, N.P., Rao, R.R. and Hajra, P.K. 1993. Flora of India I: 1-461.

Sharma, B.D., Balakrishnan, N.P. and Sanjappa, M. 1993.Flora of India II: 1-594.

Sharma, B.D. and Sanjappa, M. 1993. Flora of India. III: 1-617.

Siddhuraju, P., Vijayakumari, K. and Janardhanan, K. 1992. Nutritional and chemical evaluation of raw seeds of the tribal pulse *Vigna trilobata* (L.) Verdc. *Inter. J. Food Sci. Nutr.* 43: 97-103.

Siddhuraju, P., Vijayakumari, K. and Janardhanan, K. 1993. Genetic resources of tribal pulses. *FAO/IPGRI Plant Genet. Resour. Newslett.* 96: 47-49.

Siddhuraju, P., Vijayakumari, K. and Janardhanan, K. 1994. Chemical analysis and nutritional assessment of the less known pulses, *Vigna aconitifolia* (Jacq.) Marechal. and *V. vexillata* (L.) A. Rich. *Plant. Food Hum. Nutr.* 45: 103-111.

Singh, V. 2001. Monographic on Indian sub tribes *Cassiinae* pp. 1-270.

Singh, P.B. 1996. Wild edible plants of Mandi district in Northwest Himalaya. In: *Ethnobiology in human welfare.* (Eds.) Jain, S.K. Deep Publications, New Delhi, India. pp. 22 – 25.

Singh, U. and Singh, B. 1992. Tropical grain legume as important human foods. *Econ. Bot.* 46: 310-361.

Singh, V. and Singh, P. 1981. Edible wild plants of Eastern Rajasthan. *J. Econ. Taxon. Bot.* 2: 197-207.

Sinha, S.C. 1996. Wild edible plants of Manipur, India. In: *Ethnobiology in human welfare.* (Eds.) Jain, S.K. Deep Publications, New Delhi, India. pp. 42 – 47.

Thothathri, K.1980. Plant resources and their utilization in the Andaman and Nicobar Islands. *J. Econ. Taxon. Bot.* 1: 111-114.

Vadivel, V. and Janardhanan, K. 2000a. Preliminary agrobotanical traits and chemical evaluation of *Mucuna pruriens* (itching beans): A less-known food and medicinal legume. *J. Medi. Arom. Plant Sci.* 22: 191 – 199.

Vadivel, V. and Janardhanan, K. 2000b. Nutritional and anti-nutritional composition of velvet bean: A lesser - known food legume in South India. *Inter. J. Food Sci. Nutr.* 51: 279 – 287.

Vadivel, V. and Janardhanan, K. 2001. Nutritional and anti-nutritional attributes of the under-utilized legume, *Cassia floribunda* Cav. *Food Chem.* 73: 209 – 215.

Vartak, V.D. and Kulkarni, D.K. 1987. Monsoon wild leafy vegetables from hilly regions of Pune and neighboring districts, Maharashtra State. *J. Econ. Taxon. Bot.* 11: 331-333.

Vartak, V.D. and Suryanarayana, M.C. 1995. Enumeration of wild edible plants from Susala Island. Mulshi Reservoir, Pune district. *J. Econ. Taxon. Bot.* 19: 555-569.

Vijayakumari, K., Siddhuraju, P. and Janardhanan, K. 1993. Chemical composition and nutritional potential of the tribal pulse (*Bauhinia malabarica* Roxb). *Plant Food Hum. Nutr.* 44: 291-298.

Young, V.R. and Pellett, P.L. 1994. Plant proteins in relation to human protein and amino acid nutrition. *Amer. J. Clin. Nutr.* 59: 1203S-1212S.

2018, Ethnomedicinal Plants: A Biodiversity Treasure Pages 521–529
Editors: V.R. Mohan, A. Doss, P.S. Tresina and V. Sornalakshmi
Published by: ASTRAL INTERNATIONAL PVT. LTD., NEW DELHI

Chapter 21

Microscopic Studies of *Cadaba indica* Lam. (Capparaceae)

***A. Saravana Ganthi*[1] *and M. Padma Sorna Subramanian*[2]**

[1]*Department of Botany,*
Rani Anna Govt. College for Women, Tirunelveli, Tamil Nadu
[2]*Siddha Medicinal Plants Garden,*
CCRS, Mettur Dam, Tamil Nadu
E-mail: saran_gan@rediffmail.com

Introduction

Botanical identify of crude drugs in fragmentary form possess a problem especially in the plant based pharmaceutical arena. Many crude drugs such as bark and root simulate the fake materials so deceivingly that adulteration of the dupes with genuine ones is rendered easily. The organoleptic characters such as smell, taste, texture and some other parameters that could be retrieved with the help of the sensory organs have been one of the methods of diagnosis. However this technique has its own limitations. The more dependable method for the diagnosis of the phyto-drug will be microscopic analysis and anatomical parameters. Literature survey did not provide sufficient information about anatomical studies of this plant. The current work aims to contribute in solving the problems of controversial drugs prevalent in Ayurveda besides helping in laying down pharmacopoeial standards. In the present investigation anatomical studies of leaf, stem, root and petiole of the medicinal plant *Cadaba indica* Lam. was studied and diagnostic characters were noted.

Cadaba indica commonly known as "Vili" or "Kattatagatti" in Tamil (Eng: Indian Cadaba) is a shrub, growing in dry localities (900′ – 1700′) throughout India (Fischer, 1921). The leaves and roots are used as purgative, anthelmintic, antisyphilic (Uma Rao Singh *et al.*, 1996), emmenagogue (Chopra *et al.*, 1956), aperient (Rustomjee and Nanabhai,1984), gouts, boils, laxative, antispasmodic (Seetharam *et al.*, 2005), uterine obstructions (Kiritkar and Basu, 1987) and to treat gastrointestinal problems

(Jain,1991). An alkaloid is present in the leaves used as poultice on sores and decoction of leaves is used in uterine obstructions (Useful Plants of India, 2000). In Africa infusion of leaves is inhaled to clear from colds (Asolkar *et al.*, 2002).

Materials and Methods

The medicinal plant *Cadaba indica* Lam. was collected from Tirunelveli District, Tamil Nadu, India. The required samples of different organs *viz.*, leaf, petiole, young internodes, young root and stem, were cut and removed from the plant and then fixed in FAA. After 24 hours of fixing, the specimens were dehydrated with graded series of tertiary – butyl alcohol as per the schedule given by Sass (1940). Infiltration of the specimens was carried out by gradual addition of paraffin wax (melting point 58-60°C) until tertiary – butyl alcohol solution attained super saturation. The specimens were cast into paraffin blocks by usual method (Johanson, 1940). Paraffin embedded specimens were sectioned with the help of Rotary Microtome. The thickness of the sections was 10 to 12 µm. Dewaxing of the sections was done by customary procedure (Johansen, 1940). The sections were stained with toludine blue as per the method published by O' Brein *et al.* (1964). For studying the stomatal morphology, venation pattern and trichome distribution, paradermal sections as well as clearing of leaf with 5 per cent sodium hydroxide or epidermal peelings by partial maceration employing Jeffrey's maceration fluid (Sass, 1940) were prepared. Microphotographs of different magnifications were taken with Nikon Labphot 2 Microscopic Unit to reveal the anatomical characters. Measurement of cells was made with micrometer. For each element, 15 – 20 measurements were taken and average is presented.

Results and Discussion

Microscopical Characters of *Cadaba indica* Lam.

Venation Pattern (Figure 21.1a)

Venation pattern of the lamina has been studied from both the paradermal sections, and from cleared leaf. The lateral veins are thin forming distinct vein–islets. The vein-islets are variable in shape. The vein–terminations are present in all vein–islets. The vein terminations are mostly branched once or twice. The ends of the vein – terminations have a punch of terminal sclereids.

Epidermis and Stomata

Stomata occur mostly on the lower side. They are anisocytic type with three subsidiary cells of unequal size. The epidermal cells are amoeboid in shape and their anticlinal walls are wavy in out line. Cuticular markings are not evident.

Leaf (Figure 21.1b)

The leaf is thick and leathery, midrib prominent with thick cuticularized epidermis. The midrib is plano-convex with more or less flat adaxial side and hemispherical abaxial sides. It is 470 µm in vertical axis and 400 µm in horizontal axis. The midrib has thick adaxial layer of epidermis with prominent cuticle and tabular cells. The abaxial epidermis has circular thick walled cells with prominent echinate

Figure 21.1a-f: Anatomy of *Cadaba indica.*

VI: Vein Islet; VT: Vein termination; X: Xylem; Ph: Phloem; SC: Sclerenchyma; GT: Ground Tissue; Pe: Periderm; Co: Cortex; Cr: Crystal: Ep: Epidermis; SPh: Secondary Phloem; SX: Secondary Xylem; Ve: Vessel; PX: Primary Xylem; XF: Xylem Fibre.

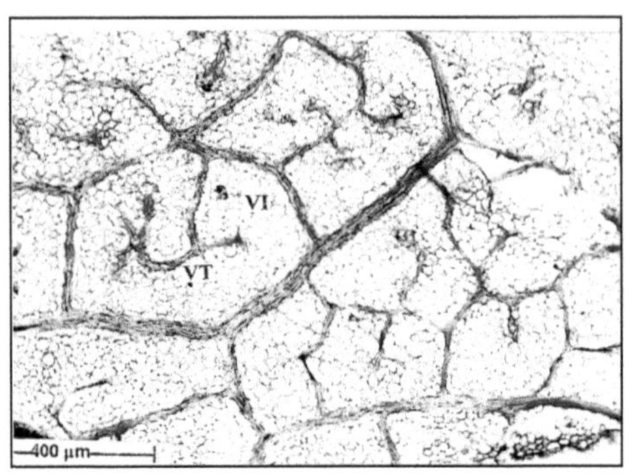

a) Paradermal section showing venation pattern

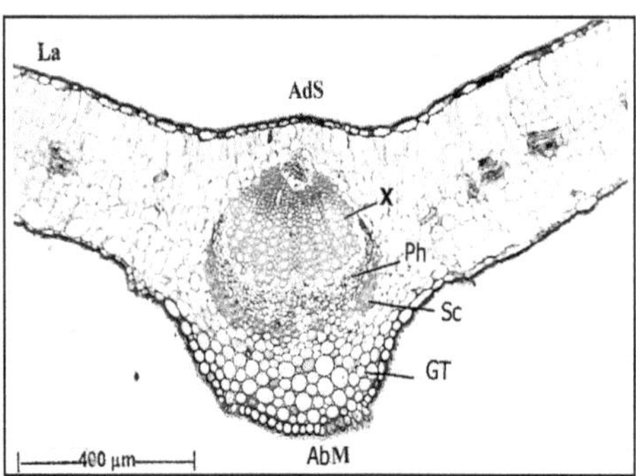

b) T.S. of leaf through midrib

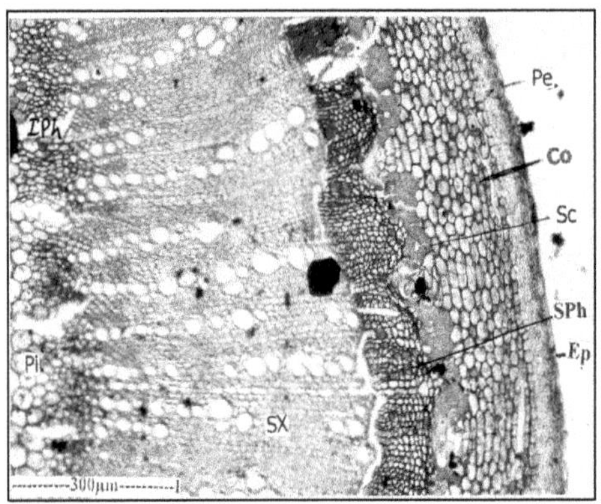

c) T.S. of stem – a portion enlarged

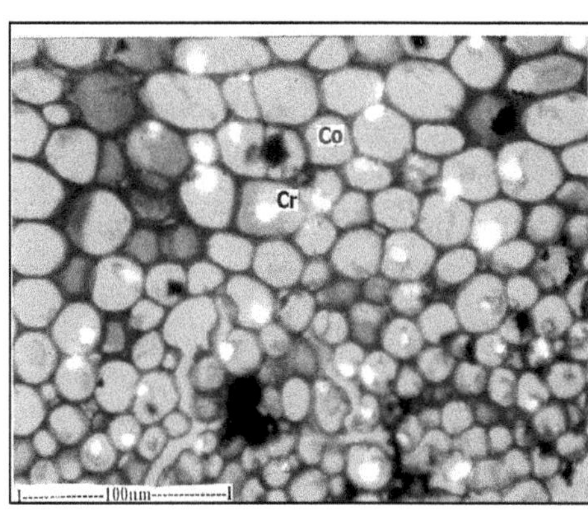

d) Cortical cells with calcium oxalate crystals

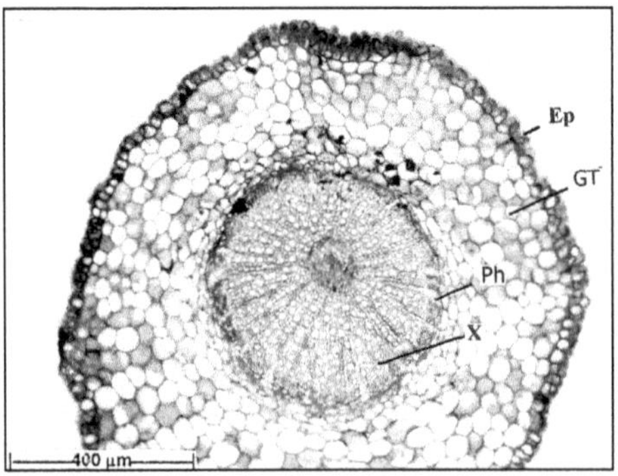

e) T.S. of Petiole

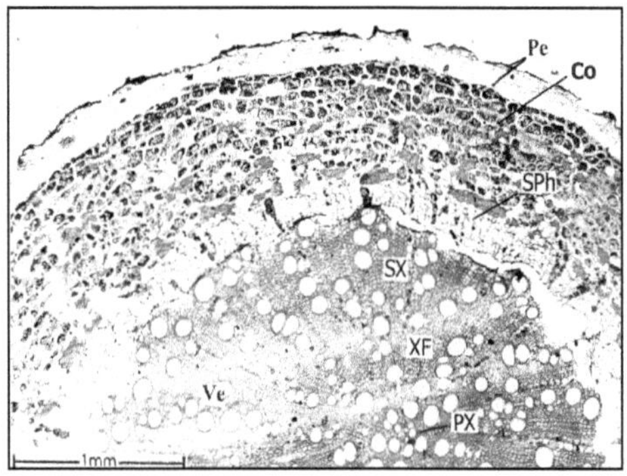

f) T.S. of root – a portion enlarged

cuticle. The ground tissue of the midrib has thick walled, circular, less compact, parenchyma cells in the abaxial part and a small mass of polygonal parenchyma cells on the adaxial part.The vascular bundle is large and broadly top shaped. It has a dense, semicircular zone of xylem comprising of radial multiples of vessels and intervening thick walled ray cells. Phloem occurs as a broad arc beneath the xylem mass, adjoining the phloem is a prominent sheath of heavily thick walled sclerenchyma cells.

The lamina has smooth and even surface with narrow epidermal layers of tabular cells and thick cuticle. The lamina is 330 µm thick. The mesophyll tissue has less defined palisade mesophyll and spongy parenchyma. The palisade cells are two layered wide and compact. The spongy mesophyll tissue has a central zone of circular cells and abaxial zone of two layered, short and wide compact parenchyma cells. The lateral veins are placed in the medium part and each vein has collateral vascular bundle with small patches of sclerenchyma caps on both sides.

Stem (Figure 21.1c)

The stem has smooth circular outline. The epidermis comprises of intact narrow thick walled tabular cells. A narrow, compact zone of thin walled cells of periderm is seen inner to the epidermis. The periderm is 40 µm in wide. Cortical zone is uniformly wide and parenchymatous. The cells are thick walled elliptical, compact having intercellular spaces. The cortex is 150 µm in wide. Inner to the cortex, large discrete masses of sclereids are closely adhering in phloem. The sclereids are brachy-sclereids with narrow lumen, thick lignified walls. Secondary phloem is wide and continuous all around the xylem. Phloem consists of narrow, undilated rays of radially elongated cells. The sieve-elements and parenchyma cells occur in regular radial files. The parenchyma cells are dilated and contain dense, amorphous inclusions. The sieve elements are rectangular narrow lumened and possess prominent companion cells on their lateral side. Secondary xylem is a wide and dense cylinder of uniform thickness. The inner boundary of the secondary xylem cylinder has a band of primary xylem elements which are surrounded by medullary phloem or internal phloem. Secondary xylem consists of wide, circular or elliptical, thick walled vessels. They are in their radial multiples or solitary. The vessels are 15 – 40 µm in diameter. Xylem fibres are thick walled and lignified with a narrow lumen. Xylem rays are narrow, straight and thick walled. Calcium oxalate crystals are diffusely distributed in the cortical cells (Figure: 1d). The crystals are druses and only one occurs per cell. The crystals are surrounded by cell contents, the cytoplasmic material. The crystals are 5 – 10 µm wide.

Petiole (Figure 21.1e)

The petiole is circular in outline along the basal part and shortly two winged along the terminal part. It has thick and prominent epidermal layer having heavy cuticle. The petiole is 750 to 800 µm in diameter. The ground tissue is homogeneous with circular, thin walled parenchyma cells. The vascular system has a thick hallow cylinder measuring 400 µm in diameter. It consists of a thin peripheral continuous layer of phloem and several radial multiples of xylem elements and xylem rays. The central core is narrow and parenchymatous.

Root (Figure 21.1f)

Thick root with well developed secondary xylem has been observed. The root has uniformly wide, continuous periderm, wide cortex dispersed with scattered masses of sclerenchyma elements, wide secondary phloem and central, solid, dense secondary xylem. The periderm is 150 µm wide, it has wide shallow fissures. It consists of tangentially oblong thin walled, suberised phellem cells and phelloderm is not evident. Cortex is wide measuring 700 µm in radial plane. It consists of large tangentially elongated, thin walled compact parenchyma cells which are densely loaded with starch grains. Dispersed at random in the cortex are thick tangential blocks of sclerenchyma cells. Secondary phloem is wide and occurs in continuous sheath around the xylem. The phloem elements are in regular radial files, phloem rays are narrow with dense contents. The sieve elements are wide, angular and thin walled. The secondary xylem has no growth rings; the vessels are wide, circular thick walled, mostly solitary or in radial multiples of two. The vessels are distributed in a characteristic pattern. They occur in several, independent, radial segments extending from the centre towards periphery. The space in-between the vessel segment is wide and is occupied by xylem fibres. The vessels in the central part are narrow and they become progressively wider towards periphery. The narrow vessels are 30 µm in diameter; the wider ones are 130 µm in diameter. Xylem fibres are fairly thick walled with wide lumen. The walls are lignified. Xylem rays are fairly prominent, thin and straight.

Powder Microscopy of the Root (Figure 21.2)

The root powder shows the vessel elements, tracheids and fibres. The vessel elements are short, wide and drum – shaped. They have wide, circular, horizontal perferation Figure. Some other vessel elements are cylindrical and slightly wide. They are 300 µm long (Figure 21.2a). They also have wide, circular horizontal perferation Figure (Figure 21.2b). The third type of vessel elements are narrowly cylindrical measuring 210 µm long (Figures 21.2b, c). These narrow vessel elements have simple, circular, oblique perforation. All the vessel elements have circular, dot – like dense bordered pits. The vessels elements may be tailed or tailless. Trachieds are more abundant, they are narrow 350 – 400 µm long, thick walled and densely pitted (Figure 21.2a). The fibres are needle – like, thick walled narrow elements. They range from 300 – 500 µm long, some of the fibres are wide and thin walled (Figures 21.2a, b). The fibres have no lateral wall pits.

Powder Microscopy of the Stem (Figure 21.3)

The powder of the stem shows the vessel elements, tracheids and xylem fibres. Vessel elements are long and cylindrical with dense, circular, lateral wall pits. The vessel elements are 200 – 230 µm long, with simple, oblique perforation Figure. Tracheids elements are similar to vessel elements. They are 400 µm long. They have bordered pits on the lateral walls similar to vessel elements. Xylem fibres are long, needle – like elements with pointed tips. Some of the fibres are narrow and thick walled, other fibres are wide and thin walled (Figures 21.3a, b). The powder has also shown small, rectangular or square shaped xylem parenchyma cells.

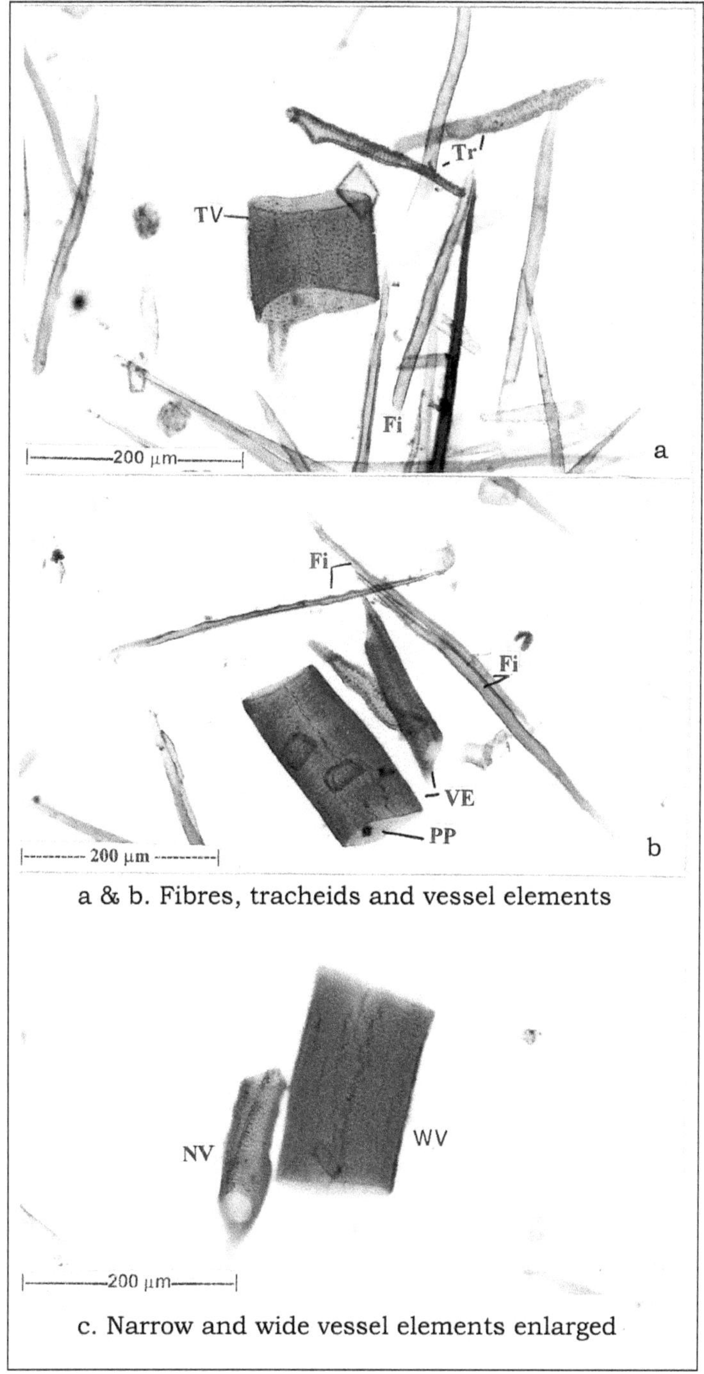

a & b. Fibres, tracheids and vessel elements

c. Narrow and wide vessel elements enlarged

Figure 21.2: Powder Microscopy of Root.

TV: Tailed Vessel; Tr: Tracheid; Ve/VE: Vessel; Fi: Fibre; PP: Perforation Plate; WFi: Wide Fibre; NFi: Narrow fibre; NV: Narrow vessel; WV: Wide Vessel; PC: Parenchyma.

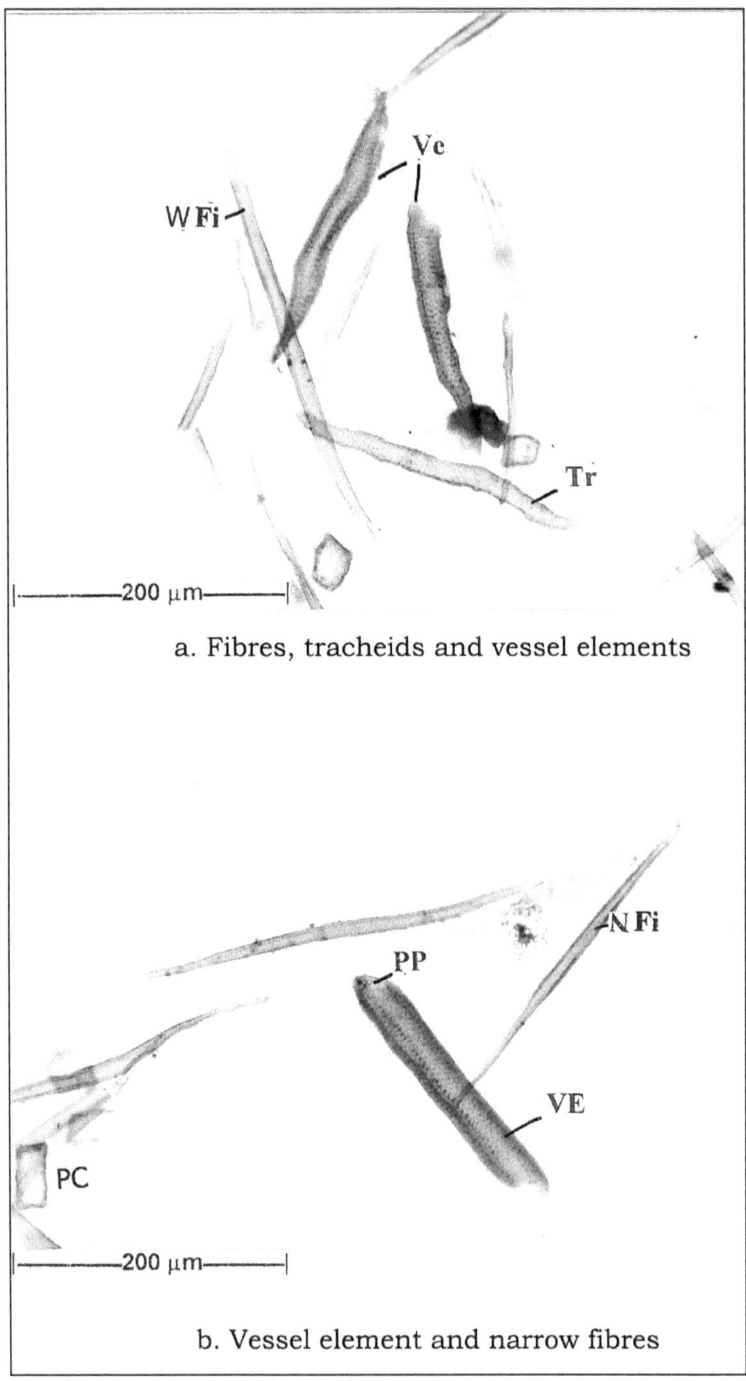

Figure 21.3: Powder Microscopy of Stem.

TV: Tailed Vessel; Tr: Tracheid; Ve/VE: Vessel; Fi: Fibre; PP: Perforation Plate; WFi: Wide Fibre; NFi: Narrow fibre; NV: Narrow vessel; WV: Wide Vessel; PC: Parenchyma.

Cell Inclusions

Starch grains are abundant in the phloem parenchyma, xylem fibres and xylem rays. The starch grains are simple with circular concentric central hilum. The cells have one or more starch grains filling the cell lumen. The starch grains are up to 10 μm in diameter.

Conclusion

Resurgence of faith in traditional herbal medicines and resilience of belief in the principle of Ayurveda and Siddha systems of medicine have promising prospects. This trend has to be sustained at global level approbation for which meticulous standardization of the phytodrugs is prerequisite. This study is helpful in the identification of crude drug *Cadaba indica.*

Acknowledgement

We thank the Principal, St. Xavier's College, Palayamkottai, Tirunelveli, Tamil Nadu, India for providing laboratory facilities and guidance.

References

Asolkar LV Kakkar KK and Chakre OJ 2000. *Cadaba fruticosa* In: Second supplement to Glossary of Indian medicinal plants with active principles (Part – I). *Publication and Information Directorate, CSIR New Delhi*: pp. 150.

Chopra RN Nayar SL and Chopra IC 1956. *Cadaba* In: Glossary of Indian Medicinal Plants. *CSIR New Delhi*: pp. 112 -114.

Fischer CEC 1921. A survey of the Flora of the Anaimalai Hills in the Coimbatore District Madras Presidency Record of the BSI Vol IX No. 1. *Superintendent Govt. Printing India Calcutta*: pp. 25.

Jain SK 1991. *Cadaba fruticosa* In: Dictionary of Indian Folk Medicine and Ethnobotany Deep publications, New Delhi: pp. 41.

Johansen DA 1940. Plant Microtechnique. *McGraw – Hill Book Company Inc. New York:* pp. 523.

Kirtikar KR and Basu BD 1987. *Cadaba* In: Indian Medicinal Plants, *International Book distributors, Dehradun*: pp. 193 – 195.

O'Brein TP Feder N and Mcull ME 1964. Polychromatic staining of plant cell walls by toludine blue – O. *Protoplasma* 59: 364-– 373.

Rustomjee Naserwanjee Khorg and Nanabhai Navrosji 1984. *Cadaba indica* In: Materia Medica of India and their therapeutics. *Neeraj Publishing House Delhi*: pp. 61 – 63.

Sass J E (1940). Elements of Botanical Microtechnique.*McGraw Hill Book Co INC New York and London*: pp. 222.

Seetharam Reddi TVV Prasanthi S and Ram Rao Naidu BVA 2005. *Cadaba fructicosa* in: Medicinal and Aromatic plants of India. *Ukaaz publications, Andhra Pradesh.*: pp. 79 – 194.

Uma Rao Singh Wadhwari AM and Johri M 1996. *Capparaceae* In: Dictionary of Economic Plants of India. *ICAR, New Delhi.*

Useful Plants of India 2000. *Cadaba indica* In: *National Institute of Science Commission, New Delhi:* pp. *93* – 94.

2018, Ethnomedicinal Plants: A Biodiversity Treasure Pages 531–549
Editors: V.R. Mohan, A. Doss, P.S. Tresina and V. Sornalakshmi
Published by: ASTRAL INTERNATIONAL PVT. LTD., NEW DELHI

Chapter 22

Pharmacognostic and Phytochemical Evaluation of the Whole Plant of *Catharanthus pusillus* (Murr.) G. Don

P. Yokeswari Nithya[1], S. Mary Jelastin Kala[2] and V.R. Mohan[3]

[1]*Department of Chemistry, A.P.C. Mahalakshmi College for Women, Tuticorin, Tamil Nadu*
[2]*Department of Chemistry, St.Xaviers College, Palaymkottai, Tamil Nadu*
[3]*Ethnopharmacology Unit, PG and Research Department of Botany, V.O. Chidambaram College, Tuticorin – 628 008, Tamil Nadu*

Introduction

Pharmacognosy literally means knowledge of drugs or pharmaceuticals, which deal with the drugs of vegetable, animal and mineral origin. It may be defined as "an applied science that deals with biological, biochemical and economical features of natural drugs and their constituents". Pharmacognosy helps to study the structural, physical, chemical and sensory characteristics of crude drugs which also include their history, cultivation, collection and other particulars relating to the treatment they receive during their passage from the procedure to the distributors of pharmacists (Wallis, 1985).

The pharmacognostical studies of the herbal drugs have become imperative for several reasons. As per the WHO norms, every drug has to undergo botanical standarization, particularly macroscopic and microscopic characterization which constitutes the major part of pharmacognosy. This primary step enables the researcher in phytodrugs to affirm the botanical identity, genuineness and purity of the samples that she is working. Pharmacognostic standardization is based upon the lenet that certain microscopic characters are specific and resticted in distribution (Metealf and Chalk, 1979). Microscopic parameter, though limited in their application under certain circumstances, have still highly reliable diagnostic values and play appreciable role in the herbal drug.

A perusal of literature on the much valued medicinal plant *Catharanthus pusillus* of Apocynaceae, reveals a lacuna in the pharmacognostic and other parameters. This fact induced the researcher to investigate the micro morphological standardization of the folklore drug to putforth a protocol of anatomical features, which may enable the pharmacological scientists to identify the raw drugs prior to their studies.

C. pusillus belonging to family Apocynaceae is known with various names in India and all over the world. It is widely used as various treatments of diseases and traditionally used as herbal medicine (Don, 1999). The roots, leaves and latex of these plants are used to treat skin and liver diseases, leprosy, dysentery, worms, ulcers, tumor and ear aches. The leaf powder of *C. pusillus* were mixed with coconut oil and used for treating the antidandruff activity and also used to kill the lice (Balajirao, 1996). The first and foremost step, is the characterization of different pharmacognostical parameters, botanical identification, microscopic study, powder characteristics and fluorescence study has been included here, A preliminary phytochemical screening of whole plant has also been carried out.

Materials and Method

Collection of Plant Material

The whole plant of *Catharanthus pusillus* (Murr.) G.Don were collected from Pechiparai, Kanayakumari District, Tamil Nadu. With the help of local flora, voucher specimens were identified and preserved in the Ethnopharmacology unit, Research department of Botany, V.O.Chidambaram College, Thoothukudi, Tamil Nadu for further references.

Microscopic Studies

Care was taken to select healthy plants. The required samples of different organs were cut and removed from the plants and fixed in FAA solution (Formalin - 10 ml, Acetic acid - 5 ml and 70 per cent Ethyl alcohol - 85 ml). After 24 hrs of fixing, the specimens were dehydrated with graded series of Tertiary Butyl Alcohol (TBA) as per the schedule given by Sass (1940). Infiltration of the specimens was carried by gradual addition of paraffin wax (melting point 58° - 60°C) until TBA solution attained super saturation. The plant materials were cast into paraffin blocks.

Sectioning

The paraffin embedded specimens were sectioned with the help of rotary microtome. The thickness of the section was 10 to 12 μm. Dewaxing of the section was done by customary procedure (Johansen, 1940). The sections were stained with Toluidine blue as per the method published by O' Brien *et al.* (1964). Since Toluidine blue is a polychromatic stain, the staining results were remarkably good. The dye rendered pink colour to the cellulose walls, blue to the lignified cells, dark green to suberin, violet to the mucilage, blue to the protein bodies *etc.* Wherever necessary, sections were also stained with Safranin and Fast - green. For studying the stomatal morphology, venation pattern and trichome distribution, paradermal sections (sections taken parallel to the surface of leaf) as well as clearing of leaf with 5 per cent sodium hydroxide or epidermal peeling by partial maceration employing Jeffrey's maceration fluid (Sass, 1940) were prepared. Glycerin mounted temporary preparations were made for macerated/cleared materials.

Photomicrographs

Microscopic descriptions of tissues were supplemented with micrographs wherever necessary. Photographs of different magnifications were taken with Nikon Labphot 2 microscopic unit. For normal observations bright field was used, for the study of crystals polarized light was employed. Since these structures have briefringent property, under polarized light they appear bright against dark background. Magnifications of the figures are indicated by the scale-bars. Descriptive terms of the anatomical features are as given in the standard books (Esau, 1964).

Preparation of Plant Sample

The collected whole plants were cut into small fragments and dried under shade until the fracture is uniform and smooth. The dried plant materials were granulated or powdered by using a blender and sieved to get uniform particles by using sieve No. 60. The final uniform powder of the plant was used for various experimental studies.

Physico-chemical Constants and Fluorescence Analysis

These studies were carried out as per the standard procedures (Lala, 1993). In the present study, the powder of the whole plant was treated with 1N aqueous sodium hydroxide and 1N alcoholic sodium hydroxide acids like 1N hydrochloric acid, 50 per cent sulphuric acid, nitric acid, acetic acid, nitric acid with ammonia, ferric chloride, ammonia, benzene, petroleum ether, acetone, chloroform, methanol and ethanol. These extracts were subjected to fluorescence analysis in visible/ daylight and UV light (254nm and 365nm). Various ash types and extractive values were determined by following standard methods (African Pharmacopoeia.1986).

Preliminary Phytochemical Analysis

Shaded dried and powdered whole plant was successively extracted with petroleum ether, benzene, ethyl acetate, methanol and ethanol and water. The extracts were filtered and concentrated using vacuum distillation The different extracts were subjected to qualitative tests for the identification of various phytochemical constituents as per standard procedure (Lala, 1993; Brindha, 1981).

Results

Exomorphic Features (Figure 22.1)

Distribution: South Part of Western Ghats of Tamil Nadu.

Status: Rare and endemic

Figure 22.1: *Catharanthus pusillus* (Murr.) G. Don.

Habit: Perennial decumbent herb grows up to 1 m tall

Leaves: Opposite, simple and entire; stipules 1–3 at each side of the leaf base; petiole 1–3 mm long; blade oblong to narrowly ovate, 1–4.5 cm × 3–13 mm, base cuneate, apex acuminate to rounded, herbaceous to thinly leathery, shiny on both sides, glabrous.

Flowers: Flowers in leaf axils, bisexual, 5-merous, regular, fragrant; pedicel 5– 25 mm long

Sepals: Slightly fused at base, 5–10 mm long

Corolla: Cylindrical, 15–22 mm long, widening near the insertion of the stamens, throat constricted, inside with a ring of hairs just below the throat and a ring of hairs below the insertion of the stamens, green, pinkish at base, lobes ovate to obovate, 11–22 mm long, dense short hair inside, spreading, pink, reddish violet or pale pink-magenta, white to cream at the base.

Stamens: Inserted just below the corolla throat, included, filaments very short

Ovary: Superior, consisting of 2 very narrowly oblong carpels, style slender, 10–16 mm long, pistil head cylindrical with a reflexed transparent frill and with rings of woolly hairs at base and apex, stigma minute.

Fruit: Composed of 2 free cylindrical follicles 1.5–5 cm long, striate, glabrous, green, dehiscent.

Seed: Oblong, 1–3 mm long, grooved at one side, black

Microscopic Features of C. *pusillus*

Leaf

In transactional view the leaf exhibits biconvex midrib, and thick lamina. The midrib is 400µm thick. The adaxial cone part is 200µm wide, abaxial part is 350µm wide (Plate 22.1 a). The midrib includes thick layer of epidermis which is uneven in outline. On the adaxial part of the midrib there is a single layer of thick walled cells beneath the epidermis. On the abaxial part there are 2 layers of thick walled cells inner to the epidermis.

The vascular strand is single, flat and bicollateral. The xylem strand consists of horizontal band of compact layers of xylem element (Plate 22.1b). The xylem elements are in vertical short rows, the individual elements being angular, thick walled with wide lumen. The xylem strand is 200µm wide and 40µm thick. Phloem elements occur both on the adaxial part of the xylem strand. The phloem elements are in small discrete units.

Lateral Vein

The lateral vein also consists of short, narrow, adaxial cone and small semicircular abaxial part. The palisade cells are horizontally transcurrent between the vascular strand and adaxial epidermis. The vascular strand is single, more or less circular with a cluster of xylem elements and a few phloem elements located on the adaxial and abaxial ends. The lateral vein is 220μm thick.

Lamina

The lamina is dorsiventral and smooth on either side. The adaxial epidermis is thick and the cells are broadly rectangular with their cuticle. It is apostomatic. The abaxial epidermis is thin, stomatiferous and the cells are squarish or rectangular. The mesophyll tissue consists of single adaxial row of vertically oblong cylindrical palisade cells with wide gaps inbetween the cells. The spongy parenchyma cells are spherical or lobed cells which form a network of wide air spaces. Lateral veins are small, circular and located in the median part. The lamina is 120μm thick (Plate 22.1b).

Leaf Margin

The marginal part of the lamina is slightly bent below and conical in shape. The epidermal cells of the leaf margin are thick walled. The mesophyll tissue and the epidermal cells are similar to middle part of the leaf. The marginal part is 40μm thick.

Epidermal Cells and Stomata

In surface view of the epidermal tissue, the epidermal cells appear polygonal with thin, straight, anticlinal walls. Stomata are fairly dense and diffuse in distribution. The stomatal type is anisocytic of the three subsidiary cells. One is smaller than the other two (Plate 22.1c).

At certain regions of the lamina, the epidermal cells are thick walled with wavy anticlinal walls. The stomata are broadly elliptical measuring 15x30μm in size (Plate 22.1e).

Venation Pattern

The venation system consists of pattern more or less similar to monocot leaves. There is a thick and straight median vein, slightly thinner, sub marginal lateral vein and quite thin marginal vein (Plate 22.1d). From these major veins arise thin, straight, horizontal veins that run half way between the midrib vein and lateral vein. Some of the lateral veinlets directly run upto the later vein. These horizontal veins less frequently produce vein terminations which are also horizontal and straight. The vein termination are invariably unbranched and are parallel to the lateral veins (Plate 22.1f). The veinlets are horizontally stretched rectangular in shape.

Petiole

The petiole consists of a bowl shaped abaxial part and semicircular adaxial part with 2 thick vertical lateral wings (Plate 22.2a). The petiole is 170μm thick. The adaxial part is 100μm wide and the abaxial part is 250μm wide. The wings are 50μm in height and 40μm in thickness.

The epidermal layer is thick with squarish, thin walled cells with thin cuticle. The ground tissue consists of small, angular, thin walled compact homogeneous parenchyma cells. The vascular strand consists of a wide, shallow, arc shaped xylem elements and adaxial and abaxial phloem strands. The xylem strand consists of

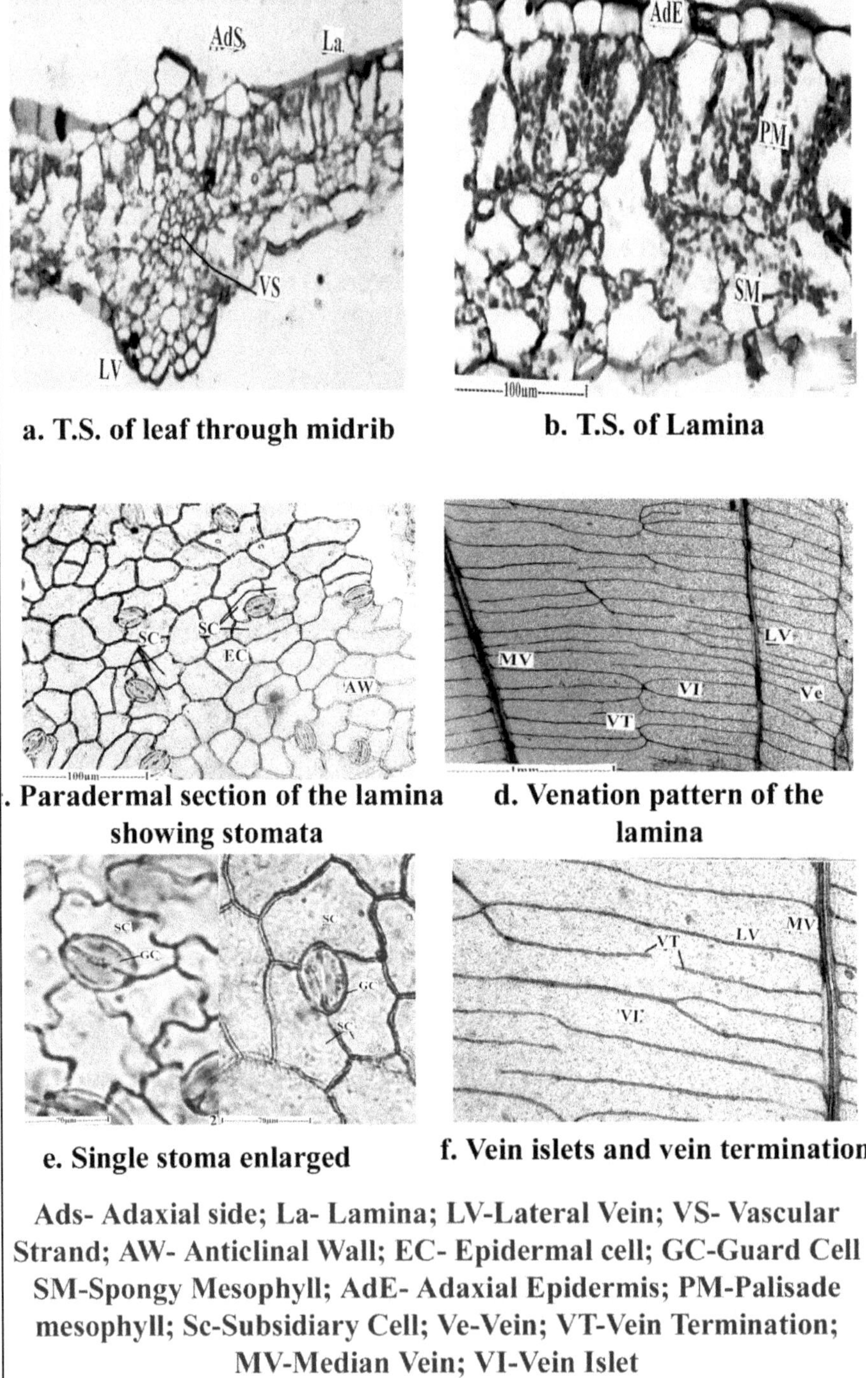

a. T.S. of leaf through midrib

b. T.S. of Lamina

. Paradermal section of the lamina showing stomata

d. Venation pattern of the lamina

e. Single stoma enlarged

f. Vein islets and vein termination

Ads- Adaxial side; La- Lamina; LV-Lateral Vein; VS- Vascular Strand; AW- Anticlinal Wall; EC- Epidermal cell; GC-Guard Cell SM-Spongy Mesophyll; AdE- Adaxial Epidermis; PM-Palisade mesophyll; Sc-Subsidiary Cell; Ve-Vein; VT-Vein Termination; MV-Median Vein; VI-Vein Islet

Plate 22.1: *Catharanthus pusillus* (Murr.) G. Don. Anatomy of Leaf.

vertical rows of 5 or 6 xyelm elements which are angular, thick walled with wide lumen. In between the xylem row occur fairly broad parallel lines of parenchyma cells. The phloem elements are in small clusters of sieve elements. The sieve elements are small, thick walled and darkly stained (Plate 22.2b).

Stem

In cross sectional view of the young stem exhibits circular outline with a pair of long, thick conical wings on either lateral side. The stem consists of a distinct layer of epidermis, the cells being squarish, small and thick walled. The ground tissue is parenchymatous, thin walled, angular and compact. There is a thin, circular cylinder of primary xylem elements around a wide pith. The xylem elements consists of 2 or 3 radially arranged, thick walled, angular wide elements. Phloem occurs on both inner and outer sides of the xylem cylinder.

The inner phloem (medullary phloem) and outer phloem consists of small, isolated groups of sieve elements. The stem is 600µm in diameter.

The old stem is perfectly circular with short, conical wings at four points around the stem (Plate 22.2c). The stem is 2.25 mm in diameter. The epidermal layer is intact and consists of small, thin walled rectangular cells. The cortical zone includes outer about 4 layers of tangentially compressed rectangular cells.

Inner to the cortex occurs a thick discontinuous cylinder of cortical fibres. The vascular cylinder is 800µm thick. The secondary phloem occurs in thin cylinder of discrete strands at the outer periphery of the xylem cylinder. Medullary phloem also occurs along the inner boundary of the secondary xylem cylinder. The secondary xylem consists of fairly dense, angular mostly solitary thin walled vessels and radial rows of wide thick walled and lignified fibres (Plate 22.2c).

Primary xylem occurs along the inner end of secondary xylem. The primary xylem consists of long, radial rows of angular xylem elements and parenchyma cells. The vessels of the secondary xylem are 40µm wide.

Root

The thin root is 280µm thick (Plate 22.3a). It includes thin broken epidermis 2 or 3 layers of shrunken cortical cells. The vascular cylinder consists of thick secondary xylem and thin layer of secondary phloem.

Secondary xylem includes wide, solitary, diffusely distributed vessels and xylem fibres. The vessels are angular, thick walled and measure 40µm in diameter (Plate 22.3c).

Secondary phloem consists of thin continuous layer of sieve elements. Xylem fibres are angular, thick walled and lignified. The primary xylem consists of diarch, thin, less prominent cells.

The thick root is 1.2mm in diameter (Plate 22.3b). It includes broken and distorted epidermal layer, inner tangentially stretched thin walled parenchyma cells. The vascular cylinder is dense, compact and circular (Plate 22.3d). The xylem cylinder consists of central, darkly stained portion with crowded vessels and thick walled narrow fibres.

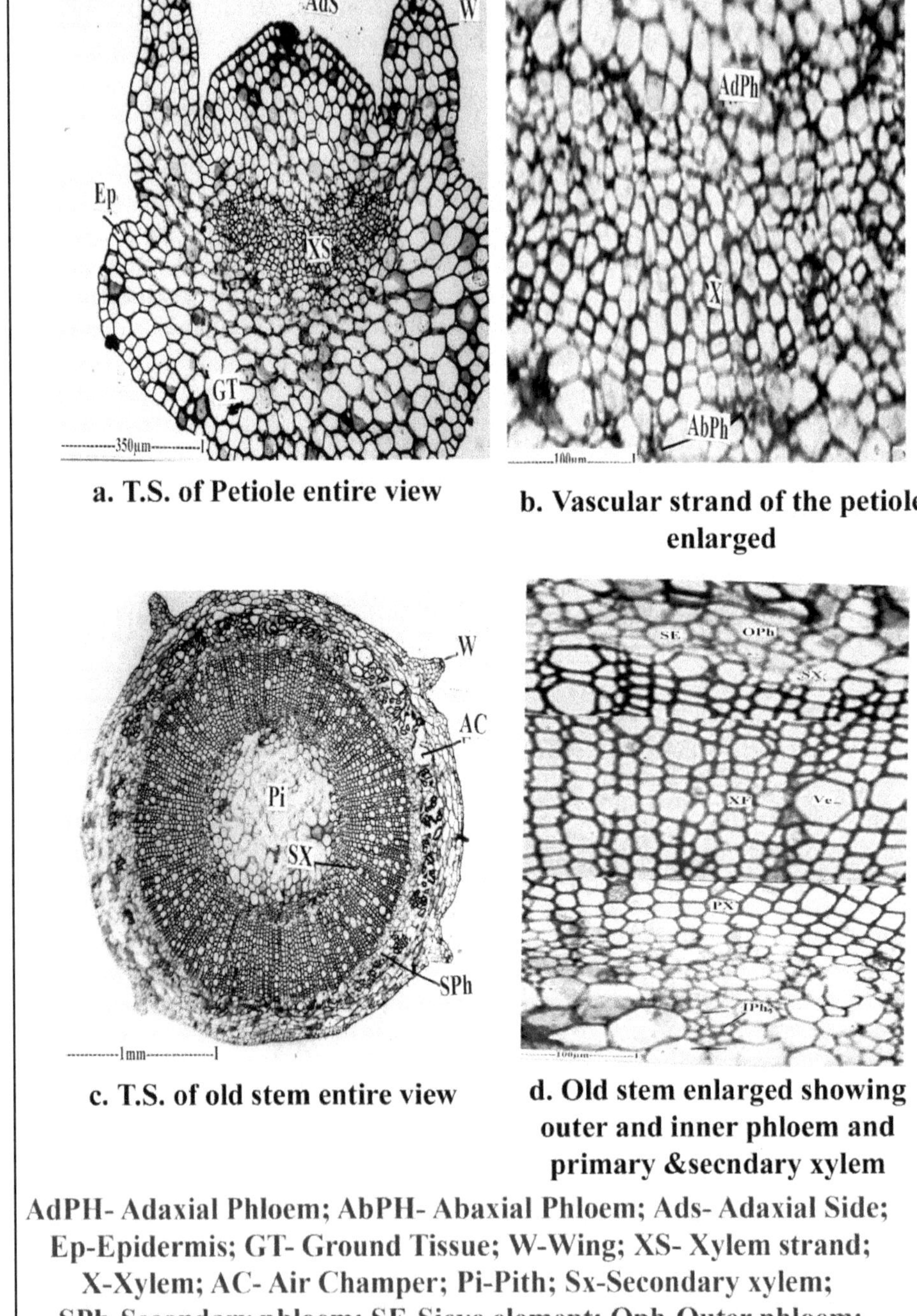

a. T.S. of Petiole entire view

b. Vascular strand of the petiole enlarged

c. T.S. of old stem entire view

d. Old stem enlarged showing outer and inner phloem and primary &secndary xylem

AdPH- Adaxial Phloem; AbPH- Abaxial Phloem; Ads- Adaxial Side; Ep-Epidermis; GT- Ground Tissue; W-Wing; XS- Xylem strand; X-Xylem; AC- Air Champer; Pi-Pith; Sx-Secondary xylem; SPh-Secondary phloem; SE-Sieve element; Oph-Outer phloem; XF-Xylem Fibre; Ve-vessels; Px-Primary xylem; Iph-Inner phloem

Plate 22.2: ***Catharanthus pusillus*** **(Murr.) G. Don. Anatomy of the Petiole and Stem.**

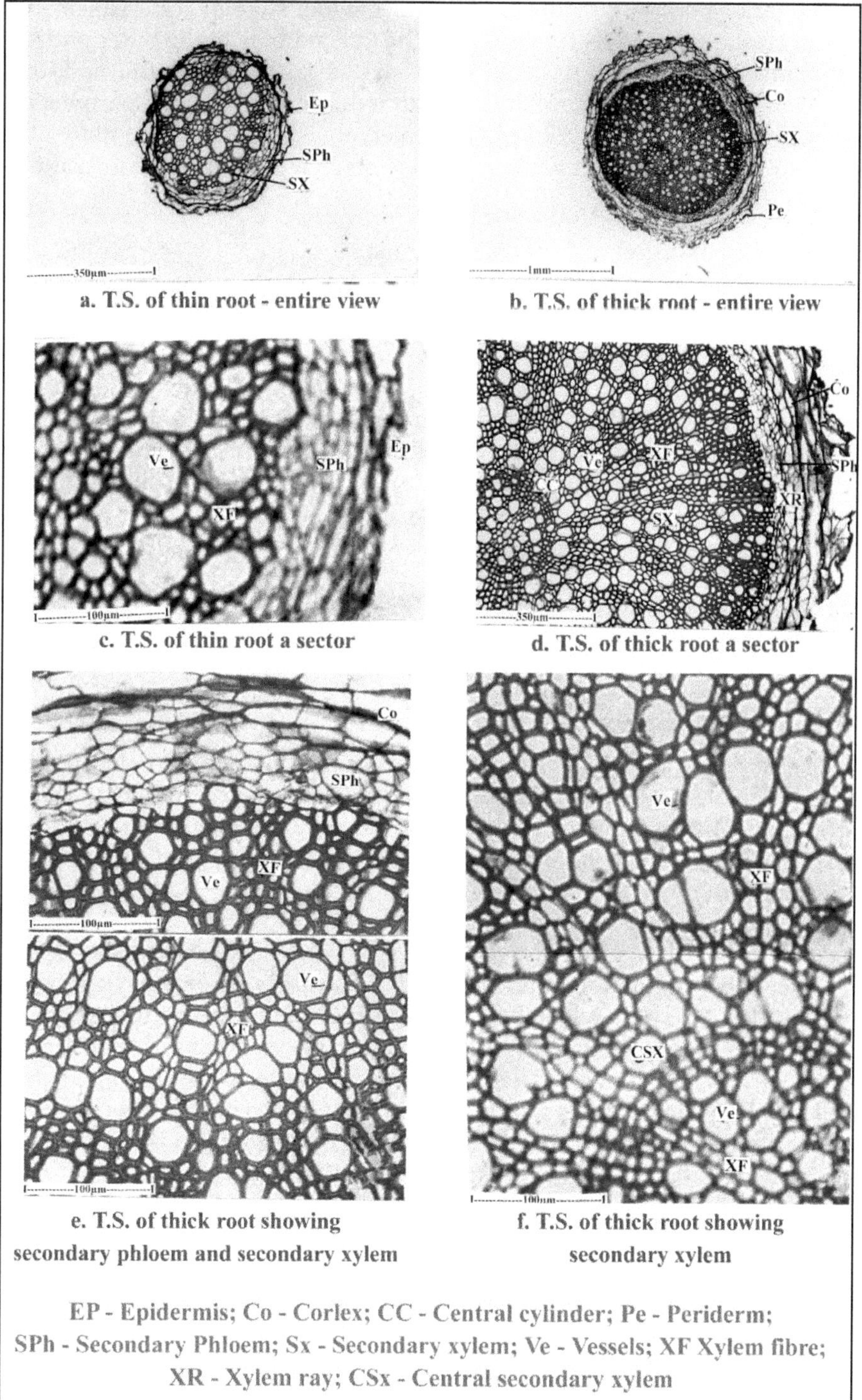

Plate 22.3: *Catharanthus pusillus* (Murr.) G. Don. Anatomy of of the Root.

The outer portion is wider and includes wide, angular thin walled solitary or small clusters of vessels (Plate 22.3e). The xylem fibres in the outer portion are comparatively wide and thin walled xylem rays are fairly distinct, thin and straight. The xylem fibres become more thicker with reduced lumen in the peripheral part of the xylem cylinder (Plate 22.3e). The vessels are 30µm wide. Secondary phloem includes wide, angular thin walled sieve elements with fairly distinct companion cells.

Powder Microscopy

The powder preparation of the sample includes the following elements:

(i) Trichomes

Thick cylindrical, multicellular, club shaped epidermal trichomes are seen all along the margins as well as on the middle part of the lamina. The trichome consists of a thick, short and conical stalk, 2 or 3 celled wide and thick cylindrical trichome part. The cells occur in vertical row and the cells are wide and thick. They measure 150µm long and 50µm thick (Plate 22.4a, c, d).

(ii) Adaxial Epidermis

The adaxial epidermal cells are seen in surface view. The cells are polygonal in outline, their anticlinal walls are thick and straight (Plate 22.4b). The cells have no cuticular structure.

(iii) Abaxial Epidermis

The abaxial epidermis was in epidermal peeling. The epidermal cells are thin walled with, highly wavy anticlinal walls. The stomata are dense and diffuse in distribution. They are anamocytic type, (*i.e.*) no definite subsidiary cells associated with these stomata are evident (Plate 22.4e). The guard cells are elliptical measuring 20x12µm in size.

(iv) Xylem Fibres

Xylem fibres are abundant in the powder. The fibres are of two types. Some are wide and short, they are wide fibres. These are also thin, narrow, longer, fibres are called narrow fibres. The wide fibres are thin walled and their lumen is very wide. Most of the wide fibres possess single, vertical row of slit-like simple pits. The wide fibres are 280 to 400µm long and upto 30µm wide. The narrow fibres have thicker walls and reduced lumen. The slit-like pits are not evident in the narrow fibres. They are upto 550µm long and 20µm wide (Plate 22.4f, g).

(v) Vessel Elements

Vessel elements are also equally abundant in the powder. The vessel elements vary in shape and size. Most of the vessel elements are narrow, long and cylindrical. Some vessel elements have fairly long tails. The long, narrow, vessel elements have circular, oblique and wall perforations.

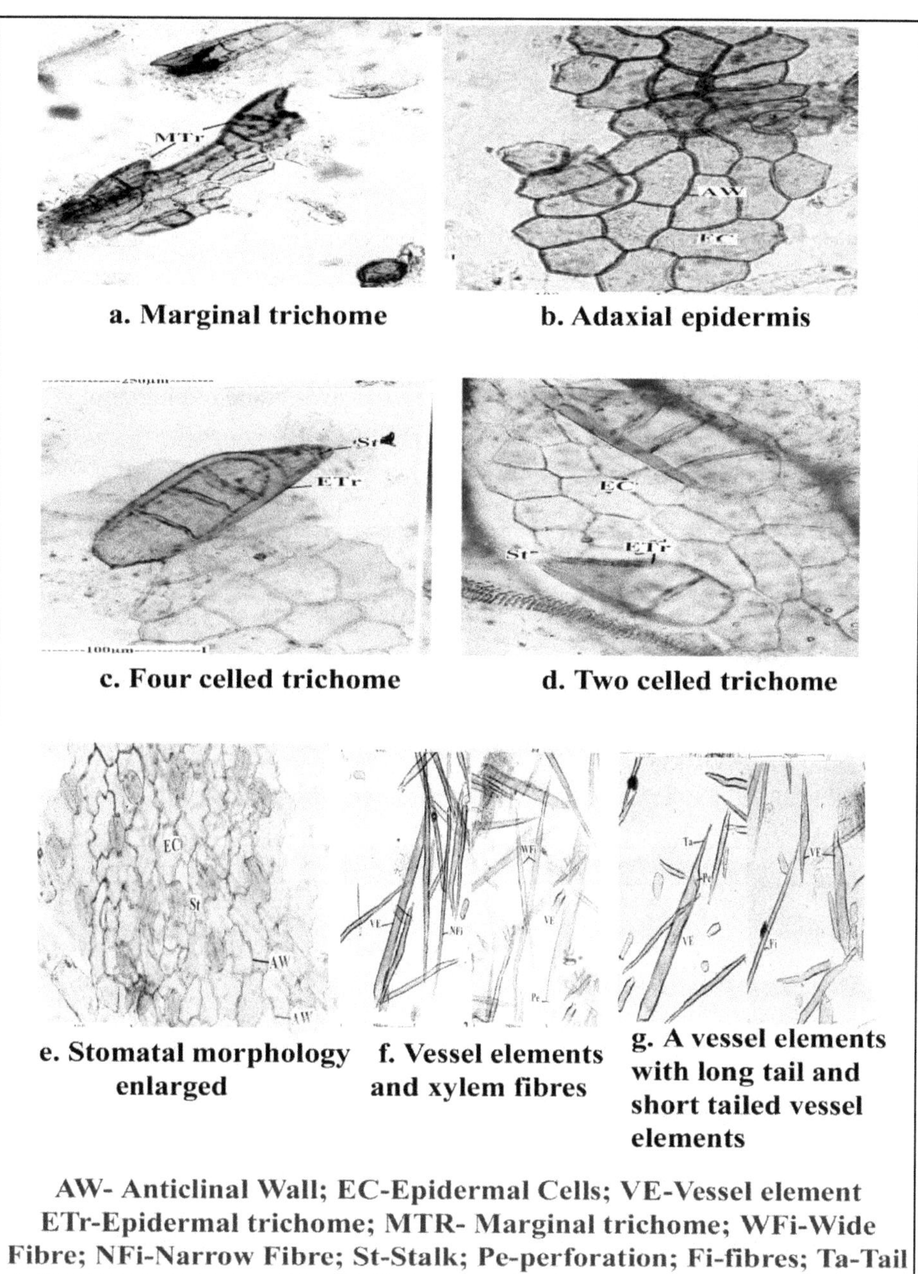

a. Marginal trichome

b. Adaxial epidermis

c. Four celled trichome

d. Two celled trichome

e. Stomatal morphology enlarged

f. Vessel elements and xylem fibres

g. A vessel elements with long tail and short tailed vessel elements

AW- Anticlinal Wall; EC-Epidermal Cells; VE-Vessel element ETr-Epidermal trichome; MTR- Marginal trichome; WFi-Wide Fibre; NFi-Narrow Fibre; St-Stalk; Pe-perforation; Fi-fibres; Ta-Tail

Plate 22.4: ***Catharanthus pusillus*** **(Murr.) G. Don. Powder Microscopy.**

There are also short and wide vessel elements with tail or without tail. The short wide barrel shaped vessel elements are 110 to 260 µm long. On the lateral walls multiseriate, dense bordered pits are seen (Plate 22.4g).

Powder Analysis

Physico-chemical Evaluation

The physico-chemical parameters like ash and extractive values, fluorescence analysis of whole plant of *C.pusillus* were determined. Preliminary phytochemical screening was also performed and results are presented below.

Ash and Extractive Values

The results of the ash and extractive values of the powdered whole plant of *C.pusillus* is depicted in Tables 22.1a and b. The total ash content of the powdered whole plant of *C. pusillus* is 9.8 per cent.The extractive value of methanol is more than the other solvents investigated in the present study.

Table 22.1a: Ash Values of the Powdered Whole Plant of *C. pusillus* *

Type of Ash	*Per cent of Ash*
Total ash	9.8± 0.05
Water soluble ash	1.96±0.01
Acid insoluble ash	2.42±0.02
Sulphated ash	8.00±0.03

Table 22.1b: Extractive Values of the Powdered Whole Plant of *C. pusillus* *

Nature of the Extract	*Per cent of Extractive Values*
Petroleum ether	4.80 ± 0.02
Benzene	4.70 ± 0.05
Chloroform	4.67 ± 0.06
Acetone	5.34 ± 0.04
Methanol	11.21 ± 0.08
Ethanol	8.23 ± 0.03
Water	8.10 ± 0.11

* All values are the mean of triplicate determination ± standard error.

Fluorescence Analysis

The results of fluorescent analysis of powdered whole plant of *C. pusillus* are shown in Table 22.2. The powder from the whole plant of *C. pusillus* emitted greenish yellow under day light, green under short and long UV light. The powder shows the characteristic fluorescent green colour when treated with 1N HCl, 50 per cent

H_2SO_4, $FeCl_3$ Conc. H_2SO_4, Conc. HCl, acetic acid, benzene, acetone, ethanol and petroleum ether.

Table 22.2: Fluorescence Analysis of the Powdered Whole Plant of *C. pusillus*

Experiment	*Visible Light/Day*	*UV Light*	
		Short Wave Length	*Long Wave Length*
Powder	Greenish yellow	Green	Green
Powder+1N HCl	Pale green	Fluorescent green	Fluorescent green
Powder+50 per cent H_2SO_4	Brown	Fluorescent green	Fluorescent green
Powder+1N aqNaOH	Greenish yellow	Yellow	Yellow
Powder+$FeCl_3$	Green	Fluorescent green	Fluorescent green
Powder+40 per cent NaOH +10 per cent Lead acetate	Yellow	Yellow	Yellow
Powder+HNO_3 +NH_3	Orange	Orange	Orange
Powder+Conc. H_2SO_4	Brown	Fluorescent green	Fluorescent green
Powder+Conc. HNO_3	Pale yellow	Pale yellow	Fluorescent green
Powder+Conc. HCl	Green	Fluorescent green	Fluorescent green
Powder+Acetic acid	Yellow	Fluorescent green	Fluorescent green
Powder+NH_3	Yellow	Yellow	Greenish yellow
Powder+Benzene	Green	Fluorescent green	Fluorescent green
Powder+Acetone	Yellow	Fluorescent green	Fluorescent green
Powder+Chloroform	Greenish yellow	Yellow	Fluorescent green
Powder+Ethanol	Green	Fluorescent green	Fluorescent green
Powder+Petroleum ether	Brown	Fluorescent green	Fluorescent green
Powder+50 per cent HNO_3	Yellow	Yellowish green	Yellowish green
Powder+Alcoholic NaOH	Yellow	Yellowish green	Yellow

Preliminary Phytochemical Screening

Petroleum ether, benzene, ethyl acetate, methanol and ethanol extracts of whole powder of *C. pusillus* are qualitatively analysed for the presence of different phytoconstituents and the results are presented in Table 22.3. The methanol and ethanol extracts of *C. pusillus* whole plant shows the presence of alkaloids, coumarins, flavonoids, saponins, steroids, tannins, glycosides, phenols, quinines and terpenoids.

Discussion

According to the World Health Organization (WHO, 1998), the macroscopic and microscopic description of a medicinal plant is the first step towards establishing the identity and the degree of purity of such materials and should be carried out before any tests are undertaken. In present study all the parameters are evaluated successfully as per ayurveda pharmacopeia (Haider, 2013).

Salient Features

- ☆ The leaf consists of biconvex midrib with thick and blunt adaxial hump and thin bifacial lamina.
- ☆ The vascular strand of the midrib consists of a thin flat layer of xylem elements and adaxial and abaxial layer of phloem elements (bicollateral vascular bundles).
- ☆ The lateral vein also has biconvex midrib with small circular vascular bundle.
- ☆ Lamina consists of thick rectangular epidermal cells, single adaxial row of loosely arranged palisade cells and 4 or 5 layers of lobed and loosely arranged spongy parenchyna cells.
- ☆ The leaf margin is conical and slightly bent down.The internal structure of the leaf margin remains unaltered.
- ☆ The epidermal cells of the lamina small polygonal with thick straight anticlinal walls.
- ☆ The stomata are anisocytic type with three uniequal subsidiary cells.
- ☆ The venation pattern is unique in being the lateral veins run horizontally parallel to each other; the vein terminations also run horizontally; they are straight and long.
- ☆ Petiole has convex wide abaxial part, thick and wide adaxial part and two thick conical and vertical wings.The vascular strand of the petiole is bicollateral, slightly and curved.
- ☆ The young stem is circular with four long and thick lateral wings and uneven outline. The vascular system includes thin cylinder of xylem with outer and inner phloem strands.
- ☆ Old stem is circular with short, conical peg like wings at four places. The vascular system includes thick and depse hollow xylem cylinder with phloem strands located in the inner and out places of the xylem cylinder.
- ☆ Xylem consists of wide, solitatory or radical multiples angular vessels, radical files of thin walled squarish fibres and straight narrow xylem rays
- ☆ Both thin and thick roots were studied. The root has less distinct periderm, thin layer of cortex and continuous layer of secondary phloem enclosing the thick and dense secondary xylem cylinder.
- ☆ The secondary xylem includes dense and diffuse angular wide vessels and thick walled and lignified fibres.
- ☆ The powder preparation of the plant shows adaxial epidermal cells of the leaf in surface which are wide with slight anticlinal walls, short started, cylindrical thick four called epidermal trichomes, abaxial epidermal fragments with anisocytic stomata, long narrow cylindrical vessel elements and wide and narrow fibres.

Pharmacochemical Characterization

Physio-chemical Constituents

Ash and Extractive Values

Evaluation of the physical constant of the drug is an important parameter in detecting adulteration or improper handling of drugs (African Pharmacopoeia, 1986). Ash is one of the components in the proximate analysis of biological materials, consisting mainly of salty, inorganic constituents. It includes metal salts which are important for processes requiring ions such as Na^+ (Sodium), K^+ (Potassium), and Ca^{2+} (Calcium). The Ash value is the residue remaining after the incineration, which consists mostly of metal oxides.

The total ash is particularly important in the evaluation of purity of drugs, that is, the presence or absence of foreign organic matter such as metallic salts and/or silica. The ash value of whole plant of *C.pusillus* is 9.80 per cent.The ash value is indicative of the impurities present in the drug and the value is also one of the diagnostic parameters of the drug. Total ash usually consists of carbonates, phosphates, oxidases, silicates and silica. In certain drugs, the percentage variation of ash from sample to sample is very small and any marked difference indicates the change in quality. A high value is indicative of contamination, substitution, adulteration or carelessness in preparing the crude drug for marketing. Samples have more water soluble ash than acid soluble ash. The ash values are generally the index of the purity as well as identity of the drug. Ash value is a criterion to judge the identity or purity of the crude drug. Extractive values are primarily useful for the determination of exhausted or adulterated drugs.

Fluorescence Analysis

Many drugs are fluorescent when their powder is exposed to ultraviolet radiation. It is important to observe all materials on reaction with different chemical reagents under UV light. The phytoconstituents present in the crude drug interact with the chemical reagents and may produce certain products which may be present inside the cell or may come out of the cell and react in the medium, thus resulting in a specific fluorescence pattern. This is the basis of the fluorescence analysis. Many phytocompounds are fluorescent when suitably illuminated (Musa *et al.*, 2006). The fluorescent colour is specific for each compound. A non-fluorescent compound may fluoresce if mixed with impurities that are fluorescent. The powder from the bark of *C.pusillus* fluoresced light green under day light, short UV light and dark green under UV light.

Phytochemical Studies

Presence or absence of certain important compounds in an extract is determined by colour reactions of the compounds with specific chemicals which act as dyes. This procedure is a simple preliminary prerequisite before going for detailed phytochemical investigation. Various tests have been conducted qualitatively to find out the presence or absence of bioactive compounds such as flavonoids, terpenoids, tannins, saponins, steroids and alkaloids are detected in *C.pusillus* which could make the plant useful for treating different ailments as having a potential of providing useful drugs of human use.

Alkaloids act as antioxidant and immunomodulatory agent. Alkaloids are known to exhibit emetic amoebicides, expectorant, anaesthetics, antipyretics, analgesics, anthelmintic and can be used for the treatment of stomach problems (Farnsworth, 1975) Flavonoids elicit a wide range of therapeutic activities such as antihypertensive, antibiotic, antimicrobial, antitumor activities and recently for immune regulatory functions. Flavonoids are known to regenerate the damaged beta cells in the alloxan diabetic rats (Chakravarthy *et al.*, 1980). Several authors reported that flavonoids, sterols/terpenoids, phenolic acids are known to be bioactive antidiabetic principles (Oliver-Bever, 1986: Rhemann and Zaman, 1989). Flavonoids act as insulin secretagogues (Geetha *et al.*, 1994). Most of the plants have been found to contain certain substances like glycosides, alkaloids, terpenoids, flavonoids, *etc.*, which are frequently implicated as having antidiabetic effects (Loew and Kaszhin, 2002).

Therapeutically terpenoids exert wide spectrum of activities such as antiseptic, stimulant, diuretic, anthelmintic, analgesic and counter-irritant (Gokhale *et al.*, 2003). Many tannin containing drugs are used in medicine as astringent. They are used in the treatment of burns as they precipitate the proteins of exposed tissues to form a protective 116 covering (Handa and Kapoor, 1992). They are also medically used as healing agents in inflammation, leucorrhoea, gonorrhoea, burns, piles and antidote (Ali, 1994).

In plants, the presence of steroidal saponins like, cardiac glycosides appear to be confined to many families and these saponins have great pharmaceutical importance because of their relationship to compounds such as the sex hormones, cortisones, diuretic steroids, vitamin D *etc.* (Evans and Saunders, 2001). From plant saponins a synthetic steroid is prepared and to treat a wide variety of diseases such as rheumatoid arthritis, collagen disorders, allergic and asthmatic conditions (Claus, 1956).

Tannins are complex moieties produced by majority of plants as protective substances. They have wide pharmacological activities and have been used since past as tanning agents and they possess astringent, antiinflammatory, antidiarrhoeal, antioxidant and antimicrobial activities.

Conclusion

The pharmacognostic characters highlighted in this paper can use for the correct identification of the drug. Phytochemical screening shows the presence of valuable secondary metabolites, so there is no doubt about that this traditional medicinal plant will give clues for the preparation of new drugs. This information will enrich the data base of Indian Pharmacopoeia by incorporating the pharmacognostic information of these selected ethnomedicinal plants and the uses of these respective plants will also be included in the Indian national ethnomedicinal inventory.

Acknowledgement

The first author Mrs.P.Yokeswari Nithya gratefully acknowledges and expresses her sincere thanks to University Grants Commission, New Delhi for providing financial assistance to this Minor research project, she is also thankful to Dr. P. Jayaraman, Plant Anatomy Research Centre, Chennai for extending their help in the anatomical studies.

References

African Pharmacopoeia.1986. General Methods for Analysis Ist ed. 2: (OAU/STRC) Lagos.123.

Ali, M. 1994. Text book of Pharmacognosy. CBS Publishers and Distributors, New Delhi, p. 405

Alvarez, E., Leiro, J.M., Rodrigue, Z.M. and Orallo, F. 2005. Inhibitory effect of leaf extracts of *Stachytarpheta jamaicansis* on the respiratory burst of of rat macrophages. *Phototherapy research.* 18: 457-462.

Balaji rao, N.S., Rajasekhar, D. and Chengal raju, D. 1996. Folklore Remedies for Dandruff from Tirumala hills of Andhra Pradesh. *Ancient Science of life.* XV (4):296 -300.

Balaji rao, N.S., Rajasekhar, D. and Chengal raju, D. 1996. Folklore Remedies for Dandruff from Tirumala hills of Andhra Pradesh. *Ancient Science of life.* XV (4): 296 -300.

Brindha, P., Sasikala, P. and Purushothaman, K.K. 1981. Pharmacognostic studies on merugan kizhangu. Bull. *Med. Eth. Bot. Res.* 3: 84 - 96.

Chakravarthy, B.K., Gupta, S., Gambir, S.S. and Gode, K.D. 1980. Pancreatic beta cell regeneration. A novel antidiabetic mechanism of *Pterocarpus marsupium* Roxb. *Ind. J. Photochem. Photo. Biol.* 12: 123-127

Claus, E.P. 1956. Pharmacognosy. (Ed.) Lea and Febiger, Philadelphia. pp. 1-697

Don, G. 1999. Catharanthus Roseus In: Ross I.A (Ed) Medical Plants of The World. Human Press, Totowa, NJ: 109-118.

Esau, K. 1964. Plant anatomy. John Wileyand sons, New York. pp. 1-767

Evans, W. C. and Saunders, W.B (Eds.) 2001.Trease and Evan's Pharmacognosy Tokyo. 1-579.

Farnsworth NR, 1988. Preliminary phytochemical screening and HPLC Analysis of Flavonoid from Methanolic Extract of Leaves of *Annona squamosa.* Screening plants for new National Academy Press, Washington, DC, 83-97.

Geetha, B.S., Mathew, B.C. and Augusti, K.T. 1994. Hypoglycemic effects of leucodelphinidin derivative isolated from *Ficus bengalensis* Linn. *Indian J. Physiol. Pharmacol.* 38: 220-222.

Gokhale, S.G., Kotate, C.K. and Purohit, A.P. 2003. A Text Book of Pharmacognosy, Nirali Prakashan, Pune-21.

Haider, M.Z. 2013. Ethnobotanical studies of potential wild medicinal plants of Ormara, Gawader, Pakistan. Emir. *J. Food Agric.* 25(10): 751-759.

Handa, S.S. and Kapoor, V. K. 1992. (Eds): Pharmacognosy. Vallabah Prakashan Puplications, New Delhi.

Johansen, D.A. 1940. Plant Microtechnique. Mc Graw Hill Book Co; New York. :523.

Lala, P.K. 1993. Lab Manuals of Pharmacognosy, CSI Publishers and Distributors, Calcutta, 5 th Edition. London.: 38-48.

Loew, D. and Kaszkin, M. 2002. Approaching the problem of bioequivalence of herbal medicinal products. *Phytother. Res.* 16: 705-711.

Metcalfe, C.R. and Chalk, L. 1979. Anatomy of Dicotyledons I. London: Oxford University Press.

Musa, K.Y., Katsayal A.U., Ahmed A., Mohammed Z. and Danmalam U.H. 2006. Pharmacognostic investigation of the leaves of *Gisekia pharmacioides*. *African J. Biotech.* 5: 956-957.

O' Brien J.P., Feder N. and Mc Cull M.E. 1964. Polychromatic staining of plant cell walls by toluidine blue-O. protoplasma. (59): 364-373.

Oliver-Bever, B. 1986. Medicinal plants in tropical West Africa, Cambridge University press, London. pp. 245-267.

Rhemann, A.V. and Zaman, K. 1989. Medicinal plants with hypoglycemic activity. *J. Ethnopharmacol.* 26: 1-5.

Sass, J.E. 1940. Elements of Botanical Microtechnique. Mc Graw Hill Book Co; New York. pp.222

Wallis, T.E. 1985. Text book of Pharmacognosy, New Delhi: 5th ed. CBS Publishers. 566.

World Health Organization (WHO), *The Global Burden of disease: 2004 Update*, WHO, Geneva, Switzerland, 2008.

2018, Ethnomedicinal Plants: A Biodiversity Treasure *Pages* **551–568**
Editors: ***V.R. Mohan, A. Doss, P.S. Tresina and V. Sornalakshmi***
Published by: **ASTRAL INTERNATIONAL PVT. LTD., NEW DELHI**

Chapter 23

Pharmacognostical, Physico-chemical and Phytochemical Assessment of *Barleria courtallica* Nees (Acanthaceae)

A. Ponmathi Sujatha[1], R. Michael Evanjaline[1], S. Muthukumarasamy[2] and V.R. Mohan[1*]

[1]Ethnopharmacology Unit, PG and Research Department of Botany, V.O. Chidambaram College, Tuticorin – 628 008, Tamil Nadu
E-mail: vrmohanvoc@gmail.com
[2]Department of Botany, Sri K.G.S. Arts College, Srivaikuntam – 628 619, Tamil Nadu

Introduction

Medicinal plants are playing very active role in traditional medicines for the treatment of various ailments (Mohammad Saleem *et al.*, 2008). However a key obstacle, which has hindered the promotion in use of alternative medicines in the developed countries have no evidence of documentation and absence of stringent quality control measures. There is a need for the record of all the research work carried out on traditional medicines in the form of documentation. According to World Health Organization (WHO) more than 80 per cent of the world's population relive on herbal medicine for their primary healthcare needs. Use of herbal medicines in Asia represents a long history of human interactions with the environment. Herbal formulation involves use of fresh or dried plant part. Therefore proper and correct identification of the plant material is very much essential. Correct identification of the starting material is an essential prerequisite to ensure reproducible quality and will contribute immensely to its safety and efficacy (Thomas *et al.*, 2008).

The term "pharmacognosy" is derived from two Greek words 'Pharmacon' meaning drug or medicine and 'gnosis' meaning knowledge. This term was first coined by C.A. Seydler in his dissertation entitled 'Analectapharmacognosia' in 1895. Adulteration or substitution is nothing but replacement of original plant with another plant material or intentionally adding any foreign substance to increase the weight or potency of the product or to decrease its cost. Therapeutic efficacy of medicinal plants depends upon the quality and quantity of chemical constituents. The misuse of herbal medicine or natural products starts with wrong identification. The most common error is one common vernacular name is given to two or more entirely different species (Dineshkumar, 2007). All these problems can be solved by pharmacognostic studies of medicinal plants. It is very important and in fact essential to lay down pharmacognostic specifications of medicinal plants which are used in various drugs. Pharmacognosy is the study of medicines derived from natural sources, mainly from plants. It basically deals with standardization, authentication and study of natural drugs. Most of the research in pharmacognosy has been done in identifying controversial species of plants, authentication of commonly used traditional medicinal plants through morphological, phytochemical and physico-chemical analysis. The importance of pharmacognosy has been widely felt in recent times. Unlike taxonomic identification, pharmacognostic study includes parameters which help in identifying adulteration in dry powder form also. This is again necessary because once the plant is dried and made into powder form, it loses its morphological identity and easily prone to adulteration. Pharmacognostic studies ensures plant identity, lays down standardized parameters which will help and prevents adulterations. Such studies will help in authentication of the plants and ensures reproducible quality of herbal products which will lead to safety and efficacy of natural products.

Genus *Barleria* belongs to the family Acanthaceae. The whole plant extract *Barleria* contains a number of active compounds like alkaloids, terpenes, flavonoids, glycosides, lignins, phenolics *etc.*, which have shown potent therapeutic activities against several diseases (Saadabi *et al.*, 2006; Mukherjee *et al.*, 2009; Agarwal *et al.*, 2011; Gantait *et al.*, 2011). *Barleria* also shows various pharmacological effects such as antimicrobial, anthelminthic, antifertility, antioxidant, antidiabteic, antiarthritic, hepatoprotective, diuretic, cytoprotective, antidiarrhoeal, analgesic, antileukemic, antiinflammatory and hypoglycemic properties without any toxic effects (Singh *et al.*, 2005; Amoo *et al.*, 2009). However, available literature revealed that no pharmacognostic study has been carried out on the plant *Barleria courtallica*, hence the present investigation was undertaken. The object of the present study is to evaluate various pharmacognostical parameters such as macroscopic, microscopic, physico-chemical, fluorescence and phytochemical studies of the above said plant.

Methodology

Whole plant of *Barleria courtallica* Nees was collected from Agasthiarmalai Biosphere Reserve, Western Ghats, Tamil Nadu. The plant samples were identified with the help of local flora and authenticated by Botanical Survey of India, Southern Circle, Coimbatore, Tamil Nadu, India. A voucher specimen (VOCB3336) of collected plants was deposited in the Ethnopharmacological Unit, PG and Research

Department of Botany, V.O. Chidambaram College, Thoothukudi District, Tamil Nadu.

Macroscopical Studies

The macroscopic characters like surface, shape, size, venation, phyllotaxy, length of the petiole, length of the leaf etc were noted.

Anatomical Studies

For anatomical studies, the required samples of stem and leaf were cut and removed from the plant and immediately fixed in FAA (formalin- 5 ml + acetic acid- 5 ml + 70 per cent Ethyl alcohol- 90 ml). The specimens were left in the preservative for two days; then the materials were washed in water and processed further. Standard microtome techniques were followed for anatomical investigation (Johanson, 1940). Transverse sections of the materials were made. The microtome sections were stained with 0.25 per cent aqueous Toluidine blue (Metachromatic stain) adjusted to pH 4.7 (O'Brien, 1964). Photomicrographs were taken with NIKON trinocular photo micrographic unit.

Physico-chemical and Fluorescence Analysis

These studies were carried out as per the standard procedures (Lala, 1993). In the present study, the powered stem and leaf were treated with various chemical reagents like aqueous 1N sodium hydroxide, alcoholic 1N sodium hydroxide, 1N hydrochloric acid, 50 per cent sulphuric acid, concentrated nitric acid, picric acid, acetic acid, ferric chloride and concentrated HNO3+NH3 These extracts were subjected to fluorescence analysis in day light and UV light (254nm and366nm). Various ash types and extractive values were determined by following standard methods (Anonymous 1996).

Preliminary Phytochemical Analysis

Shade dried and powdered stem and leaf samples were successively extracted with Petroleum ether, benzene, ethyl acetate, methanol and ethanol. The extracts were filtered and concentrated using vacuum distillation. The different extracts were subjected to qualitative tests for the identification of various phytochemical constituents as per the standard procedure (Lala, 1993; Brindha *et al.*, 1981).

Results

Exomorphic Features (Plate 23.1a)

A shrub 0.9-1.2m height: Stems and branches glabrous. Leaves 12.5-18.25 by 4.5-6.3cm,elliptic- lanceolate, long-acuminate, glab-rous, shining, lincolate on both sides. Base acutely tapering into the petiole: Main nerves 6-9 pairs. Slender: Petioles 13-38mm long flowers subsecund. In dense axillary and terminal glandular-hairy narrow spikes 2.5-10cm. Long: bracts and bracteoles 6-8mm long. Linear-lanceolate, glandular-hairy. Calyx glandular-hairy outside: Outer sepals subequal,2-2.5 by 0.6-0.8cm, Ovate-lanceolate bluntly long-acuminate, 7-9 nerved from the base, densely glandular. Hairy outside, appressedly silky-hairy inside. One of the pair

a. habit

b. T.S of midrib of the leaf

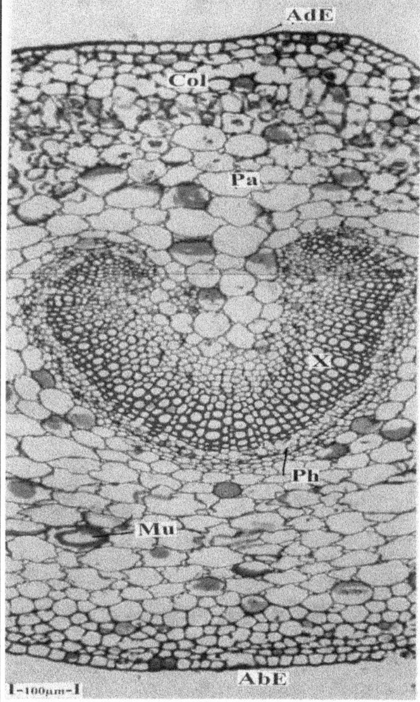

c. T.S. of midrib-a sector enlarged

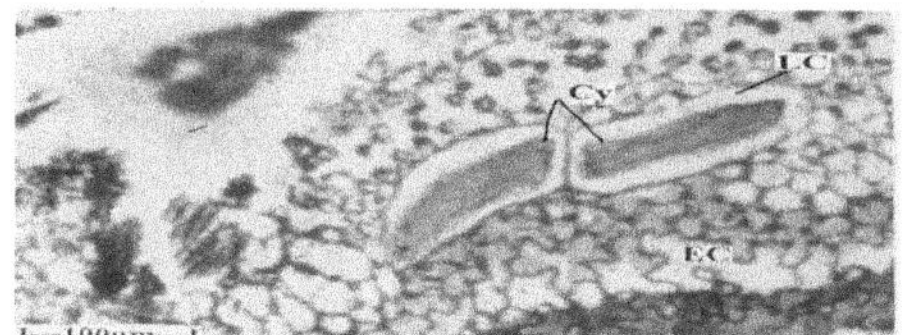

d. T.S of leaf margin of the lamina

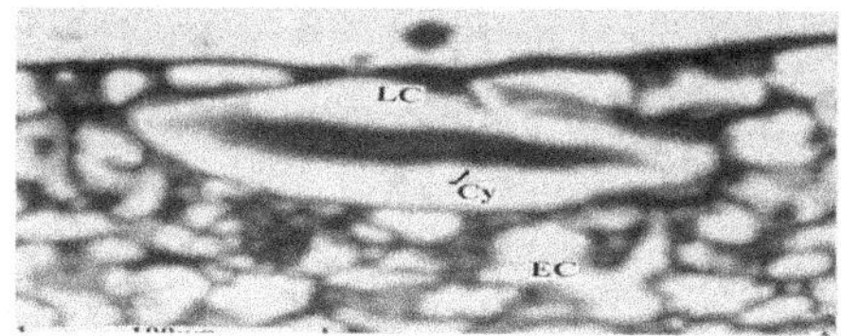

e. Double cystolith in the lamina

f. Single cystolith

Abs- Abaxial side; Ads- Adaxial side; Col- Collenchyma; Mu- Mucilage cell; Par-Parenchyma;Ph- Phloem; VS- Vascular System; WB- Wing Bundle; X-Xylem; AbE- Abaxial Epidermis; AdE- Adaxial Epidermis; Cy- Cystolith; EC- Epidermal cells; LC- Lithocyst; LM- Leaf Margin; MT- Mesophyll Tissue; Sc-Sclerenchyma cells.

Plate 23.1

vely slightly 2-toothed at the apex: inner sepals 16-20 by 2.5mm linear-subulate, usually spreading outwards and not enclosed within the larger sepals ciliate. Corolla 3.8-5cm long with a blue limb and yellow tube, glabrous outside; tube hairy inside at the insertion of the filaments; lobes 16mm long, obovate. Oblong rainded. Ovaly pubescent at the apex; style pubescent at the vely base. Capsules 2-2.5cm long, oblong, pointed, pubescent at the tip,4-seeded. Seeds 6mm diam broadly ellipsoid, compressed, silky-hairy.

Leaf

The leaf consists of thick midribs and uniformly thin and smooth lamina. The midrib has thick and wide adaxial hump and wide semicircular abaxial part (Plate 23.1b). The midrib is 1.7 mm thick and 2 mm wide. The epidermal layer of the midrib consists of fairly wide squarish cells with prominent cuticle. Inner to the epidermis, the ground tissue includes 2-3 layers of collenchymas and rest of ground tissue is parenchymatous; the cells are circular to angular, thin walled and less compact.

Midribs

The vascular system consists of a wide and deep "horse shoe" shaped main strand and two small less prominent lateral bundle and a wing bundle (Plate 23.1c). The main vascular strand consists of several, vertical lines of compact angular, thick walled xylem elements. The xylem lines are 200μm long. Along the outer zone of the xylem occurs a thin continuous layer of phloem elements. The ground parenchyma cells possess dense accumulation of mucilaginous substance.

Lamina

The lamina is smooth on either surfaces. The adaxial epidermal cells and abaxial epidermal cells are fairly wide with prominent cuticle. Stomata occur on the abaxial surface of the lamina. The mesophyll tissue is less distinctly differentiated between palisade and spongy mesophyll tissues. The cells along adaxial part are slightly vertically elongated into palisade cells. Spongy mesophyll tissue is well defined with spherical or lobed cells with wide interacellular spaces (Plate 23.1d).

Some of the mesophyll cells are dilated into wide lithocyst and possess the calcium carbonate crystal (Plates 23.1e, f). Stomata occur on the abaxial epidermis. The stoma has prominant stomatal ledges which short beak shapes.

Glandular trichomes are also seen on the abaxial epidermis. The gland is located within a shallow pit. The gland consists of a short stalk and dilated secondary head which is darkly stained.

Marginal Part of the Lamina

The leaf margin is semicircular and blunt. It is slight bent down. It is 150μm thick. The epidermal cells of the marginal part are wide with cuticle. The mesophyll cells are small, thick walled and compact (Plate 23.1d).

Cystoliths

Calcium carbonate crystals called cystoliths are common in the mesophyll tissue. The crystolith crystals are cylindrical elongated bodies with tapering ends.

The cystolith is 20µm thick and150µm long. Frequently the cystoliths are seen in pairs. The paired cystoliths are located in two independent cells and the cystoliths are seen end to end.The double cystoliths are as common as solitary cystoliths (Plate 23.1e).

Epidermal Tissues

The epidermal cells and the stomata were studied in surface view of the paradermal sections. The adaxial epidermis consists of wide and hihtly wavy anticlinal walls (Plate 23.2a). The epidermis is apostomatic (without stomata).

Stomata

The abaxial epidermis is stomatiferous. The epidermal cells have undulate anticlinal walls. The stomata are diacytic type i.e; there are two subsidiary cells for each stomata; the cells occur at polar ends of the guard cells. Their common walls are at right angles to the ling axis of the guard cells (Plate 23.2b). The stoma is broadly elliptical measuring 15×30µm in size. The guard cells have prominent spherical polar nodules (Plate 23.2c). The stomatal aperture is wide and elliptical.

Venation Pattern

Lamina was cleared to study venation type of the leaf. The venation is reticulate with wide vein islets. The vein islets are polyhedral in outline. Within the vein islets are seen vein terminatious. The vein terminatious arise from the inner part of the vein and extend into the vein islets. The terminatious are either simple (unbranched) or forked into two unequal branches (Plates 23.2 d, e, f).

Petiole

The petiole is semicircular on the abaxial side and raised into a flat adaxial side. These two short thick wings on the lateral part of the petiole (Plate 23.3a). The petiole is 2mm in vertical plane and 2.1 mm in horizontal plane. The peripheral part of the petiole consists of a thin epidermal layer of small cells followed by about six layer of collenchymas cells. The remaining major portion includes large, circular thin walled ground parenchyma cells with prominent intercellular spaces (Plate 23.3b).

The vascular system of this petiole includes a deeply curved vascular strand with a small opening on the adaxial side. The vascular strand consists of several vertical rows of xylem elements and parenchyma cells. The xylem elements are highly and thick walled. The cell lumen of the xylem elements is circular. The metaxylem elements are 20µm wide. The parenchyma cells that run along parallel to the xylem rows are rectangular and thin walled. Phloem occurs along the outer boundary of the xylem arc. The phloem elements are in 3-6 layers of thick walled darkly stained (Plate 23.3c).

Apart from main vascular system, there are two pairs of accessory vascular bundles located each pair towards the wings of the petiole. One bundle forms the marginal bundle and the other forms the wing bundle. These bundles are circular in outline with two or three groups of xylem and a small nest of phloem.

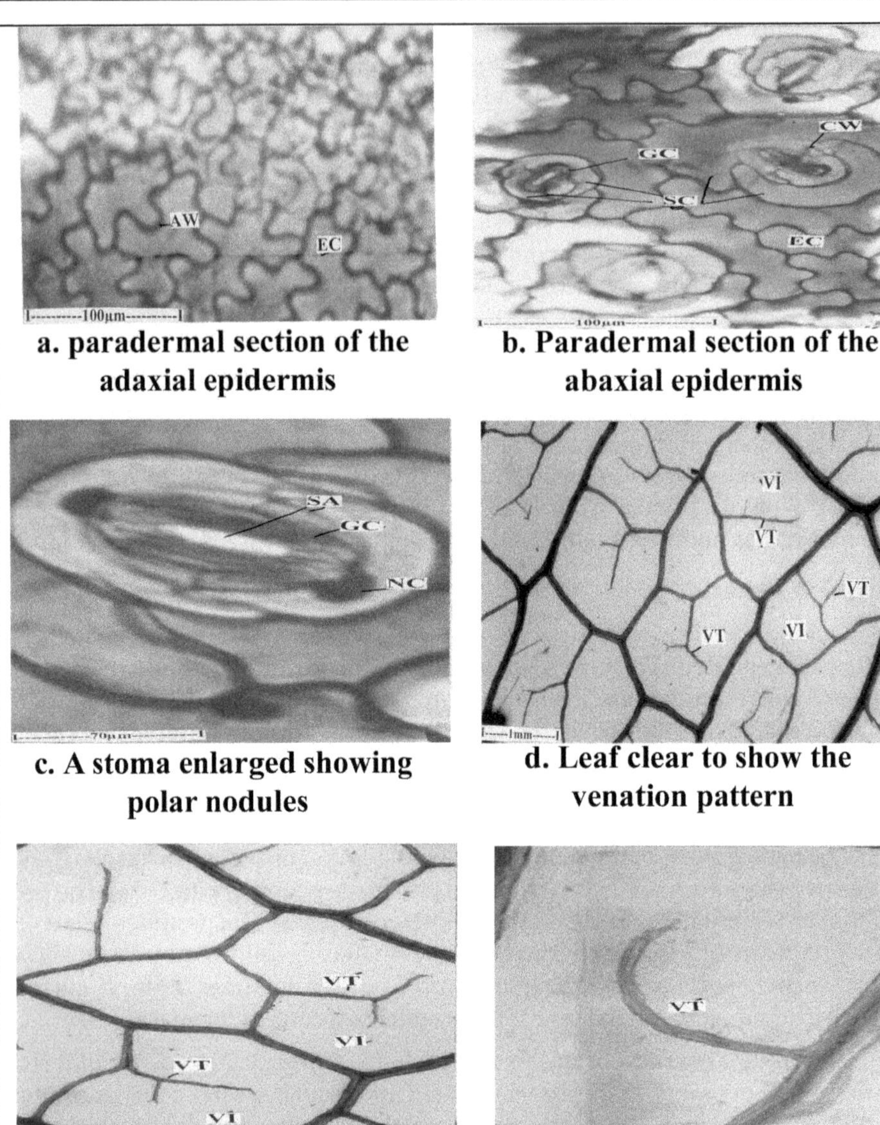

a. paradermal section of the adaxial epidermis

b. Paradermal section of the abaxial epidermis

c. A stoma enlarged showing polar nodules

d. Leaf clear to show the venation pattern

e. Vein islets and vein termination

f. A single vein termination

AW- Anticlinal wall; EC- Epidermal cells; GC- Gall cells; CW- Common wall of the subsidiary cells; NC- Nodular cells; SA- Stomatal aparture; SC- Subsidiary cells; VI- Vein islets; VT- Vein termination.

Plate 23.2

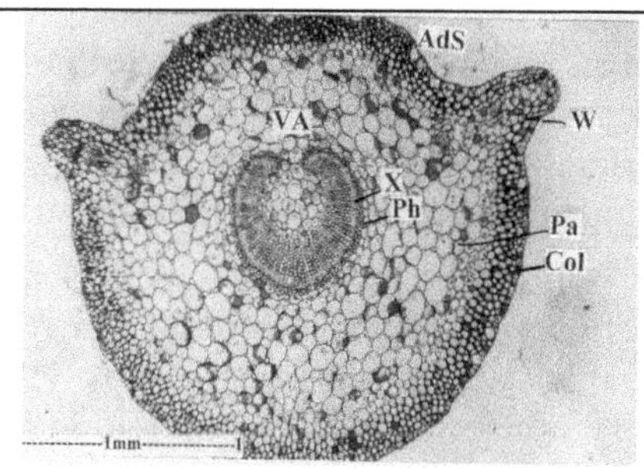

a. T.S of petiole- Over view

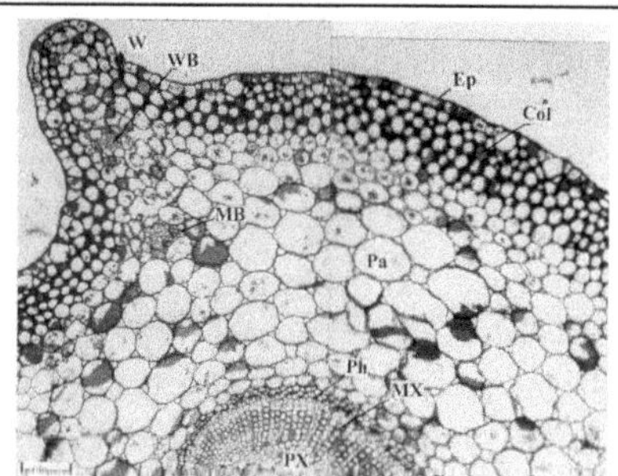

b. T.S. of petiole- a sector enlarged

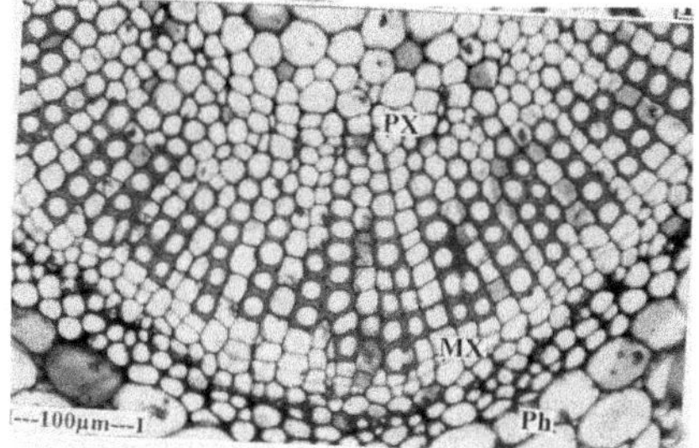

c. Main vascular arc- a sector enlarged

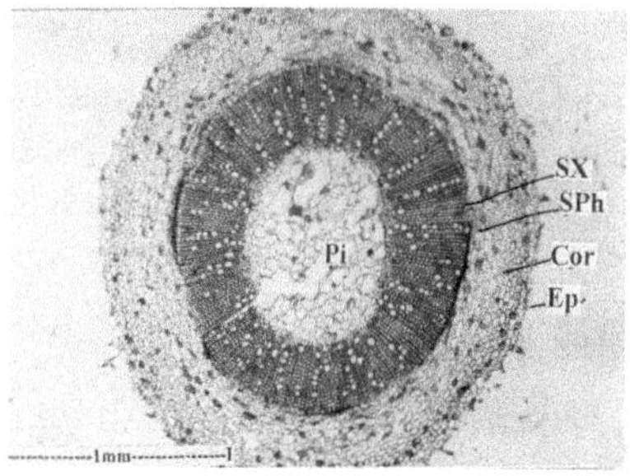

d. T.S. of stem- entire view

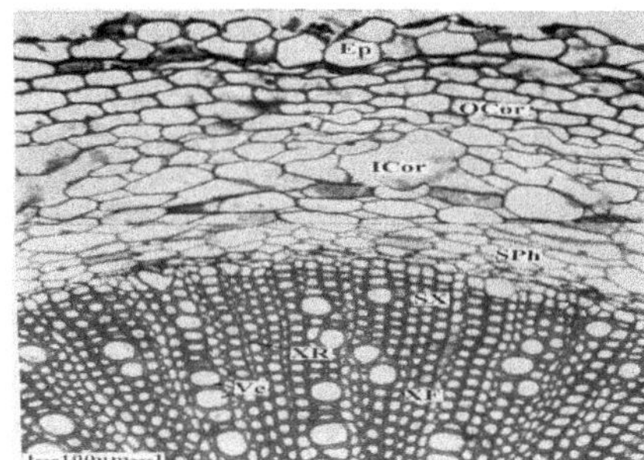

e. T.S. of stem- a secor enlarged

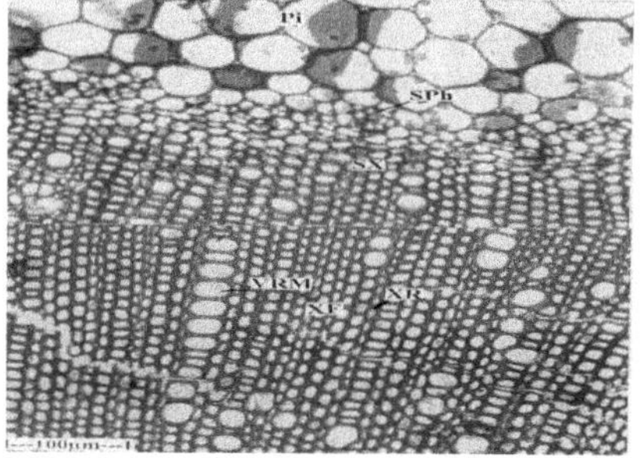

f. T.S of stem- Secondary xylem and secondary phloem enlarged

Ads- Adaxial side; Col- Collenchyma; Ep- Epidermis; MB- Marginal Bundle; MX- Metaxylem; Pa- Parenchyma; Ph- phloem; PX- Protoxylem; VA- Vascular Arc; W- Wing; WB- Wing Bundle; Co- Cortex; ICor- Inner cortex; OCor- Outer cortex; Ve- Vessel; Pi- Pith; SPh- Secondary phloem; SX- Secondary Xylem; XF- Xylem Fibres; XR- Xylem ray; VRM- Vessels in Radial Multiplies.

Plate 23.3

Stem

The stem is circular in outline measuring 2 mm in diameter. The stem consists of an epidermal layer of wide angular thick walled cells. Inner to the epidermis is thick zone of tangentially stretched rectangular thick walled outer cortex. The inner cortex consists of angular, thin walled compact parenchyma cells (Plate 23.3d).

The vascular system consists of a wide, hollow, thick cylinder which encloses a wide central pith. The vascular cylinder includes dense xylem cylinder; the xylem comprises wide circular or angular vessels; thick walled fibrous and narrow; less prominent xylem rays. The vessels are either solitary or more commonly in long radial multiples (Plate 23.3e).

The vessels are 30μm in diameter. The xylem cylinder is ensheathed by thin continuous layer of secondary phloem. The phloem elements include small angular sieve elements intermixed with phloem parenchyma (Plate 23.3f).

The pith cells are parenchymatous; the cells are angular and the cells include dense accumulation of starch grains. In the pith, there are small clusters of circular, highly thick walled and lignified sclereids. The cells have narrow and small cell lumen.

Powder Microscopy

The powder preparation exhibits the following cell types when viewed under microscope.

Fibres

Long, thick fibres are common in the powder. The fibre has thick, lignified walls and wide lumen. The fibres are 470μm long and 15μm thick. **Fibre tracheids** are occasionally seen in the powder. They are long, narrow cells with pointed ends. The cell wall is thin; cell lumen is wide; there are minute circular pits (Plate 23.4c,d).

Epidermal Cells of the Stem

Broken pieces of the epidermal cells of the stem are commonly seen in the powder. The cells are vertically elongated, polygonal and thick walled. The cells are 30μm wide and 60μm long (Plate 23.4e).

Sclereid

Rectangular thick sclereid is characteristic of the powder. It is more or less isodiametric and hence it is known brachesclereid. It consists of thick secondary wall and narrow lumen. The walls have well developed canal like simple pits (Plate 23.4f).

Mesophyll Cells of the Leaf

Large parenchyma cells with cylindrical shape are common in the powder. The cells are thin walled and dilated. They have large spherical chloroplasts. The cells are 80μm long and 50 μm thick (Plate 23.4g).

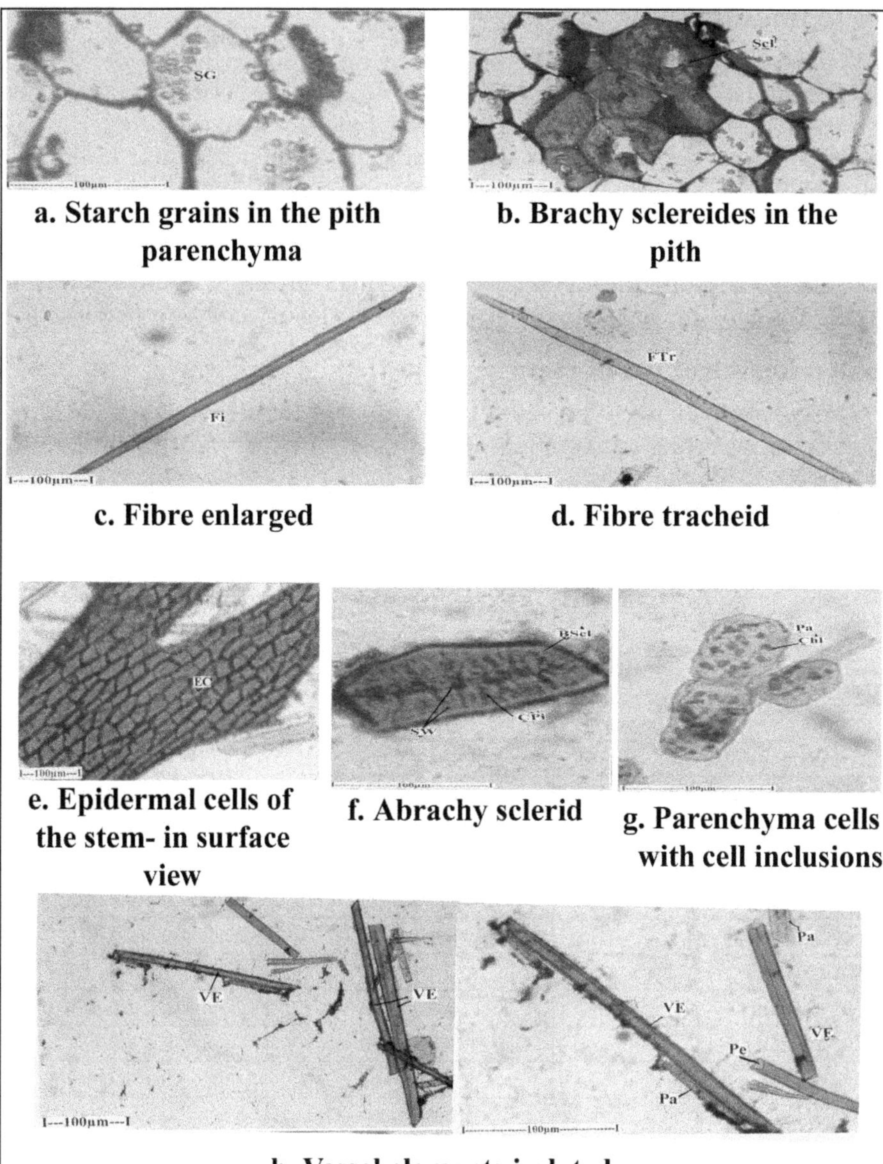

a. Starch grains in the pith parenchyma

b. Brachy sclereides in the pith

c. Fibre enlarged

d. Fibre tracheid

e. Epidermal cells of the stem- in surface view

f. Abrachy sclerid

g. Parenchyma cells with cell inclusions

h. Vessel elements isolated

SG- Starch grains; Scl- Sclereids; Fi- Fibre; FTr- Fibre Tracheid CPi- Canal like simple pith; Chl- Chloroplast; BScl- Brachy Sclereid; Ec-Epidermal cells; Pa- Parenchyma; SW- Secondary wall; VE- Vessel elements; Pe- perforation.

Plate 23.4

Vessel Elements

Long, narrow cylindrical vessel elements are common in the powder. The side walls are parallel and the end wall is element with simple perforation. The lateral walls are thick and numerous minute bordered pits are seen on the lateral walls. The pits are multi seriate. The vessel elements are 310 μm long and 15 μm thick (Plate 23.4b).

Parenchyma Cells

Rectangular short thick walled parenchyma cells are sparsely seen in the powder. The cells are 30 X40 μm in size. No cell inclusions are seen in the cells.

Physico-chemical Parameters

The physico-chemical constant evaluation of the drugs is an important parameter In detecting adulteration or improper handling of drugs. Ash values and extractive values of stem and leaf of *B.courtallica* were determined. The results are depicted in Table 23.1.

Table 23.1: Ash and Extractive Values of the Powdered Stem and Leaf of *B. courtallica*

Sl.No.	*Parameters*	*Value per cent*	
		Stem	*Leaf*
1.	**Ash values**		
	Total ash value of powder	9.8 ± 0.06	8.8 ± 0.03
	Water soluble ash	2.4 ± 0.03	2.1 ± 0.02
	Acid insoluble ash	3.2 ± 0.01	3.4 ± 0.01
	Sulphated ash	6.8 ± 0.04	7.4 ± 0.05
2.	**Extractive values**		
	Petroleum ether	5.6 ± 0.02	6.1 ± 0.03
	Benzene	6.8 ± 0.04	7.4 ± 0.02
	Chloroform	7.4 ± 0.03	7.8 ± 0.04
	Acetone	7.8 ± 0.03	8.2 ± 0.06
	Methanol	9.4 ± 0.06	9.6 ± 0.05
	Ethanol	10.2 ± 0.05	10.8 ± 0.03
	Water	9.6 ± 0.04	9.2 ± 0.08

Ash Values

The analytical results for total ash of stem and leaf were found to be 9.8 per cent and 8.8 per cent respectively. The amount of acid insoluble ash present in stem and leaf were 3.2 per cent and 3.4 per cent respectively. The water soluble ash of stem and leaf were found to be 2.4 per cent and 2.1 per cent respectively. The amount of sulphated ash present in stem and leaf were 6.8 per cent and 7.4 per cent respectively.

Extractive Values

Percentage of the extractive values of various extracts is given in Table 23.1. The results showed that various extracts of leaf contain greater proportion by mass of the extractive values than the various extracts of stem. Petroleum ether, benzene, chloroform, acetone, methanol, ethanol and aqueous soluble extractive values of stem were 5.6 per cent, 6.8 per cent, 7.4 per cent, 7.8 per cent, 9.4 per cent, 10.2 per cent and 9.6 per cent respectively and leaf were 6.1 per cent 7.4 per cent, 7.8 per cent, 8.2 per cent, 9.6 per cent, 10.8 per cent and 9.2 per cent respectively. In stem and leaf extracts, ethanol soluble extractive values were higher while the least amount of extractive value was observed in petroleum ether extract.

Table 23.2: Fluorescent Analysis of the Powdered Stem of *B. courtallica*

Sl.No.	*Experiments*	*Visible/Day light*	*UV-light*	
			254 nm (Short wave length)	*365 nm (Long wave length)*
1.	Powder as such	Green	Dark Green	Black
2.	Powder + 1N aqueous NaoH	Dark Green	Fluorescent green	Dark Blue
3.	Powder + 1N alcoholic NaoH	Yellowish Green	Dark Green	Dark Blue
4.	Powder + 1N Hcl	Green	Greenish Yellow	Dark Brown
5.	Powder + con.Hcl	Green	Fluorescent green	Brown
6.	Powder + con. H_2So_4	Brown	Fluorescent green	Dark Green
7.	Powder +50 per cent H_2So_4	Brown	Fluorescent green	Dark Green
8.	Powder + con. HNo_3	Light Green	Light Green	Dark Green
9.	Powder + 40 per cent NaoH+ 10 per cent lead acetate	Pale Green	Fluorescent green	Dark Green
10.	Powder + Acetic acid	Yellowish green	Light Green	Dark Green
11.	Powder + Ferric chloride	Green	Dark Green	Brown
12.	Powder + Chloroform	Green	Yellowish Green	Dark Green
13.	Powder + Benzene	Green	Fluorescent green	Dark Green
14.	Powder + Petroleum ether	Dark Green	Yellowish Green	Dark Brown
15.	Powder + Methanol	Yellowish green	Pale yellow	Brown
16.	Powder + Ethanol	Yellowish green	Fluorescent green	Dark Brown
17.	Powder + Acetone	Yellowish green	Fluorescent green	Dark Green
18.	Powder + NH_3	Green	Fluorescent green	Blue
19.	Powder + HNo_3 + NH_3	Green	Dark Green	Brown
20.	Powder + 50 per cent HNo_3	Pale green	Dark Green	Dark Brown

Fluorescence Analysis

The results of fluorescence analysis of powdered stem and leaf were studied at day light and UV light (245 nm and 365 nm) and the observations are presented in Tables 23.2 and 23.3. Fluorescence studies of powdered stem revealed the presence of fluorescence green when treated with 1N aqueous NaOH, conc. HCl, conc. H_2SO_4, 50 per cent H_2So_4, 40 per cent NaoH+10 per cent lead acetate, benzene, ethanol,acetone and NH_3. The leaf powder treated with conc. H_2So_4, 50 per cent H_2So_4, 40 per cent NaoH+10 per cent lead acetate, benzene, petroleum ether, acetone, NH_3, 80 per cent HNO_3 revealed the presence of fluorescence green under UV light of shorter wavelength.

Table 23.3: Fluorescent Analysis of the Powdered Leaf of *B. courtallica*

Sl.No.	*Experiments*	*Visible/Day light*	*UV-light*	
			254 nm (Short wave length)	*365 nm (Long wave length)*
1.	Powder as such	Green	Green	Dark Green
2.	Powder + 1N aqueous NaoH	Yellowish Green	Greenish Yellow	Dark Blue
3.	Powder + 1N alcoholic NaoH	Yellowish Green	Dark Green	Dark Brown
4.	Powder + 1N Hcl	Green	Light Green	Dark Brown
5.	Powder + con.Hcl	Green	Light Green	Brown
6.	Powder + con. H_2So_4	Light Brown	Fluorescent green	Dark Brown
7.	Powder +50 per cent H_2So_4	Brown	Fluorescent green	Brown
8.	Powder + con. HNo_3	Light Green	Light Green	Brown
9.	Powder + 40 per cent NaoH+ 10 per cent lead acetate	Pale Green	Fluorescent green	Dark Green
10.	Powder + Acetic acid	Yellowish Green	Light Green	Dark Green
11.	Powder + Ferric chloride	Green	Dark Green	Dark Green
12.	Powder + Chloroform	Pale Green	Dark Green	Dark Brown
13.	Powder + Benzene	Yellowish Green	Fluorescent green	Brown
14.	Powder + Petroleum ether	Yellowish Green	Fluorescent green	Bluish Green
15.	Powder + Methanol	Dark Green	Yellowish Green	Dark blue
16.	Powder + Ethanol	Yellowish Green	Yellowish Green	Blue
17.	Powder + Acetone	Dark Green	Fluorescent green	Brown
18.	Powder + NH_3	Yellowish Green	Fluorescent green	Dark Brown
19.	Powder + HNo_3 + NH_3	Yellowish Green	Yellowish Green	Dark Brown
20.	Powder + 50 per cent HNo_3	Dark Green	Fluorescent green	Dark Green

Preliminary Phytochemical Analysis

The results of preliminary phytochemical screening of stem and leaf of *Barleria courtallica* were presented in Table 23.4. The methanol and ethanol extracts of stem and leaf shows the presence of alkaloids, terpenoids, catechin, coumarin, tannin, saponin, flavonoids, quinines, anthraquinones, phenols, carbohydrate and glycosides.

Discussion

The standardization of crude drugs has become very important for identification and authentication of a drug. Owing to certain problems the importance was not up to the mark. Thus, the lack of standardization techniques fails to identify the drug

Table 23.4: Preliminary Phytochemical Screening

Bioactive Components	*Petroleum Ether*		*Benzene*		*Ethyl Acetate*		*Methanol*		*Ethanol*	
	Stem	*Leaf*	*Stem*	*Leaf*	*Stem*	*Leaf*	*Stem*	*Leaf*	*Stem*	*Leaf*
Alkaloid	-	-	-	-	+	-	+	+	+	+
Anthraquinone	+	-	-	+	+	+	+	+	+	+
Catechin	-	+	+	-	+	-	+	+	+	+
Coumarin	+	-	-	+	-	+	+	+	+	+
Flavonoid	-	-	+	+	+	+	+	+	+	+
Phenol	+	-	+	+	+	+	+	+	+	+
Quinine	-	+	-	+	-	+	+	+	+	+
Saponin	-	-	-	-	-	+	+	+	+	+
Steroid	-	-	-	-	+	-	+	+	+	+
Tannin	-	-	-	+	+	-	+	+	+	+
Terpenoid	-	+	+	-	-	+	+	+	+	+
Sugar	-	-	-	-	-	-	+	+	+	+
Carbohydrate	-	-	-	-	+	+	+	+	+	+
Glycoside	+	+	-	+	+	+	+	+	+	+
Xanthoprotein	-	-	-	-	+	+	+	+	+	+
Fixed oil	-	-	-	-	-	-	+	+	-	-

+ Present - Absent.

from its originality which thereby exploits the usage of drug from its traditional system of medicine (Charkaborthy and Ghorpade, 2009). A large variety of medicinal plants are growing in India but the trade of crude drugs always remained in the hands of unqualified and unskilled persons and causes the collection of the premature or incorrect drug and thus often lead to adulteration or substitution. Thus pharmacognosy appears to be of great value in identification of commercial samples of the market to find their authenticity and establishing identity of adulterant or substituent.

It is globally accepted that herbal based drugs have many advantages over the synthetic drugs. However, one of the major problems in utilization of phytodrugs is correct diagnosis of the medicinal plants that are used either in the traditional systems or modern systems of preparation of the drugs. It is regrettable to note that most of the people involved in the manufacture or preparation of herbal drugs lack the basic background of botanical knowledge of the drugs. Consequently adulteration or substitutions of plants in the place of original ones permeate the pharmaceutical industries, rendering the herbal drugs undependable and invalid. This will lead unpopularity of phytodrugs among the people. So, it is most essential that a medicinal plant, when found to be of high pharmacological potentials, should be subjected to thorough botanical standardization so that there will be no ambiguity with respect to botanical identify of the plants. Identifications of plants involve the study of the external features of vegetative and floral parts. This study must be complimented with anatomical parameters which are very often useful to identify the fragmentary plant specimens. Raw drugs pose problem of identifications and to establish their genuineness when they lack any external diagnostic features or organoleptic clues. During such situation, the microscopic analyses of the specimen will offer a helping hand to establish to identify the phytodrugs. The anatomical features of the plant that are reliable for diagnosis and that are least changed due to environmental stresses include:

- ☆ Structure of midrib.
- ☆ Structure of lamina and its epidermal outgrowth.
- ☆ Surface view of foliar epidermis- epidermal cells and stomatal morphology.
- ☆ Petiolar anatomy.
- ☆ Venation pattern of the lamina
- ☆ Gross anatomy of the stem and root
- ☆ Ergastic cell inclusion such as; starch, crystal, tannin, mucilage *etc.*

Literature dealing with the anatomy of *Barleria courtallica* is lacking. The present study may be claimed as the first comprehensive investigation of the stem and leaf of *Barleria courtallica.* The present investigation has laid down a set of anatomical features of stem and leaf, which can be employed for botanical diagnosis. The following are the salient features of identification of stem and leaf of *Barleria courtallica.*

Salient Features of *Barleria courtallica*

- ☆ Midrib of the leaf is semicircular on the abaxial side and wide and flat on the adaxial side.
- ☆ Vascular system of the midrib includes a wide and deep main vascular strand and pairs circular vascular bundles on the wings.
- ☆ Mucilagenous cells are common on the ground parenchyma.
- ☆ Lamina shows bifacial differentiation with abaxial stoma.
- ☆ Short stalked, multicellular secretory head bearing glandular trichomes are sparsely seen on the lamina.
- ☆ Long cylindrical double cystoliths ot solitary cystoliths seen within specialized lithocyst cavities.
- ☆ Epidermal cells of the lamina have highly wavy anticlinal walls.
- ☆ Adaxial epidermis is apostomatic; abaxial epidermis have diacyclic stomata; the guard cells have prominent polarnodules.
- ☆ Venation of the leaf is not dense; vein islets are wide with simple or forked vein teerminations.
- ☆ Petiole is semicircular with adaxial flat surface with short thick wings on either side.

- ✰ Main vascular strand of the petiole is deeply cup shaped with small adaxial opening.
- ✰ Small circular in sectional view with wide cortical and pith zones and hollow vascular cylinder.
- ✰ Xylem cylinder includes mostly long radial multiples and cells with angular thick walled liquefied walls.
- ✰ The pith tissue contains small cluster of brachysclereids and cells possessing starch grains.
- ✰ Powder preparation of the plant shows fibres, fibre tracheids, thick walled polygonal epidermal cells, wide rectangular parenchyma cells with prominent chloroplast, large prominent brachysclereids and long, narrow vessel elements with simple and wall perforations.

Physico-chemical Parameters

The physical constant evaluation of the drug is an important parameter in detecting adulteration or improper handling of drugs by African Pharmacopoedia, (1986). Equally important in the evaluation of crude drugs, is the ash value and acid insoluble ash value determination. The total ash is particularly important in the evaluation of purity of drugs, *i.e.*, the presence or absence of foreign organic matter such as metallic salts and/or silica (Musa *et al.*, 2006). The ash value of *B. courtallica* stem and leaf is 9.8 per cent and 8.8 per cent respectively. Ash values are used to determine quality and purity of crude drug. It indicates the presence of various impurities like carbonate, oxalate and silicate. The water soluble ash is used to estimate the amount of inorganic compound present in drugs. The acid insoluble ash consist mainly silica and indicate contamination with earthly material. Moisture content of drugs should be at minimal level to discourage the growth of bacteria, yeast or fungi during storage. Estimation of extractive values determines the amount of the active constituents in a given amount of plant material when extracted with a particular solvent. The extractions of any crude drug with a particular solvent yield a solution containing different phytoconstituents. The compositions of these phytoconstituents depend upon the nature of the drug and the solvent used. It also gives an indication whether the crude drug is exhausted or not (Tatiya *et al.*, 2012).

Fluorescence Analysis

Some constituents show fluorescence in the visible range in daylight. The ultra violet light produces fluorescence in many natural products which do not visibly fluoresce in daylight. If substance themselves are not fluorescent, they may often be converted into fluorescent derivatives or decomposition products by applying different reagents. Hence crude drugs are often assessed qualitatively in this way and it is an important parameter for pharmacognostic evaluation of crude drugs (Zhao, 2011).

In the present study, the stem powder of *B.courtallica* florescent green under day light, dark green under short UV-light (254 nm) and black under long UV-light

(365 nm). Similarly the leaf powder of *B. courtallica* florescent green under day light and short UV-light (254 nm) and darkgreen under long UV-light (365 nm).

Preliminary Phytochemical Analysis

Presence or absence of certain important compounds in an extract is determined by colour reactions of the compounds with specific chemicals which act as dyes. This procedure is a simple preliminary prerequisite before going for detailed phytochemical investigation. Various tests have been conducted qualitatively to find out the presence or absence of bioactive compounds. Different chemical compounds such as alkaloids, terpenoids, catechin, coumarin, tannin, saponin, flavonoids, quinines, anthraquinones, phenols, carbohydrate and glycosides are detected in *B.courtallica* methanol and ethanol extracts of stem and leaf which could made the plant useful for treating different ailments as having a potential of providing useful drugs of human use.

Thus the process of standardization can be achieved by stepwise pharmacognostic studies as stated above. These studies help in identification and authentication of the plant material. Such information can act as reference information for correct identification of particular plant and also will be useful in making a monograph of the plant. Further, it will act as a tool to detect adulterants and substituent and will help in maintaining the quality, reproducibility and efficacy of natural drugs.

References

African Phaemacopoeia, General methods for analysis 1st ed.2: (OAU/STRC) Lagos. (1986) 123.

Agrawal B, Das S and Pandey A 2011. Boerhaavia Diffusa Linn.: A review On Its Phytochemical and Pharmacological profile. *Asian Journal of Applied Science,* 4: 663-684.

Amoo SO, Finnie JF and Van Staden J 2009. In vitro pharmacological evaluation of three *Barleria* species. Journal of Ethnopharmacology 121: 274-277.

Anonymous 1996. Indian Pharmacopoeia, Government of India, Ministry of Health and Family Welfare, the Controller of Publications, Civil Lines, Delhi – 110 054. Vol.I and II.

Brindha P., Sasikala P and Purushothaman KK. 1981. Pharmacognostic studies on merugan kizhangu. *Bulletin of Medico Ethnobotanical Research.* 3: 84-96.

Charkarborthy Guno Sindhu and Ghorpade Prashant M, 2009. Pharmacognostical and Phytochemical Evaluation of Stem of *Abutilon indicum* (Linn)," *International Journal of Pharmaceutical and Clinical Research,* 1(3):188-190.

Dineshkumar C. Pharmacognosy can help minimize accidental misuse of herbal medicine. Curr Sci 2007; 3:1356-1358.

Gantait A, Maji A, Barman T, Banerji P, Venkatesh P and Mukherjee PK 2011. Estimation of Capsaicin Through Scanning Densitometry and Evaluation of Differnet Varities of Capsicum in India. *Natural Product Research* 26: 216-222.

Johanson DA. 1940. Plant Microtechnique. Mc Graw Hill Book Co, New York, 523.

Lala PK. 1993. Lab Manuals of Pharmacognosy. Edn 5, CSI Publishers and Distributors, Calcutta.

Mohammad Saleem TS, Christina AJ, Chidambaranathan N, Ravi V, Gauthaman K. Hepatoprotective activity of *Annona squamosa* Linn. on experimental animal model. *Int J Appl Res Nat Prod* 2008; 1: 1-7.

Mukherjee PK, Mukherjee A, Maji K, Rai S and Heinrich 2009. The Sacred Lotus (*Nelumbo nucifera*)- Phytochemical And Therapeutic Profile. *Journal of Pharmacy and Pharmacology,* 61: 407-422.

Musa KY, Katsayal AU,Ahmed A, Mohammed Z, UH Danmalam, pharmacognostic investigation of the leaves of *Gisekia pharmacioides*. *African Journal of Biotechnology,* 5: 956-957 (2006).

O'Brien TP., Feder N and Mc Cull ME. 1964. Polychromatic staining of plant cell walls by toluidine blue-O. *Protoplasma.* 59: 364-373.

Saadabi AMA, Sehemi AG and AL-Zailaie KA, 2006. In Vitro Antimicrobial activity of Some Saudi Arabian Plants used in Folkoric Medicine. *International Journal of Botany,* 2: 201-204.

Singh B, Chandan BK, Prabhakar A, Taneja SC, Singh J and Qazi GN 2005. Chemistry and hepatoprotective activity of an active fraction from *Barleria prionitis* Linn. In experimental animals. *Phytotherapy Research,* 19(5): 391–404.

Tatiya A, Surana S, Bhavsar S, Patil D, Patil Y. Pharmacognostic and preliminary phytochemical investigation of *Eulophia herbacea* Lindl. Tubers (Orchidaceae). *Asian Pac J Trop Disease,* 2012; 2(Suppl 1):S50-55.

Thomas S *et al.,* Pharmacognostic evaluation and physico-chemical Analysis of *Averrhoa carambola* L. fruit. *J Herbal Med Toxicol,* 2008; 2: 51-4.

Zhao Z, Liang Z, Guo P. Macroscopic identification of Chinese medicinal materials: Traditional experiences and modern understanding. *J Ethnopharmacol,* 2011; 131:556-561.

2018, Ethnomedicinal Plants: A Biodiversity Treasure Pages 569–587
Editors: V.R. Mohan, A. Doss, P.S. Tresina and V. Sornalakshmi
Published by: ASTRAL INTERNATIONAL PVT. LTD., NEW DELHI

Chapter 24

Pharmacognostic, Physico-chemical Standardization and Phytochemical Analysis of Stem Bark of *Ailanthus excelsa* Roxb.

V. Sornalakshmi[1], P.S. Tresina[2], K. Paulpriya[2] and V.R. Mohan[2]

[1]Department of Botany, A.P.C. Mahalaxmi College for Women, Tuticorin – 628 002, Tamil Nadu
[2]Ethnopharmacology Unit, Research Department of Botany, V.O. Chidambaram College, Tuticorin – 628 008, Tamil Nadu

Introduction

Plants play a major role in providing basic needs like food, timber and medicine. These plants play important roles in curing disease for long time (Aseefa *et al.*, 2010). Many medicinal plants are used by men against diseases and to improve health. Secondary metabolites produced from plants mostly include alkaloids, phenolic compounds, tannins, phytosterols and terpenoids have been exploited by men for their useful roles in various ways (Balandrin *et al.*, 1985). These compounds have been extracted with various solvents by different screening techniques (Neube *et al.*, 2008). It is believed that modern medicines have actually emerged from traditional medicines. Indeed, most of the medicines that are used to treat bacterial and other infections are isolated from plants and other natural resources (Sarwat *et al.*, 2012). Most plants found in nature have received scientific and commercial attention.

Nowadays, because of the development of modern and new sophisticated methods, scientists are taking more attention to explore new drugs from natural and biological active compounds of the plants, which may serve as an endless resource for pharmaceutical industries. It is therefore necessary to establish a worldwide documentated guiding principle for the evaluation and standardization of the quality of plants (Rajesh *et al.*, 2010; Lalitharani *et al.*, 2013). The procedure of standardization of plants can be achieved by stepwise pharmacognostic studies. Simple pharmacognostic techniques used in standardization of plant material include its morphological, anatomical, physic-chemical and biochemical characteristics (Anonymous, 1998). According to the World Health Organisation, the macroscopic and microscopic description of a medicinal plant is the first step towards establishing the identity and the degree of purity of such materials and should be carried out before any tests are under taken (WHO, 1998). Accurate identification of the plant and quality assurance is an integral part to ensure reproductive quality of crude drug before including in the pharmacopoeia. Diagnostic characters and quality standard for various plants species have already been investigated by different workers (Mohan *et al.*, 2010; Sarada *et al.*, 2014).

Ailanthus excelsa Roxb. (Simaroubiaceae) commonly known as Mahanimb. *A. excelsa* is a large tree originally from China, is known as the 'tree of heaven'. Different parts of this plant are used widely in traditional medicine for a variety of diseases. The bark is used as bitter, refrigerant, astringent, appetizer, anthelmintic, febrifuge, to treat dysentery, ear ache, skin diseases, troubles of the rectum, and fever due to tridosha and allay thirst. It is also used in gout, rheumatism, dyspepsia, bronchitis and asthma. In Ayurveda it is used to remove the bad taste of mouth (Nadkarni, 2002 and Kirtikar Basu, 2003). *Ailanthus* is used to cure wounds and skin eruptions as mentioned in traditional medicine (Asolkar *et al.*, 1992). The root bark possess cytotoxic and antitumor activity both in mice and in cell cultures (Ogura *et al.*, 1977). Stem bark extracts showed potent antibacterial and antifungal activities (Shrimali *et al.*, 2001). The alcohol extract from leaf and stem bark exhibits remarkably high anti-inflammation and early abortifacient activity (Joshi *et al.*, 2003). A recent study reveals that ethanol extracts of *A. excelsa* leaves have a significant hepatoprotective effect on experimental liver damage in rats (Hukkeri *et al.*, 2002) and antidiabetic activity (Cabrera *et al.*, 2008). The plant is reported to contain flavonoids, quassinoides, alkaloids, terpenoids, sterols and saponins (Mehta and Patel, 1959; Rahman *et al.*, 1997; Kapoor *et al.*, 1971). Based on ethnobotanical practice *A. excelsa* is a rich source of different chemical compounds with a variety of potential biological activities. The vast ethnomedical uses inspired us to investigate the pharmacognostic evaluation of stem bark of *A. excelsa*.

Methodology

The plant specimens for the proposed study were collected from a road side near Chennai. The plant samples were identified with the help of local flora and authenticated by Botanical Survey of India, Southern Circle, Coimbatore, Tamil Nadu, India. A voucher specimen of collected plants was deposited in the Ethnopharmacological Unit, PG and Research Department of Botany, V.O. Chidambaram College, Thoothukudi District, Tamil Nadu.

Macroscopical Studies

The macroscopic characters like surface, shape, size, venation, phyllotaxy, length of the petiole, length of the leaf *etc.* were noted.

Anatomical Studies

For anatomical studies, the required samples of stem bark were cut and removed from the plant and immediately fixed in FAA (formalin- 5 ml + acetic acid- 5 ml + 70 per cent Ethyl alcohol- 90 ml). The specimens were left in the preservative for two days; then the materials were washed in water and processed further. Standard microtome techniques were followed for anatomical investigation (Johanson, 1940). Transverse sections of the materials were made. The microtome sections were stained with 0.25 per cent aqueous Toluidine blue (Metachromatic stain) adjusted to pH 4.7 (O'Brien, 1964). Photomicrographs were taken with NIKON trinocular photo micrographic unit.

Physico-chemical and Fluorescence Analysis

These studies were carried out as per the standard procedures (Lala, 1993). In the present study, the powered stem bark was treated with various chemical reagents like aqueous 1N sodium hydroxide, alcoholic 1N sodium hydroxide, 1N hydrochloric acid, 50 per cent sulphuric acid, concentrated nitric acid, picric acid, acetic acid, ferric chloride and concentrated HNO_3+NH^3 These extracts were subjected to fluorescence analysis in day light and UV light (254nm and366nm). Various ash types and extractive values were determined by following standard methods (Anonymous 1996).

Preliminary Phytochemical Analysis

Shade dried and powdered stem bark samples were successively extracted with Petroleum ether, benzene, ethyl acetate, methanol and ethanol. The extracts were filtered and concentrated using vacuum distillation. The different extracts were subjected to qualitative tests for the identification of various phytochemical constituents as per the standard procedure (Lala, 1993; Brindha *et al.*, 1981).

3. Results

Ailanthus excelsa Roxb

Family: Simaroubaceae

Regional Names

Hindi and Marathi: Maharukha

Sanskrit: Mahanimba

Tamil: Pee Maram; Peruppi

Telugu: Peedamanu

Malayalam: Mattipongiliyam

Occurrence

The tree occurs in plains and often grown as avenue tree or for its soft wood.

Distribution

India, Srilanka.

The genus *Ailanthus* of Simaroubaceae (= Ailanthaceae) includes 10 species which are distributed mostly in Asia and Australia. In India, only two species are available, which are characterised as below;

1. *Ailanthus excelsa* Roxb: leaflets coarsely toothed tomentose when young, samara narrow and twisted.
2. *Ailanthus triphysa* (Denst.) Altson (A. malabarica D.C): Leaflets entire, glabrous, samara broad and not twisted.

A. excelsa is a beautiful tree with softwood used for catamarans and sword-sheath.

External Features

The plant is a moderate sized tree growing up to 20-25m high. The old branches have transverse leaf-scars. The leaves are odd-pinnate, the leaflets are 8-14 pairs, sub-opposite, oblong lanceolate, unequal at the base, margins irregularly toothed. The rachis is 10cm long and the petiole is 2.5cm long. The inflorescence is an axillary panicle. The flowers are polygamous; calyx is 5 lobed; corolla 5 free, imbricate. Stamens 10, free, filaments short and anthers oblong. Ovary is 5-lobed with one ovule per cell, style 5, connate at the base. Fruit is a samara with single seed in the middle and membranous wings.

Anatomical Features - Surface Features

The bark surface is greyish – brown; there are irregular, narrow vertically oriented, shallow fissures of varying size and shape. The bark scales off in the form of fine powder (Plate 24.1a).

Physical Features

The bark is fairly hard, fibrous in the inner part and granular in the outer part. The outer and inner barks do not separate early.

Organoleptic Characters

The fresh bark emits an agreeable odour, the taste is slightly bitter.

Microscopic Characters

The trunk bark is 1.6mm thickness. The bark can be differentiated into outer bark or the periderm and inner bark or secondary phloem (Plates 24.1b,c). The outer bark is wavy in outline with shallow irregular fissures. It consists of a single broad zone of periderm at certain regions; in other regions an inner zone of periderm is formed in the form of a shallow cup which enclosed a mass of phloem tissue including sclerenchyma; they included non-peridermous tissue occurs inbetween and arch of inner periderm and wavy outer periderm. This portion of the bark is called "Shell-Bark". The shell-bark occurs at several places around the trunk of the tree. They include sclerenchyma cells as sclerids or fibres (Plate 24.1d).

***Ailanthus excelsa* Roxb.**

a. Surface appearance of the trunk bark

b. Periderm forming shell bark

c. Outer part of the collapsed phloem

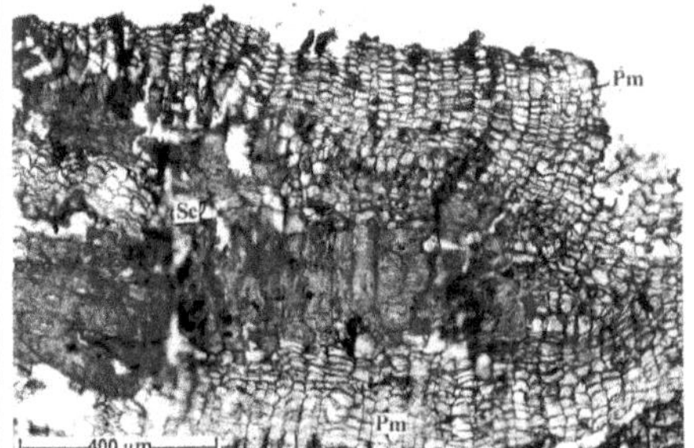

d. Shell bark enlarged showing two phellem zones with included sclerenchyma

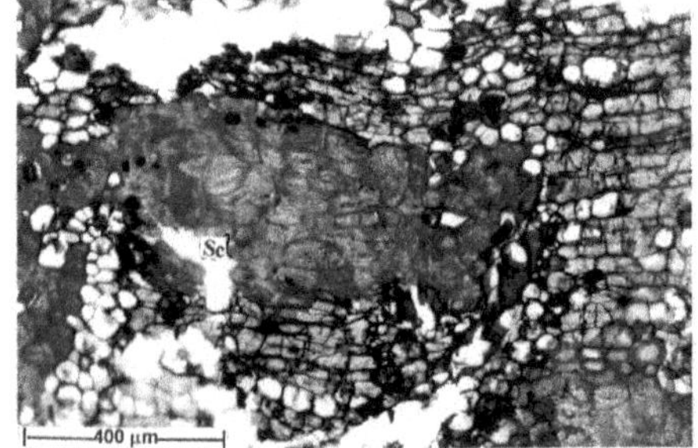

e. Sclereides mass in the collapsed phloem

Fi - Fissure; DR - Dilated rays; Pd - Phelloderm; Pe - Periderm;
SC - Sclerenchyma; PM - Phellem

Plate 24.1

The periderm has uniformly thick zone of phellem and narrow zone of phelloderm (Plate 24.1b). The phellem cells are thick walled, suberised and tabular in shape (Plate 24.1d). The shell bark region is nearly 1mm thick, the outer narrow region of periderm is 600 μm thick. The phelloderm is narrow measuring about 20 μm in width. The cells are squarish to rectangular and thin walled.

Inner to the periderm is a broad region of the collapsed phloem. This region consists of compact, tangenetially aligned, thin walled parenchymatous tissue; this tissue is formed by phloem rays which undergo extensive dilation growth as they reach outer part of the bark. The collapsed phloem also consists of large, tangential irregular masses of phloem sclerenchyma cells. The sclerenchyma may be only of brachysclereid (stone cells) type or it may be mixture of fibrous and sclereids (Plates 24.1d,e). The collapsed phloem is characteristic of crushed and obliterated sieve elements which can be seen as thin, black tangential lines occurring in parallel bands (Plate 24.2a). The sclerenchyma component of the collapsed phloem is the libriform fibres which occur in irregular masses of tangential hands (Plate 24.2b). The phloem rays starts dilating and as they reach the outer bark, they dilate extensively and loose their identity. The phloem parenchyma (axial parenchyma) has undergone hyperplasy and dilation and appears conspicuous under the microscope (Plate 24.2b).

The collapsed phloem gradually transits into a narrow zone of non collapsed phloem (Plate 24.2c). The non collapsed phloem occurs immediately next to the cambial zone (Plate 24.2d). The non collapsed phloem lacks the sclerenchyma elements of any type. It consists of radial parallel lines of sieve elements, phloem parenchyma and broad (undiluted) rays (Plate 24.2d). The sieve tube members are polyhedral in transectional view and occur in groups of four or in radial lines (Plate 24.2e). The axial parenchyma cells are squarish, thick walled and occur intermixed with the sieve tube members. The sieve plate is compound and oblique (Plate 24.2f). The sieve pores are wide and open. The phloem rays consist of radially elongated parenchyma cells.

In tangential and radial longitudinal sections of the bark, the structure and organisation of the phloem rays are evident. The rays are mostly multiseriate, occasionally uniseriate (Plate 24.3c); they are non steroid, broad and high. The individual rays are homocellular; the cell components are angular, thin walled and compact (Plate 24.3a). The sieve plate as seen in TLS view is oblique with simple plate (Plate 24.3b). The lateral wall and the sieve area show wide area and free sieve pores. The axial parenchyma are in fusiform strands with vertically oblong cells (Plates 24.3a-d).

In RLS view, the phloem rays show radial files of oblong or squarish cells. The cells are similar in height and length and the rays are homocellular (Plates 24.3e, f). The multiseriate rays range from 500μm to 1mm in height and 75-180μm in breadth; the uniseriate rays are 180-200 μm in height and 40 μm in breadth.

Powder Microscopy

Coarse powder of the bark shows two major cell types; libriform fibres and sclereids (Plate 24.4a, b).

Ailanthus excelsa **Roxb.**

a. Surface appearance of the trunk bark

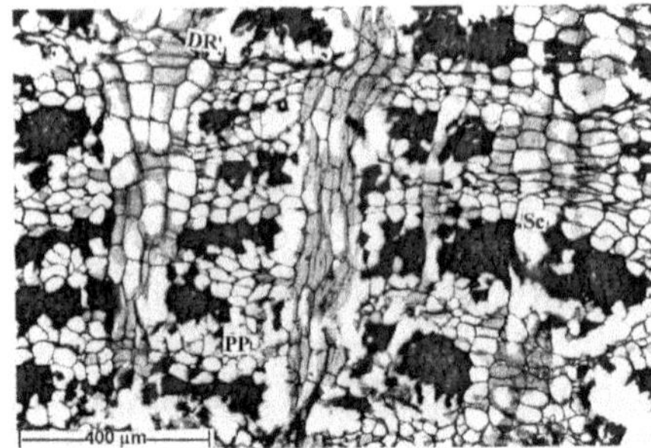

b. Surface appearance of the trunk bark

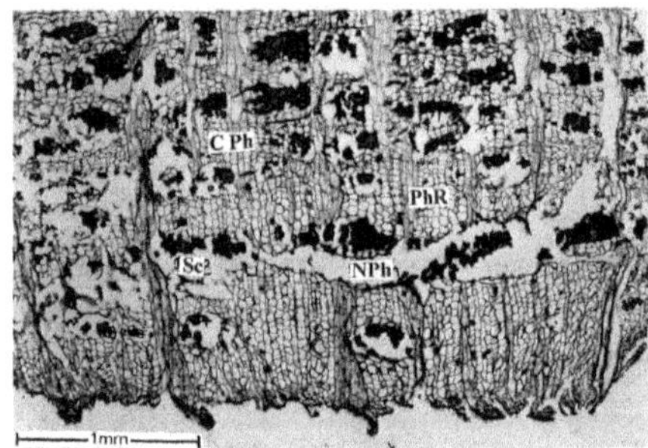

c. T.S. of inner bark showing non-collapsed phloem and collapsed phloem

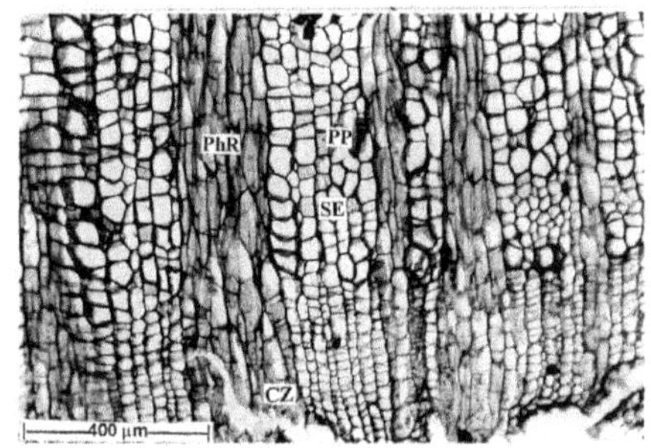

d. Non-collapsed phloem and cambial zone

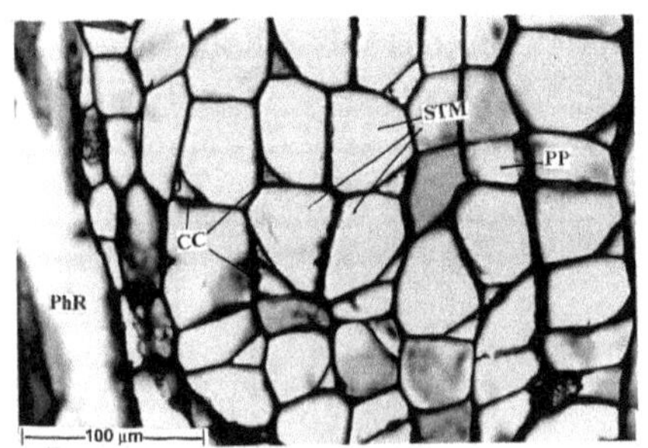

e. Sievetube members and companian cell

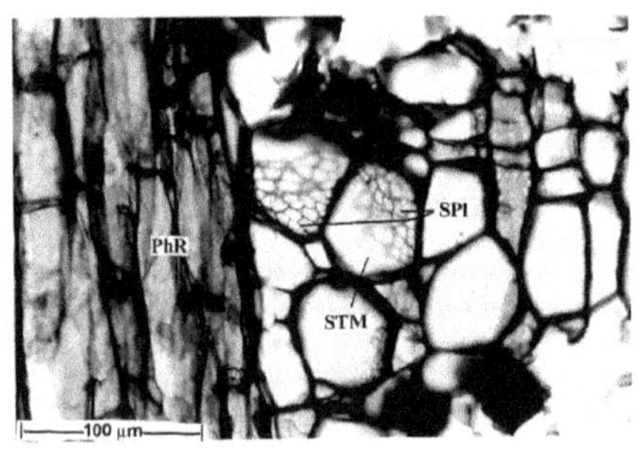

f. Sievetube members with sieve plate

Cph - Collapsed phloem; DR - Dilated rays; CZ - Cambial zone;
NPh - Non-collapsed phloem; SC - Sclerenchyma; SE - Sieve elements;
CC - Companian cell; PhR - Phloem ray; PP - Phloem parenchyma;
SP - Sieve plate; STM - Sievetube members

Plate 24.2

***Ailanthus excelsa* Roxb.**

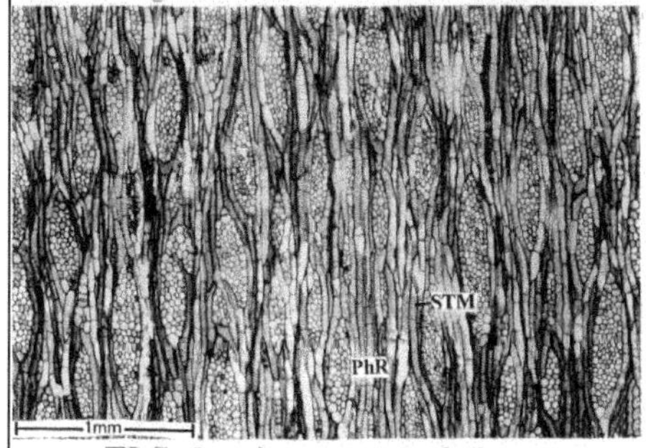

a. TLS showing non-storied rays

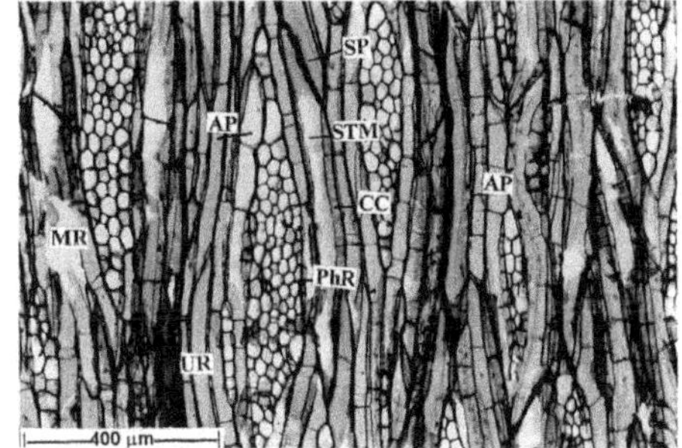

b. TLS showing different elements of phloem

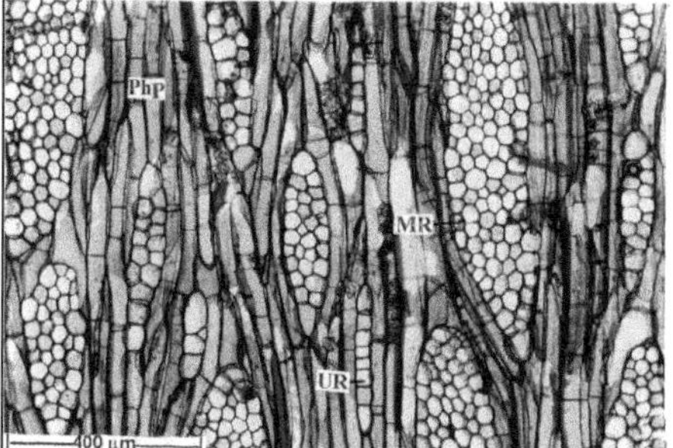

c. Multiseriate and uniseriate rays

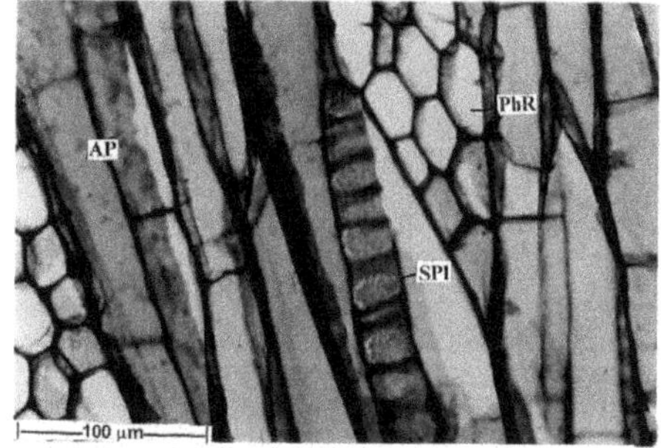

d. Axial parenchyma and sieve plate

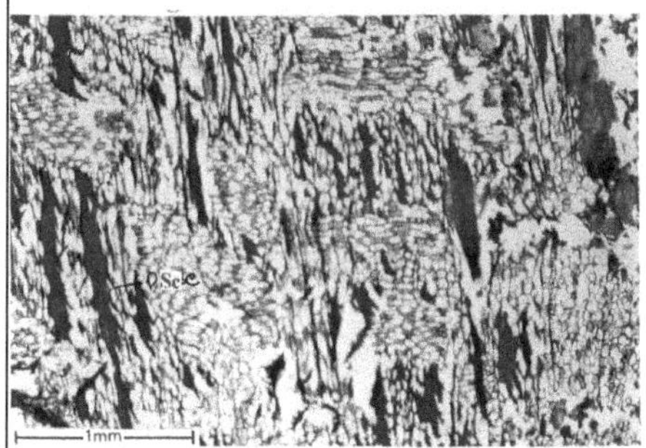

e. RLS under low magnification

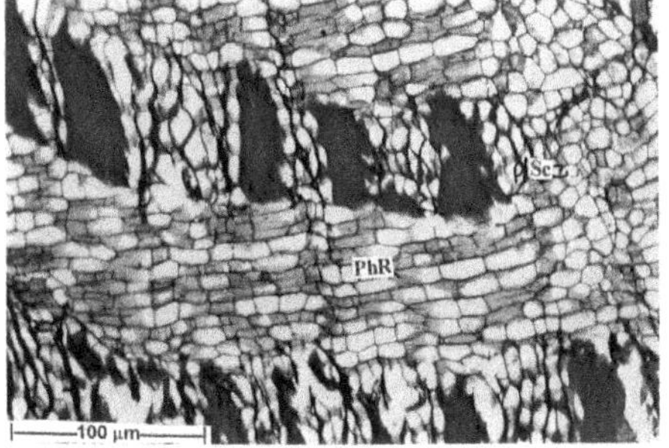

f. RLS under high magnification

PP - Phloem parenchyma;
CC - Companian cell; SP - Sieve plate; STM - Sievetube members
AP - Axial parenchyma; MR - Multiseriate rays; PhP - Phloem parenchyma;
SPl - Sieve plate; UR - Uniseriate ray; PhR - Phloem ray; SC - Sclerenchyma

Plate 24.3

***Ailanthus excelsa* Roxb.**

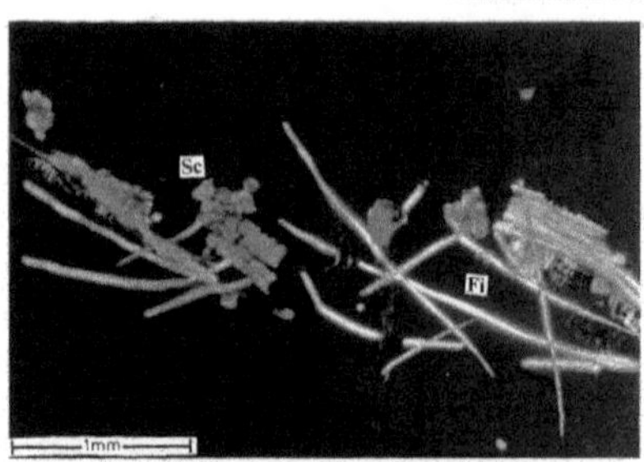

a. Fibres and sclereides

b. Fibres and sclereides

c. Starch grains and crystals in the delated parenchyma cells

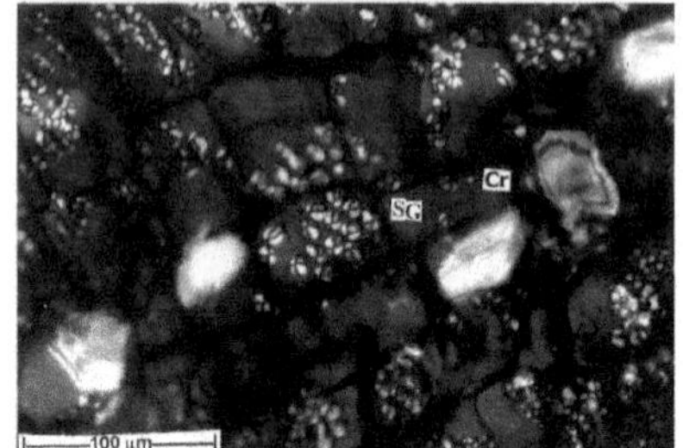

d. Starch grains and prismatic crystal enlarged

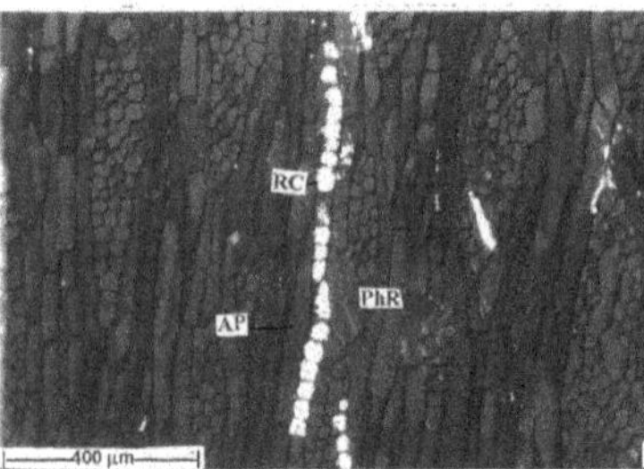

e. A vertical row of rosette crystals in the axial parenchyma

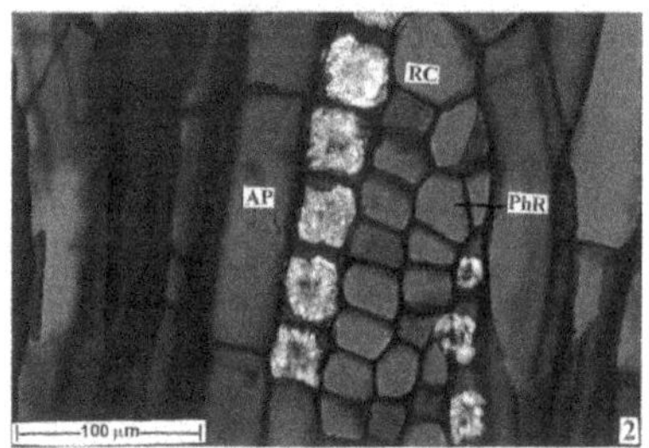

f. Rosette crystal enlarged

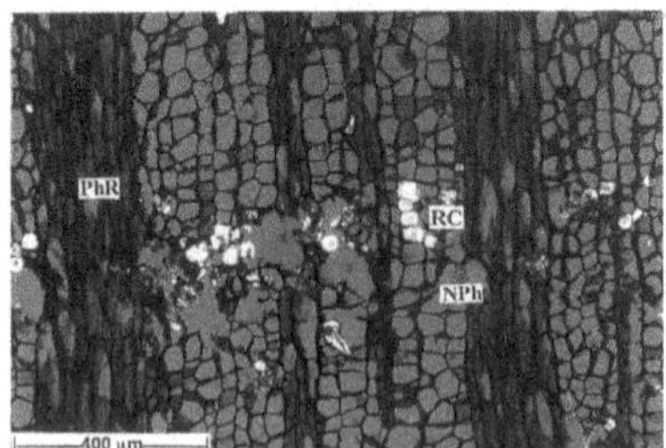

g. T.S. at low magnification

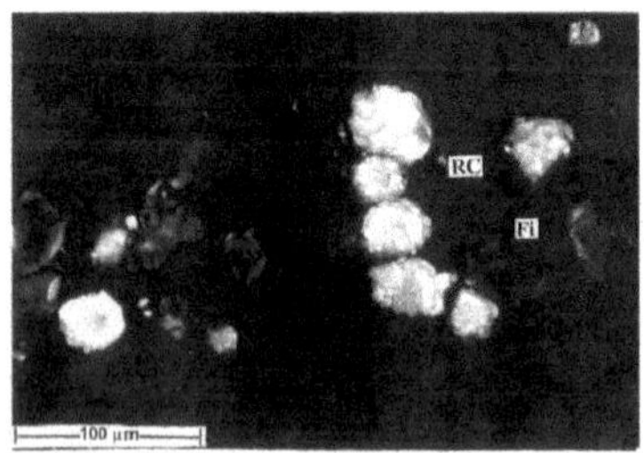

h. Rosette crystal enlarged (In T.S. view)

Fi - Fibre; Sc - Sclereids; Cr - Crystal; SG - Starch grains; SC - Sclerenchyma; AP - Axial parenchyma; PhR - Phloem ray; RC - Rosette crystal; Nph - Non-collapsed phloem

Plate 24.4

Libriform Fibres

These are major components; the fibres are thick walled, with narrow lumen; no pits are evident. The walls are lignified.

Sclereids

The sclereids are of brachysclereid type; they are squarish or oblong; the walls are thick and lignified and have narrow canal like pits. The fibres are up to 2 mm long, 25 μm broad.

Cell Inclusions

The bark contains abundant starch and calcium oxalate crystal. These cell inclusions are clearly distinguishable under the polarized light.

1. **Starch grains** - The starch grains are heavily loaded in the ray and axial parenchyma cells of the outer collapsed phloem zone (Plates 24.4e, d). The starch grains are spherical and have central hilum appearing as concentric type (Plate IV d).
2. **Crystals** – two types of crystals are seen in the bark
   (a) **Prismatic crystals** – These are seen mostly in the outer zone of the phloem parenchyma cells along with the starch grains. The crystals may be cubical, rectangular or pyramidal (Plate 24.4d).
   (b) **Roestte crystals** – These crystals occur in the axial parenchyma cells of the outer zone of non collapsed phloem. They occur mostly in vertical row (Plates 24.4e, f, g, h). The rosette crystal has a central dark part and outer radiating conical crystals; the crystal is up to 150μm in diameter.

Table 24.1: Ash and Extractive Values of Powdered Stem Bark of *Ailanthus excelsa*

	Ash values	
Sl.No.	*Type of Ash*	*Per cent of ash*
1	Total ash value of powder	9.50±0.07
2	Water soluble ash	2.85±0.03
3	Acid soluble ash	1.51±0.01
4	Sulphated ash	10.36±0.06
	Extractive Values	
Sl.No.	*Name of the Extract*	*Extractive Value (per cent)*
1	Petroleum ether	7.81±0.03
2	Benzene	6.38±0.02
3	Chloroform	5.68±0.02
4	Acetone	6.67±0.04
5	Methanol	7.98±0.03
6	Ethanol	8.36±0.05
7	Water	12.14±0.11

a: All values are means of triplicate determinations ± denotes standard error

Physico-chemical Constants Ash and Extractive Values

The results of ash and extractive values of stem bark of *A. excelsa* are depicted in Table 24.1. The total ash content of the powdered stem bark of *A. excelsa* is 9.50 per cent. The present study revealed that the extractive values of the water extract are more than the other solvent extracts studied.

Fluorescence Analysis

The results of fluorescence analysis of the experimental plant of *A. excelsa* stem bark is shown in Table 24.2. The stem bark powder of *A. excelsa* as such fluoresced brown under day and short UV light (254 nm) and dark blue under long UV light (365 nm). The powdered stem bark of *A. excelsa* emitted the characteristic fluorescent green colour when treated with 1N aqueous NaOH, 50 per cent H_2SO_4, ferric chloride, NH_3 + NH_3, HNO_3, benzene, acetone and chloroform.

Table 24.2: Fluorescent Analysis of Powdered Stem Bark of *A. excelsa*

Sl.No.	*Experiments*	*Visible/Day Light*	*UV-light*	
			254nm (Short wave length)	*365 nm (Long wave length)*
1	Powder	Brown	Brown	Dark blue
2	Powder + 1NNaOH (Aqueous)	Yellowish green	Fluorescent green	Dark brown
3	Powder + 1NNaOH (Alcohol)	Yellowish green	Green	Dark brown
4	Powder + 1N NH_4Cl	Yellow	Green	Dark violet
5	Powder + Conc. H_2SO_4,	Dark brown	Dark green	Dark brown
6	Powder + 50 per cent H_2SO_4,	Yellowish green	Fluorescent green	Dark brown
7	Powder + Conc.HNO_3	Dark brown	Greenish yellow	Dark brown
8	Powder + Conc.HCL	Dark brown	Greenish yellow	Dark violet
9	Powder + 50 per cent HNO_3	Dark brown	Yellowish green	Dark blue
10	Powder + 40 per cent NaOH+10 per cent Lead acetate	Yellow	Yellowish green	Dark blue
11	Powder + Acetic acid	Yellowish green	Yellow	Black
12	Powder + Ferric chloride	Dark green	Fluorescent green	Dark blue
13	Powder + HNO_3+ NH_3	Pale brown	Fluorescent green	Dark brown
14	Powder + HNO_3	Yellow	Fluorescent green	Fluorescent green
15	Powder + Benzene	Yellow	Fluorescent green	Fluorescent green
16	Powder + Petroleum ether	Yellow	Yellowish green	Fluorescent green
17	Powder + Acetone	Yellowish green	Fluorescent green	Dark brown
18	Powder + Chloroform	Yellowish green	Fluorescent green	Fluorescent green
19	Powder + Methanol	Yellowish green	Yellowish green	Fluorescent green
20	Powder + Ethanol	Yellowish green	Green	Fluorescent green

Preliminary Phytochemical Screening

The distribution of different phytochemical constituents in petroleum ether, benzene, ethyl acetate, methanol and ethanol extracts of stem barkpowder of *A. excelsa* was evaluated qualitatively and the results are presented in Table 24.3. The presence of phytocompounds such as alkaloids, anthraquinone, catechin, coumarin, flavonoid, phenol, quinone, saponin, steroid, tannin, terpenoid, sugar, glycoside, xanthoprotein and fixed oil have been confirmed in the methanol and ethanol extracts of the selected plant.

Table 24.3: Preliminary Phytochemical Screening of Stem Bark of *A. excelsa*

Test	*Petroleum Ether*	*Benzene*	*Ethyl Acetate*	*Methanol*	*Ethanol*
Alkaloid	+	+	+	+	+
Anthroqui-none	+	+	+	+	+
Catechin	–	–	–	+	+
Coumarin	–	–	–	+	+
Flavonoid	+	+	+	+	+
Phenol	+	+	+	+	+
Quinone	–	–	–	+	+
Saponin	+	+	+	+	+
Steroid	+	–	–	+	+
Tannin	+	+	+	+	+
Terpenoid	+	+	–	+	+
Sugar	–	+	+	+	+
Glycoside	–	–	+	+	+
Xanthoprotein		–		+	+
Fixed oil	+	+	–	+	+

+ Presence; – Absence.

Discussion

Nature always stands as a golden mark to exemplify the outstanding phenomena of symbiosis. In the developed and developing countries, as the people are becoming aware of the potency and side effects of synthetic drugs, there is an increasing interest in the natural product remedies with a basic approach towards the nature. Throughout the history of mankind, many infectious diseases have been treated with herbals. Herbal preparations called "Phytopharmaceuticals" are preparations made from different parts of plants. They come in different formulations and dosage forms including tablets, capsules, powder, extract, tincture and cream (Samantha *et al.*, 2000). The misuse of herbal medicine or natural products starts with wrong identification. Hence, standardization of herbal raw material is very important today before subjecting the plant material to biological screening.

Pharmacognostical Studies

Pharmacognostic study is the initial step to confirm the identity and assess the quality and purity of the crude drug. Pharmacognostic techniques used in plant standardization include macroscopical, microscopical and physico-chemical parameters (Pramod and Jayaraj, 2011; Ravichandra and Parakh, 2011). According to World Health Organization (WHO), the macroscopic and microscopic description of a medicinal plant is the first step towards establishing its identity and purity and should be carried out before any tests are undertaken (Anonymous, 2002). It is globally accepted that herbal based drugs have many advantages over the synthetic drugs. However, one of the major problems in utilization of phytodrugs is correct diagnosis of the medicinal plants that are used either in the traditional systems or modern systems of preparation of the drugs. It is regrettable to note that most of the people involved in the manufacture or preparation of herbal drugs lack the basic background of botanical knowledge of the drugs. Consequently adulteration or substitutions of plants in the place of original ones permeate the pharmaceutical industries, rendering the herbal drugs undependable and invalid. This will lead unpopularity of phytodrugs among the people. So, it is most essential that a medicinal plant, when found to be of high pharmacological potentials, should be subjected to thorough botanical standardization so that there wouldn't be any ambiguity with respect to botanical identity of the plants. Identifications of plants involve the study of the external features of vegetative and floral parts.This study must be complimented with anatomical parameters which are very often useful to identify the fragmentary plant specimens. Raw drugs pose problem of identifications and to establish their genuineness when they lack any external diagnostic features or organoleptic clues. During such situation, the microscopic analyses of the specimen will offer a helping hand to establish or identify the phytodrugs. The anatomical features of the plant that are reliable for diagnosis and that are least changed due to environmental stresses include:

- Structure of the midrib
- Structure of lamina and its epidermal outgrowth.
- Surface view of foliar epidermis - epidermal cells and stomatal morphology
- Petiolar anatomy
- Venation pattern of the lamina
- Gross anatomy of the stem and root
- Ergastic cell inclusion such as; starch, crystal, tannin, mucilage *etc.*

Literature dealing with the anatomy of *A. excelsa* is lacking. The present study may be claimed as the first comprehensive investigation of the bark of *A. excelsa*. The present investigation has laid down a set of anatomical features of stem bark, which can be employed for botanical diagnosis. The following are the salient features of identification of stem of *A. excelsa.*

Salient Anatomical Features of *A. excelsa* Stem Bark

- ☆ The bark surface is almost smooth with narrow irregular, vertically oriented fissures.
- ☆ The surface colour is light brown or yellowish brown.
- ☆ The bark is up to 1.6mm thick.
- ☆ The bark is aromatic and bitter in taste.
- ☆ A wide distinct outer bark comprising of periderm and broad inner bark of secondary phloem constitute the bark.
- ☆ The periderm consists of broad wavy, thin walled, homogenous phellem and narrow zone of phelloderm.
- ☆ At frequent intervals, characteristic shell – bark is seen which consists of two concave bands of elliptical phellem with sclerotic tissue sandwiched in between.
- ☆ The secondary phloem is differentiated into outer zone of collapsed phloem and inner zone of non collapsed phloem.
- ☆ The collapsed phloem comprises of much dilated rays, large, irregular masses of brachysclereids and tangential lines of crushed and obliterate sieve elements.
- ☆ The non-collapsed phloem has intact sieve elements, companion cells, phloem parenchyma and undilated rays standardisation of the fragmentary phytodrugs.

Physico-chemical Parameters

Physico-chemical standardization is a prerequisite in quality control of herbal drugs. The efficacy of herbal drug mainly depends upon its physical and chemical properties. Therefore, the determination of physico-chemical characters for the authenticity of the drug is necessary before subjected to pharmacological activities. The qualitative and quantitative analysis of major bioactive chemical components of crude drug constitute important and reliable part of quality control protocol as any change in the quality of the drug directly affects the constituents (Mukherjee *et al.*, 2008). Evaluation of ash and extractive values of crude drugs help in the identification and determination of its purity and quality (Kokate *et al.*, 2010). The commonly applied parameter for the detection of impurities and adulteration of drug is the estimation of ash value, which establishes the quality and the purity of drug. Ash value can also detect the nature of the material added to the drug for the purpose of adulteration. In the present study, the total ash value is 9.50 per cent. These ash values are generally considered the index of the purity as well as identity of the drug. The extractive values in different organic solvents are based on the quantity, which are soluble in them. It makes a valuable test to check the quality of drug and any variation in the chemical constituents may cause a change in the extractive values. Thus, it helps in the determination of the adulteration and is an index of the purity of drug. The extractive values of stem bark of *A. excelsa* were determined by successive extraction in different solvents. Since, the extractive percentage of the drug has not been reported in the literature, it may be taken as an addition to the existing stock of knowledge. The variation in the extractive values may be possible due to the presence of specific compound, solubility, soil condition, atmospheric condition and water content of the same (Nasrin *et al.*, 2008).

Fluorescence Analysis

Modern methods like powder analysis and fluorescence drug analysis are very useful in standardization of plant material. Fluorescent is the phenomenon exhibited by various chemical constituents present in the plant material. Fluorescence studies of stem powder revealed the presence of fluorescent green with 1N aqueous NaOH, 50 per cent H_2SO_4, Conc. HNO_3, ferric chloride, conc.HNO_3 + NH_3, benzene, acetone and chloroform under UV light of shorter wavelength. Some constituents show fluorescent green in the visible range in daylight. The ultraviolet light produces fluorescent green in many natural products which is not visibly fluorescent in daylight. If substance themselves are not fluorescent, they may often be converted into fluorescent derivatives or decomposition products by applying different reagents. The organic molecules absorb light over a specific range of wavelength and re-emit radiations and hence it can be used for the identification of the powdered drug, extract or fractions of herbs (Rashida *et al.*, 2012). Crude drugs are often assessed qualitatively in this way and it is an important parameter for pharmacognostic evaluation of crude drugs (Zhao *et al.*, 2011).

Preliminary Phytochemical Analysis

The screening of plants for medicinal value has been carried out by number of workers with the help of preliminary phytochemical analysis (Ram, 2001). Phytochemical analysis of the stem bark extracts of *A. excelsa* revealed the presence of phytochemicals such as alkaloid, anthraquinone, catechin, flavonoid, coumarin, phenol, quinone, saponin, steroid, glycoside, tannin, sugar, terpenoid and xanthoprotein in them. These compounds make the plants useful for treating different ailments and having potential of providing useful drug of human use. This is because, the pharmacological activity of any plant is usually traced to a particular compound. These components were well known to have curative activity against several human problems such as diuretic, choleretic, spasmodic, chronic eczema, diarrohea, dysentery and menstrual disorders (Brinkhaus *et al.*, 2005). Phytochemicals such as saponins, terpenoids, flavonoids, tannins, steroids and alkaloids have antiinflammatory effects (Orhan *et al.*, 2007). Presence of phenols indicate the plant ability for antimicrobial activities (Parekh and Chanda, 2007a). Alkaloids have been associated with medicinal uses for centuries and one of their common biological properties is their cytotoxicity (Nobori *et al.*, 1994). Several workers have reported the analgesic (Antherden, 1969), antispasmodic and antibacterial (Okwu, 2004) properties of alkaloids. Saponins help in controlling cholesterol and diabetes (Ong, 2004). Saponins are also responsible for central nervous system activities (Rupasinghe *et al.*, 2003). Steroids have been reported to have antibacterial properties (Raquel, 2007) and they are very important compounds especially due to their relationship with compounds such as sex hormones (Okwu, 2001). Flavonoids are thought to play a role in protection of plants from microbial and insect attack. Moreover, flavonoids have remarkable health promoting effects, such as antiinflammatory (Yamamoto and Gaynor, 2001), antimicrobial (Tim Cushnie and Lamb, 2005), antioxidant (Shahidi and Wanasundara, 1992), anticancer (Wei *et al.*, 1990) activity as well as the prevention of osteoporosis (Migliaccio and Anderson, 2003). Tannins decrease the bacterial proliferation by blocking key enzymes at microbial metabolism. Tannins play an important role as potent antioxidant (Trease and Evans, 1992). Herbs that have tannins as their main component are astringent in nature and are used for treating intestinal disorders such as diarrhea and dysentery (Dharmananda, 2003). Terpenoids exhibit various important pharmacological activities *i.e.*, antiinflammatory, anticancer, antimalarial, inhibition of cholesterol synthesis, antiviral and antibacterial activities (Mahato and Sen, 1997). Terpenoids are very important in attracting useful mites and consume the herbivorous insects (Kappers *et al.*, 2005). Present investigation showed that this plant is a warehouse of chemodiversity.

The pharmacognostic evaluation which comprises of macromorphology and microscopic characters, the estimation of physico-chemical parameters and the phytochemical screening are constant features of a plant which are highly essential for raw drugs or plant parts used for preparation of phytomedicine. Therefore, the results generated from this would be useful in identification and standardization of the plant material towards quality assurance and also for preparation of a monograph on the plant.

References

Anonymous 1996. Indian Pharmacopoeia, Government of India, Ministry of Health and Family Welfare, the Controller of Publications, Civil Lines, Delhi – 110 054. Vol.I and II.

Anonymous 1998. Macroscopic and microscopic examination: quality control methods for medicinal plant materials, WHO, Geneva.

Anonymous. 2002. Quality Control methods for medicinal plant material. An authorized publication of WHO. First Edition, New Delhi, India. A.I.T.B.S. Publishers and Distributors. pp. 18-21.

Antherden LM. 1969. Textbook Of Pharmaceutical Chemistry, 8th edn., Oxford University Press, London, pp. 813-814.

Asolkar LV., Kakkar KK and Chakre OJ. 1992. Glossary of Indian Medicinal Plant with active principle, Part I:34

Assefa B., Glatzel G and Buchmann C. 2010. Ethnomedicinal uses of *Hagenia abyssinica* among rural communities of Ethiopia. *J. Ethbiol. Ethmed.* 6:20.

Balandrin MF., Lclocke JA., Wartele ES and Bollinges WH. 1985. Natural plant chemicals. Sources of industries and medicinal materials. Science. 228:1154-1160.

Brindha P., Sasikala P and Purushothaman KK. 1981. Pharmacognostic studies on Merugan Kizhangu. *Bulletin of Medico Ethnobotanical Research.* 3: 84-96.

Brinkhaus B., Hentschel C. and Scand J. 2005. Herbal medicine with curcuma and fumitory in the treatment of irritable bowel syndrome: a randomized, placebocontrolled, double-blind clinical trial. *Gastroenterol*. 40: 936-43.

Cabrera S., Genta A., Said A., Farag K., Rashed S and Sanchez. 2008. Hypoglycemic activity of *Ailanthus excelsa* leaves in normal and streptozotocin-induced diabetic rats *Phytother. Res.* 22, 303-307 (2008).

Hukkeri VI., Jaiprakash B., Lavhale MS., Karadi RV and Kuppast IJ. 2002. Hepatoprotective activity of *Ailanthus excelsa* Roxb. Leaf extract on experimental liver damage in rats. *Indian J. Pharm. Educ*. 37:105-106.

Johanson DA. 1940. Plant Microtechnique. Mc Graw Hill Book Co, New York, 523.

Joshi BC., Pendey A., Chaurasia L., Pal M., Sharma RP and Khare A. 2003. Antifungal activity of the stem bark of *Ailanthus excelsa*. *Fitoterpia* 74:689-691.

Kapoor SK., Ahmed PL and Zaman A. 1971. Chemical constituents of *Ailanthus excelsa*. *Phytochemistry*. 1971; 10: 3333-3335.

Kappers IF., Aharoni A., Van Herpen TW., Luckerhoff LL. Dicke M. and Bouwmeester HJ. 2005. Genetic engineering of terpenoid metabolism attracts bodyguards to *Arabidopsis*. *Sci.* 309: 2070-2072.

Kirtikar Basu BD Indian medicinal plant, vol. 1, 2nd edition (2003) pp. 506-507.

Kokate CK., Purohit AP and Gokhale SB. 2010. Pharmacognosy 45th Edition. Nirali Prakashan, Mumbai India.2010.

Lala PK. 1993. Lab Manuals of Pharmacognosy. Edn 5, CSI Publishers and Distributors, Calcutta.

Lalitharani S., Kalpanadevi V and Mohan VR. 2013. Pharmacognostic studies on the spine of *Zanthoxylum stetsa* (Roxb). DC. *Bioscience Discov*. 4:5-11.

Mahato SB and Sen S. 1997. Advances in triterpenoid research, 1990-1994. *Phytochem*. 44: 1185-1236.

Mehta CR and Patel CN. 1959. Chemical examination of the bark of *Ailanthus excelsa*, Roxb. Part I. *Indian J. Pharm*. **21**, 143-145.

Migliaccio S and Anderson JB. 2003. Isoflavones and skeletal health: Are these molecules ready for clinical application. *Osteoporos Int*. 14: 361-368.

Mohan VR., Chendurpandy P and Kalidass C. 2009. Pharmacognostic and phytochemical investigation of *Elephantopus scaber* L. (Asteraceae). *J. Pharmaceu. Sci. Tech*. 2:191-197.

Mukherjee PK., Kumar V., Satheesh kumar N and Heinrich M. 2008. The Ayurvedic medicine *Clitoria ternatea*-from traditional use to scientific assesment. *J. Ethnopharm*. 120: 291-301.

Nadkarni KM 2000. Indian Materia Medica. Popular Prakashan Private Limited Publication, Mumbai: 3rd edition, Vol 1: 56-57.

Nasrin AB., Ma AN., Choo YM, Mohamad S., Rohaya MH., Azali A and Zainal Z. 2008. Oil palm biomass as potential substitution raw materials for commercial biomass briquettes production. *Amer. J. App. Sci.* 5: 179-183.

Neube NS., Afolayan AJ and Okoh AI. 2008. Assessment techniques of antimicrobial properties of natural compounds of plant origin: current methods and future trends. *Afr. J. Biotech.* 7:1797-1806.

Nobori T., Miurak K., Wu DJ., Takabayashik LA. and Carson DA. 1994. Deletion of cyclin-dependent kinase-4 inhibitor gene in multiple human cancers. *Nat.* 46: 753-756.

O'Brien TP., Feder N and Mc Cull ME. 1964. Polychromatic staining of plant cell walls by toluidine blue-O. *Protoplasma.* 59: 364-373.

Ogura M., Cordell MGA., Kinghorn., AD and Farnsworth NR. 1977. Potential anticancer agents vi. Constituents of *Ailanthus excelsa* (Simaroubaceae) *Lloydia.* 40(6):579-84.

Okwu DE. 2001. Evaluation of chemical composition of medicinal plants belonging to Euphorbiaceae. *Pak. Vet. J.* 14: 160-162.

Okwu DE. 2004. Phytochemicals and vitamin content of indigenous species of southeastern Nigeria. *J. Sustain. Agric. Environ.* 6: 30-37.

Ong HC. 2004. Tumbuhan liar. Khasiat ubatan dan kegunaan lain. Some Indian medicinal plants. *J. Ethnopharmacol.* 19: 425-428.

Orhan I., Kupeli E., Sener B. and Yesilada E. 2007. Appraisal of anti –inflammatory potential of the clubmoss, *Lycopodium clavatum* L. *J.Ethnopharmacol.* 109:146-150.

Parekh J. and Chanda SV. 2007. *In vitro* antimicrobial activity and phytochemical analysis of some Indian medicinal plants. *Turk. J. Biol.* 31:53-58

Pramod V. and Jayaraj PM. 2011. Pharmacognostical and phytochemical investigation of *Sida cordifolia* L.A threatened medicinal herb. *Int. J. Pharm. Pharmceut. Sci.* 4: 114-117.

Rahman S., Fukamiya N., Ohno N., Tokuda H., Nishino H., Tagahara K., Lee KH and Okano M. 1997. Inhibitory effects of quassinoid derivatives on Epstein-Barr virus early antigen activation. *Chem Pharm Bull* (Tokyo). 45(4):675-677.

Rajesh P., Latha S., Selvamani P and Kannan R. 2010. *Capparis sepiaria* Linn. Pharmacognostical standardization and toxicity profile with chemical compounds identification (GC MS). *Int. J. Phytomed.* 271-272.

Ram RL. 2001. Preliminary phytochemical analysis of medicinal plants of South Chotanagpur used against dysentery. *Advan. Plant Sci.* 14: 525-530.

Raquel FE. 2007. Bacterial lipid composition and antimicrobial efficacy of cationic steroid compounds (Ceragenins). *Biochemica et Biophysica Acta.* 2500-2509.

Rashida GJN., Venkatarathnakumar T., Aruna AD., Gowri R., Parameshwari R., Shanthi M. and Raadhika K. 2012. Pharmacognostic and preliminary phytochemical evaluation of the leaves of *Dalbergia sissoo* Roxb. *Asian J. Pharm. Clin. Res.* 5: 115-119.

Ravichandra VD and Paarakh PM. 2011. Pharmacognostic and phytochemical investigation on leaves of *Ficus hispida*. *Int. J. Pharm. Pharmaceut. Sci.* 3:131-134.

Rupasinghe HP., Jackson CJ., Poysa V., Di Berado C., Bewley JD and Jenkinson J. 2003. Soyasapogenol A and B distribution in Soybean (*Glycine max* L.Merr) in relation to seed physiology, genetic variability and growing location. *J. Agric. Food Chem*. 51: 5888-5894.

Samantha MK., Pulok K., Mukherjee., Prasad MK and Suresh B. 2000. Development of Natural Products, The Eastern Pharmacist, August. pp. 23-27.

Sarada K., Mohan VR and Sakthidevi G. 2014. Pharmacognostic and phytochemical studies of leaves of *Naringi crenulata* (Roxb.) Nicolson. *Int. J. Pharmacog. Phytochem*. Res. 6:153-156.

Sarwat ZK., Shinwari and Ahmad N. 2012. Screening of potential medincinal plants from district swat specific for controlling women diseases. *Pak. J. Bot.* 44:1193-1198.

Shahidi F and Wanasundara PK. 1992. Phenolic antioxidants. *Crit. Rev. Food Sci. Nutr*. 32: 67-103.

Shrimali M., Jain DC., Darokar MP and Sharma RP. 2001. Antibacterial activity of *Ailanthus excelsa* (Roxb) *Phytother. Res*. 15, 165.

Tim Cushnie TP and Lamb AJ. 2005. Antimicrobial activity of flavonoids. *Int. J. Antimicrob. Agent*. 26: 343-356.

Trease GE and Evans WC. 1992. Pharmacognosy 13th ed. ELBS/Bailliere Tindal, London UK. Dharmananda, S. 2003. Gallnuts and the uses of Tannins in Chinese Medicine. In:Proceedings of Institute for Traditional Medicine, Portland, Oregon.

Wei H., Tye L., Bresnick E and Birt DF. 1990. Inhibitory effect of epigenin, a plant flavonoid, on epidermal ornithine decarboxylase and skin tumor promotion in mice. *Can. Res*. 50: 499-502.

World Health Organization. Quality control methods for medicinal plant material. WHO Library 1998 pp. 110-115.

Yamamoto, Y. and Gaynor, R.B. 2001. Therapeutic potential of inhibition of the NFJB pathway in the treatment of inflammation and cancer. *J. Clin. Invest*. 107:135-142.

Zhao Z., Liang Z and Guo P. 2011. Macroscopic identification of Chinese medicinal materials: Traditional experiences and modern understanding. *J.Ethnopharmacol*. 131: 556-561.

2018, Ethnomedicinal Plants: A Biodiversity Treasure *Pages 589–605*
Editors: V.R. Mohan, A. Doss, P.S. Tresina and V. Sornalakshmi
Published by: **ASTRAL INTERNATIONAL PVT. LTD., NEW DELHI**

Chapter 25

Pharmacognostic Study and Establishment of Quality Parameters of Whole Plant of *Beloperone plumbaginifolia* (Jacq.) Nees.

G. Sathiyabalan[1], K. Paulpriya[2], P.S. Tresina[2], S. Muthukumarasamy[3], V.R. Mohan[2*]

[1]*Department of Pharmacognosy, College of Pharmacy, Madurai Medical College, Madurai, Tamil Nadu*
[2]*Ethnopharmacology Unit, Research Department of Botany, V.O. Chidambaram College, Tuticorin – 628008, Tamil Nadu*
[3]*Department of Botany, KGS College for Arts and Science, Srivaikundam, Tamil Nadu*
**Corresponding author E-mail: vrmohanvoc@gmail.com*

Introduction

Medicinal plants have been used in traditional healing practices for treating various human ailments since time immemorial. Such traditional practices have provided the basis of scientific investigation of medicinal plants which led to the discovery of many potential drug molecules of today's modern medicine. Herbal medicine has therefore become the most reliable form of alternative medicine for treating human disorders around the world. In recent years, there has been a dramatic rise in use of herbal drugs/preparations in the developed countries because of their easy availability and cost effectiveness besides having desired pharmacological effectiveness with high level of safety/low toxicity profile. It is estimated that world's one-forth population *i.e.* 1.42 billion people are dependent on traditional herbal medicines for the treatment of various ailments (Kadam *et al.*, 2012). However, the lack of documentation and stringent quality control procedures has hindered the easy acceptance of such plant drugs (crude preparations) to be used as herbal medicine.

Pharmacognostical study is the preliminary step in the standardization of crude drugs. The detailed pharmacognostical evaluation gives valuable information regarding the morphology, microscopical and physical characteristics of the crude drugs. Pharmacognostic studies have been done on many important drugs and the resulting observations have been incorporated in various pharmacopoeias (Sharma, 2004). Despite the modern techniques, identification of plant drugs by pharmacognostic studies is more reliable. According to the World Health Organization (WHO), the macroscopic and microscopic description of a medicinal plant is the first step towards establishing the identity and the degree of purity of such materials and should be carried out before any tests are undertaken (WHO, 1998)

Beloperone plumbaginifolia (Jacq.) Nees is a traditional medicinal plant used in Siddha system of Indian medicine are being used by tribals in Kerala as an antidote for snake bites. This plant is known by Visapachilai in Tamil. It is a flowering plant and belongs to the family Acanthaceae. In Tamil Nadu, it has been used to treat various ailments like antidote for scorpion bites, psoriasis, *etc.*, as a folklore medicine. Despite its rich pharmacological potential, so far no study has been conducted on this medicinal plant except a shoot multiplication and callogenesis technique developed by Shameer *et al.* (2008). The objective of the present study is to evaluate various pharmocognostic standards like microscopy, physico-chemical constant, fluorescence analysis and qualitative preliminary phytochemical analysis of *B. plumbaginifolia* (Jacq.) Nees, these findings would be helpful for authentication, purification, quality control and for better use in pharmaceutical herbal formulations.

Materials and Methods

The plant specimens for the proposed study were collected from Agasthiarmalai Biosphere Reserve, Western Ghats, Tamil Nadu. The plant samples were identified with the help of local flora and authenticated by Botanical Survey of India, Southern Circle, Coimbatore, Tamil Nadu, India. A voucher specimen of collected plants was deposited in the Ethnopharmacological Unit, PG and Research Department of Botany, V.O. Chidambaram College, Thoothukudi District, Tamil Nadu.

Macroscopical Studies

The macroscopic characters like surface, shape, size, venation, phyllotaxy, length of the petiole, length of the leaf *etc.*, were noted.

Anatomical Studies

For anatomical studies, the required samples of root, stem and leaf were cut and removed from the plant and immediately fixed in FAA (formalin- 5 ml + acetic acid- 5 ml + 70 per cent Ethyl alcohol- 90 ml). The specimens were left in the preservative for two days; then the materials were washed in water and processed further. Standard microtome techniques were followed for anatomical investigation (Johanson, 1940). Transverse sections of the materials were made. The microtome sections were stained with 0.25 per cent aqueous Toluidine blue (Metachromatic stain) adjusted to pH 4.7 (O'Brien, 1964). Photomicrographs were taken with NIKON trinocular photo micrographic unit.

Physico-chemical and Fluorescence Analysis

These studies were carried out as per the standard procedures (Lala, 1993). In the present study, the powered stem bark was treated with various chemical reagents like aqueous 1N sodium hydroxide, alcoholic 1N sodium hydroxide, 1N hydrochloric acid, 50 per cent sulphuric acid, concentrated nitric acid, picric acid, acetic acid, ferric chloride and concentrated HNO_3 + NH_3 These extracts were subjected to fluorescence analysis in day light and UV light (254nm and366nm). Various ash types and extractive values were determined by following standard methods (Anonymous 1996).

Preliminary Phytochemical Analysis

Shade dried and powdered plant samples weres successively extracted with Petroleum ether, benzene, ethyl acetate, methanol and ethanol. The extracts were filtered and concentrated using vacuum distillation. The different extracts were subjected to qualitative tests for the identification of various phytochemical constituents as per the standard procedure (Lala, 1993; Brindha *et al.*, 1981).

Results

Subshrubs 0.7 – 1.5 cm tall, branched. Stems tetrangular, swollen at nodes, glabrous. Petiole 3-10 mm; leaf blade lanceolate, 6-10 X 1-1.5 cm, glabrous, secondary viens 5-7 on each side of mid-vein, base cuneate to attenuate, margin subsinuate, apex acute to shortly acuminate. Spikes terminal or axillary, 3-12 cm, usually in a leafy panicle; peduncle 0.5 – 1.5 cm; bracts triangular, 2-6 x 1-2.5 mm, basal ones longer than calyx then gradually smaller with apical most ones shorter than calyx, margin ciliate, apex cute; bracteoles elliptic to linear–lanceolate, ca. 3 x 1 mm, margin ciliate, apex acute. Calyx ca. 5 mm, 5 – lobed; lobes linear lanceolate, 3-4 x ca. 0.5 mm, sub-equal, apex acuminate. Corolla variously coloured, usually reddish, 1.2 -1.5 cm; tube basally cylindric and ca. 2 mm wide for 8-9 mm; lower lip 3 lobes, cuneate-obovate, 6-10 mm broad, lobes oblanceolate and 3-5 x ca. 3.5 mm; upper lip triangular, ca. 7 x 3.5 mm, 2 –cleft. Stamens exerted; filaments 3-6 mm, glabrous, anther thecae oblong, ca. 1.2 mm, superposed, lower one spurred at base, upper one muticous. Ovary glabrous; style ca. 1 cm, glabrous; stigma capitate, shortly 2 – lobed. Capsule clavate, ca. 1.2 cm.

Anatomy of Root

The root exhibits hydromorphic features. It is circular in sectional view and measures 1.6 mm in thickness. The root consists of an aerenchymatous cortical zone and a stele with primary vascular tissues of xylem and phloem (Plate 25.1b). The cortex consists of an outer zone of three or four layers of compact parenchyma cells. The middle zone has wide, radially stretched air chambers, separated by lateral,

Plate 25.1a-e: Anatomical Features of *B. plumbaginifolia*

(Rh: Rhizodermis, Ac: Air chamber, Oc: Outer cortex, Ic: Inner cortex, MX: Metaxylem, Pc: Pericycle, Pi: Pith, PF: Partition filament, Xy: Xylem, Ph: Phloem, Ec: Epidermal cell, Co: Cortex, Chl: Chlorenchyma, Col: Collenchyma, Cy: Cystolith, Pi: Pith cavity, SPh: Secondary Phloem, SX: Secondary Xylem).

a. Habit

T.S. of Root

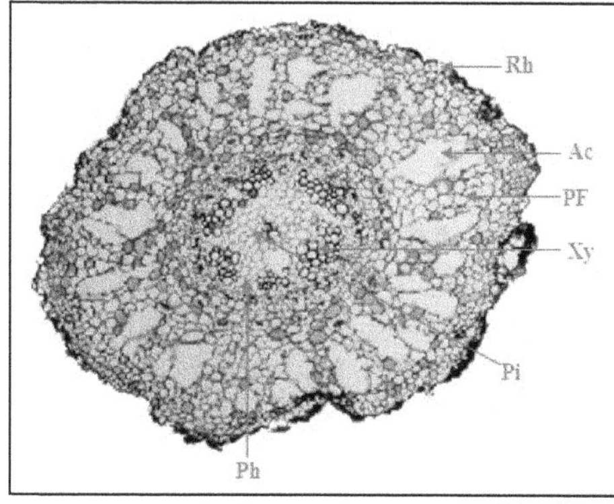

b. Enter view

c. Central stele with xylem, phloem and pith

T.S. of Young (Thin) Stem

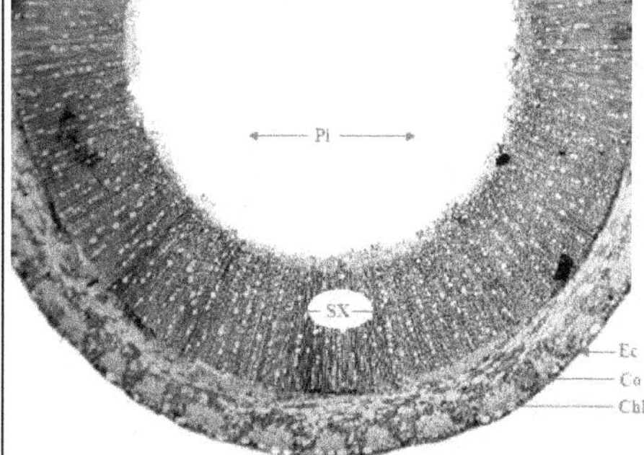

d. Enter view

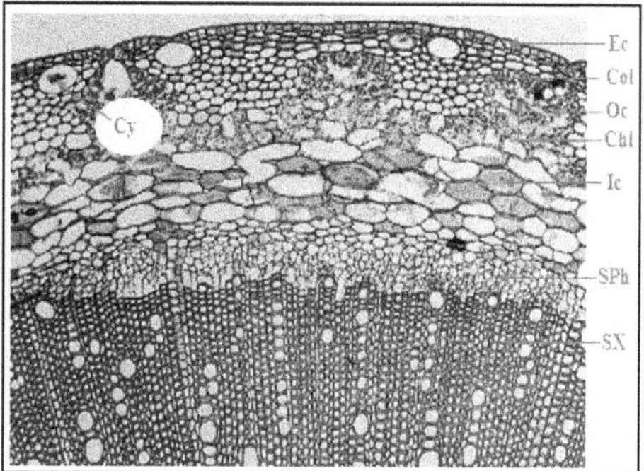

e. A sector enlarged

two or three cells thick partitions. The inner cortex includes four or five layers of angular compact parenchyma cells (Plate 25.1b). The endodermis is fairly distinct and consists of rectangular thin walled cells. There are six, fairly large, triangular mass of exarch xylem strands, alternating with small, less prominent phloem masses. The metaxylem elements are wide, angular and thick walled. The center of the stele has narrow parenchymatous pith (Plate 25.1c).

Anatomy of Young (Thin) Stem

The young stem is circular in outline and is smooth. It is hollow, cylindrical and measures 800 µm thick (Plate 25.1d). The epidermal layer is thin and intact and the cells are rectangular in shape. The epidermal cells have undergone periclinal divisions at several places producing thin and incipient periderm (Plate 25.1e). The cortex is distinguished into an outer zone and an inner zone. The outer zone is made up of smaller cells of collenchyma in which are embedded small masses of chlorenchyma at regular intervals. The inner zone is made up of four or five layers of tangentially elongated elliptical, large and thin walled parenchyma cells (Plate 25.1e).

The vascular cylinder consists of an outer thick continuous cylinder of secondary phloem and an inner dense and thick cylinder of secondary xylem. The secondary phloem elements are angular in outline, thin walled and are arranged in compact radial rows. The phloem rays are prominent and are extensions of the xylem rays. The secondary phloem is followed by a thick cylinder of secondary xylem which includes several radial multiples of long or short vessels and a few solitary vessels. The vessels are circular or angular, thin walled and are 30 µm wide (Plate 25.1e).

Epidermal Cells of the Stem

The epidermal layer of the stem is viewed from the surface. The epidermal cells are small, polyhedral, thick walled and have straight anticlinal walls (Plate 25.1e). Stomata are sparsely seen on the epidermis. The stomata have elliptical guard cells and the stomatal pore is not evident. The guard cells are surrounded by two circles distillation. The different cells and thus and they are cyclocytic type.

Anatomy of Old (Thick) Stem

The old stem is 2.1 mm in diameter. The epidermal layer is crushed into dark, thick fissures. There is a thin superficial layer of periderm which comprises of three or four layers of tubular cells. The cortical zone is 350 µm thick and is aerenchymatous in nature (Plate 25.2a,b). The median part of the cortical zone has wide air chambers and the air chambers are arranged in a single circle. The air chambers are separated from each other by vertical, thick, three or four cells thick short portion (Plate 25.2b).

The vascular cylinder of the old stem is hallow, dense and comprises of an outer secondary phloem and an inner secondary xylem. The secondary phloem elements are distributed diffusely and the sieve elements are angular and thin walled. The companion cells are prominent and they are seen at one corner of the sieve element.

Plate 25.2a-f: Anatomical Features of *B. plumbaginifolia*.

(Ep: Epidermis, Ac: Air chamber, Oc: Outer cortex, Ic: Inner cortex, Pc: Pith cavity, SPh: Secondary Phloem, SXy: Secondary Xylem, Cy: Cystolith, SE: Sieve Elements, Ve: Vessel, Ads: Adaxial side, Col: Collenchyma, Gp: Ground parenchyma, LB - Lateral Bundle, MB: Median Bundle).

T.S. of Old (Thick) Stem

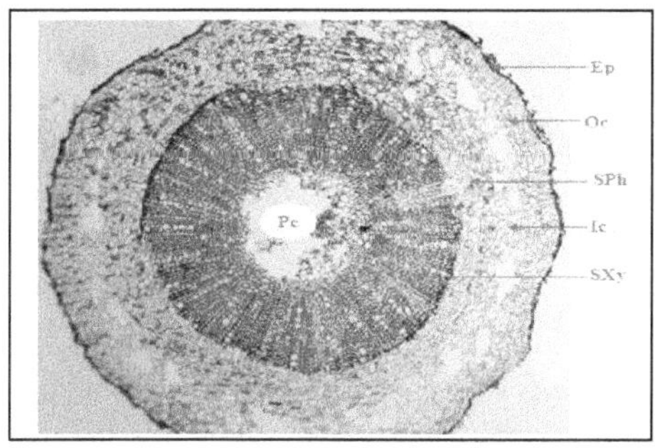

a. Enter view

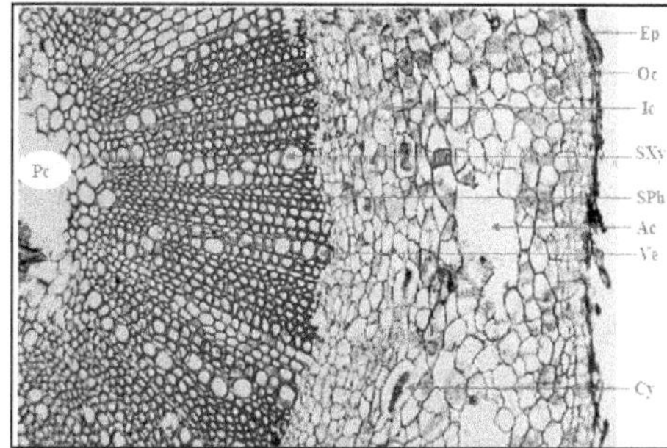

b. A sector enlarged

T.S. of Petiole

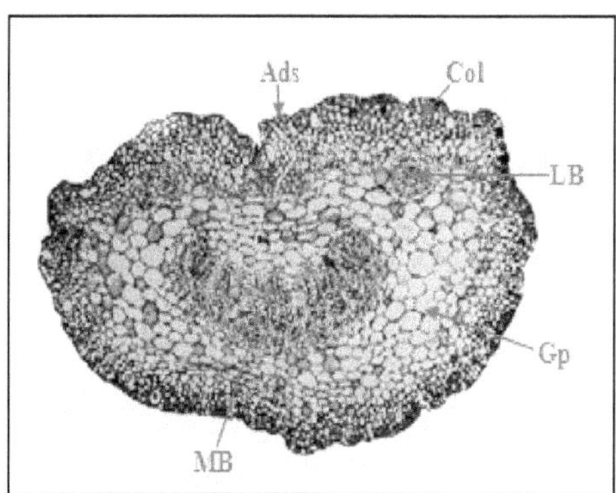

c. Enter view

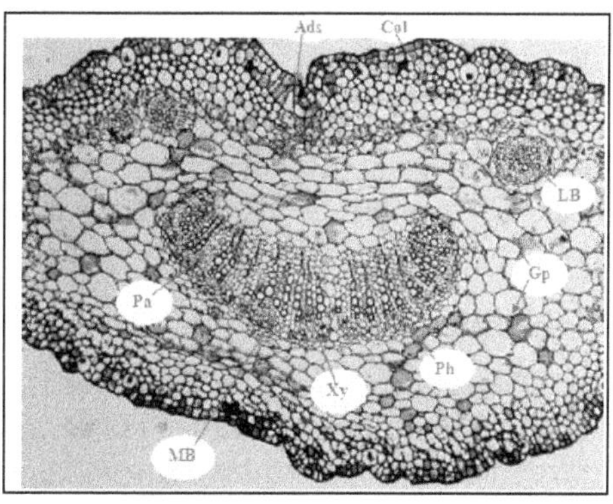

d. Central part of the petiole enlarged

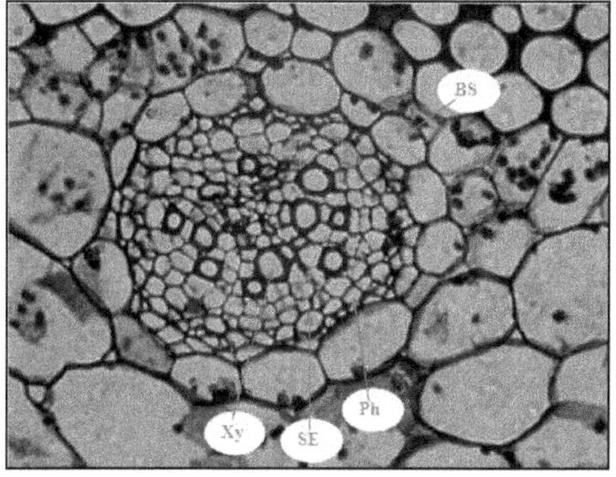

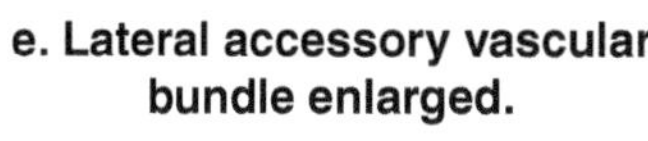

e. Lateral accessory vascular bundle enlarged.

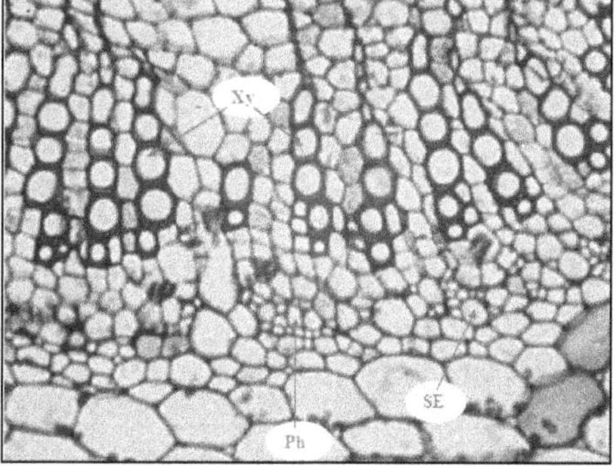

f. Vascular elements of the Main vascular bundle.

The secondary xylem elements include vessels, fibres and ray cells. The vessels are angular or circular in outline, narrow and highly thick walled. The vessels are in long, radial multiples or solitary. The vessels are up to 30 µm wide. The xylem fibres are squarish or radially oblong, highly thick walled and lignified. They are in regular radial rows. The xylem rays are thin and straight and the ray cells are radially elongated and thick walled.

Anatomy of Petiole

The petiole is semicircular in sectional view with shallow depressions on the adaxial side. It is 900 µm thick and 1.5 µm wide. The epidermal layer of the petiole is thin and the epidermal cells are small, squarish and fairly thick walled. The ground tissue is differentiated into an outer zone of four or five layers of small thick walled collenchyma cells and an inner zone of large polyhedral, thin walled and compact parenchyma cells (Plate 25.2e,d).

The vascular system consists of a wide bowl shaped main strand of xylem and phloem and small circular, one or two accessory vascular bundles located in the adaxial lateral part of the petiole. The main strand is 550 µm wide and 150 µm thick (Plate 25.2d). The strand consists of long vertical lines of angular, thick walled wide xylem elements. Each line ends in thin narrow, compressed or crushed protoxylem cells. Phloem occurs in discrete groups at the lower part of the xylem strands (Plate 25.2f).

The accessory adaxial bundles are 150 µm in diameter. Each bundle consists of a few xylem elements which occur in groups of two or in solitary thick walled angular cells. The phloem is located along the peripheral part of the bundle. The bundle is surrounded by a single layer of sclerenchymatous bundle sheath (Plate 25.2e).

Anatomy of Leaf

The leaf consists of a thick and wide midrib and several smaller lateral veins which are similar in structure as the main midrib. Both the midrib and lateral veins have short and broad adaxial hump and wide and thick abaxial midrib (Plate 25.3a).

Midrib

The midrib is 800 µm thick and the adaxial hump is 400 µm wide. The abaxial midrib is 900 µm wide. The midrib consists of a thick and prominent intact epidermal layer of thin walled cells. The ground tissue in the abaxial part has angular or circular, thin walled compact parenchyma cells. The adaxial cone includes small, slightly thick walled collenchyma cells. The palisade layer of the lamina forms a horizontally transcurrent palisade layer, passing through the region between the adaxial hump and vascular strand of the abaxial midrib (Plate 25.3a,b).

The vascular system consists of a wide and deep bowl shaped main vascular strand and two small less prominent circular vascular strands, one on either side of the main strand (Plate 25.3). The main vascular structure consists of several, uniseriate long xylem elements, with wide parenchymatous gaps, in between the xylem lines. The xylem elements are angular, thick walled and wide. Phloem

Plate 25.3a-f: Anatomical Features of *B. plumbaginifolia*

(Ep: Epidermis, La: Lamina, Adh: Adaxial hump, Lv: Lateral Vein, MR: Midrib, MVS: Median Vascular Strand, Col: Collenchyma, Pa: Parenchyma, Gp: Ground parenchyma, Xy: Xylem, Ph: Phloem, SE: Sieve Element, TPC: Transcurrent layer of palisade cells, VS: Vascular Strand, Adc: Adaxial cone, VB: Vascular Bundle, AdE: Adaxial Epidermis, AbE: Abaxial Epidermis, Cy: Cystolith, Lc: Lithocyst, MT: Mesophyll Tissue, PM: Palisade Mesophyll, SM: Spongy Mesophyll, AW: Anticlinal wall, Gc: Guard cell, Sc: Subsidiary cell, St: Stomata).

Anatomy of Leaf

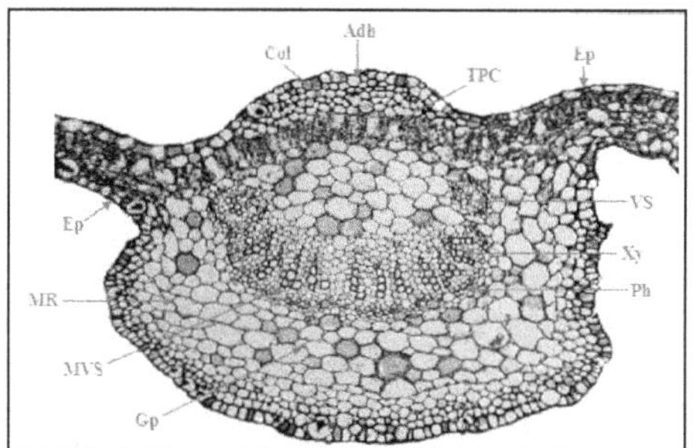

a. Midrib region enlarged

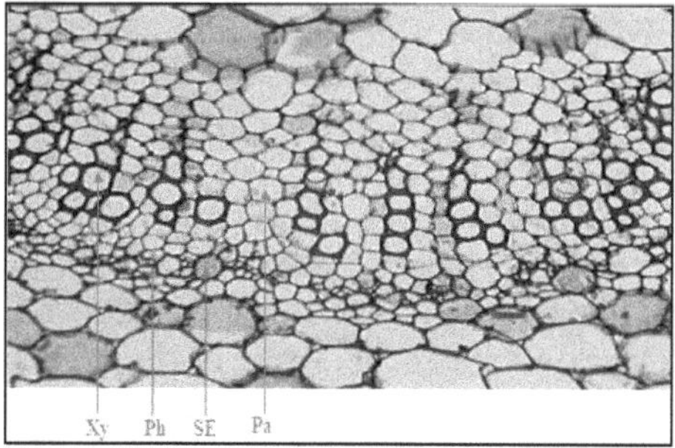

b. Vascular strand of midrib enlarged

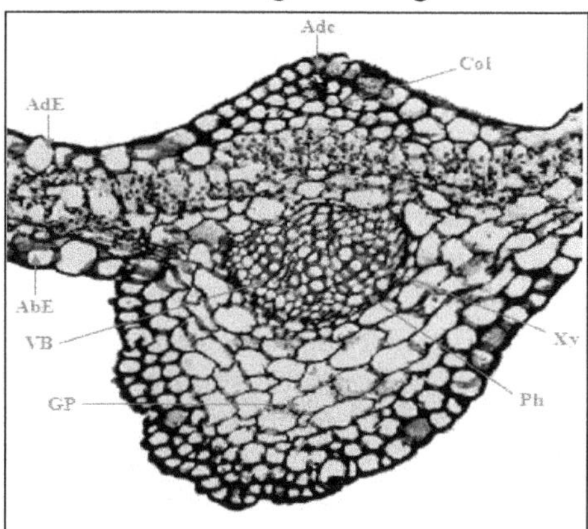

c. Lateral vein region enlarged

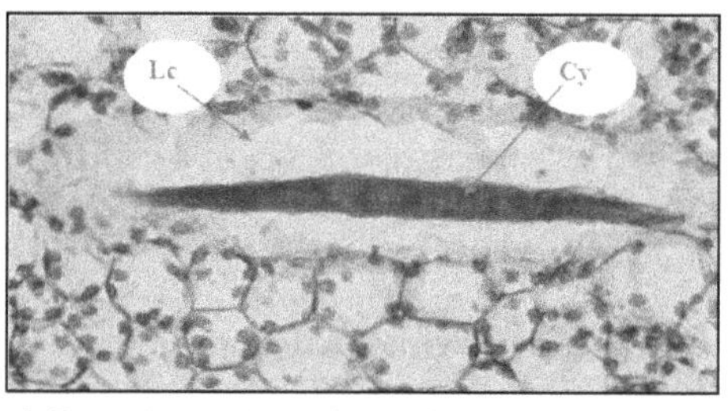

d. Paradermal section of lamina showing Lithocyst and Cystolith

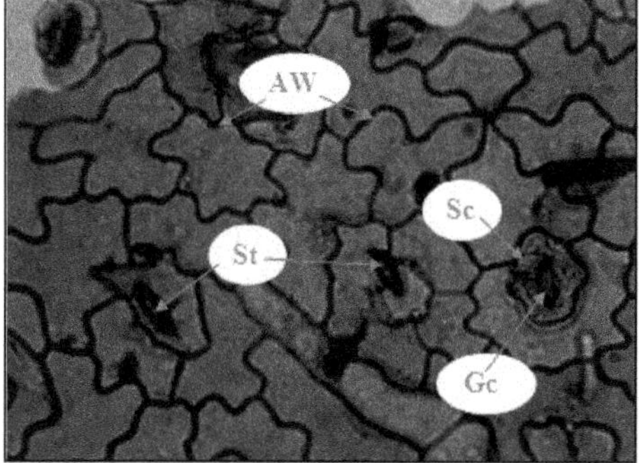

e. Paradermal section of the Abaxial Epidermis showing stomata

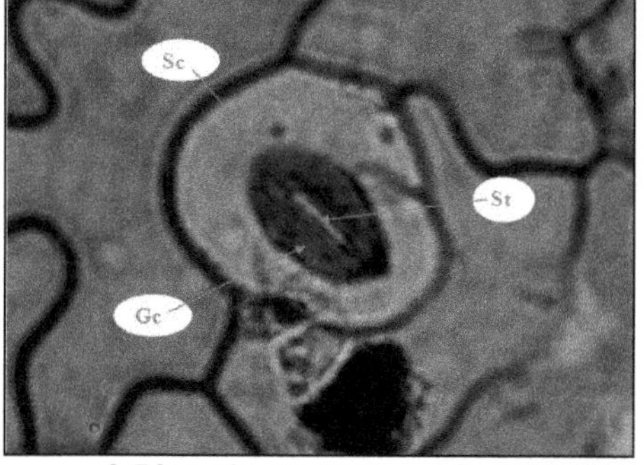

f. Diacytic stomata enlarged

elements occur on the lower end of the xylem strands and they are in small discrete groups embedded in the parenchymatous ground tissue.

Lateral Vein

The lateral vein is also prominent, having an adaxial cone and an abaxial semicircular part. It is 700 μm thick. The epidermal layer consists of wide squarish cells with thick walls. The ground cells in the adaxial cone include collenchyma cells. The cells in the abaxial part are angular, thin walled with wavy walls. The vascular strand is single, bowl shaped and collateral. It consists of five or six vertical short lines of thick walled, angular xylem elements and thin lines of phloem elements situated along the lower part of the xylem strands (Plate 25.3e).

Lamina

The lamina is dorsiventral, thin and smooth on both sides. The lamina is 250 μm thick. The adaxial epidermal cells are large and 30 - 40 μm thick. The cuticle is prominent. The abaxial epidermal cells are spindle shaped, thin walled and have thin cuticle (Plate 25.3c). Some of the epidermal cells are highly dilated into elongated wide cells and possess the calcium carbonate cystoliths. The cells containing cystoliths are called lithocysts (Plate 25.3d). The cystoliths may be smooth or warty or echinate and are 200 μm long and 15 μm thick.

Epidermal Cells and Stomata

The epidermal tissue was studied from paradermal sections of the lamina and it was viewed from the surface. The adaxial epidermal cells are wide and have thick, highly wavy anticlinal walls. The abaxial epidermal cells are slightly smaller; their anticlinal walls are fairly thick and highly wavy. The abaxial epidermis is stomatiferous (Plate 25.3e). The stomata are broadly elliptical and darkly stained. They have narrow slit-like aperture. The stomata are diacytic type. They have two subsidiary cells, one on each polar end of the guard cells. Their common wall is at right angles to the guard cells. The guard cells are 40 x 70 μm in size (Plate 25.3f).

Venation of the Lamina

The veins are uniformly thin and straight. The veins form wide vein-islets which are squarish, rectangular, triangular or polyhedral in outline (Plate 25.4a). They have distinct vein boundaries. The vein terminations are well developed and are either branched into dendroid outline (Plate 25.4b) or unbranched (Plate 25.4c). The vein terminations are mostly curved or wavy.

Powder Microscopy

The powdered preparation of the sample includes vessel elements, fibres, parenchyma cells and glandular trichomes (Plate 25.4d).

(i) Vessel Elements

The vessel elements are 350 to 560 μm long and 30 to 40 μm wide. They are mostly long, narrow and with thin long tails at the one end or at both the ends. The

Plate 2.4a-h: Anatomical Features of *B. plumbaginifolia*

(BVT: Branched vein termination, VI: Vein islet, VT: Vein termination, Fi: Fibre, RPa: Rectangular Parenchyma, SPa: Spherical Parenchyma, Pe: Perforation, Ta: Tail, VE: Vessel element, NFi: Narrow Fibre, WFi: Wide Fibre, LGtr: Large glandular trichome, SGTr: Small glandular trichome, BC: Body cell, SC: Stalk cell).

Venation Pattern of Leaf and Powder Microscopic Features

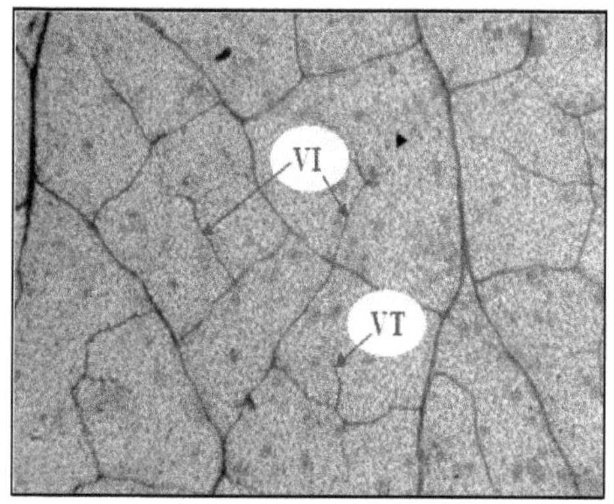

a. Venation pattern

b. Dendroid type of vein termination

c. Unbranched, curved vein termination

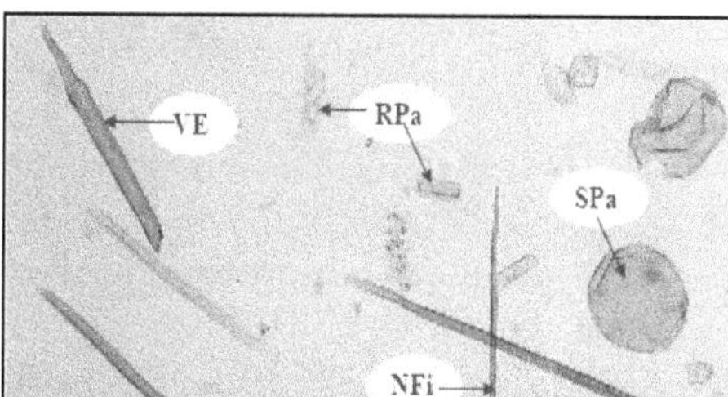

d. Fibres, parenchyma and vessel element

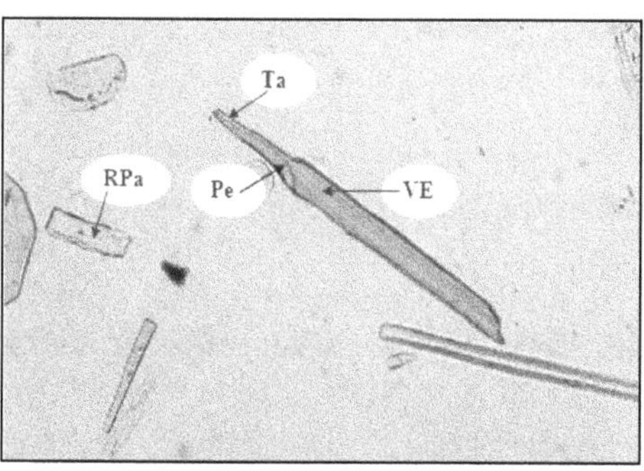

e. Vessel element and rectangular parenchyma

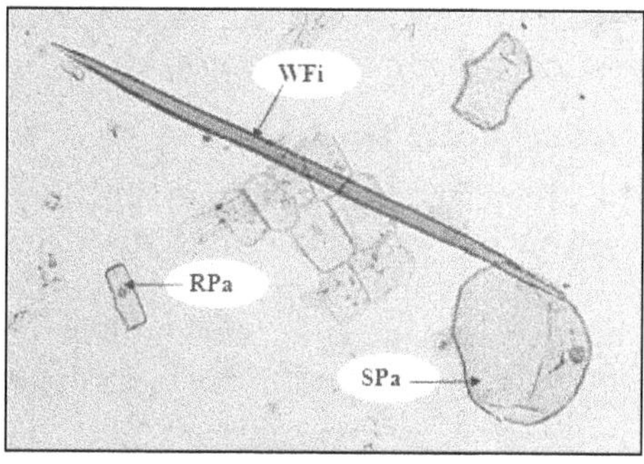

f. Wide fibre and parenchyma

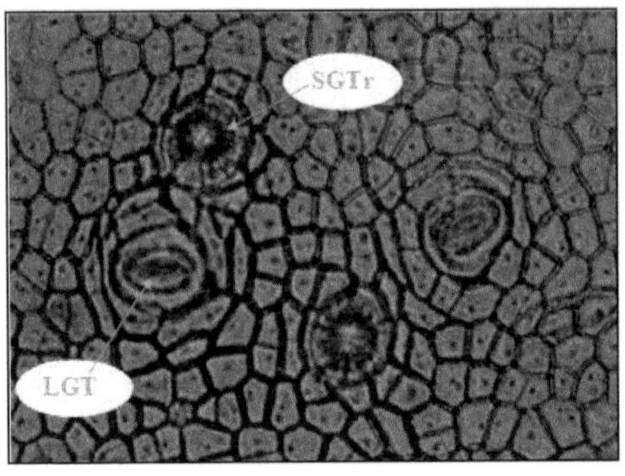

g. Venation pattern

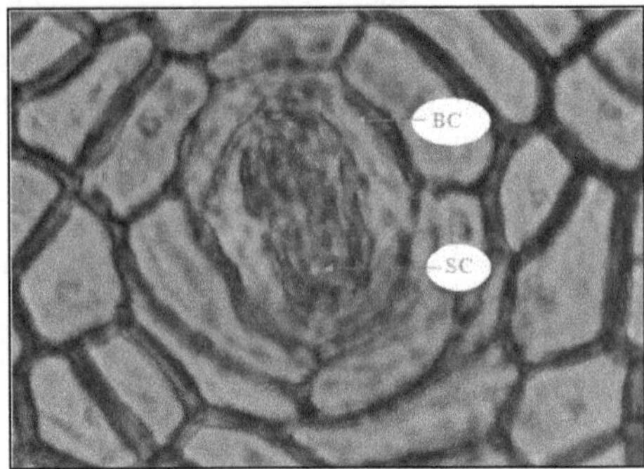

h. Venation pattern

end wall perforation is elliptical and oblique (Plate 25.4d). The lateral wall pits are minute, circular and multiseriate.

(ii) Fibres

The fibres are abundant in the powdered preparation. There are two types of fibres seen in the powder (Plate 25.4d).

(a) Narrow fibres: These fibres are 450 μm to 1mm long and 10 to 20 μm wide. They have reduced lumen, thick lignified walls and with gradually tapering ends on both sides (Plate 25.4e).

(b) Wide fibres: The wide fibres are as long as the narrow fibres but their walls are very thin and with wider lumen than the narrow fibres. The wide fibres have bluntly conical ends and they do not have tapering pointed tips (Plate 25.4f).

(iii) Parenchyma

The parenchyma cells are also equally abundant in the powdered sample. Most of the parenchyma cells are large spherical bodies with thin walls and with prominent nuclei. These spherical parenchyma cells are 140 to 240 μm in diameter (Plate 25.4e,f). There are also elongated, rectangular and thin walled parenchyma cells in the powdered sample. The rectangular parenchyma cells are 16 μm long and 20 to 40 μm wide (Plate 25.4e,f).

(v) Glandular Trichomes

The glandular trichomes are abundant in the powder sample. They may be seen in the surface view attached on the fragments of epidermal peeling. There are two types of glandular trichomes which are distinctly different in size. Some of the trichomes are large, circular with central stalk cell and with terminal spherical body of 4 cells. The terminal body appears circular in outline. The small trichomes are 10 μm in diameter and are also circular in outline. They have a short and wide stalk and with a thick neck cell. The neck bears a spherical body which is 30 μm in diameter and is made up of 4 cells (Plate 25.4g,h). The glands are of capitates type.

Powder Analysis of the Whole Plant

Physico-chemical Constant

The physico-chemical parameters like ash and extractive values, fluorescence analysis of whole plant of *B. plumbaginifolia* were determined. Preliminary phytochemical screening was also performed and results are presented below.

The powdered whole plant of *B. plumbaginifolia* was investigated for physic-chemical constant like total ash value, water soluble ash, acid insoluble ash, sulphated ash and extractive values (Table 25.1). The total ash content of the powdered whole plant of *B. plumbaginifolia* is 9.84 per cent. The extractive value of methanol is more than the other solvents investigated in the present study.

Table 25.1a: Ash Values of the Powdered whole Plant of *B. plumbaginifolia**

Sl.No.	Type of Ash	Per cent of Ash Values
1.	Total ash value of powder	9.84 ± 0.04
2.	Water soluble ash	3.24 ± 0.03
3.	Acid insoluble ash	2.86 ± 0.02
4.	Sulphated ash	10.32 ± 0.11

* All values are mean of implicate determination.

Table 25.1b: Extractive Values of the Powdered whole Plant of *B. plumbaginifolia**

Sl.No.	Nature of the Extract	Per cent of Extractive Values
1.	Petroleum ether	6.58 ± 0.05
2.	Benzene	5.76 ± 0.03
3.	Chloroform	6.94 ± 0.02
4.	Acetone	6.58 ± 0.04
5.	Methanol	8.98 ± 0.05
6.	Ethanol	8.54 ± 0.03
7.	Water	7.56 ± 0.04

* All values are mean of implicate determination.

Fluorescence Analysis

The results of fluorescent analysis of whole plant of *B. plumbaginifolia* are shown in Table 25.2. The powder form of the whole plant of *B. plumbaginifolia* emitted pale green under day light, dark green under short and long UV light. The whole plant powder shows the characteristic fluorescent green colour when treated with 1NH_4Cl, HCl, H_2SO_4, HNO_3+NH_3, 50 per cent HNO_3, Petroleum ether and acetone.

Preliminary Phytochemical Screening

Petroleum ether, benzene, chloroform, methanol and ethanol extracts of whole plant of *B. plumbaginifolia* are qualitatively analysed for the presence of different phytoconstituents and the results are presented in Table 25.3. The methanol and ethanol extracts of whole plant of *B. plumbaginifolia* shows the presence of alkaloid, anthraquinone, catechin, flavonoid, phenol, quinone, saponin, steroid, tannin, tepernoid, sugar, glycoside and xanthoprotein.

Table 25.2: Fluorescence Analysis of the Powdered whole Plant of *B. plumbaginifolia*

Treatment	*Under Day Light*	*Under UV light*	
		245 nm	*365 nm*
Powder as such	Pale green	Dark green	Dark green
Powder + 1N Aqueous NaOH	Yellowish green	Greenish yellow	Dark green
Powder + 1N Alcoholic NaOH	Yellowish green	Greenish yellow	Dark green
Powder + 1N HCL	Green	Fluorescent green	Brown
Powder + Conc. HCL	Light green	Fluorescent green	Dark green
Powder + Conc. H_2SO_4	Green	Fluorescent green	Dark green
Powder + 50 per cent H_2SO_4	Green	Light green	Black green
Powder + Con.HNO_3	Light green	Green	Pink
Powder + 40 per cent NaOH + 10 per cent Lead Acetate	Pale green	Dark green	Violet
Powder + Acetic acid	Green	Light green	Dark green
Powder + Ferric Chloride	Dark green	Dark green	Dark green
Powder + Chloroform	Yellowish green	Yellowish green	Brown
Powder + Benzene	Green	Pale yellow	Dark green
Powder + Petroleum ether	Yellowish green	Fluorescent green	Dark green
Powder + Methanol	Yellowish green	Pale Yellow	Dark green
Powder + Ethanol	Yellowish green	Green	Dark green
Powder + acetone	Dark green	Fluorescent green	
Powder + HNO_3 + NH_3	Yellowish green	Yellowish green	Dark brown
Powder + 50 per cent HNO_3	Yellowish green	Yellowish green	Dark brown

Discussion

The macroscopic and microscopic evaluations of any plant drug are considered to be the preliminary steps for establishing their quality control profile. According to WHO, botanical standards should be proposed as a protocol for the diagnosis of the herbal drugs.

Literature dealing with anatomy of whole plant of *B. plumbaginifolia* is minimal. The present study attempts a modest comprehensive investigation of whole plant of *B.plumbaginifolia* claims has therapeutic qualities. The present investigation has laid dawn set of anatomical features of the whole plant which can be employed for its botanical diagnosis. The salient features of identification of the whole plant sample are as follows.

Table 25.3: Preliminary Phytochemical Screening of Powdered whole Plant of *B. plumbaginifolia*

Tests	*Petroleum Ether*	*Benzene*	*Ethyl acetate*	*Methanol*	*Ethanol*
Alkaloid	–	–	+	+	+
Anthraquinone	–	–	+	+	+
Catechin	+	+	–	+	+
Coumarin	–	+	–	–	–
Flavonoid	–	–	+	+	+
Phenol	+	+	+	+	+
Quinone	+	+	+	+	+
Saponin	–	–	–	+	+
Steroids	+	–	–	+	+
Tannin	–	+	+	+	+
Terpenoids	+	+	+	–	–
Sugar	–	–	+	+	+
Glucoside	–	–	+	+	+
Xanthoprotein	+	+	+	+	+
Fixed oil	+	–	–	–	+

+ Present ; – Absent.

Salient Diagnostic Features

- ☆ Root is circular with slight wavy outline. It consists of single ring of radially oblong cortical air chambers and the stele with a ring of about six collateral xylem and phloem elements. Xylem elements are wide, angular and thick walled.
- ☆ Both young and old stems are circular with even outline. The young stem has intact epidermis and heterogenous cortex comprising of an outer collenchymatous cortex, a middle zone with discrete masses of thin walled fibres and an inner zone of dilated, elliptical, tangentially elongated parenchyma cells.
- ☆ The vascular cylinder of the stem consists of thick, continuous layer of secondary phloem encircling thick secondary xylem cylinder which consists of long radial multiples of vessels and thick walled and lignified fibres.
- ☆ Mature stem has narrow periderm layers, a parenchymatous outer cortex, a middle zone of air chambers and a parenchymatous inner cortex. The vascular tissues are similar to those of the young stem.
- ☆ The petiole is planoconvex with narrow adaxial groove. There is a wide bowl shaped, collateral main vascular strand and one or two

accessory, circular small bundles located in the adaxial lateral part. The accessory strands have central xylem and outer circle of phloem and sclerenchymatous bundle sheath.

- ☆ The leaf is dorsiventral with smooth lamina and with prominent midrib possessing wide, short adaxial hump and a wide semi-circular abaxial part. The midrib is 800 μm thick.
- ☆ The ground tissue in the abaxial part is parenchymatous and in the adaxial cone it is collenchymatous.
- ☆ The vascular system consists of a wide bowl shaped main strand, which is collateral and comprises of several uniseriate xylem elements and phloem elements occur abaxially in small discrete units.
- ☆ The palisade tissue is horizontally transcurrent along the adaxial part of the midrib.
- ☆ Lamina is 250 μm thick. It is dorsiventral with single short conical palisade cell layer and two or three lobed spongy parenchyma cell layers.
- ☆ Some of the epidermal cells are modified into large cylindrical lithocysts possessing elongated cylinder of calcium carbonate crystals of cystolith.
- ☆ The adaxial and abaxial epidermal cells have highly wavy anticlinal walls. The abaxial epidermis is stomatiferous consisting of diacytic stomata.
- ☆ Venation of the lamina is densely reticulate. The vein-islets are well defined and possess either simple unbranched vein-terminations or dendroid type of terminations.
- ☆ Powdered preparation of the plant sample shows narrow fibres and wide fibres; narrow, cylindrical and long vessel elements with prominent tails and with oblique end wall perforations and abundant parenchyma cells.
- ☆ There are also abundant glandular, capitate types of trichomes. The gland may be either small or much larger. Both the small and large trichomes have a shot stalk cell, a narrow neck cell and large spherical body cells.

Physio-chemical Constant

Ash values are used to determine quality and purity of crude drug. The water soluble ash is used to estimate the amount of inorganic compound present in drugs. The acid insoluble ash consist mainly silica and indicate contamination with earthy material. The ash content gives an idea about the inorganic content of powdered whole plant under investigation and thus the quality of the drugs can be assessed.

Ash value determination is a very important tool to access the quality of herbal raw material since higher ash value is an indication of adulteration and or improper processing of raw material. It indicates presence of various impurities like carbonate, oxalate and silicate. The percentage variation of the weight of ash in certain drugs from sample to sample is very small and any marked difference indicates a change in quality.

Fluorescence Analysis

The fluorescent character of powdered drug plays a vital role in the determination of quality and purity of the drug material. In the present study, whole plant powder treated with various reagents showed characteristic fluorescence at 254nm and 366 nm wavelength. Some constituents show fluorescence in the visible range in day light. The ultra violet light produces fluorescent in many natural products which do not visibly fluoresce in day light. If substance themselves are not fluorescent, they may often be converted into fluorescent derivatives or decomposition products by applying different reagents. Hence crude drugs are often assessed qualitatively in this way and it is an important parameter for pharmacognostic evaluation of crude drugs (Kumar and Kumar, 2012; Zhao *et al.*, 2011).

Phytochemical Analysis

The plants are considered as biosynthetic laboratory for a multitude of compounds that exert physiological effects. Secondary metabolites are the compounds which are responsible for imparting therapeutic effects. The preliminary phytochemical anlysis will give an idea about the chemical nature of the drug. The information obtained will be useful in further structural characterization of the nature of constituents present in the plant material under investigation. It could also be helpful to extract out particular constituents by a particular solvent. The preliminary qualitative phytochemical investigation of methanol and ethanol extracts of *B. plumbaginifolia* whole plant was performed which shows the presence of alkaloid, anthraquinone, catechin, favonoid, phenol, quinone, seponin, steroid, tannin, sugar, glycoside and xanthoprotein. Some secondary metabolites or phytochemicals have been reported in the literature to have pharmacological activities. For instance, alkaloids may be responsible for the anticancer, antidiabetics, antiaging and antiviral activities. (Evans and Trease, 2002). The presence of tannins may be responsible for ability to cure diseases such as diabetes, diarrhea, sore throat, skin ulcer and dysentery. The presence of flavonoids may be responsible for its uses to cure cancer, inflammations and allergies (Cushine and Lamb, 2005). Flavonoids are effective water soluble antioxidants and free radical scavengers, which can prevent oxidative cell damage and exhibit anticancer effect (Doss and Phil, 2009) presence of phenolic compounds in the plants indicate that the plant may posseses antimicrobial properties (Seow *et al.*, 2013). The occurrence of majority of secondary metabolites in the whole plant of *B. plumbaginifolia* is a promising sign of discovery of new medicinal principles.

Standardization is an essential analytical aspect for the study of identity, purity and quality of crude drug sample of plant origin. Chemical and physico-chemical analysis reveal useful information which is of utmost importance for the quality control of *B.plumbaginifolia* whole plant to be used as crude drugs. Since *B.plumbaginifolia* whole plant is known for its various medicinal properties, the study could be useful to supplement information with respect to its identification, authentication and standardization. The information generated can also be useful for preparation of monograph of the plant, which could be incorporated in the preparation of Indian Herbal Pharmacopoeia.

References

Anonymous 1996. India pharmacopoea vol. I and II. Government of India, Ministry of Health and Family Welfare. The controller of Publications. Civil Lines, Delhi – 110054.

Brinda, P., Sasikala, P. and Purushothaman, K.K. 1981. Pharmacognostic studies on *Merugan Kizhangu. Bull-Med. Ethnobot*. Res. 3: 84 – 96.

Cushine T.P.T. and Lamb A.J. 2005. Antimicrobial activity of flavonoids. *Int. J. Antimicro. Age*. 343 – 356.

Doss A and Phil M. 2009. Preliminary phytochemical screening of some Indian medicinal plants. *Anc. Sci. Life*. 29: 12-16.

Evans W C and Trease G E. 2002. Pharmacognosy. 15th Edition W.R. Saunders. 214 – 314.

Johanson, D.A. 1940. Plant Microtechnique, 1st edition. McGraw Hill Book Co. Inc., New York and London. p. 104–106.

Kadam, P.V., Deoda R.S., Shivatare, R.S., Yadav, K.N. and Patil, M.J. 2012. Pharmacognostic, phytochemical and physico-chemical studies of *Mimusops elengi* Linn stem bark (Sapotaceae). *Der. Pharm. Lettr*. 4: 607 – 613.

Kumar, D and Kumar, A.O.P. 2012. Pharmacognostic evaluation of stem bark of *Pongamia pinnata* (L.) Pierre. *Asian Pac. J. Trop. Biomed*. 2: S543 – S546.

Lala PK. 1993. Lab manuals of Pharmacognosy CSI publications and Distributers, Kolkatta.

O'Brien, T.P., Feder, N. and McGull. M.E. 1964. Polychromatic staining of plant cell walls by toluidine blue – O. *Protoplas*. 59: 368 – 373.

Seow, L.J., Beh, H.K., Sadikun A and Asmawi, M.Z. 2013. Preliminary phytochemical and physico-chemical characterization of *Gynura segetum* (Lour) Merr. (Compositae) leaf. *Trop. J. Pharmaceu. Res*. 12: 777 - 782.

Shameer, M.C., Saeeda, V.P, Madhusoodanan, P.V. and Benjamin, S. 2008. Direct organogenesis and somatic embryogenesis in *Beloperone plumbaginifolia* (Jacq.) Nees. *Indian J. Biotech*. 8: 132 – 155.

Sharma, S.K, 2004. Recent approach to herbal formulation development and standardization, http//pharmainfo.net.

Who, 1998. World Health Organization. Quality control method for medicinal plant material. WHO Library. 110 – 115.

Zhao Z, *et al.*, 2011. Macroscopic identification of Chinese medicinal materials: Traditional experiences and modern understanding *J. Ethnopharmacol*. 131: 556 – 561.

2018, Ethnomedicinal Plants: A Biodiversity Treasure Pages 607–615
Editors: V.R. Mohan, A. Doss, P.S. Tresina and V. Sornalakshmi
Published by: ASTRAL INTERNATIONAL PVT. LTD., NEW DELHI

Chapter 26

GC-MS Analysis and Antibacterial Activity of *Myristica fragrans* Seed Extracts against the Lower Respiratory Tract Pathogen *Klebsiella pneumoniae*

T.P. Kumari Pushpa Rani and A. Doss

Department of Microbiology,
Kamaraj College, Thoothukudi
E-mail: androdoss@gmail.com

Introduction

Several plants were reported for their therapeutic and pharmaceutical virtues, especially antioxidant, anti-tumor and anti-infectious activities. A big part of the world's population still relies on the benefits of food for the treatment of common illnesses (Zhang, 2004). These benefits are due to their big content of bioactive compounds (Cheruvank, 2004). Since the introduction of antibiotics there has been tremendous increase in the resistance of diverse bacterial pathogens (Cohen, 1992; Gold, 1996). *Myristica fragrans*, commonly known as nutmeg, belongs to the Myristicaceae family. *M. fragrans* is a common flavoring agent in Indian cooking. It also possesses medicinal and aromatic properties. It serves as an antidiarrhoeal, stomachic stimulant, carminative, intestinal catarrh and colic to stimulate appetite. The chemical composition of the seed of *Myristica fragrans* contains sugars, phenols, proteins, *etc.* (Abdurrasheed and Janardanan, 2009).

Materials and Methods

Collection of Plant Material

The fresh fruits of *M. fragrans* were collected from the local areas. It was then cut into small pieces and dried for 7 days at room temperature (25°C). The dried samples were ground into fine powder and kept away from heat, moisture, and sunlight.

Preparation of Plant Extracts

500g dry powder of *Myristica fragrans* was sequentially extracted with hexane, toluene, tetrahydrofuran, methanol and water using the Soxhlet apparatus on the water bath for 12 h each. Each of the mixtures was carefully filtered using filter paper (Whatman No. 1) and concentrated using a rotary evaporator. The extracts were stored in sterile bottles at -18 °C kept as aliquots until further evaluation.

Test Organism

Twenty five isolates of *Klebsiella pneumoniae* were recovered from sputum samples and were used in this study. All the samples were collected from different patients in Kanyakumari District. The isolates were sub-cultured at 37 °C for 24 h and maintained on nutrient agar slants. The pathogenic cultures were inoculated into sterile nutrient broth and incubated at 37°C for 3 h.

Preparation of Discs for Antibacterial Activity

The sterile discs (10 mm) were soaked in the diluted extracts. The prepared discs were dried in controlled temperature to remove excess of solvent and used for study.

Antibacterial Activity Using Disc Diffusion Method

The modified paper disc diffusion method was employed to determine the antibacterial activity of the plant extracts. The prepared *K. pneumoniae* inoculum was spread over the Muller Hinton agar plate using a sterile cotton swab in order to obtain uniform microbial growth. Then the prepared antibacterial discs were kept over the lawn and pressed slightly along with positive and negative control. Rifampicin 10 mcg/disc (Hi-Media) was used as positive control. The plates were incubated for 18-24 h at 37°C. The antibacterial activity was evaluated for 5 mg/disc and diameter of inhibition zones were measured (Bauer, 1966).

Phytochemical Screening

Phytochemical screening of plant extracts was carried out qualitatively in the presence of carbohydrates, terpenoids, tannins, flavonoids, phenolic compounds, saponins, phlobactanin, quinones and alkaloids (Harborne, 1998).

GC - MS Analysis

GC-MS technique was used in this study to identify the components present in the methanol extract. GC-MS technique was carried out at The *Cashew Export Promotion Council* (*CEPC*), Kollam, Kerala. GC-MS analysis was carried out on a GC clarus 500 Varian, USA system comprising a AOC-20I auto sampler and gas chromatograph interfaced to a mass spectrophotometer instrument employing the following conditions: Column Elite-1 fused silica capillary column (30mm×0.25mm I.D ×1 μ M df, composed of 100 per cent Dimethyl poly siloxane), operating in electron impact mode at 70 eV; helium (99.999 per cent) was used as carrier gas at a constant flow of 1ml/min and an injection volume (0.5 μl) was employed (split ratio of 10:1) injector temperature 250 °C; ion-source temperature 280 °C. The oven temperature was programmed from 110 °C (isothermal for 2 min), with an increase of 10 °C/min, to 200 °C, then 5 °C/min to 280 °C, ending with a 9 min isothermal at 280 °C. Mass spectra were taken at 70 eV; a scan interval of 0.5 seconds and fragments from 45 to 450 Da. Total GC running time is 46min.

Result and Discussion

The medicinal properties of plant species have made an outstanding contribution in the origin and evolution of many traditional herbal therapies. Over the past few years, medicinal plants have regained a wide recognition due to an escalating faith in herbal medicine (Pooja *et al.*, 2012).

In this study, we have tested the hexane, toluene, tetrahydrofuran, methanol and water extracts of *M. fragrans* for their antibacterial activity against *K. pneumoniae* strains. Phytochemical analysis of all extracts showed the presence of alkaloids, carbohydrates, flavonoids, terpenoids, phlobatanins and quinones, but all extracts showed absence of phenolic compounds and tannins. Saponins were observed except in methanol and water extracts. The results are shown in Table 26.1. Our studies were similar to the observations made by other researchers on the phytochemical compounds elucidated in the hexane extract of *M. fragrans* for alkaloids, tannins, flavonoids, saponins and phenolic compounds. In one of the earlier studies, methanol extracts showed presence of only alkaloids and other tests showed negative result (Reena Saxena and Pramod Patil, 2012). Earlier studies reported positive phytochemical analysis of phenols in both hexane and methanol extracts (Pooja *et al.*, 2012). In another study, methanol extract of *M. fragrans* showed the presence of tannins, flavonoids, alkaloids, terpenoids, phenolic compounds

Table 26.1: Phytochemical Tests for *M. fragrans* using different Solvents

Phytochemical Tests	*Hexane*	*Toluene*	*Tetrahydro-furan*	*Methanol*	*Water*
Alkaloids	+	+	+	+	+
Carbohydrates	+	+	+	+	+
Saponins	–	–	–	+	+
Phenolic compounds	–	–	–	–	–
Flavanoids	+	+	+	+	+
Terpenoids	+	+	+	+	+
Phlobatanins	+	+	+	+	+
Tannins	–	–	–	–	–
Quinones	+	+	+	+	+

+: Present; –: Absent.

and absence of saponins (Essam *et al.*, 2012). However in the present investigation the methanol extract of *M. fragrans* revealed the presence of alkaloids, flavonoids, terpenoids, tannins, saponins, carbohydrates and phenolic compounds.

The antibacterial activity of *M. fragrans* using toluene, tetrahydrofuran and methanol solvents against twenty five strains of *K. pneumoniae* was initially assessed by disc diffusion method. Rifampicin was used as standard antibiotic. Methanol was found to be the best extract and showing high activity against *K.pneumoniae* strains (Krishnan, 2010). The results were shown in Table 26.2.

Table 26.2 Antimicrobial Activity of *M. fragrans* against Strains of *K. pneumoniae*

Isolates of K. pneumoniae	*Zone of inhibition in mm*					
	Rifampicin	*Toluene*	*Tetrahydro-furan*	*Methanol*	*Hexane*	*Water*
K1	22	10	9	16	5	-
K2	24	11	10	18	6	-
K3	22	10	10	19	8	-
K4	23	12	9	18	-	-
K5	25	12	12	13	-	-
K6	22	11	11	14	-	6
K7	26	12	11	12	-	-
K8	20	10	9	11	-	-
K9	22	9	10	13	-	-
K10	21	9	9	10	-	-
K11	20	9	9	14	7	-
K12	23	11	11	13	-	-
K13	26	12	12	11	-	-
K14	24	10	12	11	-	-
K15	28	12	10	15	-	-
K16	22	10	10	16	-	-
K17	23	10	9	15	-	-
K18	23	11	9	20	-	-
K19	25	12	11	17	-	-
K20	27	11	12	16	-	-
K21	26	11	9	12	-	-
K22	28	11	10	12	-	-
K23	25	10	12	10	-	-
K24	25	10	12	13	-	-
K25	22	9	10	15	-	-

Seventeen compounds were identified in methanolic extract of *M. fragrans* by GC-MS analysis. The identified bioactive compounds possess many biological properties. The active principle Molecular Weight (MW), Molecular Formula

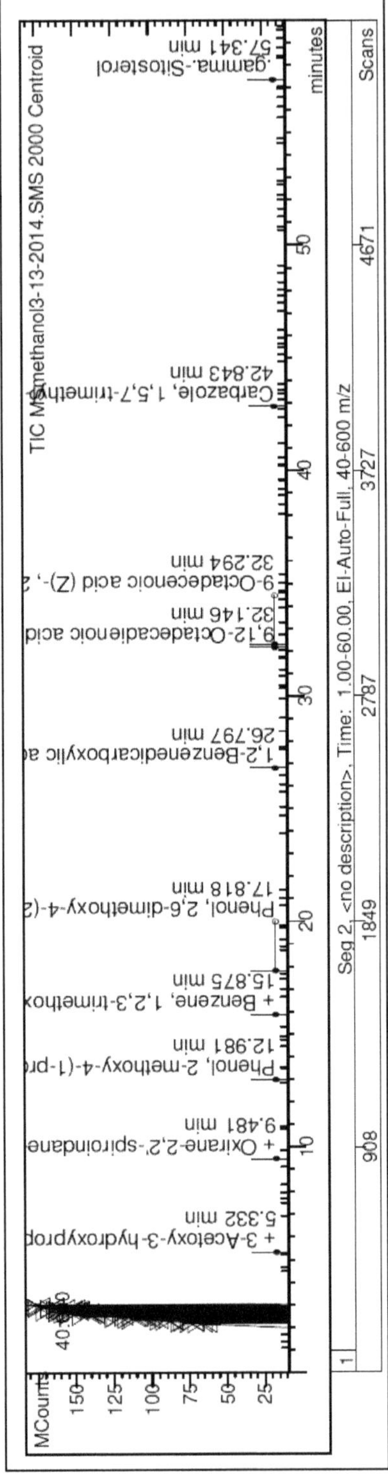

Figure 26.1: Phytochemical Compounds of the Methanolic Extract of the whole Plant of *M. fragrans* using GC-MS Analysis.

(MF), Retention Time (RT) and their bioactivity are presented in Figure 26.1 and Table 26.3. The compounds detected in the GC MS analysis namely, Safrole and Isoeugenol were used as the pesticide and flavoring agent respectively as listed in earlier literature. In an earlier study, ethanolic extract of *Vitex altissima* characterized by GC-MS analysis revealed the presence of Eugenol which possess analgesic, anesthetic, allergenic, antibacterial, anticonvulsant, anti-inflammatory, antioxidant, antipyretic, antisalmonella, antistaphylococcus and antiseptic activities. The similar compound Eugenol was also identified in the GC-MS analysis in the present study. Octadeca-9, 12-dienoic acid identified in the GC MS analysis of *Vitex altissima* revealed hypocholesterolemic, 5-Alpha reductase inhibitor, antihistaminic, insectifuge, and antiacne activities (Sahaya Sathish *et al.*, 2012).

Table 26.3: Phytochemical Compounds of the Methanolic Extract of the whole Plant of *M. fragrans* using GC-MS Analysis

Sl. No.	*RT*	*Peak Area*	*Name of the Compound*	*Chemical Formula*	*Mol. Wt*	*Chemical Structure*
1	5.332	14.5%	3- Acetoxy-3-hydroxy propyl	$C_5H_9O_3$	117.12	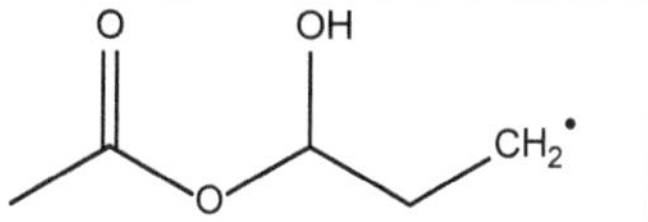 O OH O $CH_2^{\bullet}$
2.	5.805	1.7%	Naphthalene-2-yl tetradecanoate	$C_{24}H_{34}O_2$	354.24	 O O
3.	5.942	1.6%	(2,4-dinitro-phenyl) hydrazine	$C_6H_6N_4O_4$	198.04	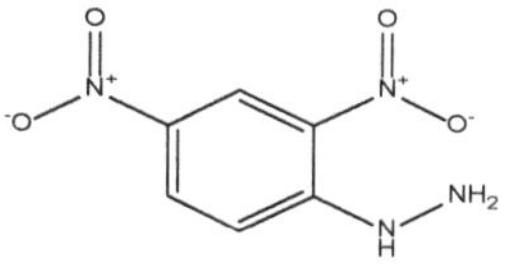 O O $^-$O N$^+$ N$^+$ O$^-$ N H NH_2
4	7.584	0.4%	1-tert-butyl-3-isopropyl-5-methylbenzene	$C_{14}H_{22}$	190.17	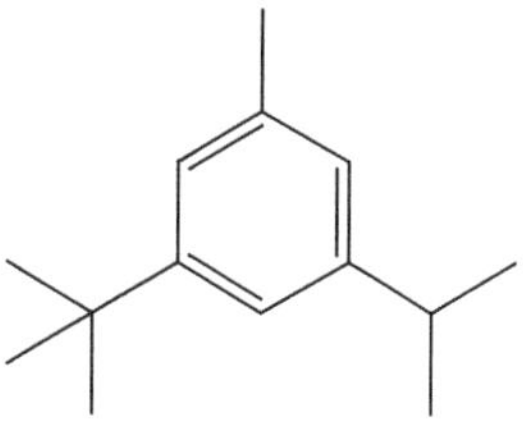
5	8.912	0.8%	5-allylbenzo[d] [1,3]dioxole	$C_{10}H_{10}O_2$	162.19	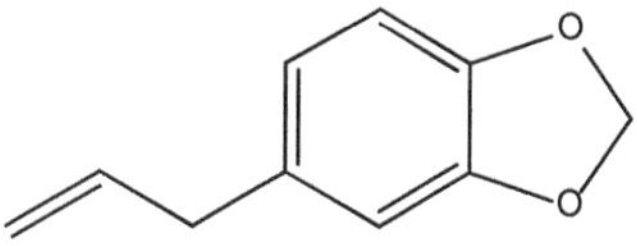 O O
6	9.481	6.2%	1,3-dioxo-1,3-dihydrospiro [indene-2,2'-oxirane]-3,3'-dicarbonitrile	$C_{12}H_4N_2O_3$	224.17	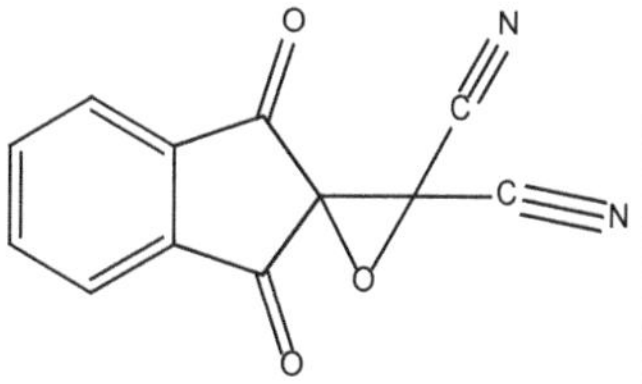 O N C C N O O

Sl. No.	RT	Peak Area	Name of the Compound	Chemical Formula	Mol. Wt	Chemical Structure
7	10.418	0.4%	4-allyl-2-methoxyphenol	$C_{10}H_{12}O_2$	164.08	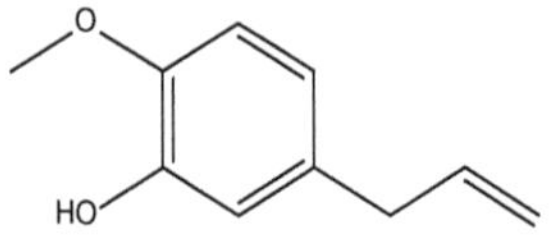
8	12.981	6.5%	2-methoxy-4-(prop-l-enyl) phenol	$C_{10}H_{12}O_2$	164.20	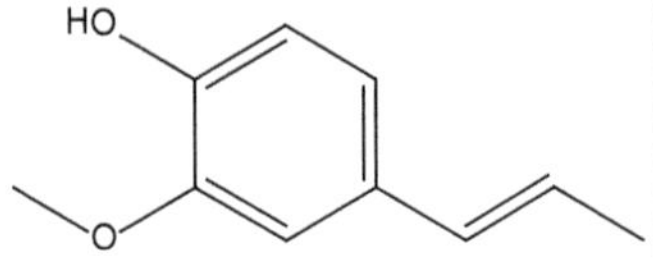
9	13.721	1.2%	2-(4-hydroxy-2,5-dioxo-2,5-dihydrofuran-3-yl) Acetic acid	$C_{12}H_{16}O_5$	240.25	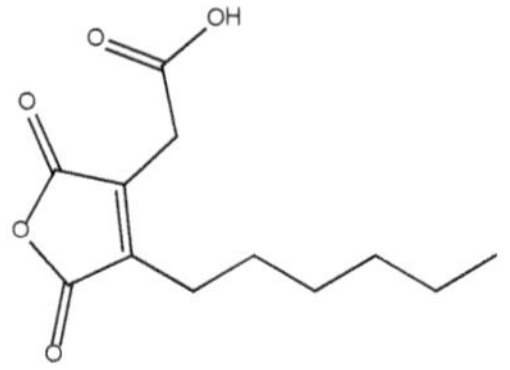
10.	15.074	11%	6-allyl-4-methoxy-benzo[d][1,3] dioxole	$C_{11}H_{12}O_3$	192.08	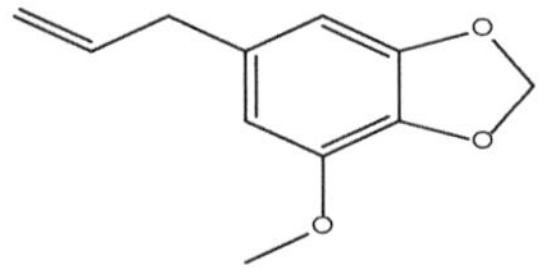
11	15.875	13.2%	5-allyl-1,2,3-trimeth-oxybenzene	$C_{12}H_{16}O_3$	208.25	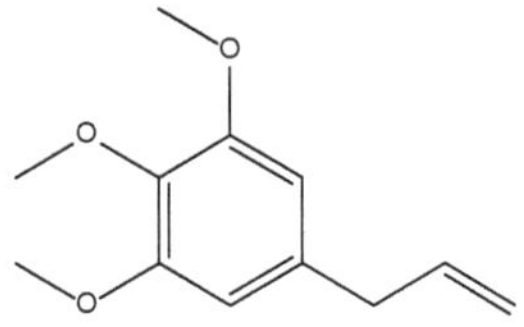
12	17.818	9.8%	4-allyl-2,6-dimeth-oxyphenol	$C_{11}H_{14}O_3$	194.23	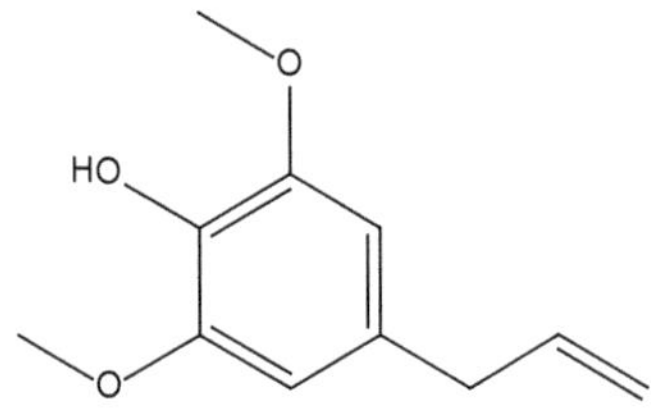
13	26.798	5%	Diisobutyl phthalate	$C_{16}H_{22}O_4$	278.34	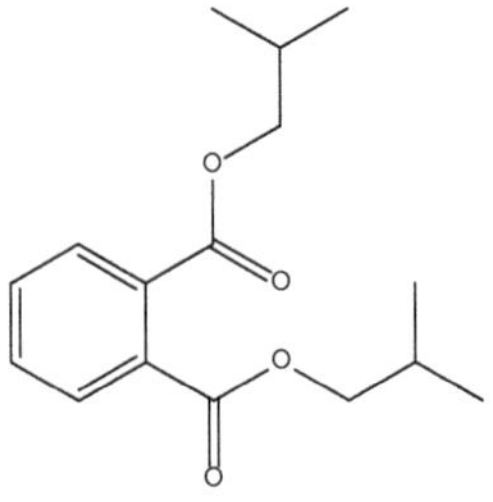

Sl. No.	RT	Peak Area	Name of the Compound	Chemical Formula	Mol. Wt	Chemical Structure
14	32.146	0.9%	Octadeca-9,12-dienoic acid	$C_{18}H_{32}O_2$	280.24	O, OH
15	32.294	0.4%	Heptadec-9-enoic acid	$C_{18}H_{34}O_2$	282.26	O, OH
16	42.843	0.5%	1,5,7-trimethyl-9H-carbazole	$C_{15}H_{15}N$	209.29	H N
17	57.341	0.07%	Gamma sitosterol	$C_{29}H_{50}O$	414.71	HO

Conclusion

Therefore, *M. fragrans* fruit extract possesses a significant inhibitory effect towards the potentially serious *Klebsiella pneumoniae* isolates. These results therefore support the traditional use of this plant product in pain and related conditions. However, further studies are necessary to examine the underlying mechanisms of the above mentioned phytochemical constituents responsible for these pharmacological activities.

References

Bauer A.N., Kirby W.M.M., Sherries J.C and Truck M., 1966. Antibiotic susceptibility testing by a standardized single disc method. *American Journal of Clinical Pathology*, 45(4): 493-436.

Cheruvank, H., 2004. Method for Treating Hepercholerolemia and Atherosclerosis. United States *Journal of Pathology*, 6(4): 733-799.

Cohen, M.L., 1992. Epidemiology of drug resistance, implications for a post antimicrobial era. *Science*, 257: 1050-1055.

Essam F, Al-Jumaily and Maytham H. A, Al-Amiry, 2012. Extraction and Purification of Terpenes from Nutmeg (*Myristica fragrans*). *Journal of Al-Nahrain University*, 15 (3): 151-160.

Gold SG. and Moellering RC, 1996. Antimicrobial drug resistance. *N. Engl. J. Med*,

335: 1445-1453.

Harborne J.B.Phytochemical methods: A guide to modern technique of plantanalysis, Champman and Hall, London, 1998.

K. M. Abdurrasheed, C. Janardanan, J. *Spices Aromatic Crops*, 2009, 18(2), 108.

Krishnan.N., 2010. Antimicrobial activity evaluation of *Cassia spectabilis* leaf extracts. *International Journal of Pharmaclogy*, 4 (5): 1-5.

Pooja V, Sanwal H, Goyal A, Bhatnagar S and Ashwani K. Srivastava, 2012. Activity of *Myristica fragrans* and its effect against filamentous and non-filamentous fungus, *International Journal of Pharmacy and Pharmaceutical Sciences*,4 (1): 1-3

Reena Saxena and Pramod Patil, 2012. Phytochemical Studies on *Myristica fragrans* Essential Oil, *Biological Forum – An International Journal of Pharmacology*, 4(2): 62-64.

Sahaya Sathish S, Janakiraman N, Johnson M, 2012. Phytochemical Analysis of *Vitex altissima* L. using UV-VIS, FTIR and GC-MS, *International Journal of Pharmaceutical Sciences and Drug Research*, 4(1): 56-62.

Zhang X. Traditional Medicine and it's Importance and Protection. In: S. Twarog and P. Kapoor, (eds), "Protecting and Promoting Traditional Knowledge. Part 1, New York: United Nations; 2004, pp: 3-6.

Index

A

Abortifacient 51, 56, 57, 59, 109, 114, 117, 154, 331, 455

Abrus precatorius 9, 21, 52, 53, 56, 57, 58, 73, 111, 128, 151, 154, 155, 159, 176, 195, 196, 206, 210, 266, 295, 304, 313, 317

Acalypha fruticosa 9, 21, 52, 128, 158, 159, 347

Acalypha indica 9, 21, 58, 59, 73, 115, 117, 128, 148, 176, 241, 251, 255, 266, 295, 347, 353, 366, 376, 436, 445, 452, 472, 475, 477

Acanthaceae 9, 10, 14, 17, 47, 49, 75, 76, 77, 82, 86, 88, 106, 107, 129, 130, 134, 135, 138, 143, 173, 177, 180, 183, 199, 200, 204, 207, 233, 234, 241, 251, 252, 253, 254, 255, 277, 297, 298, 313, 315, 347, 349, 354, 367, 371, 383, 412, 428, 439, 441, 442, 443, 446, 447, 448, 449, 473, 474, 551, 552, 590

Acanthus ilicifolius 383

Achyranthes aspera 9, 21, 56, 58, 59, 73, 128, 149, 151, 154, 161, 207, 210, 265, 266, 295, 438, 451, 452, 458, 461, 473

Acicennia marina 383

Acorus calamus 9, 21, 58, 61, 73, 118, 128, 176, 177, 223, 251, 267, 347, 374, 472

Aegle marmelos 9, 21, 53, 54, 55, 59, 128, 151, 154, 155, 159, 204, 210, 223, 242, 251, 267, 295, 304, 495

Agasthiarmalai Biosphere Reserve 67, 68, 70, 72, 186, 552, 590

Agricultural knowledge 189

Ailanthus excelsa 198, 569, 570, 571, 572, 578, 585, 586, 587

Aloe vera 9, 23, 56, 129, 149, 151, 156, 177, 202, 206, 209, 251, 255, 267, 295, 304, 348, 366, 475, 476, 477, 488

Alpinia calcarata 20, 38, 45, 74, 177, 348

Alpinia galanga 177

Alstonia scholaris 9, 23, 55, 56, 57, 59, 74, 114, 115, 117, 118, 367

Alternanthera sessilis 129, 162, 267, 452, 461

Amaranthus viridis 74, 177, 199, 208, 268, 295, 348, 367, 438, 457, 488, 514

Ammannia baccifera 74, 129, 159, 162

Amorphophallus sylvaticus 9, 23, 485, 488

Androgarphis paniculata 177

Andrographis paniculata 9, 23, 55, 56, 59, 60, 61, 75, 85, 113, 114, 115, 117, 129, 151, 154, 156, 161, 176, 207, 210, 233, 241, 255, 257, 258, 367, 371, 376, 412, 413, 473, 474, 477

Angiosperm diversity 425

Anisomeles indica 9, 23, 75, 111, 112, 113, 115, 129, 177, 252, 313, 319, 331, 333, 334, 336, 440, 452

Anisomeles malabarica 9, 23, 58, 59, 75, 117, 129, 159, 206, 210, 241, 268, 313, 319, 331, 335, 336, 348, 367, 376, 440, 452, 458, 459

Annona muricata 177

Anti-bacterial plants 415

Anti-cancer plants 413

Anti-diabetic plants 409

Anti-hepatotoxicity plants 411

Anti-Inflammatory plants 418

Antimicrobial activities 381

Antioxidant plants 420

Anti-viral plants 417

Apocynaceae 9, 11, 14, 18, 19, 47, 49, 73, 74, 78, 91, 92, 94, 106, 107, 130, 131, 134, 138, 140, 141, 143, 178, 181, 182, 203, 204, 205, 235, 241, 251, 253, 254, 271, 294, 296, 299, 313, 316, 351, 358, 359, 367, 372, 374, 409, 439, 441, 443, 446, 447, 449, 450, 492, 495, 532

Araceae 9, 49, 73, 75, 90, 106, 108, 128, 132, 177, 264, 267, 268, 347, 352, 367, 375, 485, 488, 489

Argemone mexicana 75, 113, 114, 129, 148, 149, 156, 159, 348, 367, 376, 436, 452, 458

Argyreia pilosa 9, 23, 61, 485, 514

Aristolocaceae 106

Aristolochia indica 9, 23, 57, 58, 59, 60, 61, 75, 93, 111, 114, 115, 117, 177, 296, 367, 368, 373

Asclepias curassavica 9, 24, 204, 348

Asparagus racemosus 10, 24, 53, 55, 57, 58, 151, 154, 156, 177, 233, 242, 349, 367, 485

Asteraceae 10, 12, 19, 49, 82, 92, 94, 106, 107, 129, 133, 140, 141, 143, 180, 184, 200, 233, 234, 241, 251, 252, 253, 254, 255, 264, 267, 272, 274, 275, 285, 294, 296, 298, 301, 314, 347, 353, 358, 366, 370, 373, 374, 440, 441, 449, 450, 585

Asthma 51, 56, 112, 113, 114, 115, 116, 146, 150, 152, 153, 154, 156, 159, 160, 172, 179, 183, 243, 333, 336, 393, 455, 458

Astringent 51, 109, 115, 331, 456

Avicennia alba 383, 387

Avicennia officinalis 382, 383, 384, 385, 439, 444

AYUSH systems 422

Azadirachta indica 10, 24, 53, 55, 59, 75, 76, 83, 85, 88, 90, 91, 93, 94, 111, 113, 114, 115, 117, 129, 149, 151, 152, 154, 156, 159, 177, 203, 204, 205, 206, 207, 209, 210, 233, 242, 269, 296, 313, 319, 320, 324, 327, 328, 330, 331, 332, 333, 334, 335, 336, 349, 367, 369, 376, 413, 438, 452, 458, 459, 460, 472, 474, 477, 495

Azima tetracantha 177, 269

B

Bacillus subtilis 415

Bacopa monnieri 76, 130, 162, 177

Bambusa arundinacea 90, 195, 196, 201, 202, 203, 204, 205, 207, 208, 209, 210, 252

Barleria acuminata 10, 24

Barleria courtallica 551, 552, 563, 565

Bauhinia acuminata 178

Bauhinia purpurea 130, 162, 198, 208, 367

Bauhinia tomentosa 10, 24, 130, 270

Begonia malabarica 10, 24, 76, 115, 313, 320, 331, 367, 488

Beloperone plumbaginifolia 589, 590, 605

Bhils 53, 54, 61, 64, 111, 112, 118, 120, 149, 161, 164, 332, 333, 339, 376

Bidens pilosa 10, 24, 58, 252

Biodiversity conservation 64, 120, 163, 186, 339, 517

Biodiversity related knowledge 189

Biophytum sensitivum 77, 113, 178, 367

Bischofia javanica 24

Blepharis maderaspatensis 10, 25, 77, 130

Bone fracture 51

Botanical Survey of India 62, 63, 64, 72, 143, 185, 225, 259, 289, 290, 305, 311, 337, 338, 339, 340, 363, 366, 377, 378, 465, 484, 552, 570, 590

Bowel disorder 51, 109

Breathing problem 294

Breynia vitis-idaea 10, 25

Bupleurum wightii 10, 25, 498

C

Cadaba indica 521, 522, 523, 528, 529

Caesalpiniaceae 106

Calotropis gigantea 10, 25, 53, 54, 55, 57, 58, 59, 77, 112, 113, 114, 116, 117, 118, 130, 148, 150, 154, 156, 158, 159, 206, 265, 270, 296, 313, 320, 333, 335, 350, 368, 376, 439, 453, 477

Canarium strictum 4, 10, 25, 78, 368, 483, 498

Canavalia gladiata 492, 498, 515, 518

Canthium dicoccum 10, 25

Capparis sepiaria 10, 25, 78, 586

Capsicum annum 10, 26, 58, 130, 178, 195, 196, 271, 304

Caralluma adscendens 11, 26, 59, 350, 487

Caralluma umbellata 178, 233

Cardiospermum canescens 11, 26

Cardiospermum halicacabum 11, 26, 52, 55, 58, 60, 78, 111, 113, 114, 116, 117, 118, 271, 296, 303, 313, 320, 331, 334, 335, 336, 489

Carissa carandas 11, 26, 78, 130, 241, 351, 492, 495

Cassia klenni 178

Cassia tora 131, 149, 150, 152, 160, 199

Catharanthus pusillus 178, 531, 532, 534, 537, 539, 540, 542

Catharanthus roseus 131, 152, 154, 156, 160, 178, 204, 210, 241, 271, 296, 351, 439, 444, 453, 458, 459, 460

Cayratia pedata 178, 351

Celtis philippensis 11, 26, 498

Centella asiatica 11, 26, 53, 57, 58, 79, 178, 241, 252, 271, 296, 303, 351, 419, 420, 423

Ceriops decandra 383, 384, 387

Chenopodium album 178

Chest pain 51, 53, 57, 58, 109

Chlorophytum heynei 11, 28, 79

Chloroxylon swietenia 195, 196, 201, 208

Cissampelos pareira 11, 28, 55, 56, 57, 79, 112, 113, 114, 116, 131, 153, 154, 156, 178, 351, 368

Cissus quadrangularis 11, 28, 53, 60, 79, 131, 149, 150, 152, 155, 156, 178, 272, 297, 304, 313, 320, 352, 368, 436, 447, 453, 474, 476, 477, 487, 489

Citrus aurantifolia 179

Cleome viscosa 11, 28, 131, 155, 156, 368, 376, 438, 453, 463, 489

Clerodendrum phlomidis 179, 272, 297

Clinical trials 360

Clitoria ternatea 11, 28, 53, 58, 60, 80, 112, 114, 115, 116, 131, 155, 157, 160, 223, 272, 297, 369, 376, 437, 585

Cnidoscolus chayamansa 179

Cocculus hirsutus 11, 28, 53, 60, 131, 489

Cocos nucifera 4, 11, 39, 40, 45, 73, 76, 77, 78, 79, 80, 81, 82, 85, 86, 87, 88, 90, 91, 93, 94, 96, 125, 131, 142, 181, 203, 209, 273, 297, 317, 322, 348, 351, 352, 359, 369, 435, 453, 458, 459, 474, 475, 483

Coleus aromaticus 242, 258

Colocasia esculenta 132, 162, 273, 352, 485, 489

Commelina benghalensis 11, 28, 53, 54, 80, 112, 113, 114, 116, 132, 162, 352, 369, 376, 435, 453, 459, 460, 489

Commelina ensifolia 11, 28, 489

Commiphora pubescens 11, 28, 492, 495

Comparative ethnomedicobotany 53, 55, 56, 59, 111, 112, 114, 117, 149, 151, 153, 154, 155, 158, 159, 333, 334, 335, 336

Crinum defixum 80, 176, 179, 369

Cross-cultural ethnomedicobotanical studies 52, 148

Crotalaria calycina 12, 28

Cryptolepis buchananii 12, 30

Cucurbitaceae 15, 16, 48, 80, 82, 87, 88, 106, 131, 132, 133, 136, 143, 179, 197, 241, 252, 253, 254, 272, 273, 277, 279, 285, 294, 297, 347, 352, 355, 370, 372, 374, 410, 437, 441, 442, 447, 448, 450, 489, 490, 492, 493, 495, 496

Cuminum Cyminum 26

Curculigo orchioides 12, 30, 53, 55, 57, 58, 59, 81, 112, 113, 114, 116, 132, 149, 150, 151, 152, 154, 157, 158, 160, 161, 314, 322, 325, 331, 333, 334, 335, 336, 369, 376, 485, 514

Curcuma angustifolia 179

Curcuma longa 12, 35, 36, 38, 45, 52, 53, 96, 111, 132, 142, 148, 152, 155, 157, 160, 177, 179, 184, 243, 317, 328, 355, 367, 369, 376, 409

Cyanotis tuberosa 12, 30

Cycas circinalis 176, 179, 489, 498, 515

Cylista scariosa 12, 30

Cymbopogan flexuosus 179

Cymbopogon polyneurus 483

Cynodon dactylon 12, 30, 132, 200, 204, 208, 210, 242, 265, 274, 297, 352, 435, 454, 459, 460, 462

Cyperus difformis 132, 162

D

Dalbergia latifolia 195, 196, 201, 202, 208, 209

Datura metel 133, 151, 152, 179, 258, 298, 314, 322, 334, 440, 451, 454, 459

Dengu 146, 151, 154, 156

Diabetes 51, 54, 55, 56, 57, 60, 84, 109, 114, 118, 146, 150, 151, 152, 153, 154, 155, 157, 158, 159, 160, 161, 172, 233, 331, 333, 334, 393, 455, 459

Dichrostachys cinerea 12, 30, 133, 151, 152, 162

Digestive disorder 146

Digestive troubles 223

Dioscorea alata 176, 179

Dioscorea bulbifera 12, 30, 81, 180, 485, 514

Dioscorea oppositifolia 12, 30, 70, 169, 309, 353, 485, 514

Dioscorea pentaphylla 4, 12, 30, 81, 483, 485, 514

Dioscorea tomentosa 12, 31, 81, 485, 514

Dodonaea angustifolia 195

Dodonaea viscosa 12, 31, 243, 474

Dog-bite 51, 56, 59, 109, 155

Dry deciduous forests 70, 126, 311

Drypetes sepiaria 12, 31, 498

Dry teak forests 5, 70, 311, 481

Dyspepsia 51, 58, 60, 109, 151, 452

E

Ear ailments 51, 109

Eclipta alba 133, 180, 234, 353

Eclipta prostrata 12, 31, 53, 54, 57, 58, 59, 60, 61, 82, 112, 113, 114, 115, 116, 117, 118, 200, 234, 241, 252, 265, 274, 298, 314, 322, 331, 332, 333, 335, 353, 356, 370, 376

Eczema 51, 56, 114, 146, 150, 151, 152, 154, 156, 158, 159, 160, 161, 172, 335

Ehretia ovalifolia 12, 31, 198

Endangered plants 61, 118, 119, 337

Enicostema axillare 13, 31, 52, 53, 133, 242

Ensete superbum 13, 31, 61, 82, 498

Entada phaseoloides 515

Erythrina variegata 83, 89, 112, 133, 162, 274, 314, 324, 327, 331, 370

Escherichia coli 415

Estuariane ecosystem 425

Ethnoveterinary practices 469, 472

Eupatorium triplinerve 222

Euphorbiaceae 9, 10, 12, 13, 14, 16, 20, 47, 49, 73, 77, 81, 83, 85, 87, 89, 96, 106, 108, 128, 133, 135, 137, 138, 140, 142, 143, 173, 176, 179, 180, 181, 182, 196, 201, 202, 203, 207, 233, 234, 241, 251, 252, 253, 254, 255, 264, 265, 266, 274, 275, 277, 278, 281, 282, 294, 295, 300, 315, 347, 350, 353, 354, 356, 357, 366, 369, 371, 372, 374, 428, 436, 441, 442, 443, 445, 447, 448, 449, 472, 475, 493, 495, 497, 498, 586

Euphorbia hirta 13, 31, 57, 83, 133, 160, 234, 252, 265, 275, 290, 353, 436, 454, 458

Euphorbia rosea 13, 31

Euphorbia tirucalli 133, 162, 436, 444, 454, 458, 459

Evolvulus alsinoides 13, 31, 38, 43, 45, 54, 55, 59, 83, 112, 113, 116, 117, 134, 155, 157, 206, 210, 252, 275, 355, 370, 440, 451

F

Fabaceae 9, 11, 12, 14, 16, 17, 18, 19, 47, 48, 73, 80, 82, 83, 85, 88, 90, 91, 93, 96, 106, 128, 130, 131, 135, 138, 140, 141, 142, 143, 176,

178, 179, 180, 182, 184, 196, 197, 199, 200, 201, 202, 203, 206, 233, 234, 235, 242, 251, 252, 253, 254, 255, 264, 266, 272, 273, 274, 282, 283, 284, 285, 294, 313, 314, 316, 317, 347, 350, 352, 353, 354, 369, 370, 371, 372, 374, 416, 428, 437, 441, 442, 443, 445, 446, 447, 448, 449, 450, 475, 485, 490, 492, 494, 498, 499, 500, 515, 516

Family-wise distribution 48, 106, 232, 254, 256, 442

Ficus benghalensis 195

Ficus microcarpa 13, 32, 84, 353

Ficus racemosa 13, 32, 54, 55, 84, 134, 154, 180, 207, 275, 353, 496

Ficus religiosa 13, 32, 84, 134, 150, 152, 204, 210, 275, 298, 370, 496

Fiscus benghalensis 32

Flacourtia indica 13, 32, 496, 514

Foeniculum vulgare 20, 32, 370

G

Garcinia gummiguta 180

Gardenia resinifera 13, 32, 496

GC-MS analysis 400, 607, 611, 612

Givotia rottleriformis 13, 32, 498

Globba marantina 13, 32

Gloriosa superba 13, 34, 59, 61, 84, 134, 157, 162, 234, 370, 376

Gonds 53, 55, 56, 61, 111, 112, 113, 114, 149, 151, 161, 165, 166, 260, 332, 333, 334, 376, 514, 515

Gymnema sylvestre 13, 34, 54, 55, 56, 60, 61, 84, 134, 223, 234, 241, 245, 252, 258, 409, 410, 423

H

Heart disorders 223

Hedyotis puberula 14, 34, 47, 94

Hemidesmus indicus 14, 34, 47, 54, 55, 56, 57, 58, 60, 61, 85, 94, 112, 113, 115, 116, 117, 118, 119, 134, 150, 152, 154, 157, 180, 234, 253, 255, 298, 303, 315, 325, 331, 333, 334, 335, 336, 370, 376, 486

Herbal remedies 2, 51, 68, 109, 146, 163, 168, 223, 422

Hibiscus rosasinensis 180

Hibiscus surattensis 85, 180

Hibiscus vitifolius 14, 34, 134

Hot spots 2, 68, 168, 307, 469, 480

Hugonia mystax 14, 34, 85, 315, 325

Human ailments 389

Hybanthus enneaspermus 14, 34, 243, 389, 390, 391, 394, 395, 396, 397, 398, 400, 401, 402, 403, 404, 405, 489

Hygrophila auriculata 134, 162, 439, 448, 454, 458, 459

I

Ichnocarpus frutescens 14, 34, 56, 57

Indian medicinal plants 409, 528

Indigenous ethnomedicinal plants 1

Indigenous wild edible plants 479

Indigofera tinctoria 135, 155, 157, 158, 160, 180, 354

Indigofera wightii 14, 34, 58, 85

Indonesiella echinoides 180

Inflammations 54, 144, 157, 257, 300, 309, 452, 455, 456, 604

Ipomoea barlerioides 14, 35

Ipomoea staphylina 14, 35, 195, 196, 199, 208

Irulas Tribes 53, 123

IUCN list 162

Ixora coccinia 180

J

Jasminum roxburghianum 14, 35

Jatropha curcas 14, 35, 54, 56, 57, 60, 135, 149, 150, 151, 152, 155, 157, 160, 180, 222, 234, 277, 371

Jatropha glandulifera 14, 35, 277, 436, 454, 459, 460, 461

Jatropha gossypifolia 14, 35, 57, 85, 115, 135, 315, 325, 335

Jaundice 51, 54, 56, 58, 59, 60, 109, 112, 113, 115, 116, 117, 147, 150, 151, 152, 153, 154, 156, 158, 159, 161, 172, 233, 241, 242, 243, 258, 335, 419, 452, 458

Justicia adhatoda 14, 35, 86, 112, 115, 116, 117, 135, 157, 277, 315, 325, 330, 331, 333, 335, 336, 371

Justicia gendrussa 180

Justicia glauca 14, 35, 135, 440

Justicia simplex 180

K

Kaempferia galanga 85, 86, 180, 222, 223, 371

Kalakad-Mundanthurai tiger reserve sanctuary 307

Kalanchoe pinnata 14, 35, 86, 116, 371, 489

Kalanchoe tubiflora 14, 36

Kanikkars Tribe 53, 58, 59, 61, 64, 68, 70, 72, 96, 106, 109, 110, 111, 112, 114, 115, 117, 118, 121, 149, 155, 161, 163, 165, 166, 168, 169, 171, 176, 186, 307, 308, 309, 311, 312, 317, 332, 333, 334, 335, 336, 363, 364, 366, 376, 377, 515

Kani Tribes 167, 175

Kanyakumari wildlife sanctuary 167

Kidney stone 51, 53, 60, 109, 287

Klebsiella pneumoniae 607, 608, 614

L

Lamiaceae 9, 15, 16, 17, 47, 49, 75, 86, 89, 106, 107, 129, 131, 134, 135, 137, 138, 143, 177, 181, 182, 202, 203, 205, 206, 207, 234, 252, 253, 254, 264, 268, 277, 278, 280, 294, 297, 299, 313, 347, 348, 356, 367, 371, 372, 375, 440, 441, 442, 443, 446, 448, 449, 473, 474

Lantana camara 4, 15, 36, 86, 125, 135, 153, 154, 180, 201, 208, 243, 277, 298, 420, 421, 440, 496

Lantana wightiana 15, 36

Lawsonia inermis 36, 54, 86, 91, 112, 116, 117, 118, 135, 157, 181, 200, 201, 208, 277, 298, 371, 373, 376, 419, 437, 448, 454

Leucas biflora 15, 36, 55

Leucas lavandulifolia 418

Liliaceae 9, 11, 13, 50, 74, 79, 84, 94, 96, 106, 108, 129, 139, 142, 197, 202, 206, 233, 242, 251, 313, 347, 348, 366, 367, 370, 375, 475, 485, 488

Lindernia crustacea 15, 36, 87

Lippia nodiflora 150, 181

Lobelia nicotianifolia 15, 36, 54, 55

Luminitzera racemosa 383

M

Madras herbarium 143

Malayali Tribals 239, 247, 260

Malvaceae 14, 16, 18, 47, 48, 73, 85, 89, 92, 93, 106, 107, 128, 134, 136, 137, 139, 140, 143, 180, 182, 197, 220, 235, 251, 253, 254, 264, 266, 276, 283, 284, 294, 295, 298, 301, 347, 354, 357, 366, 373, 375, 428, 438, 441, 442, 445, 447, 448, 449, 450, 485, 492

Manakudy estuary 425, 430, 431, 433, 435, 442, 444, 445, 452, 462

Mangifera indica 15, 36, 135, 151, 153, 157, 160, 185, 195, 201, 205, 208, 210, 253, 278, 299, 355, 371, 474, 496

Mangrove species 381, 383

Manihot esculenta 70, 135, 169, 181, 278, 309, 429

Medicinal knowledge 189

Melia azadirachta 10, 181, 242, 349

Menispermaceae 11, 18, 48, 79, 80, 92, 93, 106, 107, 131, 140, 178, 183, 220, 233, 254, 316, 347, 351, 358, 368, 369, 370, 373, 374, 473, 489

Menstrual disorder 51

Michelia champaca 136, 153, 181

Microscopic studies 521, 532

Mimosa pudica 15, 36, 54, 57, 58, 60, 87, 112, 115, 116, 117, 136, 153, 158, 162, 242, 253, 278, 299, 304, 315, 327, 331, 333, 335, 336, 355, 371, 376, 437, 444, 455, 459, 460

Mimusops elengi 15, 38, 55, 496, 605

Mitragyna parvifolia 15, 37, 38, 56

Moist deciduous forests 70, 311

Mollugo cerviana 15, 38, 45, 355

441, 442, 447, 449, 450, 488, 490, 492, 493, 495, 497

Solanum surattense 18, 43, 58, 59, 92, 116, 117, 255, 265, 283, 287, 373, 440, 456, 458, 459, 460

Solanum torvum 18, 44, 92, 139, 183, 254, 283, 493

Solanum trilobatum 18, 26, 43, 58, 59, 78, 79, 92, 139, 183, 235, 243, 254, 255, 283, 301, 440, 451, 456, 464, 490, 493, 497

Sonneratia apetala 382, 383, 385

Sonneratia caseolaris 383

Sorghum vulgare 195, 197, 205, 209

Souropsis androgynus 183

Staphylococcus aureus 385, 399, 415

Sterculia urens 18, 44, 54, 56, 486, 487, 499

Stomachache 51, 53, 54, 55, 56, 57, 58, 60, 110, 114, 116, 150, 154, 155, 157, 158, 159, 160, 161, 331, 336

Strychnos nux-vomica 44, 203, 209, 499

Sustainable development 163

Swertia chirata 417

Swertia corymbosa 18, 44, 61

Syzygium aromaticum 183, 403

Syzygium cumini 18, 44, 54, 57, 60, 61, 84, 92, 139, 151, 153, 155, 161, 198, 284, 355, 497, 514

T

Tabernaemontana divaricata 18, 44, 61, 140, 284

Tamarindus indica 18, 23, 38, 44, 74, 93, 140, 153, 158, 161, 195, 196, 198, 223, 254, 284, 301, 319, 320, 352, 358, 437, 473, 490, 491, 493, 497, 499

Taxonomical categories 251

Tephrosia purpurea 18, 44, 56, 60, 61, 93, 140, 203, 235, 254, 284, 287, 437, 456, 459, 461

Terminalia bellirica 18, 38, 44, 54, 56, 58, 59, 61, 93, 198, 254, 500

Terminalia chebula 18, 38, 45, 56, 57, 60, 61, 93, 140, 153, 155, 198, 201, 202, 207, 209, 210, 235, 349, 358, 359, 475, 500

Thespesia populnea 18, 45, 93, 116, 117, 140, 235, 284, 301, 373, 376, 438, 456, 459

Throat pain 223, 266

Tibetan Medicine 390

Tinospora cordifolia 18, 45, 54, 56, 60, 93, 112, 114, 140, 150, 153, 161, 183, 254, 285, 316, 330, 331, 332, 333, 334, 358, 473

Traditional knowledge 123, 211, 212, 225, 227, 289, 290, 291, 305, 338, 340, 360, 377, 378, 401, 615

Traditional medicine 167, 386, 587, 615

Trianthema decandra 19, 45

Tribulus terrestris 45, 184

Trichodesma zeylanicum 19, 45, 94, 254

Trichopus zeylanicus 94, 184, 223, 373, 376

Tridax procumbens 19, 45, 54, 56, 57, 58, 59, 60, 61, 94, 111, 112, 114, 115, 117, 118, 140, 149, 150, 151, 153, 155, 158, 161, 184, 241, 285, 287, 301, 303, 374, 376, 440, 444, 456, 459

Tropical riparian fringe forests 70, 311

Tropical wet evergreen forests 70, 311

Tylophora indica 19, 45, 140, 439, 456, 461, 473

U

Umbrella thorn forest 70, 311

Urinary disorder 114, 147

V

Vaccinium oxycoccos 421

Vachellia leucophloea 416

Vachellia nilotica 415, 416

Vetan community 263, 266

Vetiveria zizanioides 19, 47, 94, 374

Vigna bourneae 494, 500, 515

Vigna radiata 184, 356, 494, 500, 515, 516

Vitex negundo 19, 47, 54, 57, 60, 94, 114, 115, 118, 141, 148, 149, 150, 155, 158, 184, 207, 235, 243, 254, 265, 285, 301, 316, 330, 331, 334, 335, 336, 414, 415, 472, 477

W

Wet temperate forests 70

World Health Organization 123, 167, 187, 248, 263, 407, 422, 544, 549, 551, 581, 587, 590, 605

World Intellectual Property Office (WIPO) 189

Wrightia laevis 141, 162

Wrightia tinctoria 19, 47, 94, 112, 117, 150, 153, 158, 205, 210, 254, 359, 374, 376

X

Xylocarpus granatum 383, 384

Z

Zingiberaceae 12, 13, 20, 74, 86, 95, 96, 106, 108, 132, 141, 142, 173, 177, 179, 180, 243, 273, 286, 317, 347, 348, 359, 369, 371, 375, 412, 472, 485

Zingiber officinale 20, 38, 40, 45, 54, 60, 78, 83, 84, 92, 96, 141, 142, 150, 153, 243, 286, 317, 320, 324, 330, 350, 352, 353, 354, 355, 359, 412

Ziziphus xylopyrus 19, 47, 57, 95

www.ingramcontent.com/pod-product-compliance
Ingram Content Group UK Ltd.
Pitfield, Milton Keynes, MK11 3LW, UK
UKHW021435280726
14060UKWH00001BA/96